Army Techniques Publication
No. 3-21.20

MW00934467

Washington, DC, 28 December 2017

Infantry Battalion

Contents

DISTRIBUTION RESTRICTION: Approved for public release; distribution is unlimited.

*This publication supersedes FM 3-21.20, dated 13 December 2006 and FM 3-21.12, dated 1 July 2008.

Figures

Tables

Preface

ATP 3-21.20 provides doctrine for the Infantry battalion of the Infantry brigade combat team (IBCT). ATP 3-21.20 describes relationships, organizational roles and functions, capabilities and limitations, and responsibilities within the Infantry battalion. *Techniques,* nonprescriptive ways or methods used to perform missions, functions, or tasks (CJCSM 5120.01), are discussed in this publication and are intended to be used as a guide. The techniques are not prescriptive. ATP 3-21.20 publication supersedes FM 3-21.20 and FM 3-21.12.

Readers must first understand the principles of the Army profession and the Army ethic as described in ADRP 1 to comprehend the doctrine contained in this publication. Readers must understand the principles of joint operations, the nature of unified land operations, and the links between the operational and tactical levels of war described in JP 3-0, ADP 3-0, and ADRP 3-0; FM 3-94, ATP 3-91, and FM 3-96. In addition, readers should understand the fundamentals of the operations process found in ADP 5-0 and ADRP 5-0 associated with offensive and defensive tasks contained in FM 3-90-1 and reconnaissance, security, and tactical enabling tasks contained in FM 3-90-2. The reader must comprehend how stability tasks described in ADP 3-07 and ADRP 3-07 carry over and affect offensive and defensive tasks and vice versa. Readers must understand how the operation process fundamentally relates to the Army's design methodology, military decision-making process, troop-leading procedures, and the principles of mission command as described in ADP 6-0, ADRP 6-0, FM 6-0, and ATP 6-0.5.

The principal audience for ATP 3-21.20 is the commanders, staff, officers, and noncommissioned officers within the Infantry battalion. The audience also includes the United States Army Training and Doctrine Command institutions and components, and the United States Army Special Operations Command. This publication serves as an authoritative reference for personnel developing doctrine, materiel and force structure, institutional and unit training, and standard operating procedures for the Infantry battalion.

Commanders, staffs, and subordinates ensure their decisions and actions comply with United States, international, and in some cases, host-nation laws and regulations. Commanders at all levels ensure their Soldiers operate within the law of war and the rules of engagement. (See FM 27-10.)

ATP 3-21.20 implements STANAG 2020.

ATP 3-21.20 uses joint terms where applicable. Selected joint and Army terms and definitions appear in both the glossary and the text. Terms for which ATP 3-21.20 is the proponent publication (the authority) are marked with an asterisk (*) in the glossary. Definitions for which ATP 3-21.20 is the proponent publication are boldfaced in the text and the term is italicized. For other definitions shown in the text, the term is italicized and the number of the proponent publication follows the definition.

ATP 3-21.20 applies to the Active Army, the Army National Guard/Army National Guard of the United States, and the United States Army Reserve unless otherwise stated.

The proponent for ATP 3-21.20 is the United States Army Training and Doctrine Command. The preparing agency is the United States Army Maneuver Center of Excellence. Send comments and recommendations on DA Form 2028, (*Recommended Changes to Publications and Blank Forms*) to Commander, Maneuver Center of Excellence, Directorate of Training and Doctrine, Doctrine and Collective Training Division, ATTN: ATZK-TDD, 1 Karker Street, Fort Benning, GA 31905-5410; by email to usarmy.benning.mcoe.mbx.doctrine@mail.mil; or submit an electronic DA Form 2028.

Introduction

The Army provides readily available and trained regionally aligned and globally responsive forces to prevent conflict, shape the security environment, and win wars. Army forces maintain proficiency in the fundamentals of decisive action, and possess capabilities to meet specific geographic combatant command requests. Regionally aligned forces provide combatant commanders with an Army headquarters tailored to missions from tactical level to joint task force capable. (See FM 3-94.) The Infantry battalion of the Infantry brigade combat team shapes the security environment and wins across the range of military operations.

ATP 3-21.20 addresses the tactical application of techniques associated with the offense, the defense, and operations focused on stability. ATP 3-21.20 does not discuss defense support of civil authorities. (Refer to ADRP 3-28 and ATP 3-28.1 for information about defense support of civil authorities.) Employing the techniques addressed in ATP 3-21.20 requires using and integrating the tactics and procedures found in FM 3-96, FM 3-90-1, and FM 3-90-2. *Tactics* are the employment and ordered arrangement of forces in relation to each other (CJCSM 5120.01). *Procedures* are standard, detailed steps that prescribe how to perform specific tasks (CJCSM 5120.01).

The techniques addressed in ATP 3-21.20 includes the movement and *maneuver*—the employment of forces in the operational area through movement in combination with fires to achieve a position of advantage in respect to the enemy (JP 3-0)—of units in relation to each other, the terrain, and the enemy. Techniques vary with terrain and other circumstances; they change frequently as the enemy reacts and friendly forces explore new approaches. Applying techniques usually entails acting under time constraints with incomplete information. Techniques always require judgment in application; they are always descriptive, not prescriptive.

Fictional scenarios, used as discussion vehicles throughout this publication, illustrate different ways an Infantry battalion can accomplish its mission regardless of which element of decisive action (offense, defense, or stability) currently dominates. Scenarios focus on potential challenges confronting the Infantry battalion commander in accomplishing a mission, but are not intended to be prescriptive of how the Infantry battalion performs any particular operation.

Note. These same scenarios drive the techniques used to develop ATP 3-21.10, *Infantry Rifle Company*. ATP 3-21.10 will replace FM 3-21.10, *The Infantry Rifle Company,* and focus on the techniques used to perform missions, functions, or tasks in support of the Infantry battalion.

ATP 3-21.20 incorporates the significant changes in Army doctrinal terminology, concepts, constructs, and proven tactics developed during recent operations. It also incorporates doctrinal terms and changes based on Doctrine 2015.

The following is a brief introduction and summary of changes by chapter:

Chapter 1 – Organization

Chapter 1 provides a brief overview of the operational environment, the Army's operational concept of unified land operations through decisive action, and the eight elements of combat power. In addition, chapter 1—

- Discusses combined arms, hasty versus deliberate operations, close combat, and operations structure.
- Describes the organizations, missions, capabilities, and limitations of the IBCT, the Infantry battalion, and the forward support company supporting the Infantry battalion.
- Discusses the exercise of mission command, specific to the foundational mission command philosophy together with the mission command warfighting function, guided by the principles of mission command.
- Discusses information operations, specific to degrading the enemy's decision-making ability and the enemy's ability to degrade friendly decision-making and mission command systems.

Chapter 2 – Offense

Chapter 2 discusses offensive actions to destroy the enemy. It describes the four offensive tasks: movement to contact, attack, exploitation, and pursuit. It also discusses—

- The doctrinal basis for the offense.
- Offensive considerations for the Infantry battalion.
- Subordinate forms of attack.
- Tactical movement.
- Tactical enabling tasks.
- Transitions.

Chapter 3 – Defense

Chapter 3 addresses the defensive tasks—area defense, mobile defense, and retrograde; and subordinate forms of the defense—defense of a linear obstacle, perimeter defense, and a reverse-slope. This chapter also describes in detail the—

- The doctrinal basis for the defense.
- Defense considerations for the Infantry battalion.
- Forms of maneuver.
- Tactical enabling tasks.
- Transitions.

Chapter 4 – Stability

Chapter 4 discusses operations in support of stability-focused tasks, and various military missions, tasks, and activities conducted in support of stabilization. In addition, chapter 4 addresses—

- The foundation for operations focused stability.
- Organization of forces.
- Operational area security.
- Security force assistance.
- Transitions.

Nine appendixes complement the body of this publication. They include—

- Appendix A: Command Post Operations and Organizations.
- Appendix B: Planning and Preparation.
- Appendix C: Fires.
- Appendix D: Infantry Weapons Company (replaces FM 3-21.12).
- Appendix E: Battalion Sniper Squad.
- Appendix F: Combined Arms Breaching Operations.
- Appendix G: Chemical, Biological, Radiological, and Nuclear Defense and Countering Weapons of Mass Destruction.
- Appendix H: Sustainment.
- Appendix I: Base Operations.

Chapter 1
Organization

The Infantry battalion of the Infantry brigade combat team (IBCT) organizes to conduct *decisive action*—the continuous, simultaneous combinations of offensive, defensive, and stability or defense support of civil authorities tasks (ADRP 3-0). The Infantry battalion can deploy rapidly, execute entry operations, and execute missions throughout the range of military operations. The Infantry battalion can conduct effective combat or other operations immediately upon arrival in an *operational area*—an overarching term encompassing more descriptive terms (such as area of responsibility and joint operations area) for geographic areas in which military operations are conducted (JP 3-0). Chapter 1 provides the doctrinal foundation for the Infantry battalion during decisive action. The chapter addresses briefly key doctrinal concepts on how the Army fights, but focuses primarily on the mission, capabilities, limitations, and internal organization of the Infantry battalion.

SECTION I – OPERATIONAL OVERVIEW

1-1. While an Infantry battalion's operation predominant characteristic is offense, defense, or operations in support of stability, different units involved in that operation conduct different types and subordinate forms of operations, and often transition rapidly from one element of decisive action or subordinate task to another. The battalion commander rapidly shifts emphasis from one element or task to another, continually keeping the enemy off balance, while positioning available forces for maximum effectiveness. Flexibility in transitioning contributes to a successful operation. The commander and staff use their *situational understanding,* the product of applying analysis and judgment to relevant information to determine the relationship among the operational and mission variables to facilitate decision making (ADP 5-0), to choose the right combinations of combined arms to place the enemy at the maximum disadvantage. The section briefly covers key doctrinal concepts on how the Army fights. For a complete discussion, refer to ADP 3-0 and ADRP 3-0.

OPERATIONAL ENVIRONMENT

1-2. An *operational environment* is a composite of the conditions, circumstances, and influences that affect the employment of capabilities and bear on the decisions of the commander (JP 3-0). Operational environments shape the nature and affect the outcome of military operations. The complex and dynamic nature of an operational environment and the threats that exist within an operational environment make determining the relationship between cause and effect difficult and contributes to the uncertainty of the military operation. An understanding of the operational environment in which the unit fights is required.

1-3. The operational environment encompasses physical areas and factors of the air, land, maritime, space, and cyberspace domains, and the information environment. *Information environment* is the aggregate of individuals, organizations, and systems that collect, process, disseminate, or act on information (JP 3-13). Although an operational environment and information environment are defined separately, they are interdependent and integral to each other. The information environment is comprised of three dimensions: physical, informational, and cognitive (for additional information about each dimension, see Chapter 1, FM 3-13).

1-4. The information element of combat power (see paragraph 1-22) is integral to optimizing combat power, particularly given the increasing relevance of operations in and through the information environment to achieve decisive outcomes. Information operations and the information element of combat power are related but not the same. Information is a resource. As a resource, it must be obtained, developed, refined, distributed, and protected. Information operations, along with knowledge management and information management, are ways the Infantry battalion can harness this resource and ensure its availability, as well as operationalize and optimize it. *Knowledge management* is the process of enabling knowledge flow to enhance shared understanding, learning, and decision making (ADRP 6-0). *Information management* is the science of using procedures and information systems to collect, process, store, display, disseminate, and protect data, information, and knowledge products (ADRP 6-0).

1-5. The operational environment of the battalion includes all enemy, adversarial, friendly, and neutral systems across the range of military operations and is part of the higher commander's operational environment. The battalion's operational environment includes the physical environment, the state of governance, technology, and local resources, and the culture of the local populace. As the operational environment for each operation is different, it also evolves as the operation progresses. Commander and staff continually assess and reassess the operational environment as they seek to understand how changes in the nature of threats and other variables affect not only their force but other actors as well. The commander and staff use the Army design methodology, operational variables, and mission variables to analyze an operational environment in support of the operations process.

Army Design Methodology

1-6. The *Army design methodology* is a methodology for applying critical and creative thinking to understand, visualize, and describe unfamiliar problems and approaches to solving them (ADP 5-0). The Army design methodology is particularly useful as an aid to conceptual thinking about unfamiliar problems and to gain a greater understanding of the operational environment. To produce executable plans, the commander integrates this methodology with the detailed planning typically associated with the military decision-making process. (Refer to ATP 5-0.1 and FM 3-96 for additional information.)

Operational and Mission Variables

1-7. When alerted for deployment, redeployment within a theater of operations, or assigned a mission, the Infantry battalion's assigned higher headquarters provides an analysis of the operational environment. That analysis includes *operational variables,* a comprehensive set of information categories used to define an operational environment (ADP 1-01). Information categories are political, military, economic, social, information, infrastructure, physical environment, and time, commonly referred to as PMESII-PT.

1-8. Upon receipt of a mission, commander and staff filter information categorized by operational variables into relevant information with respect to the assigned mission. Commanders use *mission variables*, which are the categories of specific information needed to conduct operations (ADP 1-01), to focus on specific elements of an operational environment during mission analysis. This analysis enables the commander and staff to combine operational variables and tactical-level information with knowledge about local conditions relevant to their mission. Mission variables are mission, enemy, terrain and weather, troops and support available, time available and civil considerations (METT-TC). (Refer to FM 6-0 for additional information.)

Threat

1-9. The commander must understand threats, criminal networks, enemies, and adversaries, to include both state and non-state actors, in the context of the operational environment. When the commander understands the threat, the commander can visualize, describe, direct, lead, and assess operations to seize, exploit, and retain the initiative. A *threat* is any combination of actors, entities, or forces that have the capability and intent to harm United States forces, United States national interests, or the homeland (ADRP 3-0). Threats may include individuals, groups of individuals (organized or not organized), paramilitary or military forces, nation-states, or national alliances. When threats execute their capability to do harm to the United States, they become enemies.

1-10. In general, the various actors in any area of operation can qualify as a threat, an enemy, an adversary, a neutral, or a friend. An *enemy* is a party identified as hostile against which the use of force is authorized (ADRP 3-0). An *adversary* is a party acknowledged as potentially hostile to a friendly party and against which the use of force may be envisaged (JP 3-0). *A neutral* is a party identified as neither supporting nor opposing friendly, adversary, or enemy forces (ADRP 3-0). Land operations often prove complex because a threat, an enemy, an adversary, a neutral, or a friend intermix, often with no easy means to distinguish one from another.

1-11. A *hybrid threat* is the diverse and dynamic combination of regular forces, irregular forces, terrorist forces, or criminal elements unified to achieve mutually benefitting threat effects (ADRP 3-0). The term hybrid threat evolved to capture the seemingly increased complexity of operations, the multiplicity of actors involved, and the blurring among traditional elements of conflict. (Refer to FM 3-96 and ADRP 2-0 for more information.)

SPECIFIC OPERATIONAL ENVIRONMENTS

1-12. Specific operational environments include urban, mountain, desert, and jungle. Subsurface areas are conditions found in all four operational environments. Offensive and defensive tasks in these environments follow the same planning process as operations in any other environment, but they do impose specific techniques and methods for success. The uniqueness of each environment may affect more than their physical aspects but also their informational systems, flow of information, and decision-making. As such, mission analysis must account for the information environment and cyberspace within each specific operational environment (see appendix B). Each specific operational environment has a specific manual because of their individual characteristics.

Urban Terrain

1-13. Operations in urban terrain are infantry-centric combined arms operations that capitalize on the adaptive and innovative leaders at the squad, platoon, and company level. Plans must be flexible to promote disciplined initiative by subordinate leaders, characterized by a simple scheme of maneuver and detailed control measures for interaction with civilian population and/or noncombatants. In the offense, task organizing the battalion combined arms team at the right place and time is key to achieving the desired effects. In the defense, the combined arms team turns the environment's characteristics to its advantage. Urban areas are ideal for the defense because they enhance the combat power of defending units. (Refer to Army tactics, techniques, and procedures (ATTP) 3-06.11 for additional information.)

Mountainous Terrain

1-14. Operations in mountainous terrain are conducted for three primary purposes: to deny an enemy a base of operations; to isolate and defeat enemy; and to secure lines of communication. Enemy tactics commonly involve short violent engagements followed by a hasty withdrawal through preplanned routes. The enemy often strikes quickly and fights only as long as the advantage of the initial surprise is in their favor. Attacks may include direct fires, indirect fires, or improvised explosive devices and may be against stationary or moving forces. The design of the landscape, coupled with climatic conditions, creates a unique set of mountain operations characteristics that are characterized by close fights with dismounted infantry, decentralized small-unit operations, degraded mobility, increased movement times, restricted lines of communications, and operations in thinly populated areas. (Refer to ATTP 3-21.50 for additional information.)

Desert Terrain

1-15. Operations in desert terrain require adaptation to the terrain and climate. Equipment must be adapted to a dusty and rugged landscape with extremes in temperature and changes in visibility. The battalion orients on primary enemy approaches but prepares for an attack from any direction. Considerations for operations in desert terrain include lack of concealment and the criticality of mobility; use of obstacles to site a defense, which are limited; strong points to defend choke points and other key terrain; and mobility and sustainment. (Refer to FM 90-3 for additional information.)

Jungle Terrain

1-16. Operations in jungle terrain combine dispersion and concentration. For example, a force may move out in a dispersed formation to find the enemy. Once the force makes contact, its subordinate forces close on the enemy from all directions. Operations are enemy-oriented, not terrain-oriented. Forces should destroy the enemy wherever found. If the force allows the enemy to escape, the force will have to find the enemy again, with all the risks involved. Jungle operations use the same defensive fundamentals as other defensive operations. Considerations for offensive and defensive tasks in a jungle environment include limited visibility and fields of fire, ability to control units, and limited and restricted maneuver. (Refer to FM 90-5 for additional information.)

Subsurface Areas

1-17. A subsurface area is a condition found in all four operational environments described above. Subsurface areas are areas below ground level that may consist of underground facilities, passages, subway lines, utility corridors or tunnels, sewers and storm drains, caves, or other subterranean spaces. This dimension includes areas both below the ground and below water. Additional subterranean areas include drainage systems, cellars, civil defense shelters, mines, and other various underground utility systems. In older cities, subsurface areas include ancient hand-dug tunnels and catacombs.

1-18. Subsurface areas may serve as secondary and, in fewer instances, primary avenues of approach at lower tactical levels. Subsurface areas are used for cover and concealment, troop movement, command functions, and engagements, but their use requires intimate knowledge of the area. When thoroughly reconnoitered and controlled, subsurface areas offer excellent covered and concealed lines of communications for moving supplies and evacuating casualties. Attackers and defenders can use subsurface areas to gain surprise and maneuver against the rear and flanks of an enemy and to conduct ambushes. However, these areas are often the most restrictive and easiest to defend or block. The commander may need to consider potential avenues of approach afforded by the subsurface areas of rivers and major bodies of water that border urban areas.

1-19. Knowledge of the nature and location of these subsurface areas is of great value to both friendly and enemy forces. The effectiveness of subsurface areas depends on superior knowledge of their existence and overall design. A thorough understanding of the environment is required to exploit the advantages of subsurface areas. Maximizing the use of these areas could prove to be a decisive factor while conducting offensive and defensive tasks. (Refer to FM 3-06, TC 2-91.4, and ATP 3-34.81 for additional information on subsurface areas.)

DECISIVE ACTION

1-20. Army forces demonstrate the Army's core competencies through decisive action. In unified land operations, a commander seeks to seize, retain, and exploit the initiative to gain and maintain a position of relative advantage over an enemy or threat while synchronizing unit actions to achieve the best effects possible. Decisive action begins with the commander's intent and concept of operations. *Commander's intent* is a clear and concise expression of the purpose of the operation and the desired military end state that supports mission command, provides focus to the staff, and helps subordinate and supporting commanders act to achieve the commander's desired results without further orders, even when the operation does not unfold as planned (JP 3-0). *Mission command* is the exercise of authority and direction by the commander using mission orders to enable disciplined initiative within the commander's intent to empower agile and adaptive leaders in the conduct of unified land operations. (ADP 6-0). *Concept of operations* is a statement that directs the manner in which subordinate units cooperate to accomplish the mission and establish the sequence of actions the force will use to achieve the end state (ADRP 5-0).

Note. *Mission* is the task, together with the purpose, that clearly indicates the action to be taken and the reason therefore (JP 3-0).

1-21. As a single, unifying idea, decisive action provides direction for the entire operation. Mission command requires commanders to convey a clear commander's intent and concept of operations. These become essential in operations where multiple operational and mission variables interact with the lethal application of ground combat power. Such dynamic interaction often compels subordinate commanders to make difficult decisions in unforeseen circumstances. Based on a specific idea of how to accomplish the mission, commander and staff refine the concept of operations during planning and adjust the concept of operations throughout the operation as subordinates develop the situation or conditions change. Often, subordinates acting on the higher commander's intent develop the situation in ways that exploit unforeseen opportunities. (Refer to ADRP 3-0 for additional information.)

COMBAT POWER

1-22. Commanders conceptualize capabilities in terms of combat power. *Combat power* is the total means of destructive, constructive, and information capabilities that a military unit or formation can apply at a given time (ADRP 3-0). The eight elements of combat power are leadership, information, mission command, movement and maneuver, intelligence, fires, sustainment, and protection. Commanders apply leadership and information throughout to multiply the effects of the other six elements of combat power. The other six elements—mission command, movement and maneuver, intelligence, fires, sustainment, and protection—are collectively known as warfighting functions. A *warfighting function* is a group of tasks and systems united by a common purpose that commanders use to accomplish missions and training objectives (ADRP 3-0).

1-23. *Leadership* is the process of influencing people by providing purpose, direction, and motivation to accomplish the mission and improve the organization (ADP 6-22). Army professionals are expected to act and apply force ethically and in accordance with shared national values and Constitutional principles, which are reflected in the law, oaths, and the Army Ethic. Refer to ADRP 6-22 and ADRP 1 for more information.

1-24. Information is the meaning that a human assigns to data by means of the known conventions used in their representation. Information enables the commander to make informed decisions on how to apply combat power. Information operations is the commander's primary means to optimize the information element of combat power and supports and enhances all other elements in order to gain an operational advantage over an enemy or adversary. *Information operations* is the integrated employment, during military operations, of information-related capabilities in concert with other lines of operation to influence, disrupt, corrupt, or usurp the decision making of adversaries and potential adversaries while protecting our own (JP 3-13). Refer to FM 6-0 and FM 3-13 for more information.

1-25. The *mission command warfighting function* is the related tasks and systems that develop and integrate those activities enabling a commander to balance the art of command and the science of control in order to integrate the other warfighting functions (ADRP 3-0). Refer to ADRP 6-0 for more information.

1-26. The *movement and maneuver warfighting function* is the related tasks and systems that move and employ forces to achieve a position of relative advantage over the enemy and other threats (ADRP 3-0). Refer to ADRP 3-90 and FM 3-96 for more information.

1-27. The *intelligence warfighting function* is the related tasks and systems that facilitate understanding the enemy, terrain, weather, civil considerations, and other significant aspects of the operations environment. (ADRP 3-0). Refer to ADRP 2-0 and FM 2-0 for more information.

1-28. The *fires warfighting function* is the related tasks and systems that provide collective and coordinated use of Army indirect fires, air and missile defense, and joint fires through the targeting process (ADRP 3-0). Refer to ADRP 3-09 and FM 3-09 for more information.

1-29. The *protection warfighting function* is the related tasks and systems that preserve the force so the commander can apply maximum combat power to accomplish the mission (ADRP 3-0). Refer to ADRP 3-37 for more information.

1-30. The *sustainment warfighting function* is the related tasks and systems that provide support and services to ensure freedom of action, extend operational reach, and prolong endurance (ADRP 3-0). Refer to ADRP 4-0 for more information.

1-31. Commanders employ three means to organize combat power: force tailoring, task-organizing, and mutual support, which are described below:

- *Force tailoring* is the process of determining the right mix of forces and the sequence of their deployment in support of a joint force commander (ADRP 3-0).
- *Task-organizing* is the act of designing a force, support staff, or sustainment package of specific size and composition to meet a unique task or mission (ADRP 3-0).
- *Mutual support* is that support which units render each other against an enemy, because of their assigned tasks, their position relative to each other and to the enemy, and their inherent capabilities (JP 3-31).

Note. *Task organization* is a temporary grouping of forces designed to accomplish a particular mission (ADRP 5-0).

1-32. Commanders consider mutual support when task-organizing forces, assigning areas of operations, and positioning units. The two aspects of mutual support are supporting range and supporting distance. *Supporting range* is the distance one unit may be geographically separated from a second unit, yet remain within the maximum range of the second unit's weapons systems (ADRP 3-0). *Supporting distance* is the distance between two units that can be traveled in time for one to come to the aid of the other and prevent its defeat by an enemy or ensure it regains control of a civil situation (ADRP 3-0). (Refer to ADRP 3-0 and FM 3-96 for additional information.)

COMBINED ARMS

1-33. Applying combat power depends on combined arms to achieve its full destructive, disruptive, informational, and constructive potential. *Combined arms* is the synchronized and simultaneous application of all elements of combat power that together achieve an effect greater than if each element was used separately or sequentially (ADRP 3-0). Combined arms integrates leadership, information, and each of the warfighting functions and their supporting systems. Used destructively, combined arms integrates different capabilities so that counteracting one makes the enemy vulnerable to another. Used constructively, combined arms multiplies the effectiveness and efficiency of Army capabilities used in operations in support of stability.

1-34. Combined arms uses the capabilities of each warfighting function and information in complementary and reinforcing capabilities. Complementary capabilities protect the weaknesses of one system or organization with the capabilities of a different warfighting function. For example, commanders use artillery (fires) to suppress an enemy bunker complex pinning down an infantry unit during tactical movement (movement). The infantry unit then closes with (maneuver) and destroys the enemy. In this example, the fires warfighting function complements the movement and maneuver warfighting function.

Note. Avoid confusing tactical movement with maneuver. Tactical movement is movement in preparation for contact; maneuver is movement while in contact. Actions on contact is the process by which a unit transitions from tactical movement to maneuver. (See chapter 2.)

1-35. Reinforcing capabilities combine similar systems or capabilities within the same warfighting function to increase the function's overall capabilities. In urban operations, for example, infantry, aviation, and armor (movement and maneuver) often operate close to each other. This combination reinforces the protection, maneuver, and direct fire capabilities of each. The infantry protects tanks from enemy infantry and antitank systems; tanks provide protection and firepower for the infantry. Army aviation attack and reconnaissance units maneuver above buildings to observe and fire from positions of advantage, while other aircraft may help sustain the ground elements. Army space-enabled capabilities and services such as communications and global positioning satellites enable communications, navigation, situational awareness, protection, and sustainment of land forces.

1-36. Joint capabilities—such as close air support (see ATP 3-09.32) and special operations forces (see FM 6-05)—can complement or reinforce Army capabilities throughout both the generating force and the operating force. The generating force consists of those Army organizations whose primary mission is to generate and sustain the operational Army's capabilities for employment by joint force commanders. Operating forces consist of those forces whose primary missions are to participate in combat and the integral supporting elements thereof. Often, commanders in the operating force and commanders in the generating force subdivide specific responsibilities. Army generating force capabilities and organizations are linked to operating forces through collocation and reachback.

1-37. Combined arms multiplies Army forces' effectiveness in all operations. Units operating without support of other capabilities generate less combat power and may not accomplish their mission. Employing combined arms requires highly trained Soldiers, skilled leadership, effective staff work, and integrated information systems. Commanders synchronize combined arms through mission command to apply the effects of combat power to the best advantage. They conduct simultaneous combinations of offensive, defensive, and stability tasks to defeat an opponent on land and establish conditions that achieve the commander's end state.

HASTY VERSUS DELIBERATE OPERATIONS

1-38. Army forces are task organized specifically for an operation to provide a fully synchronized combined arms team. That combined arms team conducts extensive rehearsals while also conducting shaping operations to set the conditions for the conduct of the force's decisive operation. Most operations lie somewhere along a continuum between two extremes—hasty operations and deliberate operations. A *hasty operation* is an operation in which a commander directs immediately available forces, using fragmentary orders, to perform activities with minimal preparation, trading planning, and preparation time for speed of execution (ADRP 3-90). A *deliberate operation* is an operation in which the tactical situation allows the development and coordination of detailed plans, including multiple branches and sequels (ADRP 3-90). Determining the right choice involves balancing several competing factors.

1-39. The decision to conduct a hasty or deliberate operation is based on the commander's current knowledge of the enemy situation and assessment of whether the assets available (to include time) and the means to coordinate and synchronize those assets are adequate to accomplish the mission. If they are not, the commander takes additional time to plan and prepare for the operation or bring additional forces to bear on the problem. The commander makes that choice in an environment of uncertainty, which always entails some risk. Ongoing improvements in mission command systems continue to assist in the development of a common operational picture of friendly and enemy forces while facilitating decision-making and communicating decisions to friendly forces. These improvements can help diminish the distinction between hasty and deliberate operations; they cannot make that distinction irrelevant.

1-40. The commander may have to act based only on available *combat information*—unevaluated data, gathered by or provided directly to the tactical commander which, due to its highly perishable nature or the criticality of the situation, cannot be processed into tactical intelligence in time to satisfy the user's tactical intelligence requirements (JP 2-01)—in a time-constrained environment. The commander must understand the inherent risk of acting only on combat information, since it is vulnerable to enemy deception operations and can be misinterpreted. The commander's intelligence staff helps assign a level of confidence to combat information used in decision-making.

1-41. A commander cannot be successful without the capability of acting under conditions of uncertainty while balancing various risks and taking advantage of opportunities. Although a commander strives to maximize knowledge of available forces, the terrain and weather, civil considerations, and the enemy, a lack of information cannot paralyze the decision-making process. A commander who chooses to conduct hasty operations must mentally synchronize the employment of available forces before issuing fragmentary orders. This includes using tangible and intangible factors, such as subordinate training levels and experience, a commander's own experience, perception of how the enemy will react, understanding of time-distance factors, and knowledge of the strengths of each subordinate and supporting unit to achieve the required degree of synchronization. (Refer to ADRP 3-90 for additional information.)

CLOSE COMBAT

1-42. Only on land do combatants routinely and in large numbers come face-to-face with one another. *Close combat* is warfare carried out on land in a direct-firefight, supported by direct and indirect fires, and other assets (ADRP 3-0). Close combat encompasses all actions that place friendly forces in immediate contact with the enemy where the commander uses fire and movement in combination. It can be initiated by our forces or by the enemy.

1-43. The primary mission of the Infantry battalion is to close with the enemy by means of fire and movement in order to destroy, defeat, or capture the enemy, to repel the enemy assault by fire, close combat, and counterattack, or all of these. An Infantry battalion in close combat may—

- Be receiving effective direct fire.
- Have no or only a limited ability to maneuver.
- Be receiving indirect fire.
- Have the entire battalion or one or more of its rifle companies decisively engaged

1-44. Close combat places a premium on leadership, positive face-to-face control, and clear and concise orders. During close combat, leaders have to think clearly, give concise orders, and lead under great stress. Key terms used within this section and throughout this publication include the following:

- *Destroy*—a tactical mission task that physically renders an enemy force combat-ineffective until it is reconstituted. Alternatively, to destroy a combat system is to damage it so badly that it cannot perform any function or be restored to a usable condition without being entirely rebuilt (FM 3-90-1).

- *Defeat*—a tactical mission task that occurs when an enemy force has temporarily or permanently lost the physical means or the will to fight. The defeated force's commander is unwilling or unable to pursue that individual's adopted course of action, thereby yielding to the friendly commander's will and can no longer interfere to a significant degree with the actions of friendly forces. Defeat can result from the use of force or the threat of its use (FM 3-90-1).
- *Direct fire*—fire delivered on a target using the target itself as a point of aim for either the weapon or the director (JP 3-09.3).
- *Fire and movement*—the concept of applying fires from all sources to suppress, neutralize, or destroy the enemy, and the tactical movement of combat forces in relation to the enemy (as components of maneuver, applicable at all echelons). At the squad level, it entails a team placing suppressive fire on the enemy as another team moves against or around the enemy (FM 3-96).
- *Fires*—the use of weapon systems or other actions to create specific lethal or nonlethal effects on a target (JP 3-09).
- *Indirect fire*—fire delivered at a target not visible to the firing unit (TC 3-09.81).
- *Neutralize*—a tactical mission task that results in rendering enemy personnel or materiel incapable of interfering with a particular operation (FM 3-90-1).
- *Suppress*—a tactical mission task that results in the temporary degradation of the performance of a force or weapon system below the level needed to accomplish its mission (FM 3-90-1).
- *Suppression*—the temporary or transient degradation by an opposing force of the performance of a weapons system below the level needed to fulfill its mission objectives (JP 3-01).

OPERATIONS STRUCTURE

1-45. The operations structure—the operations process, warfighting functions, and operational framework—is the Army's common construct for operations. It allows Army leaders to rapidly and effectively organize effort in a manner commonly understood across the Army. The operations process provides a broadly defined approach to developing and executing operations. The warfighting functions provide an intellectual organization for common critical functions (see paragraphs 1-25 to 1-30). The operational framework provides Army leaders with basic conceptual options for visualizing and describing operations. (Refer to ADRP 3-0 for additional information.)

OPERATIONS PROCESS

1-46. The Army's framework for exercising mission command is the *operations process*—the major mission command activities performed during operations: planning, preparing, executing, and continuously assessing the operation (ADP 5-0). The operations process is a commander-led activity, informed by the mission command approach to planning, preparing, executing, and assessing military operations. These activities may be sequential or simultaneous. In fact, they are rarely discrete and often involve a great deal of overlap. The battalion commander, assisted by the battalion staff, uses the operations process to drive the conceptual and detailed planning necessary to understand, visualize, and describe the operational environment, make and articulate decisions; and direct, lead, and assess military operations. (Refer to ADRP 5-0 for additional information.)

1-47. *Planning* is the art and science of understanding a situation, envisioning a desired future, and laying out effective ways of bringing that future about (ADP 5-0). Planning consists of two separate but interrelated components: a conceptual component and a detailed component. Successful planning requires the integration of both these components. Commanders employ three methodologies for planning: the Army design methodology, the military decision-making process (battalion echelon and above), and troop leading procedures (company echelon and below). Commanders determine how much of each methodology to use based on the scope of the problem, their familiarity with it, and the time available. Planning helps the commander create and communicate a common vision between the staff, subordinate commanders, their staffs, and unified action partners. Planning results in an order that synchronizes the action of forces in time, space, and purpose to achieve objectives and accomplish missions.

Note. *Unified action partners* are those military forces, governmental and nongovernmental organizations, and elements of the private sector with whom Army forces plan, coordinate, synchronize, and integrate during the conduct of operations (ADRP 3-0).

1-48. *Preparation* is those activities performed by units and Soldiers to improve their ability to execute an operation (ADP 5-0). The military decision-making process (battalion echelon and above) and troop leading procedures (company echelon and below) drive preparation. Since time is a factor in all operations, the commander and staff conduct a time analysis early in the planning process. This analysis helps them determine what actions they need to take and when to begin those actions to ensure forces are ready and in position before execution. The plan may require the commander to direct subordinates to start necessary movements; conduct task-organization changes; begin reconnaissance, surveillance, and security operations; and execute other preparation activities before completing the plan.

1-49. *Execution* is putting a plan into action by applying combat power to accomplish the mission (ADP 5-0). The battalion commander positions where best to exercise command during execution. This may be forward of the main command post to provide command presence, sense the mood of the unit, and to make personal observations. A position forward of the main command post and near the main effort or decisive operation facilitates an assessment of the situation and decision-making. Staffs synchronize actions, coordinate actions, inform the commander, and provide procedural control to support the commander's ability to assess, use judgment, and make decisions. FM 6-0 describes decision making during execution and describes the rapid decision-making and synchronization process.

1-50. A*ssessment* is the determination of the progress toward accomplishing a task, creating a condition, or achieving an objective (JP 3-0). Assessment is continuous; it precedes and guides every operations process activity and concludes each operation or phase of an operation. Commander and staff conduct assessments by monitoring the current situation to collect information, evaluating progress towards achieving end state conditions or objectives, and recommending or directing action to modify or improve the existing course of action. The commander establishes priorities for assessment in planning guidance; commander's critical information requirements (CCIR) (priority intelligence requirements and friendly force information requirements), essential element of friendly information, and decision points. By prioritizing the effort, the commander avoids excessive analyses when assessing operations.

1-51. Throughout the operations process, the commander integrates their assessments with the staff, subordinate commanders, and other unified action partners. Primary tools for assessing progress of the operation include the operation order, the common operational picture, personal observations, running estimates, and the assessment plan. The latter includes measure of effectiveness (MOE), measure of performance (MOP), and reframing criteria. The commander's visualization forms the basis for the commander's personal assessment of progress. Running estimates provide information, conclusions, and recommendations from the perspective of each staff section. ADRP 5-0 addresses the assessment process during the operations process. (See FM 6-0 for doctrine on developing assessment plans.)

Operational Framework

1-52. The commander and staff use an operational framework, and associated vocabulary, to help conceptualize and describe the concept of operations in time, space, purpose, and resources. An *operational framework* is a cognitive tool used to assist commanders and staffs in clearly visualizing and describing the application of combat power in time, space, purpose, and resources in the concept of operations (ADP 1-01). An operational framework establishes an area of geographic and operational responsibility for the commander and provides a way to visualize how the commander will employ forces against the enemy. To understand this framework is to understand the relationship between the area of operations and operations in *depth*—the extension of operations in time, space, or purpose to achieve definitive results (ADRP 3-0). Proper relationships allow for simultaneous operations and massing of effects against the enemy.

1-53. The operational framework has four components. First, the commander is assigned an area of operation for the conduct of operations. Second, the commander can designate a deep, close, and support areas to describe the physical arrangement of forces in time and space. Third, within this area, the commander conducts decisive, shaping, and sustaining operations to articulate the operation in terms of purpose. Finally, the commander designates the main and supporting efforts to designate the shifting prioritization of resources.

Area of Operation

1-54. An *area of operations* is an operational area defined by a commander for land and maritime forces that should be large enough to accomplish their missions and protect their forces (JP 3-0). In operations, the commander uses *control measures*—a means of regulating forces or warfighting functions (ADRP 6-0)—to assign responsibilities, coordinate maneuver, and control combat operations. Within the area of operation, the commander integrates and synchronizes combat power. To facilitate this integration and synchronization, the commander designates targeting priorities, effects, and timing within the assigned area of operation. Responsibilities within an assigned area of operation include—

- Terrain management.
- Information collection, integration, and synchronization.
- Civil affairs operations.
- Movement control.
- Clearance of fires.
- Security.
- Personnel recovery.
- Airspace control of assigned airspace.
- Minimum-essential stability tasks.

1-55. The commander considers the battalion's area of influence when assigning an area of operation to subordinate commanders. An *area of influence* is a geographical area wherein a commander is directly capable of influencing operations by maneuver or fire support systems normally under the commander's command or control (JP 3-0). Understanding the area of influence helps the commander and staff plan branches to the current operation in which the force uses capabilities outside the area of operations. An area of operation should not be substantially larger than the unit's area of influence. Ideally, the area of influence would encompass the entire area of operations. An area of operations that is too large for a unit to control can allow sanctuaries for enemy forces and may limit joint flexibility.

1-56. An *area of interest* is that area of concern to the commander, including the area of influence, areas adjacent thereto, and extending into enemy territory (JP 3-0). An area of interest for operations in support of stability tasks (see chapter 4) may be much larger than that area associated with the offense and defense.

1-57. Areas of operations may be contiguous or noncontiguous. When they are contiguous, a boundary separates them. When areas of operations are noncontiguous, subordinate commands do not share a boundary. The higher headquarters retains responsibility for the area not assigned to subordinate units. (Refer to ADRP 3-0 for additional information.)

Deep, Close, and Support Areas

1-58. A *deep area* is the portion of the commander's area of operations that is not assigned to subordinate units (ADRP 3-0). Operations in the deep area involve efforts to prevent uncommitted enemy forces from being committed in a coherent manner. The commander's deep area generally extends beyond subordinate unit boundaries out to the limits of the commander's designated area of operations. The purpose of operations in the deep area is frequently tied to other events distant in time, space, or both time and space. Operations in the deep area might disrupt the movement of reserves; cannon, rocket, or missile; and follow-on forces. In an operational environment where the enemy recruits insurgents from a population, deep operations might focus on interfering with the recruiting process, disrupting the training of recruits, or eliminating the underlying factors that enable the enemy to recruit.

1-59. The *close area* is the portion of a commander's area of operations assigned to subordinate maneuver forces (ADRP 3-0). Operations in the close area are operations that are within a subordinate commander's area of operations. The commander plans to conduct decisive operations using maneuver in the close area, and positions most of the maneuver force within it. Within the close area, depending on the echelon, one unit may conduct the decisive operation while others conduct shaping operations. A close operation requires speed and mobility to rapidly concentrate overwhelming combat power at the critical time and place and to exploit success.

1-60. In operations, the commander may refer to a support area. The *support area* is the portion of the commander's area of operations that is designated to facilitate the positioning, employment, and protection of base sustainment assets required to sustain, enable, and control operations (ADRP 3-0). The commander assigns a support area as a subordinate area of operations to support functions. It is where most of the echelon's sustaining operations occur. (See appendix H for additional information on support area operations.)

Decisive, Shaping, and Sustaining Operations

1-61. Decisive, shaping, and sustaining operations lend themselves to a broad conceptual orientation. The *decisive operation* is the operation that directly accomplishes the mission (ADRP 3-0). The decisive operation determines the outcome of a major operation, battle, or engagement. The decisive operation is the focal point around which the commander designs an entire operation. Multiple subordinate units may be engaged in the same decisive operation. Decisive operations lead directly to the accomplishment of a commander's intent. The commander typically identifies a single decisive operation, but more than one subordinate unit may play a role in a decisive operation.

1-62. A *shaping operation* is an operation that establishes conditions for the decisive operation through effects on the enemy, other actors, and the terrain (ADRP 3-0). In combat, synchronizing the effects of aircraft, artillery fires, and obscurants to delay or disrupt repositioning forces illustrates shaping operations. Information operations, for example, may integrate Soldier and leader engagement tasks into the operation to reduce tensions between subordinate units within the battalion and different ethnic groups through direct contact between subordinate leaders and local leaders. Shaping operations may occur throughout the area of operations and involve any combination of forces and capabilities. Shaping operations set conditions for the success of the decisive operation. The commanders may designate more than one shaping operation.

1-63. A *sustaining operation* is an operation at any echelon that enables the decisive operation or shaping operations by generating and maintaining combat power (ADRP 3-0). Sustaining operations differ from decisive and shaping operations in that they focus internally (on friendly forces) rather than externally (on the enemy or environment). Sustaining operations include personnel and logistics support, support area security, movement control, terrain management, and infrastructure development. Sustaining operations occur throughout the area of operations, not just within a support area. Failure to sustain may result in mission failure. Sustaining operations determine how quickly the force can reconstitute and how far the force can exploit success.

1-64. Throughout decisive, shaping, and sustaining operations, the commander and staff ensure that—

- Forces maintain positions of relative advantage.
- Operations are integrated with unified action partners.
- Continuity is maintained throughout operations.

Position of Relative Advantage

1-65. A *position of relative advantage* is a location or the establishment of a favorable condition within the area of operations that provides the commander with temporary freedom of action to enhance combat power over an enemy or influence the enemy to accept risk and move to a position of disadvantage (ADRP 3-0). Positions of relative advantage provide the commander with an opportunity to compel, persuade, or deter an enemy decision or action. The commander maintains the momentum through exploitation of opportunities to consolidate gains and continually assess and reassess friendly and enemy effects for further and future opportunities. The commanders seek positions of relative advantage before combat begins, and exploits success throughout operations.

1-66. The battalion commander understands that positions of advantage are temporary, seeks positions of relative advantage before combat begins, and exploits success throughout operations. As they recognize and gain positions of relative advantage, enemy forces will attempt to gain a position of advantage over the battalion. As such, subordinate units of the battalion leverage terrain to their advantage and pit their strength against a critical enemy weakness. Subordinate units maneuver to a position that provides either positional advantage over the enemy for surveillance and targeting, or a position from which to deliver fires in support of continued movement towards an advantageous position or to break contact.

Integration in Operations

1-67. The commander integrates battalion operations within the larger effort. The commander, assisted by the staff, integratcs numerous processes and activities (see appendix B) within the headquarters and across the force. Integration involves efforts to operate with unified action partners and efforts to conform battalion capabilities and plans to the larger concept. The commander extends the depth of operations through joint integration and multi-domain battle.

> ***Note.*** Army forces conduct multi-domain battle, as part of a joint force, to seize, retain, and exploit control over enemy forces. For example, Army forces use aviation and unmanned aircraft systems in the air domain, and protect vital communications networks in cyberspace, while retaining dominance in the land domain. (Refer to ADRP 3-0 for additional information.)

1-68. When determining an operation's depth, the commander consider the battalion's own capabilities as well as available joint capabilities and limitations. The commander sequences and synchronizes operations in time and space to achieve simultaneous effects throughout an area of operations. The commander seeks to use capabilities within the battalion that complement those of unified action partners. Effective integration requires the staff to plan for creating shared understanding and purpose through collaboration with unified action partners.

Maintaining Continuity in Operations

1-69. Decision making during operations is continuous; it is not a discrete event. The commander balances priorities carefully between current and future operations. The commander seeks to accomplish the mission efficiently while conserving as many resources as possible for future operations. To maintain continuity of operations, the commander and staff ensure they—

- Make the fewest changes possible.
- Facilitate future operations.

1-70. The commander makes only those changes to the plan needed to correct variances. The commander keeps as much of the current plan the same as possible. This presents subordinates with the fewest possible changes. The fewer the changes, the less resynchronization needed, and the greater the chance that the changes will be executed successfully.

1-71. When possible, the commander and staff ensure that changes do not preclude options for future operations. The staff develops options during planning, or the commander infers them based on the staff assessment of the current situation. Developing or inferring options depends on validating earlier assumptions and updating planning factors and staff estimates. The concept of future operations may be war-gamed using updated planning factors, estimates, and assumptions. (See appendix B.) Commanders project the situation in time, visualize the flow of battle, and project the outcomes of future operations and consolidating gains.

Main and Supporting Efforts

1-72. The commander designates main and supporting efforts to establish clear priorities of support and resources among subordinate units. The *main effort* is a designated subordinate unit whose mission at a given point in time is most critical to overall mission success (ADRP 3-0). The main effort is usually weighted with the preponderance of combat power. Typically, the commander shifts the main effort one or more times during execution. Designating a main effort temporarily prioritizes resource allocation. When the commander designates a unit as the main effort, it receives priority of support and resources in order to maximize combat power. The commander establishes clear priorities of support, and shifts resources and priorities to the main effort as circumstances and the commander's intent require. The commanders may designate a unit conducting a shaping operation as the main effort until the decisive operation commences. However, the unit with primary responsibility for the decisive operation then becomes the main effort upon the execution of the decisive operation.

1-73. A *supporting effort* is a designated subordinate unit with a mission that supports the success of the main effort (ADRP 3-0). The commander resources supporting efforts with the minimum assets necessary to accomplish the mission. The force often realizes success of the main effort through the success of the supporting effort(s). (Refer to ADRP 3-0 and FM 3-96 for additional information.)

KEY DOCTRINAL TERMS AND DEFINITIONS

1-74. The following key doctrinal terms and definitions are used throughout this and other chapters and appendixes. Refer to referenced publications for additional information.

- *Commander's critical information requirements*—an information requirement identified by the commander as being critical to facilitating timely decision making (JP 3-0).
- *Decision point*—a point in space and time when the commander or staff anticipates making a key decision concerning a specific course of action (JP 5-0).
- *Essential element of friendly information*—a critical aspect of a friendly operation that, if known by the enemy, would subsequently compromise, lead to failure, or limit success of the operation and therefore should be protected from enemy detection (ADRP 5-0).
- *Friendly force information requirement*—information the commander and staff need to understand the status of friendly and supporting capabilities (JP 3-0).
- *Priority intelligence requirement*—an intelligence requirement, stated as a priority for intelligence support, that the commander and staff need to understand the adversary or other aspects of the operational environment (JP 2-01).

SECTION II – INFANTRY BATTALION

1-75. The Infantry battalion within the IBCT can deploy rapidly and can be sustained by an austere support structure. It conducts operations against conventional and unconventional enemy forces in all types of terrain and climate conditions. The battalion's composition and training uniquely equip it to conduct its mission. In addition to its primary warfighting missions, the Infantry battalion may be tasked to perform other types of operations semi-independently or as an integral part of a larger force. The Infantry battalion within the IBCT can be task-organized as part of an armored brigade combat team (ABCT), a Stryker brigade combat team, a multifunctional brigade, or supporting functional brigade. This section addresses the Infantry battalion's mission, capabilities, limitations, and organization and the exercise of mission command within the battalion. (Refer to FM 3-96 for additional information.)

MISSION

1-76. The Infantry battalion fights and wins engagements to support operational and strategic objectives. The mission of the Infantry battalion is to close with the enemy using fire and movement to destroy or capture enemy forces, or to repel enemy attacks by fire, close combat, and counterattack to control land areas, including populations and resources. The Infantry battalion commander exercises mission command, and directs the operation of the battalion and attached units while conducting decisive action throughout the depth of the battalion's area of operations. Battalion missions, although not inclusive, may include reducing fortified areas, infiltrating and seizing objectives in the enemy's rear, eliminating enemy force remnants in restricted terrain, securing key facilities and activities, and conducting operations in support of stability tasks in the wake of maneuvering forces. Reconnaissance and surveillance (R&S) tasks and security operations remain a core competency of the Infantry battalion and Infantry rifle company, platoon, and squad.

CAPABILITIES

1-77. The Infantry battalion is an expeditionary formation optimized for dismounted operations in *complex terrain*—a geographical area consisting of an urban center larger than a village and/or of two or more types of restrictive terrain or environmental conditions occupying the same space (ATP 3-34.80). The Infantry battalion conducts entry operations by ground, airland, air assault, or amphibious assault (via surface and vertical) into austere areas of operations with little or no advanced notice. Airborne Infantry battalions can conduct vertical envelopment by parachute assault. The Infantry battalion is particularly effective in urban terrain, where subordinate Infantry units can infiltrate and move rapidly to the rear of enemy positions. The commander can enhance tactical mobility using rotatory and fixed wing airlift. The Infantry battalion's capabilities include—

- Strategic and operational deployability (may be the dominant arm during initial entry phase).
- Entry operations to gain the initiative early, seize and hold ground, and mass fires to stop the enemy.

- Forcible entry operations, through airborne assault (airborne Infantry battalions), air assault, and amphibious operations.
- Offensive and defensive tasks and tactical enabling tasks in all types of environments.
- Screen and guard missions against similarly equipped enemy forces.
- Dismounted operations in restrictive or severely restrictive terrain.
- Transportable by Army aviation brigades.
- Enhanced situational awareness, including a common operational picture down to company commander level (and platoon leaders assigned wheeled vehicles).
- Reduced logistics requirement compared to ABCT and Stryker brigade combat team.
- Sustainment provided by its supporting forward support company.

LIMITATIONS

1-78. Combat power and protection, relative to the Stryker brigade combat team and ABCT vehicle platforms, are the battalion's primary limitations. While insertion means vary, all Infantry battalions are comprised mostly of foot-mobile Soldiers, and require organic or supporting unit vehicles for increased mobility of troops. However, the command posts, weapons company, scout platoon, mortar platoon, sustainment assets, and other battalion elements have wheeled vehicles. Other limitations include—

- Vulnerability to enemy armor, artillery, and air assets in open terrain.
- Limited decontamination capability.
- No assigned information operations expertise.

ORGANIZATION OF THE INFANTRY BATTALION

1-79. The Infantry battalion of the IBCT is the Army's most numerous, versatile, and adaptive combat formation. The Infantry battalion can deploy rapidly and strategically because of its relatively light organization. As a result, the battalion often conducts the initial entry into an operational area as part of a mission-tailored joint initial entry force. The Infantry battalion has a headquarters and headquarters company; three Infantry rifle companies, each with three Infantry rifle platoons; and an Infantry weapons company. The combination of rifle companies, a weapons company, and battalion scouts, mortars, and snipers, allow the commander to internally task-organize as needed. Figure 1-1 shows the internal organization of the Infantry battalion.

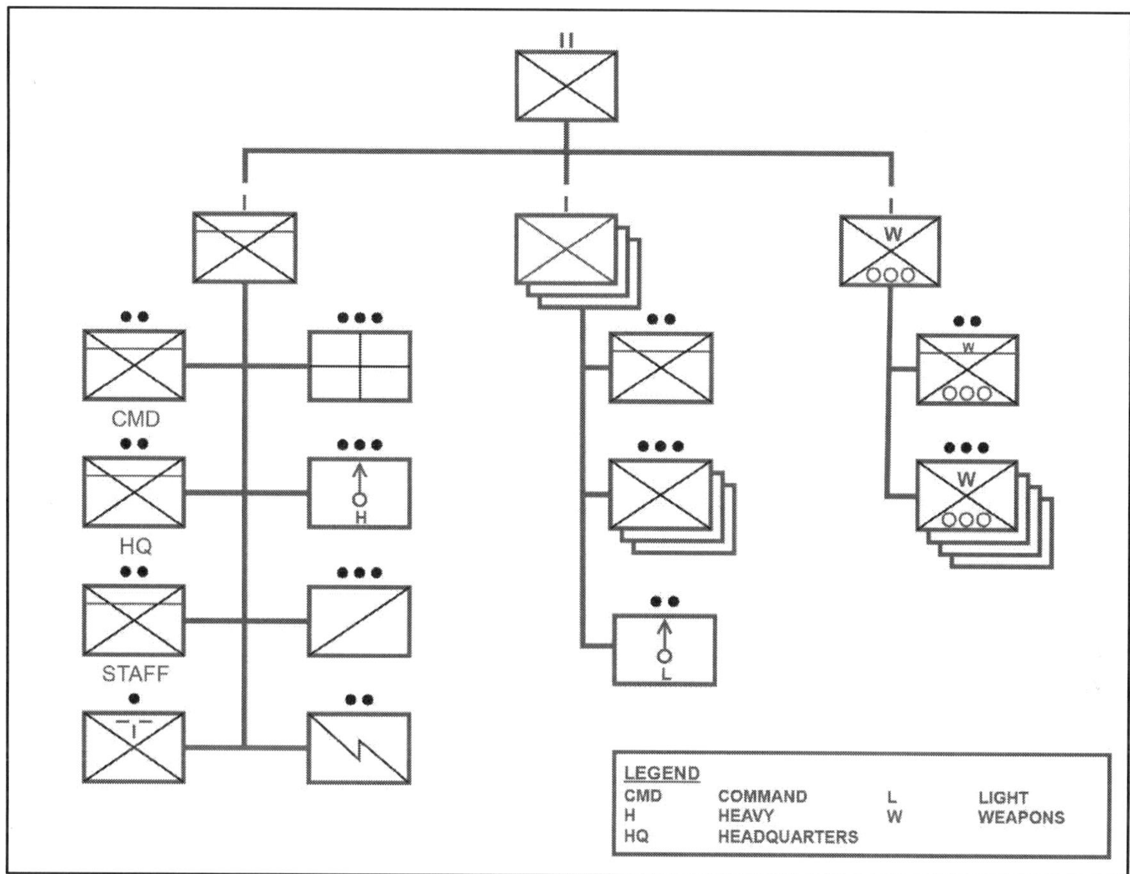

Figure 1-1. Infantry battalion

HEADQUARTERS AND HEADQUARTERS COMPANY

1-80. The headquarters and headquarters company consists the battalion command section, battalion staff sections, the company headquarters, the battalion's medical, mortar, and scout platoons, signal and retransmission sections, and sniper squad. The headquarters and headquarters company enables the exercise of mission command by the commander through the headquarter sections, staff sections and cells, specialty platoons and squad, and other attached support elements.

Headquarters

1-81. The battalion command section consists of the battalion commander, the battalion executive officer (XO), and the battalion command sergeant major. The battalion command section has several wheeled vehicles to assist with mission command and transportation of command section personnel, but during execution of their duties, they are often on foot.

1-82. The battalion echelon is the first level of command that includes an assigned staff. The basic staff structure includes various staff sections, by area of expertise, under a coordinating, special, or personal staff officer. The staff supports the commander in the exercise of mission command, assists subordinate units, and informs units and organizations outside the battalion.

Headquarters Company

1-83. The headquarters company consists of the company headquarters and the battalion's medical, mortar, and scout platoons, the signal and retransmission sections, and sniper squad. Company headquarters personnel, and specialty platoons, sections, and squads provide the commander with flexibility across the range of military operations.

Company Headquarters

1-84. The company headquarters of the headquarters company provides leadership, human resources and supply support to headquarters and headquarters company. The headquarters company includes the headquarters and headquarters company commander, executive officer, first sergeant, and supporting supply and signal personnel. The headquarters and headquarters company commander, first sergeant, and executive officer do not have a set location to conduct their duties; they are positioned to best support the battalion's mission.

Battalion Medical Platoon

1-85. The mission of the medical platoon is to provide Role 1 Army Health System support to the Infantry battalion. The medical treatment platoon is organic to the battalion and is the unit's level Role 1 medical treatment facility, usually referred to as the battalion aid station. Role 1 medical treatment is provided by the combat medic or by the physician, the physician assistant, or the health care specialist in the battalion aid station. (Refer to FM 4-02 for additional information.) Roll 1 includes the following:

- Tactical combat casualty care (immediate far forward care) consists of those lifesaving steps that do not require the knowledge and skills of a physician. The combat medic is the first individual in the medical chain that makes medically substantiated decisions based on medical military occupational specialty-specific training.
- At the battalion aid station, the physician and the physician assistant are trained and equipped to provide advanced trauma management to the combat casualty. The aid station supports casualty collection and medical evacuation from supported units to the battalion aid station or forward aid station. Aid station elements also conducts routine sick call when the operational situation permits.

1-86. The medical platoon habitually establishes the battalion aid station where it can best support the battalion and has the capability to split into a battalion aid station and a forward aid station for wider area coverage. The medical platoon is dependent on the Army Health System for direct support and augmentation/reinforcement. The battalion surgeon or, in the battalion surgeon's absence, the physician assistant, is the medical advisor to the battalion commander and, at the commander's discretion, the medical platoon leader. The medical platoon is dependent upon the maneuver elements to which it is assigned for all logistic support, with the exception of Class VIII (medical) supplies. The battalion medical platoon is configured with a headquarters element, medical treatment squad, ambulance squad (ground), and combat medic section (figure 1-2).

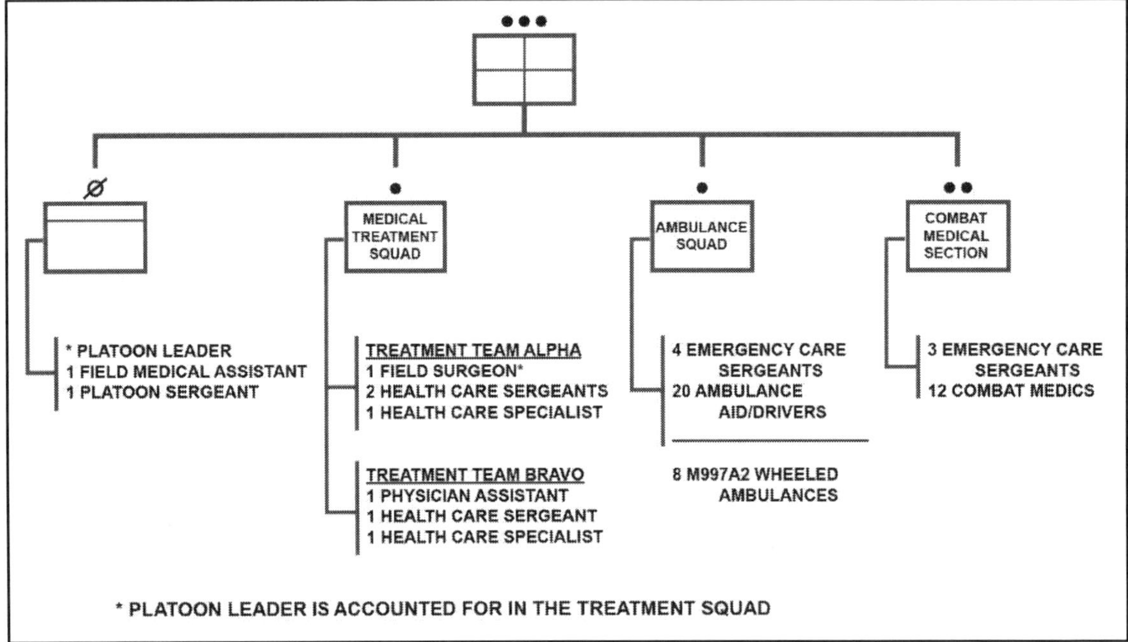

Figure 1-2. Infantry battalion medical platoon

1-87. The headquarters element provides mission command and coordinates resupply for the medical platoon. The platoon headquarters consists of the field surgeon/medical platoon leader, the field medical assistant, and the platoon sergeant. It normally collocates with the treatment squad to form the battalion aid station. The battalion surgeon, assisted by the field medical assistant and the platoon sergeant, is responsible for the force health protection plan for the Infantry battalion. The field medical assistant is the operations/readiness officer and plans, coordinates, and executes the force health protection plan. During periods where the Infantry battalion is not deployed, the battalion surgeon is normally attached to the local medical care facility. As a result, the platoon's field medical assistant/operations/readiness officer often serves as the medical platoon leader.

1-88. The treatment squad consists of two teams (treatment team alpha and team bravo). The treatment squad operates the battalion aid station and provides Role 1 medical care and treatment (to include sick call, tactical combat casualty care, and advance trauma management). When established, team alpha is clinically staffed with the battalion surgeon while team bravo is clinically staffed with the physician assistant. Each team is supported by a health care sergeant and two health care specialists and can operate for a limited amount of time in split-based operations in direct support of battalion subordinate units for wider area coverage. Each team employs treatment vehicles with two medical equipment sets: one trauma field and one sick call field.

1-89. Medical platoon ambulances provide medical evacuation and en route care from the Soldier's point of injury, the casualty collection point, or an ambulance exchange point to the battalion aid station. The ambulance squad is four teams of two ambulances composed of one emergency care sergeant and two ambulance aide/drivers assigned to each ambulance. The ambulance team in support of the maneuver company works in coordination with the trauma specialists supporting the platoons. In mass casualty situations, nonmedical vehicles may be used to assist in *casualty evacuation*—nonmedical unit use this to refer to the movement of casualties aboard nonmedical vehicles or aircraft without en route medical care (FM 4-02)—as directed by the supported commander. Plans for the use of nonmedical vehicles to perform casualty evacuation should be included in the Infantry battalion's tactical standard operating procedures or operations order.

1-90. Combat medics are normally allocated to the supported maneuver companies on a basis of one emergency care sergeant per company plus one combat medic per platoon. The medical platoon's emergency care sergeants normally locate with, or near, the maneuver company commander or first sergeant to provide guidance and direction to the subordinate platoon combat medics. The platoon's combat medic locates with, or near, the platoon leader or platoon sergeant. (Refer to ATP 4-02.3 for additional information.)

Battalion Mortar Platoon

1-91. The battalion mortar platoon's primary role is to provide immediate, responsive, indirect fires in support of maneuver companies or the battalion. The battalion mortar platoon consists of a mortar platoon headquarters, a fire direction center section, and four mortar squads (figure 1-3, page 1-18). The platoon's fire direction center controls and directs a mortar platoon's fires. Infantry battalion mortar squads are equipped with 120-mm and 81-mm mortars, but are only authorized enough personnel to operate one of the systems at any one time (arms room concept). The battalion mortar platoon carries the four medium mortars (81-mm) for dismounted operations. The mortar platoon is equipped with trucks and trailers to carry mortars (For more information, refer to the fires appendix of this publication and ATTP 3-21.90.)

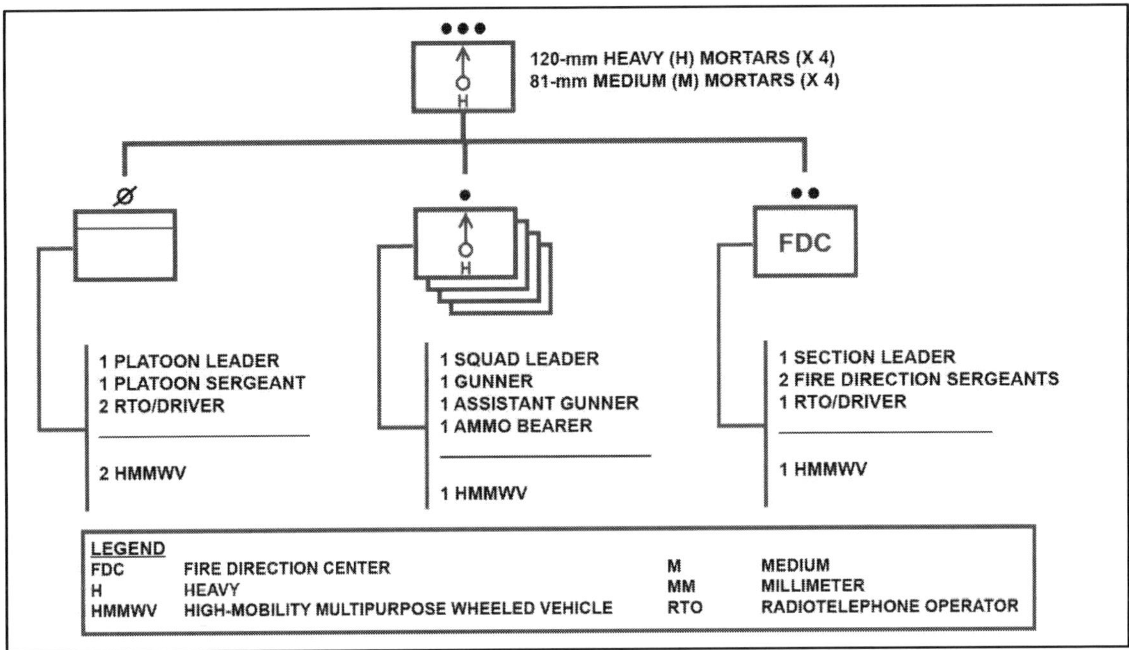

Figure 1-3. Infantry battalion mortar platoon

1-92. The mortar platoon provides the commander with the ability to shape the Infantry's close fight with indirect fires that—

- Provide close supporting fires for assaulting Infantry forces in any terrain.
- Destroy, neutralize, suppress, or disrupt enemy forces and force armored vehicles to button up.
- Fix enemy forces or reduce the enemy's mobility.
- Canalize enemy assault forces into engagement areas.
- Deny the enemy the advantage of defile terrain and force the enemy into areas covered by direct fire weapons.
- Optimize indirect fires in urban terrain.
- Significantly improve the Infantry's lethality and survivability against a close dismounted assault.
- Provide obscuration for friendly movement.

1-93. Each mortar system can provide three primary types of mortar fires as follows:

- High explosive rounds are used to suppress or destroy enemy dismounted Infantry, mortars, and other supporting weapons and to interdict the movement of men, vehicles, and supplies in the enemy's forward area. Bursting white phosphorus rounds are often mixed with high explosive rounds to enhance their suppressive and destructive effects. [See ATP 3-09.32 for information on precision guided mortar munition (120-mm only)]
- Obscuration rounds are used to conceal friendly forces as they maneuver or assault, and to blind enemy supporting weapons. Obscurants can also be used to isolate a portion of the enemy force while it is destroyed piecemeal. Some mortar rounds use bursting white phosphorus to achieve this obscuration. Bursting white phosphorus may be used to mark targets for engagement by other weapons, usually aircraft, and for signaling.
- Illumination rounds, to include infrared illumination, are used to reveal the location of enemy forces hidden by darkness. They allow the commander to confirm or deny the presence of the enemy without revealing the location of friendly direct fire weapons. Illumination fires are often coordinated with high explosive fires both to expose the enemy and to kill or suppress the enemy.

Battalion Scout Platoon

1-94. The battalion scout platoon is the battalion commander's primary reconnaissance asset. The scout platoon is a specialty platoon comprised of the most tactically and technically proficient Infantry Soldiers in the battalion.

Unlike traditionally Infantry rifle platoons whose primary mission is to kill or capture the enemy; the scout platoon's primary mission is to provide the battalion commander information about the enemy. The Infantry battalion commander uses the scout platoon to gather critical information regardless of which element of decisive action (offense, defense, or stability) currently dominates. The battalion commander and staff use this information throughout the operations process (plan, prepare, execute, and assess) during the conduct of operations. The following paragraphs discuss the role, organization, missions, and employment of the scout platoon. (Refer to ATP 3-20.98 for additional information.)

1-95. While the battalion commander determines the role of the scout platoon, the leaders within the scout platoon must understand how a battalion operates in the tactical environment. Leaders utilize stealth in every mission; exercise initiative in the absence of guidance; and are knowledgeable, resourceful, dependable, and disciplined. They must know their duties and responsibilities to the battalion—to provide accurate and timely information. They know the enemy's tactics, techniques, and equipment and understand the importance of their mission to the battalion—and what is required to accomplish that mission.

1-96. *Reconnaissance* is a mission undertaken to obtain, by visual observation or other detection methods, information about the activities and resources of an enemy or adversary, or to secure data concerning the meteorological, hydrographic, or geographic characteristics of a particular area (JP 2-0). *Surveillance* is the systematic observation of aerospace, surface, or subsurface areas, places, persons, or things, by visual, aural, electronic, photographic, or other means (JP 3-0). By performing R&S tasks, the scout platoon enables the battalion commander to maneuver companies, concentrate combat power, and prevent surprise by providing the commander with current and continuous combat information. Refer to FM 3-96, chapter 4 for a detailed discussion of reconnaissance and surveillance missions, to include reconnaissance fundamentals and the five forms of reconnaissance.

1-97. *Security operations* are those operations undertaken by a commander to provide early and accurate warning of enemy operations, to provide the force being protected with time and maneuver space within which to react to the enemy, and to develop the situation to allow the commander to effectively use the protected force (ADRP 3-90). By performing security operations (specific to screen and guard missions) the scout platoon protects the Infantry battalion from surprise and reduce the unknowns in any situation. A *screen* is a security task that primarily provides early warning to the protected force (ADRP 3-90). A *guard* is a security task to protect the main force by fighting to gain time while also observing and reporting information and preventing enemy ground observation of and direct fire against the main body. Units conducting a guard mission cannot operate independently because they rely upon fires and functional and multifunctional support assets of the main body (ADRP 3-90). As a shaping operation, economy of force is often a condition of tactical security operations. Refer to FM 3-96, chapter 4 for a detailed discussion of security operations, to include the five fundamentals of security operations and the five security operations tasks.

Note. The main difference between the conduct of security operations, and reconnaissance and surveillance missions, is that the conduct of security orients on the force or facility being protected, while reconnaissance and surveillance are enemy and terrain oriented. Reconnaissance and surveillance missions and security operations techniques are addressed throughout this publication.

1-98. The Infantry battalion scout platoon (figure 1-4, page 1-20) is organized into a platoon headquarters and three squads of six men each. The scout platoon is equipped with individual weapons, night vision devices, and communications equipment specific to their R&S missions. The platoon's mission and geographic location may require a modified table of organization and equipment.

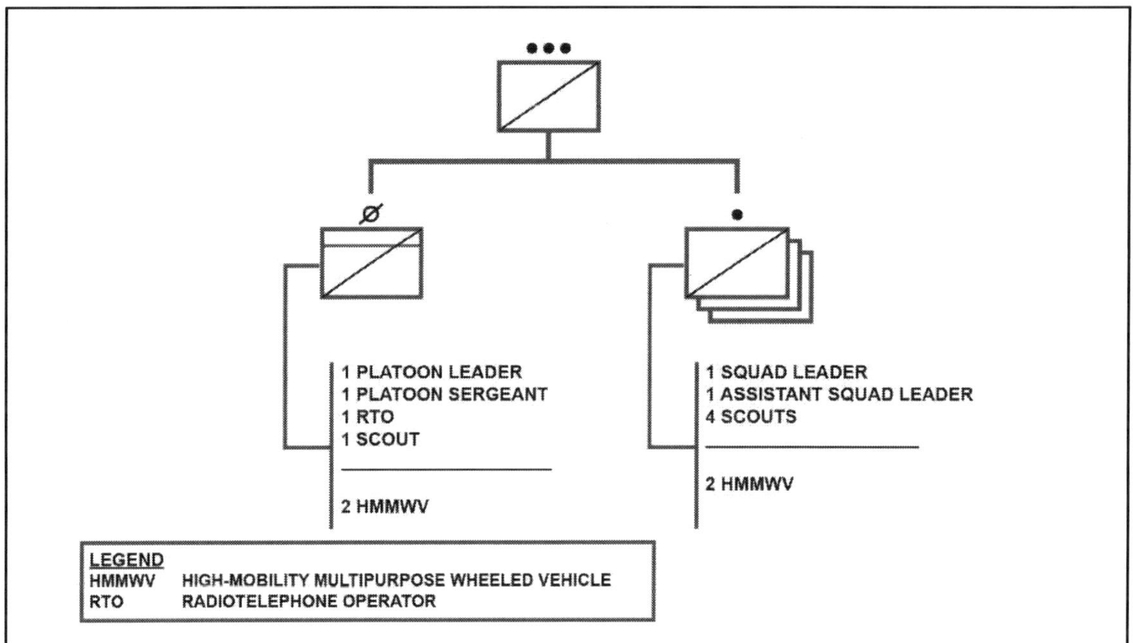

Figure 1-4. Infantry battalion scout platoon

1-99. The scout platoon headquarters provides leadership and control of platoon operations. It consists of the platoon leader, platoon sergeant, and a radiotelephone operator. The platoon headquarters controls and employs attachments provided by battalion.

1-100. Each squad consists of a squad leader, assistant squad leader, and Soldiers. The squads perform R&S missions and security operations as directed by the platoon leader. Squads also assist in tactical control coordination.

1-101. The Infantry battalion scout platoon is organized, equipped, and trained to conduct R&S, and limited security operations for its parent battalion. The platoon's primary mission is to provide combat information. The scout platoon also assist in the tactical control, movement, and positioning/repositioning of the battalion's companies and platoons. The scout platoon is employed under battalion control, but it may be detached for a specific operation. The scout platoon can be tasked to—

- Conduct zone reconnaissance.
- Conduct area reconnaissance.
- Conduct route reconnaissance.
- Screen (and to a lesser degree guard) within the platoon's capabilities.
- Conduct surveillance of critical areas of interest.
- Linkup and conduct liaison.
- Guide maneuver forces.
- Conduct chemical detection and radiological survey and monitoring.

1-102. The platoon leader has overall responsibility for ensuring that the platoon accomplishes its mission. The platoon leader must know the tactical strengths and weaknesses of the platoon, and must determine the most effective and efficient method of employing the platoon. The scout platoon plans, prepares, and executes its assigned missions with the assistance of the battalion staff. Refer to ATP 3-21.8 for information on reconnaissance patrolling. Primary and special staff officers provide expertise for a particular warfighting function as follows:

- Battalion intelligence staff office (S-2) provides information on the enemy and terrain.
- Battalion operations staff officer (S-3) assigns missions and integrates the scout platoon into the battalion plan.
- Battalion fire support officer ensures that aviation, artillery, and mortar fires support the scout platoon's plan.

- Battalion logistics staff officer (S-4) ensures that the logistical requirements of the platoon are met.
- Battalion signal officer ensures that the platoon's communications requirements are met.

Battalion Sniper Squad

1-103. The battalion sniper squad, like the scout platoon, is comprised of the most tactically and technically proficient Infantry Soldiers in the battalion. The Infantry battalion sniper squad consists of a squad leader, three three-man sniper teams, three long-range sniper rifle systems, and three standard sniper rifle systems. The primary mission of the sniper squad in combat is to support combat operations by delivering precise long-range fire on selected targets. The secondary mission of the sniper squad is collecting and reporting combat information.

1-104. The sniper squad employs in support of offensive and defensive tasks, and operations in support of stability tasks in which precision fire is delivered at long range. Sniper teams engage and destroy high payoff targets, they create casualties among enemy troops, slow enemy movement, frighten enemy Soldiers, lower morale, and add confusion to the enemy's operation. Employment includes combat patrols, ambushes, counter sniper operations, forward observation elements, operations in urban terrain, and retrograde operations in which snipers are part of forces left in contact or as stay-behind forces.

1-105. The squad leader is the primary advisor to the battalion commander on sniper employment. If the commander does not directly control sniper employment, the commander should designate a sniper employment officer over the sniper squad. The HHC commander or executive officer could provide this role. Each of the three sniper teams in the sniper squad organize with three personnel; the sniper, the observer, and security. As a result, the sniper squad can effectively employ three sniper teams at any one time, although the commander could employ up to five ad hoc sniper teams for limited duration missions by employing two man teams. Sniper teams can be task organized to any unit in the battalion or employed directly under battalion control. Snipers are most effective when leaders in the supported unit understand capabilities, limitations, and tactical employment of sniper teams. (See appendix E for a detailed discussion.)

INFANTRY RIFLE COMPANY

1-106. The composition and training of the Infantry rifle company uniquely enable it to conduct missions against conventional and unconventional enemy forces in all types of terrain and climate conditions. The Infantry rifle company has three Infantry rifle platoons, a mortar section, and a headquarters section (figure 1-5, page 1-22). Each rifle platoon has three Infantry rifle squads and a weapons squad. The mortar section has two squads, each with a 60-mm mortar. Habitual attachments to the Infantry rifle company includes: a fire support officer at company level and a forward observer at platoon level, and combat medics assigned to rifle platoons and an emergency care sergeant (commonly referred to as the senior medic) assigned to the rifle company.

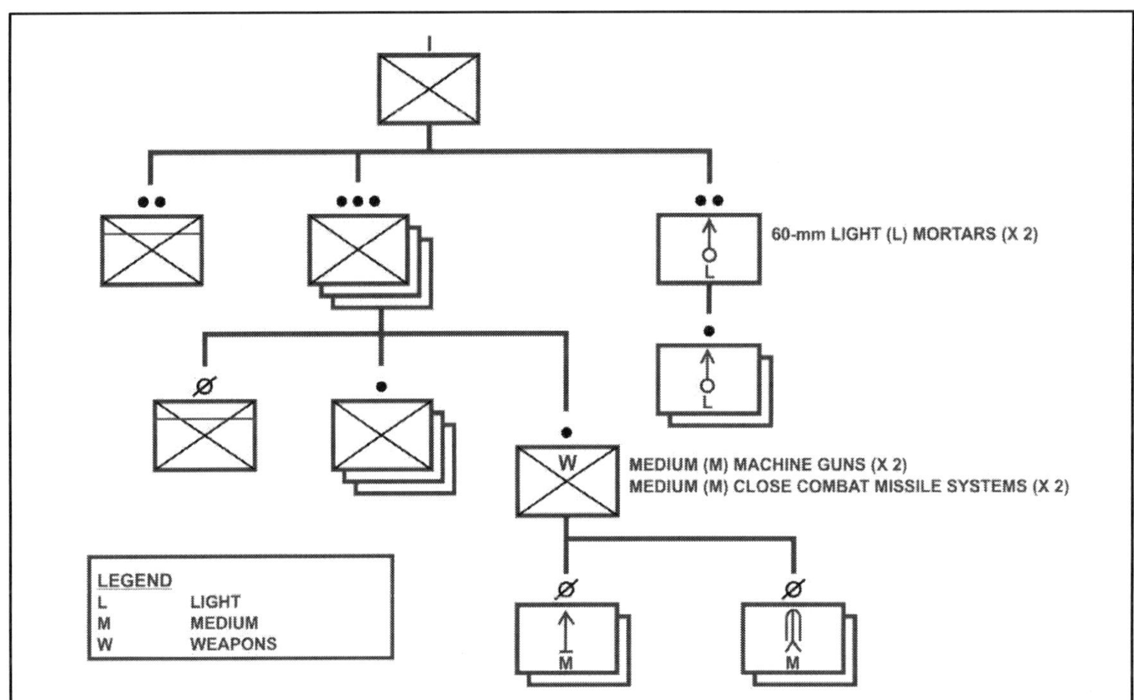

Figure 1-5. Infantry rifle company

1-107. The Infantry rifle company can be task-organized based on the mission variables of METT-TC. Its effectiveness increases through the synergy of combined arms including elements from the Infantry weapons company, tanks, Bradley fighting vehicles and Stryker vehicles, engineers, and other support elements. Effective application of the Infantry rifle company as a combined arms force can capitalize on the strengths of the team's elements while minimizing its respective limitations. The commander can enhance the Infantry rifle company's tactical mobility by providing available transportation assets, such as trucks and mine resistant ambush protected vehicles (MRAPs), and by using helicopters and tactical airlift. (Refer to FM 3-21.10 for additional information.)

INFANTRY WEAPONS COMPANY

1-108. The Infantry weapons company provides the Infantry battalion with mobile heavy weapons. Heavy weapons units can suppress, fix, or destroy enemy forces, allowing other Infantry units or combined arms teams to maneuver. The Infantry weapons company has a company headquarters and four assault platoons. Each assault platoon has two sections of two squads and a leader's vehicle. Each squad contains four Soldiers and a vehicle mounting the heavy weapons. The organization of the weapons company permits its flexible employment. Assault platoons of the weapons company are controlled by the weapons company commander or attached to rifle companies. (See appendix D for a detailed discussion.)

FORWARD SUPPORT COMPANY

1-109. The Infantry battalion receives its logistics support through a forward support company (FSC), less medical supplies. The FSC commander is the senior logistician for the Infantry battalion for general supply, distribution, and maintenance. The FSC commander assists the battalion S-4 with the battalion concept of support and is responsible for executing logistics support according to the Infantry battalion and brigade support battalion (BSB) commanders' guidance. Integrating the logistics plan early into the battalion S-3's operational plan will help to mitigate logistic shortfalls, and support the commander to seize, retain, and exploit gains.

1-110. The FSC has a headquarters section, a distribution platoon, and a maintenance platoon (figure 1-6). The headquarters' food service section provides class I support. This section provides food service and food preparation for the battalion. The food service section prepares, serves and distributes the full range of operational rations. The distribution platoon consists of a platoon headquarters, and four squads that can be task organized to distribute class II, III, IV, V, and VII. The maintenance platoon varies based upon the assigned or attached equipment and major weapon systems within the Infantry battalion. Generally, the maintenance platoon consists of a platoon headquarters, maintenance control section, field maintenance section, service and recovery section and the field maintenance teams. (Refer to ATP 4-90 for additional information.)

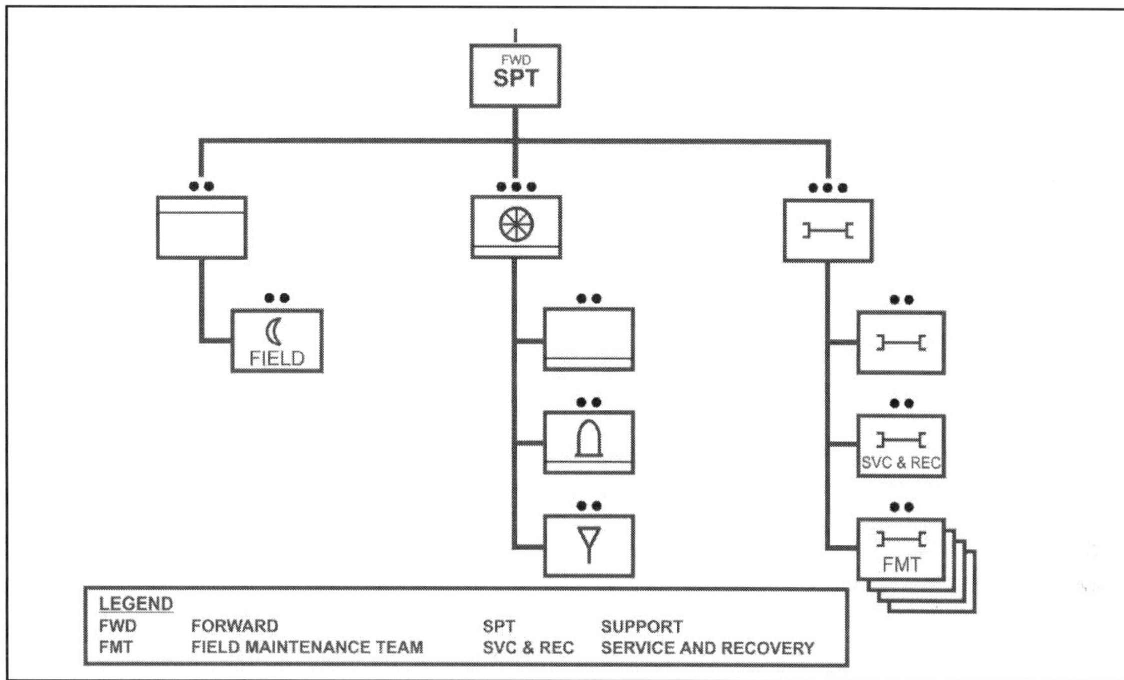

Figure 1-6. Forward support company supporting the Infantry battalion

Notes:

1. The BSB has six organic FSCs that provide direct support to each of the IBCT maneuver battalions and squadron, the field artillery battalion, and the brigade engineer battalion. The FSCs are the link from the BSB to the supported battalions and squadron and are the organizations that provide the IBCT the greatest flexibility for providing logistics support. The BSB provides administrative support, some logistic support, and technical oversight to the FSCs.

2. FSCs receive technical logistic directions from the BSB commander. This allows the BSB commander and the BSB support operations officer to task organize FSCs and cross-level assets amongst FSCs when it is necessary to weight logistics support within the IBCT. The task organization of the FSCs is a collaborative, coordinated effort that involves analysis by the staff and consensus amongst all commanders within the IBCT.

3. The IBCT commander may attach or place a FSC under operational control of its supported battalion or squadron. Upon the advice of the BSB commander, the IBCT commander decides to establish these types of command relationships. All commanders must understand that these types of command relationships limit the BSB commander's, and ultimately the IBCT commander's, flexibility to support the IBCT.

4. The FSC attachment or operational control to its supported battalion or squadron is generally limited in duration and may be for a specific mission or phase of an operation. Regardless of what command relationship is determined for the FSCs, they must retain their technical relationship with the BSB commander. (Refer to FM 3-96 for additional information.)

EXERCISE OF MISSION COMMAND

1-111. To function effectively and have the greatest chance for mission accomplishment, the battalion commander, supported by the staff, exercises mission command throughout the conduct of operations. In this discussion, the "exercise of mission command" refers to an overarching idea that unifies the mission command philosophy of command and the mission command warfighting function. The exercise of mission command encompasses how the commander, supported by the staff, applies the foundational mission command philosophy together with the mission command warfighting function, guided by the principles of mission command.

1-112. The commander uses the mission command philosophy to exploit and enhance uniquely human skills within complex and ever-changing operational environments. Understanding the fundamental nature and philosophy of mission command is essential to the effective conduct of operations. The commander's ability to visualize relationships among opposing human wills is essential to understanding the fundamental nature of operations. To account for the uncertain nature of operations, mission command (as opposed to detailed command) tends to be decentralized and flexible. This uncertain nature requires an environment of mutual trust and shared understanding among commanders, subordinates, and partners.

1-113. The mission command warfighting function enables the commander to balance the art of command and the science of control in order to integrate other warfighting functions. The exercise of mission command enables a shared understanding of an operational environment and the commander's intent. Information operations focus on protecting information, information systems, and decision-making, enhances the commander's ability to integrate other elements of combat power and create necessary shared understanding. At the same time, information operations seek to degrade the enemy's decision-making ability and the enemy's ability to degrade friendly decision-making.

1-114. Information operations supports the accomplishment of several mission command warfighting tasks, including inform and influence audiences inside and outside an organization, conduct knowledge management and information management, synchronize information-related capabilities, and conduct cyberspace electromagnetic activities. Informing and influencing are effects that occur in the cognitive dimension of the information environment. By effectively synchronizing an *information-related capability*—a tool, technique, or activity employed within a dimension of the information environment that can be used to create effects and operationally desirable conditions (JP 3-13)—and, when appropriate, conducting cyberspace electromagnetic activities, the commander can tailor the battalion's influence and manner of informing to the situation and audience at hand. Information and knowledge management support the commander and staff's ability to access information quickly and completely, as well as segment and protect information and mission command systems, thereby enhancing decision-making and gaining an advantage over adversaries and enemies.

Note. Appendix A and B address various factors that degrade the efficiency of mission command systems within the battalion and the techniques and procedures to operate in a degraded environment specific to enemy and friendly capabilities and environmental factors such as terrain and weather.

1-115. This section describes the exercise of mission command, to include duties and responsibilities, for the battalion commander, executive officer, and command sergeant major, and the staff within the Infantry battalion headquarters. See ADRP 6-0, FM 6-0, and FM 3-96 for further information on the exercise of mission command as it refers to the overarching idea that unifies the mission command philosophy of command and the mission command warfighting function.

Note. Refer to FM 3-21-10 or ATP 3-21.8 for duties and responsibilities of key leaders and organizations within the Infantry rifle company or Infantry rifle platoon and squad, respectively.

BATTALION COMMANDER

1-116. The battalion commander uses mission command, the exercise of authority and direction, to seize, retain, and exploit the initiative through mission orders. *Mission orders* are directives that emphasize to subordinates the results to be attained, not how they are to achieve them (ADP 6-0). The commander focuses the order on the purpose of the operation through mission orders. The commander, assisted by the staff, uses the guiding principles of mission command to balance the *art of command,* which is the creative and skillful exercise of authority through timely decision making and leadership (ADP 6-0), with the *science of control,* which is the systems and procedures used to improve the commander's understanding and support accomplishing missions (ADP 6-0). The six principles of mission command are—

- Build cohesive teams through mutual trust.
- Create shared understanding.
- Provide a clear commander's intent.
- Exercise disciplined initiative.
- Use mission orders.
- Accept prudent risk.

1-117. Mission command—as a warfighting function—assists the commander in balancing the art of command with the science of control, while emphasizing the human aspects of mission command. The commander encourages disciplined initiative through a clear commander's intent while providing enough direction to integrate and synchronize the battalion at the decisive place and time. To this end, the commander performs three primary mission command warfighting function tasks. Commander tasks are—

- Drive the operations process through the activities of understanding, visualizing, describing, directing, leading, and assessing operations.
- Develop teams, both within the battalion and with joint, interagency, and multinational partners.
- Inform and influence audiences, inside and outside the battalion.

1-118. The battalion staff supports the commander in the exercise of mission command by performing four primary mission command warfighting function tasks. Staff tasks are—

- Conduct the operations process: plan, prepare, execute and assess.
- Conduct knowledge management and information management.
- Synchronize information-related capabilities.
- Conduct cyberspace electromagnetic activities.

1-119. Five additional tasks reside within the mission command warfighting function. These commander-led and staff-supported additional tasks are—

- Conduct military deception.
- Conduct civil affairs operations.
- Install, operate, and maintain the network.
- Conduct airspace control.
- Conduct information protection.

1-120. Command presence, as a philosophy, requires the commander to lead from a position that allows timely decisions based on an operational environment assessment of the operational environment and application of judgment. Depend on the situation, the battalion commander may find it necessary to locate forward of the main or tactical command posts. For example, the commander may position with the main effort to gain understanding, prioritize resources, influence others, and mitigate risk or support by fire positon to better control assault forces. To do this, the battalion commander must understand how the fundamental principles of mission command guide and help balance the art of command with the science of control.

Battalion Executive Officer

1-121. The battalion executive officer is the commander's principal assistant and directs staff tasks, manages and oversees staff coordination, and special staff officers. The battalion commander normally delegates executive management authority to the executive officer. The executive officer provides oversight of sustainment planning and operations for the battalion commander. As the key staff integrator, the executive officer frees the commander from routine details of staff operations and the management of the headquarters and ensures efficient and prompt staff actions. The executive officer exercises the duties and responsibilities of the second in command.

Note. The executive officer is the senior knowledge management officer in the battalion and advises the commander on knowledge management policy. The executive officer is responsible for directing the activities of each staff section and subordinate unit to capture and disseminate organizational knowledge. When staffed, a knowledge management officer, working through the executive officer, is responsible for developing the knowledge management plan that integrates and synchronizes knowledge and information management within the battalion. (Refer to FM 3-96 and ATP 6-01.1 for additional information.)

BATTALION COMMAND SERGEANT MAJOR

1-122. The battalion command sergeant major is the senior noncommissioned officer within the Infantry battalion. The command sergeant major carries out policies and enforces standards for the performance, training, and conduct of enlisted Soldiers. The command sergeant major provides advice and initiates recommendations to the commander and staff in matters pertaining to enlisted Soldiers. In operations, the battalion commander employs the command sergeant major throughout the battalion's area of operations to extend command influence, assess the morale of the force, and assist during critical events.

BATTALION STAFF

1-123. The battalion staff supports the commander, assist subordinate units, and informs units and organizations outside the headquarters. The staff supports the commander's understanding, making and implementing decisions, controlling operations, and assessing progress. The battalion staff makes recommendations and prepares plans and orders for the commander. The staff establishes and maintains a high degree of coordination and cooperation with staffs of higher, supporting, supported, and adjacent units. The battalion staff does this by actively collaborating and communicating with commanders and staffs of other units to solve problems. The staff keeps civilian organizations informed with relevant information according to their security classification as well as their need to know. The staff structure for the battalion includes the executive officer and a grouping of staff members by area of expertise under a coordinating, personal, or special staff officer. (Refer to FM 6-0 for additional information.)

Note. The basis for staff organization depends on the mission and the activities to accomplish. These activities determine how the commander organizes, tailors, or adapts individual staffs to accomplish the mission. The mission also determines the size, composition, and location of a staff, including the establishment of integrating and functional cells, elements, and staff augmentation. Within command posts, the commander cross-functionally organize the staff into command post cells and staff sections to assist in the exercise of mission command. Regardless of mission, every staff section, cell, or element has common broad areas of expertise that determine how the commander organizes within each command post (see appendix A) to perform essential staff functions to aid with planning and controlling operations.

Coordinating Staff Officers

1-124. Coordinating staff officers are the battalion commander's principal assistants who advise, plan, and coordinate actions within their area of expertise or warfighting function. Coordinating staff officers may also exercise planning and supervisory authority over designated special staff officers. Coordinating staff officers within the battalion are the personnel staff officer (S-1), intelligence staff officer (S-2), operations staff officer (S-3), logistics staff officer (S-4), and signal staff officer (S-6).

Personnel Staff Officer, S-1

1-125. The battalion S-1 is the principal staff officer for all matters concerning human resources support. Specific responsibilities include manning, personnel services, and personnel support. (See appendix H.) The S-1 prepares a portion of Annex F (Sustainment) to the operation order. (Refer to FM 1-0, FM 1-04, and ATP 1-0.1 for additional information.)

Note. Legal support—the paralegal noncommission officer must be supervised by a judge advocate and works in the consolidated legal office, generally within the IBCT headquarters. (Refer to FM 1-04 for additional information.)

Intelligence Staff Officer, S-2

1-126. The battalion S-2 is the principal staff officer responsible for providing intelligence to support current operations and plans. The S-2 gathers and analyzes information on enemy, terrain, weather, and civil considerations for the commander and participates, with the staff, in performing intelligence preparation of the battlefield. The S-2, together with the S-3, helps the commander coordinate, integrate, and supervise information collection planning and operations, and targeting. The S-2 gives the commander and the S-3 the initial intelligence synchronization plan, which facilitates R&S integration. The S-2 assists the S-3 to develop the initial R&S plan. The S-2 is responsible for the preparation of Annex B (Intelligence) and assists the S-3 in the preparation of Annex L (Information Collection). (Refer to FM 2-0 for additional information.)

Operations Staff Officer, S-3

1-127. The battalion S-3 is the principal staff officer for integrating and synchronizing the operation as a whole for the commander. The S-3 integrates R&S during plans and operations. The S-3, together with the S-2, helps the commander coordinate, integrate, and supervise information collection planning and operations, and targeting. The S-3 synchronizes R&S with the overall operation throughout the operations process. The S-3 develops plans and orders, and determines potential branches and sequels. The S-3 coordinates and synchronizes warfighting functions in all plans and orders. Additionally, the S-3 is responsible for and prepares Annex L (Information Collection). The S-3 prepares Annex A (Task Organization), Annex C (Operations), Annex E (Protection), and Annex M (Assessment) to the operation order. In conjunction with the executive officer (or when staffed the knowledge management officer), the S-3 prepares Annex R (Reports) and Annex Z (Distribution). (Refer to FM 6-0 for additional information.)

Logistics Staff Officer, S-4

1-128. The battalion S-4 is the principal staff officer for logistics planning (see appendix H) and operations, supply, maintenance, transportation, services, field services, distribution, and operational contract support. The S-4 develops the battalion's concept of support and prepares Annex F (Sustainment), Annex P (Host-Nation Support) and Annex W (Operational Contract Support) to the operation order. (Refer to FM 6-0 for additional information.)

Signal Staff Officer, S-6

1-129. The battalion S-6 is the principal staff officer for all matters concerning network operations. The signal staff officer provides network transport, information services, and information management, conducts network operations to operate and defend the network, enables knowledge management (see FM 3-96), manages LandWarNet and combat net radios assets in area of operation, and performs spectrum management operations. The S-6 prepares Annex H (Signal) and participates in preparation of Appendix 12 (Cyberspace Electromagnetic Activities) to Annex C (Operations) with input from the S-2 and in coordination with the S-3, to the operation order. (Refer to FM 6-02, ATP 6-02.70, and FM 6-02.71 for additional information.)

Note. The Infantry battalion may also be authorized a battalion civil affairs operations officer (S-9) to aid in civil-military operations. If an S-9 is not authorized the attached civil affairs team can perform these functions. Additionally, the battalion commander can assign this responsibility to an officer or senior NCO to assist with relations between the civilian population and military operations during the conduct of stability tasks. (Refer to FM 3-57 or ATP 3-57.60 for more information.)

Personal Staff Officers

1-130. Personal staff officers work under the immediate control of, and have direct access to, the battalion commander. They advise the commander, provide input to orders and plans, and interface and coordinate with entities external to the battalion headquarters. Personal staff officers to the battalion commander include the command sergeant major (see paragraph 1-100), the battalion surgeon, and the battalion chaplain.

Battalion Surgeon

1-131. The battalion surgeon is responsible for coordinating health service support (see appendix H) and operations within the command. The surgeon provides force health protection mission planning to support battalion operations. The surgeon provides and oversees medical care to Soldiers, civilians, and enemy prisoners of war. (See ATP 4-02.3.) The surgeon prepares a portion of Annex E (Protection) and Annex F (Sustainment) of the operation order. (Refer to FM 4-02 and FM 3-96 for additional information.)

Battalion Chaplain

1-132. The chaplain is responsible for religious support operations. The chaplain advises the commander and staff and subordinate command on potential operational impact of religion, morals, ethics, and morale within the battalion and throughout its area of operation. The chaplain prepares a Portion of Annex F (Sustainment) to the operations order. (Refer FM 1-05 for additional information.)

Special Staff Officers

1-133. Special staff officers, within the Infantry battalion, provide specific areas of expertise to assist the exercise of mission command. These areas of expertise vary with authorizations, mission requirements, and the desires of the commander. When a special staff officer is not assigned or attached, the officer with coordinating staff responsibility for the area of expertise assumes those functional responsibilities. For example, the operations officer or fire support officer are typically assigned responsibility for information operations, to include preparation of Appendix 15 (Information Operations) to Annex C (Operations). Special staff officers or noncommissioned officers, common to the Infantry battalion, include the fire support officer (assigned or attached), the engineer noncommissioned officer (NCO), the air liaison officer, the chemical, biological, radiological, and nuclear (CBRN) officer, the electronic warfare NCO, and the liaison officer. (Refer to FM 6-0 for additional information.)

Fire Support Officer

1-134. The fire support officer serves as the special staff officer for fires and integrates fires into the scheme of maneuver for the commander. The fire support officer leads the targeting process and fire support planning for the delivery of fires to include preparation fires, harassing fires, interdiction fires, suppressive fires, and destruction fires, as well as, the generation of non-lethal effects, such as those generated by information operations. The fire support officer leads the fire support cell (when established) and prepares Annex D (Fires) of the operation order. The fire support officer also coordinates with the electronic warfare noncommissioned officer and the air liaison officer. The battalion S-3 coordinates this position. (Refer to ADRP 3-09 for additional information.)

Engineer Noncommissioned Officer

1-135. The engineer NCO is the engineer on staff responsible for coordinating engineer support to combined arms operations. The engineer (in coordination with the battalion S-3) integrates specified and implied engineer tasks into the maneuver force plan. The engineer ensures that mission planning, preparation, execution, and assessment activities integrate supporting engineer units. The engineer prepares Annex G (Engineer) to the operation order. (Refer to ATP 3-34.22 for additional information.)

Air Liaison Officer

1-136. The air liaison officer is the senior Air Force officer (or noncommissioned officer) in the tactical air control party. The air liaison officer plans and executes close air support in accordance with the battalion commander's guidance and intent. The air liaison officer is responsible for coordinating aerospace assets and operations such as close air support, air interdiction, air reconnaissance, airlift, and joint suppression of enemy air defenses. The air liaison officer supports the fire support cell (when established) and assists in preparing Annex D (Fires) of the operation order. (Refer to JP 3-09.3, FM 3-52, and ATP 3-52.1 for additional information.)

Chemical, Biological, Radiological, and Nuclear Officer

1-137. The CBRN officer is responsible for CBRN operations, obscuration operations, and CBRN asset used. When established, the CBRN officer leads the CBRN working group. The CBRN officer prepares a portion of Annex E (Protection) and a portion of Annex C (Operations) of the operation order. (Refer to ATP 3-11.36 for additional information.)

Electronic Warfare Noncommissioned Officer

1-138. The electronic warfare NCO serves as the designated staff officer for the planning, integration, synchronization, and assessment of electronic warfare, to include cyberspace electromagnetic activities. The electronic warfare NCO coordinates through other staff members to integrate electronic warfare or/and cyberspace electromagnetic activities into the commander's concept of operations. The electronic warfare NCO prepares Appendix 12 (Cyberspace Electromagnetic Activities) to Annex C (Operations) to the operation order and contributes to any section that has a cyberspace electromagnetic activities subparagraph in the operations order. (Refer to ATP 3-36 and FM 3-12 for additional information.)

Liaison Officer

1-139. Liaison officer(s) are the commander's representative at the headquarters or agency to which they are sent. Liaison activities augment the commander's ability to synchronize and focus combat power. Liaison activities promote coordination, synchronization, and cooperation between the battalion and higher headquarters, interagency, multinational, host-nation, adjacent, and subordinate organizations as required. When embedded as subject matter experts, liaison officers provide face-to-face coordination.

Augmentation

1-140. The battalion staff can receive augmentation teams by specialty to assist the exercise of mission command. For example, teams may be air and missile defense planners, liaison officers from joint or multinational support agencies, or additional augmentation to the fire support cell, which may include a naval surface fire support team, air force weather team, and a space support team. When received the commander integrates these elements into the planning process as early as possible.

This page intentionally left blank.

Chapter 2
Offense

Offensive action is the critical part of any engagement. The primary purpose of the offense for the Infantry battalion is to decisively defeat, destroy, or neutralize the enemy force, or to seize key terrain. The commander also may take offensive action to collect information, deceive the enemy, deprive the enemy of resources or decisive terrain, or fix the enemy in position. Even in the defense, offensive action normally is required to destroy an attacker and exploit success. The key to a successful offense is to identify the enemy's most vulnerable point; choose a form of maneuver that avoids the enemy's strength while exploiting enemy weakness and one that masses overwhelming combat power. This chapter discusses the doctrinal basis for the offense and introduces a fictional scenario used as a discussion vehicle for illustrating one of many ways that an Infantry battalion conducts the offensive element of decisive action. The scenario focuses on potential challenges confronting the Infantry battalion commander in accomplishing a mission but is not intended to be prescriptive of how the Infantry battalion performs any particular operation.

SECTION I – DOCTRINAL BASIS FOR THE OFFENSE

2-1. Offensive techniques cannot be discussed in isolation. There must be a seamless continuity and understanding between the fundamental doctrinal principles, tactics and procedures, covered in Army doctrinal reference publications and field manuals, and the techniques covered in this Army techniques publication. This section briefly discusses offensive tasks, tactical movement, and supporting doctrinal terms. The reader should refer to FM 3-96, FM 3-90-1, FM 3-90-2, and ADRP 3-90 for additional information.

OFFENSIVE TASKS

2-2. *Offensive tasks are* tasks conducted to defeat and destroy enemy forces and seize terrain, resources, and population centers (ADRP 3-0). The four primary offensive tasks are movement to contact, attack, exploitation, and pursuit. See the specific following sections for tasks discussions.

CHARACTERISTICS OF THE OFFENSE

2-3. Audacity, concentration, surprise, and rapid tempo characterize the offense. See ADRP 3-90 for a detailed discussion of each characteristic listed below:

- Audacity**.** Audacity means boldly executing a simple plan of action.
- Concentration. Concentration is the massing of overwhelming effects of combat power to achieve a single purpose.
- Surprise. Commanders achieve surprise by attacking the enemy at a time or place the enemy does not expect or in a manner for which the enemy is unprepared.
- Tempo. A rapid tempo allows attackers to quickly penetrate barriers and defenses and destroy enemy forces in-depth before they can react.

PLANNING CONSIDERATIONS

2-4. Understanding, visualizing, describing, directing, leading, and assessing are aspects of leadership common to all commanders. The commander begins with an assigned area of operation, identified mission, and available forces. The commander develops and issues planning guidance based on their visualization in terms of the means to accomplish the mission. The six-warfighting functions are the framework for discussing planning considerations that apply to all primary and subordinate offensive tasks. (See FM 3-96 and ADRP 3-90 for a detailed discussion, by warfighting function, of offensive planning considerations.)

FORMS OF MANEUVER

2-5. *Forms of maneuver* are distinct tactical combinations of fire and movement with a unique set of doctrinal characteristics that differ primarily in the relationship between the maneuvering force and the enemy (ADRP 3-90). The battalion uses the six basic forms of maneuver during an attack: envelopment, turning movement, frontal attack, penetration, infiltration, and flank attack. When the battalion executes a form of maneuver, subordinate units may execute different forms of maneuver in executing the battalion's concept of operation. Forms of maneuver are conducted in relation to or relative to an enemy force. The six forms of maneuver are addressed in the following paragraphs. (Refer to FM 3-90-1 for addition information.)

Note. Each form of maneuver has a virtual equivalent in the informational and cognitive dimensions of the information environment. For example, infiltration is not only possible in physical space, but also in cyberspace. Frontal and flank attacks can be both physical and psychological. Through the integration of information operations and cyber electromagnetic activities into its operations, the battalion is able to out-maneuver the enemy or adversary physically, informationally, and psychologically. (Refer to FM 3-13 for additional information.)

ENVELOPMENT

2-6. *Envelopment* is a form of maneuver in which an attacking force seeks to avoid the principal enemy defenses by seizing objectives behind those defenses that allow the targeted enemy force to be destroyed in their current positions (FM 3-90-1). Envelopments focus on seizing terrain, destroying specific enemy forces, and interdicting enemy withdrawal routes. An envelopment avoids the enemy's front; where the enemy is strongest, where the enemy's attention is focused, and where the enemy's fires are most easily concentrated. During an envelopment, the battalion shaping operation fixes the defender, while the battalion decisive operation maneuvers out of contact around the enemy's defenses to strike at assailable flanks, the rear, or both. If no assailable flank is available, the attacking force creates one through the conduct of a penetration.

2-7. Envelopments may be conducted against a stationary or moving enemy force. Sometimes the enemy exposes their flank by their own forward movement, when unaware of their opponent's location. In noncontiguous areas of operation, the combination of air and ground fires may isolate the enemy on unfavorable terrain and establish conditions for maneuver against an assailable flank or rear. Attacking forces need to be agile enough to concentrate and mass combat power before the enemy can reorient their defense. Fixing forces must have sufficient combat power to keep the enemy engaged, while the enveloping force maneuvers to close with the enemy. The four varieties of envelopment are the single envelopment, double envelopment, encirclement, and vertical envelopment (figure 2-1). Infantry battalions do not possess the resources to execute a double envelopment or encirclement independently, but rather, as a part of an IBCT or higher operation. The four varieties of envelopment are defined below:

- A *single envelopment* is a form of maneuver that results from maneuvering around one assailable flank of a designated enemy force (FM 3-90-1). For information on a single envelopment, see FM 3-90-1 and FM 3-96.
- A *double envelopment* results from simultaneous maneuvering around both flanks of a designated enemy force (FM 3-90-1). For information on a double envelopment, see FM 3-90-1 and FM 3-96.
- *Encirclement operations* are operations where one force loses its freedom of maneuver because an opposing force is able to isolate it by controlling all ground lines of communication and reinforcement (ADRP 3-90). For a discussion of offensive encirclement operations, see FM 3-90-2 and FM 3-96.
- *Vertical envelopment* is a tactical maneuver in which troops that are air-dropped, air-landed, or inserted via air assault, attack the rear and flanks of a force, in effect cutting off or encircling the force (JP 3-18). For a discussion of airborne and air assault operations, see FM 3-99.

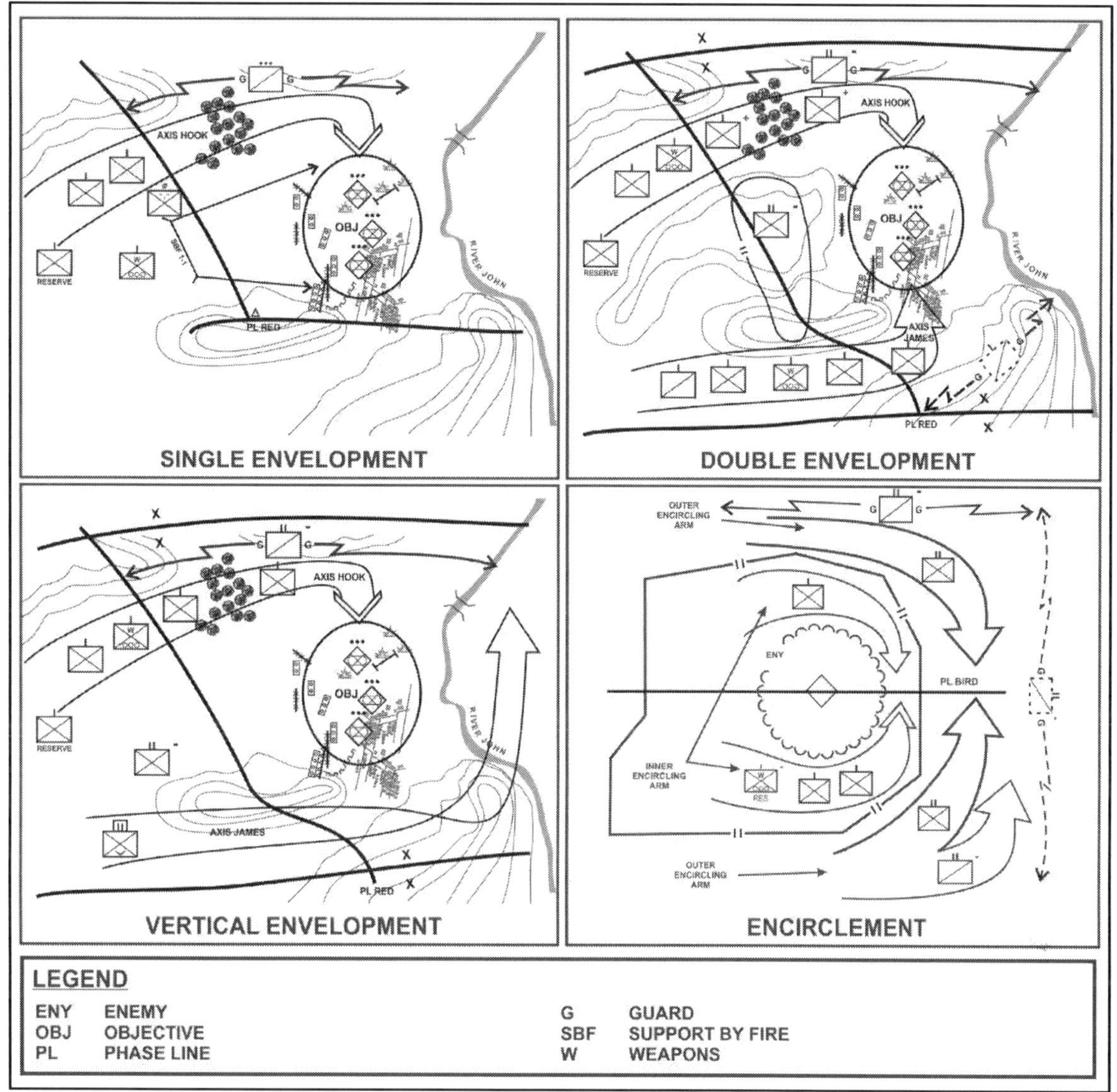

Figure 2-1. Envelopments

Note. A unit can conduct offensive encirclement operations designed to isolate an enemy force (addressed above) or conduct defensive encirclement operations (addressed in chapter 3) because of the unit's isolation by the actions of an enemy force. (Refer to FM 3-90-2 for additional information.) See paragraphs 3-96 through 3-116 for information on linkup during offensive and defensive encirclement operations.

TURNING MOVEMENT

2-8. A *turning movement* is a form of maneuver in which the attacking force seeks to avoid the enemy's principle defensive positions by seizing objectives behind the enemy's current positions thereby causing the enemy force to move out of their current positions or divert major forces to meet the threat (FM 3-90-1). In a turning movement, the turning force passes around and avoids the enemy's main force. The force then secures an objective that causes the enemy to move out of its current position or divert forces to meet the threat. The objective of the turning movement is to make contact with the enemy, but at a location of the commander conducting the turning movement's advantage and out of the enemy established kill zones. A turning movement differs from envelopment because the force conducting a turning movement seeks to make the enemy forces displace from their current locations, whereas an enveloping force seeks to engage the enemy forces in their current locations from an unexpected direction.

2-9. A turning movement is particularly suited for division-sized or larger forces possessing a high degree of tactical mobility. The Infantry battalion conducts a turning movement as part of that larger force, most likely as a shaping operation or fixing force as opposed to the decisive operation involving the turning force. The commander directing the turning movement can employ a vertical envelopment using airborne or air assault forces to effect a turning movement. (See FM 3-99.) The commander can also conduct a turning movement using amphibious means. (See JP 3-02.) The commander uses this form of offensive maneuver to seize vital areas in the enemy's support area before the main enemy force can withdraw or receive support or reinforcements.

2-10. The commander directing a turning movement task organizes available resources to conduct three main tasks: conduct a turning movement, conduct shaping operations (fixing force), and conduct reserve operations. (Figure 2-2). Normally the force conducting the turning movement conducts the echelon's decisive operation given the appropriate mission variables of METT-TC. It is not until a commander has access to the resources of these echelons that the commander normally has the combat power to resource a turning force that can operate outside supporting range of the main body to allow the turning force-to-force enemy units out of their current positions. The commander bases the task organization of these forces on the mission variables of METT-TC and the concept of operations for the turning movement. A commander frequently transition this form of offensive maneuver from the attack into an exploitation or pursuit.

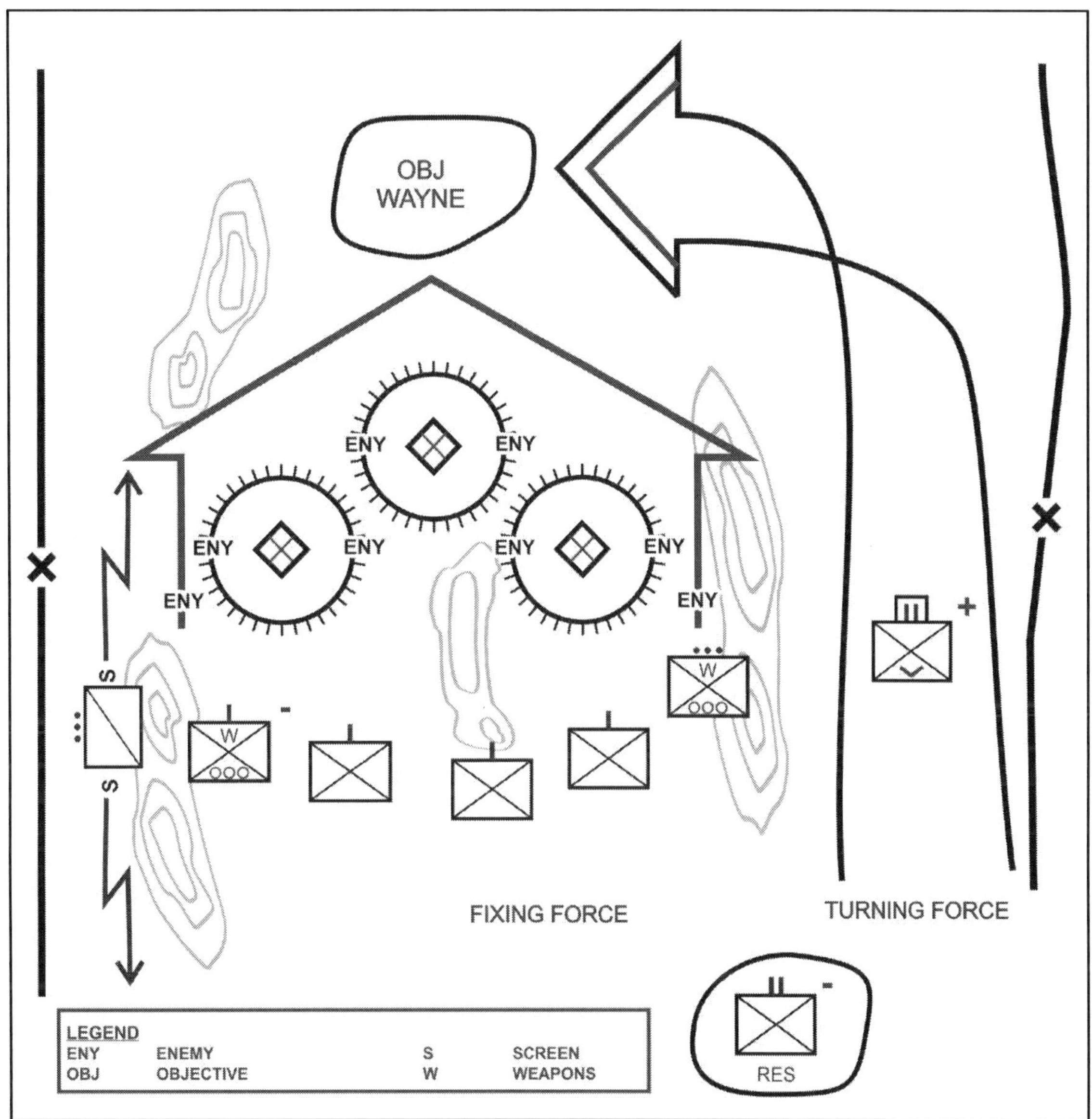

Figure 2-2. Turning movement

Frontal Attack

2-11. A *frontal attack* is a form of maneuver in which an attacking force seeks to destroy a weaker enemy force or fix a larger enemy force in place over a broad front (FM 3-90-1). The frontal attack is usually the least desirable form of maneuver because it exposes the majority of the offensive force to the concentrated fires of the defenders. The battalion normally conducts a frontal attack as part of a larger operation against a stationary or moving enemy force (figure 2-3, page 2-6). Unless frontal attacks are executed with overwhelming and well synchronized speed and strength against a weaker enemy, they are seldom decisive. The battalion attacks the enemy across a wide front and along the most direct approaches. It uses a frontal attack to overrun and destroy a weakened enemy force or to fix an enemy force. Frontal attacks are used when commanders possess overwhelming combat power and the enemy is at a clear disadvantage or when fixing the enemy over a wide front is the desired effect and a decisive defeat in that area is not expected. The frontal attack may be appropriate in an attack or meeting engagement where speed and simplicity are paramount to maintain *tempo*—the relative speed and rhythm of military operations over time with respect to the enemy (ADRP 3-0)—and, ultimately, the initiative; or in a shaping attack to fix an enemy force.

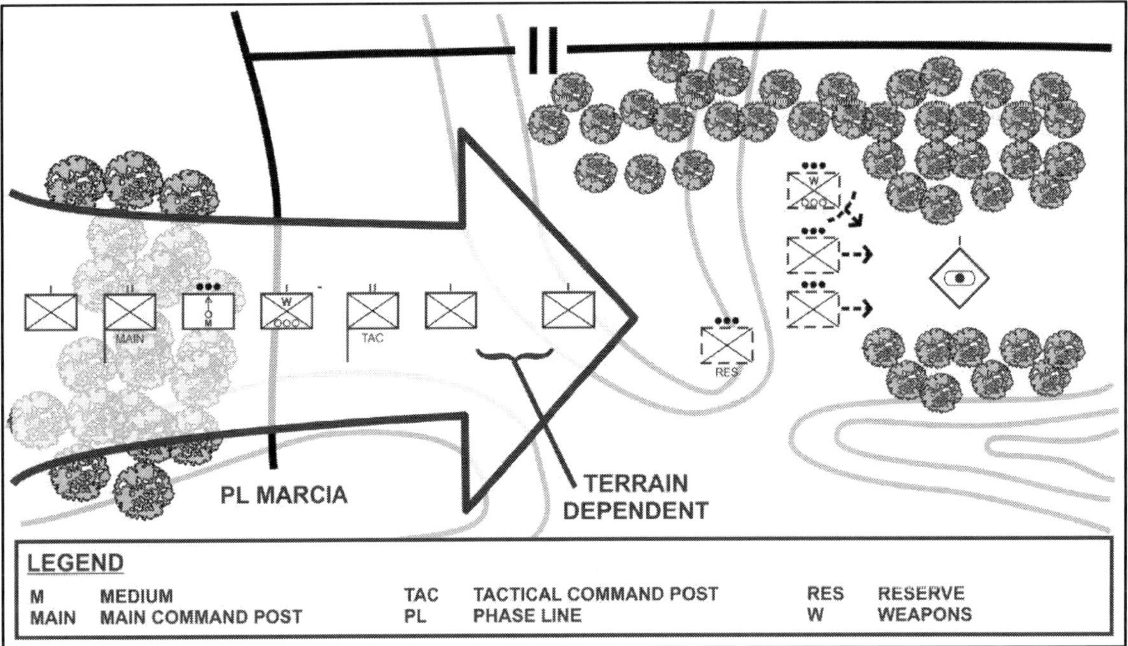

Figure 2-3. Frontal attack

Penetration

2-12. A *penetration* is a form of maneuver in which an attacking force seeks to rupture enemy defenses on a narrow front to disrupt the defensive system (FM 3-90-1). In a penetration, the battalion commander concentrates forces to strike at an enemy's weakest point, rupture the defense, and break up its continuity to create an assailable flank (figure 2-4). Penetration of an enemy position requires a concentration of combat power to permit continued momentum of the attack. The commander uses the breach (see appendix F) created to pass forces through to defeat the enemy through attacks into the enemy's flank and rear. The attack should move rapidly to destroy the continuity of the defense since, if it is slowed or delayed, the enemy is afforded time to react.

2-13. A successful penetration depends on the attacking force's ability to suppress enemy weapons systems, to concentrate forces to overwhelm the enemy defender at the point of attack, and to pass sufficient forces through the gap to defeat the enemy quickly. If the attacker does not make the penetration sharply and secure objectives promptly, the penetration is likely to resemble a frontal attack. This may result in high casualties and permit the enemy to fall back intact, thus avoiding destruction.

2-14. Normally, a penetration is tried when enemy flanks are unassailable or when conditions permit neither envelopment nor a turning movement such as an attack against the enemy's main defensive belt. To allow a penetration, the terrain must facilitate the maneuver of the penetrating force. The concentration of the battalion is planned to penetrate the defense where the continuity of the enemy's defense has been interrupted such as gaps in obstacles and minefields or areas not covered by fire. Multiple penetrations are normally only conducted at echelons above battalion level. When essential to the accomplishment of the mission, intermediate objectives should be planned for the attack. Usually, when the penetration is successfully completed, the battalion will transition to another form of maneuver.

2-15. After the initial breach of the enemy's main line of resistance, the sequence of the remaining two phases is determined by the situation. If the enemy is in a weak position, it may be possible for the lead attacking force to seize the penetration's final objective while simultaneously widening the initial breach. An example would be penetration at battalion level, which is described below:

- Phase 1. Breaching the enemy's main defensive positions. A reinforced company can execute the initial penetration. The weapons company has the potential to play a significant role in this initial breach.

- Phase 2. Widening the breach to secure flanks. The battalion seizes enemy positions behind the obstacles and widens the shoulders of the penetration to allow assaulting forces room to attack deep objectives.
- Phase 3. Seizing the objective and subsequent exploitation. Exploitation of the penetration is made as companies complete the destruction of the enemy and attack to secure deeper objectives. Objectives for the assaulting force are deep enough to allow an envelopment of the rest of the enemy position and should facilitate attack by fire against second echelon enemy positions and enemy counterattack routes.

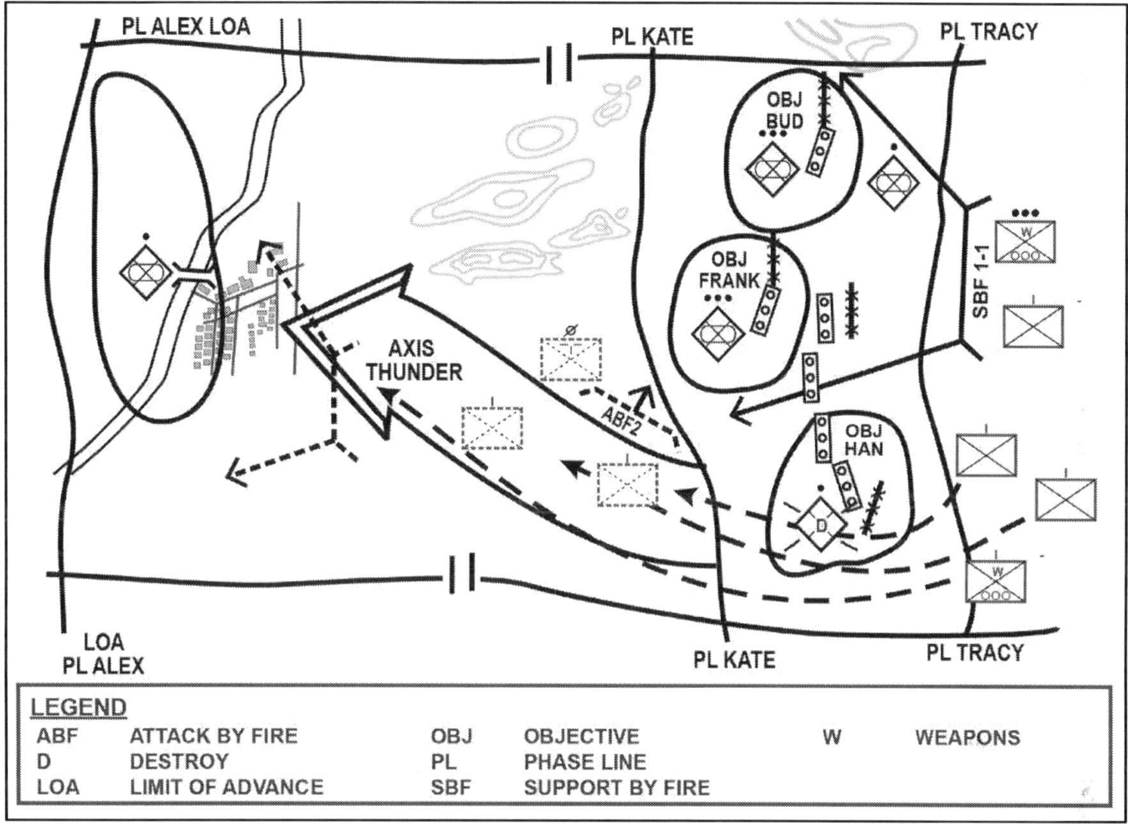

Figure 2-4. Penetration

INFILTRATION

2-16. An *infiltration* is a form of maneuver in which an attacking force conducts undetected movement through or into an area occupied by enemy forces to occupy a position of advantage behind those enemy positions while exposing only small elements to enemy defensive fires (FM 3-90-1). The commander uses infiltration to—

- Attack lightly defended positions or stronger positions from the flank and rear.
- Secure key terrain in support of the decisive operation.
- Disrupt or harass enemy defensive preparations/operations.
- Relocate the battalion by moving to battle positions around an engagement area.
- Reposition to attack vital facilities or enemy forces from the flank or rear.

2-17. The commander can impose measures to control the infiltration including checkpoints, linkup point, phase lines, and assault positions on the flank or rear of enemy positions. If it is not necessary for the entire infiltrating unit to reassemble to accomplish its mission, the *objective area*—a geographical area, defined by competent authority, within which is located an objective to be captured or reached by the military forces (JP 3-06)—may be broken into smaller objectives or positions. Each infiltrating element would then move directly to its objective or position to conduct operations (figure 2-5, page 2-8). A ***checkpoint*** **is a predetermined point on the ground used to control movement, tactical maneuver, and orientation. Also called a CP**. The commander designates checkpoints along the route to assist marching units in complying with the timetable. A *linkup point* is a point where two infiltrating elements in the same or different infiltration lanes are scheduled to meet to consolidate before proceeding on with their missions (FM 3-90-1). A linkup point should be an easily identifiable point on the ground, large enough for all infiltrating elements to assemble, and offer cover and concealment. Additional control measures for an infiltration may include—

- An area of operation for the infiltrating unit.
- One or more infiltration lanes.
- A line of departure or point of departure.
- Movement routes with their associated start points and release points, or a direction or axis of attack.
- Rally point, including an objective rally point.
- Support by fire, attack by fire, or assault positions.
- A limit of advance.

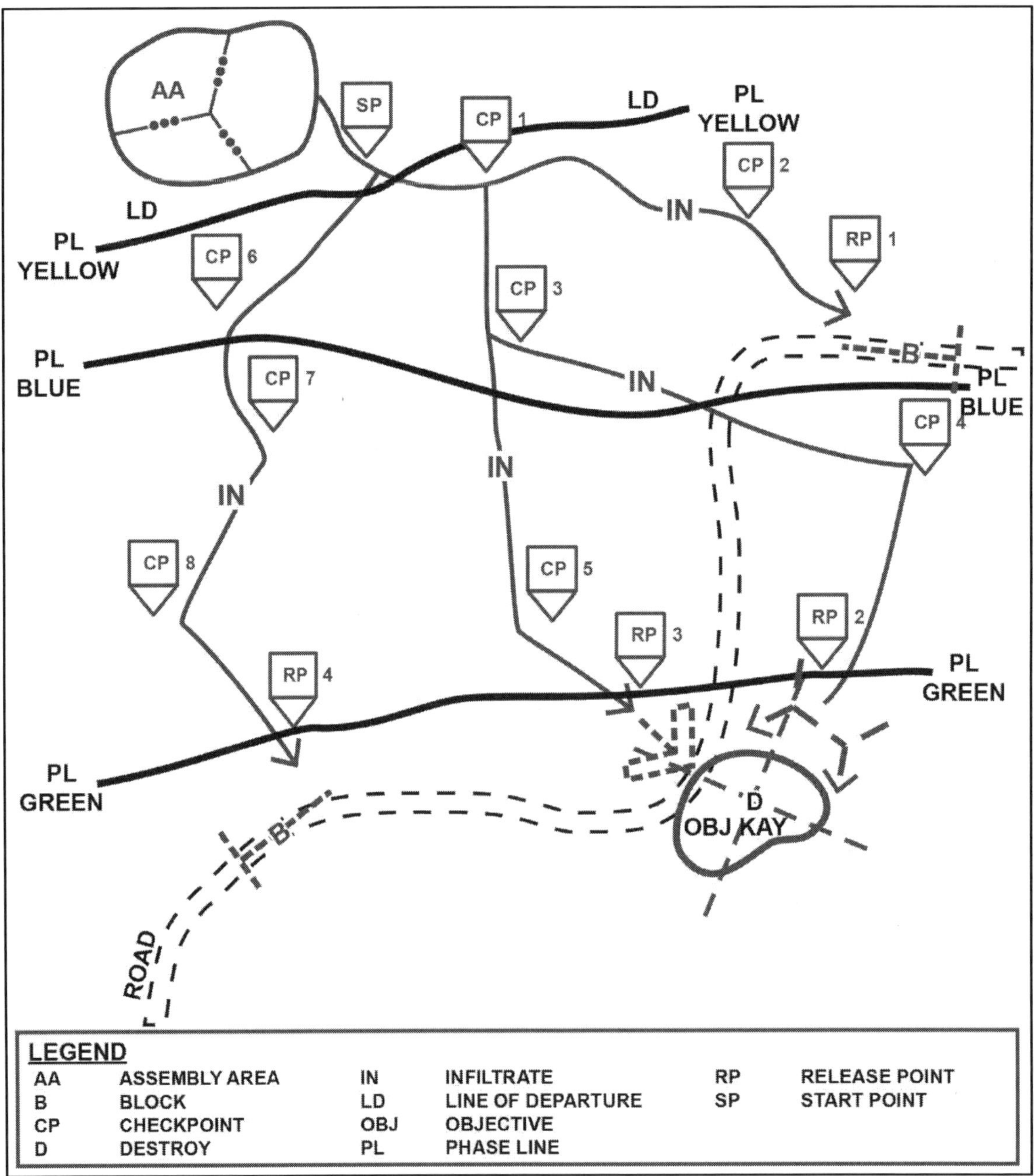

Figure 2-5. Infiltration route

2-18. Mission analysis will dictate the location of infiltration routes, the number of infiltration lanes, the size of the infiltration lane(s), the anticipated speed of movement, and the time of departure. An infiltration should be planned during limited visibility through areas the enemy does not occupy or cover by surveillance and fire. Planning should incorporate infiltration lanes, rally points along the route or axis, and contact points.

2-19. Companies and platoons usually conduct infiltrations, but it is possible to execute at a battalion or squad level also. Careful planning considerations for the integration of the weapons company throughout an infiltration must be exercised. Although the weapons company may provide a larger and louder signature, the speed, mobility, shock, and firepower of the weapons company may provide an overwhelming advantage to either set conditions for or execute a decisive operation. The weapons company may have its own infiltration lane, follow another unit on a lane, or may be held as a reserve element. Another consideration for the infiltration means is the vertical infiltration. Due to limited resources, this means of infiltration may be limited to reconnaissance and weapons company assets.

2-20. An *infiltration lane* is a control measure that coordinates forward and lateral movement of infiltrating units and fixes fire planning responsibilities (FM 3-90-1). Single or multiple infiltration lanes can be planned. Using a single infiltration lane—facilitates navigation, control, and reassembly, reduces susceptibility to detection, reduces the area requiring detailed intelligence, and increases the time required to move the force through enemy positions. Using multiple infiltration lanes—reduces the possibility of compromise, allows more rapid movement, and makes control more challenging (figure 2-6).

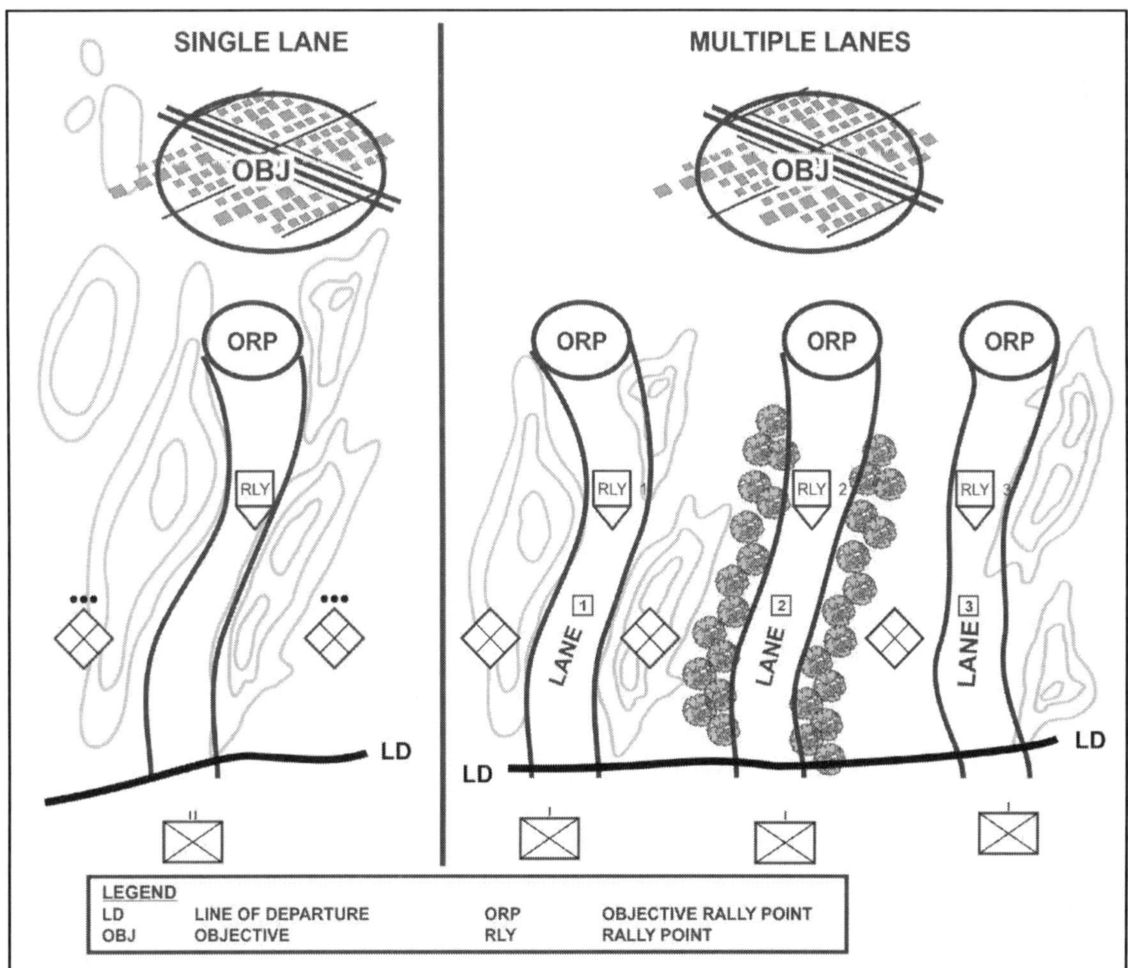

Figure 2-6. Infiltration lane

2-21. Forces conducting an infiltration or a patrol commonly use a rally point as a control measure. **A *rally point* is an easily identifiable point on the ground at which units can reassemble and reorganize if they become dispersed. Also depicted in graphics as RLY.** An *objective rally point* is a rally point established on an easily identifiable point on the ground where all elements of the infiltrating unit assemble and prepare to attack the objective (ADRP 3-90). This rally point is typically near the infiltrating unit's objective; however, there is no standard distance from the objective to the objective rally point. It should be far enough away from the objective so that the enemy will not detect the infiltrating unit's attack preparations.

FLANK ATTACK

2-22. A flank *attack* is a form of offensive maneuver directed at the flank of an enemy (FM 3-90-1). The primary difference between a flank attack and an envelopment is one of depth. A flank attack is an envelopment delivered squarely on the enemy's flank. Conversely, an envelopment is an attack delivered beyond the enemy's flank and into the enemy's support areas, but short of the depth associated with a turning movement. It is designed to defeat the enemy force while minimizing the effect of the enemy's frontally-oriented combat power. Flanking attacks are normally conducted with the main effort directed at the flank of the enemy. Usually, a supporting effort engages the enemy's front by fire while the main effort maneuvers to attack the enemy's flank. This supporting effort diverts the enemy's attention from the threatened flank. It is often used for a hasty attack or meeting engagement where speed and simplicity are paramount to maintaining battle tempo and, ultimately, the initiative (figure 2-7).

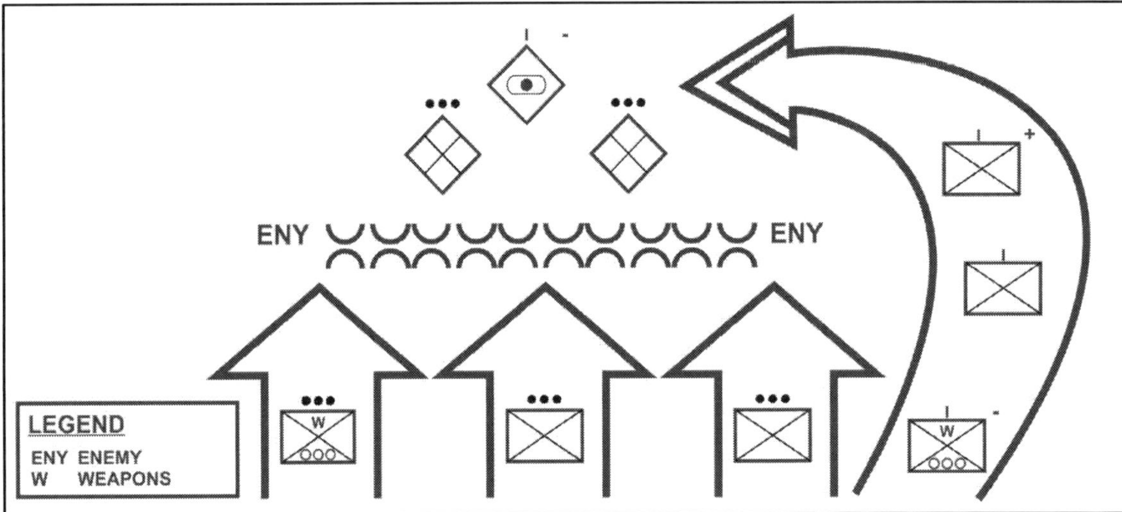

Figure 2-7. Flank attack

COMBAT FORMATIONS

2-23. A *combat formation* is an ordered arrangement of forces for a specific purpose and describes the general configuration of a unit on the ground (ADRP 3-90). Terrain characteristics and visibility determine the actual arrangement and location of the unit's personnel and vehicles within a given formation. There are seven different combat formations: column, line, echelon (left or right), box, diamond, wedge, and vee. The Infantry battalion general moves in any one of the following five combat formations on the battlefield—column, line, wedge, vee, or echelon. (See FM 3-90-1 for information on the box and diamond combat formation.) The commander may direct subordinate maneuver companies to move in a certain formation or allow them to determine the formation to use. Maneuver companies may move in different formations at any one time.

2-24. The battalion commander may use more than one formation in a given movement, especially if the terrain changes during the movement. For example, the commander may elect to use the column formation during a passage of lines and then change to another formation such as a wedge. Companies within a battalion formation may conduct combat formations different from that of the battalion. Although the battalion may be moving in a wedge formation, one company may be in a wedge, another in an echelon right, and yet another in a column. Other factors, such as the distance of the move or the enemy dispositions, may also prompt the commander to use more than one formation. Distances between units depend on the mission variable of METT-TC.

COLUMN FORMATION

2-25. The Infantry battalion moves in a column formation when early contact is not expected and the objective is far away (figure 2-8). The battalion's lead element normally uses traveling overwatch while the following units travel. The column formation—

- Speeds movement, eases control, and increases usefulness in close terrain.
- Allows quick transition to other formations.
- Requires flank security.
- Places most of the firepower on the flanks.

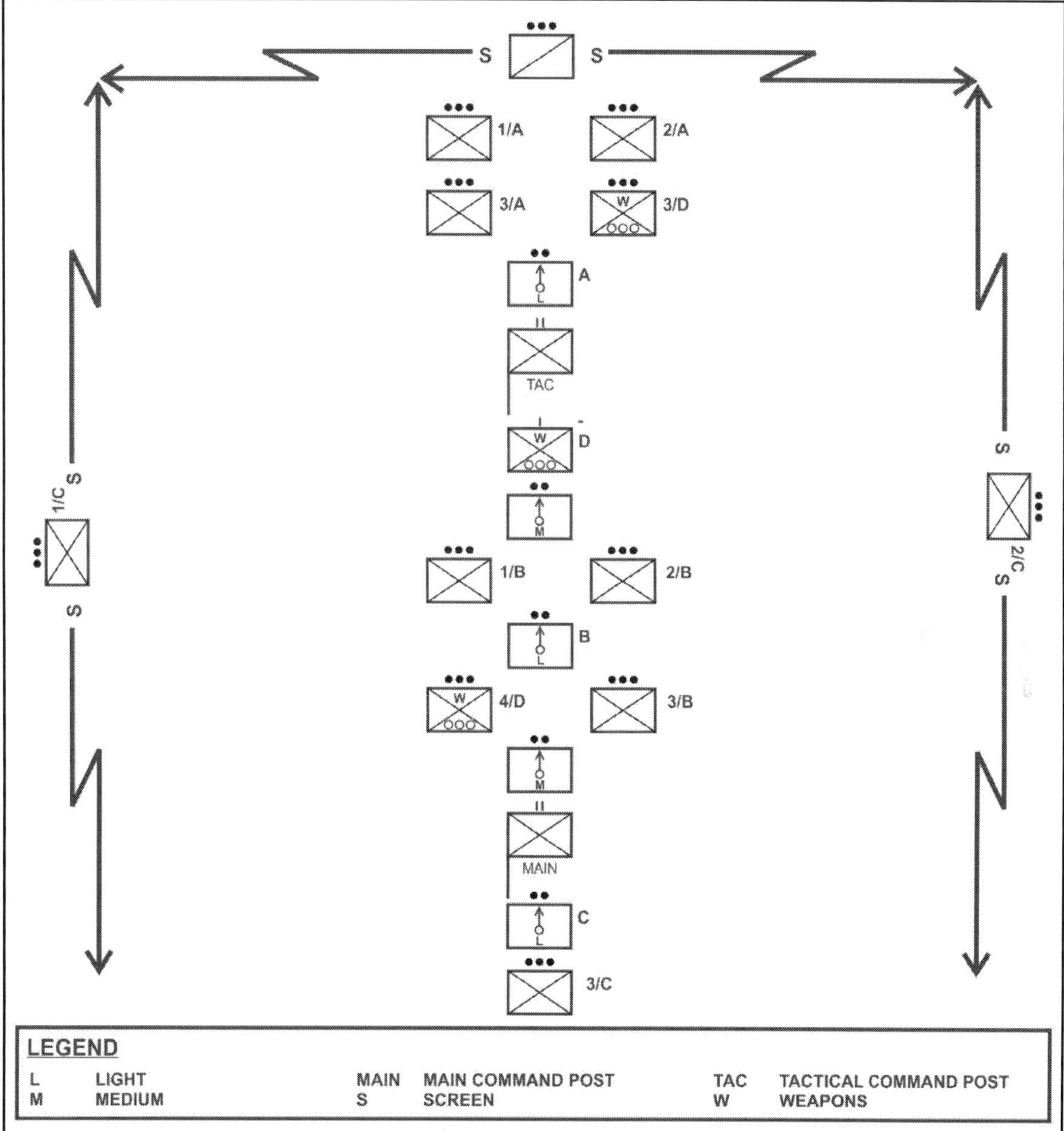

Figure 2-8. Battalion in company column, example

Line Formation

2-26. The line formation postures the battalion with companies on line and abreast of one another (figure 2-9). The line formation provides less flexibility of maneuver than other formations because it does not dispose companies in-depth. The battalion uses the line when it requires continuous movement with maximum firepower to the front in an assault.

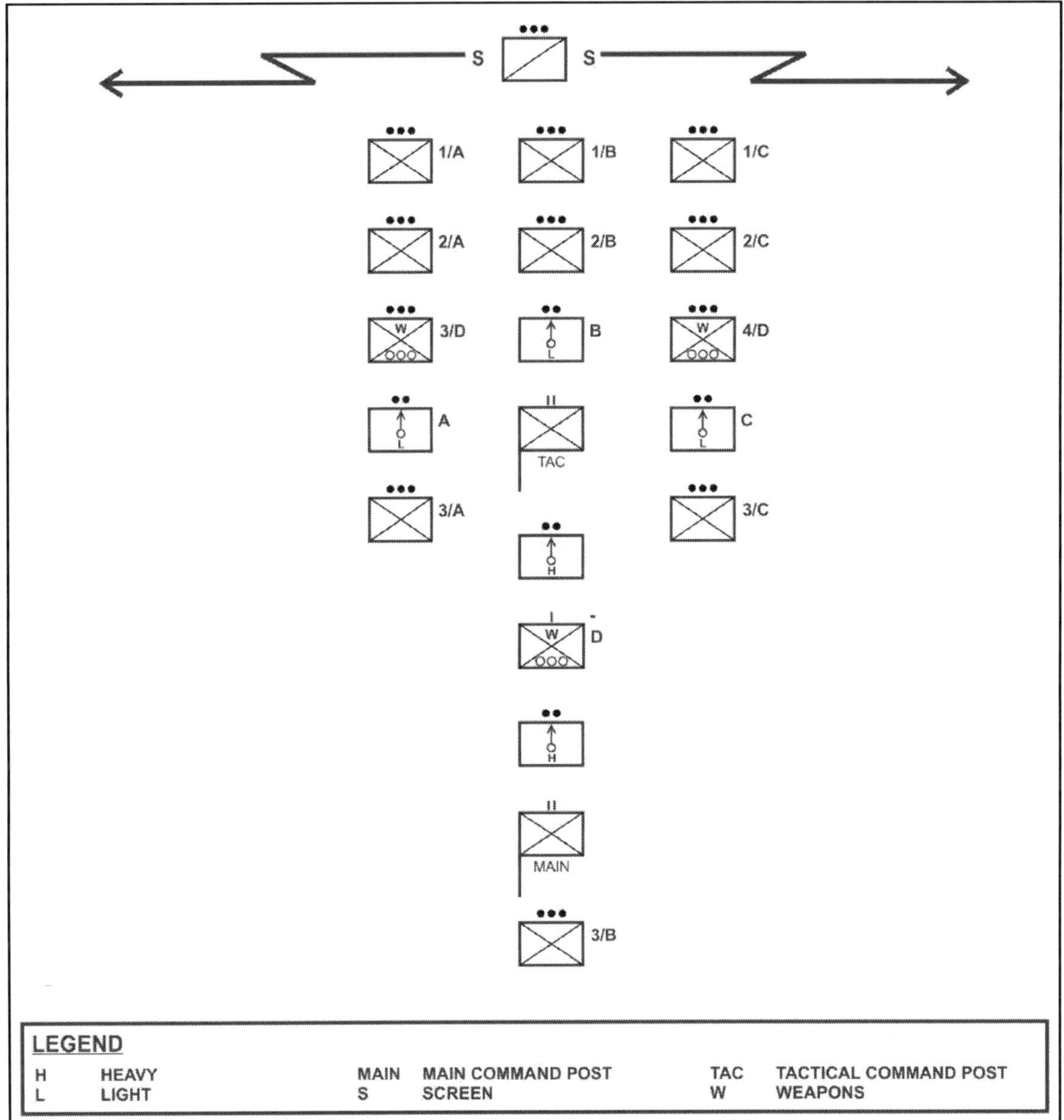

Figure 2-9. Battalion in company line, example

Wedge Formation

2-27. The wedge formation postures the battalion for enemy contact on its front and flanks (figure 2-10). The force uses the wedge when enemy contact is possible or expected but the location and disposition of the enemy is vague. When not expecting enemy contact, it may use the wedge to cross open terrain rapidly. The wedge formation—

- Facilitates control and transition to the assault.
- Provides for maximum firepower forward and good firepower to the flanks.
- Requires sufficient space to disperse laterally and in-depth.

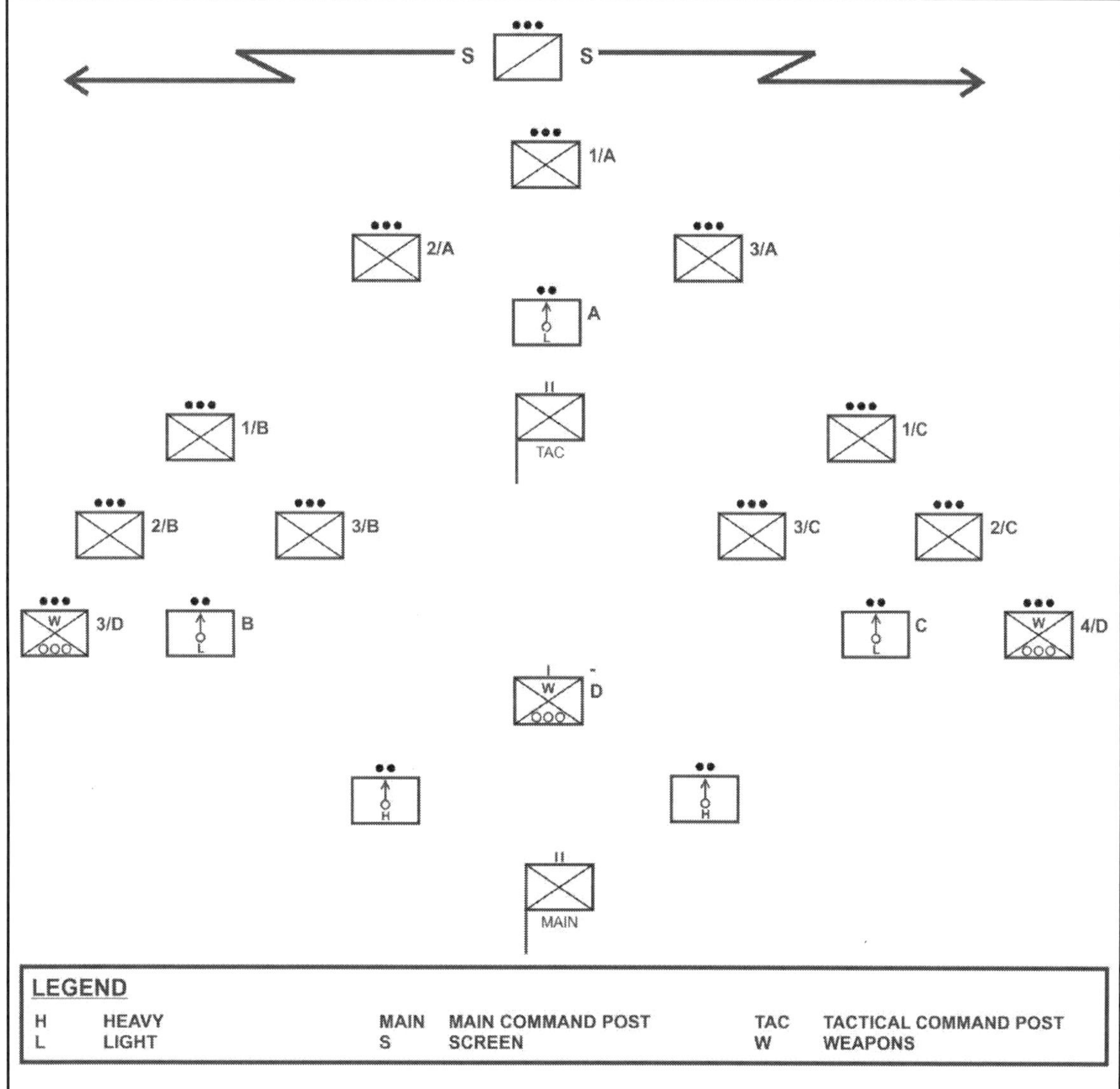

Figure 2-10. Battalion in wedge formation, example

VEE FORMATION

2-28. The vee formation postures the battalion with two companies abreast and one trailing (figure 2-11). This arrangement is most suitable to advance against an enemy known to be to the front of the battalion. The battalion may use the vee when enemy contact is expected and the location and disposition of the enemy is known. The following planning considerations apply:

- Formation is hard to orient and control is more difficult in close or wooded terrain.
- Formation provides for good firepower forward and to the flanks.

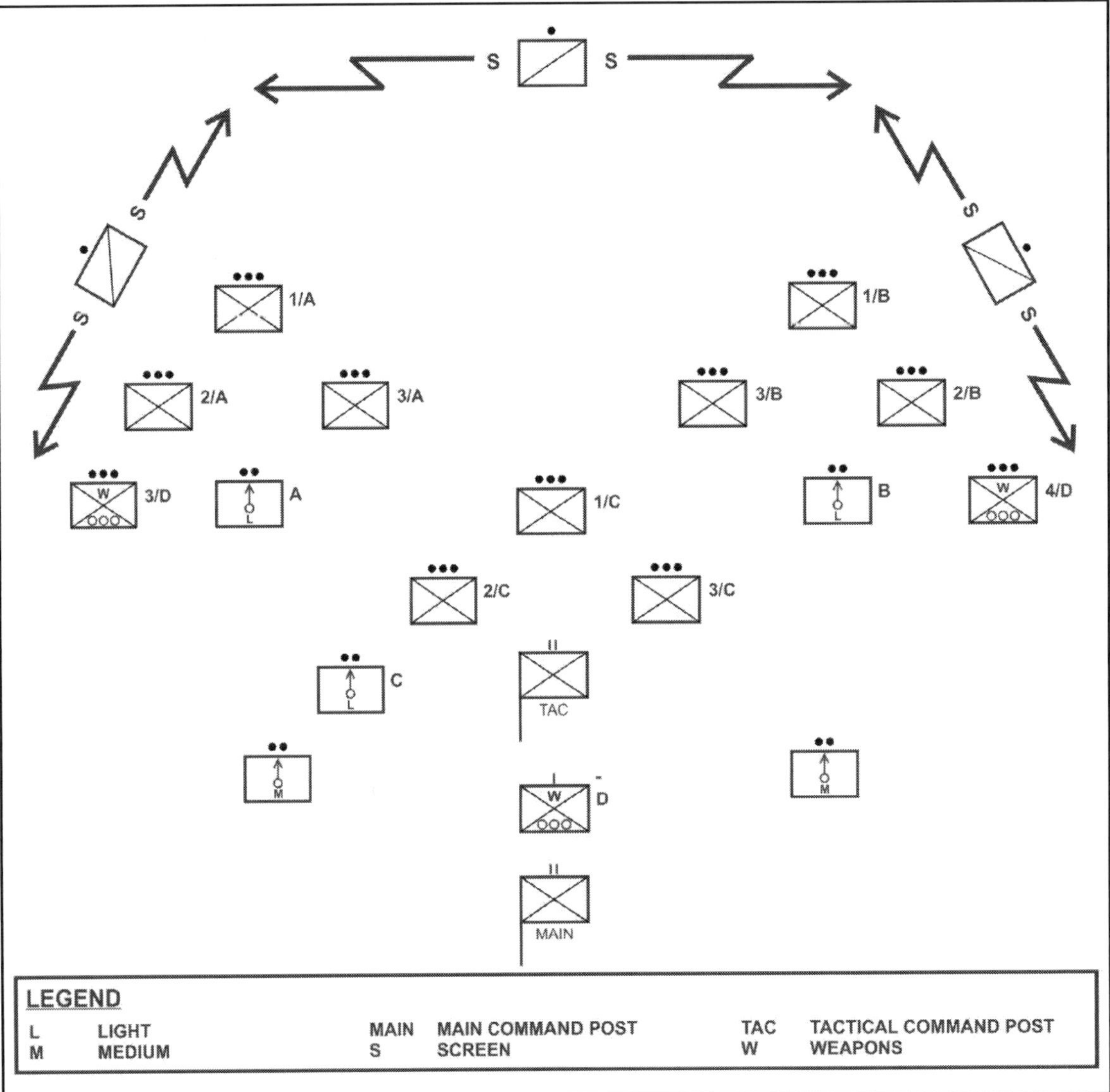

Figure 2-11. Battalion in vee formation, example

ECHELON FORMATION

2-29. The echelon formation arranges the battalion with the companies in column formation in the direction of the echelon (right or left) (figure 2-12). The battalion commonly uses the echelon when providing security to a larger moving force. The echelon formation—

- Provides for firepower forward and in the direction of echelon.
- Facilitates control in open areas but makes it more difficult in heavily wooded areas.

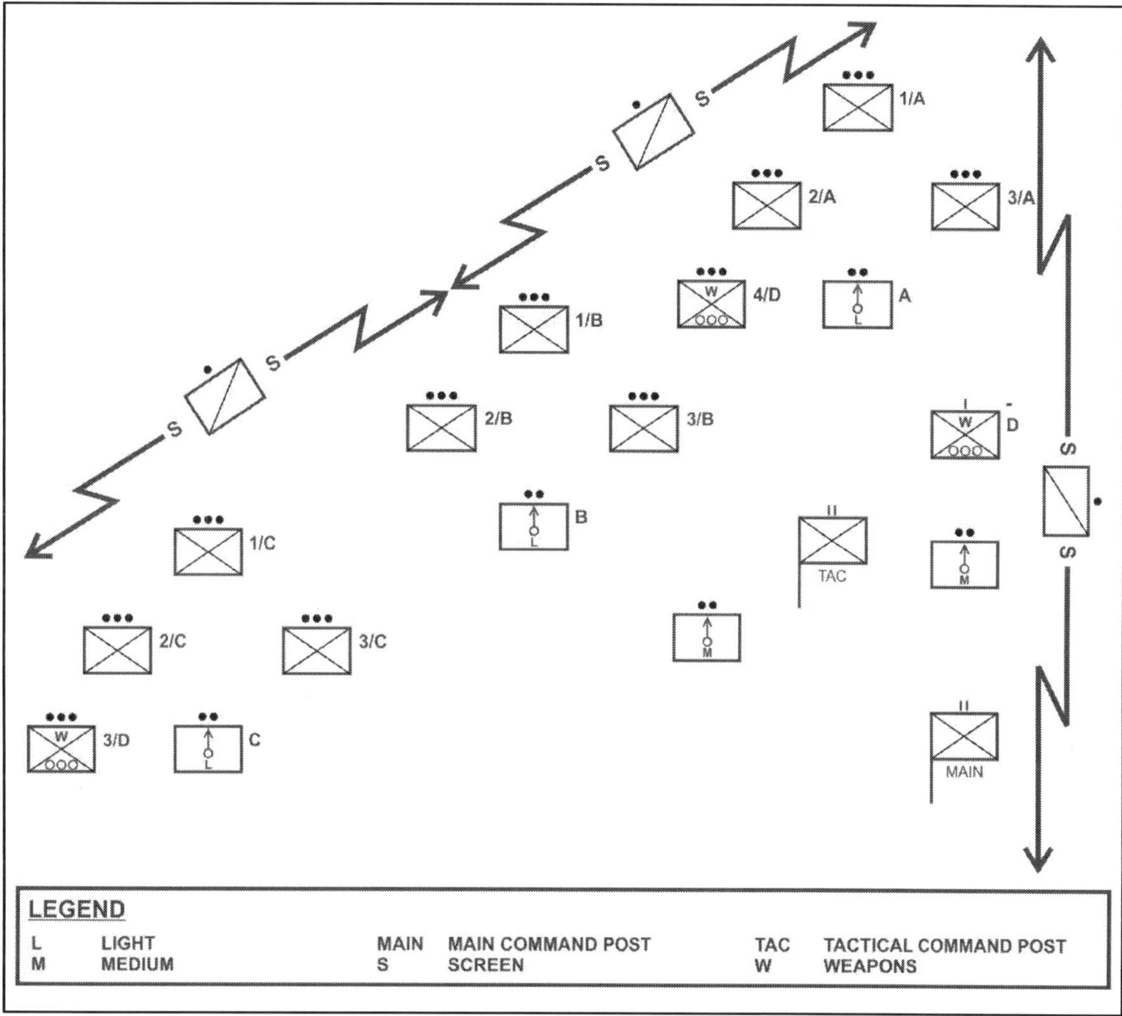

Figure 2-12. Battalion in echelon left, example

MOVEMENT TECHNIQUES

2-30. The Infantry battalion uses the combat formations described above in conjunction with three movement techniques: traveling, traveling overwatch, and bounding overwatch. Based on the chance of enemy contact, commanders select the appropriate movement technique to limit the unit's exposure to enemy fire and to position the unit in a good formation to react to enemy contact. Contact with the enemy is made with the smallest force possible to allow the majority of the battalion freedom to maneuver against the enemy. For example, the lead company in a battalion column may have its lead platoon conduct bounding overwatch with the following platoons in travelling. There also may be situations where the lead company is conducting company bounding overwatch with the lead platoon itself conducting bounding overwatch. This is secure but also slow, but may be required in certain situations.

TRAVELING

2-31. ***Traveling* is a movement technique used when speed is necessary and contact with enemy forces is not likely. All elements of the unit move simultaneously. The commanders or small-unit leader locates where best to control the situation. Trailing elements may move in parallel columns to shorten the column and reaction time.** Interval between units is based on visibility, terrain, and weapon ranges (figure 2-13).

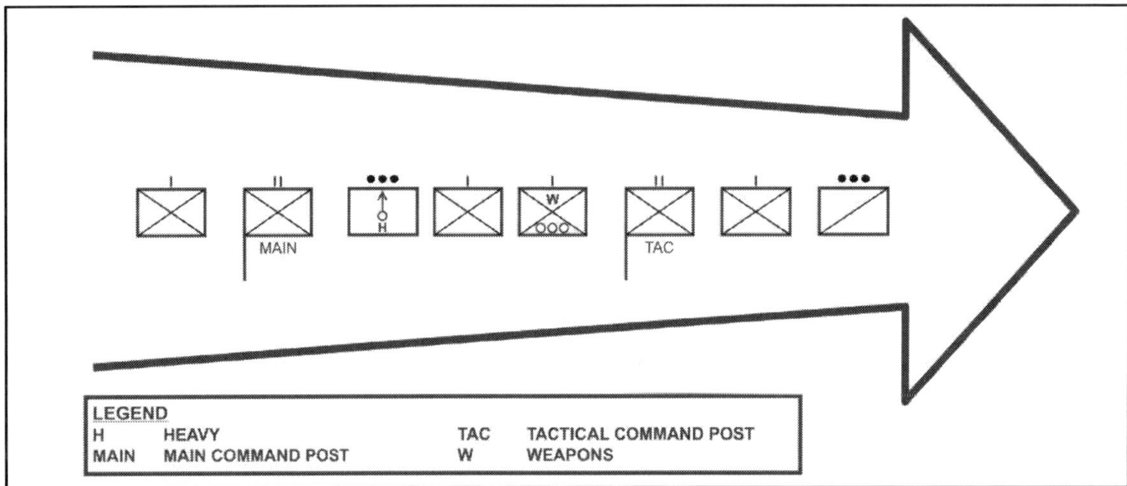

Figure 2-13. Traveling, example

TRAVELING OVERWATCH

2-32. *Traveling overwatch* is a movement technique used when contact with enemy forces is possible. The lead element and trailing element are separated by a short distance, which varies with the terrain. The trailing element moves at variable speeds and may pause for short periods to overwatch the lead element. It keys its movement to terrain and the lead element. The trailing element overwatches at such a distance that, should the enemy engage the lead element, it will not prevent the trailing element from firing or moving to support the lead element (FM 3 90-2). The overwatch unit displaces as necessary, moving at a variable speed (figure 2-14).

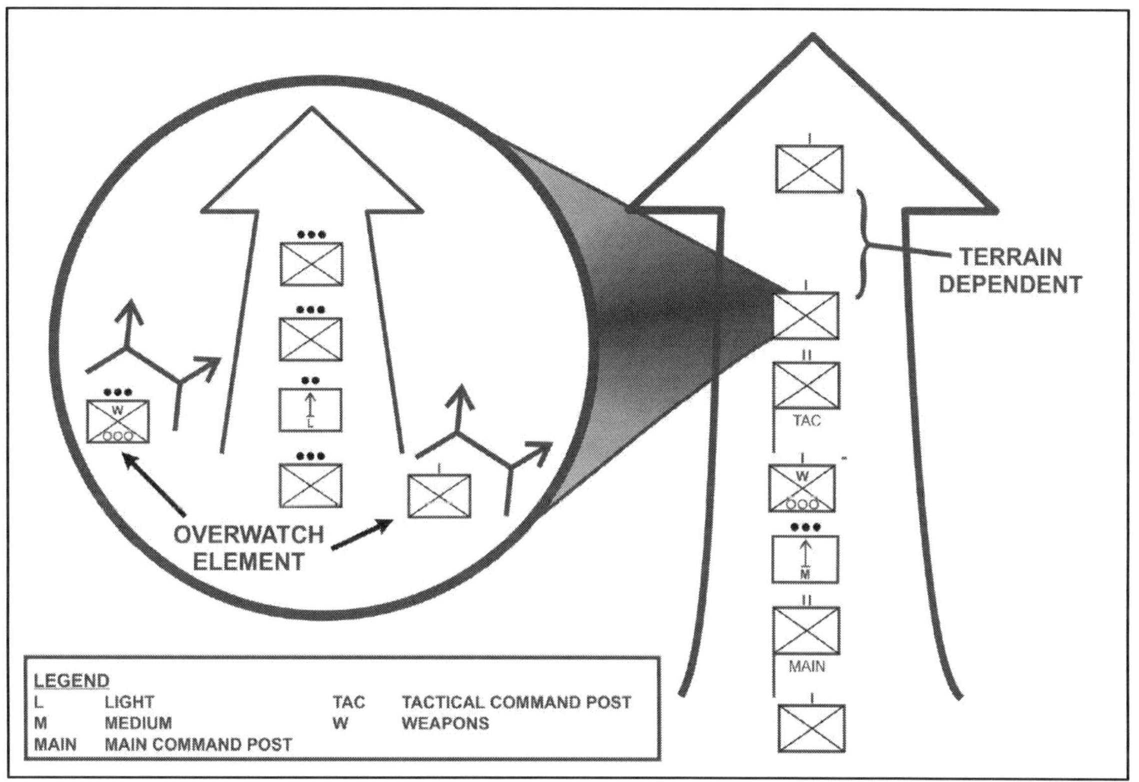

Figure 2-14. Traveling overwatch, example

BOUNDING OVERWATCH

2-33. *Bounding overwatch* is a movement technique used when contact with enemy forces is expected. The unit moves by bounds. One element is always halted in position to overwatch another element while it moves. The overwatching element is positioned to support the moving unit by fire or fire and movement (FM 3-90-2). There are two variations of this technique: alternate bounds and successive bounds (figure 2-15). In both cases, the overwatching elements cover the bounding elements from covered, concealed positions with good observation and fields of fire against possible enemy positions. They can immediately support the bounding elements with maneuver or fires, if the bounding elements make contact. Unless they make contact en route, the bounding elements move via covered and concealed routes into the next set of support by fire positions. The length of the bound is based on the terrain and the range of overwatching weapons.

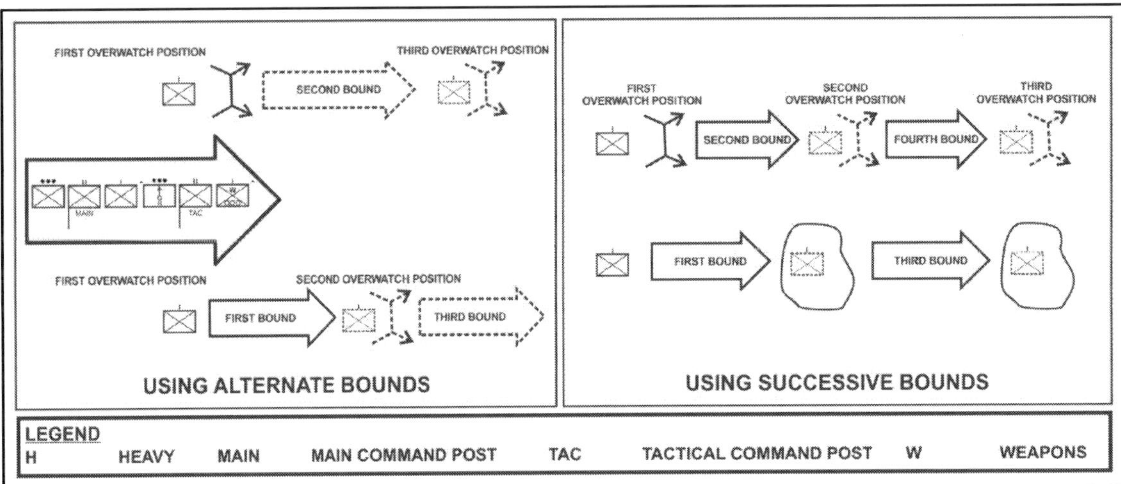

Figure 2-15. Bounding overwatch, example

TROOP MOVEMENT

2-34. *Troop movement* is the movement of troops from one place to another by any available means (ADRP 3-90). Troop movement can be administrative or tactical. Successful movement places troops and equipment at their destination at the proper time and ready for combat. Troop movements are made by different methods, such as dismounted and mounted marches using organic combat and tactical vehicles and motor transport air, rail, and water means in various combinations (see FM 3-90-2). The method employed depends on the situation, the availability of transport, the size and composition of the moving unit, the distance the unit must cover, the urgency of execution, and the condition of the troops. The three types of troop movement are administrative movement, tactical road march, and approach march.

2-35. An *administrative movement* is a movement in which troops and vehicles are arranged to expedite their movement and conserve time and energy when no enemy interference is anticipated (FM 3-90-2). The commander conducts administrative movements only in secure areas. Normally once units deploy into a theater of war, administrative movement is not employed. When conducting administrative movement in secure areas, units should maintain integrity practicing security techniques pertaining to tactical marches.

2-36. The Infantry battalion conducts tactical movement when contact with the enemy is possible or anticipated. The battalion maintains integrity throughout the movement and plans for enemy interference en route to or shortly after arrival at its destination. During a tactical road march (see paragraphs 2-38 to 2-80) or an approach march (see paragraphs 2-196 to 2-206), the battalion uses combat formations and movement techniques consistent with the mission variables of METT-TC. During tactical movements, the commander must be prepared to maneuver against an enemy force. Once deployed in its assigned area of operations, the battalion moves using proper techniques for assigned missions. When contact is made fire and movement is executed.

KEY DOCTRINAL TERMS AND DEFINITIONS

2-37. The following key doctrinal terms and definitions are used throughout this and other chapters and appendixes. Refer to referenced publications for additional information.

- *Assault position* is a covered and concealed position short of the objective, from which final preparations are made to assault the objective (ADRP 3-90).
- *Assured mobility* is a framework—of processes, actions, and capabilities—that assures the ability of a force to deploy, move, and maneuver where and when desired, without interruption or delay, to achieve the mission (ATP 3-90.4).
- *Attack by fire* is a tactical mission task in which a commander uses direct fires, supported by indirect fires, to engage an enemy force without closing with the enemy to destroy, suppress, fix, or deceive that enemy (FM 3-90-1).
- *Attack position* is the last position an attacking force occupies or passes through before crossing the line of departure (ADRP 3-90).
- *Countermobility operations* are those combined arms activities that use or enhance the effects of natural and man-made obstacles to deny an adversary freedom of movement and maneuver (FM 3-34).
- *Fire superiority* is that degree of dominance in the fires of one force over another that permits that force to conduct maneuver at a given time and place without prohibitive interference by the enemy (FM 3-90-1).
- *Follow and assume* is a tactical mission task in which a committed force follows a force conducting an offensive task and is prepared to continue the mission if the lead force is fixed, attrited, or unable to continue (FM 3-90-1).
- *Follow and support* is a tactical mission task in which a committed force follows and supports a lead force conducting an offensive task (FM 3-90-1).
- *Interdiction* is an action to divert, disrupt, delay, or destroy the enemy's military surface capability before it can be used effectively against friendly forces, or to achieve enemy objectives (JP 3-03).
- *Local security* is a security task that includes low-level security activities conducted near a unit to prevent surprise by the enemy (ADRP 3-90).
- *Mobility* is a quality or capability of military forces which permits them to move from place to place while retaining the ability to fulfill their primary mission (JP 3-17).
- *Named area of interest* (NAI) is a geospatial area or systems node or link against which information that will satisfy a specific information requirement can be collected. NAIs are usually selected to capture indications of adversary courses of action, but also may be related to conditions of the operational environment (JP 2-01.3).
- *Observation post* is a position from which military observations are made, or fire directed and adjusted, and which possesses appropriate communications. While aerial observers and sensor systems are extremely useful, those systems do not constitute aerial observation posts (FM 3-90-2).
- ***Reconstitution* is actions that commanders plan and implement to restore units to a desired level of combat effectiveness commensurate with mission requirements and available resources**.
- *Support by fire* is a tactical mission task in which a maneuver force moves to a position where it can engage the enemy by direct fire in support of another maneuvering force (FM 3-90-1).

SECTION II – MOVEMENT TO CONTACT

2-38. *Movement to contact* is an offensive task designed to develop the situation and establish or regain contact (ADRP 3-90). The commander conducts a movement to contact when the enemy situation is vague or not specific enough to conduct an attack. This section discusses the doctrinal basis for the conduct of a movement to contact and introduces a fictional scenario illustrating one of many ways that an Infantry battalion might conduct a movement to contact as part of an IBCT. The scenario includes tasks that the Infantry battalion may conduct prior to the start of a movement to contact or as a follow on task. This section concludes with a discussion of the cordon and search, and search and attack techniques. Both techniques are subordinate tasks to a movement to contact. When involved in operations in support of stability tasks, unit offensive actions normally are closely related to these subordinate tasks of movement to contact.

Note. Tasks illustrated within the movement to contact scenario include—conduct a tactical road march, occupy an assembly area, conduct a passage of lines, conduct battle handover, and conduct an approach march.

CONDUCT A TACTICAL ROAD MARCH

2-39. The movement of troops from one location to another is inherent in any phase of a military operation. Mission accomplishment relates directly to the ability to arrive at the proper place, at the proper time, in effective condition, and in the formation best suited for the assigned mission. Units conducting road marches may or may not be organized into a combined arms formation.

DOCTRINAL BASIS

2-40. A *tactical road march* is a rapid movement used to relocate units within an area of operations to prepare for combat operations. (ADRP 3-90.) Though the primary consideration of the tactical road march is rapid movement, the moving force maintains *security,* which is the measures taken by a military unit, activity, or installation to protect itself against all acts designed to, or which may, impair its effectiveness (JP 3-10). Even when contact with enemy ground forces is not expected the moving force is prepared to act upon enemy contact. When contact is expected, the commander uses a mix of combat formations and movement techniques discussed earlier in section I of this chapter.

2-41. The complexity of operational environments mean that tactical road marches will often occur in and through population centers. As such, commanders must proactively shape or influence interactions with indigenous peoples through the effective conduct of information operations. Integrating and synchronizing information-related capabilities such as presence, posture, and profile; operations security; public affairs; and operations security will significantly improve cooperation and security, among other outcomes.

Tactical March Techniques

2-42. When conducting a tactical road march the Infantry battalion employs three tactical march techniques—open column, close column, and infiltration. The mission variables of METT-TC require adjustments in the standard distances between dismounted Soldiers and vehicles during the conduct of a tactical road march.

Open Column

2-43. In an open column, the commander increases the distance between dismounted Soldiers and vehicles for greater dispersion. The distance between dismounted Soldiers varies from two to five meters to allow for dispersion. Any distance that exceeds five meters between dismounted Soldiers increases the length of the column and hinders control. The ***vehicle distance,* which is the clearance between vehicles in a column, which is measured from the rear of one vehicle to the front of the following vehicle,** varies from 50 to 100 meters, and may be greater if required. Normally, the open column technique is used during daylight. It may also be used at night with infrared lights, blackout lights, or passive night-vision equipment. Using an open column roughly doubles the column's length and thereby doubles the time it takes to clear a point when compared to a close column moving at the same speed. The open column is the most common movement technique because it offers the most security while still providing the commander with a reasonable degree of control. In an open column, a single infantry company, with intervals between its platoons, occupies roughly a kilometer of road or trail with vehicle density varying from 15 to 20 vehicles per kilometer.

Close Column

2-44. In a close column, the dismounted equivalent is a limited visibility march. The distance between individual Soldiers is reduced to one to three meters to help maintain contact and facilitate control. Limited visibility marches are characterized by close formations, reconnaissance, a slow rate of march, and good concealment from enemy observation and air attack. When mounted in a close column, the commander spaces vehicles about 20 to 25 meters apart. At night, vehicles are spaced so each driver can see the two lights in the blackout marker of the vehicle ahead. The commander normally employs a close column for marches during darkness under blackout driving conditions or for marches in restricted terrain. This method of marching takes maximum advantage of the traffic capacity of a route but provides little dispersion. Normally, vehicle density is from 40 to 50 vehicles per kilometer along the route in a close column.

Infiltration

2-45. During troop movement by infiltration, the commander dispatches Soldiers and vehicles in small groups, or at irregular intervals, at a rate that keeps the traffic density down and prevents undue massing of vehicles during a move by infiltration. Infiltration provides the best possible passive defense against enemy observation and attack. It is suited to tactical road marches when there is enough time and road space and when the commander desires the maximum security, military deception, and dispersion. The disadvantages of an infiltration are that more time is required to complete the move, column control is nearly impossible, and recovery of broken-down vehicles by the trail party is more protracted when compared to vehicle recovery in close and open columns. Additionally, unit integrity is not restored until the last group of Soldiers and vehicles arrive at the destination, thus complicating the unit's onward deployment to some degree.

> ***Note.*** Infiltration during troop movement should not be confused with infiltration as a form of maneuver in which an attacking force conducts undetected movement through or into an area occupied by enemy forces as discussed in section I of this chapter.

Extended Tactical Road March

2-46. During an extended tactical road march, halts are necessary to rest Soldiers, service vehicles, and adjust movement schedules. The march order or unit standard operating procedures regulate when to take halts. When halted units establish security and take other measures to protect the force. Once a unit stops moving, there is a natural tendency for Soldiers to let their guard down and relax their vigilance. The commander addresses this by defining in standard operating procedures unit actions for various types of halts, such as maintenance halts, security halts, and unexpected halts. Unit leaders promptly notify commanders of the time and approximate length of unscheduled halts. In mounted movement, the commander schedules short halts for every two to three hours of movement and halts may last up to an hour. Long halts occur on marches that exceed 24 hours and last no more than 2 hours. Long halts are not scheduled at night, which allows maximum time for night movement. During halts, each unit normally clears the march route and moves to a previously selected assembly areas to prevent route congestion and avoid being a lucrative target.

Organization for a Tactical Road March

2-47. The organization for a tactical road march is the march column. A *march column* consists of all elements using the same route for a single movement under control of a single commander (FM 3-90-2). The commander organizes a march column into four elements: reconnaissance, quartering party, main body, and trail party (figure 2-16, page 2-22). Commanders conducting a tactical road march can organize their columns for administrative convenience by similar type, speed, and cross-country capabilities.

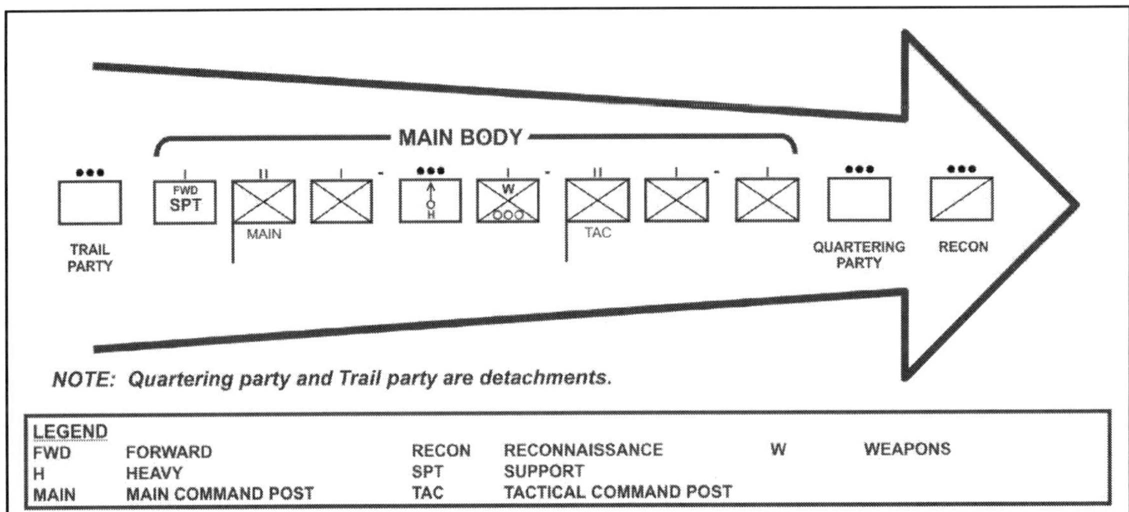

Figure 2-16. Organization of a tactical road march, example

2-48. A march column provides excellent speed, control, and flexibility, but sacrifices flank security. The commander uses a march column when speed is essential and enemy contact is unlikely. March columns are organized to maintain unit integrity and to maintain a task organization consistent with mission requirements. All march units provide their own security. The four elements of a march column, which are reconnaissance, quartering party, main body, and trail party, are discussed in the following paragraphs. (Refer to ATP 3-21.18 and FM 3-90-2 for additional information.)

Reconnaissance Element

2-49. Each march plan is based on a thorough ground reconnaissance if time permits. Map reconnaissance and aerial reconnaissance help formulate a plan but are not substitutes for ground reconnaissance. A reconnaissance element performs route reconnaissance and usually consists of a reconnaissance element, an engineer element (if available) from an attached or supporting engineer unit, and a traffic control element. When the situation dictates, CBRN survey teams may be included in the reconnaissance element. The battalion's tactical standard operating procedures generally establish the reconnaissance element's composition, which can be modified to meet the specific march requirements.

Quartering Party

2-50. A *quartering party* is a group of unit representatives dispatched to a probable new site of operations in advance of the main body to secure, reconnoiter, and organize an area before the main body's arrival and occupation (FM 3-90-2). Based on the order of march, a plan is prepared to guide each unit over a designated route. This route begins at the release point and extends to the unit's new site of operations. Guides must understand and rehearse the plan. This prevents congestion or delays near release points. The actual dispatch of the quartering party can follow the issuance of the movement order.

Main Body

2-51. The main body includes the march commander and majority of Soldiers and march vehicles. The march commander positions within this section for effective mission command. Battalion march units of the main body consist of individual maneuver units with their mortars, trains, command posts, and attachments.

Trail Party

2-52. The *trail party* is the last march unit in a march column and normally consists of primarily maintenance elements in a mounted march (FM 3-90-2). The assistant march commander usually leads a battalion-size unit trail party or the battalion maintenance officer when vehicles are included in the march. The party may consist of elements of the maintenance and medical sections.

Control and Scheduling within a March Column

2-53. To facilitate control and scheduling within a march column, units are organized into subordinate elements, a march serial and a march unit, and are given an order of march (see figure 2-19, page 2-30). A *march serial* is a major subdivision of a march column that is organized under one commander who plans, regulates, and controls the serial (FM 3-90-2). An example is a battalion serial formed from a brigade-sized march column. A *march unit* is a subdivision of a march serial. It moves and halts under the control of a single commander who uses voice and visual signals (FM 3-90-2). An example of a march unit is a company from a battalion-sized march serial. (Refer to ATP 3-21.18 and FM 3-90-2 for additional information.)

March Planning

2-54. Tactical road marches require extensive planning and coordination, especially when multiple units are using the same route. The commander and staff determine how best to execute a move from one point to another. Refer to ATP 3-21.18 and FM 3-90-2 for a detailed discussion of movement planning considerations, terms, and movement time computation.

Key March Considerations

2-55. Key march considerations include the following:

- Requirements for the movement. (Refueling time, the distance between march units factoring in the time required at the refuel site, and time to clear choke points.)
- Organic and nonorganic movement capabilities; determine external movement requirements.
- Unit movement priorities.
- Enemy situation and capabilities, terrain and weather, and civil considerations.
- Actions on contact and likely engagements.
- Positioning recovery vehicles throughout the march unit's flow to provide flexibility and timely push or pull recovery to close all vehicles on the destination.
- Ensure all march units have security and firepower.
- Security measures before and during the movement and at the destination.
- Assembly of the march units.
- Fire support coverage during movement and at the destination.
- Communications; particularly units without Blue Force Tracker.
- Movement of civilians along the same or intersecting routes.
- Actions at the destination.

Sequence of March Planning

2-56. When preparing for a tactical road march, the battalion uses the following sequence of march planning, as time permits:

- Prepare and issue a warning order (WARNORD) as early as possible to allow subordinates time to prepare for the march.
- Analyze routes designated by higher headquarters and specify organization of the march serial.
- Prepare and issue the march order.
- Prepare a detailed movement plan and site destination plan (for example, assembly area plan).
- Organize and dispatch reconnaissance and quartering parties as required.

Dismounted March and Mounted March

2-57. *Dismounted march* is the movement of troops and equipment, mainly by foot, with limited support by vehicles. Also called foot march (FM 3-90-2). *Mounted march* is the movement of troops and equipment by combat and tactical vehicles (FM 3-90-2). The headquarters and headquarters company, weapons company, and assigned forward support company can move all their Soldiers with organic vehicles while the Infantry rifle companies are primarily foot-mobile. This can become more complicated if elements of the weapons company are attached to the rifle companies. To compensate, the battalion commander may do one or a combination of the following:

- Request vehicles to move dismounted Infantry. This permits the march to move much faster and, more importantly, keeps Soldiers rested and better prepared for combat.
- Move the mounted and dismounted elements as separate march columns. Depending on the separation, this may reduce the Infantry weapons company and mortar platoon ability to support forward elements of the march.
- Use the weapons company to establish overwatch positions along the route of march. Have the mortars displace by sections to ensure immediate fires.
- Move the mounted element as a separate march column while using available vehicles to shuttle the Infantry companies forward. Vehicles may include those from the forward support company that have off-loaded their normal loads and returned to provide transport. The rifle companies may move dismounted and be picked up en route or remain in the previous assembly area to wait for vehicles.

Movement Order and Movement Table

2-58. The movement order and movement table provide clear and concise information and instructions to subordinates to accomplish movement within the framework of the commander's intent. The movement order clearly states all required information for units to perform their assigned tasks. Tasks must be understood for a movement order since it may be preceded by a tactical operation or follow after an operation or mission. The movement table, as an attachment to the movement order, is a convenient means of transmitting time schedules and other essential march details to subordinate units. (Refer to ATP 3-21.18, appendix B for additional information.)

Control Measures

2-59. The commander uses control measures to assist in controlling the battalion during the tactical road march to include speed and rate of march. The commander directing a tactical road march often uses a strip map (figure 2-17) or overlay (figure 2-18, page 2-29) to graphically depict critical information about the route to subordinates. Common march control measures include route of march, start point, release point, checkpoints, critical points (such as bridges, major cities and towns), and traffic control posts. Unless the commander directs them not to do so for security reasons, march units report when they have crossed each control point. Other control measures and information may include assembly areas, scheduled halts, distance between checkpoints (strip map), and north orientation (strip map). (Refer to FM 3-90-2 for additional information.)

Note. Additional control measures, commonly established above battalion echelon, may include a light line (during periods of limited visibility, designated line requiring the use of blackout lights) and movement corridor (designated area established to set the conditions to protect and enable movement of traffic along a designated surface route). Refer to FM 3-90-2 and ATP 3-91, respectively, for additional information.

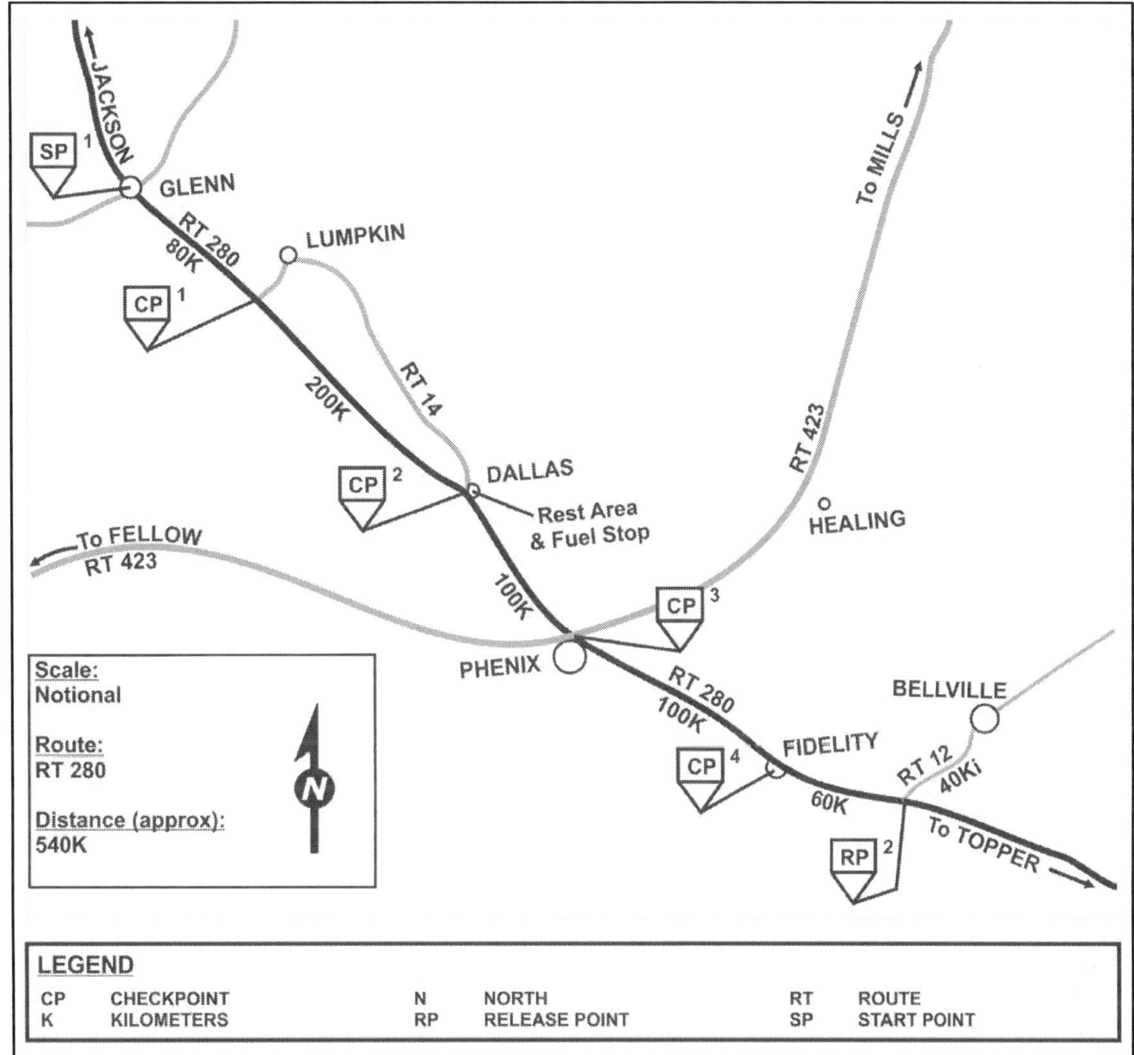

Figure 2-17. Strip map, tactical road march (mounted march)

Start Point

2-60. The *start point* is a location on a route where the marching elements fall under the control of a designated march commander (FM 3-90-2). At this point the column is formed by the successive passing, at an appointed time, of each of the elements comprising the column. The start point should be an easily recognizable point on the map and on the ground. It should be far enough from the assembly area to allow units to be organized and moving at the prescribed interval and rate when the start point is reached.

Release Point

2-61. A *release point* is a location on a route where marching elements are released from centralized control (FM 3-90-2). At the release point, each element continues its movement toward its own destination. Multiple movement routes from the release point enable units to disperse rapidly and navigate to their assembly areas or area of operation.

Scheduled Halts

2-62. Scheduled halts are used to control and sustain the march. Scheduled halts are preplanned along the march route for maintenance and rest, or to follow higher echelon movement orders. They should be located on defensible, covered, and concealed terrain. During scheduled halts, vehicles and Soldiers pull to the side of the road while maintaining march dispersion. Local security, including at least one observation post for each platoon, is established immediately, and drivers perform during operation maintenance checks. Observation posts should not be established outside small arms range and should be readily retrievable so the unit is ready to move at a moment's notice. Unit leaders promptly notify the march commander of the time and approximate length of unscheduled halts.

Critical Points

2-63. Critical points on a route are places used for information references, places where obstructions or interference with movement might occur, or places where timing may be a critical factor. The commander may positions traffic control posts along the route to prevent congestion and confusion at these critical points. During mounted movements, maintaining and improving routes, and creating bypass or alternate routes at critical points are major engineering tasks because movement routes are subject to fires from enemy artillery and air support systems. Obscurant curtains, blankets, and haze may protect withdrawing columns at this critical points along the route.

Communications

2-64. The ability to communicate during march operations is essential. Radio nets must be established to link the march commander with higher headquarters, fire support, element commanders, reconnaissance forces, weapons trucks, medics, and the quartering and trail party. Within the column, each march element may have its own control net with the march element commander and quartering and trail party. Other communications techniques such as signals must be established and rehearsed.

2-65. March units must be prepared to operate with degraded communications and digital networks. Messengers and visual signals are also excellent means of communication during marches. The battalion generally moves under radio silence and uses radio only in emergencies or when it can use no other means of communication. The battalion also can use road guides to pass messages from one march unit to a following march unit. Road guides are also important in controlling march units and the interval between them.

Traffic Control

2-66. The headquarters controlling the march may post road guides and traffic signs at designated traffic control posts. A *traffic control post* is a manned post that is used to preclude the interruption of traffic flow or movement along a designated route (FM 3-39). At critical points, guides assist in creating a smooth flow of traffic along the march route. Attached military police (if available) or designated elements from the quartering party may serve as guides. They should have equipment or markers that will allow march elements to identify them in darkness or other limited visibility conditions.

Security

2-67. During the movement, march units maintain security through observation, weapons orientation, dispersion, and concealment. Commanders assign sectors of observation to their personnel to maintain 360-degree observation. Main weapons are oriented on specific sectors throughout the column. The lead elements cover the front, following elements cover alternate flanks, and the trail element covers the rear. Air guards also are designated if there is an air threat.

Halts

2-68. While taking part in a tactical road march, the march elements must be prepared to conduct both scheduled and unscheduled halts. In either case, Soldiers and vehicles should move to the side of the road while maintaining vehicle dispersion. Security at halts is always a priority.

Air Defense

2-69. Planning for air defense and implementing all forms of air defense security measures are imperative to minimize the battalion's vulnerability to enemy air attack. The commander must integrate the battalion's fire support plan effectively with any attached or supporting air defense assets. Furthermore, the commander must ensure the battalion plans and uses passive and active air defense measures.

2-70. Should hostile aircraft attack the battalion during the march, the march unit under attack moves off the road into a defensive posture and immediately engages the aircraft with all available automatic weapons. The rest of the battalion moves to covered and concealed areas until the engagement ends.

Chemical, Biological, Radiological, and Nuclear

2-71. If a CBRN threat exists, units should conduct monitoring activities during movement. This may include having mounted elements checking upwind of the march columns during the march. A CBRN reconnaissance should be conducted prior to the movement of the main body and areas of contamination marked. Units will move around CBRN-contaminated areas along designated routes.

Obstacles

2-72. The battalion should bypass reported obstacles, if possible. If it cannot bypass obstacles, the lead march unit establishes a hasty defense, provides overwatch, then breaches the obstacle, working with engineers, if available. As lead march elements conduct the breach (see appendix F), the other march units move at a decreased speed or move off the road and monitor the battalion command net. The commander of the breaching element communicates the posted location of the obstacle as soon as possible.

Enemy Indirect Fire

2-73. Should the battalion come under attack by enemy indirect fire during the march, the unit in contact continues to move. The remainder of the battalion tries to bypass the impact area.

Restrictions

2-74. Restrictions are points along the route of march where movement may be hindered or obstructed. The march planner should stagger start times or adjust speeds to compensate for restrictions, or he should plan to halt the column en route until the restriction ends.

Limited Visibility

2-75. Units must be able to operate routinely under limited visibility conditions caused by darkness, smoke, dust, fog, heavy rain, or heavy snow. Limited visibility decreases the speed of movement and increases the difficulty in navigating, recognizing checkpoints, and maintaining proper interval between units. To overcome control problems caused by limited visibility, commanders may position themselves just behind lead elements. More restrictive control measures, such as additional checkpoints, phase lines, and use of a single route, may become necessary.

March Preparation

2-76. Before starting a march, each march unit of a serial reconnoiters and rehearses the route to the start point and determines the exact time to reach it. The movement order states the unit's arrival and clearance times at the start point. The serial commander then determines and announces the times for march units to arrive at and clear the start point. Arrival time at the start point is critical. Each march unit must arrive at and clear the start point on time; otherwise, movement of other elements may be delayed. Additionally each march unit, and the serial as a whole, rehearse actions along the entire road march route, specifically actions on contact with the enemy. (Refer to ATP 3-21.18 for additional information.) Finally, each march unit must anticipate and rehearse possible interactions with indigenous populations and plan for acceptable (culturally-attuned) and unacceptable (insensitive) actions upon encountering them. Consideration should be given to having interpreters in the formation. (See FM 3-13, Chapter 9, for further information.)

March Execution

2-77. March execution depends upon organizations, tasks organized to accomplishment critical tasks and that are flexible enough to adjust to changing conditions to ensure mission success. During movement, march units move at the constant speed designated in the order, maintaining proper interval and column gap. Elements in a column of any length may simultaneously encounter many different types of routes and obstacles, resulting in different parts of the column moving at different speeds at the same time. This can produce an undesirable accordion-like action. March units report crossing each control point as directed by the march order and maintain air and ground security during the move. (Refer to ATP 3-21.18 for additional information.) When contact is made with indigenous populations, the serial commander executes planned information operations activities. (See FM 3-13.)

Note. The following illustration introduces a fictional scenario as a discussion vehicle for illustrating one of many ways that an Infantry battalion can conduct a tactical road march. Prior to the tactical movement below, elements of the IBCT occupied assembly areas to the rear of the division's deployed forces. The IBCT conducts troop movement to stage units forward from the division's rear boundary along Route Bear (Infantry battalion 2) and Route Lion (Infantry battalion 1) to Assembly Area Eagle and Assemble Area Falcon, respectively, to prepare for future combat operations (IBCT movement to contact). Discussion within the illustration focuses on Infantry battalion 2.

Illustration of a Tactical Road March

2-78. In this scenario, Infantry battalion 2 conducts a tactical road march (mounted march) along Route Bear to Assembly Area Eagle. After initial occupation of the assembly area, the battalion continues to prepare for offensive action (battalion movement to contact). On order, the battalion conducts a second tactical road march (dismounted) along Route Fox to Release Point 4 located north of Attack Position 2 (figure 2-18).

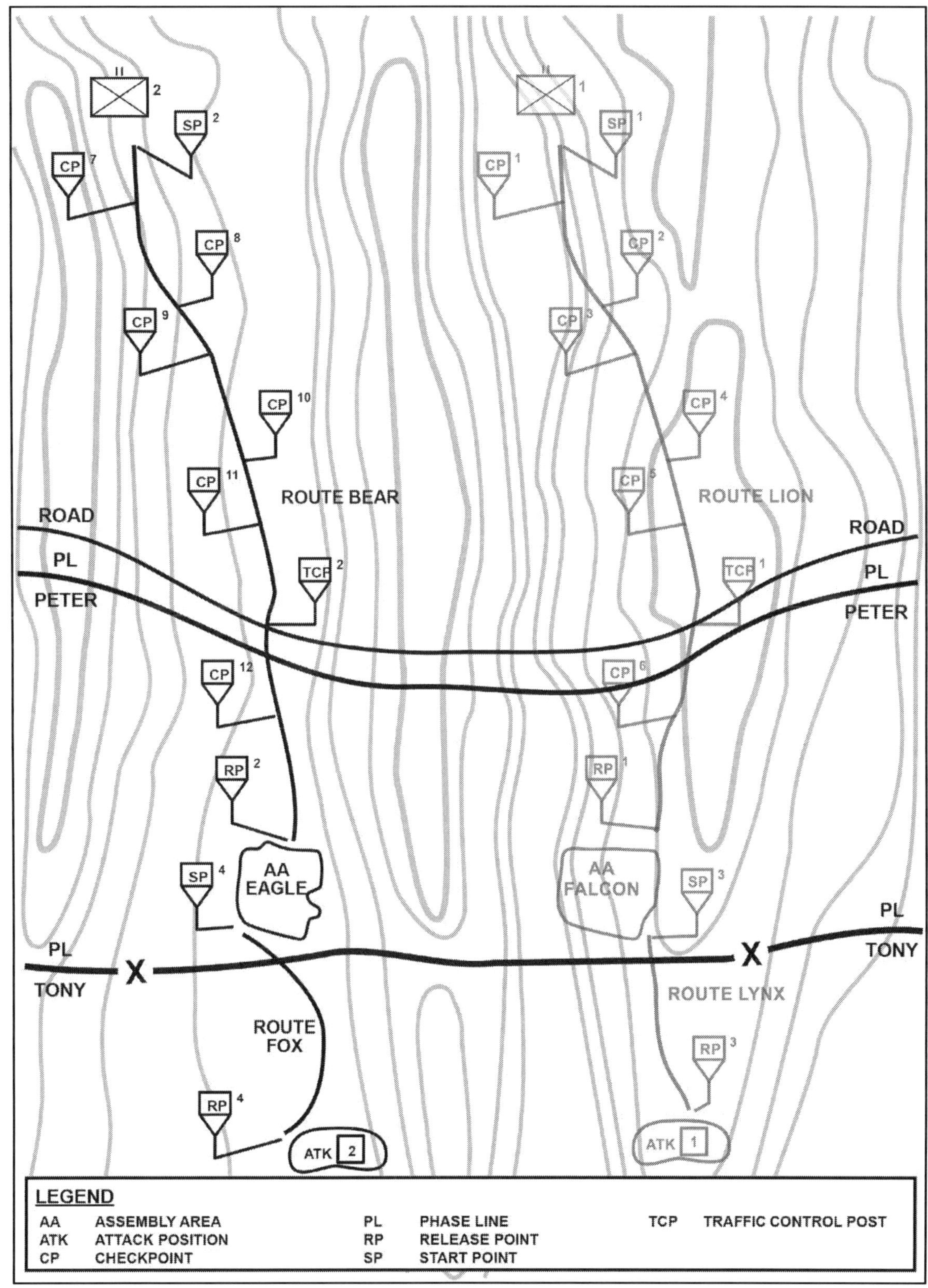

Figure 2-18. Overlay with route control measures, example

Infantry Battalion 2—March Serial

2-79. Staging forward to prepare for combat operations, Infantry battalion 2 conducts tactical movement by mounted march to the rear of an established area defense. Speed is vital, and security requirements are minimal. Infantry battalion 2 coordinates the movement of march units and elements with the IBCT and the unit(s) responsible the route and for the area of operation where the assembly area is located. March units and elements within the battalion mounted march serial include reconnaissance, quartering party, main body, and trail party (figure 2-19). During tactical movement, march elements prepare for actions on contact against likely enemy contact along the route.

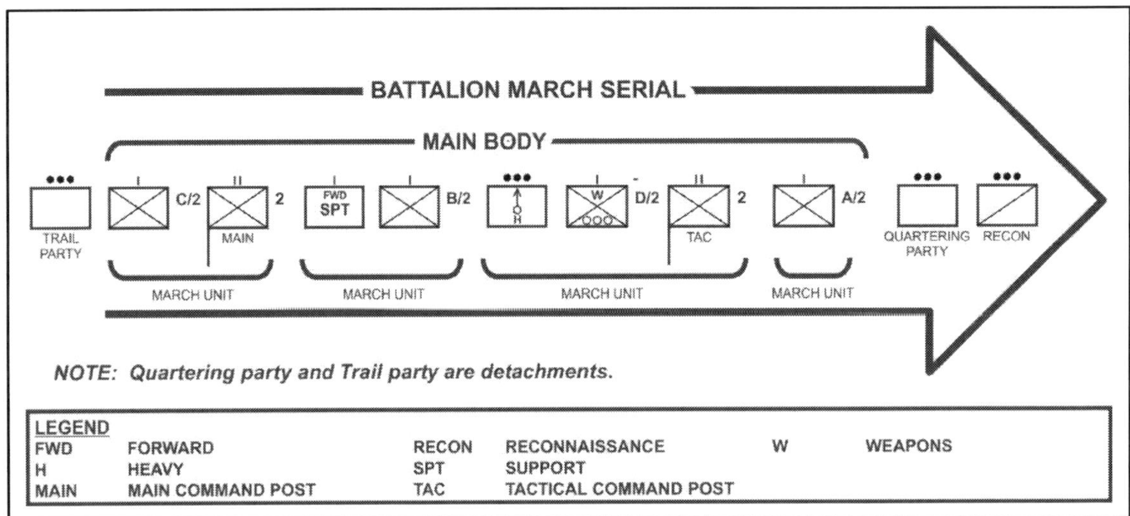

Figure 2-19. Battalion march serial (mounted march), example

> ***Note.*** March serials and units are organized to maintain unit integrity and task-organized as determine by the mission variables of METT-TC.

Reconnaissance

2-80. The reconnaissance element (scout platoon) conducts a route reconnaissance moving ahead of the quartering party. The reconnaissance element does not have attached engineer support; the scout platoon enables information requirements regarding technical reconnaissance–route classification, as required. The reconnaissance element does not have attached CBRN reconnaissance support; the scout platoon monitors its chemical and radiological detectors, as required. Reconnaissance element makes contact with responsible unit(s) at Release Point 2 in preparation for quartering party's arrival then confirms their linkup and arrival at Assembly Area Eagle with the main body and the quartering party.

Quartering Party

2-81. Prior to the tactical movement, the quartering party leader and reconnaissance leader, meet and are briefed by the responsible unit leader(s) regarding the route, release point, and occupation of the assembly area. During this movement, the battalion's quartering party (detachment) is led by the headquarters and headquarters company commander and consists of—

- An assault platoon.
- One mortar squad with a fire direction center element, led by the mortar squad leader.
- Representatives from each company and attached unit in the battalion. Each company sends the company executive officer, one representative from each of its platoons, and a representative from the headquarters section.
- Representatives from the battalion tactical and main command posts, battalion trains, and communications section.

2-82. The quartering party takes the same route as the battalion, Route Bear. Led by the scout platoon conducting a route reconnaissance, the quartering party moves to Assembly Area Eagle. The assault platoon is the primary security element for the quartering party.

2-83. In route, prior to Release Point 2, responsible unit leader(s) link up with the quartering party, separate at Release Point 2 and move to their designated areas. Quartering party members identify and mark the exact positions for their units and ensure they are tied into the units to their flanks. Subordinate unit command posts are marked. Other actions include—

- Tactical and main command posts (if not combined) locations within the assemble area are identified and marked.
- The mortar squad establishes a firing position and firing positions for the other mortars are identified.
- Sustainment representatives identify and mark positions for their units.

2-84. The main body of the battalion conducts the march while the quartering party establishes the assembly area. Unit guides are sent from the quartering party to Release Point 2 to meet their units, and without stopping the unit, lead their unit into position. Subordinate unit release points are established to assist in the occupation of the assembly area.

Main Body

2-85. Infantry battalion 2 main body conducts mounted march, crosses the start point at the correct time and moves along Route Bear to Assembly Area Eagle. The main body includes rifle companies, weapons company, battalion command posts, battalion mortar platoon, and battalion unit and company trains. The battalion conducts the move under radio-listening silence. Each vehicle commander has a route overlay or strip map of Route Bear. The order of march by march unit is as follows:

- Company A with its company trains.
- Company D (Weapons company-minus), battalion mortar platoon, and tactical command post.
- Company B with its company trains and the battalion unit trains.
- Company C with its company trains and main command post.

Trail Party

2-86. Trail party with one assault platoon, one medical treatment team and one ambulance squad, one field maintenance team, and recovery vehicles.

Battalion March Units

2-87. Battalion march units conduct mounted march in accordance with the movement order and the battalion's tactical standard operating procedures. The reconnaissance element crosses the start point 90 minutes prior to the main body. The quartering party crosses the start point 60 minutes prior to the main body. The trail party crosses the start point 30 minutes after the last element of the main body crosses the start point. Mounted march procedures include the following:

- Move in an open column at the designated speed.
- Under radio silence, march unit leaders monitor the battalion command net.
- Orient vehicle weapons systems outward, mounted Infantry face outward.
- Post air guards.
- March unit leaders report passing checkpoints through the digital system.
- Monitor and be prepared for situational updates reported by the reconnaissance element (to include obstacle markings, changes to route, and other factors).
- Vehicle break down procedures—move to the side, provide local security, and wait for the trail party to repair or recover the vehicle.
- Infantry on vehicles that break down—leave a fire-team sized element to help secure the vehicle, cross-load remaining Infantry to other vehicles within the march unit.

- During scheduled or unscheduled halts, march unit moves off the road and assumes a herringbone formation (see ATP 3-21.8). Drivers and crew conduct maintenance. Infantry dismount to provide local security.

March Unit Release Point

2-88. Release point—march units linkup with guides and move into area of operations within Assembly Area Eagle. March units do not stop at the release point. Units within each march unit are organized to ease its movement into an area of operations within the assembly area.

Note. Within this scenario, Infantry battalion 2 moves to Assembly Area Eagle to make final preparation for a forward passage of lines and movement to contact.

OCCUPY AN ASSEMBLY AREA

2-89. An assembly area is a location where a unit prepares or regroups for future action. Units move with as much secrecy as possible, normally at night and along routes that prevent or degrade the enemy's capabilities to observe or detect during the occupation of an assembly area. Units avoid congesting in assembly areas and occupy them for the minimum possible time. While in the assembly area, each unit is responsible for its own protection activities, such as local security. Designation and occupation of an assembly area may be directed by a higher headquarters or by the unit commander during relief or withdrawal operations or unit movements.

DOCTRINAL BASIS

2-90. An *assembly area* is an area a unit occupies to prepare for an operation (FM 3-90-1). Assembly areas are areas occupied by forces where enemy contact is likely and commitment of the unit directly from the assembly area to combat is possible or anticipated. Units likely to occupy assembly areas include units designated as tactical reserves, units completing a rearward passage of lines, units preparing to move forward to execute a forward passage of lines, units performing tactical movements, and units conducting reconstitution. An assembly area should provide—

- Concealment from air and ground observation.
- Adequate entrances, exits, and internal routes.
- Space for dispersion; with enough distance from other assembly areas to preclude mutual interference.
- Cover from direct fire.
- Good drainage and soil conditions that can sustain unit vehicles and individual Soldier movements.
- Terrain masking of electromagnetic signatures.
- Terrain allowing observation of ground and air avenues into the assembly area.
- Sanctuary from enemy medium-range artillery fires.
- Sufficient space for sustainment operations.

Organization

2-91. Assembly areas may be organized using one of three methods. The battalion may be part of a larger unit's assembly area, occupy an assembly area on its own, or assign companies their separate assembly areas.

Occupy a Portion of an Assembly Area

2-92. The battalion may occupy a portion of the perimeter of an assembly area. It occupies the area of operation assigned by the higher commander and by arraying companies, generally on a line oriented on avenues of approach into the assembly area. Leftmost and rightmost units tie in their fires and areas of observation with adjacent units. Depending on the tactical situation and width of the area assigned to it, the battalion may maintain a reserve. Battalion trains are located to the rear of the companies. Centrally located in the assembly area; the main command provides mission command, battalion mortar platoon provide fire support. The battalion scout platoon screens and establishes observation post along the most likely or most dangerous avenues of approach into the assembly area.

Occupy a Separate Assembly Area

2-93. The battalion may assign areas to subordinate companies and require them to tie in fires and observation with each other. The main command post, trains, and mortar platoon are located near the center of the assembly area. Company area of operations are assigned based on their future mission, their combat power and the presence of enemy avenues of approach within their area of operation. The scout platoon occupies observation posts at key points around the entire perimeter of the battalion or screens along the most dangerous or likely enemy avenues of approach. This method configures the battalion in a perimeter defense. This is the most common organization of a battalion assembly area.

Assign Companies to Separate Assembly Areas

2-94. The battalion may assign separate individual assembly areas to subordinate companies, which establish their own perimeter defense. Areas between companies are secured through surveillance and patrolling and is usually under the control of the battalion or higher headquarters. The main command post, trains, and mortar platoon establish positions central to outlying companies. When the battalion is dispersed over a large area, air defense assets (if available) may need to collocate with companies.

Quartering Party

2-95. A quartering party is a group of unit representatives dispatched to a new site of operations to secure, reconnoiter, and organize an area before the main body's arrival and occupation. The battalion's tactical standard operating procedures establish the general composition of the quartering party and its transportation, security, communications equipment, and specific duties. Quartering parties typically reconnoiter, to include CBRN reconnaissance, and confirm the route and tentative locations previously selected from map reconnaissance. Quartering parties also serve as a liaison between their parent headquarters and the quartering party of their higher headquarters to change unit locations in the assembly area based on their reconnaissance.

2-96. The intelligence staff officer (S-2) routinely receives intelligence information from higher headquarters throughout the battalion's deployment and operations. From this information, the S-2 determines the characteristics and likelihood of the air and ground threat to the quartering party during its movement to and occupation of the assembly area. This information assists the battalion staff and the quartering party officer in charge in determining the security required and the desirability of maintaining the quartering party in the assembly area during the movement of the rest of the battalion.

2-97. The quartering party may move with another unit's quartering party, depending on the likelihood of enemy contact. Ideally, the quartering party moves over the routes to be used by the battalion and executes a route reconnaissance and time distance check. The quartering party typically includes an officer in charge or noncommissioned officer in charge and representatives from the battalion main command post, battalion trains, and the battalion's subordinate units. The battalion S-3 air, S-3 sergeant major, and S-1, headquarters and headquarters company commander and executive officer, and command sergeant master are potential quartering party leaders.

2-98. Composition of the maneuver company quartering parties usually is determined by the company commander but may be specified by the battalion commander. Headquarters and headquarters company representatives typically include noncommissioned officers from key support sections such as communications or supply. Representatives from the mortar platoon and the scout platoon may also be represented in the quartering party.

2-99. The main command post quartering party identifies potential command post locations based on tactical requirements such as cover and concealment and the line-of-sight signal requirements. When planning time is short, key members of the staff or the tactical command post may move with the quartering party. This enables the staff to begin detailed planning immediately upon arrival in the assembly area. This technique also facilitates transitions to new missions by prepositioning key staff members so planning can occur concurrently with the movement of the main body. When the battalion moves and occupies its assembly area as part of the IBCT, the IBCT makes all coordination for fire support. Otherwise, the battalion conducts its own planning and coordination.

2-100. During its planning, the staff determines sustainment requirements for the quartering party. The estimate of necessary supplies and equipment covers the entire quartering party, including accompanying staff section representatives and fire support, protection, and sustainment assets. The quartering party may move under radio silence or other emission restrictive postures, especially during movement into the assembly area.

2-101. The quartering party leader briefs the quartering party after the plan is completed. The briefing follows the standard five-paragraph field order format. In it, the quartering party leader emphasizes actions at halts and critical areas, actions of the quartering party in the assembly area, contingency plans, and procedures to request and receive fire support, protection, and sustainment. The leader covers in detail medical evacuation procedures, actions on contact, and actions to take if separated from the quartering party. Rehearsal and backbriefs (see appendix B) times are scheduled.

2-102. Prior to and after rehearsals and during final preparations the S-2 ensures the quartering party leader is aware of any changes to the current enemy situation, probable enemy course of actions, the weather forecast, and the terrain and vegetation likely en route to and in the assembly area. The quartering party leader coordinates with the S-3 to determine any mission changes, for example, whether or not the quartering party is to remain in the assembly area and await the remainder of the battalion, or a change to the route and movement restrictions to be used by the quartering party.

2-103. During rehearsals the quartering party leader ensures subordinate unit quartering parties know where and when the battalion quartering party will be located in the assembly area. The battalion S-3 determines whether it is required to send engineer personnel, if available, with the quartering party after final reconnaissance and evaluations of routes, bridges, and cross-country mobility.

2-104. Air defense units, when available, may move with the quartering party en route to and within the new tactical assembly area. If air defense assets move with the quartering party, the air defense unit leader ensures he knows both the current and projected weapons control status and air defense warning. If a CBRN threat is present, CBRN reconnaissance is conducted in conjunction with the route reconnaissance. The route is adjusted around any CBRN contamination sites and guides may be required to re-direct the main column onto the adjusted route.

2-105. The quartering party navigates to the assembly area, generally along one route. If the quartering party moves along a route to be used by the main body and the main body has not yet sent a reconnaissance element forward, the quartering party conducts a route reconnaissance during its movement. The quartering party also may execute a time distance check of the designated route. The quartering party reports these times and distances to the main command post after moving through the release point.

2-106. Upon arrival in the assembly area, the quartering party moves to its assigned positions and executes the required reconnaissance. The quartering party also has the following responsibilities at the assembly area:

- Determines locations for units.
- Identifies unit left and right limits of fire, records this information, and sends updates to the unit's commander.
- Determines the location for the main command post. This may include establishing communications equipment, laying wires, and so forth.
- Verifies subordinate unit locations and sectors of fire to ensure there are no gaps in coverage.
- Transmits changes or updates to the main command post to alert the main body to changes in the route and assembly area.

2-107. If the proposed location for the assembly area is unsuitable, the quartering party leader attempts to adjust the assigned areas. If adjustment is not possible, he immediately notifies the S-3 or commander.

2-108. If an element of the main command post has accompanied the quartering party, it moves to the location reconnoitered by its representative and establishes the tactical command post. If air defense assets have accompanied the quartering party, they occupy firing positions oriented on air avenues of approach. Representatives organize their respective areas by selecting and marking positions for vehicles and support facilities.

2-109. If the battalion quartering party is not going to remain in the assembly area, it does not depart the assembly area until all subordinate unit quartering parties have reported. The unit quartering parties should provide the results of their reconnaissance and identify requested changes to their tentative locations.

2-110. Guides move to the battalion release point to meet and guide their units. Guides are especially needed during periods of limited visibility. Sustainment assets may accompany the quartering party. Sustainment elements generally conduct resupply operations for the quartering party at scheduled halts or in the new assembly area.

Note. The following illustration introduces a fictional scenario as a discussion vehicle for illustrating one of many ways that an Infantry battalion can occupy an assembly area.

ILLUSTRATION OF AN ASSEMBLY AREA

2-111. The following scenario illustrates an Infantry battalion occupying an assembly area. In this scenario, Infantry battalion 2 completes a tactical road march along Route Bear and occupies Assembly Area Eagle. During the battalion's main body movement to the assembly area the quartering party secured, reconnoitered, and organized the area. Quartering party elements identified and marked exact positions for units and ensured positions within the assembly area provide mutual support. The quartering party identified and marked command post positions, firing positions, and unit train locations. Unit guides from the quartering party went to release point 2, meet their units, and without stopping the units, lead them to their areas within the assembly area. Subordinate unit release points were established to ease congestion. Figure 2-20 illustrates the occupation of Assembly Area Eagle.

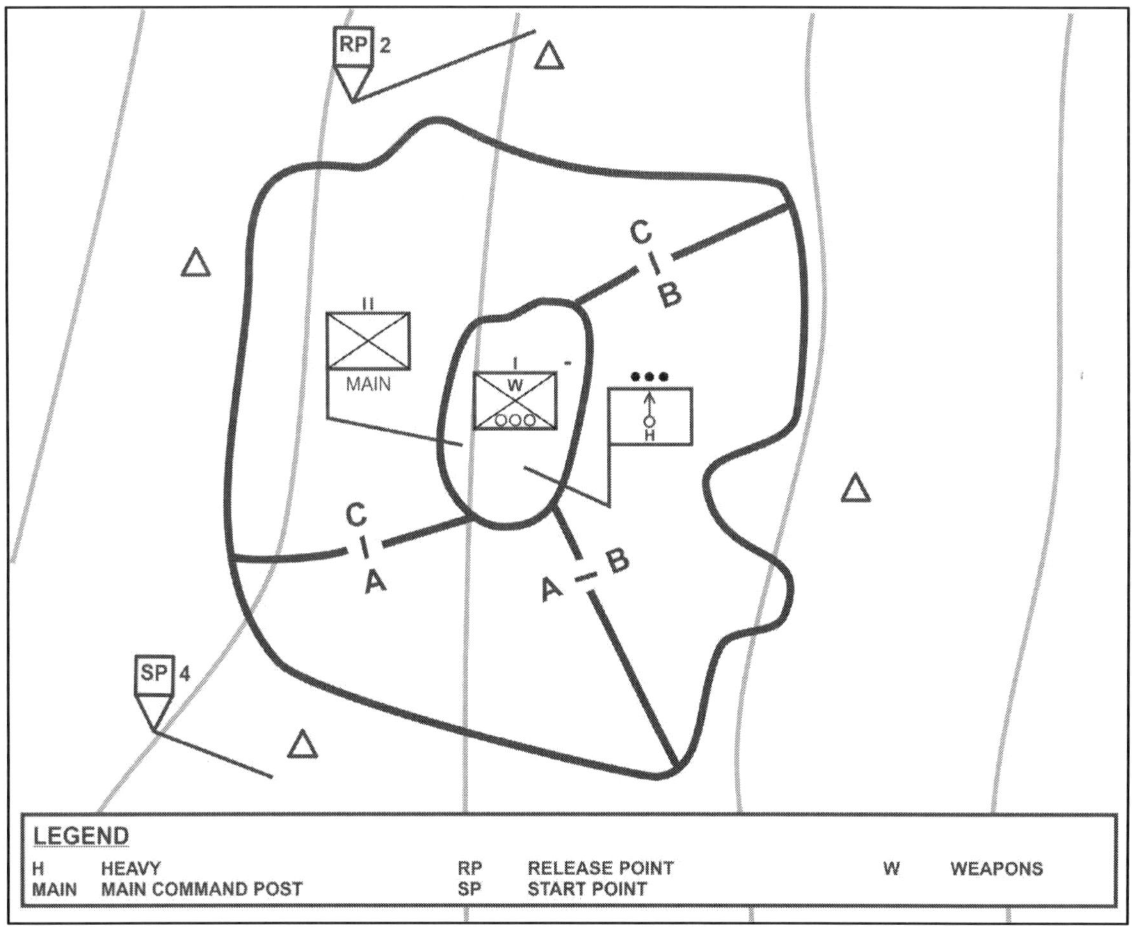

Figure 2-20. Occupy Assembly Area Eagle, example

Occupation of the Assembly Area

2-112. Once clear of the Release Point 2, march units follow guides into assigned areas within the assembly area. For example Company A, the lead serial of the main body, has an assigned area on the south side of the assembly area. Each march unit moves to its area of operation, dismounts Soldiers, forms its portion of the perimeter, and immediately establishes local security. The guide shows the commander and subordinate leaders their initial areas of operation. The commander make changes as required as units move into position. Units establish communications with higher headquarters and adjacent units. Crew-served weapons, heavy weapons, and mortars are positioned. Depending on the priorities work, fighting positions are built.

Actions at the Assembly Area

2-113. In the assembly area, units prepare for the forward passage of lines and movement to contact, which includes the following:

- Receive required supplies.
- Conduct rehearsals.
- Conduct inspections.
- Maintain weapon, vehicles, and other equipment.
- Check communications.
- Receive orders.

Departure from Assembly Area

2-114. Planning considerations for occupying the assembly area are based largely on the anticipated future missions of units. Units position in the assembly area so they can depart the assembly area en route to their assigned tactical missions without countermarching or moving through another unit.

Start Point

2-115. Units departing the assembly area must reach the start point at the correct interval and time. The start point is located outside the assembly area to allow units to maneuver out of their positions and configure for the tactical road march before reaching the start point. The start point for the battalion's tactical movement is located an adequate distance from the assembly area to permit the march units (companies) to attain proper speed and intervals before crossing it. Units do not halt at the start point rather they have measured the travel time between their positions within the assembly area to the start point and start their movements accordingly.

Contact With the Next Unit

2-116. When units are dispersed or terrain in the assembly area prohibits visual contact, subordinate units maintain contact with the unit they will follow in the march to ensure their movement is coordinated and do not bunch up or become intermingled. March units must make their start point time but also need to know and compensate for any changes to the march order.

Note. Within this scenario, Infantry battalion 2 occupied Assembly Area Eagle and made final preparation for a forward passage of lines and movement to contact.

PASSAGE OF LINES AND BATTLE HANDOVER

2-117. A passage of lines is the coordinated movement of one or more units through another unit. Battle handover is a coordinated operation executed to sustain continuity of the combined-arms fight and to protect the combat potential of both forces involved. A passage of lines and battle handover often are conducted sequentially. The moving unit passes through a staionary unit to either initiate or relinquish contact with the enemy.

DOCTRINAL BASIS FOR A PASSAGE OF LINES

2-118. A *passage of lines* is an operation in which a force moves forward or rearward through another force's combat positions with the intention of moving into or out of contact with the enemy (JP 3-18). The primary purpose of a passage of lines is to transfer responsibility for an area from one unit to another. A commander conducts a passage of lines to continue an attack or conduct a counterattack, pass through security or main battle forces, and anytime one unit cannot bypass another unit's position. The battalion or its subordinate units may execute a forward or rearward passage of lines. A passage of lines may involve engagement with the enemy shortly before or after the completion of the task, and usually involves a battle handover.

Planning Considerations

2-119. A passage of lines is a complex operation requiring close supervision and detailed planning, coordination, and synchronization between the commander of the unit conducting the passage and the unit being passed. Terrain management is critical to successful completion of a passage of lines. At least two units are occupying and concentrated on the same terrain. The commander and subordinate leaders at all levels have to understand their respective commander's plan and be flexible in its execution. Terrain is controlled through the sharing of a common operational picture and overlays that contain—

- Primary and alternate routes.
- Checkpoint data.
- Friendly and enemy unit locations and status.
- Passage points and lanes.
- Fire support coordination measures.
- Friendly and enemy obstacle types and locations.
- Sustainment locations and descriptions.
- Contact points.

2-120. A passage of lines may require either the reduction of some obstacles or the opening and closing of lanes through friendly obstacles. The passing engineer officer must coordinate with the stationary unit engineer. At a minimum, this coordination must address the following:

- Location and status of friendly and enemy tactical obstacles.
- Routes and locations of lanes and bypasses through friendly and enemy obstacles.
- Transfer of obstacle and passage lane responsibilities.

2-121. The battalion fire support officer reviews the fire support plan of the stationary unit and conducts direct coordination to ensure that a clear understanding exists between the passed and passing units on the established fire support coordination measures. The fire support officer does so through the transfer of digital fire support overlays between the two fire support teams via the Advanced Field Artillery Tactical Data System. Procedures to establish fire support battle handover or transfer of control also are identified and approved by the passing and passed commanders. Terrain and route management for fire support assets and their support assets are especially important due to potential terrain limitations. Sufficient fire support assets must be positioned to support the passage if enemy contact is possible during the operation.

2-122. During the conduct of a passage of lines, units participating in the operation present a lucrative target for air attack. The passing commander coordinates air defense protection with the stationary force commander during the passage of lines. This method allows the passing force supporting air defense assets to conduct a move at the same time. If the passing force requires static air defense, then it coordinates the terrain with the stationary battalion S-3.

2-123. The battalion trains normally will be one of the first battalion elements through the passage lanes during a rearward passage of lines and one of the last units through the passage lanes during a forward passage of lines. During a passage of lines, the Infantry battalion commander may choose to keep unit trains positioned with the stationary unit or in an assembly area to the rear of the stationary unit until maneuver units are well clear of the passage points.

Rehearsal

2-124. During rehearsals for a passage of lines, the battalion commander ensures that each organization knows when and where to move as well as how to execute the required coordination. Rehearsal items include—

- Fire support observation plan, target execution, communication linkages, and mutual support.
- Confirm fire support coordination measures.
- Review unit routes and positioning.
- Locations and descriptions of obstacles, lanes, bypasses, and markings.

- Locations of any stockpiles, especially engineer stockpiles.
- Responsibility by unit for closing passage lanes after the passage of lines is complete.
- Air defense weapons locations, early warning communications, air threat, and weapons control status.
- Passage point recognition procedures.
- Route management, contact points, checkpoints, and use of guides.
- Locations for and movement of sustainment units.
- Locations of aid stations, ambulance exchange points, and casualty evacuation procedures.

Forward Passage of Lines

2-125. In a forward passage of lines, conducted as part of an attack, both the stationary and passing commanders must be aware of the passing battalion's objective. This awareness is especially important if the stationary battalion must provide supporting fires.

2-126. On receipt of an order, the passing battalion commander begins preparing the passage of lines plan by conducting a reconnaissance while concurrently updating the information received from the stationary battalion. For example, the passing battalion receives an operations overlay from the stationary battalion that delineates routes to the contact points as well as the location of the actual linkup site (figure 2-21).

2-127. During the physical reconnaissance, the S-3 from the passing battalion updates the initial operations overlay, incorporating information received from the stationary battalion by adding pertinent control measures. Upon completion, the S-3 forwards this overlay to the main command post. Based on this information; the staff completes development of the plan.

2-128. The main command post forwards the approved operations overlay update from the stationary and passing battalion, higher quarters, and subordinate units to the liaison teams. This technique allows the S-3 and battalion commander to develop their scheme of maneuver for the passage of lines overlay concurrent with reconnaissance. At the conclusion of the reconnaissance and subsequent coordination with the stationary battalion, the revised battalion plan is distributed to subordinate units and higher headquarters.

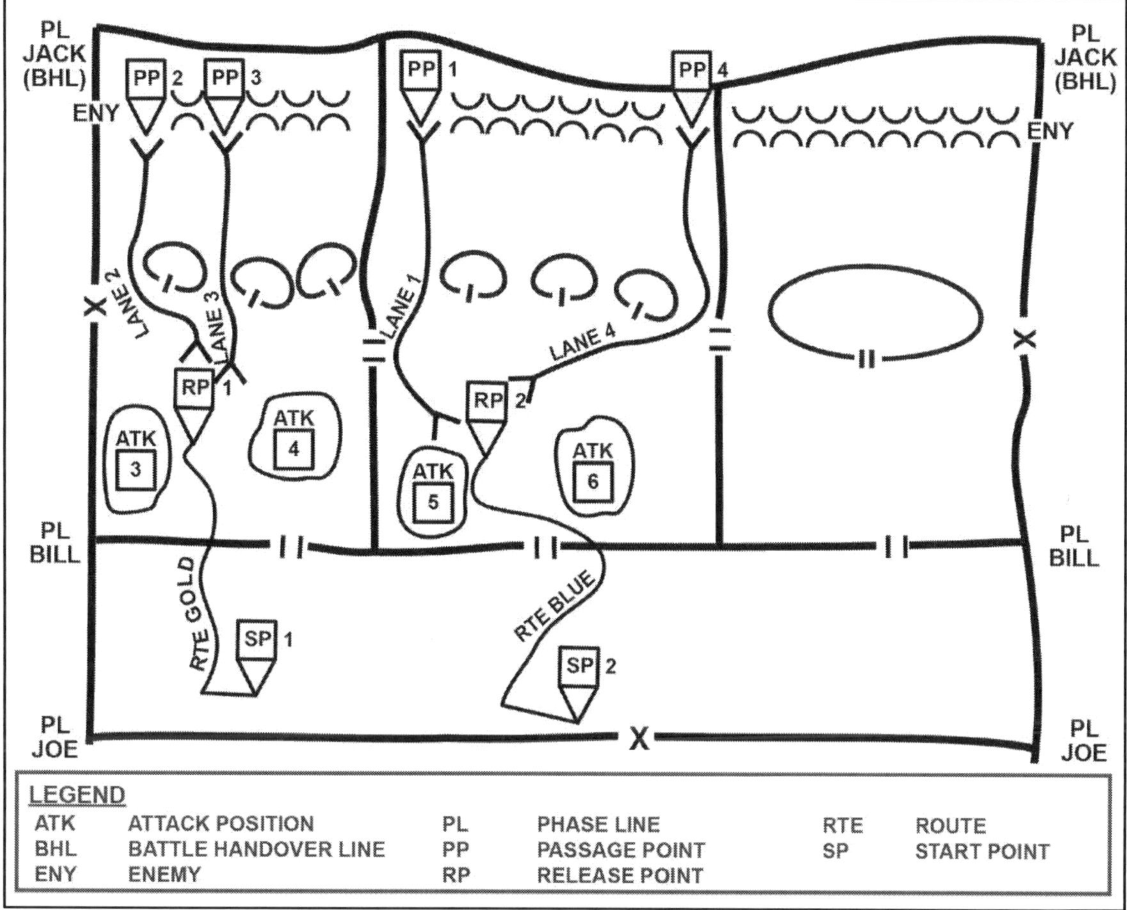

Figure 2-21. Control measures, forward passage of lines

Rearward Passage of Lines

2-129. Typically, a rearward passage of lines occurs within a defensive framework in which elements of the security force operate forward of the main battle area. The main battle area forces are the stationary unit in a rearward passage of lines. Security forces withdraw through main battle area forces handing off control of the fight at the battle handover line (figure 2-22, page 2-40).

2-130. To facilitate a rearward passage of lines, the stationary force commander designates—

- The battle handover line.
- Contact points forward of the battle handover line.
- Passage points along the forward edge of the battle area.
- Lanes through the main battle area.

2-131. Once the overlay is prepare, the stationary commander provides it and any amplifying information to the passing force commander. The stationary and passing commanders determine the best method of exercising mission command to avoid slowing the tempo of the operation and fratricide.

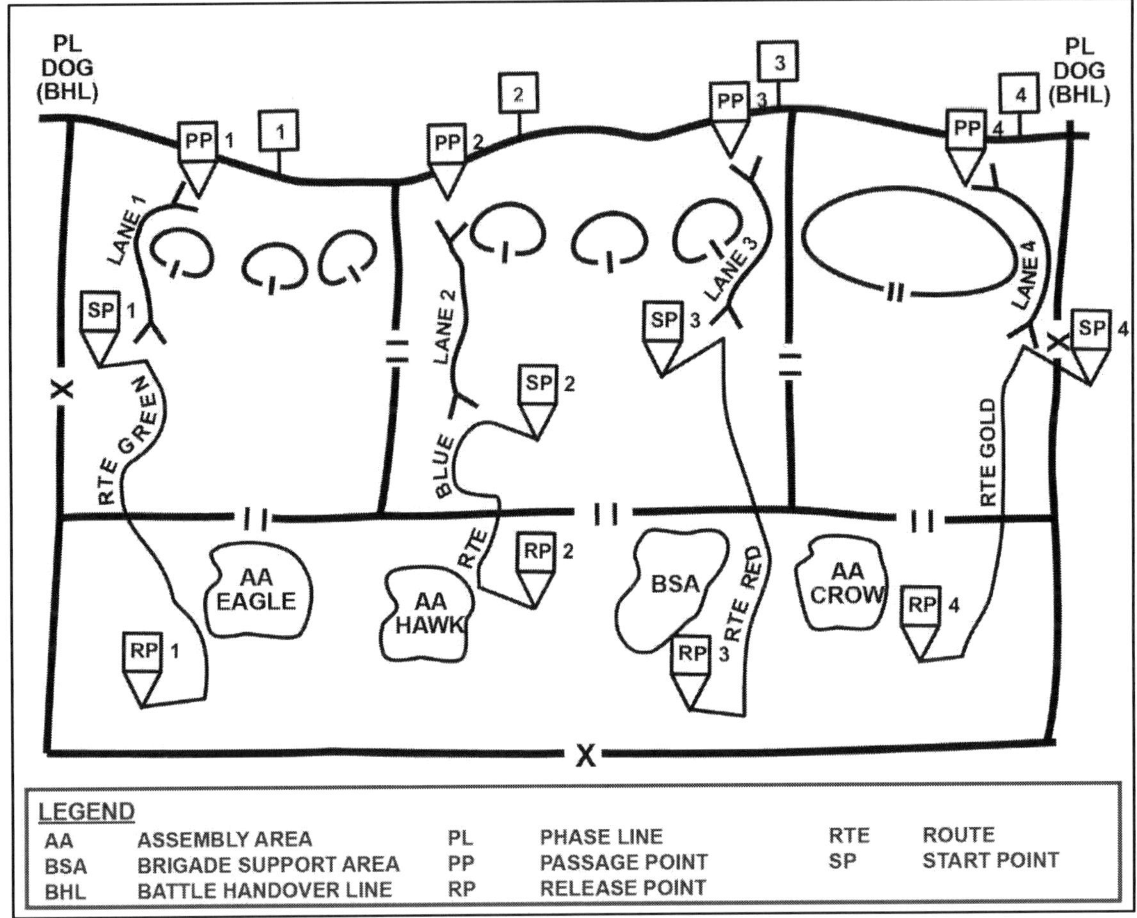

Figure 2-22. Control measures, rearward passage of lines

DOCTRINAL BASIS FOR BATTLE HANDOVER

2-132. Battle handover is the act of transitioning responsibility from the stationary force to a moving force and vice versa and is designated by a line. A *battle handover line* is a designated phase line on the ground where responsibility transitions from the stationary force to the moving force and vice versa (ADRP 3-90). The common higher commander of the two forces establishes the battle handover line after consulting both commanders. The stationary commander determines the exact location of the line.

2-133. The battle handover line is forward of the forward edge of the battle area in the defense or the forward line of troops (FLOT) in the offense. The commander draws it where elements of the passing unit can be supported effectively by the direct fires of the forward combat elements of the stationary unit until passage of lines is complete. The area between the battle handover line and the stationary force belongs to the stationary force commander. The stationary force commander may employ security forces, obstacles, and fires in the area.

2-134. During the defense, the battle handover is normally planned and coordinated in advance to facilitate execution and usually involves a rearward passage of lines. Battle handover during the offense can also be planned, such as when a unit seizes an objective and follow-on forces pass through to continue the attack. Battle handover during the offense can also be situational dependent and initiated by an operations order or fragmentary order.

2-135. Physical handover normally occurs at the battle handover line. Events may dictate that a force break contact forward of or behind the battle handover line such as when a gap exists between echelons of the attacking enemy force. Close coordination—physical or by FM voice—between the battalions involved in the handover allows them to coordinate and execute this process at the small unit level.

2-136. Battle handover begins on order of the higher headquarters commander from either unit, or when a given set of conditions occurs. Defensive handover normally is complete when the passing battalion is completely clear and the stationary battalion is ready to engage the enemy. These actions may occur at the same time. Offensive handover normally is complete when the passing battalion combat elements completely cross the battle handover line. The battle handover line may be the line of departure for an attacking battalion. Until the handover is complete and acknowledged by the commanders, the battalion commander in contact is responsible for the fight.

2-137. Coordination for battle handover flows from the battalion commander out of contact to the battalion commander in contact. The coordination for a battle handover overlaps with the coordination for a passage of lines; the coordination for both is accomplished at the same time. The battalion's standard operating procedures should outline these coordination requirements to facilitate rapid accomplishment.

2-138. Each unit transmits or delivers a complete copy of their operations order or fragmentary order and overlays. Any changes made after initial distribution are updated immediately. The coordination effected between the two commanders includes—

- Establishing frequency modulation voice and digital communications.
- Providing updates of both friendly and enemy situations (voice and graphic).
- Coordinating passage points and routes and ensuring these are displayed on operational overlays.
- Collocating mission command and exchanging liaison personnel (if required).
- Coordinating fires and fire control measures (direct and indirect) and ensuring these are displayed on operational overlays and the common operational picture.
- Determining the need for and dispatching contact point representatives.
- Establishing and coordinating recognition signals.
- Exchanging locations of obstacles and related covering fires.
- Exchanging route information to include waypoints.
- Determining fire support, protection, and sustainment requirements.

Note. The following illustration introduces a fictional scenario as a discussion vehicle for illustrating one of many ways that an Infantry battalion can conduct a passage of lines and battle handover.

Illustration of a Passage of Lines and Battle Handover

2-139. The following scenario illustrates an Infantry battalion conducting a forward passage of lines and battle handover with a defending force. In this scenario, Infantry battalion 2 conducts a second tactical movement (dismounted march) from Assembly Area Eagle along Route Fox to the area of operation where the passage of lines and battle handover will occur (figure 2-23, page 2-42).

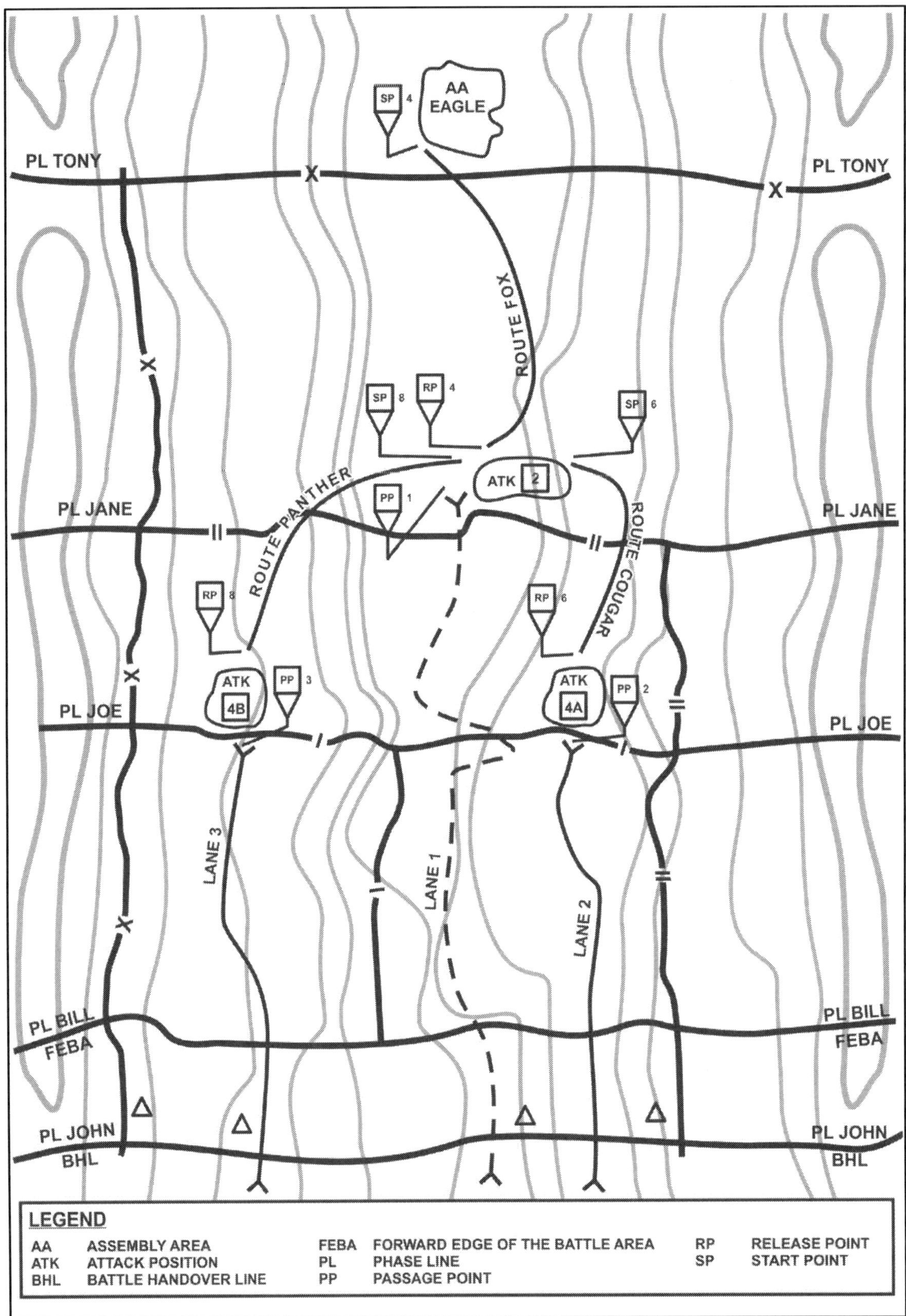

Figure 2-23. Tactical movement overlay with control measures, example

Passage of Lines and Battle Handover

2-140. The battalion commander and staff develop the best plan in the time available. They allow the subordinate maneuver companies as much time as possible to develop their own plans and to prepare for the passage of lines, battle handover, and the follow on mission—a movement to contact along Axis of Advance Red (figure 2-30 on page 2-60). The battalion commander actively participates in the military decision-making process to plan for the mission. By the commander's direct involvement and the use of WARNORDs, the commander enhances the process and allows subordinate commander's sufficient time to plan in parallel with the battalion. Prior to the battalion's movement from Assembly Area Eagle, the battalion commander, S-3, battalion rifle company commanders, weapons company commander, scout platoon leader, and an assault platoon (used as a security element) move forward to make direct coordination for the passage of lines and battle handover with the defending battalion.

Enemy Information Along the Axis of Advance Red

2-141. During direct coordination as stated in the prior paragraph, the defending battalion commander and staff provide Infantry battalion 2 with all available information on the area of operation, specifically enemy information along Axis of Advance Red, the proposed battalion route for the movement to contact. The defending battalion commander states that—

- The immediate enemy force is located just to the south of the battle handover line (Objective Kiwi).
- The enemy force is an Infantry (dismounted) platoon size element defending in position, occupying prepared individual fighting positions.
- Anti-personnel mines are located just to the north of the defending enemy force (not illustrated).
- Small enemy forces move along Axis Red, but no known enemy defensive positions exist south of Objective Kiwi.
- The defending battalion conducts daily ground and aerial reconnaissance and surveillance (R&S) missions along the forward edge of the battle area and within the battalion's forward security area.

Battalion Dismounted March

2-142. During occupation of Assembly Area Eagle, Infantry battalion 2 task organizes for tactical movement to Release Point 4 located north of Attack Position 2 within the brigade support area (BSA) of the defending IBCT. On order, Infantry battalion 2 conducts a tactical movement (dismounted march) and moves in march units from Assembly Area Eagle (figure 2-24, page 2-44). The movement starts at 2000 hours, two days prior to the movement to contact, with all units closed into attack positions no later than 0430 hours the following day. Sequence and actions of the movement are—

- Battalion scout platoon, with attached sniper squad, moves in advance of the march serial main body. The scout platoon moves along Route Fox to Release Point 4, establishes linkup with defending forces (within the BSA) at Release Point 4.
- Company A moves (main body movement starts at 2130 hours) along Route Fox to Release Point 4.
- Company D (weapons company-minus), with battalion tactical command post and attached mortar section, moves along Route Fox to Release Point 4.
- Company B, with attached engineer platoon and assault platoon, moves along Route Fox through Release Point 4 (guided by the defending force) into Attack Position 2.
- Company C, with battalion main command post and attached mortar section, moves along Route Fox through Release Point 4 (guided by the defending force) into Attack Position 2.
- Unit trains remain in Assembly Area Eagle until Infantry battalion 2 passes March Objective Snake, then on order, displaces forward to vicinity March Objective Snake. (Not illustrated.)

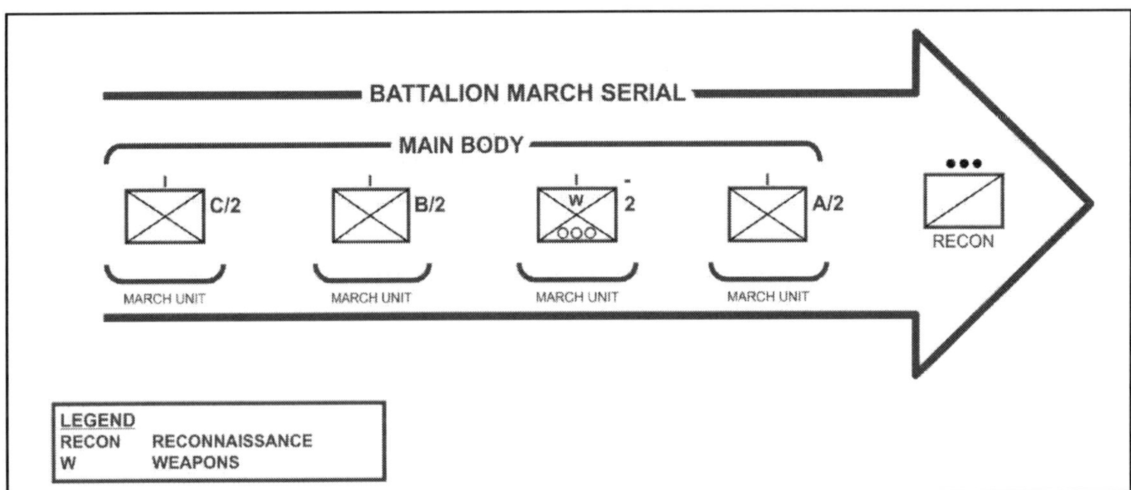

Figure 2-24. Battalion march serial (dismounted march), example

Occupation of Forward Attack Positions and Support-by-Fire Positions

2-143. Once linkup established at Release Point 4, the scout platoon moves (with guides from defending battalion) along Route Panther to Attack Position 4b (figure 2-25). The scout platoon quarters Attack Position 4b, leaving one scout squad to conduct linkup and occupation of Attack Position 4b with Company D. The scout platoon leader, with two scout squads, conducts passage of lines (Lane 3), then reconnoiters (previously reconnoitered the by defending battalion) Support-by-Fire Positions 1 and 2 and the mortar section firing position. Prior to the establishment of support by fire positions and mortar firing position, the scout platoon leader reconnoiters Objective Kiwi and inserts a surveillance team to observe the objective.

2-144. The sniper squad moves (with guides from defending battalion) along Route Cougar to Attack Position 4a (figure 2-25). The sniper squad quarters Attack Position 4a, leaving one sniper team to conduct linkup and occupation of Attack Position 4a with Company A. The sniper squad leader, with two sniper teams, conducts passage of lines (Lane 2) to conduct reconnaissance of previously reconnoitered (by defending battalion) route to the planned probable line of deployment for the Company A attack.

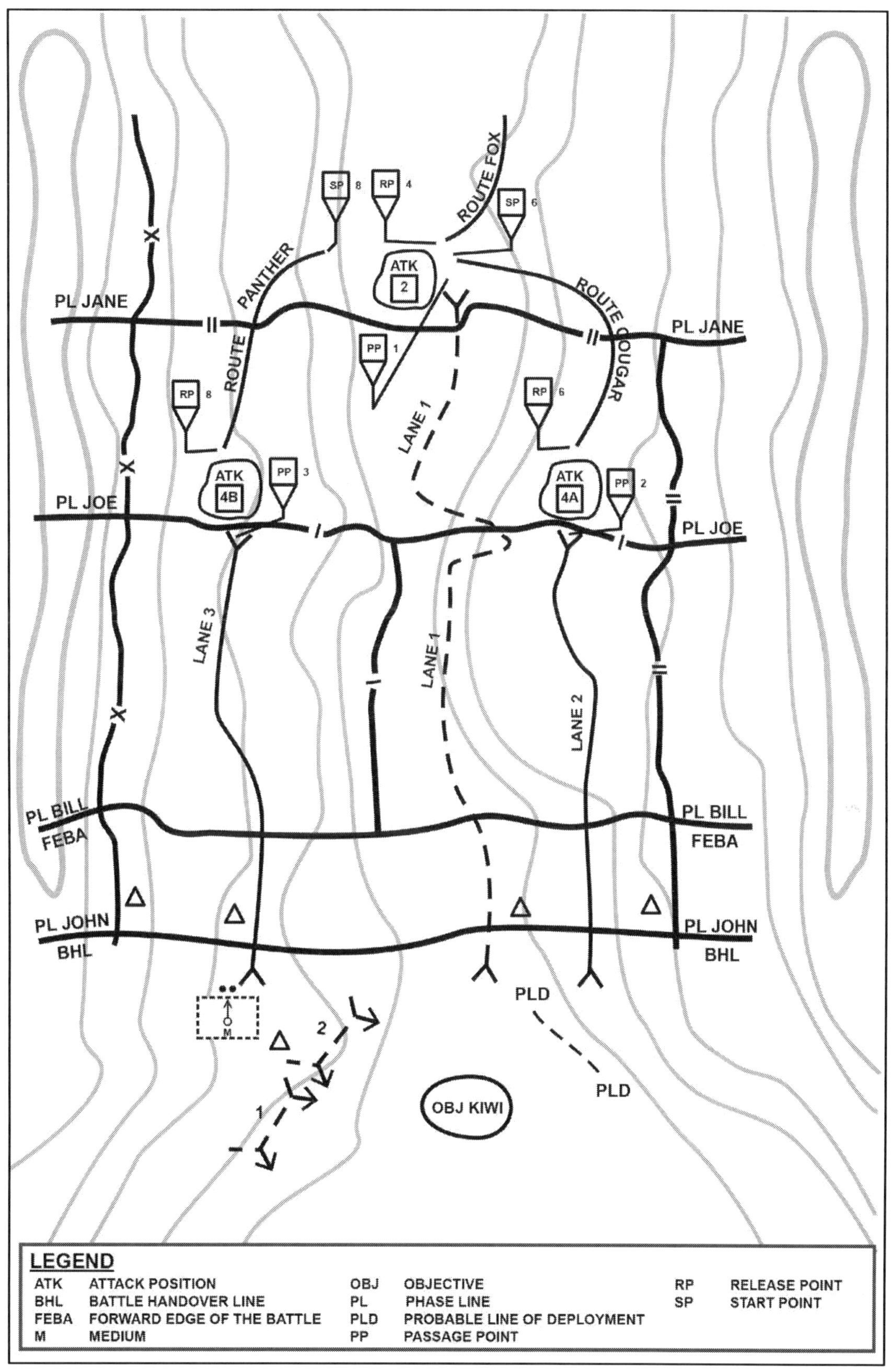

Figure 2-25. Attack positions and support-by-fire positions (occupation), example

2-145. Company D, with the battalion tactical command post and attached mortar section, conducts linkup with guides from the defending battalion at Release Point 4 (figure 2-26.) Company D moves forward along Route Panther to Attack Position 4b to the rear of the defending company of the defending battalion. Company D then conducts linkup with the scout squad to occupy Attack Position 4b. On order, Company D with the battalion tactical command post and attached mortar section conducts passage of lines (Lane 3) to previously reconnoitered Support by Fire Positions 1 and 2 and mortar firing position forward of the defending company. Company D in coordination with the defending battalion, prepares to support the seizure of Objective Kiwi and the battalion main body passage of lines by—

- Dismounting a mix of MK19 and M2 machine guns.
- Moving dismounted systems and ammunition forward into support by fire positions.
- Camouflaging positions.
- Identifying targets during daylight and make range cards.
- Laying in on designated targets.

2-146. Company A conducts linkup with guides from the defending battalion at Release Point 4 (see figure 2-25, page 2-45). Company A moves forward along Route Cougar, conducts linkup with scout squad and establishes Attack Position 4a to the rear of the defending company of the defending battalion. During occupation, Company A uses Attack Position 4a to facilitate deployment and last-minute coordination for the passage of lines and the limited visibility attack on Objective Kiwi.

Note. Whenever possible, units move through the attack position without stopping. An attacking unit occupies an attack position for a variety of reasons, including, for example, when the unit is waiting for specific results from preparation fires or when it is necessary to conduct additional coordination, such as a forward passage of lines. If the attacking unit occupies the attack position, it stays there for the shortest amount of time possible to avoid offering the enemy a lucrative target.

Limited Visibility Attack—Seize Objective Kiwi

2-147. Infantry battalion 2 conducts a limited visibility attack (see paragraph 2-295) with Company A attacking from the northeast, while avoiding anti-personnel mines to the front of the objective (figure 2-26). Company D and attached battalion mortar section support the attack from Support by Fire Positions 1 and 2 and a mortar firing position to the northwest of the objective. Company A moves forward from the northeast side of the objective in company column to the probably line of departure (Phase Line Sue), then attacks with two rifle platoons on line. Each rifle platoon has a designated breaching team. The third rifle platoon, in reserve, follows behind the two assaulting rifle platoons. Company mortars move with the reserve, on order establish firing position(s) to support the attack and follow-on mission. Company A commander designates Assault Positions 1 and 2, prior the Phase Line Sue, from which final preparations are made (if required) to assault the objective. These final preparations can involve tactical considerations, such as a short halt to coordinate the final assault, reorganize to adjust to combat losses, make final breach preparations, or make necessary adjustments in the attacking force's dispositions.

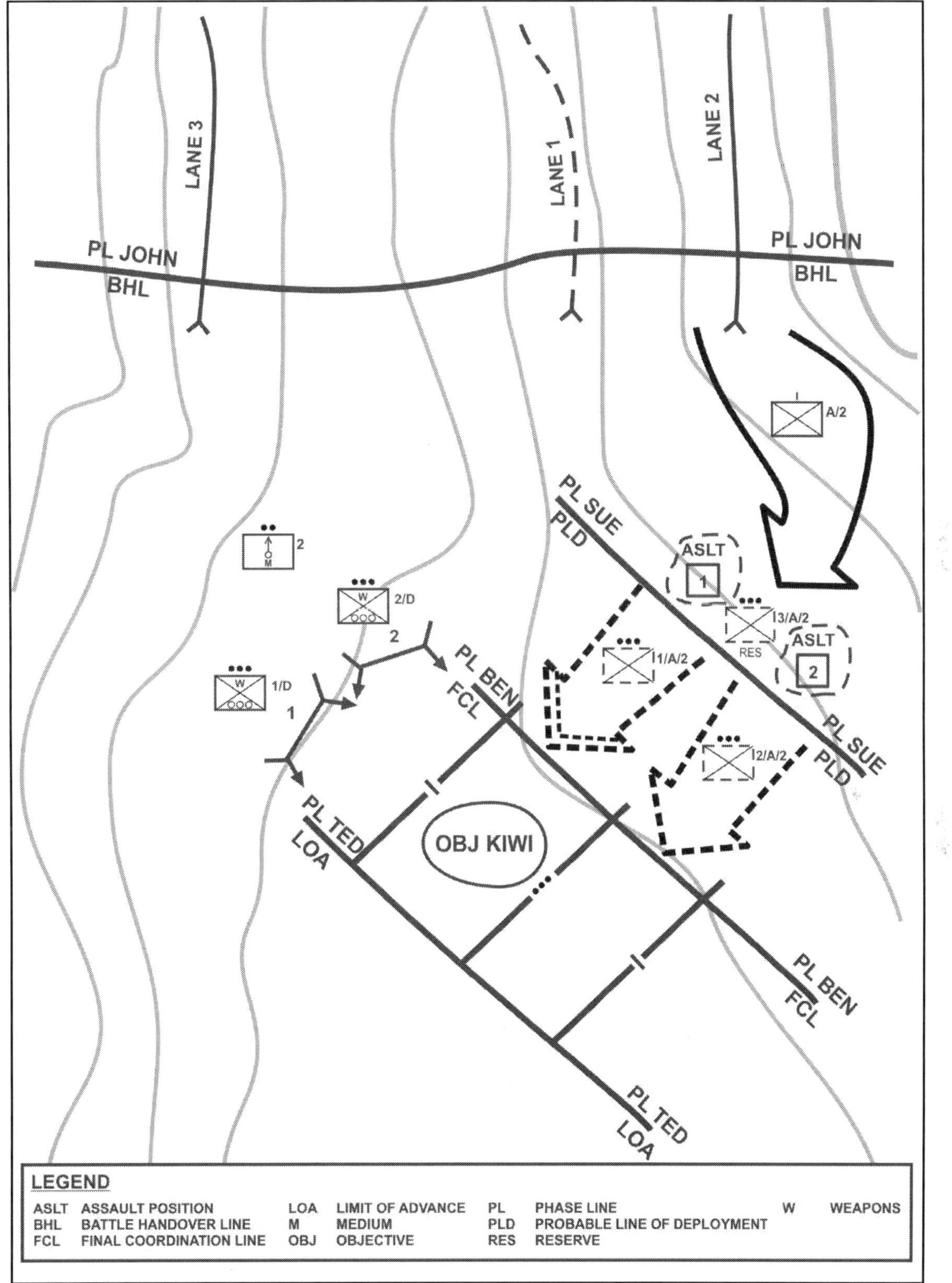

Figure 2-26. Limited visibility attack (maneuver), example

2-148. Prior to the attack and the establishment of support positions, the scout platoon leader returns to make contact with the battalion commander, vicinity Attack Position 4b, to update the commander on the current enemy situation. Once updated, the commander (with key personnel from the battalion tactical command post) and scout platoon leader conduct a passage of lines, Lane 3, guided by the defending force. The scout platoon leader then guides the commander on the leader's reconnaissance of the objective. During the leader's reconnaissance the battalion commander confirms the plan and establishes a forward position (vicinity, Support by Fire Position 2), to control the attack and follow-on mission. Scout squads provided overwatch and security throughout the commander's movement.

Note. Optimally, the commander conducts a leader's reconnaissance with key personnel to confirm or modify the plan. Depending on the enemy situation and battalion scheme of maneuver, the leader's reconnaissance may just involve the commander (with a security team). After the leader's reconnaissance, the commander modifies the plan and disseminates those changes to subordinate leaders and other affected organizations. Under the least favorable condition, the commander may not be able to conduct a leader's reconnaissance. In this case, the commander may utilize other available assets (to include IBCT or higher-level asset) to confirm or modify the plan. When these assets are unavailable and when known conditions are unchanged, the commander executes the mission according to plan.

2-149. During the battalion commander's conduct of the leader's reconnaissance, Company D begins final coordination for the movement and on order conducts a passage of lines, Lane 3, guided by the defending force. Shortly after Company D beings its passage of lines, Company A beings final coordination for movement and on order conducts a passage of lines, Lane 2, guided by the defending force. Once Company D passage of lines complete, assault platoons and attached mortar section, on order maneuver into and establish support positions, guided by the scout platoon.

2-150. As support forces maneuver into position, the sniper squad leader makes contact with the Company A commander at the passage point to update the commander on the current enemy situation. The sniper squad leader then leads the company along the previously reconnoitered route to Phase Line Sue, the probable line of deployment. Sniper teams provide overwatch and security during Company A's maneuver to and through the probable line of deployment. As Company A approaches Phase Line Sue, assault platoons in Support by Fire Positions 1 and 2 prepare to suppress the enemy, vicinity Objective Kiwi. On order, the mortar section (attached to Company D) provides preparatory fires on Objective Kiwi. The battalion plans targets on the objective and on routes into and out of the objective area. As the attack progress, the defending battalion mortars and IBCT indirect fire support assets, and Army aviation will conduct observed fires (joint fires observers) to the south of Objective Kiwi to isolate the objective.

2-151. At 0300 hours, on the day of the movement to contact, lead elements of Company A pass Phase Line Sue, the probable line of deployment (figure 2-27). The probable line of deployment is located on the reverse slope of a ridgeline, overwatched and secured by battalion sniper teams. The probable line of deployment, where the two lead platoons deploy on line, is marked by infrared (or thermal) signal, away from enemy view. On order from the commander, preparatory fires begin and the platoons assault forward while guiding on the company main effort on the right side of the assault. The assaulting rifle platoons pass through Phase Line Ben (final coordination line) and sweep through the objective, while supporting fires shift to the front of the advancing assault force. The assault force stops at the Phase Line Ted, the limit of advance. Designated parties move back across the objective, seize prisoners and ensure the objective is secure. Company A commander informs the battalion commander that Objective Kiwi is secure. (Refer to FM 3-21.10 for additional information.)

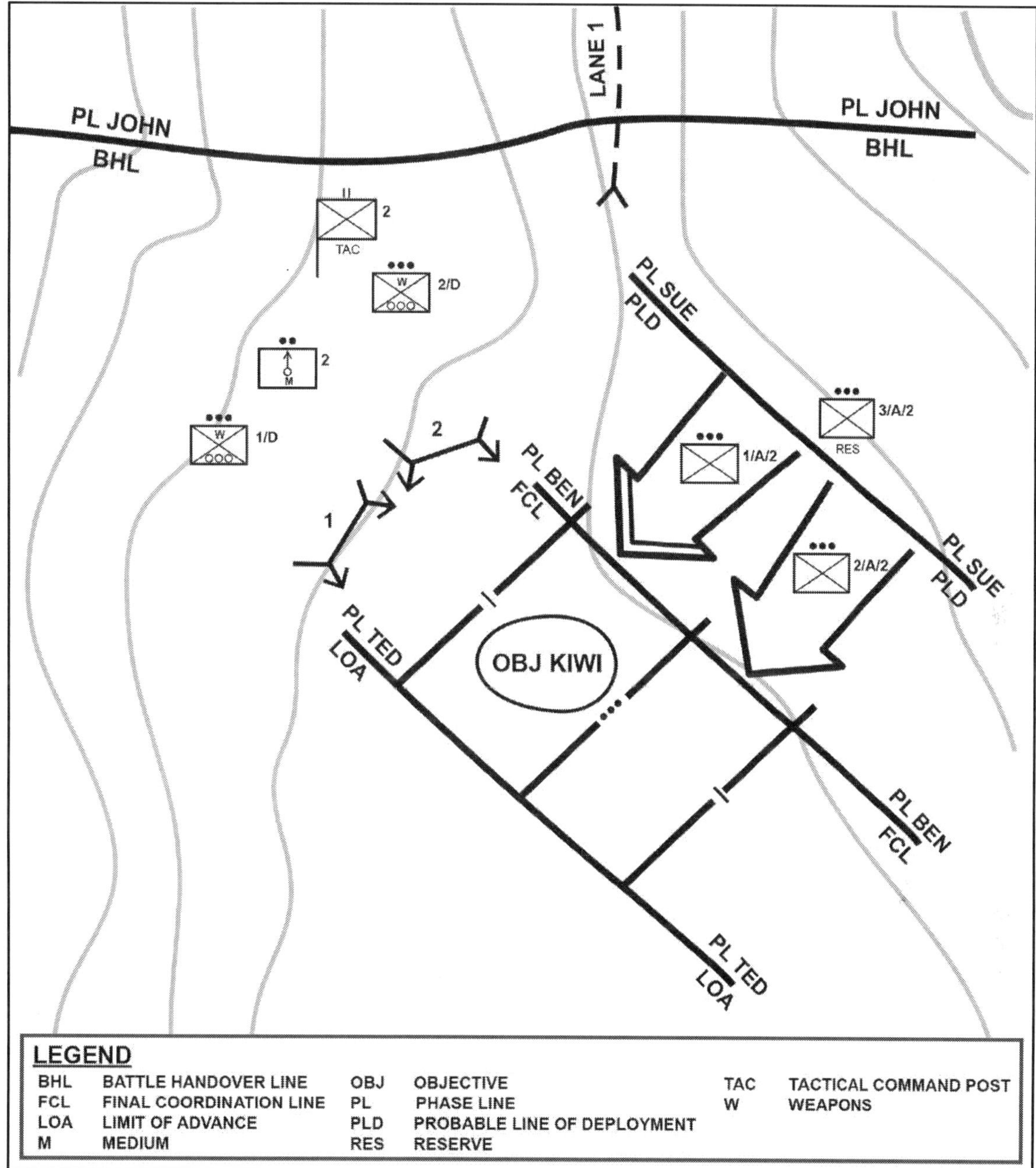

Figure 2-27. Limited visibility attack (company assault), example

Conduct Passage of Lines

2-152. Once Objective Kiwi is secure, Company A establishes a blocking position to the south of Objective Kiwi. The scout platoon conducts screen to the west of Company D. The sniper squad occupies observation posts to the east of Objective Kiwi. At 0300 hours, Infantry battalion 2 initiated tactical movement to conduct passage of lines (Line 1) by march unit from Attack Position 2 (figure 2-28, page 2-50). Company B, (lead element) with attached engineer platoon, moved from Attack Position 2 through Lane 1 to breach obstacle north of Objective Kiwi. Battalion (-) and Company C moved on order to reduce congestion and maintain operational tempo. While elements of Company B provided overwatch, engineer platoon cleared a single lane to allow lead elements of Company B to pass. One engineer squad moved with the lead elements of Company B. Engineer platoon (-) then immediately started to clear and mark a vehicles lane. Once vehicle lane cleared and marked, engineer platoon (-) attaches to Company C, in order of movement.

Note. Breaching activities require the precise synchronization of breaching fundamentals and the critical events of a breaching activity. Refer to ATP 3-90.4 for a detailed discussion of the breaching fundamentals and critical events of a breaching activity.

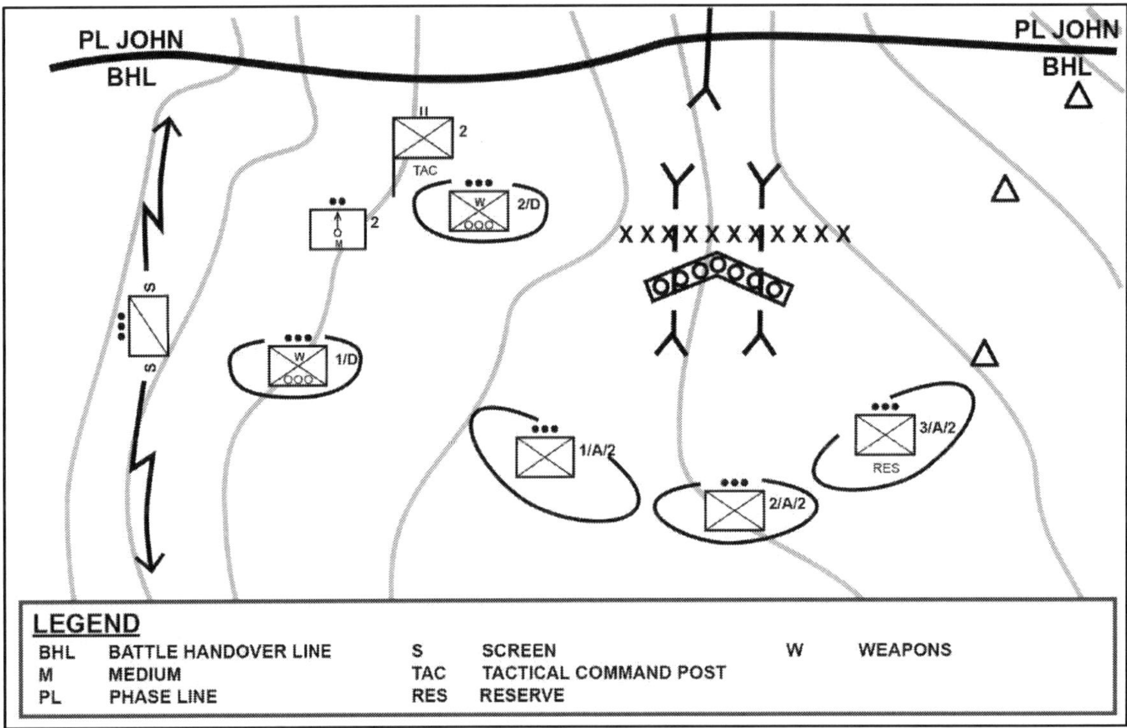

Figure 2-28. Passage of follow-on forces (breach lanes north of Objective Kiwi), example

Conduct Battle Handover

2-153. Battle handover occurs along a line (battle handover line) forward of the defending force where elements of Infantry battalion 2 (D Company) can effectively overwatch by direct fires and\or indirect fires until battle handover is complete. Once the battalion main command post clears the Passage Lane 1, the battalion tactical command post affects battle handover. Infantry battalion 2 now controls its area of operation forward of the battle handover line and the fires within its area of operation. While within range, the defending battalion may provide fire support if requested. Infantry battalion 2 mortars begin to displace by section to provide continuous fire support.

Note. Within this scenario, Company A conducted a limited visibility attack, seized Objective Kiwi then established blocking position south of Objective Kiwi. As Objective Kiwi seized, Infantry battalion 2 (-) conducted passage of lines (Lane 1) and battle handover with defending battalion, then initiated movement to contact along Axis of Advance Red. Infantry battalion 2 movement to contact order of movement—battalion scout platoon with attached sniper squad, Company B, Company C, Company D, and Company A.

DOCTRINAL BASIS FOR A MOVEMENT TO CONTACT

2-154. Commanders conduct a movement to contact to create favorable conditions for subsequent tactical tasks. A commander conducts a movement to contact when the tactical situation is not clear, or when the enemy has broken contact. A properly executed movement to contact develops the combat situation and maintains the commander's freedom of action after contact.

Fundamentals of a Movement to Contact

2-155. A movement to contact employs purposeful and aggressive movement, decentralized control, and the hasty deployment of combined arms formations from the march to conduct offensive and defensive tasks or operations in support of stability tasks. The fundamentals of a movement to contact are—

- Focus all efforts on finding the enemy.
- Make initial contact with the smallest force possible, consistent with protecting the force.
- Make initial contact with small, mobile, self-contained forces to avoid decisive engagement of the main body on ground chosen by the enemy. (This allows the commander maximum flexibility to develop the situation.)
- Task-organize the force and use combat formations and movement techniques to deploy and attack rapidly in any direction.
- Keep subordinate forces within supporting distances to facilitate a flexible response.
- Maintain contact regardless of the course of action adopted.

2-156. Whether an Infantry battalion conducts a movement to contact independently or is moving along a separate axis, it organizes its forces with a security force and a main body. If the battalion is moving as part of a higher unit, such as the IBCT, then it will either be the advance guard, with the requisite attachments, or be part of the main body.

2-157. Considerations for conducting a movement to contact along a single versus multiple axes include—

- Control. A single column is easier to control.
- Speed. The speed of the column or columns depends more on METT-TC than whether the unit travels in single or multiple columns.
- Length of column. Multiple columns reduce the length of the column. If the battalion is reinforced with other elements, it may become necessary to move on multiple columns.
- Enemy situation. A situation where the enemy is known to be deliberately defending favors moving with depth and minimum forces forward, that is, in a single column. If the mission requires finding all enemy in zone, multiple columns may be required.
- Width of zone. A wide zone favors multiple columns particularly if the zone must be cleared.
- Routes. In order to advance on multiple columns, adequate forward and lateral routes should exist. Routes not only impact on the speed and security of forward movement, but also on reaction time and mutual support between units.
- Moving as part of the brigade. If the battalion is moving as part of a brigade column or in one of its multiple columns, the basic organization does not change. Depending on the battalion's location in the column, however, certain security responsibilities may be increased, decreased, or eliminated.

Organization of Forces for a Movement to Contact

2-158. A movement to contact is organized (as a minimum) with a forward security force, either a covering force (division or corps level) and an advance guard, and a main body. Based on the mission variables of METT-TC, the commander may increase the unit's security by resourcing additional forward security forces and assets, as well as establishing flank and rear security (normally a screen or guard). The main body consist of forces not detailed to security duties and is normally the element that will conduct the decisive operation within the conduct of the movement to contact. The main body may be composed with a portion of the commander's sustaining base. (Refer to FM 3-90-1 for additional information.)

The Infantry Battalion as Part of a Larger Unit's Movement to Contact

2-159. The IBCT commander determines whether the Infantry battalion is part of the security force, such as the advance guard, and part of the main body. If time and conditions allow, the commander may consider infiltrating Infantry forces to positions of advantage to the suspected enemy's rear. The force may report and bypass enemy positions, such as roadblocks, to maintain its momentum. The weapons company; may be used to support forward movement of the Infantry rifle companies or given a more open area as its area of operation. The mortar platoon may displace by sections to ensure continuous coverage and immediate fire support when needed. Army aviation, if available, can conduct R&S, occupy support by fire positions, conduct aerial insertion, medical evacuation, and resupply, and conduct air assault to engage and destroy enemy forces or to seize and hold key terrain (see FM 3-04).

2-160. An advance guard is a task-organized combined arms unit that precedes the main body to protect it from ground observation or surprise by the enemy. The IBCT typically organizes an advance guard to lead the brigade with or without a covering force from a higher echelon. The advance guard composition is METT-TC dependent. Within the IBCT however, only Infantry battalions, their companies, or the cavalry squadron with augmentation from one of the Infantry rifle companies has sufficient combat power to serve as an advance guard. Generally, the advance guard requires fire support, anti-armor, and engineer support. The advance guard should remain within range of the main body's indirect fire weapons systems.

2-161. The main body consists of forces not detailed to security missions. The combat elements of the main body prepare to respond to enemy contact with the maneuver unit's security forces. Fire support teams may displace forward to be immediately responsive to calls for fire. The main body follows the advance guard and keeps enough distance between itself and the advance guard to maintain flexibility. The IBCT commander may designate a portion of the main body as the reserve. The combat formation the battalion uses as part of the main body is METT-TC dependent. The commander however, must be responsive to the actions of the advance guard. (Refer to FM 3-96 for additional information.)

Infantry Battalion Conducting a Movement to Contact

2-162. When an Infantry battalion conducts a movement to contact independently, the battalion is normally organized (as a minimum) with a forward security force and a main body. The forward security force is generally comprise of an R&S force, an advance guard, and flank and rear security. Forward security forces, in most cases, develop the situation using actions on contact (see paragraphs 2-170).The main body consists of forces not detailed to security duties. Combat elements of the main body prepare to respond to enemy contact with security forces. Fire support teams may displace forward to be immediately responsive to calls for fire. The main body follows the advance guard and keeps enough distance between itself and the advance guard to maintain flexibility. The Infantry battalion commander may designate a portion of the main body as the reserve (figure 2-29).

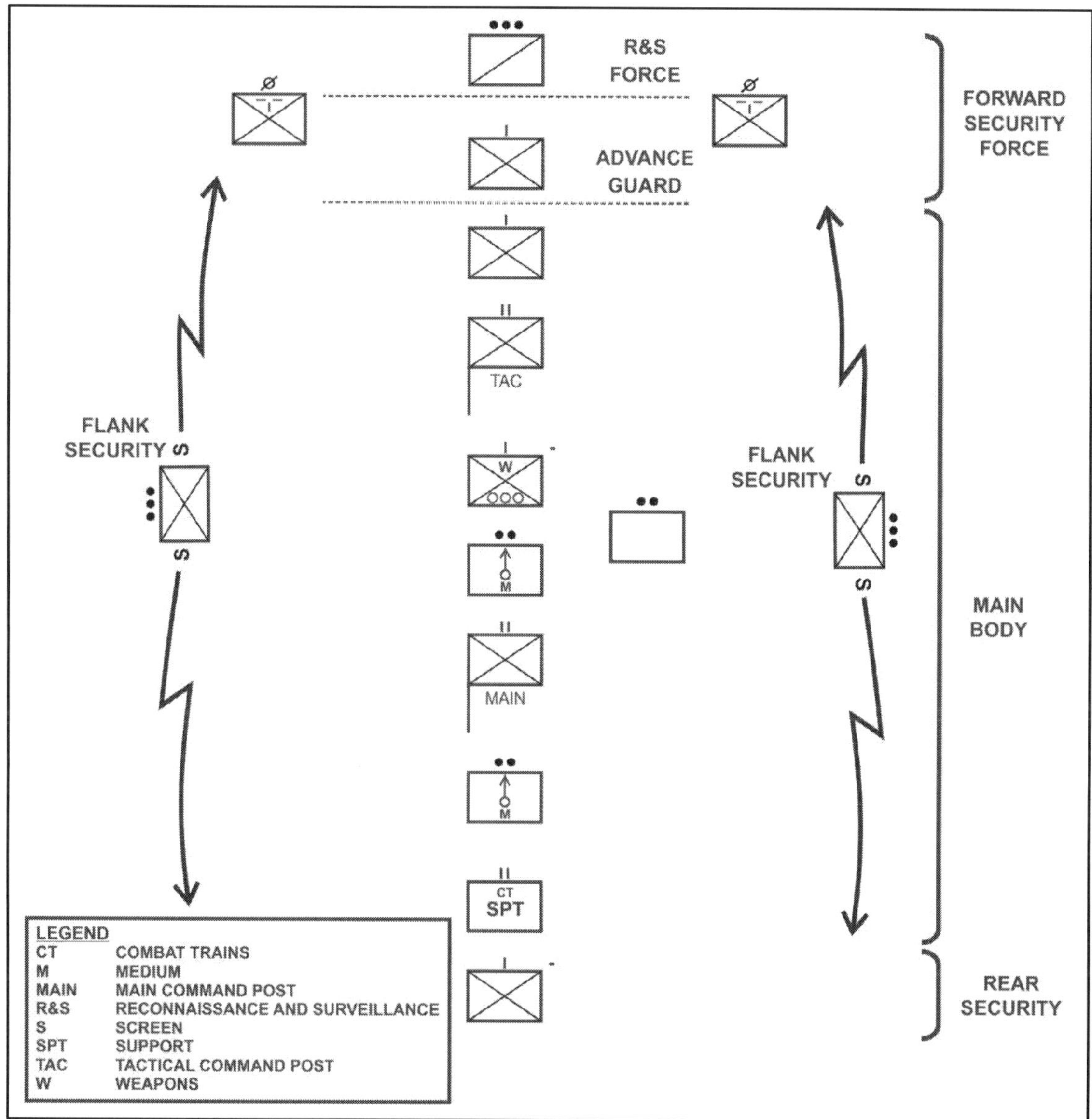

Figure 2-29. Battalion movement to contact (organization of forces), example

Reconnaissance and Surveillance Forces

2-163. The commander executes the reconnaissance missions and surveillance tasks to determine the enemy's location and intent while conducting security to protect the main body. The R&S force for the battalion is normally the scout platoon; additional augmentation may include Army aviation attack and reconnaissance units, battalion snipers, Infantry weapons company assault platoons, engineer assets, and forward observers (to include joint terminal attack controllers) to develop the situation before committing the advance guard or main body. The commander tasks R&S forces with conducting route reconnaissance of route(s) the main body will traverse. That route may be a cross-country mobility corridor. *Route reconnaissance* is a directed effort to obtain detailed information of a specified route and all terrain from which the enemy could influence the battalion's movement along that route (ADRP 3-90).

2-164. Throughout the forward security R&S effort, the force provides new or updated information on route conditions, such as obstacles and bridge classifications, and enemy and civilian activity along the route. Specifically, the R&S force must answer the priority intelligence requirements established by the commander. Other tasks, similar to an *area reconnaissance,* which is a form of reconnaissance that focuses on obtaining detailed information about the terrain or enemy activity within a prescribed area (ADRP 3-90) and a *zone reconnaissance*, which is a form of reconnaissance that involves a directed effort to obtain detailed information on all routes, obstacles, terrain, and enemy forces within a zone defined by boundaries (ADRP 3-90), normally include—

- Reconnaissance of routes, bridges, and roads.
- Reconnaissance of obstacles and restrictive terrain.
- Surveillance of critical areas, danger areas, or key terrain.

2-165. The R&S force covers the frontage of the battalion axis of advance. The force avoids decisive engagement, but once found it must keep the enemy under surveillance and report the enemy's activity. The R&S force normally remains within supporting range of the battalion's indirect fires and initially has priority of fires for those systems. Aviation maneuver and fires external to the battalion, when available, provide the force with the capability to delay, disengage, or destroy an enemy force.

2-166. The R&S force is far enough ahead of the advance guard to provide adequate warning, a detailed picture of the enemy force (size, activity, location, and depth of the enemy force), and sufficient space for the advance guard to maneuver. However, the force must not be so far ahead that the advance guard cannot rapidly assist it in disengaging from the enemy, should that become necessary. The advance guard keys its movement on the movement of the R&S force.

2-167. The R&S force must be able to receive (and provide) the latest information available from the battalion S-2, as well as information available from adjacent units, other battlefield surveillance assets, and the cavalry squadron when influence by its employment. With this information, the force can confirm information provided by these assets to reduce the risks and unknowns normally associated with a movement to contact. This information is made available to other subordinate elements of the battalion.

Advance Guard

2-168. The advance guard for the Infantry battalion is usually a company or company team. Its composition depends on the mission variables of METT-TC. Engineers, if available, follow or are attached to the lead elements to ensure mobility and provide route/bridge classification expertise. Assault platoons may be attachment from the weapons company. The two lead companies are task-organized accordingly when a battalion moves in parallel columns.

2-169. The advance guard operates forward of the main body to provide security for the main body and ensure its uninterrupted advance. The advance guard protects the main body from surprise attacks and develops the situation to allow time and space for the deployment of the main body when it is committed to action. The advance guard accomplishes this by destroying or suppressing enemy reconnaissance or ambushes, delaying enemy forces, and marking bypasses for or reducing obstacles. The advance guard—

- Remains oriented on the main body.
- Reports enemy contact to the battalion commander.
- Collects and reports all information about the enemy.
- Selects tentative fighting positions for following battalion units.
- Tries to penetrate enemy security elements and reach or identify the enemy main force.
- Destroys or repels enemy reconnaissance forces.
- Prevents enemy ground forces from engaging the main body with direct fires.
- Locates, bypasses, or breaches obstacles along the main body's axis of advance.
- Executes tactical tasks such as fix, contain, or block, against enemy forces to develop the situation for the main body.
- May conduct a passage of lines with the main body.

2-170. Until the main body is committed, the advance guard is the battalion commander's initial main effort. Priority of fires shifts to the main body once committed. In planning the movement to contact, each contingency operation should revolve around the actions of the advance guard. The lead elements must be well trained on battle drills, especially those involving obstacle reduction and actions on contact. Obstacle reduction by maneuver and engineer units reduce or negate the effects of existing or reinforcing obstacles with the objective is to maintain freedom of movement for maneuver units, weapon systems, and critical supplies. (See ATP 3-90.4.).

2-171. The enemy situation becomes clearer as forward security elements conduct actions on contact to rapidly develop the situation in accordance with the commander's plan and intent. By determining the strength, location, and disposition of enemy forces, these security elements allow the commander to focus the effects of the main body's combat power against the enemy main body. The overall force must remain flexible to exploit both intelligence and combat information. The forward security force should not allow the enemy force to break contact unless it receives an order from the commander. *Actions on contact* are a series of combat actions, often conducted simultaneously, taken upon contact with the enemy to develop the situation (ADRP 3-90). Actions on contact are—

- Deploy and report.
- Evaluate and develop the situation.
- Choose a course of action.
- Execute selected course of action.
- Recommend a course of action to the higher commander.

Note. In both the offense and defense, contact occurs when a unit encounters any situation that requires an active or passive response to a threat or potential threat. The eight forms of contact are visual; direct; indirect; non-hostile; obstacles; aircraft; chemical, biological, radiological, and nuclear (see appendix G) and electronic warfare (see appendix B). The conduct of tactical offensive and defensive tasks most often involves conduct using the visual, direct, and indirect forms.

2-172. When contact is made, the advance guard forces the enemy to withdraw or destroys small enemy groups before they can disrupt the advance of the main body. When the advance guard encounters large enemy forces or heavily defended areas, it takes prompt and aggressive action to develop the situation and, within its capability, defeat the enemy. The commander reports the location, strength, disposition, and composition of the enemy and tries to find the enemy's flanks and gaps or other weaknesses in the enemy's position. The main body then may join the attack. The battalion commander usually specifies how far in front of the main body the advance guard is to operate and reduces those distances in complex terrain and under low-visibility conditions.

Flank and Rear Security

2-173. When adjacent units are not protecting the battalion's flanks or rear, the battalion internally provides flank and rear security. Flank and rear security missions may be given to one company or to a platoon-size element from one of the companies within the main body to conduct security missions under organic company control or battalion control. These security elements remain at a distance from the main body to allow the battalion time and space to maneuver to either the flanks or the rear. Flank and rear security elements also operate far enough out to prevent the enemy from placing direct or observed indirect fires on the main body. Indirect fires are planned on major flank and rear approaches to enable security. The weapons company may conduct security missions because of its observation capabilities, firepower, and mobility. The battalion may use elements of the scout platoon for flank and rear security or may require main body forces to provide flank and rear security.

Main Body

2-174. The main body consist of forces not detailed to the security of the Infantry battalion and is normally the force that conducts the decisive operation or the force resourced as the main effort within the conduct of the movement to contact. The main body contains most of the battalion's combat elements and is arrayed to achieve all-round security throughout the movement. Combat elements of the main body are prepared to deploy and attack, giving them the flexibility to maneuver to a decisive point on the battlefield to destroy the enemy. The battalion commander designates a portion of the main body for use as the reserve. The size of the reserve is based upon the mission variables of METT-TC and the amount of uncertainty concerning the enemy.

2-175. The main body's rate of movement is dictated by the advance guard. The main body maintains situational awareness of the advance guard's progress and current enemy situation and provides responsive support when the advance guard is committed. The use of standard formations and battle drills allows the battalion commander, based on the information available to the commander, to shift combat power rapidly on the battlefield. Companies employ the appropriate formations and movement techniques within the battalion formation. Company commanders, based on their knowledge of the battalion's situation, anticipate the battalion commander's decisions for commitment of the main body and plan accordingly.

Planning

2-176. As in any type of operation, the commander plans to focus operations on finding the enemy and then delaying, disrupting, and destroying each enemy force as much as possible before direct-fire range. The commander and staff analyze the terrain to include enemy air avenues of approach and the enemy's most dangerous course of action as determined in the war-gaming portion of the military decision-making process. Because of the battalion's vulnerability, by the nature of a movement to contact, the enemy must not be underestimated.

2-177. The plan for the movement to contact addresses not only actions anticipated by the commander based on available intelligence information, but also the conduct of meeting engagements at anticipated times and locations where they might occur. A thorough intelligence preparation of the battlefield (IPB) and on-going running estimates enhance the force's security by indicating danger areas where the force is most likely to make contact with the enemy. (See ATP 2-01.3 and FM 6-0.) IPB products necessary to support planning and operations include—

- Enemy situation overlays with associated course of action statements and high-value target lists.
- Event templates and associated event matrices.
- Modified combined obstacle overlays, terrain effects matrices, and terrain assessments.
- Weather forecast charts, weather effects matrices, light and illumination tables, and weather estimates.
- Civil considerations overlays and assessments.

Note. The IPB process consists of the following four steps: define the operational environment; describe environmental effects on operations; evaluate the threat; and determine threat course of actions. A threat is any combination of actors, entities, or forces that have the capability and intent to harm United States forces, United States national interests, or the homeland. An enemy is a party acknowledged as potentially hostile to a friendly party and against which the use of force may be envisaged.

2-178. Potential danger areas are likely enemy defensive locations, engagement areas, observation posts, and obstacles. Fire support plans target these areas and they become on-order priority targets placed into effect and cancelled as the lead element can confirm or deny enemy presence. The intelligence annex of the movement to contact order must address coverage of these danger areas not only based on actions anticipated but also on available intelligence information. If determined during war-gaming that R&S forces will most likely not be able to clear these areas, more deliberate movement techniques are planned.

2-179. In a movement to contact, the battalion commander seeks to gain contact with the enemy using the smallest elements possible. Within the Infantry battalion, this element is normally the battalion scout platoon or an Infantry rifle platoon performing R&S. Army aviation attack and reconnaissance units or other R&S assets tasked organized to these forward security forces provide additional combat power, increasing their abilities to develop the situation. The main body's planned combat formation and movement technique should contribute to the goal of making initial contact with the smallest force possible.

2-180. Both the R&S force and advance guard should have sufficient uncommitted forces to develop the situation without requiring the deployment of the main body in most cases. The commander can rely on fire support assets to weight lead element combat power, but still provides lead elements with the combat multipliers they need to accomplish their mission. The frontage assigned by the battalion commander to these subordinate units in a movement to contact must allow them to apply sufficient combat power to maintain the momentum of the operation. The frontage should also provide for efficient movement of the force. Air-ground integrations enable fire superiority when organized correctly to fire immediate suppression missions to help maneuver forces get within direct-fire range of the enemy.

2-181. Within a movement to contact the commander can opt not to designate a main effort until forces make contact with the enemy, unless there is a specific reason to designate it. In this case, the commander retains resources under direct control to reinforce the main effort. The commander may designate the decisive operation during the initial stages of a movement to contact because of the presence of a key piece of terrain or an avenue of approach. The commander may designate *bypass criteria*, which is measures during the conduct of an offensive operation established by higher headquarters that specify the conditions and size under which enemy units and contact may be avoided (ADRP 3-90). Bypass criteria is clearly stated by the commander, and dependent on the mission variables of METT-TC. Criteria may also include maneuver around an obstacle or position to maintain the momentum of the operation.

2-182. The commander uses the minimal number and type of control measures possible in a movement to contact because of the uncertain enemy situation. These measures include designation of an area of operation with left, right, front, and rear boundaries, or a separate area of operation bounded by a continuous boundary (noncontiguous operations). The commander uses these control measures along with mission orders, coupled with battle drills and formation discipline, to synchronize the movement to contact. Company and company teams are not normally assigned their own areas of operations during the conduct of a movement to contact.

2-183. A movement to contact usually starts from a line of departure at the time specified in the operation order or fragmentary order. The commander controls the movement to contact by using phase lines, contact points, and checkpoints as required and controls the depth of the movement to contact by using a limit of advance or a forward boundary. March objectives (one or more) may be used to limit the extent of the movement to contact and orient the force. This movement is often terrain-oriented and used only to guide the force. Although a movement to contact may result in taking a terrain objective, the primary focus should be on the enemy force. When the commander has enough information to locate significant enemy forces, the commander should plan another type of offensive action. (Refer to FM 3-96 for additional information.)

PREPARATION

2-184. During preparation, the battalion commander and staff will receive the most current information from organic and higher echelon information collection assets. The staff must ensure that fragmentary orders are published and that plans are updated to reflect any changes. The battalion commander must ensure subordinates understand the intent and concept of operation and their individual missions even as new information becomes available. The commander uses confirmation briefs, backbriefs, and rehearsals to ensure missions are understood and all actions are integrated and synchronized. Simple plans that are flexible and rehearsed repetitively against various enemy conditions and that rely on established tactical standard operating procedures are essential to success.

2-185. Subordinate unit preparations are reviewed to ensure they are consistent with the commander's intent and concept of operations. Subordinate rehearsals should emphasize movement through danger areas, actions on contact, passage of lines, and transitions. Commanders and leaders ensure subordinate units (to include attachments) understand assigned missions during movement and maneuver options during execution. Plans are war-gamed and rehearsed against enemy course of actions that would cause the battalion to execute various maneuver options at different times and locations. The goal is to rehearse subordinate commanders on potential situations that may arise during execution to promote flexibility while reinforcing the commander's intent.

2-186. The commander seeks to rehearse the operation from initiation to occupation of the final objective or limit of advance. Rehearsals include decision points and actions taken upon each decision. Often, the commander prioritizes maneuver options and enemy course of actions to be rehearsed based on the time available. The rehearsal focuses on locating the enemy, developing the situation, executing a maneuver option, exercising direct and indirect fire control measures, and exploiting success. The rehearsal must consider the potential of encountering stationary or moving enemy forces. Other actions to consider during rehearsals include—

- Forward security force actions on contact.
- Actions to cross danger areas.
- Advance guard making contact with a small enemy force.
- Advance guard making contact with a large force beyond its capabilities to defeat.
- Advance guard making contact with an obstacle the R&S force has not identified and reported.
- Flank/rear security force making contact with a small force.

- Flank/rear security force making contact with a large force beyond its capability to defeat.
- Actions to report and bypass of an enemy force (based on bypass criteria).
- Transitions and maneuver option.

EXECUTION

2-187. The battalion moves rapidly to maintain the advantage of a rapid tempo. However, the commander must balance the need for speed with the requirement for security. The commander bases this decision on the effectiveness of the R&S effort, friendly mobility assets, effects of terrain, and the enemy's capabilities. The battalion must closely track the movement and location of subordinate companies and other battalion units. This ensures that security forces provide adequate security for the main body and that they remain within supporting range of the main body, mortars, and artillery. The movement of fire support, protection, and sustainment units is controlled by the Infantry battalion or their parent organizations (depending on command and support relationships), which adjust their movements to meet support requirements, avoid congestion of routes, and ensure responsiveness.

2-188. Obstacles pose a significant threat to the battalion's momentum due in great part to the Infantry battalion's limited breaching capability. Collaborating with the staff engineers can provide valuable information on terrain mobility, where the enemy is likely to emplace obstacles, and where the enemy could employ engineer assets. Once a battalion element detects an obstacle, it immediately disseminates the location and description throughout the battalion. The battalion seeks a secure and favorable bypass. If a bypass is available, the unit in contact with the obstacle exploits, marks, and reports the bypass. Enemy forces normally overwatch obstacles. Units should approach all obstacles and restricted terrain with the same diligence with which they approach a known enemy position.

2-189. When the battalion must conduct a combined arms breach, it maneuvers to suppress and obscure any enemy forces overwatching the obstacle and then reduces the obstacle to support its movement. The battalion's intelligence estimate includes IPB products necessary to support planning and operations, to include the modified combined obstacle overlay, (commonly referred to as MCOO) terrain effects matric, and terrain assessment. Engineer assets from the brigade engineer battalion (when attached) support the breach effort by creating lanes, marking lanes, and guiding the main body through the obstacle. R&S forces, for example the battalion sniper squad, can provide long-range observation and fires on enemy units overwatching the obstacle. (See appendix E.)

2-190. The Infantry battalion main body destroys enemy forces with a combination of fire and movement. Depending on the commander's bypass criteria and the composition of the advance guard, the advance guard may fix company or smaller size enemy forces identified by the R&S force. Once committed as the fixing force, the advance guard fixes the enemy until the main body can destroy it. The advance guard must provide the location of such a fixed enemy force to the battalion S-2, who then disseminates the information to all units in the battalion. The communication between main body and fixing force commanders is critical to coordinate actions and avoid fratricide. The fixing force directs or guides the finishing force to the best location to attack the enemy. Once the battalion destroys the enemy, all forces continue the advance.

2-191. When conducting a movement to contact as the advance guard of a larger force, the higher commander establishes bypass criteria that allow the battalion to report and bypass enemy forces of a specific size. When an enemy force meets the criteria, the battalion fixes the enemy force and leaves a small force to maintain contact while the remainder of the battalion continues the advance. Once bypassed, the destruction of the enemy force becomes the responsibility of the main body or a follow-on force. Bypassed forces present a serious threat to forces that follow the maneuver elements, especially sustainment elements. As they move around these threats, it is imperative the bypassed enemy forces' locations and strengths are disseminated throughout the battalion to enable following units to properly orient their security forces.

2-192. A *meeting engagement* is a combat action that occurs when a moving force, incompletely deployed for battle, engages an enemy at an unexpected time and place (FM 3-90-1). The enemy force may be moving or stationary. A meeting engagement is most likely during a movement to contact. Once in contact, the goal is to maneuver and overwhelm the enemy with combat power before the enemy can react. This requires the commander to keep the battalion in a posture ready to act immediately to contact and develop the situation. Subordinate companies must act on contact, develop the situation, report, and gain a position of advantage over the enemy to give the battalion time to act effectively. The battalion's success depends on its subordinate units' ability to develop the situation.

2-193. Usually in a battalion movement to contact, R&S forces makes initial contact. The force must determine the size and activity of the enemy force and avoid being fixed, or destroyed. If possible, the R&S force avoids detection. When the enemy is moving, the R&S force determines the direction of movement and the size and composition of the force. The R&S force's observers can disrupt lead enemy forces by placing indirect fires on them. Speed of decision and execution is critical when the enemy is moving. When the enemy is stationary, the R&S force determines if the enemy is occupying prepared positions and is reinforced by obstacles and minefields. The R&S force tries to identify any crew-served weapon or antitank weapon positions, the enemy's flanks, and gaps in the enemy positions.

2-194. When committed the advance guard maneuvers to overpower and destroy platoon-size and smaller security forces. Larger forces normally require the deployment of the main body. The advance guard protects the main body by fixing enemy forces, which allows the battalion main body to retain its freedom to maneuver. In developing the situation, the advance guard commander maintains pressure on the enemy by fire and movement. The advance guard probes and conducts a vigorous reconnaissance of the enemy's flanks to determine the enemy's exact location, composition, and disposition. Once contact is made with an enemy force, the battalion commander uses this information to execute one of five-planned options—attack, defend, bypass, delay, or withdraw.

2-195. During planning the commander and staff develop and war-game plans for each of the five options based on the higher commander's intent and the situation. They define the conditions in terms of the enemy and friendly strengths and dispositions that are likely to trigger the execution of each option. They identify likely locations of engagements based on known or suspected enemy locations. The commander states the bypass criteria for the advance guard recognizing the loss of tempo that is created by the lead element fighting every small enemy force it encounters. Normally, the commander makes the final decision for execution of an option based on the progress of the initial engagement of the advance guard. The movement to contact generally ends with the commitment of the main body.

Note. The following illustration introduces a fictional scenario as a discussion vehicle for illustrating one of many ways that an Infantry battalion can conduct a movement to contact. This illustration is a continuation of the scenario started earlier in this chapter focusing on Infantry battalion 2.

ILLUSTRATION OF A MOVEMENT TO CONTACT

2-196. Within this scenario, an IBCT conducts a movement to contact. The IBCT moves along two axes of advance with Infantry battalion 1, main effort moving along Axis of Advance Blue (Axis Blue) to the east and Infantry Battalion 2, supporting effort moving along Axis of Advance Red (Axis Red) to the west. The cavalry squadron conducts a guard mission to the front of the brigade main effort along Axis Blue. Remaining elements of the IBCT move, on order, behind Infantry Battalion 1 along Axis Blue (figure 2-30, page 2-60).

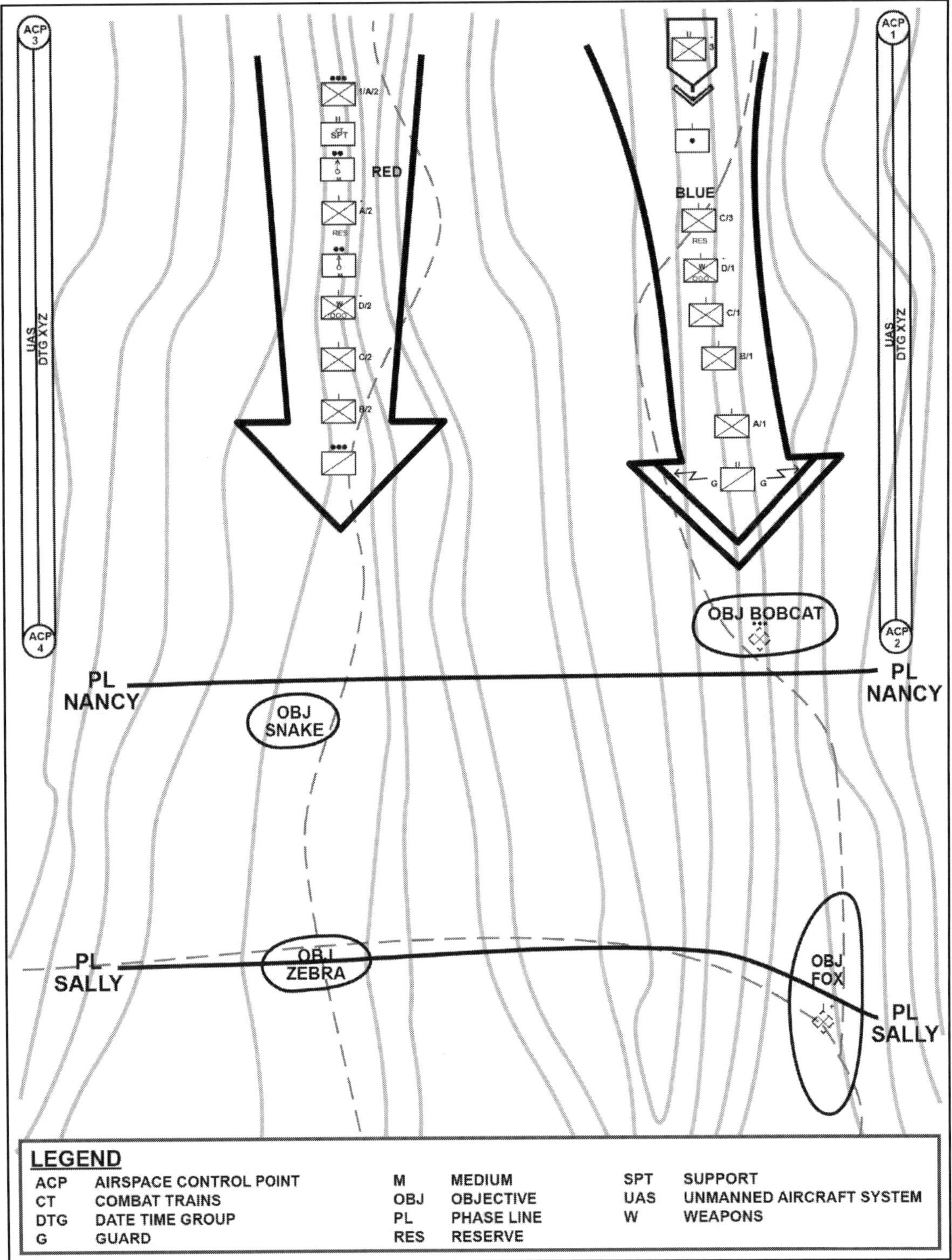

Figure 2-30. Movement to contact (axes of advance), example

INFANTRY BRIGADE COMBAT TEAM—SUBORDINATE UNIT TASK AND ORDER OF MOVEMENT

2-197. The IBCT conducts a movement to contact along axes Blue and Red, the IBCT's main effort moves along Axis Blue. Brigade limit of advance is Phase Line Dale to the south (not illustrated) with Objective Bear being the brigade's final march objective (not illustrated). IBCT subordinate unit task and order of movement are as follows:

- Cavalry squadron conducts guard mission forward of the main effort.
- Infantry Battalion 1, main effort, conducts movement to contact along Axis Blue.
- Infantry Battalion 2, supporting effort, conducts movement to contact along Axis Red.
- Infantry Battalion 3 conducts follow and assume tactical mission task along Axis Blue. Company C, detached under IBCT control.
- Company C is the IBCT reserve, follows Infantry battalion 1 along Axis Blue.
- On order, field artillery battalion moves along Axis Blue with priority of fires to Infantry Battalion 1 (brigade main effort) then to Infantry Battalion 2.
- The brigade engineer battalion priority of engineer effort, mobility to Infantry Battalion 1, then to Infantry Battalion 2. Tactical unmanned aircraft system platoon conducts surveillance to the flanks of the IBCT movement.
- Brigade support battalion establishes BSA, vicinity Assemble Area Falcon (not illustrated).

Infantry Battalion 2—Subordinate Unit Task and Order of Movement

2-198. Infantry Battalion 2 conducts movement to contact along Axis Red, seizes March Objective Snake, on-order continues movement to contact to seize March Objective Zebra. The Infantry battalion is organized with a forward security force— R&S force and advance guard—and a main body (figure 2-31, page 2-62). The limit of advance for the movement to contact is Phase Line Rita (not illustrated). Battalion unit trains, organize to establish echelon trains—battalion field trains and battalion combat trains—to support the battalion's movement to contact. Battalion subordinate unit task and order of movement are as follows:

- Battalion scout platoon (R&S force), with attached sniper squad, conducts route reconnaissance forward of Company B along Axis Red.
- Company B (advance guard), with attached engineer squad and assault platoon, conducts movement to contact forward of the main body.
- Company C (main body lead element), with attached engineer platoon (-), conducts movement to contact, battalion tactical command post moves with company.
- Company D (main body), with three assault platoons, follows Company C.
- Battalion mortar platoon (main body) moves by split section along Axis Red. Displaces by section to provide continuous fire support, priority of fires initially to forward security forces.
- Company A, battalion reserve, moves ahead of battalion combat trains. Attaches one Infantry rifle platoon to battalion control. Battalion main command post moves with Company A.
- Battalion combat trains (main body) follows Company A along Axis Red with priority of support to the main effort, initially forward security forces. This echelon support includes maintenance support and treatment team alpha (clinically staffed with the battalion surgeon), and carries mostly Class I, V, and limit Class III.
- Third Platoon Company A, under battalion control, is the rear security for the main body.
- Battalion field trains initially occupies position to the rear of defending battalion, vicinity Assembly Area Eagle. This echelon support includes all battalion sustainment not located with the combat trains, includes treatment team bravo (clinically staffed with the physician assistant). On order, conducts passage of lines and moves along Axis Red (not illustrated).

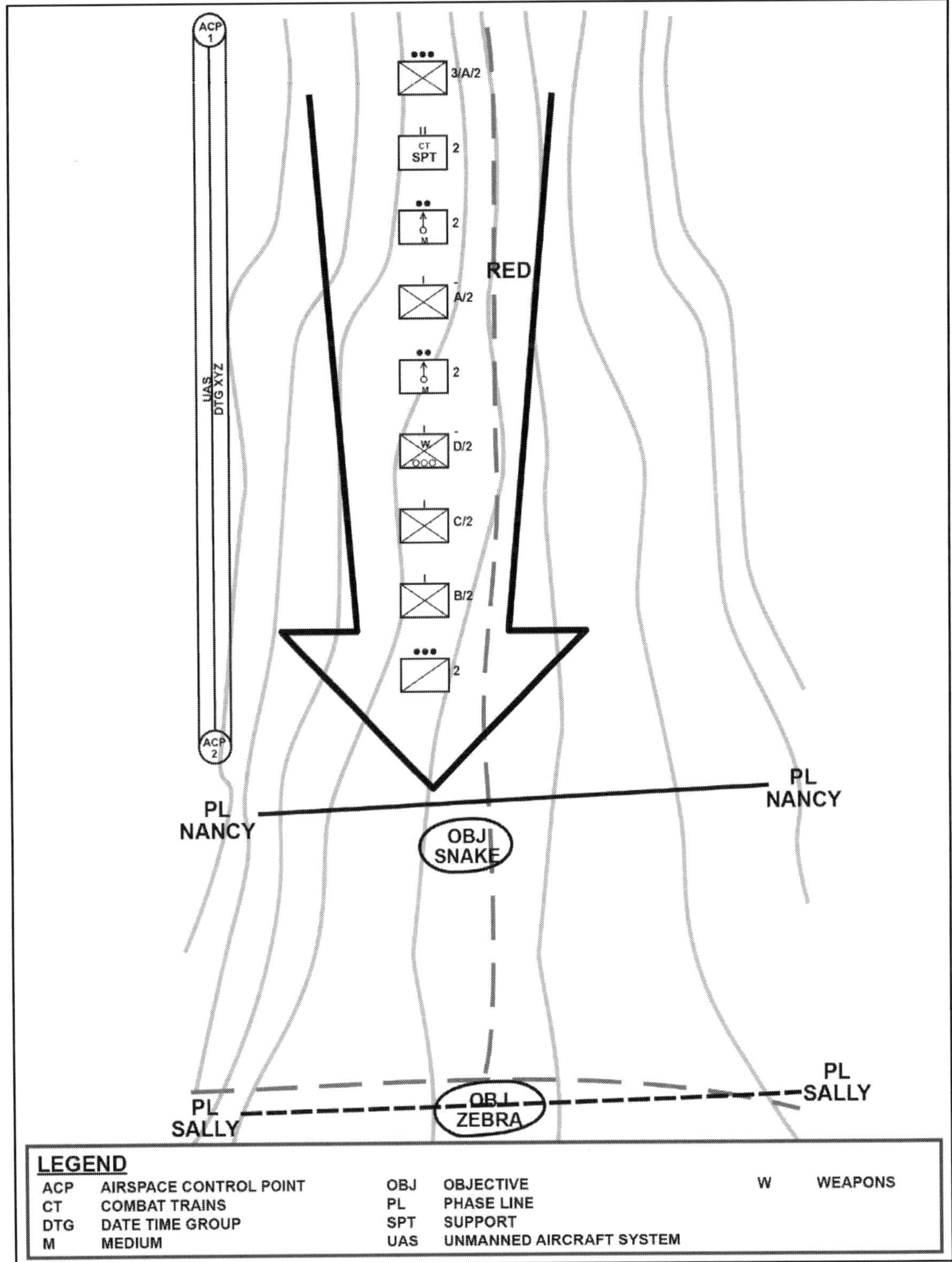

Figure 2-31. Movement to contact (Infantry Battalion 2 organization of forces), example

Gain and Maintain Contact

2-199. Within the scenario, Infantry Battalion 2 forward security forces (R&S forces and the advance guard) did not have enemy contact along Axis Red through March Objective Snake. As Company B (advance guard) crossed Objective Snake, the scout platoon, with attached sniper squad, (R&S force) continued movement to March Objective Zebra. During the movement to Objective Zebra, the scout platoon (forward of Company B) identified an enemy platoon size Infantry force (dismounted) establishing hasty defensive positions, 1 kilometer north of Phase Line Sally along Axis Red. The scout platoon passed this information to the battalion's tactical command post and Company B commander.

2-200. As the scout platoon leader developed the situation, the platoon moved to covered and concealed positions to maintain visual contact and observe further enemy activities. The sniper squad, under the control of the scout platoon leader moved to and established positions that provide for observation and precision fires. Company B continued movement while maintaining contact with the scout platoon to coordinate combat actions and exchange information prior to the meeting engagement. The scout platoon continued to maintain visual contact with the enemy and update the tactical command post and Company B maneuver forces. Regardless of the technique, actions between forward security forces are closely coordinated before engagement to prevent fratricide and confusion.

Note. No matter how the force makes contact, seizing the initiative is the overriding imperative. Prompt execution of battle drills at platoon level and below, and standard actions on contact for larger units, can give that initiative to the friendly force.

Disrupt and Fix the Enemy

2-201. The battalion's fire support plan includes allocating support for meeting engagements or hasty attacks that occur during the movement to contact. The fire support officer plans targets on and beyond the projected locations (enemy positions and march objectives). The Infantry battalion commander relies primarily on fire support assets to weight security forces during movement. The fires system helps develop fire superiority when organized correctly to fire immediate suppression missions to help maneuver forces get within direct-fire range of the enemy.

2-202. Fire support systems available to the battalion should tend to focus on suppression missions to disrupt enemy forces as they are encountered and smoke missions to obscure or screen exposed friendly forces when conducting movement. The battalion commander synchronizes battalion fire support systems with the movement and coverage of higher echelon fire support assets. This scenario includes the friendly defending force to the rear of Infantry Battalion 2 while still in range. The commander synchronizes the employment of Army aviation attack and reconnaissance units and close air support to prevent the enemy from regaining balance while ground fire support assets are repositioning.

2-203. Prior to the meeting engagement, the scout platoon provides overwatch and concentrates fires (direct and indirect) to suppress enemy weapons, and obscure enemy observation. During Company B's approach, the company commander chooses a movement technique (traveling overwatch and bounding overwatch), based on the mission variables of METT-TC, to make contact with the smallest possible force while providing flexibility for maneuver. Whatever combat formation the commander chooses, the unit must be able to deploy once the enemy's location is determined to ensure the axis of advance traveled by the main body is free of enemy forces.

Maneuver

2-204. During the movement to contact, the main body keeps enough distance between itself and its forward security forces to maintain flexibility for maneuver. This distance varies with the level of command, the unit, the terrain, and the availability of information about the enemy. The main body may advance over multiple parallel routes with numerous lateral branches to remain flexible and reduce the time needed to initiate maneuver. While it is preferred for the battalion to use multiple routes, the battalion and smaller units can move on just one route.

2-205. The main body's march dispositions must allow maximum flexibility for maneuvering during the movement to contact and when establishing contact with the enemy force. The main body may move continuously (using traveling and traveling overwatch) or by bounds (using bounding overwatch). It moves by bounds when contact with the enemy is imminent and the terrain is favorable. Command posts and trains may travel along high-mobility routes along the axis of advance (or area of operation) and occupy hasty positions as necessary. Indirect-fire assets, both organic and external to the battalion, move and position to support the movement to contact. The priority target list is continually updated throughout the movement.

2-206. Within the scenario, Infantry Battalion 2, forward security forces conducted a meeting engagement prior to March Objective Zebra along Axis Red. The scout platoon with attached snipers (R&S force) suppress the enemy in position with indirect and precision fires. The lead platoon (first platoon) of Company B was able to assault through the enemy positon. First platoon established limit of advance just south of the enemy position. Once limit of advance established, second platoon bounded right and third platoon followed by attached assault platoon bounded left around the enemy position to establish local security south of the first platoon's consolidation and reorganization line (see paragraph 2-411)/limit of advance (figure 2-32).

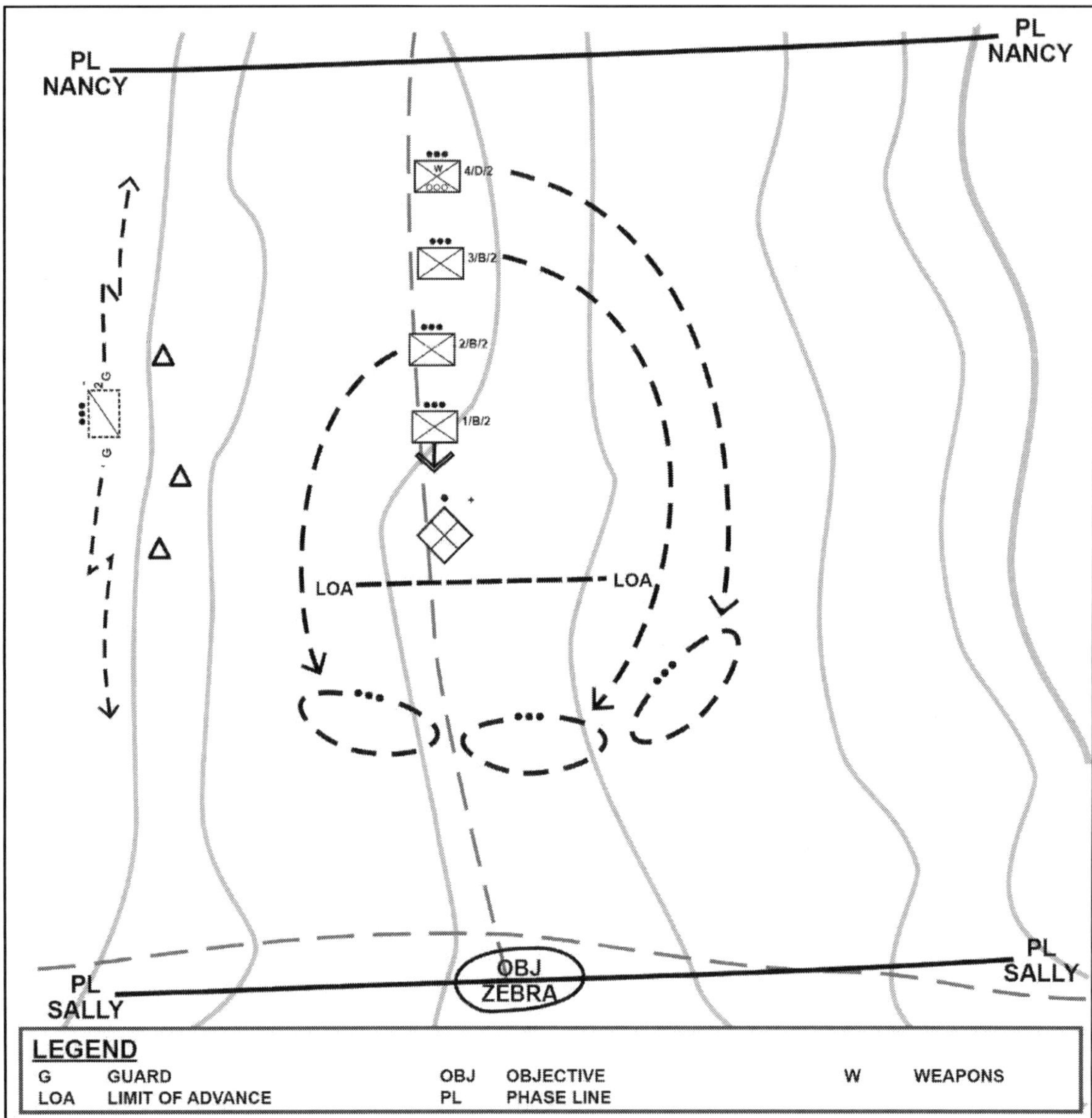

Figure 2-32. Movement to contact (meeting engagement), example

FOLLOW THROUGH

2-207. Once consolidation and reorganization (see paragraph 2-411) complete, Infantry Battalion 2 continues the movement to contact as rapidly as possible to maintain the initiative. The scout platoon and sniper squad moved from their overwatch positions to continue the R&S force mission for the movement to contact along Axis Red. Company C moves around Company B and assumes the advance guard mission for the battalion. Once consolidation and reorganization complete, Company B, the lead element now for the battalion's main body continues movement along Axis Red.

Note. This scenario for discussion purposes continues later in the chapter for the approach march. (See paragraph 2-214.)

APPROACH MARCH

2-208. The commander employs tactical movement, by means of a tactical road march (see paragraph 2-39) or approach march, when contact with the enemy is possible or anticipated. Successful tactical movement depends upon the establishment of combined arms organizations and contingencies for actions on contact. The execution of the march must be flexible to changing conditions and responsive to the commander. The commander conducts tactical movement using combat formations (described in paragraph 2-23) in conjunction with movement techniques (described in paragraph 2-30). The following paragraphs provide the doctrinal basis for an approach march for the Infantry battalion.

DOCTRINAL BASIS

2-209. An *approach march* is the advance of a combat unit when direct contact with the enemy is intended. (ADRP 3-90). The commander employs an approach march when the enemy's approximate location is known, emphasizing speed over tactical deployment, and less physical security or dispersion. When the Infantry battalion conducts an approach march it uses formations and movement techniques consistent with the mission variables of METT-TC. An approach march terminates in a march objective, such as an assembly area, attack position, or assault position, or it can be used to transition to an attack. Follow-and-assume, follow-and-support, and reserve forces may also conduct an approach march forward of a line of departure. (Refer to FM 3-90-2 and ATP 3-21.18 for additional information.)

Combined Arms Organizations

2-210. The commander arranges units conducting an approach march into combined arms organizations, task-organized before the march begins to allow for transition to an on-order or a be-prepared mission without making major organizational adjustment. The commander assigns an area of operation or an axis of advance, in combination with routes to a force conducting an approach march, based on the mission variables of METT-TC (figure 2-33, page 2-66). Area of operations, axes of advance, or routes should facilitate the force's movement and maximize its use of concealment. Within the approach march, the commander assigns the force conducting the decisive operation or main effort and forces conducting each shaping operation or supporting effort respectively, area of operations, axes of advance, and separate routes unless an individual sub-unit has the task of either follow-and-assume or follow-and-support.

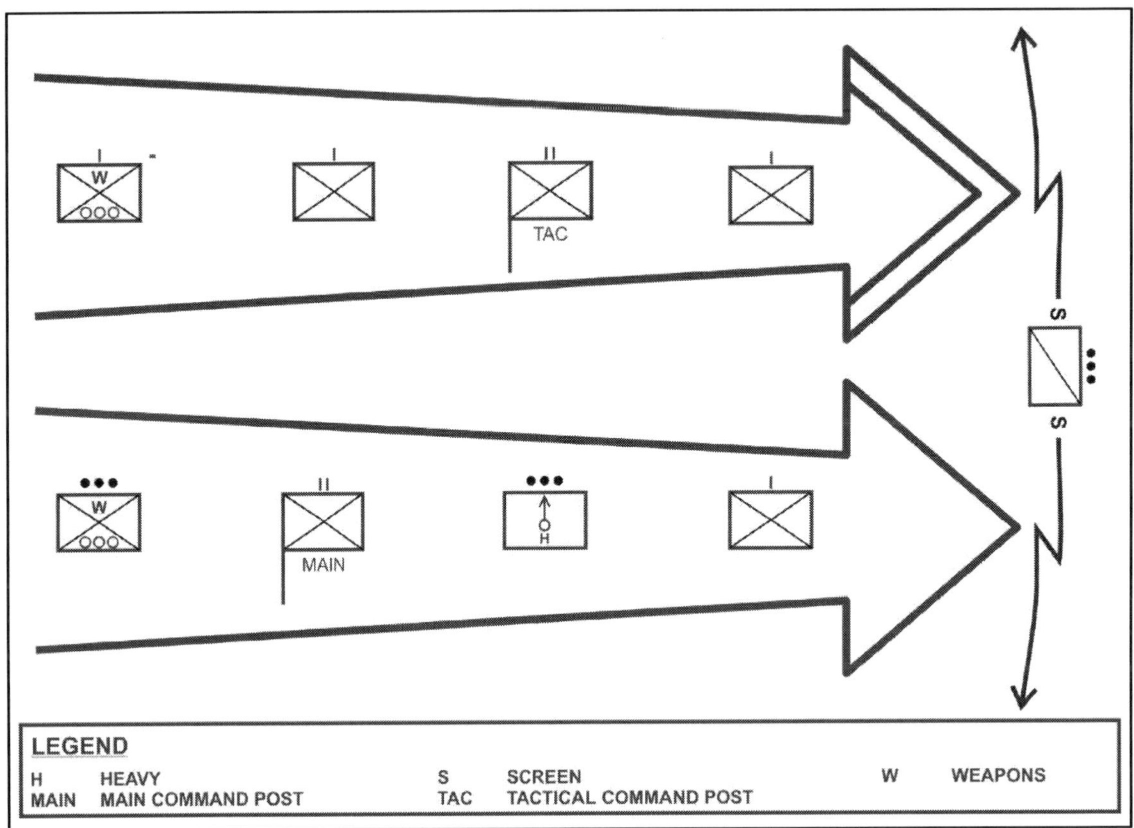

Figure 2-33. Approach march, axes of advance

2-211. As the approach march nears areas of likely enemy interference, the commander may divide the battalion main body into smaller, less vulnerable columns that move on additional multiple routes or cross-country while continuing to employ security forces (figure 2-34). The commander employs forward and flank security forces to increase the distance traveled before the main body transitions to a tactical formation. Forward and flank security forces remain within supporting distance of the main body, which stays in these smaller columns to facilitate rapid movement.

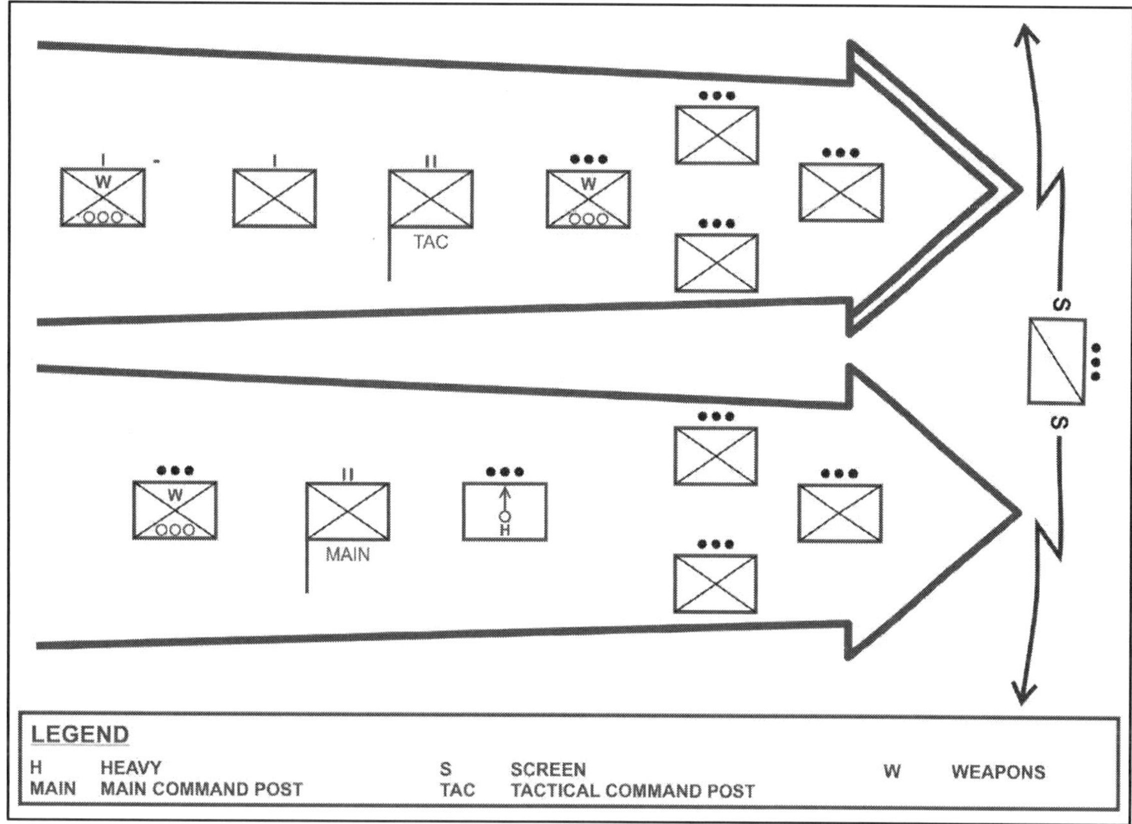

Figure 2-34. Approach march, less vulnerable columns

2-212. An approach march facilitates the commander's decision-making by allowing freedom of action and movement of the main body. The commander can choose to have all or part of the force conduct an approach march as part of the movement to contact, example—the main body may conduct movement to contact while forward security forces conduct an approach march or vice versa. The commander may execute an approach march for all or part of the tactical movement to efficiently use the available road network or reduce the time needed to move from one location to another.

Actions on Contact

2-213. Forward security forces focus on determining the enemy's composition, dispositions, and intent, and on providing the commander with relevant combat information to ensure commitment of the main body under optimal conditions. During movement, the enemy situation becomes clearer as forward security forces conduct actions on contact (see paragraph 2-170) to develop the situation in accordance with the commander's plan and intent. In determining the strength, location, and disposition of enemy forces, security forces allow the commander to focus the effects of the main body's combat power against an enemy position or the enemy's main body. As the overall force must remain flexible to exploit both combat information and intelligence, forward security forces must maintain contact with the enemy unless ordered to break contact by the commander.

2-214. When encountering an enemy force or an obstacle, forward security forces must quickly determine the threat it faces. For an enemy force, it must determine the enemy's composition, dispositions, activities, and movements and assess the implications of that information. For an obstacle, it must determine the type and extent of the obstacle and whether it is covered by fire. Obstacles can provide the attacker with information concerning the location of enemy forces, weapon capabilities, and organization of fires. Forward security forces will often use manned and unmanned aircraft systems to locate and target possible enemy units surrounding (example–enemy reaction or counterattack forces) or overwatching the engagement. When in advance of the battalion's main body, forward security forces may seize terrain that offers essential observation.

Note. The following illustration introduces a fictional scenario as a discussion vehicle for illustrating one of many ways that an Infantry battalion can conduct an approach march. This illustration is a continuation of the scenario started earlier in this chapter focusing on Infantry Battalion 2.

Illustration of an Approach March

2-215. The following scenario illustrates an approach march conducted by Infantry Battalion 2. In this scenario, started earlier in this chapter, an IBCT continues a movement to contact along two axes of advance. Infantry Battalion 1, main effort moves along Axis of Advance Blue (Axis Blue) to the east. Infantry Battalion 2, supporting effort moves along Axis of Advance Red (Axis Red) to the west (see figure 2-30, page 2-60). During the movement to contact, Infantry Battalion 1 cleared enemy resistance north of Objective Fox but encountered stiff resistance along Axis Blue from platoon size enemy elements. Because of the unrestricted terrain and the enemy's clear fields of fires from Objective Fox to the north along Axis Blue, Infantry Battalion 1 was unable to maneuver to generate sufficient combat power to seize Objective Fox. During the movement to contact, Infantry Battalion 2, forward security forces conducted a successful meeting engagement (see paragraph 2-192) prior to March Objective Zebra along Axis Red (see figure 2-35, page 2-70).

Infantry Brigade Combat Team, Subordinate Unit Task and Order of Movement

2-216. Within the scenario, Infantry Battalion 2 monitored the progress of the brigade's main effort along Axis Blue. The battalion commander, per the IBCT commander's guidance, begins to visualize how Infantry Battalion 2 might move to assist the brigade in regaining the initiative along Axis Blue. The battalion commander and staff began preliminary planning and WARNORD development to conduct an approach march to Objective Fox. As Infantry Battalion 2 continued its movement to contact, and as forward security forces of the battalion seized Objective Zebra (unopposed), the battalion received a change of mission from the IBCT commander (figure 2-35, page 2-70). The IBCT commander directed—

- Infantry Battalion 2, now the main effort, conduct approach march to Assault Positions 3 and 4. On order conduct an attack, seize Objective Fox.
- Infantry Battalion 1, now a supporting effort, establish support by fire and attack by fire positions north of Objective Fox.
- Cavalry squadron, conduct guard mission to the east of Infantry Battalion 1.
- Infantry Battalion 3 continues follow and assume tactical mission task. Company C continues under IBCT control.
- Company C continues as an IBCT reserve positioned to the rear of Infantry Battalion 1 along Axis Blue.
- On order, field artillery battalion provides priority of fires to Infantry Battalion 2.
- On order, brigade engineer battalion priority of engineer effort, mobility of Infantry Battalion 2. Tactical unmanned aircraft system platoon continues to conduct surveillance to the flanks of the IBCT movement (not illustrated).
- Brigade support battalion establishes BSA, vicinity Assemble Area Crow (not illustrated).

Infantry Battalion 2, Subordinate Unit Task and Order of Movement

2-217. Infantry Battalion 2 received change of mission to conduct an approach march to Assault Positions 3 and 4, west of Objective Fox. On order conduct an attack, seize Objective Fox. Not to slow the tactical movement to the assault positions, the battalion commander initially maintained the same combined arms organizations and order of movement task-organized for the movement to contact. The commander assigned an axis of advance and established a probable line of deployment (Phase Line Joe) for the approach march, based on the mission variables of METT-TC and the position of the assault positions (figure 2-35, page 2-70). During the approach march, the commander assigned tasks to the force conducting the main effort and forces conducting supporting efforts along with area of operations, axes of advance, phase lines, and separate routes to support the battalion scheme of maneuver for the attack. Battalion subordinate unit task and order of movement is as follows:

- Battalion scout platoon (R&S force), with attached sniper squad, conduct route reconnaissance forward of Company C along Axis Red-Alpha.
- Company C (advance guard), with attached engineer platoon (-), conduct approach march along Axis Red-Alpha forward of the main body.
- Company B (main body lead element) with attached engineer squad and assault platoon, conduct approach march along Axis Red-Alpha, battalion tactical command post moves with company.
- Company D (main body), with three assault platoons, follows Company B in main body.
- Battalion mortar platoon (main body) moves by split section along Axis Red-Alpha. Displaces by section to provide continuous fire support, priority of fires initially to forward security forces.
- Company A, battalion reserve, follows Company D in main body. Attaches one Infantry rifle platoon to battalion control. Battalion main command post moves with Company A.
- Third Platoon Company A, under battalion control, is the rear security for the main body.
- Battalion combat trains, occupy support area vicinity Objective Snake (not illustrated), with priority of support to the main effort, initially forward security forces. This echelon support includes maintenance support and treatment team alpha (clinically staffed with the battalion surgeon), and carries mostly Class I, V, and limit Class III.
- Battalion field trains, moves along Axis Red and occupies support area, vicinity Objective Kiwi (not illustrated). This echelon support includes all battalion sustainment not located with the combat trains, includes treatment team bravo (clinically staffed with the physician assistant).

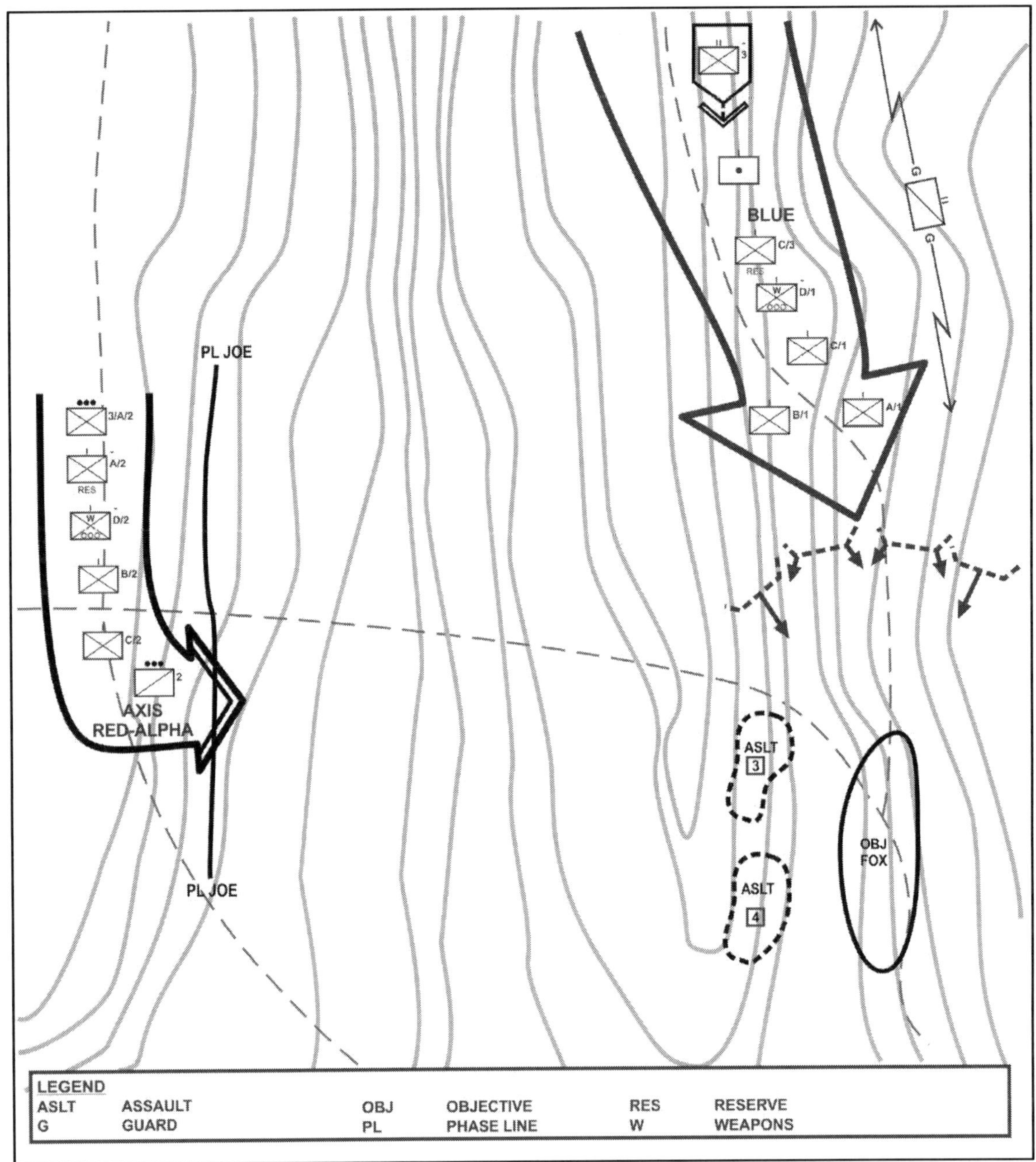

Figure 2-35. Approach march (initial axis of advance), example

2-218. As the battalion approach march neared areas of likely enemy interference from Objective Fox, vicinity Phase Line Joe (probable line of deployment), the battalion commander divided the battalion axis of advance into two company columns, Axis Green and Axis Yellow. Lead companies moved on multiple routes or cross-country while employing security forces (platoon size element) forward of the column's main body. Trail elements of each column provide rear security. Company D attaches one assault platoon to the main effort, Company C. The scout platoon moved from its security force position forward of the battalion to a position south of Axis Yellow to conduct a guard mission on battalion's southern flank. Battalion sniper squad moved from its security force location forward of the battalion to a position north of Axis Green to conduct linkup with Company D on the northern flank of the battalion. Forward and flank security forces remain within supporting distance of the battalion main body (figure 2-36, page 2-72). Battalion subordinate unit task and order of movement is as follows:

- Company C (attack-main effort), with attached engineer platoon (-) and assault platoon, conducts approach march along Axis Green to Assault Position 3.
- Company B (attack-supporting effort), with attached engineer squad and assault platoon, conducts approach march along Axis Yellow to Assault Position 4.
- Company D (attack-supporting effort), with two assault platoons, conducts approach march following Company C along Axis Green. Battalion tactical command post follows main effort (Company C), moves with Company D.
- Company A (battalion reserve) follows Company B along Axis Yellow. The rifle platoon detached to battalion control returns to Company A control. Battalion main command post follows Company B, moves with Company A.
- Battalion mortar platoon (supporting fires) moves by split section (Section A moves with Company D and Section B moves with Company A). Priority of fires to Company C.
- Battalion scout platoon conducts guard mission south of Company B along Axis Yellow.
- Battalion sniper squad conducts movement to a position north of Axis Green to conduct linkup with Company D on the northern flank of the battalion.

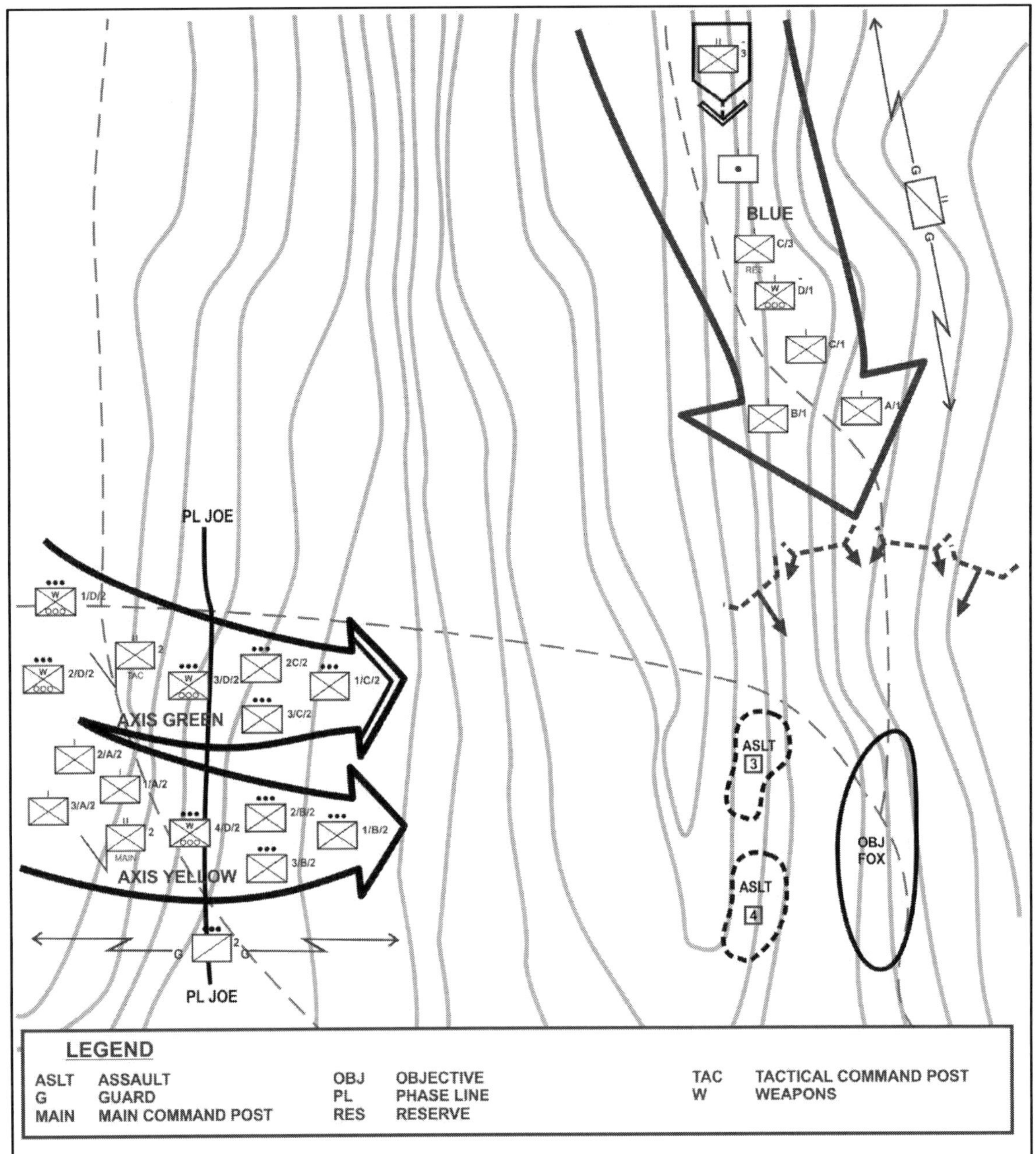

Figure 2-36. Approach march (advance in company columns), example

Note. This scenario for discussion purposes continues later in the chapter for the attack. See paragraph 2-413.

SUBORDINATE TASKS OF MOVEMENT TO CONTACT

2-219. When the Infantry battalion is involved in operations, such as peace operations, irregular warfare, and military engagement, unit offensive actions normally are closely related to the movement to contact tasks of cordon and search or search and attack. The conduct of offensive tasks in these operations will normally employ restrictive rules of engagement throughout the mission regardless of the element of decisive action (offense, defense, and stability) dominant at any specific moment. In the conduct of these operations, the emphasis on the stability element is much more dominant than the defensive element.

Cordon and Search

2-220. *Cordon and search* is a technique of conducting a movement to contact that involves isolating a target area and searching suspected locations within that target area to capture or destroy possible enemy forces and contraband (FM 3-90-1). A cordon and search may be thought of as a movement to contact, clear (tactical mission task), or area reconnaissance based on the availability or fidelity of intelligence.

Note. ATP 3-06.20 establishes multi-Service tactics, techniques, and procedures for cordon and search operations. Target area is the geographical space, containing one or more targets that must be controlled in order to permit the search to occur—is defined for the purposes of ATP 3-06.20.

2-221. While the primary purpose of a cordon and search is to find enemy or material, often parallel, is the search for and exploitation of material of evidentiary or intelligence value to use in the execution of future operations and criminal prosecutions. *Site exploitation* is a series of activities to recognize, collect, process, preserve, and analyze information, personnel, and materiel found during the conduct of operations. Also called SE (DOD) (JP 3-31). *Site exploitation* is the synchronized and integrated application of scientific and technological capabilities and enablers to answer information requirements, facilitate subsequent operations, and support host-nation rule of law (Army) (ATP 3-90.15). Additional individuals with the requisite technical expertise or equipment may be provided to the unit conducting the cordon and search. Information gained from site exploitation provides the commander with additional information to identify and target enemy personnel and materials. The following paragraphs provide the doctrinal basis for cordon and search operations conducted by the Infantry battalion.

Principles of Cordon and Search

2-222. The principles of cordon and search are comprehensive and fundamental rules guiding the Infantry battalion on the conduct of cordon and search. The principles are not a checklist. While the commander considers these principles, they do not apply in the same way to every situation, rather, they summarize characteristics of successful cordon and search operations. The value in these principles lie in assisting the commander in analyzing a pending operation while synchronizing efforts and determining if or when to deviate from the principles based on the current situation. The nine principles of cordon and search are—

2-223. Speed. Cordons and movement to a target area should occur rapidly to maintain initiative and momentum. The cordon and search force commander must carefully consider time and speed factors, especially as they relate to enemy reactions. When the target of the cordon and search force is personnel or sensitive material, the force must achieve greater relative tempo than the threat to preclude the enemy's escape or the destruction of material. Good reconnaissance, complete with guides, can greatly ease the cordon and search force's maneuver burdens and increase the speed of relative movement.

2-224. Surprise. Through tempo and deception, all efforts must be made to deny the enemy the opportunity to react. Foot movement may be combined with vehicular movement to increase speed and stealth. By itself, or with other types of movement, air mobility can provide the cordon and search force with considerable speed, some elements of surprise, and the ability to move in or remove additional resources, detainees, and caches. While waterborne movement can canalize the cordon and search force to certain, specific avenues of approach, it can also provide relative speed, surprise, and the ability to deliver additional resources and remove detainees, and caches. Special operations forces capabilities significantly enhance the speed and surprise of mission execution, decreasing the probability of compromise and enabling operational tempo.

2-225. Isolation. The target and target area must be effectively sealed off to defeat enemy reactions and free the cordon and search force to conduct actions on the objective. When possible, the Infantry battalion should isolate the objective using stealth and rapid movement in order to surprise the enemy to ensure complete control of the area before starting contact with the enemy. When isolating, the commander considers three-dimensional and in-depth isolation of the objective (front, flanks, rear, upper stories, basements, and rooftops). The commander employs all available direct and indirect fire weapons consistent with the rules of engagement. Isolating the objective is a key factor in facilitating the assault and preventing casualties.

2-226. Proper target identification. Cordon and search forces must be properly tasked and trained to identify, capture, and exploit targeted enemy personnel and material. During the execution of a cordon and search operation, specially trained special operation forces can facilitate sensitive site exploitation of an objective. Sensitive site exploitation may also provide further intelligence required for follow-on operations. Informants can provide positive identification of locations and personnel identified as high-payoff targets. Special aircraft systems are used to enable friendly-centric fire support—continuous monitoring of friendly ground forces while simultaneously sweeping the inner and outer perimeters for threats to the operation and identification of hiding and fleeing personnel.

2-227. Timeliness. Time is a driving force in cordon and search operations. The commander must strike a balance between combat information, target activities, desired end state, and execution of the cordon and search to gain the initiative and deny the enemy the ability to reposition or escape. While sudden opportunities may arise with little planning time, the nature of a search and what it may uncover can result in significant amounts of time spent in the objective area. Commanders must be prepared to execute cordon and searches with very little notice while simultaneously being prepared to spend many hours, even days, conducting the operation. The size of the objective area most directly affects the size of the security element, and the time necessary to emplace inner and outer cordons.

2-228. Accountability. Accountability is critical throughout, and until completion of the operation as cordon and search elements and sub-elements often operate dispersed. In addition to all actions associated with the target(s) and the target area, search and site exploitation, detainee handling, casualty evacuation, and interaction with the local population, actions on the objective will include internal accountability from arrival to withdrawal. Cordon and search forces reinforce accountability procedures to mitigate possible risks associated with multidirectional egress.

2-229. Minimization of collateral damage. *Collateral damage*—unintentional or incidental injury or damage to persons or objects that would not be lawful military targets in the circumstances ruling at the time (JP 3-60). While cordon and search operations focus on eliminating threats or potential threats, excessive collateral damage may constitute violations under both the law of war and the Uniform Code of Military Justice. Such violations create resentment within the local populace, embolden our enemy's, and cause damage to our credibility and good standing with the local populace. A key consideration when executing a non-permissive cordon and search is the collateral damage to property in the vicinity of the target and the target area. Proportionality will require the anticipated loss of life and damage to property incidental to attacks must not be excessive in relation to the concrete and direct military advantage expected to be gained.

2-230. Detailed search. Target areas must be searched to ensure all enemy assets are captured. This requires proper training, coordination, marking, and adherence to tactical standard operating procedures with all members of the cordon and search force understanding that clearing and searching are not the same thing. *Clear* is a tactical mission task that requires the commander to remove all enemy forces and eliminate organized resistance within an assigned area (FM 3-90-1) while *search* is a systematic reconnaissance of a defined area, so that all parts of the area have passed within visibility (JP 3-50). When time is a limitation, the detail and quality of the search process is retained by narrowing the scope of the search and increasing the manpower devoted to the search.

2-231. Legitimacy: cordon and search operations are always conducted within the law of war and according to the rules of law. Regardless of who is in charge and where a cordon and search is conducted, United States forces execute and participate in the operation in such a way as to underscore the appropriateness of the operation, and the legitimacy of the security and governmental forces associated with it. When possible, the presence of effective local police or the approval by respected local officials of a cordon and search lend it legitimacy and further the rule of law.

Phases of a Cordon and Search

2-232. A *phase* is a planning and execution tool used to divide an operation in duration or activity (ADRP 3-0). Phasing is typically used to distinguish when the decisive operation or main effort changed the tactical or operational focus. The phasing of a cordon and search describe how the battalion commander envisions the overall operation unfolding. The six phases to a cordon and search are planning and reconnaissance, movement to the objective, cordon, actions on the objective, and retrograde.

2-233. During planning (includes preparation) and reconnaissance, mission analysis is conducted through either the military decision-making process or troop leading procedures process focusing on the task and purpose, IPB (identification of target and target areas), and tentative scheme of maneuver. Support forces external the Infantry battalion are task organized and WARNORDs are issued. As planning serves to focus the information collection effort by identifying, what to look for and where to look for it, reconnaissance helps to refine the planning process by answering through collection, information requirements and priority intelligence requirements. Although the plan is continually updated, and R&S continuous throughout the operation, this phase ends with issuing the completed order, conducting final rehearsals and inspections, and crossing the line of departure.

2-234. Movement to the objective begins with the physical movement of the cordon and search force to the target and objective area. Methods of movement, timing, and scheme of maneuver depend on the mission variables of METT-TC. Movement by the cordon and search force may be as a single element on one route (single points of ingress or egress) or by a number of different elements across different routes (multidirectional points of ingress or egress) at different times. (See ATP 3-06.20 for additional information on the methods of ingress and egress.) A cordon and search may use any manner of ground tactical movement (dismounted and mounted march), air movement, or amphibious movement, and may blend types of movement. For example, the security element may deploy in wheeled vehicles while the search element conducts an air assault into the target area. Forces delivered to the objective area by air or amphibious means will still use various combat formations and movement techniques to relocate from the point of insertion to their assigned positions. The nature of the terrain and environment, and the capabilities of the threat may also dictate methods of movement. An area inaccessible to vehicles and an enemy threat possessing significant air defense capabilities might require foot movement. The movement phase normally ends when the security element reaches its release point, assault position, or similar control measure.

2-235. A cordon seals off the objective area by containing the enemy in the objective area and interdicting external attempts to influence events. A cordon consists of an outer and inner cordon, emplaced simultaneously or sequentially. Simultaneous or nearly simultaneous occupation of the inner and outer cordons can be achieved regardless of the means of movement. Multidirectional ingress lends itself to simultaneous action. Sequential occupation of the inner and outer cordon positions allows for ease of control and simplicity of maneuver while accepting some risk that an alert threat in the target area may have extra time to react. A single ingress lends itself to sequential actions. (See ATP 3-06.20 for additional information on simultaneous and sequential occupation of the inner and outer cordon.) The cordon phase ends when the cordons are in place and the security element has sealed off the objective area. Actions on the objective begin immediately after the cordon is established.

2-236. Actions on the objective includes all actions associated with the target(s) and the target area from arrival to withdrawal, from search and site exploitation to detainee handling, casualty evacuation, and interaction with the local population. The nature of the target dictates both operational and logistical considerations: operational in the sense of determining the resources needed to find the target, and logistical in the sense of determining the resources needed to meet the required end state regarding the target. Key considerations when examining the nature of a target include is the target a person(s), thing(s), or are both included, where is the target, is the target to be detained, evacuated, seized, or destroyed. The size and nature of the target area depend on the nature of the target, number of targets in that area, geographical relationship of multiple targets, and environment that makes up the area. The target area consists of that area in which the cordon and search force must establish dominance and control to enable an effective search. This phase ends with issuing the retrograde order.

2-237. Retrograde is the movement of the cordon and search force, regardless of the type of retrograde, method of movement, timing, and phasing from the objective area. Similar to a raid, the search element must depart prior to the security element displacing. The retrograde phase begins with the order to retrograde. In a manner similar to the movement and cordon phases, the cordon and search force may execute a single or multidirectional method of withdrawal, departing the target and objective areas via one or more routes. It may also include using stay behind R&S forces. The retrograde must receive proper attention during the planning and reconnaissance phase as the cordon and search force is most vulnerable to enemy action, accountability missteps, complacency risk, and other threats upon the retrograde. The retrograde phase ends when all elements of the cordon and search force (to include stay-behind R&S elements) return to the designated assembly area(s).

Task Organization

2-238. The cordon and search force includes a command element, a security element, a search element, and a support element, each with a clear task and purpose. Within the Army's operational framework, the search element is normally the decisive operation or main effort of the operation. The size and composition of the cordon and search force is based on the size of the area to be cordoned, the size of the area to be searched, and the suspected enemy situation. Host nation security forces that are dependable and competent, especially police forces, are extremely valuable during search operations in urban terrain. Assets employed during the cordon and search may include—

- Interpreters.
- Host nation or multinational forces.
- Human intelligence collection teams.
- Law enforcement professionals.
- Technical intelligence teams.
- Special advisors.
- Attack reconnaissance aviation.
- Signals intelligence enablers.
- Measurement and signature intelligence enablers.
- Military working dog teams.
- Biometrics collection efforts.
- Tactical military information support operations teams.
- Civil affair teams.

2-239. The battalion often receives additional assets to assist in a cordon and search based on availability and the mission variables of METT-TC. Assets may be included as teams in the security element or the search element, or they may remain independent and on call. Assets may be internal or external to the battalion and can include military police, engineers, civil affairs, military information support operations organizations, military intelligence, or artillery that form—

- Mine detection teams.
- Demolition teams.
- Interrogation teams.
- Documentation or biometric teams (uses a recorder with a camera).
- Scout dog teams.
- Fire support teams.
- Tactical air control parties
- Military information support operations and civil affairs augmentation teams.
- Detainee and enemy prisoner of war teams.
- Tunnel reconnaissance teams.
- Crowd control teams.
- Female search teams.
- Escort parties.
- Transportation teams.

Command Element

2-240. The command element serves as the headquarters of the cordon and search force. The commander and staff apply the same planning and decision-making processes used in other operations. While sudden opportunities may arise with little planning time, the nature of a search, and what it may uncover can result in significant amounts of time spent in the objective area. The commander must be prepared to execute cordon and searches with very little notice while simultaneously being prepared to spend many hours, even days, conducting the operation.

2-241. The command element considers numerous mission variables when planning and preparing for a cordon and search operation. The commander identifies elements and assigns units to them along with a clear task and purpose. Ideally, existing base units (companies and platoons) within the Infantry battalion are task-organized into a security element, a search element, and support elements for the cordon and search. These elements organize sub-elements as necessary to accomplish the mission. The command element normally nests within one of the other elements, rather than travel as its own entity.

Security Element

2-242. The security element is responsible for sealing off the objective through emplacement of an outer and inner cordon. The security element limits or prevents enemy or civilian influence in the objective area and prevents targets from escaping the cordon. The security element may receive the bulk of the available combat power due to multiple avenues of approach and requirements to disperse widely across the objective area to accomplish its mission. It may have to establish multiple blocking positions, observation posts, and conduct patrols in order to seal off the target area. Whether the cordon and search force travels as a single force (single point ingress) or moves in multiple elements on multiple routes (multidirectional ingress), the security element normally leads movement. It must have the inner and outer cordons in place, or nearly in place, prior to actions by other elements.

2-243. As security elements deploy to set-up the outer and inner cordons, the actions of both reinforce each other, creating an environment in which unwanted influences and actions, from both outside the objective area and from within the target area, are prevented from interfering with the success of the mission. Cordon elements (outer and inner) focus both inward and outward for security purposes. As security elements deploy, actions are dictated by the requirements of the cordons, methods by which the security element moves into the objective area, and manner in which it chooses to occupy the positions that will make up both cordons. The security element leader uses information collection assets internal and external to the Infantry battalion to observe the target area before the approach of security elements. The establishment of the cordon starts when the first security element reaches its release point position or similar control measure, and ends as the security element has sealed off the objective area.

2-244. Establishment of the outer cordon (figure 2-37, page 2-78) during the cordon phase requires detailed planning, effective coordination, and meticulous integration and synchronization to achieve the required effects. The outer cordon prevents anyone from entering the objective area and assists the inner cordon in preventing the enemy from escaping from the objective area. The element leader of the outer cordon maintains situational awareness, and within the leader's abilities, situational understanding to facilitate the progress of the operation, specifically the inner cordon and search efforts. Tactical tasks associated with the outer cordon security element include the following:

- *Block*—a tactical mission task that denies the enemy access to an area or prevents their advance in a direction or along an avenue of approach (FM 3-90-1). ***Note.*** See paragraph 3-117 for the definition of the term, block, when used as an obstacle effect.
- ***Deny*—a task to hinder or prevent the enemy from using terrain, space, personnel, supplies, or facilities.**
- *Interdict*—a tactical mission task where the commander prevents, disrupts, or delays the enemy's use of an area or route (FM 3-90-1).

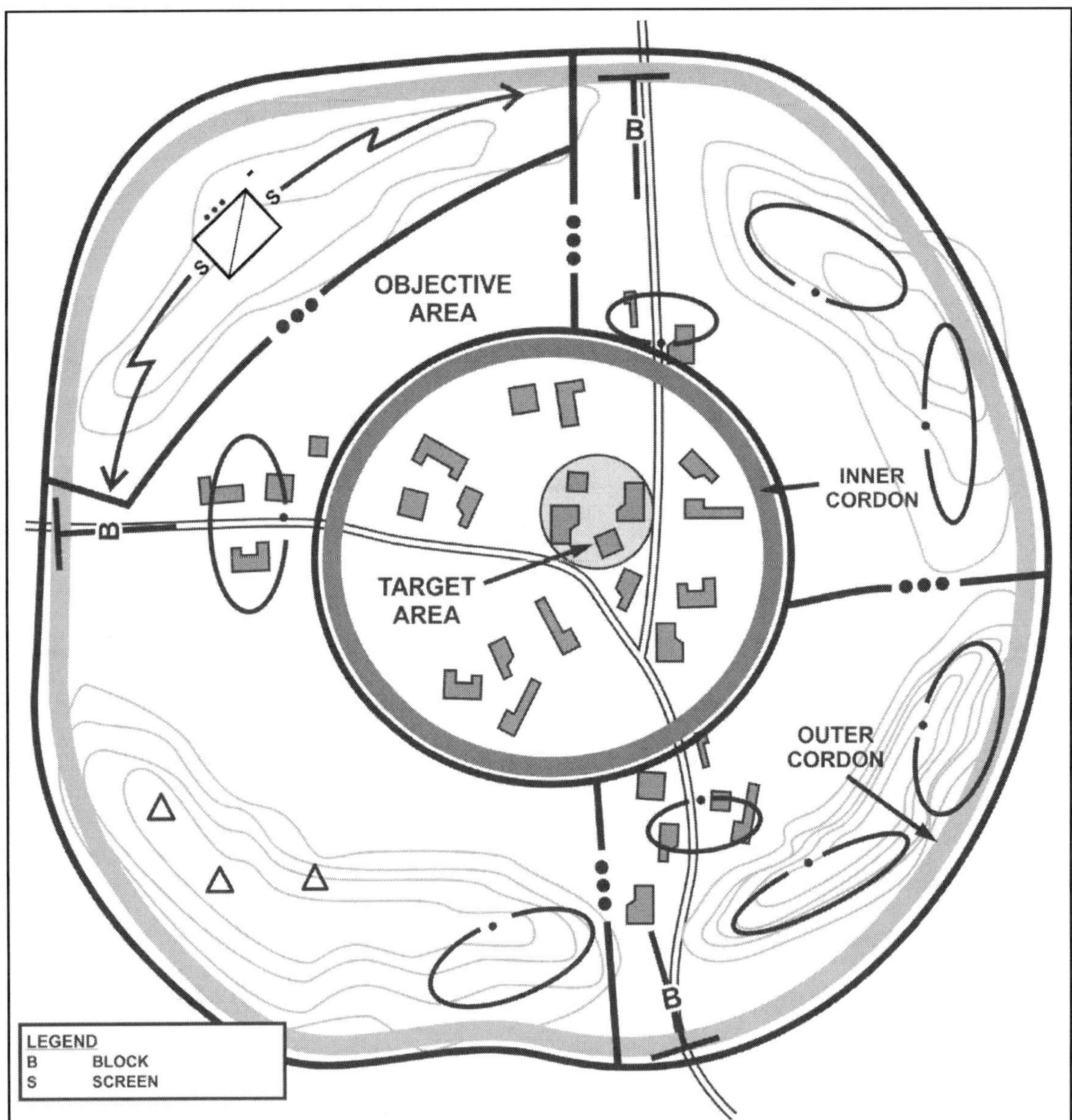

Figure 2-37. Cordon and search (outer cordon)

2-245. Establishment of the inner cordon requires the same level of planning, effective coordination, and integration and synchronization to achieve desired effects as did the outer cordon. The inner cordon seals off the target area to protect the search element from enemy activity. It prevents enemy movement within the target area and prevents enemy entry or exit. The security element is properly armed and equipped to control the ground and mitigate the most likely issues they are to face as determined in the planning and reconnaissance phase. The inner cordon's primary orientation is inward toward the target area. However, the inner cordon performs a secondary function of controlling movement into the objective area as well (figure 2-38). Tactical tasks associated with the inner cordon security element include the following:

- *Contain*—a tactical mission task that requires the commander to stop, hold, or surround enemy forces or to cause them to center their activity on a given front and prevent them from withdrawing any part of their forces for use elsewhere (FM 3-90-1).
- *Fix*—a tactical mission task where a commander prevents the enemy from moving any part of the force from a specific location for a specific period (FM 3-90-1). ***Note.*** See paragraph 3-117 for the definition of the term, fix, when used as an obstacle effect.

- Overwatch—a task that positions an element to support the movement of another element with immediate fire.
- *Suppress*—a tactical mission task that results in temporary degradation of the performance of a force or weapons system below the level needed to accomplish the mission (FM 3-90-1).

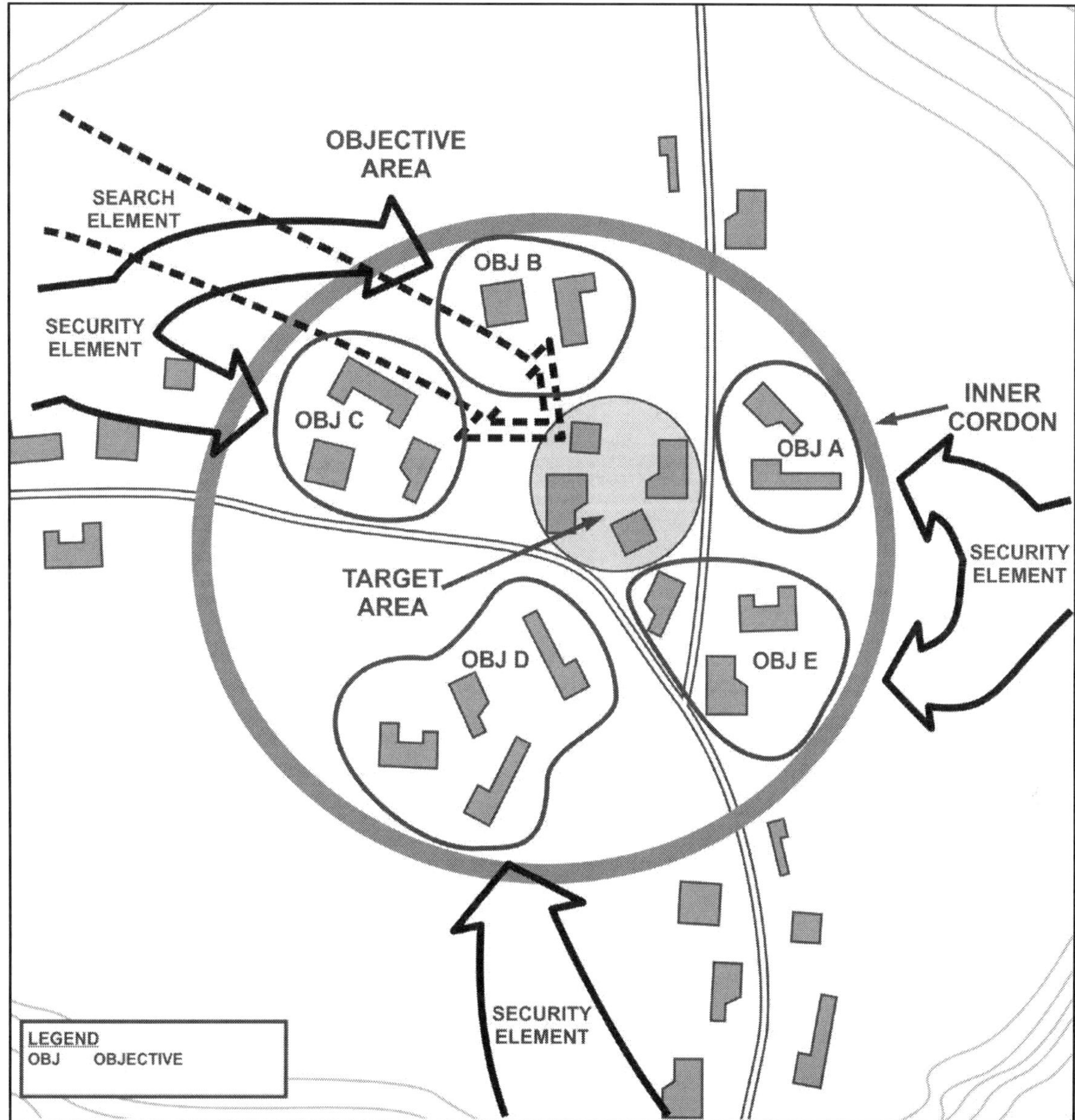

Figure 2-38. Cordon and search (inner cordon-multidirectional ingress)

2-246. During the actions on the objective phase, the security element maintains the inner and outer cordons. The inner cordon overwatches actions on the objective, prevents egress from the target area, and coordinates actions with the outer cordon. The outer cordon prevents external influences from entering the objective area, prevents or controls movement in and out of the objective area, and coordinates with the inner cordon.

2-247. Whether the cordon and search force withdraws as a single force or moves in multiple elements on multiple routes, the security element normally is the last to depart. The inner and outer cordons and their controls on the objective and target areas provide overwatch for the search and support elements. Should the withdrawal be cancelled by design or by enemy action, the cordon remains in place. Once the other cordon and search force elements have committed to withdrawal, and are safely away from the target area, the security element first collapses the inner, then the outer cordon.

Search Element

2-248. The search element's mission is to clear, search, and conduct site exploitation on the objective in order to locate and seize contraband material; and identify, search, or detain suspected insurgents. The search element moves as either the second or the third element of movement, in the order of march. When resistance is not expected, or when speed and surprise are paramount, the commander places the search element as the second march element so it can move immediately and directly into the target area. When resistance is expected or when ensuring the target area is sealed off, the search element will travel behind the security and support elements.

2-249. The search element normally divides into three types of teams: search teams, security teams, and support teams (figure 2-39). As these teams are most often-in direct contact with the local populous and potential threats, these teams train and enable accordingly. Teams within the search element initiate actions once the outer and inner cordons are in place. The following three paragraphs address the actions of each team.

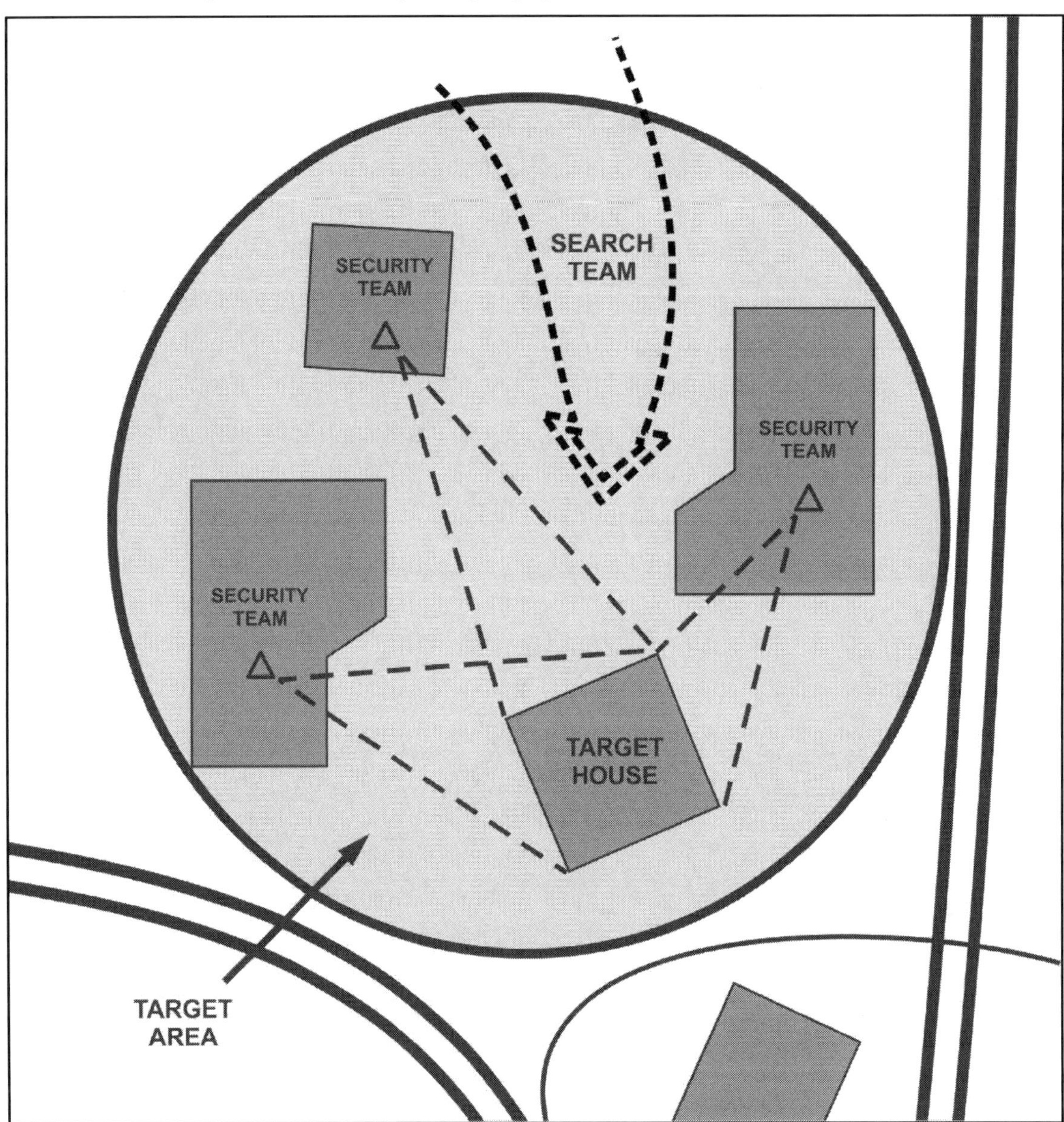

Figure 2-39. Cordon and search (target area)

2-250. Search teams focus on the actual conduct of the search and the processes of site exploitation. The search team is not normally part of the security of the target area, but can be, if required. The team relies on the support team for secondary tasks such as evidence and detainee handling.

2-251. Security teams are responsible for gaining access, clearing, and maintaining security of the immediate objective. Security functions include providing immediate security, protection, overwatch, and supervising movement in and out of the objective. The security team should not be tasks with actions that detract from security. In most cases, the existence of inner and outer cordons does not constitute local security for the search element.

2-252. Support teams are responsible for providing organized manpower to conduct continuing actions in the objective area and for providing direct assistance to the search team and security team, if required. Such actions might consist of, but are not limited to, detainee handling, evidence collection and handling, and casualty evacuation.

2-253. During the establishment of the cordon, the search element prepares to execute movement to the target and target area. Such movement may be immediate if executing arrival at the target and target area is planned to coincide with the placement of the inner and outer cordons. Conversely, the search element executes a short security halt at a designated position while it awaits deployment of the security element. Depending on the method of ingress (single or multidirectional), this may or may not be in conjunction with the support element.

2-254. During actions on the objective, the search element moves into the target area, and conducts actions necessary to clear and search the target area. The search element clears the target area and sets up a permissive environment for a systematic search. The search may be targeted (specific people, rooms, or buildings) or it may be comprehensive (everything in the target area). Tactical tasks that may be associated with the search element include the following:

- Breach—a tactical mission task in which the unit employs all available means to break through or establish a passage through an enemy defense, obstacle, minefield, or fortification.
- Clear—a tactical mission task that requires the commander to remove all enemy forces and eliminate organized resistance within an assigned area.
- Defeat—a tactical mission task that occurs when an enemy force has temporarily or permanently lost the physical means or the will to fight. The defeated force's commander is unwilling or unable to pursue that individual's adopted course of action, thereby yielding to the friendly commander's will and can no longer interfere to a significant degree with the actions of friendly forces. Defeat can result from the use of force or the threat of its use.
- Destroy—a tactical mission task that physically renders an enemy force combat-ineffective until it is reconstituted. Alternatively, to destroy a combat system is to damage it so badly that it cannot perform any function or be restored to a usable condition without being entirely rebuilt.
- Neutralize—a tactical mission task that results in rendering enemy personnel or materiel incapable of interfering with a particular operation.
- Seize—a tactical mission task that involves taking possession of a designated area by using overwhelming force (FM 3-90-1).
- Support by fire—a tactical mission task in which a maneuver force moves to a position where it can engage the enemy by direct fire in support of another maneuvering force.

2-255. The search element departs the target area under the overwatch of the security element and support element and generally serves as the lead element in the withdrawal order of movement. The search element normally retains that position throughout the rest of the withdrawal to enable mission command through ease of withdrawal.

Support Element

2-256. Support elements are designed to act as a force multiplier during a cordon and search operation and should be positioned where they can best accomplish assigned tasks. Support elements may assist the cordon and search force by serving as a designated reserve, providing additional enabling teams to support all elements of the force, and conducting continuing actions such as establishing a temporary defensive position, conducting vehicle recovery, casualty evacuation, and resupply.

2-257. The support element moves as either the second or the third element of movement in the order of march. When resistance is not expected, or when speed and surprise are paramount, the support element moves as the trail element so it does not interfere with the security and search elements' movement, but is still positioned to support, begin and continue actions in support of actions on the objective. When resistance is expected or when ensuring the target area is sealed off, the support element travels as the second element, allowing it to move into positions of support and security around the target area before the search element is committed.

2-258. As actions on the objective occur, the support element executes its assigned tasks for the cordon and search force. The support element establishes secure positions in or near the target area in which detainees, evidence, and casualties can be safely secured. These positions are established where they are readily assessable to cordon and search force, and if necessary defendable against attack. The support element may provide extra detainee and evidence handling teams to the search element, conduct casualty evacuation and provide internal resupply, vehicle recovery and hasty repair capabilities. Often the support element controls various enablers attached to the search element until they are needed. Tactical mission tasks associated with the support element include follow and assume and follow and support, which are defined below:

- Follow and assume—a tactical mission task in which a second committed force follows a force conducting an offensive task and is prepared to continue the mission if the lead force is fixed, attrited, or unable to continue.
- Follow and support—a tactical mission task in which a committed force follows and supports a lead force conducting an offensive task.

2-259. The support element, normally with the command element nested, follows in trace of the search element and is the last of the cordon and search force to depart the target area overwatched by the security element. During the withdrawal, depending on tasking, the support element may transport multiple detainees or large amounts of captured material and tow downed vehicles.

SEARCH AND ATTACK

2-260. *Search and attack* is a technique for conducting a movement to contact that shares many of the characteristics of an area security mission (FM 3-90-1). This form of a movement to contact is conducted when the enemy operates as small, dispersed elements whose locations cannot be determined to targetable accuracy by methods other than a physical search or when the task is to deny the enemy the ability to move within a given area. Normally, a search and attack is conducted by mounted or dismounted Infantry forces often supported by Stryker and armored equipped forces. However, this technique may also be necessary during major operations when the enemy establishes smaller noncontiguous defenses. These type search and attack operations are normally characterized by robust R&S, and rapidly concentrated combat power to fix and defeat or destroy the enemy once located. (Refer to FM 3-90-1 for additional information.)

Organization of Forces

2-261. A battalion search and attack is organized with a reconnaissance force, a fixing force, and a finishing force, each with a specific purpose and task to accomplish and the battalion as the appropriate level of support in terms of fire support, sustainment, and other combat enablers. Ideally, as with a cordon and search, the commander uses existing base units (companies and platoons) within the Infantry battalion to organize search and attack forces.

Reconnaissance Force

2-262. A battalion search and attack operation uses reconnaissance forces and surveillance assets within, and external to the battalion, to include scouts, snipers, unmanned aircraft systems, and electronic warfare assets (see appendix B) in finding and potentially fixing a dispersed enemy. The size of the reconnaissance force is based on the available intelligence about the size of enemy forces in the area of operation and the size of the area of operation in terms of both the geographical size and the size of the civilian population contained in that area of operation. The less known about the situation, the larger the reconnaissance force. Once the enemy is found, the reconnaissance force must determine if it has the combat power to fix the threat or if the battalion must fix the enemy with its fixing force.

Fixing Force

2-263. A battalion fixing force (usually a company-size force) prevents the enemy from moving to or from a specific location until the enemy is engaged or destroyed. The fixing force must have enough combat power to isolate the enemy forces once the enemy is located. Initial fixing positions are determined based on the enemy situation and, more appropriately, prepositioned in areas where fixing forces could most likely support the finishing force. When contact is made in other areas, fixing forces must have procedures in place to facilitate moving to where it can influence the fight. Once the threat is fixed, the fixing force determines if it is able to conduct the attack the finishing force normally would execute. If not, another unit designated as the finishing force executes the assault.

Finishing Force

2-264. A battalion finishing force engages and destroys the enemy force and is normally the decisive operation or main effort of the search and attack. Finishing forces must have enough combat power to defeat those enemy forces expected to be located within the area of operation. Finishing forces must be responsive and require synchronization with all support forces to accomplish the desired end state. Procedures must be in place to facilitate quick movement of the finishing force. Otherwise, the enemy will rapidly attrit the fixing force, and immediately break contact before engagement by the finishing force. These procedures include, but are not limited to, good graphic control measures, responsive means of transportation if required, and effective communications with the force in contact.

Planning and Preparation

2-265. A battalion search and attack is a decentralized movement to contact against an enemy operating in dispersed elements requiring multiple, coordinated patrols at company level and below. The battalion may conduct a search and attack as a coordinated battalion level operation or decentralized to independent company level operations. Whether conducting a consolidated or decentralized operation, the battalion must designate a decisive operation or main effort and shaping operations or supporting efforts that enable it to maintain the amount of flexibility the situation dictates.

2-266. As the commander and staff consider numerous operational and mission variables that influence the search and attack mission. The military decision-making process provides the battalion commander with the framework for planning population-centric distributed operations. With receipt or in anticipation of the search and attack mission the commander initiates the process alerting all participants of pending planning requirements. This in turn enables participants to determine the amount of time available for planning and preparation, decide on a planning approach (including guidance on using the Army design methodology), and how to abbreviate the military decision-making process, if required. (Refer to FM 6-0 for additional information.) This process organizes the battalion in purpose as well as space and helps to focus the search and attack for one or more of the following purposes:

- Destroy the enemy: render enemy units in the area of operation combat-ineffective.
- Deny the area: prevent the enemy from operating unhindered in a given area; for example, in any area the enemy is using for a base camp or for logistics support.
- Protect the force: prevent the enemy from massing to disrupt and destroy friendly military or civilian operations, equipment, and property such as key facilities, headquarters, or polling places.
- Collect information: gain information about the enemy and the terrain to confirm the enemy course of action predicted as a result of the IPB process.

2-267. The commander's situational understanding and knowledge of the higher echelon's concept of operation enables planning and the development of a clear task and purpose for each subordinate force. As planning and preparation continues, the commander establishes control measures that allow for decentralized actions and small-unit initiative to the greatest extent possible. Minimum control measures for a search and attack are an area of operation, target reference points, objectives, checkpoints, and contact points. The use of target reference points facilitates responsive fire support once the reconnaissance force makes contact with the enemy. The commander uses objectives and checkpoints to guide the movement of subordinate elements. Contact points indicate a specific location and time for coordinating fires and movement between adjacent units. The commander uses other control measures, such as phase lines and NAIs, as necessary.

2-268. Planning for the search and attack, based on the IPB, combines the battalion S-2's pattern and predictive analysis with available higher echelon intelligence, surveillance, and reconnaissance to determine likely enemy locations, capabilities, patterns, and actions. The battalion scheme of maneuver for the search and attack is developed to capitalize on this information along with the integration of internal and external fire support assets available to the battalion. Fire support plans must provide for flexible and rapidly delivered fires to achieve the commander's desired effects throughout the area of operation. The commander positions fire support assets to support subordinate elements throughout the area of operation and establishes procedures for rapidly clearing fires. To clear fires rapidly, command posts and small-unit commanders must track and report the locations of all subordinate elements. Because of the uncertain enemy situation, the commander is careful to assign clear fire-support relationships.

2-269. Synchronization of movement and maneuver, fire support, protection, and sustainment is difficult to achieve in this type-decentralized movement to contact. Distances between units, complex terrain, and a vague enemy situation contribute to this difficulty. For these reasons, the commander positions where best to receive information and transmit orders, and shift and commit forces during search and attack. Command posts (main and tactical command posts) are positioned to best influence the battle and to allow the commander the best vantage point to see the battlefield. This may not necessarily be with the decisive operation or main effort. Air defense capabilities, when attached in a direct support role, are employed in accordance with the commander's protection priorities. Air defense assets can increase protection for command posts located within the battalion's area of operation. In addition, air defense assets may provide security to maneuver and sustainment units and positioned to overwatch key air and ground routes or avenues of approach.

2-270. The fire support officer for the battalion prepares the fire support plan for each phase of the search and attack and contingencies. The fire support officer recommends the positioning of attached joint fires observers to support the operation, and assists in deconflicting battalion and higher echelon airspace. Tactical air control parties are positioned well forward (within reconnaissance, fixing, and finishing forces) to increase the timeliness and accuracy of close air support. Key to air-ground integration during search and attack is the means to develop *combat identification*, which is the process of attaining an accurate characterization of detected objects in the operational environment sufficient to support an engagement decision (JP 3-09). Priority of fires for battalion mortars during the search and attack is normally to the decisive operation or main effort of the battalion. Company mortars employ organic fires to support the attack of the finishing force or to block ingress/egress routes, and prevent repositioning of the enemy in support of the fixing force. Finding the enemy may have both echelons initially providing priority of fires to the reconnaissance force.

2-271. The engineer staff officer provides expertise to help identify breach points in enemy defenses and assure mobility. When engineer asset from the brigade engineer battalion are available, engineers can conduct route reconnaissance, determine bridge classifications, and find or make bypass routes where necessary. The route clearance platoon provides the detection and neutralization of explosive hazards and reduces obstacles along routes that enable force mobility within search and attack areas of operation. When attached to the Infantry battalion the route clearance platoon can sustain lines of communications during search and attack operations. When required, a breach squad from the brigade engineer battalion can be task organized to the route clearance platoon with mine-clearing line charges.

2-272. While the brigade aviation office, who leads the brigade aviation element (BAE), works directly for the IBCT commander as a permanent member of the IBCT special staff, aviation liaison teams can represent the supporting aviation task force at designated maneuver headquarters for the duration of a specific operation. If collocated with a BAE, the liaison team normally works directly with the brigade aviation officer as a functioning addition to the BAE cell. Effective employment of liaison officers is imperative for coordination and synchronization of search and attack operations. Often aviation liaison teams coordinate with the BAE and proceed to a supported ground maneuver battalion conducting the search and attack. Aviation units (attack reconnaissance, assault, and lift) can reconnoiter, guide ground forces to the enemy, provide lift and fire support assets for air movement, direct artillery fires, aid in mission command, and protect exposed flanks. Attack reconnaissance units can reinforce when anti-armor firepower is needed to block or fix the enemy.

2-273. The essential nature of communication to a battalion search and attack, characterized by dispersed decentralized independent company level operations, alludes to an additional planning problem for this type operation—communications. The limited range of communications within the type complex terrain most associate with search and attack missions requires detailed planning and rehearsal to insure that possible terrain masking effects and distances are resolved. Relays or retransmission stations may be required, and—as with the combat multipliers discussed above—additional planning becomes essential. All of these considerations add up to a more detailed preparation for combat.

2-274. The Infantry weapons company commander, in coordination with the battalion S-3, selects the tube launched, optically tracked, wire guided (TOW) missile positions to provide overwatch and long-range direct fire support. Based on the commander's analysis, the commander can use the MK 19 or the .50-caliber machine gun in place of or in combination with the TOW missile system. The weapons company can also provide mobility and additional firepower for security missions and the reserve. During limited visibility, the weapons company with advanced optics can augment security forces at key locations, monitoring areas where the enemy is expected to travel at night.

Execution

2-275. During execution, the battalion commander and staff, and subordinate commanders focus efforts on translating decisions determined during planning and preparation into actions. The commander's application of the required combat power, at the correct time and location, to gain and maintain a position of relative advantage is essential to the accomplishment of the search and attack mission. To seize, retain, and exploit the initiative, the commander ensures the synchronization of decentralized operations by specifying where each subordinate unit will operate, establishing measures to consolidate units before attacks, and establishing fire control measures for each subordinate unit. The commander seeks the most likely enemy locations, designating the forces most likely to make contact as the decisive operation or main effort and prepares to shift the decisive operation or main effort, if necessary. Search and attack forces, at battalion or company echelon, may enter the area of operation by infiltrating as an entire unit or by infiltrating as smaller units via ground, air, or water. Finally, battalion operations do not end at the destruction of the enemy. The battalion must immediately conduct consolidation and reorganization (see paragraph 2-411) and prepare for follow on offensive or defensive missions.

Search and Attack—Separate Company Areas of Operation

2-276. When each Infantry rifle company operates in separate company areas of operation (figure 2-40, page 2-86), a technique is to task each company to organize to find (reconnaissance), fix, and finish (destroy) the enemy within its capability. The battalion commander can direct each company to retain a reserve, or retain a battalion reserve. The battalion's indirect fire support is position to respond to all companies, as needed. The commander uses the reserve, priority of fire, and other available assets to weight the decisive operation or main effort. Once the company reconnaissance force finds the enemy, the company commander concentrates combat power to quickly fix and finish the enemy.

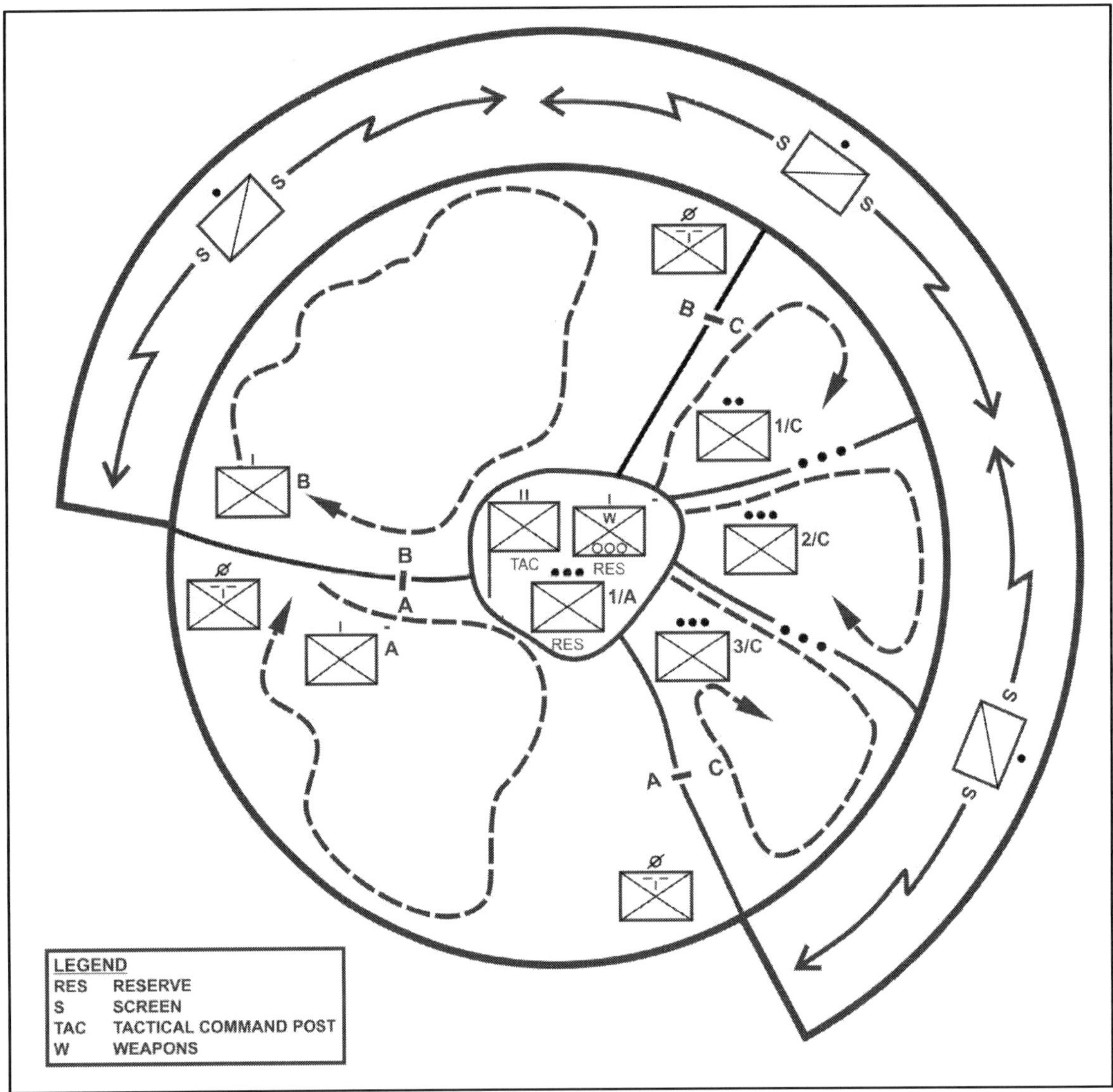

Figure 2-40. Search and attack (company areas of operation)

2-277. If the company is unable to destroy the detected enemy, the battalion commander considers means to fix or contain the enemy. The commander may divert another company to the task, use the battalion reserve or fires to contain and destroy the enemy. The commander redirects and repositions R&S assets to support any change to the plan, but also to identify enemy counterattack forces entering the area of operations. As plans change, the battalion commander provides control but allows for decentralized actions and small-unit initiative.

Search and Attack—Battalion Area of Operation

2-278. When the Infantry battalion operates in an area of operation, the commander can task organize subordinate units to find (reconnaissance), fix, and finish (destroy) the enemy, each with a specific purpose and task to accomplish. Alternatively, the battalion commander can—

- Task units with the reconnaissance effort in assigned areas of operation with individual subordinate elements tasked to perform the fixing and finishing functions based on the specifics of the situation.
- Direct subordinate companies to conduct reconnaissance and fixing in assigned areas of operation with the commander retaining direct control of the finishing force.
- Rotate subordinate elements through the reconnaissance, fixing, and finishing roles. However, rotating roles may require a change in task organization and additional time for rehearsal.

2-279. During execution, the battalion commander ensures that fire support and protection assets support the decisive operation while remaining responsive to the rest of the battalion. When the mortar platoon cannot support the entire battalion area of operation, the commander may consider splitting the platoon into sections. The commander must consider the size of the area to conduct the search and attack, the number of available forces and the time available. As the battalion staff conducts analysis to establish the duration of the mission, the staff needs to assess the battalion's ability to conduct continuous operations in order to develop and resource a plan to maintain the battalion's combat effectiveness. Subordinate unit tasks or actions, performed during search and attack may include the following:

- Locate enemy positions or routes normally traveled by the enemy.
- Destroy enemy forces within its capability.
- Fix or block the enemy until reinforcements arrive.
- Maintain surveillance of a larger enemy force through stealth until reinforcements arrive.
- Establish ambushes.
- Search towns or villages.
- Secure military or civilian property or installations.
- Act as a reserve.
- Develop the situation in a given area.

Find the Enemy

2-280. Ample time should be dedicated to determine the pattern of enemy operations. The commander decisions are most effective once the pattern has been identified; however, it may take more time than is available to accurately establish an enemy pattern. When conducting a search and attack, reconnaissance forces can expect to spend significant time reconnoitering in an area of operations. The commander may consider using one of several techniques to find the enemy. The commander may subdivide an area of operation into smaller ones and have the scout platoon reconnoiter forward of the remainder of the battalion. The scout platoon may be reinforced for this operation with snipers and unmanned aircraft systems. The commander also develops a contingency plan in the event that the reconnaissance force is compromised.

2-281. A search and attack with the scouts platoon (or other security element) forward is one technique the Infantry battalion can use to find the enemy as illustrated in figure 2-41, page 2-88. This method is initiated by the scout platoon conducting a zone reconnaissance in area of operation Green, while the remainder of the battalion conducts search and attack operations in area of operation Blue. Serving as part of the reconnaissance force, the scout platoon attempts to locate and gain information about the enemy, initially in area of operation Green, then area of operation Red.

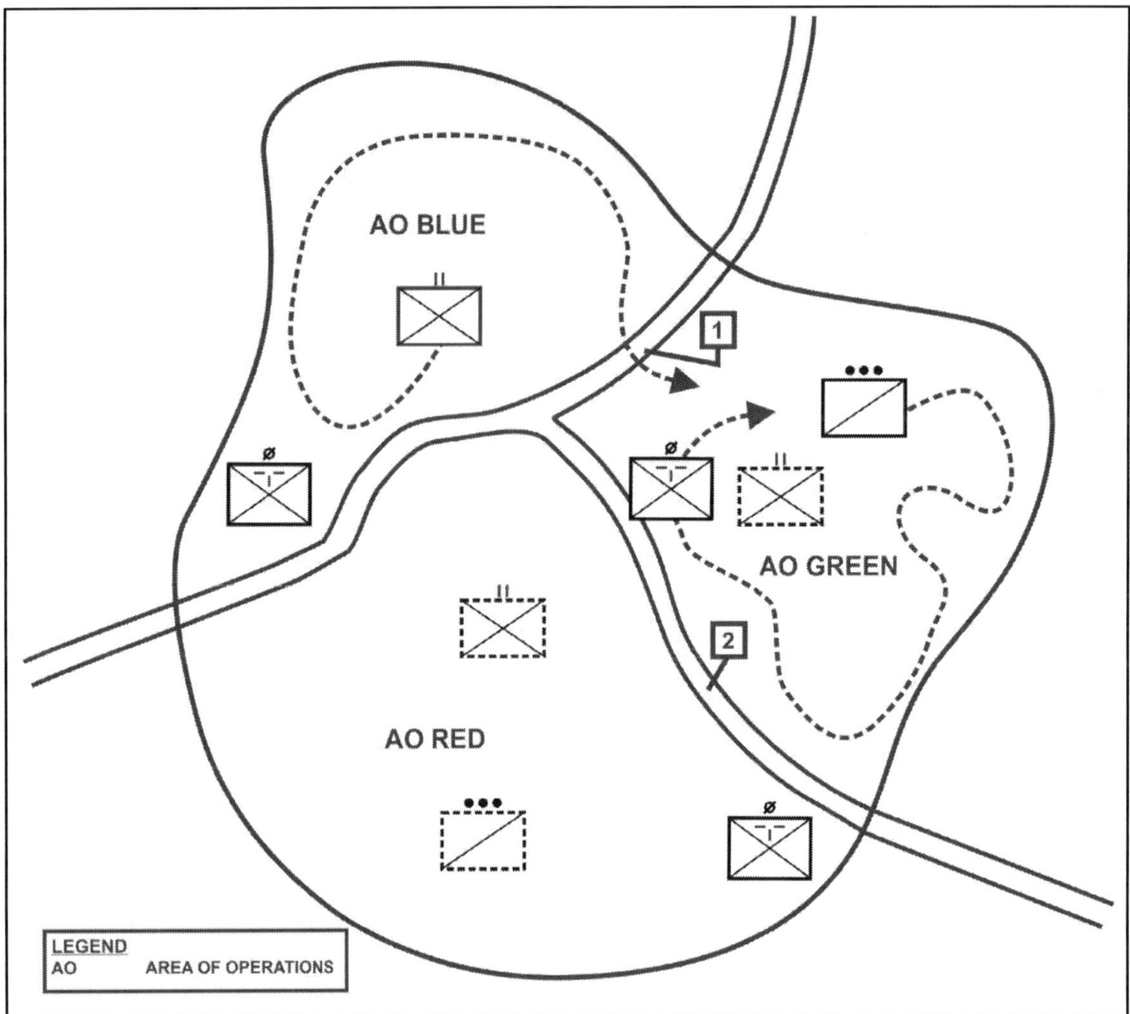

Figure 2-41. Search and attack (method with scout platoon forward)

2-282. At a designated time, the commander directs the battalion to link up with the scout platoon at contact point 1 to exchange information. If necessary, the scout platoon guides the battalion to sites of suspected or confirmed enemy activity. If the scout platoon is successful in locating the enemy, the battalion then postures itself to attack. If the enemy has not been located, the battalion then occupies area of operation Green and continues to search for the enemy. The scout platoon can then move on to reconnoiter area of operation Red.

2-283. The process is continued until the enemy force is located and destroyed, or the battalion area of operations is determined to be free of enemy activity or until the commander terminates it. If the area is determined to be free of activity, another area of operations is designated, and the process begins again. The commander may decide to emplace sensors, when available, along the border from area of operation Red to area of operation Blue to identify enemy tries to evade the battalion. An Infantry squad and a sniper team may tasked to emplace and monitor sensors.

2-284. The successive method of reconnaissance, discussed in paragraph 2-280, in which the scout platoon reaches the area of operations before the remainder of the battalion, allows the scout platoon more opportunities to gain information on enemy activity in the area. It also helps the battalion commander focus the search and attack operation when the battalion moves to the new area. Cache or airdrop most often provides logistical support for the scout platoon.

Note. The Infantry battalion rarely designates the scout platoon as the finding force or as a potential fixing force. The scout platoon is organized to find information and the enemy, not to become decisively engaged with the threat.

2-285. Enemy operations over a very large area, and in a decentralized manner, force the battalion to disperse to locate and then mass to destroy the enemy. The battalion commander can enhance (although not always possible) the units ability to mass combat power by keeping the search areas relatively small, minimizing the distance required to move the fixing and finishing force once contact with the enemy is made. Figure 2-42 provides an illustration of a battalion dispersing units to search.

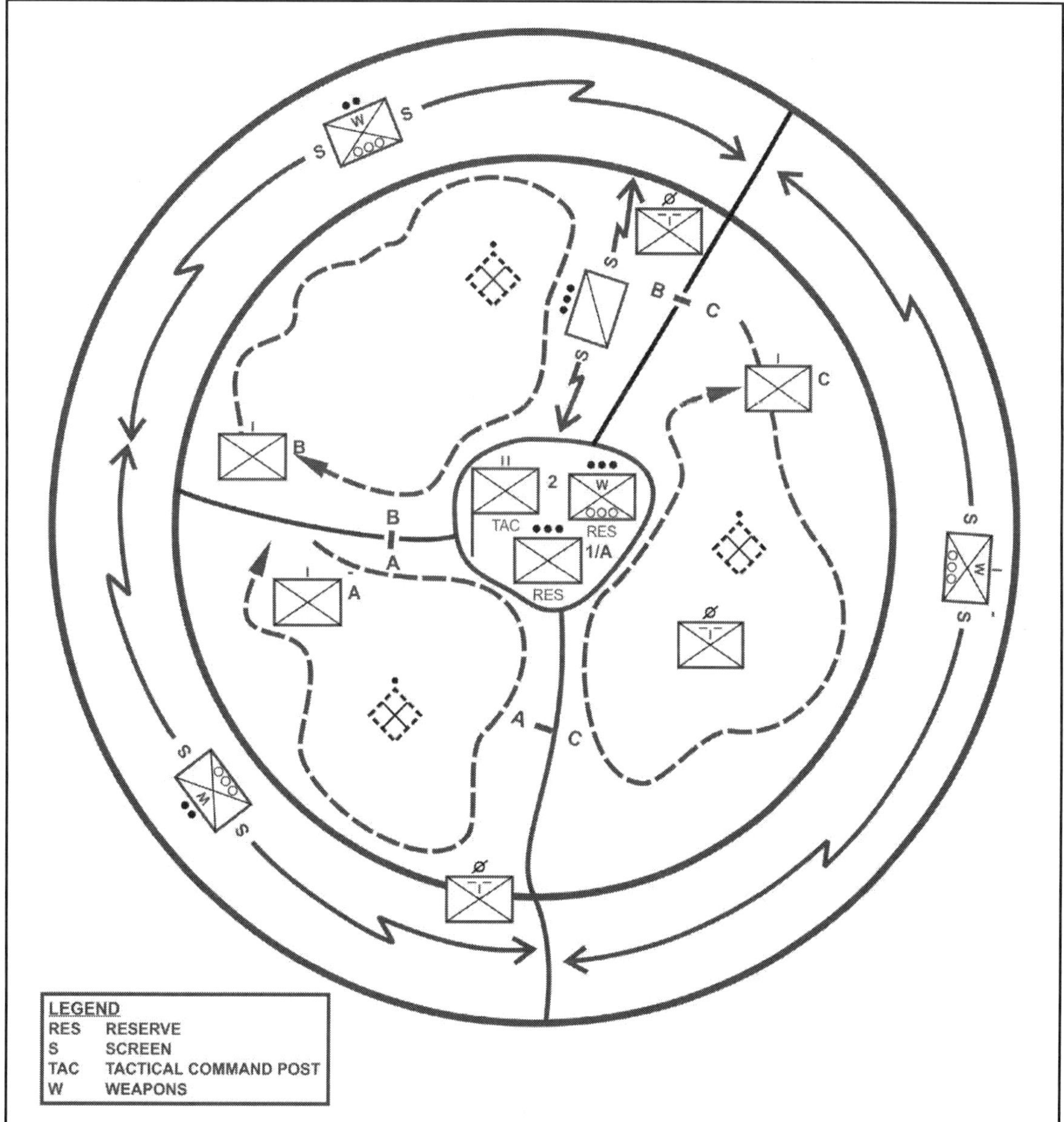

Figure 2-42. Search and attack (unit dispersing to conduct a search)

Fix the Enemy

2-286. Once the reconnaissance force finds the enemy force, the fixing force develops the situation and executes one of two options based on the commander's guidance and the mission variables of METT-TC. The first option is to block identified enemy escape and reinforcement routes. The second option is to conduct an attack to fix the enemy in its current positions until the finishing force arrives.

2-287. Block enemy escape and reinforcement routes. The fixing force maintains contact and positions its forces to isolate and prevent the enemy from moving to a position of advantage and prevent the interdiction of reinforcements. This facilitates the conduct of attack by the finishing force. Control measures and communications must be established between closing units to prevent fratricide. Unmanned aircraft systems can assist in preventing fratricide by observing forward of the moving units and identifying friendly and enemy units as they approach.

2-288. Conduct an attack. The fixing force conducts an attack to fix the enemy in its current positions until the finishing force arrives. Sniper fires can be used to disrupt the enemy and contain the enemy's movement as the finishing force approaches. The fixing force can conduct the finishing attack when it is consistent with the commander's intent with respect to tempo, and if the available forces can generate the required combat power. Depending on the enemy's mobility and the likelihood of the reconnaissance force being compromised, the commander may need to position the fixing force before the reconnaissance force enters the area of operation.

Note. If conditions are not right to use the finishing force to attack the detected enemy, the fixing force and the reconnaissance force can continue to conduct R&S activities to develop the situation further. The fixing force continues to avoid detection, reports enemy order of battle and activities. The force uses stealth in this effort, is careful to avoid an enemy ambush, and must always retain the ability to fix the enemy.

Finish the Enemy

2-289. Ultimately, the purpose of the search and attack is to destroy the enemy. Reconnaissance forces find the enemy; fixing forces rapidly fix the enemy, as finishing forces mass combat power to facilitate the enemy's destruction. Massing combat power must incorporate combined arms. The commander makes sound and timely decisions and issues mission type orders to apply combat power decisively. The commander must be able to communicate effectively with subordinate units to gather and disseminate information. During the execution of the search and attack, the commander locates where he can best obtain information and react to subordinate unit contacts with appropriate force.

2-290. During a movement to contact, there is expected to be a roughly linear arrangement between moving and stationary forces, that is to say in one general direction. This may, or may not be the case for search and attack. When a search and attack is conducted out of a centrally located command post or base, the commander's control becomes more complex during execution. The commander may remain centrally located with the reserve and mortars to ensure immediate, responsive fire support to any subordinate unit who makes contact with the enemy. Additionally, rather than moving with the decisive operation or main effort, the central position may enable a more rapid response to the decisive point where contact is made. Otherwise, the commander may not be in contact with or in communication with a shaping operation or supporting effort in contact. As these forces could quickly become the decisive operation or main effort if their "search" finds the enemy.

2-291. As the situation develops, decentralized operations can require the establishment additional fire support positions to provide responsive fires to subordinate units conducting search and attack. Fire support units would provide fire support coverage within a complete 360-degree circle. This situation would require significantly different control measures and greater planning detail because of the wide variety in observer target/ gun-target angles, which would be encountered. Decentralized operation could drive situations where finishing forces conduct point or area ambushes and use fires to drive the enemy into the ambushes. Such fires would require establishment of restrictive fire areas, coordinated fire lines, or no-fire areas around or between positions of adjacent ambush units within the area of operation.

2-292. The commander may move the finishing force behind the reconnaissance and fixing forces, or may locate the finishing force at a pickup zone to conduct an air assault into a landing zone near the enemy, once the enemy is located. The finishing force must be responsive enough to engage the enemy before the enemy can break contact with the reconnaissance force or the fixing force. The battalion intelligence officer provides the commander with an estimate of the time it will take the enemy to displace from its detected locations. The commander provides additional mobility assets, so the finishing force can respond within that timeframe. When established, the decisive operation or main effort is weighted by the commander using priority of fires and assigning priorities of support to available combat multipliers, such as engineer elements and helicopter lift support. The commander establishes control measures as necessary to consolidate units and concentrate the combat power of the force before the attack.

2-293. The commander uses the finishing force to destroy the enemy by conducting a hasty or deliberate attack, maneuvering to block enemy escape routes while another unit conducts the attack, or employing indirect fire or close air support to destroy the enemy (figure 2-43). The commander may have part of the fixing force establish an area ambush and use the reconnaissance and remaining fixing forces to drive the enemy into the ambush, establish a reserve, or screen the objective area.

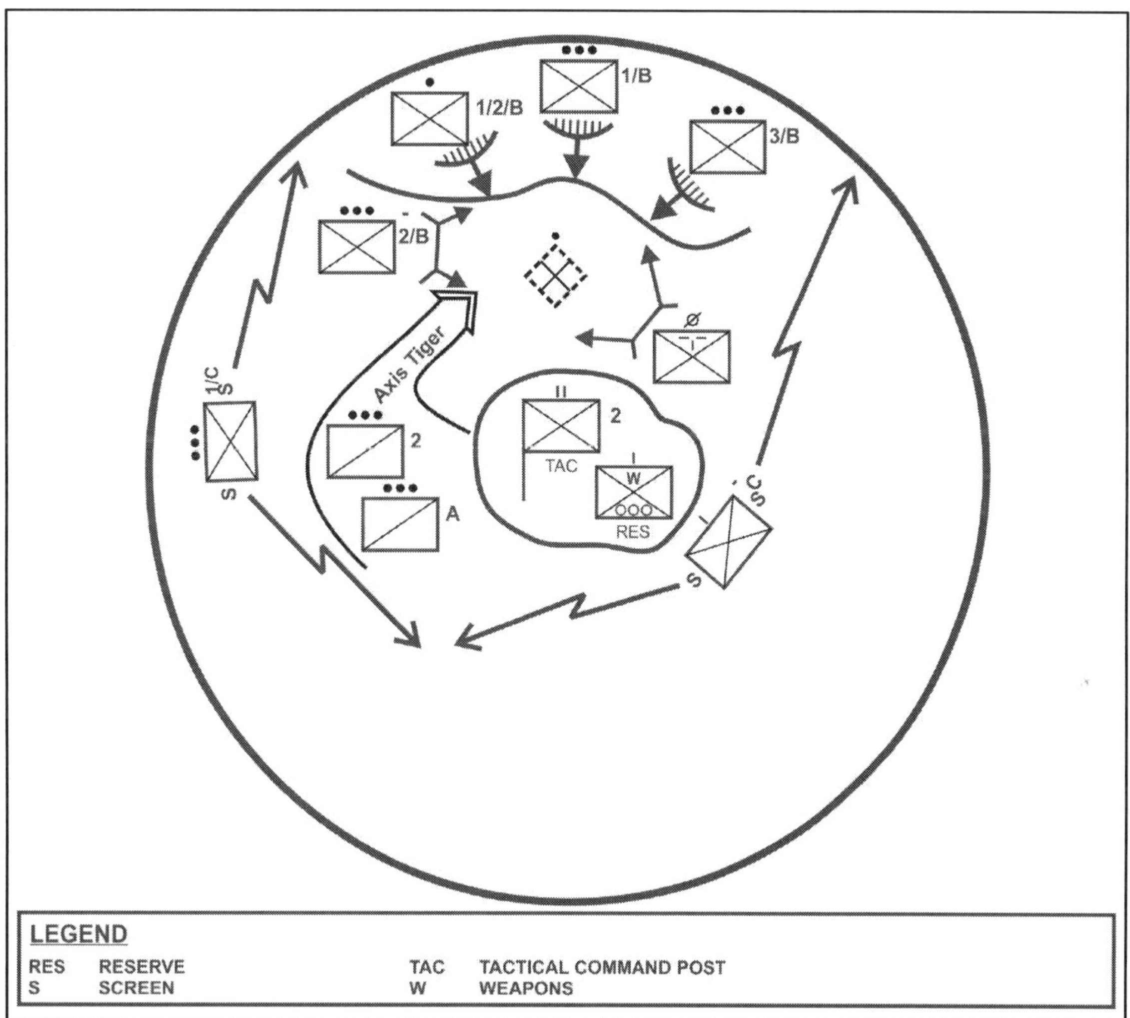

Figure 2-43. Search and attack (unit massing to attack)

SECTION III – ATTACK

2-294. An *attack* is an offensive task that destroys or defeats enemy forces, seizes and secures terrain, or both (ADRP 3-90). Attacks take place along a continuum defined at one end by fragmentary orders that direct the execution of rapidly executed battle drills by forces immediately available. For example, the Infantry battalion discovers the general enemy situation through a movement to contact and conducts an attack as a continuation of the meeting engagement to exploit a temporary advantage in relative combat power and to preempt enemy actions. The other end of the continuum includes published, detailed orders with multiple branches and sequels, detailed knowledge of all aspects of enemy dispositions, a force that has been task organized specifically for the operation, and the conduct of extensive rehearsals. Most attacks fall between the ends of the continuum as opposed to either extreme. (Refer to ADRP 3-90 for additional information.)

ORGANIZATION OF FORCES FOR AN ATTACK

2-295. Once the battalion commander determines the scheme of maneuver for the attack, the commander task organizes the battalion to accomplish the mission. Normally, the battalion attack is organized into a main body and a reserve, security forces may be utilized as addressed in paragraph 2-286. Task organized sustainment assets, from the brigade support battalion and forward support company (see appendix H), organize and perform the logistics function within the battalion echelon of support referred to as unit trains in one location or echeloned trains within an area of operation.

2-296. The commander organizes the main body into combined arms formations to conduct the decisive operation or main effort and necessary shaping operations or supporting efforts. The commander aims the decisive operation or main effort toward the decisive point, which can consist of the immediate and decisive destruction of the enemy force, its will to resist, seizure of a terrain objective, or the defeat of the enemy's plan.

2-297. The commander uses the reserve to exploit success, defeat enemy counterattacks, or restore momentum to a stalled attack. Once committed, the reserve's actions normally become or reinforce the echelon's decisive operation or main effort, and the commander makes every effort to reconstitute another reserve from units made available by the revised situation. Often the commander's most difficult and important decision concerns the time, place, and circumstances for committing the reserve. The reserve is not a committed force; it is not used as a follow and support force or a follow and assume force.

2-298. Security forces, under normal circumstances during an attack, are only resourced if the attack will uncover one or more flanks or the rear of the attacking force as it advances. In this case, the commander designates a flank or rear security force and assigns it a screen or guard mission depending on the mission variables of METT-TC. Normally an attacking unit does not need extensive forward security forces. Most attacks are launched from positions in contact with the enemy, which reduces the usefulness of a separate forward security force. An exception occurs when the attacking unit is transitioning from the defense to an attack and had previously established a security area as part of the defense. (Refer to FM 3-90-1 for additional information.)

DOCTRINAL BASIS FOR FORCE AND TERRAIN ORIENTED ATTACKS

2-299. An attack at the battalion level is an offensive action characterized by close combat. When the battalion commander decides to attack, he masses the effects of overwhelming combat power against a portion (or portions) of the enemy force with a tempo and intensity that the enemy cannot match. A force-oriented objective requires the battalion to focus its efforts on a designated enemy force. The enemy force may be stationary or moving. A terrain-oriented objective requires the battalion to seize, secure, or retain a designated geographical area. All attacks depend on the synchronization and integration of combat power for success. They require planning, coordination, and time (although time may be limited) to prepare.

2-300. When attacking against a moving enemy force the battalion commander often desires to stop the enemy's advance. In doing so, the commander makes the attack an attack against a stationary force. In this situation, during a force-oriented attack the commander can attack the enemy and force the enemy to stop. During a terrain-oriented attack, the battalion can seize key terrain and force the enemy to attempt to retake it, in effect conducting a turning movement. This section addresses techniques for conducting—

- A force-oriented attack against a stationary enemy force.
- A force-oriented attack against a moving enemy force.
- A terrain-oriented attack.

FORCE-ORIENTED ATTACK AGAINST A STATIONARY ENEMY FORCE

2-301. Normally, the Infantry battalion conducts a force-oriented attack against a stationary enemy force. The battalion may attack a stationary enemy force as part of a counterattack, spoiling attack, or as an initial attack against an enemy defense. The battalion also may attack a stationary force as part of a movement to contact or exploitation.

Planning

2-302. The focus of planning is to develop a fully synchronized plan that masses all available combat power against the enemy. A critical aspect to the commander's plan is to generate and concentrate combat power. The commander avoids using battalion resources and incurring needless casualties by attacking piecemeal. If an objective is small enough to be taken by a company attack then the battalion commander provides the necessary support to the attacking company. However, if the task requires it, the commander commits the entire battalion and all other available resources to quickly, and violently destroy the enemy.

Mission Command

2-303. The commander and staff refine the plan based on continuously updating the situation. (See appendix B.) The commander and staff develop and maintain running estimates as the situation changes throughout the operation. Subordinates conduct parallel planning as well as start their preparations for the attack immediately after the battalion issues an operation order and fragmentary order. As the situation is updated, the battalion commander revises orders and distributes those thereby giving subordinates more time to prepare for the attack. The commander issues as detailed a plan as the time available allows.

2-304. The commander assigns areas of operation to subordinate forces and normally designates (as a minimum, regardless of whether the attack takes place in a contiguous or noncontiguous area of operation): a phase line as the line of departure (which may also be the line of contact) and the point of departure; the time to initiate the operation; and the objective. The commander may designate checkpoints, additional phase lines, probable lines of deployment, assault positions, direct fire control measures, and fire support coordination measures; and can use either an axis of advance or a direction of attack to further control maneuver forces. Between the probable lines of deployment and the objective, a commander can use a final coordination line, assault positions, support by fire position, and attack by fire positions to control the final stage of the attack. Beyond the objective, the commander can impose a *limit of advance*—a phase line used to control forward progress of the attack. The attacking unit does not advance any of its elements or assets beyond the limit of advance, but the attacking unit can push its security forces to that limit. (ADRP 3-90)—if the battalion does not have a follow-on mission.

2-305. The commander uses the reverse planning process in building an effective plan. By starting with actions on the objective and working back to the line of departure or point of departure, the staff can allocate combat power, mobility assets, and indirect fires including suppression and obscuration. When possible, the commander plans avenues of approach that avoid strong enemy defensive positions, take advantage of all available cover and concealment, and place subordinate units to the flanks and rear of the defending enemy. Where cover and concealment are not available, the battalion plans obscurants to conceal movement. Any delays in establishing obscuration and suppressive fires before crossing the probable line of deployment may require the commander to establish assault positions. (Refer to FM 3-96 and FM 3-90-1 for additional information.)

Movement and Maneuver

2-306. The battalion commander directs the battalion's decisive operation or main effort against an objective, ideally an identified enemy weakness, which will cause the collapse of the enemy defense. The battalion commander seeks to attack the enemy's flanks, rear, or supporting formations causing disintegration or dislocation. By so doing, the enemy loses control of its systems and the enemy commander's options are reduced. Concurrently, the Infantry battalion retains the initiative and reduces its own vulnerabilities.

2-307. The commander seeks to identify an unobserved or covered and concealed avenue of approach to the objective. The commander attempts to attack a small unit lacking mutual support within the enemy defense or a weak flank that the commander can exploit to gain a tactical advantage. When attacking a well-prepared enemy defense, the commander normally plans to isolate and then destroy small vulnerable portions of the enemy defense in sequence. The commander incorporates plans for exploiting success and opportunities that may develop during execution. The plan emphasizes synchronization of movement, precise direct and indirect fires, and support throughout the attack.

2-308. The battalion commander is responsible for integrating and synchronizing Army aviation attack and reconnaissance units into the battalion ground scheme of maneuver. Army aviation attack and reconnaissance units, in close coordination with the Infantry battalion, attack to destroy, defeat, disrupt, divert, or delay enemy forces to enable the battalion to seize, retain, or exploit the initiative. Attacks can be either hasty or deliberate.

The battalion commander controls the distribution and de-confliction of aviation maneuver and fires through airspace coordination with the appropriate airspace control authority. Attacks can be either hasty or deliberate. (Refer to FM 3-04 for additional information.)

2-309. Assured mobility in the offense is critical. Although the rifle companies can cross almost any terrain, the supporting and sustaining forces cannot. Considerations to aid movement and maneuver include the following:

- Always search for a bypass to an obstacle.
- Maintain direct observation of the obstacle throughout the breaching operation.
- Plan for adjustment of the breach location based on the latest obstacle intelligence products.
- Ensure information on obstacles receives immediate battalion-wide dissemination, including fire support, protection, and sustainment platforms and units.
- Ensure adequate mobility support is task organized well forward during the approach to the objective to support breaching requirements.
- Retain the ability to mass engineers to support breaching operations.
- Support assaulting forces with engineers to breach enemy protective obstacles.
- Ensure adequate guides, traffic control, and lane improvements to support movement of follow on forces and sustainment vehicle traffic.
- Use situational obstacles for flank security.

2-310. When the commander is unable to bypass an obstacle, the commander and staff consider the enemy's strengths and obstacles to determine when and where the battalion may need to conduct a breach. The size of the enemy force overwatching the obstacle drives the type of breach the battalion conducts. The commander and staff consider the enemy's ability to mass combat power, reposition forces, or commit the reserve. The battalion then develops a scheme of maneuver to mass sufficient combat power at an enemy weakness. The scheme of maneuver identifies the focus of the decisive operation or main effort. The location selected for breaching and penetration depends largely on a weakness in the enemy's defense where its covering fires are limited.

2-311. The Infantry battalion's ability to fight at night and under limited-visibility conditions is a key enabler to the conduct of offensive actions during an attack. Limited visibility attacks take advantage of the night-vision and navigational superiority against most potential enemy ground forces the battalion will face. The mission variables of METT-TC will normally require limited visibility attacks to be more deliberate in nature than a daylight attack, except when it occurs as part of the follow-up to a daylight attack or as part of an exploitation or pursuit operation. Planning a limited visibility attack requires the commander to consider how limited visibility conditions complicate the control of movement and maneuver. The commander must also consider how limited visibility complicates identifying and engaging targets, navigating and moving without detection, locating, treating, and evacuating casualties, and locating, bypassing, and breaching obstacles.

Intelligence

2-312. To employ the proper capabilities and tactics, the battalion commander must have detailed knowledge of the enemy's organization, equipment, and tactics. The commander must understand the enemy's strengths and weaknesses. Ideally, this knowledge is available during the military decision-making process. The commander and staff develop enemy situational and weapons templates based on analysis of all available combat information and intelligence data.

Reconnaissance and Surveillance

2-313. R&S is performed before, during, and after operations to provide information used in the IPB process, as well as by the commander in order to formulate, confirm, or modify a course of action. Before conducting an attack, the commander through R&S identifies terrain characteristics, enemy and friendly obstacles to movement, and the disposition of enemy forces and civilian population, so commanders can maneuver their forces freely and rapidly. During hasty operations, the entire intelligence collection, analysis, processing, exploitation, and dissemination process must respond rapidly to the CCIRs.

2-314. When preparing for a deliberate operation, the commander and staff participate in the development of the information collection plan. A well-resourced and coordinated information collection effort paints a detailed picture of the enemy situation before an attack. This reconnaissance effort must include redundant information gathering systems to ensure continuous flow of information to the battalion and correspondingly from the battalion to the companies. The battalion commander uses this intelligence to decide on a course of action and make refinements to the plan. The information collection effort also provides the commander with continuous updates during the attack so he can adjust execution of the operation based on the enemy's reactions.

Note. The commander will rarely have complete information. The commander must balance the advantages of gathering more information with the advantage gained by an attack conducted with incomplete information but rapidly executed and with all available combat power. The commander that requires more complete information on the enemy may miss opportunities and allow the enemy to more thoroughly prepare the defense.

Enemy's Current Array of Forces

2-315. The first priority is to confirm information available on the enemy's strength, composition, and disposition. The next priorities are the effects of weather and terrain, and how the enemy is likely to fight. The S2 tries to identify what the enemy will do and what information the battalion needs to confirm the enemy's action. The battalion's information collection effort focuses on identifying indicators for confirming the enemy's actual course of action. This information is vital in answering the commander's information and intelligence requirements and helps the staff in developing and refining plans. Ideally, the battalion does not make final decisions on how to execute the attack until it can identify the current array of enemy forces. Key areas to identify for a defending enemy force include the—

- Composition, disposition, and strength of enemy forces along a flank or at an area selected for attack.
- Composition, strength, and disposition of security and disruption forces.
- Location, orientation, type, depth, and composition of obstacles.
- Locations of secure bypasses around obstacles.
- Composition, strength, and disposition of defending combat formations within the enemy's main battle area.
- Composition, strength, and location of reserves.
- Location of routes the enemy may use to counterattack or reinforce the defense.
- Type of enemy fortifications and survivability effort.

2-316. Reconnaissance forces patrol to collect information. As time permits, reconnaissance forces and surveillance assets observe the enemy defense from observation posts to locate gaps in the enemy's defense, identify weapons systems and fighting positions, view rehearsals and positioning, and determine the enemy's security activities and times of decreased readiness. The S-2 must discern any enemy deception efforts such as dummy obstacles and emplacements designed to confuse an attacker. If possible, the battalion positions an observation post(s) to observe the objective throughout the attack. Scouts, sniper teams, or other unit may man observation posts to report current enemy activity, conduct battle damage assessment, and call for fires, if required.

Enemy Engagement Areas

2-317. The battalion commander seeks to define the limits of the enemy engagement areas. This includes where the enemy can mass fires, weapon ranges, direct fire integration with obstacles, ability to shift fires, and mutual support between positions. This analysis requires effective terrain analysis, confirmed locations of enemy weapons systems, and a good understanding of the enemy's weapons capabilities and tactics. Reconnaissance forces and surveillance assets report locations, orientation, and composition of defending weapons systems and obstacles. The analysis of the enemy's direct and indirect fire plan assists the commander in developing the scheme of maneuver by—

- Determining the location of the probable line of contact.
- Identifying when the battalion must transition to maneuver.
- The targets and positioning of heavy direct fire weapons.
- Identifying targets for indirect fires.

Enemy Capabilities and Vulnerabilities

2-318. To understand enemy capabilities and vulnerabilities, the commander and staff require detailed, timely, and accurate intelligence produced as a result of IPB. A study of dispositions and an analysis of the terrain aid the development of conclusions concerning enemy vulnerabilities. Disrupting the enemy enables the commander to seize, retain, and exploit the initiative, maintain freedom of action, impose the commander's will on the enemy, set the terms, and select the place for battle. That disruption allows the commander to exploit enemy vulnerabilities and react to changing situations and unexpected developments more rapidly than the enemy.

2-319. The S-2 focuses information collection plans on answering commander critical information requirements and other requirements, and enables the quick retasking of reconnaissance units and surveillance assets as the situation changes. Planning requirements and assessing collection includes continually identifying intelligence gaps. Through the receipt and processing of incoming reports and messages, the S-2 staff determines the significance and reliability of incoming information and integrates incoming information with current intelligence holdings. Through analysis and evaluation the S-2 staff determines changes in threat capabilities and probable course of actions and seeks to identify enemy vulnerabilities that may include—

- Gaps in the enemy's defense.
- Exposed or weak flanks.
- Enemy units that lack mutual support.
- Unobserved or weakly defended avenues of approach to the enemy's flank or rear.
- Covered and concealed routes that allow the battalion to close on the enemy.
- Weak or poorly positioned obstacles or fortifications in an enemy defense, especially along a flank.

Fires

2-320. The battalion commander positions the battalion mortar platoon to provide continuous indirect fires to the battalion. Infantry rifle companies often have their mortars follow behind the forward platoons so they are prepared to provide immediate indirect fires. Army attack reconnaissance helicopters and close air support may be available to destroy defensive positions and interdict enemy counterattack forces. During the attack, using preparation fires, counterfire, suppression fires, and electronic warfare (see appendix B) assets provides the commander with numerous options for gaining and maintaining fire superiority. The commander uses long-range artillery systems (cannon, rocket, and air support; rotary- and fixed-wing) to engage the enemy throughout the depth of the enemy's defensive positions. Additional fire support considerations are listed below:

- Use massed fires, especially time on target fires.
- Position fire support assets to support the reconnaissance effort.
- Plan suppressive and obscuration fires at the point of penetration (POP).
- Plan fires on enemy positions supporting and overwatching the objective.
- Plan suppressive and obscuration fires in support of breaching operations.
- Plan fires in support of the approach to the objective. These fires engage enemy security forces, destroy bypassed enemy forces, and screen friendly movement.
- Plan preparation fires on the objective to suppress, neutralize, or destroy critical enemy forces that can most affect the battalion's closure on the objective.
- Plan fires beyond the objective to support an attack or defense, or to isolate the objective to prevent the egress or ingress of threat forces.
- Use indirect fires and close air support to delay or neutralize repositioning enemy forces and reserves.
- Plan locations of critical friendly zones to protect critical actions such as support-by fire positions, breaching efforts, and mortar assets.
- Use risk estimate distances to determine triggers to initiate, shift, and cease loading of rounds.
- Use echelon fires to maintain continuous suppression of enemy forces throughout the movement to and actions on the objective.

Sustainment

2-321. Commander and staff must plan for increased sustainment demands during the offense. Sustainment planners, at battalion and brigade combat team (BCT) level, synchronize and coordinate to determine the scope of the operation. Sustainment planners develop and continually refine the sustainment concept of support. Coordination between staff planners must be continuous to maintain momentum and freedom of action. The brigade support battalion and forward support company commanders anticipate where the greatest need may occur to develop a priority of support that meets the battalion commander's operational plan. Sustainment planners may consider positioning sustainment units in close proximity to operations to reduce critical support response times (see appendix H). Commanders and staff may consider alternative methods for delivering sustainment during emergencies. The following are considerations for the sustainment plan:

- Synchronize the movement and positioning of sustainment assets with the scheme of maneuver to ensure immediate support of anticipated requirements. This synchronized plan covers movement from the start point to the objective.
- Ensure adequate sustainment support to the R&S effort. The headquarters and headquarters company commander, in coordination with the battalion S-2, S-3, and S-4, plans and integrates timely resupply and evacuation support of forward R&S forces into the R&S plan.
- Plan immediate support to high-risk operations such as breaching or assaults through the forward positioning of support assets. Plan for reorganization on or near the objective once the battalion secures the objective. Articulate clear priorities of support during reorganization.

Protection

2-322. In the offense, survivability operations enhance the ability to avoid or withstand hostile actions by altering the physical environment. Conduct of survivability operations in the offense (fighting and protective position development) is minimal for tactical vehicles and weapons systems. The emphasis lies on force mobility. Camouflage and concealment typically play a greater role in survivability during offensive tasks than the other survivability operations. Protective positions for indirect fire and logistics positions, however, still may be required in the offense. The use of terrain provides a measure of protection during halts in the advance, but subordinate units of the battalion still should develop as many protective positions as necessary for key weapons systems, command posts, and critical supplies based on the threat level and unit vulnerabilities. During the early planning stages, geospatial engineer teams can provide information on soil conditions, vegetative concealment, and terrain masking along march routes to facilitate the force's survivability. (Refer to ATP 3-37.34 for additional information.)

2-323. Depending on the threat, primary protection concerns of the commander may be enemy air and CBRN threats. If these threats exist, the commander has to prepare subordinate unit and adjust the scheme of maneuver accordingly. In the face of an enemy air threat, the Infantry battalion usually has only passive and active (with its organic weapons) air defenses. Air defense units are usually not assigned below IBCT level. However, air defense assets may be located near the battalion and may provide coverage. If air defense units are assigned to the battalion, the S-2 and the air defense leader determine likely enemy air avenues of approach, and plan positions accordingly with the S-3. The commander establishes priorities for protection that may include companies, fire support, engineer elements, mission command, and logistics assets.

2-324. The commander integrates CBRN defense considerations into mission planning depending on the CBRN threat. This includes CBRN passive-defense principles, such as contamination avoidance, individual and collective protection, and decontamination. CBRN protective measures may slow the tempo, degrade combat power, and increase logistics requirements. CBRN R&S consumes resources, especially time. Personnel wearing individual protective equipment find it difficult to work or fight for an extended period. (See appendix G.) CBRN considerations include the following:

- The scout platoon should be prepared to conduct CBRN reconnaissance tasks.
- Disseminate information regarding any detected CBRN threats or hazards throughout the battalion immediately.
- Integrate and synchronize the use of obscuration to support critical actions such as breaching or assaults. Ensure artillery, mortar, and mechanical obscuration is complementary.
- Develop decontamination plans based on the commander's priorities and vulnerability assessment.
- Disseminate information regarding planned and active decontamination sites.

Preparation

2-325. The battalion uses the time available before the attack to conduct extensive R&S and rehearsals while concealing attack preparations from the enemy. The commander also will use this preparation time to conduct confirmation briefs and backbriefs with subordinate commanders to ensure they understand their task and purpose, and plan and execute within the commander's intent. (See appendix B.) Usually during the battalion's preparation phase, reconnaissance forces conduct information collection to answer the CCIRs and to confirm or deny the situation template. This allows the commander and staff time to incorporate any changes to the original course of action before executing it. The movement from the assembly area to the line of departure that precedes many attacks is troop movement (see paragraph 2-34). Additional preparation activities, although not inclusive, include sustainment preparations, finalize task organization, and perform pre-operations checks and inspections. (Refer to FM 3-96 and FM 3-90-1 for additional information.)

Reconnaissance and Surveillance

2-326. R&S is performed before (in addition to during and after) the operations to provide information used in the IPB process, as well as by the commander in order to formulate, confirm, or modify a course of action. Reconnaissance forces initially focus on the enemy's security and disruption forces forward of the main defense to locate enemy positions and obstacles along the battalion's planned routes of advance. Reconnaissance forces also locate gaps and routes that allow them to infiltrate into the enemy main defensive or rear area. The R&S effort seeks to locate enemy forces that may reposition and affect the battalion's approach to the enemy's main defense. Successful attacks are enabled when reconnaissance forces can place indirect fires on targets in the enemy's rear that isolate the enemy front line forces and prevent them from being reinforced.

Rehearsals

2-327. The battalion usually conducts rehearsals, but the type and technique may vary based on time available and the security that is required. During the combined arms rehearsal, the battalion S-2 portrays a thinking, uncooperative enemy with emphasis on enemy repositioning, employment of fires, and commitment of reserves. The primary focus of the rehearsal is actions on the objective. Each subordinate commander addresses the conduct of the mission as the rehearsal progresses. The rehearsal places special emphasis on triggers and the coordinated maneuver of forces. All subordinate commanders must accurately portray how long it takes to complete assigned tasks and how much space is required by their force. Direct and indirect fire plans are covered in detail, to include the massing, distribution, shifting, and control of fires. The commander ensures subordinate plans are coordinated and consistent with the intent. (Refer to FM 3-96 for additional information.) Additional areas to rehearse include—

- Plans to execute follow on missions or exploit success.
- Likely times and locations where a reserve is needed and its commitment criteria.
- Execution of the fire support plan, to include shifting of fires, employment of combat air support and Army attack aviation, adjusting of fire support coordination measures, and positioning of observers.
- Breaching operations.
- Passage of lines.
- Contingency plans for actions against enemy counterattacks, repositioning, commitment of reserves, or use of CBRN capabilities.
- Consolidation and reorganization.
- Execution of branches or sequels.
- Execution of the sustainment plan.
- Casualty evacuation, location, and movement.
- Activities of echeloned trains, support area movement and activities, and resupply.
- Integration of key enablers (for example CBRN R&S elements).

Attack Position and Movement to the Line of Departure

2-328. The battalion may move directly from its assembly area to the line of departure. If the distance from the assembly area to the line of departure is lengthy however, the commander may designate an attack position closer to the line of departure. An attack position is the last position an attacking force occupies or passes through before crossing the line of departure. Attack positions are used only by exception. When established the battalion conducts any final preparation and coordination in the attack position or may pass through it and proceed to the line of departure. The *line of departure* is a phase line crossed at a prescribed time by troops initiating an offensive operation (ADRP 3-90). Attacking on foot using infiltration and stealth, the commander may designate a point of departure for the attacking force instead of a line of departure. A *point of departure* is the point where the unit crosses the line of departure and begins moving along a direction of attack (ADRP 3-90). A point of departure often is used during a limited visibility attacks (see paragraph 2-310).

Note. Armored and Stryker-equipped units normally use gaps or lanes through the friendly positions to allow them to deploy into combat formations before they cross the line of departure.

Execution

2-329. The battalion commander positions reconnaissance forces and surveillance assets to maintain observation of enemy reactions to the battalion's maneuver to, and on the objective. Reconnaissance forces and surveillance assets focus on areas that the enemy is likely to use to reposition forces, commit reserves, and counterattack. As the engagement on the objective develops, reconnaissance forces report enemy reactions, repositioning, reinforcements, and battle damage assessment. Reconnaissance forces target and engage with indirect fires enemy repositioning forces, reserves, counterattacking forces, and other high-payoff targets. Early identification of enemy reactions is essential for the battalion to maintain the tempo and initiative during the attack. (Refer to FM 3-96 for additional information.)

Approach to the Probable Line of Deployment

2-330. Once pass the line of departure, the attacking force conducts tactical movement to the probable line of deployment. A *probable line of deployment* is a phase line that designates the location where the commander intends to deploy the unit into assault formation before beginning the assault (ADRP 3-90). The attacking force moves aggressively and as quickly as the terrain and enemy situation allows. The force is prepared to destroy enemy forces during tactical movement with maneuver. A reconnaissance element may lead the attacking forces if the distance from the line of departure to the probable line of deployment is long, or it may have another mission if the distance is relatively short and there is a chance that it may get in the way of the assaulting force.

2-331. The battalion commander plans to support the assault force with a base of fire, the weapons company may provide this base of fire. If the distance from the probable line of deployment to the objective is short, support by fire elements are in position prior to the assault force crossing the probable line of deployment. If the distance from the probable line of deployment to the objective is long, then the commander should consider having two support by fire positions: one to cover the movement to the probable line of deployment, and another to cover the assault on the objective. The latter support by fire element may lead the assault force and establish its position prior to the assault. As the assaulting force is committed, the battalion commander and staff ensure that information is available and current on the following:

- Locations and type of enemy contact on the objective.
- Locations of reconnaissance forces.
- Locations of lanes and obstacles, to include lane markings.
- Recognition signals and guides.
- Specific routes for the approach.
- Locations and orientation of fires from friendly forces.
- Additions or modifications of graphic control measures.

Approach to Objective

2-332. Upon reaching the probable line of deployment or before, the commander divides the battalion into assault and support forces. Supporting forces should be set in support by fire positions before the assault force crosses the probable line of deployment. The commander synchronizes the occupation of support by fire positions with the maneuver of the supported attacking units to limit the vulnerability of the forces occupying these positions. The commander uses unit tactical standard operating procedures, battle drills, prearranged signals, engagement areas, and target reference points to control the direct fires from these supporting positions. The commander normally employs restrictive fire lines between converging forces.

2-333. To further control the final stages of the attack, assault position(s) can collocate in the vicinity of the probable line of deployment. The commander ensures that the final preparations of the assault force and breach force (if required) in an assault position do not delay maneuver to the objective or breach point as soon as the conditions are set. Whenever possible, the assault force rapidly passes through the assault position. The assault force may have to halt in assault position while fires are lifted and shifted. In this case, if the enemy anticipates the assault, the assault force deploys into covered positions, screens its positions with smoke, and waits for the order to assault. As long as the assault force remains in the assault position, support forces continue their suppressive fires on the objective.

2-334. The *final coordination line* is a phase line close to the enemy position used to coordinate the lifting or shifting of supporting fires with the final deployment of maneuver elements. (ADRP 3-90). Final adjustments to supporting fires necessary to reflect the actual versus the planned tactical situation take place prior to crossing this line. The final coordination line should be easily recognizable on the ground and may be located near the assault position or the probable line of deployment. The final coordination line is not a fire support coordination measure.

2-335. Once the support force sets the conditions, the breach force (if required) reduces, proofs, and marks the required number of lanes through the enemy's tactical obstacles to support the maneuver of the assault force. (See appendix F.) To avoid confusion, the commander clearly identifies the conditions that allow the breach force to proceed. From the probable line of deployment, the assault force maneuvers against or around the enemy to take advantage of the support force's efforts to suppress the targeted enemy positions. The support force employs direct and indirect fires against the selected enemy positions to destroy, suppress, obscure, or neutralize enemy weapons and cover the assault force's movement. The assault force must closely follow these supporting fires to gain ground that offers positional advantage. During the approach, the battalion is prepared to—

- Bypass or breach obstacles.
- React to all eight forms of contact (see paragraph 2-170).
- Transition to different formations and techniques based on the terrain and enemy situation.
- Employ forces to screen or guard flanks that may become exposed or threatened during the approach.
- Avoid terrain features that are likely enemy artillery reference points, locations for CBRN attacks, or locations for situational obstacles.
- Destroy or force the withdrawal of opposing enemy security and disruption forces.
- Minimize the effects of enemy deception.

2-336. A defending enemy generally establishes a security area around their forces to provide early warning of an attack, deny friendly reconnaissance, and disrupt the friendly force's attack. The strength of the enemy's security area depends on the time available, forces available, and the enemy's doctrine or pattern of operations. The battalion must counter the effects of enemy security forces to ensure an unimpeded and concealed approach. This starts before the attack when reconnaissance forces seek to locate enemy security forces. Once located, the commander has the following options available:

- Destroy them immediately with indirect fires, attack reconnaissance helicopters, and close air support, which is the preferred option.
- Destroy them with indirect fires and close air support during the approach to the objective.
- Conduct limited objective attacks before execution of the main attack.
- Force the withdrawal of enemy security forces during the approach to the objective.
- Support by fire with sniper fire to reduce their effectiveness.

2-337. The battalion must maintain a steady, controlled movement. Speed and dispersion, facilitated by information dominance, are the norm with massing of weapons effects to destroy the enemy's defense. If the formation is too slow or becomes too concentrated, it is vulnerable to enemy fires.

Actions on the Objective

2-338. As the battalion commander concentrates available combat power on the point of attack. Before the attack initiates, the objective is isolated to prevent the ingress or egress of enemy forces. The battalion commander must set favorable conditions before committing forces. The commander uses artillery, close air support, Army attack aviation, and organic mortars, a shaping operation from another maneuver battalion and other joint, interorganizational, multinational or special operations forces assets to set conditions. The battalion commander then maneuvers combat forces and employs direct and indirect fires, situational obstacles, and obscuration to execute decisive maneuver against the enemy. The commander commits maneuver forces and fires to isolate and then rupture a small vulnerable portion of the enemy's defense to gain a flank or create a penetration.

Note. The key to a successful attack is isolation and suppression. Isolation prevents the enemy from reinforcing the objective area. Suppression of fires from the objective and from enemy locations supporting the objective allows the attacking force to maneuver and seize the objective.

2-339. Attacking from an unexpected direction and time can also limit or even change the enemy course of action. The effects of this action are fleeting and must be exploited rapidly before the enemy can recover. If the battalion is successful in locating and attacking a position that is unexpected, the enemy defense may become untenable. The battalion uses precision attacks that target key enemy command and control systems, indirect fires or reserve forces. Effective destruction of these key systems reduces the enemy control and causes enemy organizational collapse.

Set Conditions

2-340. The battalion employs fires to weaken the enemy's position and set the conditions for success before closing within direct fire range of the enemy. Initially, preparation fire focus on the destruction of key enemy forces that can most affect the scheme of maneuver. For example, during an attack to penetrate an enemy defense, the initial focus of preparation fire is to destroy the enemy positions at the selected POP. Preparation fire may also—

- Suppress or neutralize enemy reserves. Emplace artillery delivered situational obstacles to block enemy reserve routes into the objective.
- Deceive the enemy as to the battalion's actual intentions.
- Destroy enemy security and disruption forces.
- Obscure friendly movements and deployment.
- Destroy or neutralize the enemy's local command and control system.

2-341. The synchronization between indirect fires and maneuvering forces is critical. As maneuver forces approach the enemy defense, the commander uses triggers to shift fires and obscuration to maintain continuous suppression and obscuration of the enemy. Proper timing, adjustment of fires, and detailed triggers dictated by risk estimate distances enable a relatively secure closure by the maneuver force on the enemy's positions. The commander must monitor the success of the preparation fire to determine whether adequate conditions exist for commitment of the force. R&S elements provide battle damage assessment to the commander to assist in making this decision. The commander may need to adjust the tempo of the battalion's approach to the objective based on the battle damage assessment.

2-342. Prior to the assault, the battalion commander destroys the enemy or makes it ineffective through the employment of direct and indirect fires. The commander fixes the enemy in place and limits the enemy's options. One of the commander's objectives is to limit or change the options available to the enemy and to increase the enemy's uncertainty. A primary goal at the point of attack is to isolate the unit targeted for destruction by preventing the enemy from repositioning and preventing another element from reinforcing it.

2-343. The battalion may fix the enemy force by attacking an objective that isolates a portion of the enemy's defense. In open terrain, this shaping operation or supporting effort fixes the enemy with direct and indirect fire. In complex terrain, a shaping operation or supporting effort may seize terrain or destroy key enemy forces in limited objective attacks to pass the decisive operation or main effort to their objective. This ensures that the decisive operation or main effort does not have to fight its way and lose combat power en route to its objective. Demonstrations and feints may be used to fix the enemy; although, shaping operations or supporting efforts should exercise economy of force as they can take combat power from the decisive operation or main effort. The integration of indirect fires, close air support, and Army attack aviation is vital in attacking enemy forces and reserves in-depth to reduce their effectiveness or prevent their commitment against the battalion.

2-344. Before commitment, forces remain dispersed and outside the enemy's direct fire range, and they avoid exposing themselves to enemy observation. Forces not yet committed use this time to conduct final preparations and make adjustments to their plans as the situation changes. A key action during this time is the update of intelligence on the enemy locations and activities. The S-2 should have an updated intelligence summary available just before the battalion crosses the line of departure. The commander can use assault positions, phase lines, or checkpoints to control the positioning of the forces not yet committed. Commanders throughout the battalion continuously assess the situation. The commander commits subordinate forces when conditions are set, to include, being satisfied the desired levels of enemy suppression, destruction, and obscuration are achieved. Timely reporting, cross-talk, accurate assessments, and sharing of information by subordinate commanders are paramount to the success of the operation.

2-345. The coordination between the assaulting force and the support by fire force is critical. The battalion commander shifts supporting direct and indirect fires only when the assaulting force is as close to the enemy positions as possible. The assault force then can generate sufficient fire to maneuver and destroy the enemy. Properly emplaced, the weapons company provides mobile, heavy weapons platforms to destroy and fix the enemy. The company may be used in the isolation role, preventing the withdrawal of the enemy off the objective or preventing reinforcements from counterattacking.

Decisive Maneuver

2-346. The attacker concentrates and masses combat power by decisive maneuver before the enemy can reorient their defense. The destruction of a defending enemy force normally dictates an assault of the objective. A shaping operation or supporting effort shifts direct and indirect fires and repositions as required to support the decisive operation or main effort's maneuver of assaulting forces.

2-347. At the probable line of deployment, the previously dispersed assaulting force moves into its final combat formation and maneuvers to destroy the enemy forces and clear assigned objectives. The assaulting force moves along covered and concealed routes to an exposed enemy flank, created penetration, or other position of advantage. Obscuration assists with concealing the movement of assaulting forces. The assault includes destruction of defending forces and clearance of trenches and fortifications at the point of attack.

2-348. The commander focuses on maintaining the momentum and security of the assaulting force. The R&S effort continues to report enemy repositioning, battle damage assessment, and enemy counteractions to the assault. Fires reduce the ability of the enemy to reposition and mass their forces. Once the assault force has seized the objective, follow-on forces may pass through to continue the attack. Commanders and leaders throughout the battalion ensure that units remain disbursed and check to ensure all the enemy on the objective are destroyed or captured. Forces on the objective reorganize and prepare positions for an enemy counterattack.

Force-Oriented Attack Against a Moving Enemy Force

2-349. The battalion may attack a moving enemy force, especially during a counterattack, spoiling attack, exploitation, or after a movement to contact. Most of the techniques described above for the force-oriented attack against a stationary enemy equally apply to attacking a moving enemy.

Planning

2-350. The Infantry battalion in a force-oriented attack against a moving enemy force normally organizes in the same manner as a movement to contact and can be envisioned much like an ambush. Key planning considerations for a force-oriented attack against a moving enemy force include the following.

Mission Command

2-351. The decision on where to fight the enemy requires the commander to have information dominance over the enemy. The commander bases decision on a clear understanding of the effects of the terrain, the enemy situation, and what the enemy is expected to do. The commander and staff select the most advantageous location to fight the engagement and then determine other possible locations where the engagement may occur based on a slower- or faster-than-expected enemy advance or the enemy's use of an unlikely avenue of approach (figure 2-44). The commander identifies these areas as objectives, intermediate objectives, or engagement areas (EA). Example engagement area options include—

- Option EA Rain. Enemy lead elements cross phase line (PL) Nita, Infantry battalion engages enemy in EA Rain (lead battalion elements vicinity PL Sally).
- Option EA Hail. Enemy lead elements move east through named area of interest (NAI) 3, Infantry battalion engages enemy in EA Hail (lead battalion elements vicinity PL Tracy).
- Option EA Snow. Enemy lead elements move east through NAI 4, Infantry battalion engages enemy in EA Snow (lead battalion elements vicinity PL Tracy).
- Option EA Sleet. Enemy lead elements cross PL Sue, Infantry battalion engages enemy in EA Sleet (lead battalion elements vicinity PL Nita).

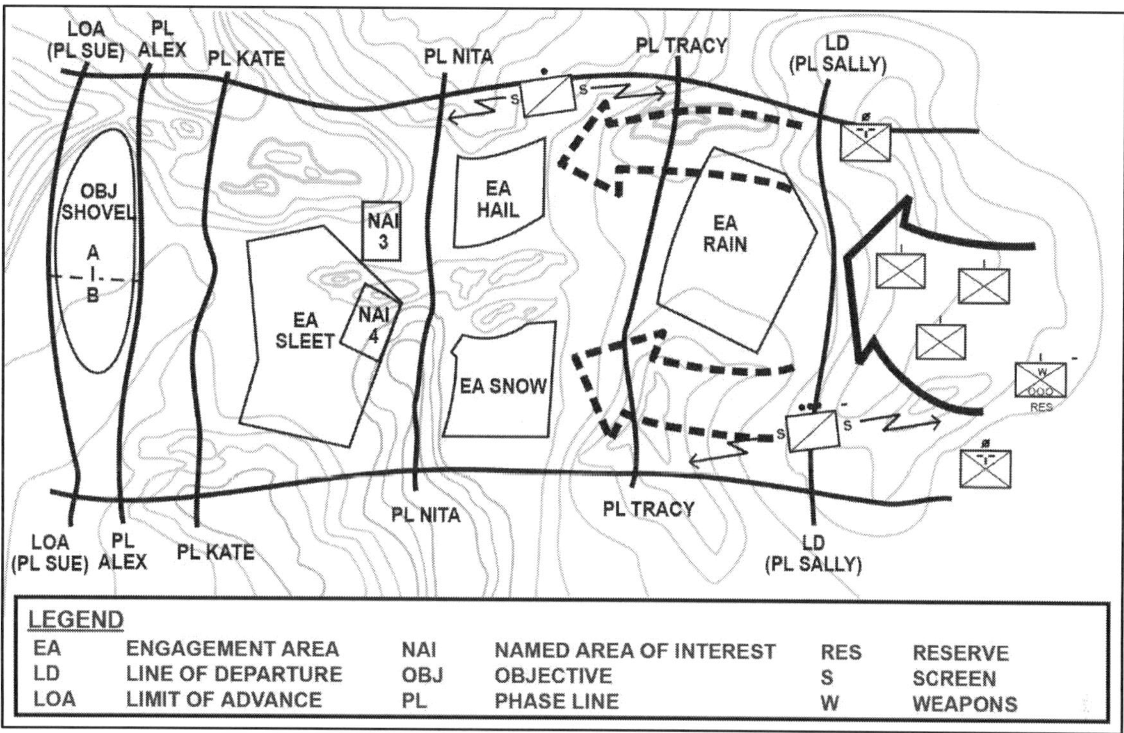

Figure 2-44. Planning options (attack against a moving enemy)

2-352. The commander and staff develop control measures to help coordinate actions throughout the battalion's area of operation. The commander, primarily assisted by the S-3 and S-2, develops decision points for the commitment of the battalion to each location based on relative locations and rates of movement of the battalion and the enemy. The S-2 selects NAIs to identify the enemy's rate and direction of movement to support the commander's decision of where to fight the engagement.

Movement and Maneuver

2-353. Reconnaissance forces move well forward of the battalion. They reconnoiter obstacles and areas that may slow the battalion's movement and disrupt the timing and planned location of the attack. Reconnaissance forces seek to detect obstacles, contaminated areas, enemy security forces, and suitable routes for the battalion's use. Once contact is established, the reconnaissance force may receive a change of mission to secure an uncovered flank or the rear of the attacking force as it advances. This security force mission will involve a screen or guard task, with a possible task organization change.

2-354. The commander uses the terrain to maximize the battalion's freedom of maneuver while limiting the freedom of maneuver available to the enemy. The commander and staff plan avenues of approach that allow the battalion to strike the enemy from a flank or the rear. One or two companies block the enemy's advance while the other companies attack into the enemy's flank. In this example, the terrain prevents the enemy from moving away from the main attack while protecting the battalion's flank from an enemy attack. *Situational obstacles,* which are obstacles that a unit plans and possibly prepares prior to starting an operation, but does not execute unless specific criteria are met (ATP 3-90.8), are employed as part of a target area of interest and provide the commander with flexibility for emplacing tactical obstacles based on battlefield development (figure 2-45).

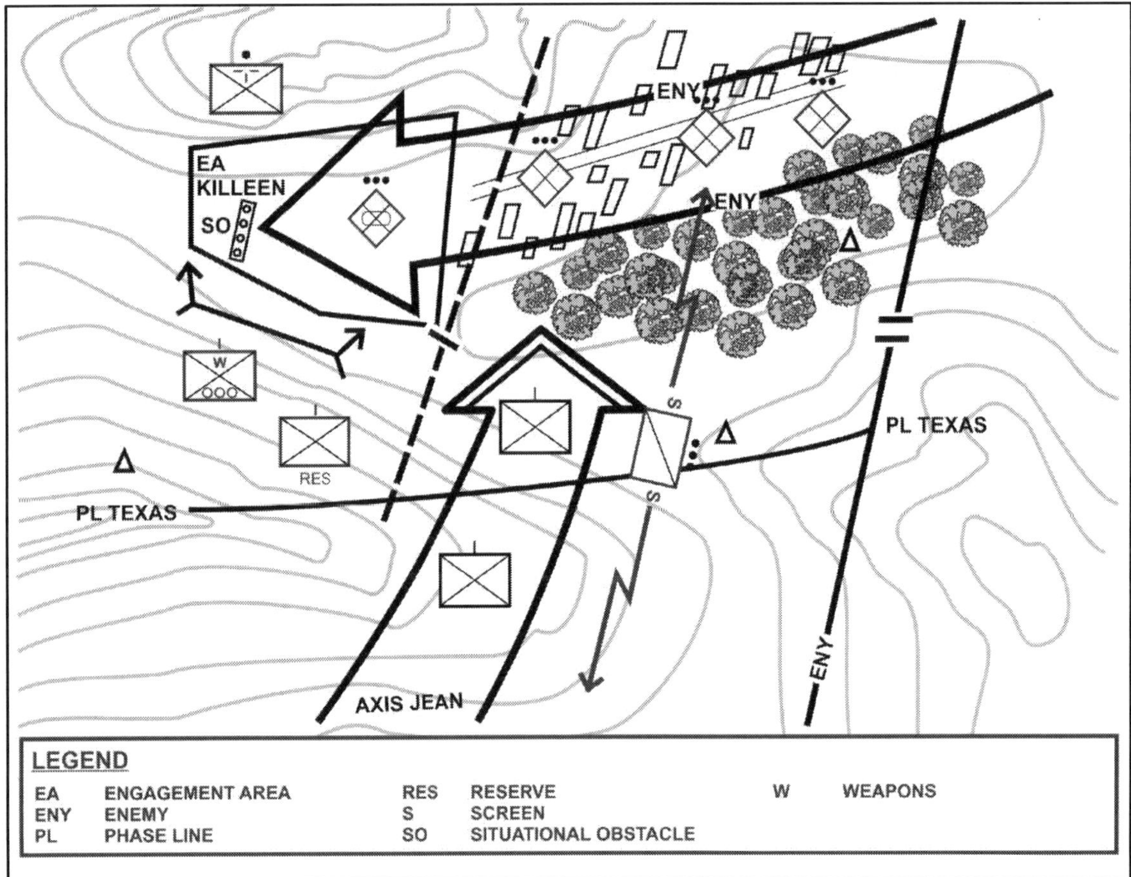

Figure 2-45. Battalion flank attack against a moving enemy

2-355. Although the commander develops the plan to fight the enemy at the most advantageous location for the battalion, the commander retains enough flexibility to attack the enemy effectively regardless of where the engagement develops. The scheme of maneuver includes provisions to fight the enemy at other possible locations. For simplicity, the commander seeks to keep the scheme of maneuver in each location as similar as possible.

2-356. In some situations, such as a movement to contact, the battalion may have constraints in the time or ability to select when and where to fight a moving enemy force. If so, the commander issues a fragmentary order for an attack based on a physical view and knowledge of the battlefield. As collection assets gather information, the commander quickly deploys and maneuvers the battalion to develop the situation and defeat the enemy.

2-357. The mobility of the Infantry weapons company of the Infantry battalion, with its heavy weapons and long range close combat missile fires, makes it well suited for a force-oriented attack against a moving enemy force. When employed in a support by fire role, to suppress, fix, or destroy enemy at long ranges, mass and depth are key consideration to allow Infantry units or combined arms teams to maneuver. The weapons company provides the Infantry battalion with a highly mobile, multi-functional assault force with the ability to task organize assault platoons to an Infantry company(s) to create combined arms teams that can execute a variety of missions during an attack on a moving enemy.

2-358. Engineers in support of the battalion focus on enabling movement and maneuver throughout the attack. Force-oriented attacks will tend to concentrate on assured mobility to attack the enemy's ability to disrupt the Infantry battalion's actions, mobility, and momentum during the attack. The battalion engineer staff officer's running estimate provides the framework to synchronize and integrate engineer support into the mission. Conducting parallel planning is vital in allowing engineer units to position critical assets (when available), establish linkup, and task-organize to their supported units. Early linkup with supported maneuver units provides critical time for combined arms planning and rehearsal.

2-359. The brigade engineer battalion or the assistant brigade engineer coordinates engineer reconnaissance to support the collection of necessary obstacle information and other technical information. They also coordinate the movement and positioning of required engineer augmentation assets (combat and general engineering). Key considerations for the scheme of engineer operations include the following:

- Task-organize engineer forces well forward to support breaching
- Assign normal priority of support to the lead company.
- Prepare to bypass or breach enemy situational obstacles.
- Integrate situational obstacles with fires and triggers to affect the movement of the enemy and support flank security in support of the commander's intent.
- To support flank security, plan obstacle belts and measures, and plan situational obstacles.
- Develop and adjust obstacles and triggers for execution based on the battalion's movement and the enemy situation.

Intelligence

2-360. The enemy's tactical deployment is the relative position of units with respect to one another or to the terrain. Enemy tactical formations are designed for executing the various tactical maneuvers. If this deployment can be predetermined, it leads to an accurate appraisal of intentions. The knowledge of how enemy units are echeloned may indicate (if the enemy assumes the offense) which units will be used in the main attack and which units will be used in supporting reserve roles.

2-361. Tactical deployment with respect to terrain is also important. A study of dispositions and an analysis of the terrain aid the development of conclusions concerning enemy capabilities, vulnerabilities, and intentions. Key terrain features are usually forward of friendly dispositions and are often assigned as objectives. Adjacent terrain features may be key terrain if their control is necessary for the continuation of the attack or the accomplishment of the mission.

2-362. The identification of avenues of approach is important because all course of actions that involve maneuver depend on available avenues of approach. During an attack, the evaluation of avenues of approach leads to a recommendation on the best avenues of approach to the battalion's objective and identification of avenues available to the enemy for counterattack, withdrawal, or the movement of reinforcements or reserves. Additional considerations for attacking a moving enemy force include the following.

Reconnaissance and Surveillance

2-363. The R&S effort focuses on answering the CCIRs to support the commander's decisions on when and where to initiate fires, where to fight the enemy, and how best to maneuver the battalion against the enemy. The battalion S-2 develops named areas of interest to identify enemy actions and decisions that indicate the enemy selected course of action.

Anticipate Enemy Course of Action

2-364. The IPB (see ATP 2-01.3) identifies the enemy's most likely course of action and most dangerous course of action. The IPB details the enemy's likely formation(s) and route(s) and how the enemy will most likely try to fight the ensuing meeting engagement. Analysis will show the enemy expected rate of movement and how the enemy force is likely to be arrayed. This information is based on a detailed terrain and time-distance analysis. The enemy normally has four course of actions listed below:

- Defend before or after initial contact to retain control of defensible terrain.
- Defend to limit the attacking force's advantage.
- Attack to defeat or penetrate the attacking force.
- Delay or bypass the attacking force.

2-365. The S-2 determines those enemy actions that may indicate the enemy's selection of a course of action and ensures observers are positioned to detect and report these indicators. The S-2 always must portray the enemy's flexibility, likely actions, and available maneuver options. The goal is to identify the enemy's most likely course of action and have the battalion anticipate and prepare for it.

Establish Contact

2-366. Preferably, the battalion establishes contact with the enemy using digital sensor platforms before it makes physical contact. The battalion, with support from the IBCT, receives information from battlefield surveillance systems. Intelligence produced from the information gathered by these sensors helps the battalion direct ground reconnaissance forces to advantageous positions to physically observe and report information on the enemy. Once made, the battalion maintains contact.

2-367. The information gained from the sensors as well as ground reconnaissance forces must be shared with all elements of the battalion and with higher as quickly as possible. Information requirements normally include—

- The enemy's rate and direction of movement.
- The enemy's formation, strength, and composition to include locations of security forces, main body, reserves, and artillery formations.
- Enemy actions and decisions that indicate a future enemy action or intention.
- Location of enemy high-payoff targets.
- Enemy vulnerabilities such as exposed flanks or force concentrations at obstacles.

Fires

2-368. Successful maneuver requires close coordination and effective employment of available fire support assets. The commander emphasizes the development of a clear, concise, and rapidly integrated fire support plan to ensure responsive fires are available when needed. Fires are planned to quickly attack enemy forces on key terrain, flanks, and in dead space to enable the commander to economize and concentrate to exploit advantages that arise. In order to be within supporting range of maneuver companies, the fire support officer for the battalion fully integrates fire support assets, external to the battalion, into the commander's scheme of maneuver for the attack. The commander positions battalion mortars forward within the attacking formation to facilitate continuous and effective fires to the battalion.

2-369. The commander employs fires to support reconnaissance forces and during preparation fires, using precision and other munitions to destroy enemy reconnaissance, security forces, and identified high-payoff targets, and to disrupt enemy maneuver. Planned precision-guided munitions to limit collateral damage permit effective engagement of point targets. For example, precision-guided munitions can be used to destroy a high-payoff target during an attack on an urban objective. Prior to and during the attack, target acquisition assets focus on identifying enemy systems that can interdict maneuver forces of the battalion as they move towards the objective. Radars and observers linked to fire support systems enable this effort. Additional considerations are listed below:

- Use fires to affect the enemy's maneuver well forward of the battalion to disrupt the enemy's formations and timetable.
- Plan triggers, observer locations, and targets to maintain flexibility and ensure achievement of required effects before contact with the enemy.
- Coordinate and synchronize with brigade the movement and positioning of artillery to support essential fire support tasks within each engagement location and to engage high-payoff targets before the enemy enters the selected area.
- Retain flexibility to mass fires at the decisive point in any location where the engagement may occur.
- Plan triggers to put targets into effect and cancel them based on the battalion's movement and the commander's decision of where to fight the enemy.
- Coordinate terrain requirements.

Sustainment

2-370. Sustainment planners within the battalion, led by the battalion S-4, synchronize and coordinate the concept of support development for the battalion's attack. Sustainment planners continually refine (running estimate) and coordinate the concept of support to maintain momentum and freedom of action. Sustainment planners may consider positioning sustainment units in close proximity to operations to reduce critical support response times. Sustainment planners echelon logistics assets to weight the effort supporting mission requirements. The commander and staff may consider alternative methods for delivering sustainment during emergencies and as the situation changes. Key considerations for the sustainment plan are listed below:

- Plan support from initiation of the operation to the final objective or limit of advance and, as required, a follow-on mission.
- Ensure the sustainment plan is responsive and flexible enough to support all maneuver options and continuously updated as the situation may change.
- Integrate resupply operations with the scheme of maneuver.
- Weigh the risk extended distances create for security of main supply routes and sustainment assets based on the potential of undetected or bypassed enemy forces.
- Use all available assets to develop and maintain an accurate enemy picture behind the lead maneuver elements.
- Develop and rehearse plans for enemy contact.
- Plan and coordinate the locations, displacements, and routes of sustainment assets to maintain responsive support.
- Develop triggers based on the battalion scheme of maneuver to activate or deactivate collection points and logistics release points.
- Plan casualty evacuation, resupply, and equipment recovery for engagements in each potential location.

Protection

2-371. If allocated, the air defense artillery supporting the IBCT operates in direct support to the battalions with the normal priority of protection to the decisive operation. The air defense artillery assets shift locations on the battlefield as required by the phase of the operation to maintain adequate air defense coverage of critical forces and events. Air defense coverage increases in areas and activities most vulnerable to air attack such as breaching operations or movements through restricted terrain. The CBRN assets are employed in a similar manner to their employment in an attack against a stationary force. Obscurants and CBRN reconnaissance assets typically support the decisive operation.

Preparation

2-372. Preparation for an attack against a moving enemy force is limited because the opportunity to attack the enemy at the appropriate time and place depends on the enemy's movement. This forces the battalion to focus the preparation on executing fires and maneuver actions within each location. The commander prioritizes each area to ensure the battalion prepares for the most likely engagements first. The commander ensures all subordinate companies and supporting forces understand their role in each area and the decision point for execution at each area. Battalion leaders rehearse actions in each area against various enemy conditions to promote flexibility and initiative consistent with the commander's intent. Repetitive rehearsals against likely enemy actions are essential for success at all levels.

Execution

2-373. Because of the battalion's vulnerability, by the nature of an attack against a moving enemy force, the enemy must not be underestimated. Only a determined attack, conducted at a high tempo and having an information dominance over the enemy, attains the enemy's total destruction. When executed correctly the battalion's principal advantage is possession of the initiative. The commander plans for, and concentrates the effects of subordinate forces and sets conditions to maintain the initiative. The battalion strikes the enemy in unexpected ways at unexpected times and places. The commander focuses on attacking the right combination of targets. Force-oriented attacks against a moving enemy force, once conditions are set, are rapidly and violently executed, and unpredictable in nature to disorient the enemy.

Approach to Objective

2-374. The commander seeks to conceal the movement of the battalion from the enemy to maintain surprise. The battalion moves in dispersed formations, and masks its movement using covered and concealed routes. By gaining contact with the enemy force through R&S, the battalion can use long-range artillery fires, close air support, and attack reconnaissance helicopters to destroy and disrupt the enemy throughout the their formation. Security forces (if employed), can detect and destroy enemy reconnaissance and security elements.

2-375. The battalion creates favorable conditions for the attack by weakening and disrupting the enemy's formation, destroying their security forces, and fixing the enemy's main body. The battalion employs fires reinforced with situational obstacles to set the conditions for the engagement area or objective fight, disrupting and weakening the enemy before they get to the area. Preparation fires should provide time for the battalion to deploy before contact. Reconnaissance elements normally control these initial fires.

2-376. The battalion deploys, attacks from unexpected and multiple directions, masses effects, and destroys the remaining enemy before he can adequately react. The commander adjusts the speed of the battalion to ensure that fires have set the appropriate conditions and that the battalion arrives at the designated location at the proper time in relation to the enemy. Effective reporting and analysis of the enemy's rate and direction of movement through R&S are critical to the timing of the attack.

Defeat Enemy Security Forces

2-377. Normally, the enemy employs security forces to protect their main body. The enemy's ability to seize the initiative often depends on their security forces. The battalion must avoid, destroy, or fight through enemy security forces to gain contact with the bulk of the enemy force. The commander employs fires in conjunction with their lead maneuver force to defeat enemy security forces. Ideally, battalion lead forces attack enemy forward and flank security forces to develop the situation. The commander initiates maneuver to destroy enemy security forces, and gain contact with the enemy's main body before the enemy can react effectively.

Fix Enemy Main Body

2-378. The battalion normally fixes the enemy's main body to create the conditions for the battalion's decisive operation or main effort. The battalion executes this task once opposing enemy security forces are effectively bypassed, fixed, or destroyed. Fires against lead enemy forces allow the battalion to deploy and gain contact with the enemy main body. Reconnaissance forces and available surveillance assets are position to keep the battalion commander informed of the enemy's strength and actions throughout the battalion's attack. It is paramount that the battalion commander receives accurate and timely reports and information on the enemy's situation. Commander critical information requirements, enemy main body's strength, disposition, and reactions, enable the commander to make final adjustments to the main body's attack.

Maneuver Main Body

2-379. As reconnaissance forces (and security forces if established, see paragraph 2-297) develop the situation, the commander begins to maneuver the remaining main body combined arms formations to favorable positions for commitment. The commander positions the battalion to attack the enemy formation from an assailable flank where the battalion's total combat power can be massed against an enemy weakness to reach a quick decision. Rapid movement and massed fires characterize the attack.

2-380. The battalion commander establishes support by fire positions (shaping operation or support effort) and uses indirect fires to suppress the enemy force that directly opposes the battalion's decisive operation or main effort. The battalion strikes the enemy force with overwhelming strength and speed. As the battalion maneuvers against the enemy, the fire support officer monitors and adjusts, as required, control measures (see appendix B) to provide continuous support and prevent fratricide.

2-381. If the commander determines the enemy force is trying to bypass or avoid contact, the commander directs fires to delay and disrupt the enemy's movement away from the battalion. The commander maneuvers forces to destroy or penetrate any enemy forces trying to fix or delay the battalion and strikes the bulk of the evading enemy force from the flank or rear.

Terrain-Oriented Attacks

2-382. Terrain-oriented attacks require the battalion to seize and retain control of a designated area to support future operations. The battalion attacks to seize terrain for many reasons. For example,—

- To seize key terrain or structures such as bridges, airfields, or public services to support follow-on operations or as part of a turning movement form of maneuver.
- To seize terrain such as a chokepoint or route to block enemy withdrawals or reinforcements.
- To block movements against a higher echelon's decisive operation.
- To facilitate friendly force passage.
- To secure an area, such as a lodgment area, for future operations.

2-383. The battalion plans, prepares, and executes terrain-oriented attacks in the same manner as attacks against enemy forces. The major distinction in a terrain-oriented attack is that the battalion focuses its efforts on the seizure and control of terrain instead of effects on the enemy (figure 2-46). The commander plans and directs the attack to gain control of the terrain as quickly as possible and only conducts necessary actions against the enemy. Success of the mission normally does not entail action against all enemy forces within the battalion's area of operation.

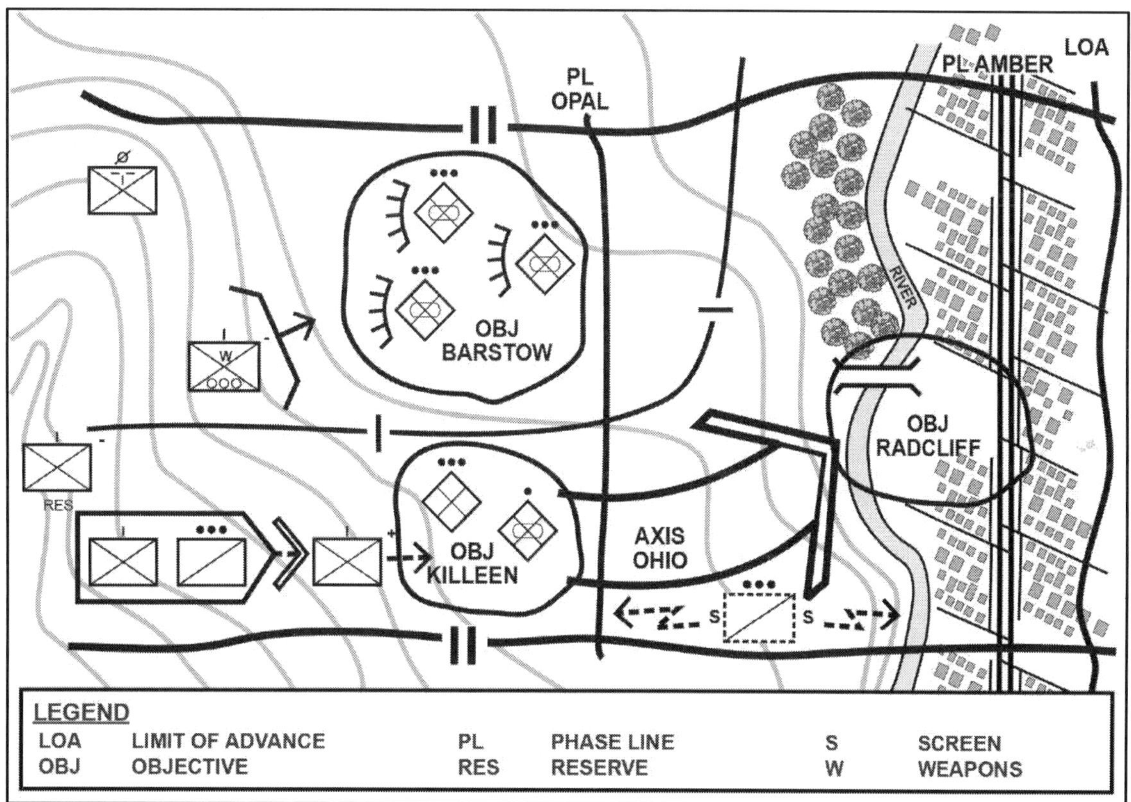

Figure 2-46. Battalion terrain-oriented attack

2-384. The Infantry battalion attacks only the enemy who directly affects the seizure of the objective or who might affect future operations. The commander understands that seizure of terrain-oriented objectives and likely key terrain probably will dislocate the enemy force. The commander also understands, after seizure, the enemy may counterattack to dislodge and defeat friendly forces occupying the objective. Other key considerations that differ from force-oriented attacks include—

Reconnaissance and Surveillance

2-385. The battalion commander and S-3, assisted by the S-2, consider enemy forces within the battalion's area of operation and area of interest for both force and terrain oriented attacks. Though, terrain oriented attacks will normally require the commander to place greater emphasis on the battalion's area of interest, specifically enemy outside the battalions area of operation that may react to the battalion's seizure of the terrain oriented objective.

2-386. Once enemy forces are located, reconnaissance forces and surveillance assets will try to determine the enemy's strength and disposition as well as possible enemy weaknesses and bypasses the battalion could exploit. This helps the S-2 develop enemy course of actions and identify indicators of the enemy's commitment to a future action. Normally, the S-2 considers enemy actions to defend in place; reinforce threatened enemy units; counterattack; delay; or possibly withdraw.

2-387. The surveillance effort, as in other attacks, capitalizes on all the battlefield surveillance assets available to the battalion, as well as those belonging to the battalion, to identify the enemy situation on the objective and any sizable enemy forces within the battalion area of operation and area of interest. Ground reconnaissance forces and surveillance assets, external to the battalion, can occupy positions to gain observation and report information on enemy activities within the battalion's area of interest.

Level of Risk

2-388. The battalion commander must determine the level of risk by leaving or bypassing enemy forces in both a terrain oriented attack and an enemy oriented attack. The commander bases this decision on the higher commander's intent and established bypass criteria, the enemy's capabilities, and the commander's assessment of the situation. The commander must recognize the potential effects that bypassed enemy forces may have on the battalion's sustainment and future operations. The commander normally employs economy of force missions to contain, destroy, or fix bypassed enemy forces. The risk imposed by these bypassed forces is to the elements moving behind the maneuver forces in the battalion's area of operation. Once the battalion secures the objective, other forces or fires can destroy bypassed enemy forces or force their surrender.

Seizure of the Objective

2-389. Once the Infantry battalion seizes the objective, the battalion immediately establishes local security, prepares hasty defensive positions, and prepares for an enemy attack to capture the objective. The commander positions forces in a manner that best defends the objective while allowing a rapid transition to follow-on operations. Security forces may establish a screening force forward of the secured objective to provide security and early warning to the battalion to prevent a surprise attack by the enemy. Reconnaissance forces may establish observation positions and surveillance assets may be deployed to provide early warning and detail information on the enemy situation forward of the battalion limit of advance. Engineers may provide countermobility and survivability support as time and resources allow. Fires assets reposition to support follow-on operations or the defense (if established) and the extended target coverage in the security area beyond the main battle area.

ATTACK—SEQUENCE OF EVENTS, EXAMPLE

2-390. Most attacks will following a sequence of events similar to the example sequence addressed below. This sequence of events is used for discussion purposes and is not the only way of conducting an attack. The battalion commander uses this or a similar sequence to plan events necessary to accomplish the mission. The commander understands events will vary depending on the mission variables of METT-TC and to some degree, events will overlap.

Assembly Area

2-391. The commander organizes the battalion for the attack, directs and supervises mission preparations in the assembly area. Assembly area activities include coordination, pre-combat checks and inspections, rehearsals, and sustainment preparations. (See paragraph 2-88 for information on occupying an assembly area.)

> ***Note.*** The commander may decide that rapid action is essential to retain a tactical advantage and decide not to use an assembly area. Detailed advance planning—combined with digital communications; standard operating procedures; and battle drills—may reduce negative impacts of such a decision.

Reconnaissance and Surveillance

2-392. R&S provides the battalion commander with the information needed to plan, prepare, execute, and assess the operation. The commander aggressively seeks information about the terrain and enemy before, during, and after the attack. The situation and available time may limit the R&S prior to crossing the line of departure. The commander understands the balance between the benefits of having reconnaissance forces on the ground providing combat information with the level of risk involved.

2-393. The R&S collection effort reports on enemy activity in the battalion's area of operation, this includes occupation of an assembly area, tactical movement, and establishment of attack position(s), line and point of departure, axis of advance, line of probable deployment, assault position(s), and the objective. R&S assets from higher echelons enable the battalion's collection effort.

2-394. The battalion commander may employ the scout platoon, unmanned aircraft systems, rifle company or rifle platoon to enable collection. Reconnaissance patrols from the scouts and rifle companies may move forward to identify enemy positions, mark and report on planned routes. The commander may deploy (dependent on available time and enemy situation) smaller elements, such as a scout and sniper team, to infiltrate to a position with direct observation of the objective, any obstacle, possible route for enemy reinforcement, or other critical area. These named areas of interest and observation positions are designated as no-fire areas.

Movement to the Line of Departure

2-395. When attacking from positions not in contact, the Infantry battalion often stages in rear assembly areas or moves to an attack position behind friendly units in contact with the enemy, then conducts passage of lines (see paragraph 2-116) to begin the attack. Lead elements of the battalion cross the line of departure at the time specified in the operations order or fragmentary order. Before movement, a patrol can be tasks to reconnoiter and mark the route and check the time it takes to move to the line of departure. When attacking from a position in direct contact, the line of departure may also be the *line of contact*—a general trace delineating the locations where friendly and enemy forces are engaged (FM 3-90-1). In certain circumstances, there may not be a line of departure, for example an operation in a noncontiguous area of operation.

2-396. Battalion support by fire force(s) may move into position prior to assault forces moving from the assembly area or may be in the lead in order to occupy an overwatch position to support the assault force as it crosses the line of departure. The commander should consider the risk of losing the element of surprise when moving vehicles close to the line of departure. Mortars move forward to a firing position near the line of departure to allow maximum coverage of the movement. The commander avoids stopping in the attack position prior to the line of departure. However, if units are ahead of schedule, are told to hold, or have to make final preparations for the attack, elements occupy the attack position, post security, and wait until time to move.

2-397. The fire support officer prepares to execute preplanned fires during tactical movement to the probable line of deployment and preparation fires on the objective (depending on the distance to the objective) prior to crossing the probable line of deployment. The fire support officer prepares for call for fire missions on enemy positions identified during movement and obscuration to cover the battalion's movement when required. During movement, any element with direct observation of the enemy may request or adjust fires.

2-398. Assault forces cross the line of departure in a manner that supports its deployment prior to the assault. Units guide on the designated element, usually the force conducting the main attack. Battalion units may cross the line of departure supported by suppressive direct and indirect fires, or wait until the appropriate time to initiate fires. The commander must consider the time of the movement and the ammunition available. The commander also should consider the effect of fires on the ability to achieve surprise. If so employed, these fires may continue until the assault force reaches the assault position(s) or final coordination line, fires then shift, as required, to allow the assault on the objective. The commander initiates indirect fires based on the battalion's scheme of maneuver.

2-399. Battalion mortars may be the most responsive to on-call missions but may be limited in range and ammunition. Mortars move forward to a firing position(s) near the line of departure to provide support during movement and, depending on the distance, to the objective. Mortars may displace by split section to provide support at the objective. The displacement provides continuous indirect fire support throughout the operation. Company mortars usually are in direct support of their company.

Movement to the Probable Line of Deployment

2-400. During movement from the line of departure to the probable line of deployment, the battalion makes the best use of cover, concealment, and supporting fires. The attacking force usually avoids known enemy positions. During movement if an element is engaged by indirect fire, the element moves quickly out of the impact area. Reaction to unexpected enemy contact, actions on contact (see paragraph 2-170) should be in accordance with the operations order or fragmentary order. Depending on the battalion plan and the location and type of resistance, enemy position may be bypassed.

2-401. When the enemy cannot be bypassed, the maneuver force in contact must take prompt and aggressive action. The commander quickly conducts an estimate of the situation and issues guidance as needed to coordinate actions and fires to attack the enemy. The commander maneuvers the force, to assault the flank or rear of the enemy position. When the enemy is destroyed, the force continues movement. When the enemy is suppressed, follow-on battalion forces pass around the enemy to continue the mission. Throughout the movement, the battalion commander continues to focus on the primary objective and continues the battalion's advance to that objective.

2-402. Maneuver forces of the battalion bypass obstacles when possible and when the terrain allows, but will breach when required. During the attack, a maneuver force may conduct the breach with organic equipment and personnel or with attached engineer or other combat enabler. (See Appendix F for information on combined arms breaching operations.) The maneuver force commander informs the battalion commander of obstacles that may affect following units. Engineers (when available) are position to provide a rapid assessment of the obstacle throughout the movement.

Actions on the Objective

2-403. As the battalion maintains the pace of its advance as it approaches the probable line of deployment. The battalion deploys into one or more assault and support forces either before or upon reaching the probable line of deployment. The battalion commander synchronizes the occupation of support and attack by fire positions with the maneuver of the assault force to limit the vulnerability of the forces occupying these positions. All means are employ to suppress and destroy the enemy to sustain the momentum of the battalion's attack.

2-404. The commander synchronizes the effects of indirect-fire systems, and available attack reconnaissance helicopters and close air support as assault forces close on the objective. Fires are planned in series or groups to support maneuver against enemy forces on or near the objective. Support forces maintain suppressive fires to isolate the objective and prevent the enemy from reinforcing or counterattacking.

2-405. Once assault forces reach the far side of the objective, selected elements clear remaining pockets of resistance while the bulk of the assault force prepares for possible enemy counterattack. When assault forces reach the limit of advance, or on-order, support forces reposition support and attack by fire positions to counter possible enemy counterattacks and to support battalion follow-on missions.

Support by Fire

2-406. Battalion support forces occupy support by fire positions that, ideally, afford unobstructed observation, clear fields of fire, and cover and concealment. Selection of support by fire positions are based on a study of the terrain, knowledge of enemy locations, or likely enemy locations. When the enemy situation is vague or unknown, positions are selected to place effective fire on terrain that dominates the area the assault force will traverse and seize. Once in position, support forces are responsible for both suppressing known enemy forces and for scanning assigned sectors of observation to identify previously unknown enemy elements to suppress them. The protection provided by support forces allow the assault force to continue movement and to retain the initiative even when under enemy observation or within range of enemy weapons.

2-407. Battalion support by fire (direct fire) and attack by fire (direct fire, support by indirect fire) positions and other supporting fires may destroy designated targets prior to the assault force crossing the probable line of deployment. At the planned time or, preferably, when the lead elements of the assault force cross a designated line, supporting fires begin suppressive fires. Supporting fires cease or shift fires when assault forces reach a designated point or when directed. Best case, supporting fires suppress the target with indirect and then direct fires until the assault force is as close to the objective as possible and then shift to other targets. Supporting fires suppress directly in front of the assault force (when conditions allow) as it moves through the objective.

Assault

2-408. The battalion's assault is short, violent, and well ordered. The assault force seizes or secures a geographic objective, or destroys, defeats, or disrupts a designated enemy force. From the probable line of deployment, the assault force maneuvers against or around the enemy to take advantage of support fires. The assault force must closely follow these supporting fires to gain ground that offers positional advantage.

Note. When required, the breach force reduces, proofs, and marks the required number of lanes through the enemy's tactical obstacles to support the maneuver of the assault force. The commander clearly identifies the conditions that allow the breach force to proceed from the probable line of deployment ahead of the assault force. (See appendix F.)

2-409. Between the probable line of deployment and the objective, the battalion can establish an assault position(s) and a final coordination line to control the final stages of the assault. Assaulting forces usually plan to pass through the assault position and deploy to their final assault formation at final coordination line. Stopping at the assault position may allow the enemy to react to the assault and may make it difficult for the assault force to regain the momentum of the assault once the force halts in a covered position. When used, the assault force pauses at the assault position only to—

- Make final equipment preparations, such as to make final preparations for demolitions.
- Ensure all assault forces are in their planned order.
- Wait for preparation fires to finish.

2-410. The objective for the assault force may vary from operation to operation. In every case, the assault force's actions on the objective are critical. Assault forces maneuver through fire and movement. Supporting fires must immediately gain fire superiority on the objective in coordination with fires around and beyond the objective to disrupt enemy reinforcement. As the assault closes on the objective, fires shift just forward of the assault force as it moves across the objective or until ordered to stop or shift to other target beyond the assault force or limit of advance.

Note. The key is to minimize the time between the shifting of fires and the maneuver of the assault force on the objective. The enemy must be suppressed during this time when the assault force is most vulnerable.

Consolidation, Reorganization, and Follow Through

2-411. On the objective, battalion assault forces move across the objective to a predetermined or, on order, limit of advance to control the forward progress of the attack. Assault forces consolidate and reorganize once the objective is seized. Assault forces then dispatch designated teams to go back through the objective to ensure all enemy forces are destroyed, incapacitated, or willing to surrender. Teams also conduct searches for any material of intelligence value. (Refer to FM 3-21.10 for additional information.)

2-412. *Consolidation* is the organizing and strengthening a newly captured position so that it can be used against the enemy (FM 3-90-1). *Reorganization* is all measures taken by the commander to maintain unit combat effectiveness or return it to a specified level of combat capability (FM 3-90-1). During consolidation and reorganization, the battalion executes follow-on missions as directed. One mission is to continue the attack against targets of opportunity in the objective area. Whether a raid, attack (as a hasty or deliberate operation), or movement to contact, the battalion postures and prepares for continued action and to defeat local counterattacks.

2-413. The battalion commander may pass follow-on forces through the assault force, or have them bypass the assault force, to continue the attack. On the objective, the assault force may be tasked to establish support by fire and attack by fire positions. Assault platoons, from the weapons company, and battalion mortars may move to the objective to establish positions. The battalion commander or assault force leader identifies initial support by fire and attack by fire positions to reinforce the battalion's limit of advance and to support follow-on missions.

Note. The following illustration introduces a fictional scenario as a discussion vehicle for illustrating one of many ways that an Infantry battalion can conduct an attack. This illustration is a continuation of the scenario started earlier in this chapter focusing on Infantry Battalion 2.

ILLUSTRATION OF AN ATTACK

2-414. The following scenario illustrates an attack conducted by Infantry Battalion 2. In this scenario, started earlier in this chapter, an IBCT conducted a movement to contact along two axes of advance. Infantry Battalion 1, initially the main effort, conducted a movement to contact along Axis of Advance Blue (Axis Blue) to the east. Infantry Battalion 2, initially a supporting effort, conducted a movement to contact along Axis of Advance Red (Axis Red) to the west. Infantry Battalion 3 conducted a follow and assume tactical mission task along Axis Blue (see figure 2-30, page 2-60). During the movement to contact, Infantry Battalion 1 cleared enemy resistance north of Objective Fox but encountered stiff resistance along Axis Blue from platoon size enemy elements. Infantry Battalion 1 was unable to maneuver to generate sufficient combat power to seize Objective Fox, because of the unrestricted terrain and the enemy's clear fields of fires from Objective Fox to the north along Axis Blue.

INFANTRY BATTALION 2 – CHANGE OF MISSION

2-415. Within the scenario, Infantry Battalion 2 commander monitored the progress of the IBCT's main effort along Axis Blue. The battalion commander, per the IBCT commander's guidance, begins visualizing how the battalion might move to assist the IBCT in regaining the initiative along Axis Blue. In coordination with the IBCT commander, the battalion commander and staff begin preliminary planning and WARNORD development to conduct an approach march to Objective Fox. As Infantry Battalion 2 continued its movement to contact along Axis Red, and as forward security forces of the battalion seized Objective Zebra (unopposed), the battalion received a change of mission (conduct an approach march) from the IBCT commander (figure 2-47). The IBCT commander directed—

- Infantry Battalion 2, main effort, conduct approach march to Assault Positions 3 and 4. On order conduct an attack, seize Objective Fox. Once objective seized, pass follow-on forces through to continue IBCT movement to contact south along Axis Blue and Axis Red.
- Infantry Battalion 1, supporting effort, establish support by fire and attack by fire positions north of Objective Fox. On order, disrupt, fix, and destroy enemy forces surrounding and on Objective Fox to support main effort.
- Cavalry squadron, establish guard mission to the east of Infantry Battalion 1. On order, conduct movement along the east side of Objective Fox to conduct guard mission forward of IBCT movement to contact south of Objective Fox along Axis Blue.
- Infantry Battalion 3 continues follow and assume tactical mission task. On order, conduct movement along east side of Objective Fox to continue IBCT's movement to contact south of Objective Fox along Axis Blue. (Follows Cavalry squadron.)
- On order, field artillery battalion provides priority of fires to Infantry Battalion 2.
- Brigade engineer battalion, priority of engineer effort (mobility) is to Infantry Battalion 1 then to Infantry Battalion 2. On order priority of engineer effort, mobility of forward security forces and Infantry Battalion 3. Tactical unmanned aircraft system platoon continues to conduct surveillance to the flanks of the IBCT movement (not illustrated).
- Brigade support battalion establishes BSA, vicinity Assemble Area Crow (not illustrated).

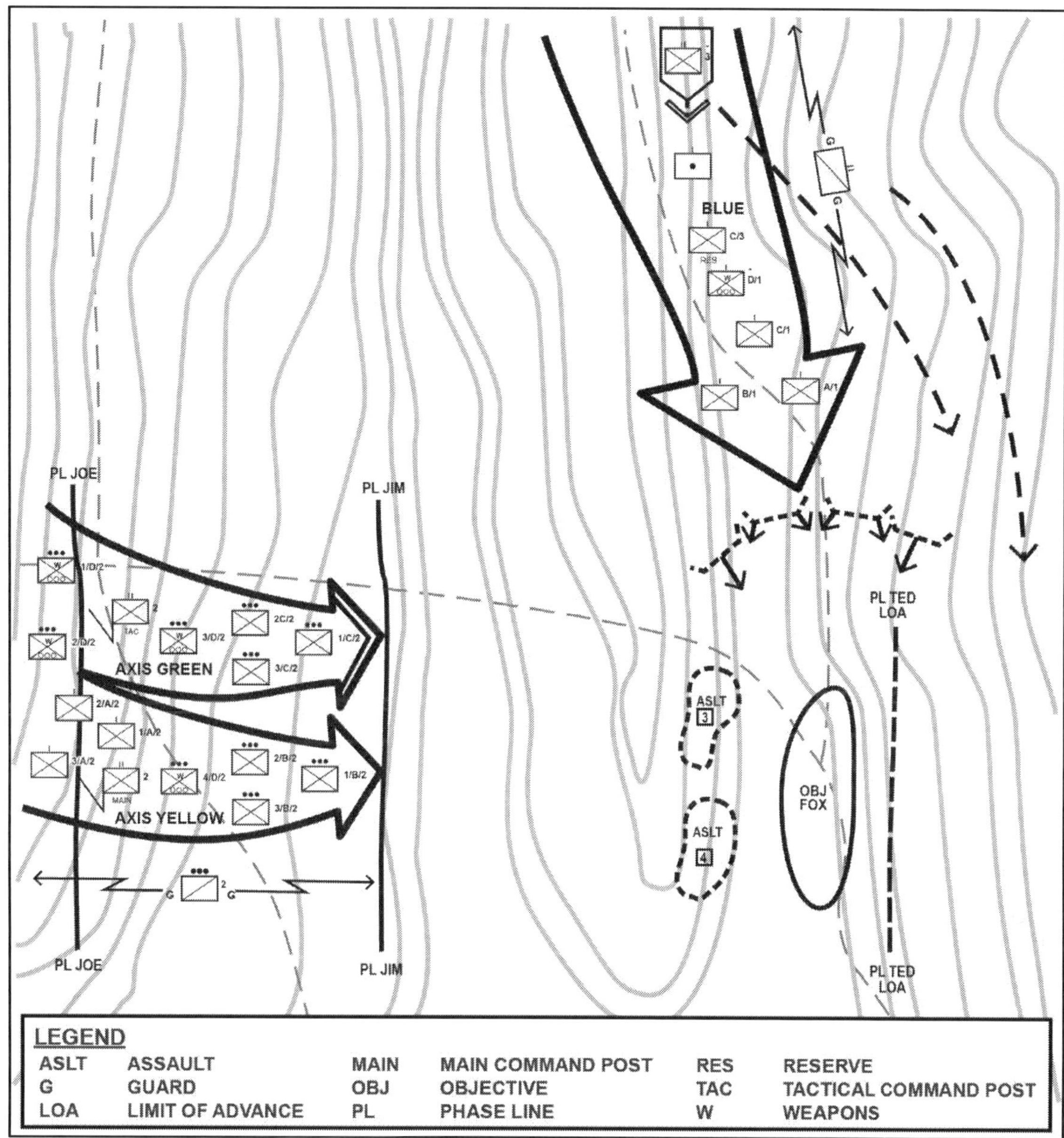

Figure 2-47. IBCT attack (scheme of maneuver), example

> ***Note.*** The IBCT establishes restrictive fire support coordination measures (not illustrated), on order, east of Phase Line Ted (Infantry Battalion 2 limit of advance) to safeguard Infantry Battalion 2's maneuver, and attack on Objective Fox. In addition to fratricide, these measures facilitated the rapidly changing operation and prevented duplication of engagement between the two maneuver battalions. Restrictive fire support coordination measures included an airspace coordination area, a no-fire area, a restrictive fire area, and a restrictive fire line. (Refer to appendix C and FM 3-09 for additional information.)

INFANTRY BATTALION 2—SUBORDINATE UNIT TASK AND ORDER OF MOVEMENT

2-416. Within the scenario, Infantry Battalion 2 conducts approach march to Assault Positions 3 and 4, on order, conducts an attack and seizes Objective Fox. Not to slow the tactical movement to the assault positions, the battalion commander maintained the same combined arms teams and order of movement task-organized for the movement to contact. The commander assigned an axis of advance for the approach march and established a probable line of deployment (Phase Line Joe) for the attack, based on the mission variables of METT-TC and the location of the assault positions (see figure 2-35, page 2-70). During the approach march, the commander assigned tasks to the force conducting the main effort and forces conducting supporting efforts respectively, along with area of operations, axes of advance, phase lines, and separate routes to support the battalion scheme of maneuver for the attack.

2-417. As the battalion approach march neared areas of likely enemy interference from Objective Fox, vicinity Phase Line Joe, the battalion commander divided the battalion axis of advance into two company columns, Axis of Advance Green (Axis Green) and Axis of Advance Yellow (Axis Yellow). Lead companies moved on multiple routes cross-country while employing security forces (platoon size element) forward of the company column's main body (see figure 2-36, page 2-72). The scout platoon moved from its security force position forward of the battalion to a position south of Axis Yellow to provide security on the battalion's southern flank. Battalion sniper squad moved from its security force location forward of the battalion to a position north of Axis Green to conduct linkup with Company D on the northern flank of the battalion. As the sniper squad crosses Phase Line Jim, the squad conducts route reconnaissance of Axis of Advance Orange (Axis Orange) to the front of Company D. Battalion security forces remain within supporting distance of the battalion main body (figure 2-48). Battalion subordinate unit task and order of movement is as follows:

- Company C (main effort), with attached engineer platoon (-) and assault platoon, conducts approach march along Axis Green to Assault Position 3.
- Company B (supporting effort), with attached engineer squad and assault platoon, conducts approach march along Axis Yellow to Assault Position 4.
- Company D (supporting effort), with two remaining assault platoons, conducts approach march following Company C along Axis Green. Battalion tactical command post follows main effort (Company C), moves with the Company D. On order, vicinity Phase Line Jim moves on Axis Orange through Assault Position 5 into Support by Fire Position 3.
- Company A (battalion reserve) conducts approach march, follows Company B along Axis Yellow. One rifle platoon detached to battalion control, detached platoon provides battalion rear security. Battalion main command post follows Company B, moves with Company A.
- Battalion mortar platoon (supporting fires) moves by split section (Section A moves with Company D and Section B moves with Company A). Priority of fires to Company C.
- Battalion scout platoon provides battalion flank security (right flank) south of Axis Yellow.
- Battalion sniper squad, on order vicinity Phase Line Joe conducts route reconnaissance along Axis Orange, establishes position to overwatch Company D movement through Assault Position 5 into Support by Fire Position 3.
- Battalion combat trains, on order occupies support area vicinity Objective Snake (not illustrated), with priority of support to the main effort, initially forward security forces. This echelon support includes maintenance support and treatment team alpha (clinically staffed with the battalion surgeon), and carries mostly Class I, V, and limit Class III.
- Battalion field trains, occupies support area, vicinity Objective Kiwi (not illustrated). This echelon support includes all battalion sustainment not located with the combat trains, includes treatment team bravo (clinically staffed with the physician assistant).

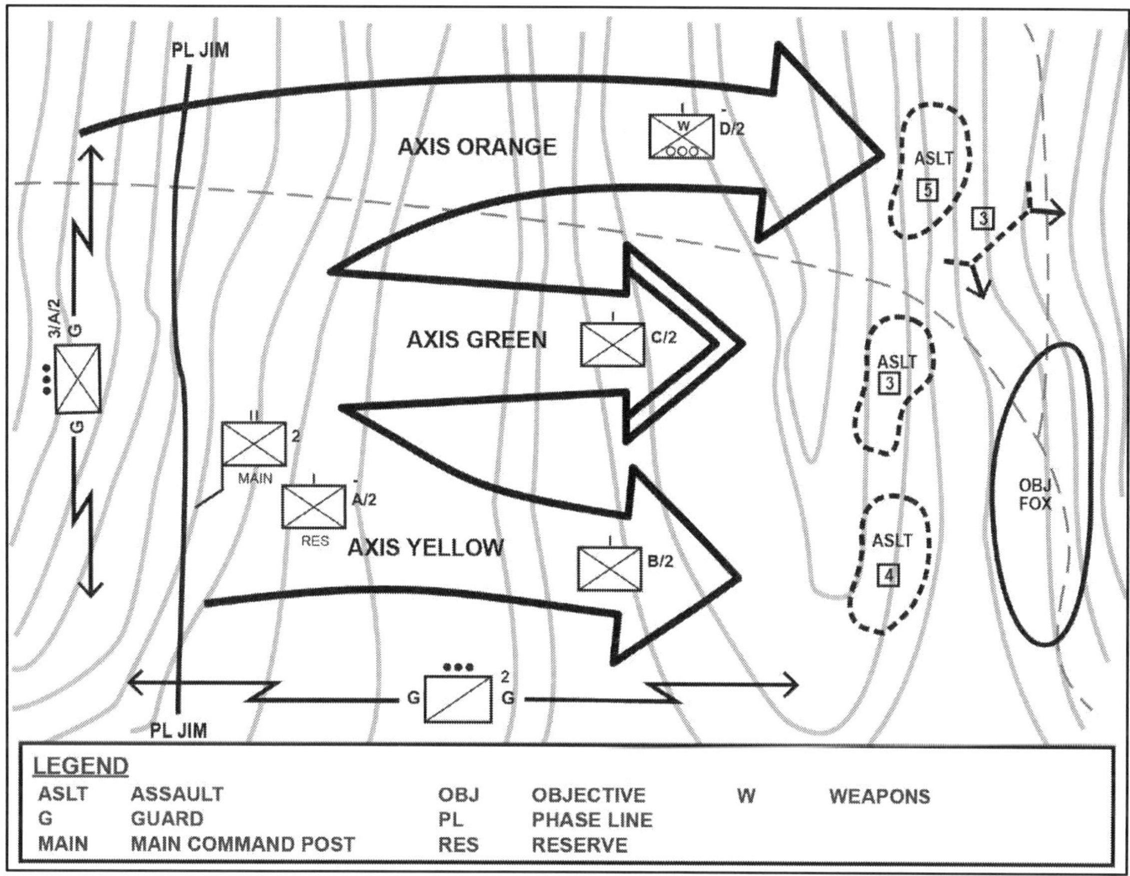

Figure 2-48. Battalion attack (axes of advance), example

Gain and Maintain Enemy Contact

2-418. As Infantry Battalion 2 forces close on the objective, the enemy situation becomes clearer. Subordinate units, through actions on contact, rapidly develop the situation in accordance with the commander's plan and intent for the attack on Objective Fox. Once battalion security forces gain contact with the enemy they maintain contact to further develop the enemy situation. Lead maneuver elements of the battalion, along axes green and yellow, employ combat power to destroy (within their capability) enemy security forces trying to hide enemy disposition, capabilities, and intent.

2-419. The battalion sniper squad moves ahead of Company D and conducts a route reconnaissance along Axis Orange. The sniper squad leader establishes one observation post (sniper team) to overwatch the movement of Company D through Assault Position 5 into Support by Fire Position 3. The sniper squad leader then moves the other two sniper teams to establish forward observation posts to make visual contact with the enemy on Objective Fox. As battalion assault and support forces advance, sniper teams report on enemy activity and maintain surveillance on Support-by-Fire Position 3 and Objective Fox.

Disrupt and Fix the Enemy

2-420. As Infantry Battalion 2 moves into position for the attack, the IBCT commander synchronizes all available elements of combat power on the battlefield to achieve a decisive massing of effects on the objective. Critical to the success of the attack is the IBCT commander's assessment and decisions regarding what to disrupt, when to disrupt, and to what end. Disrupting one or more parts of the enemy weakens the entire enemy force allowing the IBCT to isolate and attack key portions of the remaining enemy force. Field artillery counterfire, to destroy or neutralized enemy weapons, disrupts the enemy allowing attacking forces to maneuver and mass effects against the enemy.

2-421. Prior to the attack, attack by fire and support by fire positions established by Infantry Battalion 1 disrupt and fix enemy actions on (initially) and surrounding Objective Fox (see figure 2-36, page 2-72). Additional shaping operations to disrupt and fix the enemy include preparation fires and deception operations to deceive the enemy. By disrupting and fixing the enemy, the objective of the IBCTs main effort is isolated to prevent the enemy from maneuvering to reinforce the enemy targeted for destruction. Massing forces in one place by using economy of force measures in other areas allows Infantry Battalion 2 to mass the effects of overwhelming combat power against a portion of the enemy. Support by fire positions, established by Company D isolate the portion of Objective Fox that can affect the battalion's assault.

> ***Note.*** Once fire superiority on the objective is achieved by Infantry Battalion 2, the battalion controls all fires west of the restrictive fire line (not illustrated). The IBCT retains control of all fires east of the restrictive fire line. (Refer to appendix C and FM 3-09 for additional information.)

Maneuver

2-422. Seeking to avoid the enemy's defensive strength oriented north against the IBCT's attack south along Axis Blue. The IBCT commander ordered Infantry Battalion 2 to conduct a flank attack from the west, seize Objective Fox and clear enemy to Phase Line Ted (see figure 2-47, page 2-115). Prior to the establishment of Support by Fire Position 3, the sniper squad leader makes contact with the battalion commander to update the commander on the current enemy situation then guides the commander on the leader's reconnaissance of the objective. During the leader's reconnaissance the commander confirms the plan and establishes a forward position (vicinity, Support by Fire Position 3), to control the attack after assault forces cross Phase Line Ben. Sniper teams provided overwatch and security throughout the commander's movement.

> ***Note***. The battalion commander establishes command presence in a variety of ways, to include a leader's reconnaissance. Optimally, the commander conducts a leader's reconnaissance with key personnel to confirm or modify the plan. Depending on the enemy situation and battalion scheme of maneuver, the leader's reconnaissance may just involve the commander (with a security team). After the leader's reconnaissance, the commander modifies the plan and disseminates those changes to subordinate leaders and other affected organizations. Under the least favorable condition, the commander may not be able to conduct a leader's reconnaissance. In this case, the commander may utilize other available assets (BCT or higher-level asset) to confirm or modify the plan. When these assets are unavailable and when known conditions are unchanged, the commander executes the mission according to plan.

2-423. Once Company D achieves fire superiority, the battalion attacks (figure 2-49) with Company C (main effort) on the left and Company B on the right, seizes Objective Fox and clears enemy to Phase Line Ted (battalion limit of advance). Company C establishes a local support by fire position (attached assault platoon) to the left of the assault to support the company's attack. Company A (battalion reserve) be prepared mission, establish blocking position south of Objective Fox, oriented south on Route Blue. Battalion mortars operated split section. On order priority targets located vicinity Objective Fox, alternate targets enemy avenues of approach south Objective Fox and east of Phase Line Ted. As the assault reaches the limit of advance, assault forces clear remaining enemy within assigned areas, establish local security, conduct consolidation and reorganization, and prepare for follow-on mission. Company B reserve (attached assault platoon) be prepared mission, establish blocking position on the right flank of the company oriented south.

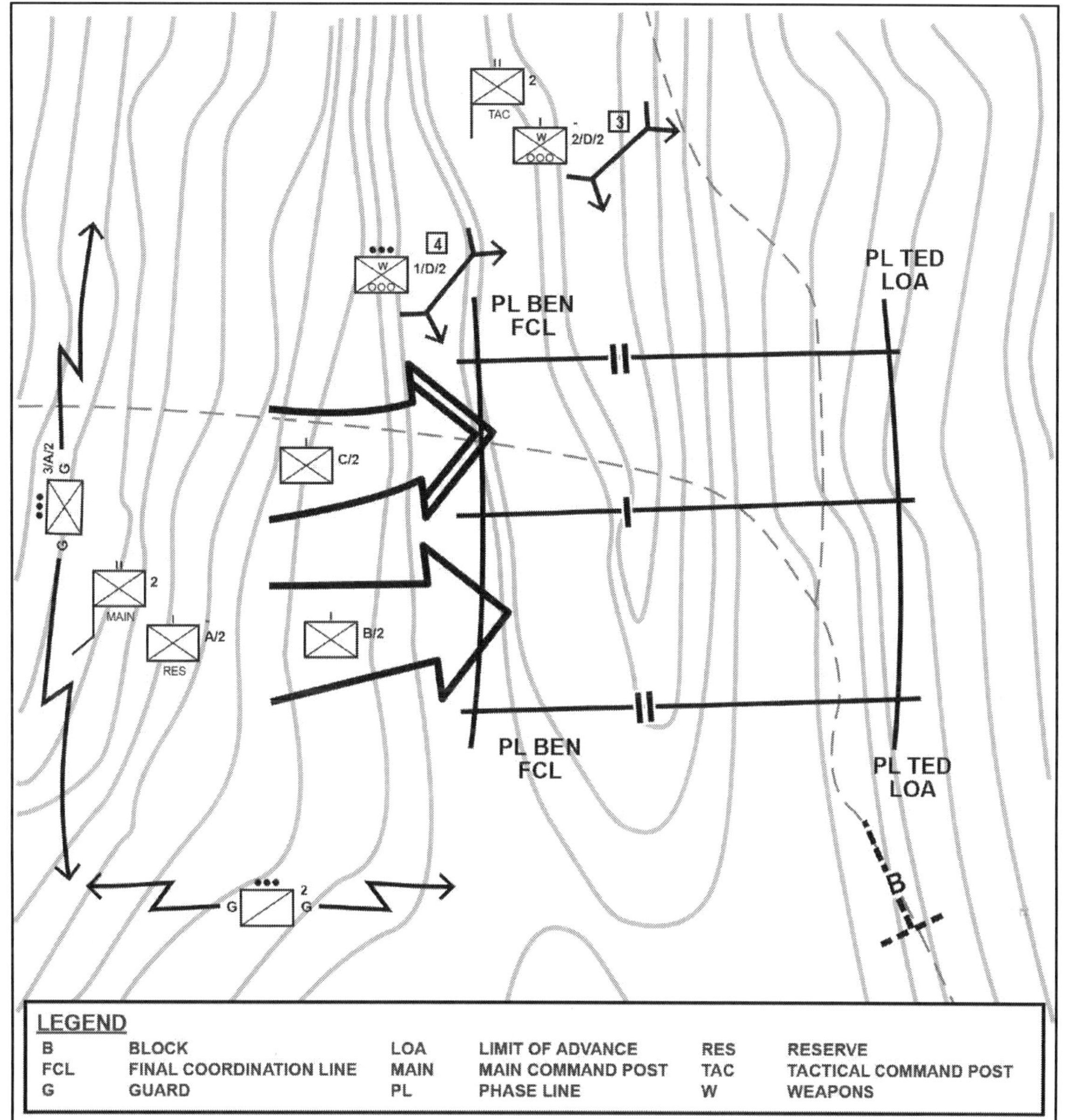

Figure 2-49. Battalion attack (scheme of maneuver), example

Follow Through

2-424. As Infantry Battalion 2 reaches Phase Line Ted, the IBCT commander issues a change of mission (fragmentary order) to subordinate units. IBCT base mission did not change—IBCT continues movement to contact along axes Blue and Red, the IBCT main effort continues to move along Axis Blue (figure 2-50, page 2-120). IBCT limit of advance is still Phase Line Dale to the south (not illustrated) with Objective Bear being the IBCT's final march objective (not illustrated). IBCT subordinate unit task and order of movement is as follows:

- Cavalry squadron, from guard location east of Infantry Battalion 1, conduct movement along the east side of Objective Fox to conduct guard mission forward of IBCT movement to contact south of Objective Fox along Axis Blue.
- Infantry Battalion 3, from follow and assume position, conduct movement along the east side of Objective Fox, conduct movement to contact (main effort) south of Objective Fox along Axis Blue. (Follows Cavalry squadron.)

- Infantry Battalion 1, from support by fire position, conduct movement along the west side of Objective Fox to conduct movement to contact (supporting effort) south of Objective Zebra along Axis Red.
- Infantry Battalion 2, from Phase Line Ted location, conduct follow and assume tactical mission task along Axis Blue. (Follows Infantry Battalion 3.) Company C, detached under IBCT control.
- Company C, IBCT reserve, follows Infantry Battalion 3 along Axis Blue.
- On order, field artillery battalion moves along Axis Blue with priority of fires to Infantry Battalion 3.
- The brigade engineer battalion priority of engineer effort, mobility to Infantry Battalion 3, then to Infantry Battalion 1. Tactical unmanned aircraft system platoon conducts surveillance to the flanks of the IBCT movement.
- Brigade support battalion establishes BSA, vicinity Assemble Area Hawk (not illustrated).

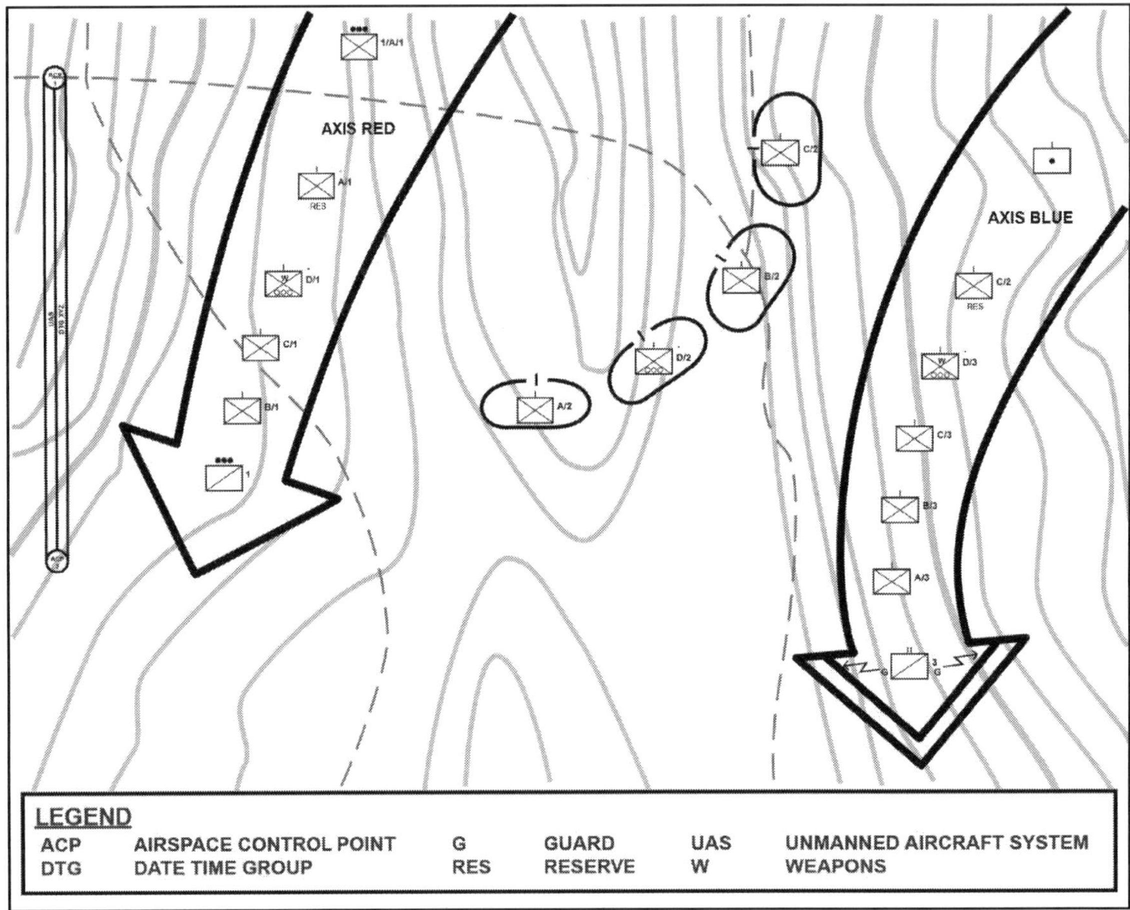

Figure 2-50. Follow-through (mission continuation), example

Note. This ends the scenario for the offense.

SUBORDINATE FORMS OF THE ATTACK

2-425. Subordinate forms of the attack have special purposes and include ambush, counterattack, demonstration, feint, raid, and spoiling attack. As subordinate attack tasks, they share many of the planning, preparation, and execution considerations of the attack. The commander's intent and the mission variables of METT-TC determine which of these forms of attack are employed. The commander can conduct each of these forms of attack, except for a raid, as either a hasty or a deliberate operation. Demonstrations and feints, while forms of attack, are also associated with the conduct of military deception operations (see JP 3-13). This section discusses the unique considerations of each subordinate attack task.

AMBUSH

2-426. An *ambush* is an attack by fire or other destructive means from concealed positions on a moving or temporarily halted enemy (FM 3-90-1). An ambush stops, denies, or destroys enemy forces by maximizing the element of surprise. An ambush can employ direct fire systems as well as other destructive means, such as command-detonated mines, indirect fires, and supporting nonlethal effects. An ambush may include an assault to close with and destroy enemy forces. Doctrine categorizes ambushes as near or far ambushes, based on the proximity of the friendly force to the enemy and that ground objectives do not have to be seized and held.

2-427. Ambushes are categorized as hasty or deliberate but take place along a continuum. (See paragraph 1-36.) A hasty ambush is an immediate reaction to an unexpected opportunity conducted using standard operating procedures and battle drill. A deliberate ambush is planned as a specific action against a specific target. Detailed information about the target; such as size, organization, and weapons and equipment carried, route and direction of movement, and times the target will reach or pass certain points on its route, may be available. During terrain analysis, leaders identify at least four different locations: the ambush site, the kill zone, security positions, and rally points. As far as possible, so-called ideal ambush sites should be avoided because an alert enemy will avoid them if possible and increase their vigilance when they must be entered. A typical ambush is organized into three elements: assault, support, and security.

2-428. The assault element's goal is to destroy the enemy force. The assault element attacks into and clears the kill zone; additional tasks may include searching for items of intelligence value, capturing prisoners, and completing the destruction of enemy equipment. The support element supports the assault element by firing into and around the kill zone, and provides the ambush's primary killing power. The support element attempts to destroy the majority of enemy combat power before the assault element moves into the objective or kill zone. The security element isolates the kill zone, provides early warning of the arrival of any enemy relief force, and provides security for the assault and support elements. The security element secures the objective rally point and blocks enemy avenues of approach into and out of the ambush site, which prevents the enemy from entering or leaving. The three forms of an ambush are the point ambush, area ambush, and anti-armor ambush.

2-429. In a point ambush, a unit deploys to attack a single kill zone. The *kill zone* is that part of an ambush site where fires are concentrated to isolate, fix, and destroy the enemy (FM 3-90-1). A point ambush usually employs a linear or an L-shaped formation. The names of these formations describe deployment of the support element around the kill zone. The linear ambush formation (deployed in a line formation) is effective in close terrain, which restricts the target's movement, and in open terrain where one flank is blocked by existing or reinforcing obstacles. An advantage of the line formation is that it is relatively easy to control under all conditions of visibility. A disadvantage of the line formation is that the target may be so dispersed that it is larger than the kill zone. The L-shaped formation is a variation of the line formation. The long leg of the "L" (assault element) is parallel to the kill zone and provides flanking fire. An advantage of the "L" formation is that the short leg (support element) is at the end of the kill zone and at a right angle to long leg and blocks the enemy's forward movement. The L-shaped formation also provides enfilading fire that interlocks with fire from the other leg. The commander can employ an L-shaped formation on a straight stretch of trail, road, stream, or at a sharp bend.

2-430. In an area ambush, a unit deploys into two or more related point ambushes. A unit smaller than a platoon does not normally conduct an area ambush. An area ambush is most effective when enemy movement is largely restricted to trails or roads. The area should offer several suitable point ambush sites. The commander selects a central ambush site around which the commander can organize outlying ambushes. Once the site is selected, the commander must determine the enemy's possible avenues of approach and escape routes. Outlying point ambush sites are assigned to subordinates to cover these avenues. Once they occupy these sites, they report all enemy traffic going toward or away from the central ambush site to the commander. These outlying ambushes allow the enemy to pass through their kill zones until the commander initiates the central ambush. Once the central ambush begins, the out-lying ambushes prevent enemy troops from escaping or entering the area.

2-431. Antiarmor ambushes focus on moving or temporarily halted enemy armored vehicles and may be part of an area ambush. The anti-armor ambush assault element will include an armor-killer element. The armor-killer element is built around close combat missile systems with additional Soldier launched munitions available to supplement the close combat missile system fires. The commander considers the mission variables of METT-TC to position all anti-armor weapons to take advantage of their best engagement aspect (rear, flank, or top). The support and security elements function in the same manner as the other forms of ambushes. (Refer to FM 3-21.10 and ATP 3-21.8 for additional information.)

COUNTERATTACK

2-432. A *counterattack* is an attack by part or all of a defending force against an enemy attacking force, for such specific purposes as regaining ground lost or cutting off, or destroying enemy advance units, with the general objective of denying to the enemy the attainment of the enemy's purpose in attacking. In sustained defensive actions, it is undertaken to restore the battle position and is directed at limited objectives (FM 3-90-1). The commander directs a counterattack—normally conducted from a defensive posture—to defeat or destroy enemy forces, exploit an enemy weakness, such as an exposed flank, or to regain control of terrain and facilities after an enemy success. A unit conducts a counterattack to seize the initiative from the enemy through offensive action. A counterattacking force maneuvers to isolate and destroy a designated enemy force.

2-433. The Infantry battalion as a counterattack force, attacks by fire into an engagement area to defeat or destroy an enemy force, to restore the original position, or to block an enemy penetration. The counterattack is often the deciding action in the defense and becomes the decisive operation upon commitment. The Infantry battalion is best suited for this role in restricted terrain. In unrestricted terrain, the battalion is vulnerable to fires and only possess, with the weapons company, the mobility and potential firepower to counterattack. The commander may plan counterattacks as part of the battalion's defensive plan, or the battalion may be the counterattack force for the brigade or division. (Refer to FM 3-96 and FM 3-90-1 for additional information.)

DEMONSTRATIONS AND FEINTS

2-434. In military deception, a *demonstration* is a show of force in an area where a decision is not sought that is made to deceive an adversary. It is similar to a feint but no actual contact with the adversary is intended (JP 3-13.4). A *feint* in military deception is an offensive action involving contact with the adversary conducted for the purpose of deceiving the adversary as to the location and time of the actual main offensive action (JP 3-13.4).

2-435. A commander uses demonstrations and feints in conjunction with other military deception activities. The commander attempts to deceive the enemy and induce the enemy commander to move reserves and shift fire support assets to locations where they cannot immediately affect the friendly decisive operation or take other actions not conducive to the enemy's best interests during the defense. Both forms are always shaping operations. The commander must synchronize the conduct of these forms of attack with higher and lower echelon plans and operations to prevent inadvertently placing another unit at risk.

2-436. The principal difference between these forms of attack is that in a feint the commander assigns the force an objective limited in size, scope, or some other measure. Forces conducting a feint make direct fire contact with the enemy but avoid decisive engagement. Forces conducting a demonstration do not seek contact with the enemy. The planning, preparing, and executing considerations for demonstrations and feints are the same as for the other forms of attack. Demonstrations support a BCT or higher-level units' plan; the battalion will not conduct demonstrations alone. Demonstrations must be clearly visible to the enemy without being transparently deceptive in nature. Demonstration forces use fires, movement of maneuver forces, smoke, electronic warfare assets, and communication equipment to support the deception plan. Planning considerations for a demonstration include the following:

- Establish a limit of advance for demonstration forces that allows the enemy to see the demonstration but not to engage it effectively with direct fires.
- Establish other security measures necessary to prevent engagement by the enemy.
- Employ demonstrations to reinforce the enemy's expectations and contribute to the decisive operation.
- Develop contingency plans for enemy contact and avoiding becoming decisively engaged.
- Issue clear follow-on missions to the demonstration force.
- Establish the means to determine the effectiveness of the demonstration and assess its effect on the enemy.

2-437. Force conducting a feint must be of sufficient strength and composition to cause the desired enemy reaction. Feints must appear real; therefore, some contact with the enemy is necessary. The feint is most effective under the following conditions:

- When it reinforces the enemy's expectations.
- When the enemy perceives it as a definite threat.

- When the enemy consistently has committed a large reserve early.
- When the attacker has several feasible course of actions.

2-438. The purposes of a feint may include the following:

- To force the enemy to employ their reserves away from the decisive operation.
- To force the enemy to remain in position.
- To attract enemy supporting fires away from the decisive operation.
- To force the enemy to reveal defensive fires or weaknesses.
- To accustom the enemy to shallow attacks in order to gain surprise with another attack.

2-439. Planning considerations for a feint include the following:

- Resource the feint so it looks like the decisive operation or at least like a significant threat.
- Establish clear guidance regarding force preservation.
- Ensure adequate means of detecting the desired enemy reaction.
- Designate clear disengagement criteria for the feinting force.
- Assign attainable objectives.
- Issue clear follow-on missions to the feinting force.

RAID

2-440. A *raid* is an operation to temporarily seize an area to secure information, confuse an adversary (Army uses the term enemy in place of adversary), capture personnel or equipment, or to destroy a capability culminating in a planned withdrawal (JP 3-0). Raids are usually small scale, involving battalion-sized or smaller forces, surprise attacks requiring detailed intelligence, planning, and preparation. Planners require precise, time-sensitive, all-source intelligence. The planning process determines how mission command, target acquisition and target servicing, and sustainment will occur during the raid. The key elements in determining the level of detail and the opportunities for rehearsal before mission execution are time, operations security, and military deception requirements.

2-441. Raids normally are conducted in five phases: In the first phase, the raiding force inserts or infiltrates into the objective area. In the second phase, the objective area is sealed off from outside support or reinforcement, to include enemy air assets. In phase three any enemy forces at or near the objective are overcome in a violently executed surprise attack using all available firepower for shock effect. In phase four, the force seizes the objective and accomplishes its assigned task quickly before any surviving enemy in the objective area can recover or be reinforced. In phase five, the raiding force extracts or exfiltrates from the objective area, usually using a different route than what was used for movement to the objective.

Note. Operations designed to rescue and recover individuals and equipment in danger of capture are normally conducted as raids.

2-442. A simplified raid chain of command is an essential organizational requirement. A raid usually requires a force carefully tailored to neutralize specific enemy forces operating vicinity of the objective and to perform whatever additional functions are required to accomplish the objective of the raid. The task organization of a raiding force based on the general objective for a raid normally consists of four base elements. Base elements within the raiding force include support, assault, breach, and security forces or some modification of these elements, see FM 3-21-10 and appendix D respectively. The raiding force operates within or outside the battalion's supporting range and moves to its objective by land, air, or water for a quick, violent attack.

2-443. The commander and staff develop as many alternative course of actions as time and the situation permit and carefully weigh each alternative. Techniques and procedures for conducting operations across the forward line of own troops are also developed, given the specific mission variables of METT-TC expected to exist during the conduct of the raid. In addition to those planning considerations associated with other offensive actions, the commander and staff must determine the risks associated with conducting the mission and possible repercussions. Typical raid missions are—

- Capture prisoners, installations, or enemy materiel.
- Destroy enemy materiel or installations.
- Obtain specific information on an enemy unit such as its location, disposition, strength, or operating scheme.
- Deceive or harass enemy forces.
- Liberate captured friendly personnel.

2-444. Reconnaissance and constant surveillance of the raid objective, external to and within the capability of the raid force, provide the information and intelligence required for planning and updates to the enemy situation on the objective. Fire support systems are positioned to provide immediate responsive fires during the approach, actions on the objective, and withdrawal. Interdiction fires, deception fires, counterfires, and situational obstacles reduce the enemy's ability to react to the raid. The following are additional planning considerations:

- Security in all directions throughout the raid.
- Clear abort criteria for the raid based on the CCIRs. Criteria may include loss of personnel, equipment, or support assets, and changes in the enemy situation.
- Contingency plans for contact before and after actions on the objective.
- Casualty evacuation and raiding force extraction throughout the entire depth of the operation.
- Rally points for units to assemble during movement, to prepare for the attack, and to assemble after the mission is complete and when the force is ready to withdraw.

2-445. Logistical considerations include the type and number of weapons that the raid force will have, movement distance, length of time the raid force will operate in enemy territory, and expected enemy resistance. Aircraft and ground linkup procedures for casualty evacuation or resupply, if required, en route, on the objective, and during the withdrawal.

SPOILING ATTACK

2-446. A *spoiling attack* is a tactical maneuver employed to seriously impair a hostile attack while the enemy is in the process of forming or assembling for an attack (FM 3-90-1). The objective of a spoiling attack is to disrupt the enemy's offensive capabilities and timelines while destroying targeted enemy personnel and equipment, not to seize terrain and other physical objectives. A commander conducts a spoiling attack to—

- Disrupt the enemy's offensive preparations.
- Destroy key assets that the enemy requires to attack, such as fire support systems, fuel and ammunition stocks, and bridging equipment.
- Gain additional time for the defending force to prepare its positions.
- Reduce the enemy's current advantage in the correlation of forces.

2-447. A commander conducts a spoiling attack whenever possible during the conduct of friendly defensive tasks to strike an enemy force while it is in assembly areas or attack positions preparing for its own offensive operation or is stopped temporarily. The commander synchronizes the conduct of the spoiling attack with other defensive actions. A spoiling attack usually employs armored, attack helicopter, or fire support elements to attack enemy assembly positions in front of the friendly commander's main line of resistance or battle positions. (Refer to FM 3-90-1 and FM 3-96 for additional information.)

SECTION IV – EXPLOITATION AND PURSUIT

2-448. During the offense, combined arms maneuver involves taking the fight to the enemy and never allowing enemy forces to recover from the initial shock of the attack. *Exploitation*, which is an offensive task that usually follows a successful attack and is designed to disorganize the enemy in-depth (ADRP 3-90) and *pursuit*, which is an offensive task designed to catch or cut off a hostile force attempting to escape, with the aim of destroying it (ADRP 3-90). Often involve pushing all available forces to the limit of their endurance to capitalize on momentum and retain the initiative. Commanders maintain momentum by anticipating and transitioning rapidly as the situation develops. Retaining the initiative pressures enemy commanders into abandoning their preferred course of actions, accepting too much risk, or making costly mistakes. As these conditions occur, friendly forces seize opportunities and create new avenues for exploitation or pursuit to break the enemy's will through relentless and continuous pressure.

EXPLOITATION

2-449. Exploitation is the primary means of translating tactical success into operational advantage. Exploitation can occur regardless of the operational theme or point along the range of operations in which the exploitation occurs. All units, regardless of their size, conduct exploitation, although discussions tend to focus on the activities of large units during conduct of major operations. Small tactical units also conduct exploitations. For example, during counterinsurgency operations, an Infantry battalion could conduct a company level raid on a particular civilian residence during the night to exploit the information and intelligence gathered during the conduct of a company level cordon and search operation that occurred earlier in the day. In this example, effective search procedures, tactical site exploitation, tactical questioning, and the use of R&S assets are keys to the company being able to conduct exploitation.

2-450. During the conduct of major operations, exploitation often follows a successful attack to take advantage of a weakened or collapsed enemy. The purpose of exploitation can vary, but it generally focuses on capitalizing on a temporary advantage or preventing the enemy from establishing an organized defense or conducting an orderly withdrawal. To accomplish this, the brigade combat team or higher-level unit attacks rapidly over a broad front to prevent the enemy from establishing a defense, organizing an effective rear guard, withdrawing, or regaining balance. The Infantry battalion as part of the IBCT secures objectives, severs escape routes, and destroys enemy forces. Failure to exploit success aggressively gives the enemy time to reconstitute an effective defense or regain the initiative by a counterattack.

2-451. The conditions for exploitation develop quickly. Often the lead battalion or subordinate unit in contact identifies the collapse of the enemy's resistance. The higher-level commander must receive accurate assessments and reports of the enemy situation to capitalize on the opportunity for exploitation. Typical indications of good conditions for exploitation include—

- A significant increase in enemy prisoners of war.
- An increase in abandoned enemy equipment and material.
- The overrunning of enemy artillery, command and control facilities, and logistics sites.
- A significant decrease in enemy resistance or in organized fires and maneuver.
- A mixture of support and combat vehicles in formations and columns.
- An increase in enemy movement rearward, especially of reserves and fire support units.

2-452. Should the Infantry battalion conduct exploitation as part of a larger operation, it might receive the mission to seize a terrain-oriented objective. In this case, the battalion avoids decisive engagement and moves to the objective as quickly as possible. If assigned a force-oriented objective, the battalion seeks and destroys enemy forces anywhere within its area of operation. Air assaults by the battalion or part of the battalion are an effective method to seize blocking positions in the enemy's rear. The exploitation ends when the enemy reestablishes its defense, all organized enemy resistance breaks down, or the friendly force culminates logistically or physically. (Refer to FM 3-90-1 and FM 3-96 for additional information.)

PURSUIT

2-453. A pursuit differs from the exploitation in that it always focuses on completing the destruction of fleeing enemy forces by destroying their ability and will to resist. Unlike an exploitation, which may focus on seizing key or decisive terrain instead of the enemy force, pursuit operations begin when an enemy force attempts to conduct retrograde operations. At that point, it becomes most vulnerable to the loss of internal cohesion and complete destruction. An aggressively executed pursuit leaves the enemy trapped, unprepared, and unable to defend, faced with the options of surrendering or complete destruction.

2-454. Pursuits include the rapid shifting of units, continuous day and night movements, hasty attacks, containment of bypassed enemy forces, large numbers of prisoners, and a willingness to forego some synchronization to maintain contact with and pressure on a fleeing enemy. Pursuits require swift maneuver and attacks by forces to strike the enemy's most vulnerable areas. A successful pursuit requires flexible forces, initiative by commanders at all echelons and a high tempo during execution.

2-455. Two options exist when conducting a pursuit. Both pursuit options involve assigning a subordinate the mission of maintaining direct-pressure on the rearward moving enemy force. The first option is a frontal pursuit that employs only direct-pressure. The second is a combination that uses one subordinate element to maintain direct-pressure and one or more other subordinate elements to encircle the retrograding enemy. Thc combination pursuit is more effective, generally. The subordinate applying direct-pressure or the subordinate conducting the encirclement can conduct the decisive operation in a combination pursuit.

2-456. During the pursuit, the commander exerts unrelenting pressure to keep the enemy force from reorganizing and preparing its defenses. The Infantry battalion or brigade combat team may be a part of a corps or division pursuit, either functioning as the direct-pressure or encircling force. Although the Infantry battalion may pursue a physical objective, the mission is the destruction of the enemy's main force.

2-457. A mobility advantage over the enemy is vital to the effectiveness of the pursuit. A combination of Armored and Stryker forces, combined with Infantry conducting air assaults, can be extremely effective when cutting off the enemy forcing them to either surrender or be destroyed. The range, speed, and weapons load of attack reconnaissance units makes them uniquely useful in an exploitation or pursuit to extend the ground commander's reach. Dismounted movement over difficult terrain allows infantry units to seize blocking positions. (Refer to FM 3-90-1 and FM 3-96 for additional information.)

Chapter 3

Defense

The Infantry battalion conducts defensive tasks to defeat enemy attacks, gain time, control key terrain, protect critical infrastructure, secure the population, and economize forces. Most importantly, the battalion sets conditions to transition to the offense or operations focused on stability. Defensive tasks alone are not decisive unless combined with offensive tasks to surprise the enemy, attack enemy weaknesses, and pursue or exploit enemy vulnerabilities. Even within the conduct of the Infantry brigade combat team (IBCT) defense, the Infantry battalion exploits opportunities to conduct offensive actions within its area of operation to deprive the enemy of the initiative, and create the conditions to assume the offense. The chapter discusses the doctrinal basis for the defense and introduces a fictional scenario as a discussion vehicle for illustrating one of many ways that an Infantry battalion conducts the defensive element of decisive action. The scenario focuses on potential challenges confronting the Infantry battalion commander in accomplishing a mission but is not intended to be prescriptive of how the Infantry battalion performs any particular operation.

SECTION I – DOCTRINAL BASIS FOR THE DEFENSE

3-1. As with the offense, defensive techniques cannot be discussed in isolation. There must be a seamless continuity and understanding between the fundamental doctrinal principles, tactics, and procedures, covered in Army doctrinal reference publications and field manuals, and the techniques covered in this Army techniques publication. This section briefly discusses defensive tasks, forms of the defense, and supporting doctrinal terms. The reader should refer to FM 3-96, FM 3-90-1, FM 3-90-2, and ADRP 3-90 for additional information.

DEFENSIVE TASKS

3-2. *Defensive tasks* are tasks conducted to defeat an enemy attack, gain time, economize forces, and develop conditions favorable for offensive or stability tasks (ADRP 3-0). There are three basic defensive tasks—area defense, mobile defense, and retrograde. These three tasks have significantly different concepts and pose significantly different problems. Each defensive task is dealt with differently when planning and executing the defense. Although, the names of these defensive tasks convey the overall aim of a selected defense, each typically contains elements of the other and combines static and mobile elements. As with offensive tasks, defensive tasks can result in non-physical effects, such as those generated in the information environment. For example, the use of deception in support of operations security can be highly effective at gaining time and tactical deception can support the economization of forces.

CHARACTERISTICS OF THE DEFENSE

3-3. Characteristics of the defense include disruption, flexibility, maneuver, mass and concentration, operations in-depth, preparation, and security. See ADRP 3-90 for a detailed discussion of each characteristic. These characteristics are defined below:

- Disruption. The defender disrupts enemy tempo and synchronization, ability to mass fires, reconnaissance and security forces, and main body formation.
- Flexibility. The defense requires preparation in-depth, use of reserves, the ability to shift the battalion's main effort, supplementary positions within the defense, and the ability to counterattack.
- Maneuver. Maneuver allows the commander to achieve a position of advantage over the enemy, mass and concentrate combat power, and to take full advantage of terrain.

- Mass and concentration. The defender shapes and decides the engagement by massing the effects of combat power in time and space and accepting risk in some areas to mass effects elsewhere.
- Operations in-depth. Simultaneous application of combat power throughout the depth of the defender's area of operation allows for the destruction of the enemy with attacks to its flanks, as that enemy force is most exposed and vulnerable.
- Preparation. Preparation, an inherent strength of the defense, provides the defender time to study the ground and select positions that allow the massing of fires on likely approaches. Defenders use available time to combine natural and manmade obstacles to canalize attacking forces into engagement areas, coordinate and rehearse actions on the ground, gaining intimate familiarity with the terrain, place security, intelligence, and reconnaissance forces throughout the area of operations, and continue defensive preparations in-depth, even as the close engagement begins.
- Security. Measures taken to protect the defender against all acts designed to, or which may, impair the defender's effectiveness to deceive the enemy as to friendly locations, strengths, and weaknesses, inhibit or defeat enemy reconnaissance, provide early warning, or to disrupt enemy attacks early and continuously.

PLANNING CONSIDERATIONS

3-4. The product of planning is an order—a directive for future action. The commander begins with an assigned area of operation, identified mission, and available forces. The commander develops and issues planning guidance based on their visualization in terms of the means to accomplish the mission. The commander controls the defense by using control measures to provide the flexibility needed to respond to changes in the situation, and to allow the commander to concentrate combat power rapidly at the decisive point. The six-warfighting functions are the framework for discussing planning considerations that apply to all primary and subordinate defensive tasks. (See FM 3-96 and FM 3-90-1 for a detailed discussion of defensive planning considerations and control measures.)

FORMS OF THE DEFENSE

3-5. The three subordinate forms of the defense (defense of a linear obstacle, perimeter defense, and reverse-slope defense) have special purposes and unique consideration associated with each. When conducting a subordinate form of the defense, proper evaluation and organization of the battalion's area of operation are essential to maximize the effectiveness of the defending force. The battalion commander exploits the advantages of occupying the terrain where the battle will occur and positions the battalion to engage the attacker from locations that give the defending force an advantage. These locations may include defiles, rivers, thick woods, swamps, cliffs, canals, built-up areas, and reverse slopes. In all three forms, the commander uses existing and reinforcing obstacles and other key terrain to impede the enemy's movement. The commander selects terrain that allows massing friendly fires but forces the enemy to commit forces piecemeal into friendly engagement areas, exposing portions of the enemy force for destruction without giving up the advantages of fighting from protected positions. The three forms of the defense provide distinct advantages to the battalion and its subordinate units during an area defense and the operations of the fixing force during a mobile defense. (Refer to FM 3-90-1 for additional information.)

DEFENSE OF A LINEAR OBSTACLE

3-6. The defense of a linear obstacle is similar to a forward defense (see paragraphs 3-193) with the intent being to limit the terrain over which the enemy can gain influence or control. A linear obstacle adds to the strength of the defense and can be a river, a stream with steep embankments or a manmade obstacle such as a highway or embankment. The key to success in a defense of a linear obstacle is maintaining the integrity of the defense by preventing the enemy from securing a foothold on the friendly side of the obstacle. When the enemy is able to gain and maintain a foothold, the battalion must contain it and prevent its expansion. The battalion commander should have a plan to conduct a delay if the enemy gains sufficient strength to attack out of the bridgehead. Defending units integrate additional obstacles to stop enemy forces, channel them into planned engagement areas, and to further, enable the integrity of the linear obstacle. The defense of a linear obstacle usually forces the enemy to deploy, concentrate forces, and conduct breaching operations (figure 3-1). When attacked, the defending force isolates the enemy, conducts counterattacks, and delivers fires onto the concentrated force to defeat attempts to breach the obstacle.

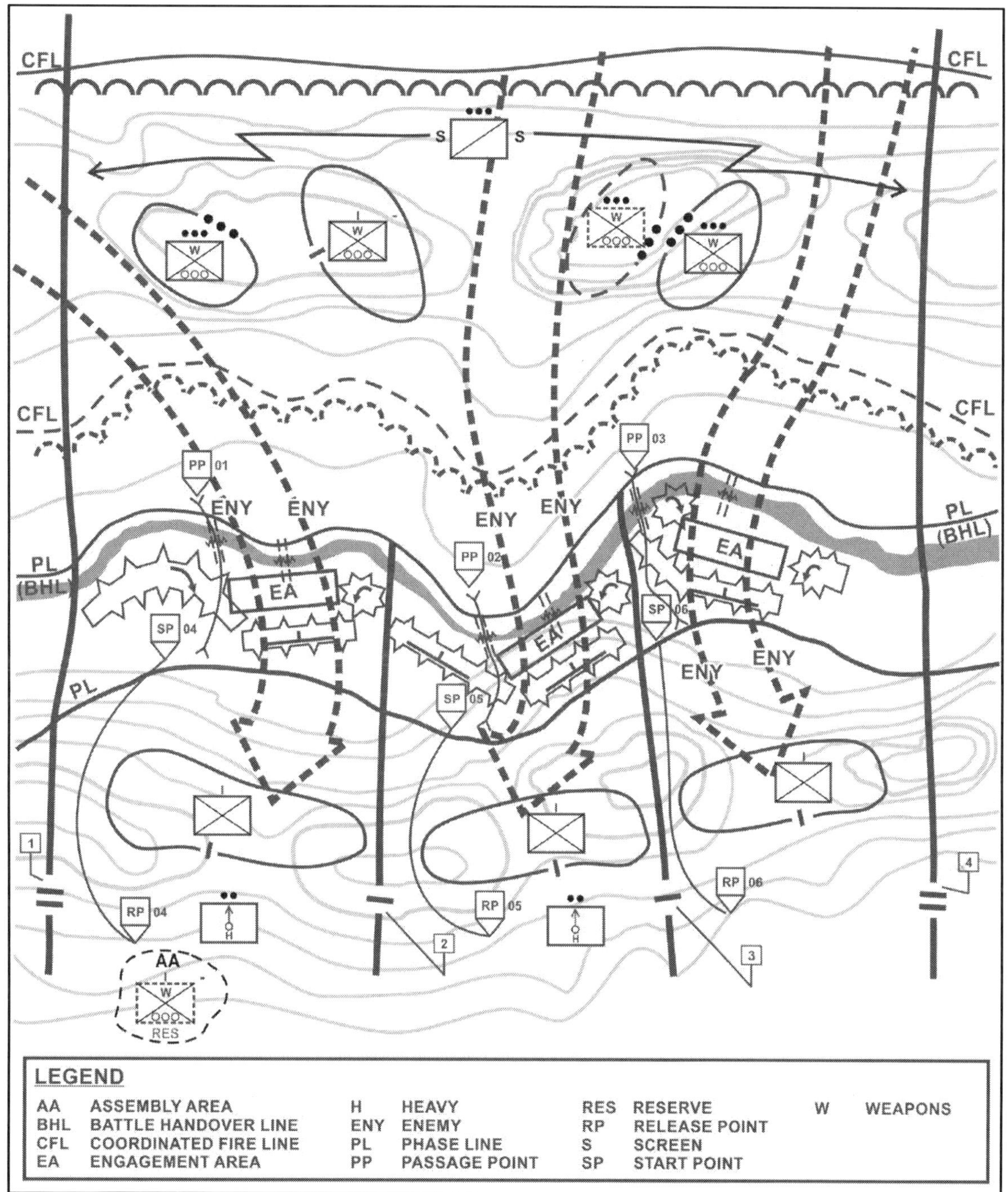

Figure 3-1. Defense of a linear obstacle (mutual support between rifle companies)

3-7. The main purpose of the defense of linear obstacle, as with any defense, is to force or deceive the enemy into attacking under unfavorable circumstances. The defending commander seeks to dictate where the fight will occur, preparing the terrain and other conditions to their advantage while simultaneously denying the enemy adequate intelligence on the battalion's defense. During the planning process, the commander uses intelligence products to identify probable enemy objectives and approaches. From those probable objectives and approaches, named areas of interest and target areas of interest are developed. The commander considers the mission variables of METT-TC to determine how best to concentrate efforts and economize forces. A detailed terrain analysis might be the most important process that the commander and staff complete. A successful defense relies on a complete understanding of terrain in order to determine likely enemy course of actions and the best positioning of battalion assets to counter them.

3-8. During preparation, the commander and staff monitor preparatory actions, and track higher and adjacent unit situations and the enemy situation. Commander and staff update and refine plans based on additional reconnaissance and updated intelligence information. The staff continues to disseminate these modifications through fragmentary orders and conducts much of the preparation phase simultaneously with security operations, continuing even as forward-deployed forces gain contact with the enemy. Throughout the preparation phase, the battalion commander, company commanders, and key staff members physically inspect preparatory activities. Weapons positioning, obstacle emplacement, direct and indirect fire plans and associated triggers, sustainment operations, and Soldier knowledge of their missions are all critical checks. The battalion, in coordination with the IBCT or conducted by the IBCT or higher headquarters, conducts spoiling attacks throughout the preparation phase to disrupt the enemy's offensive preparations.

3-9. During execution, a battalion's defense of a linear obstacle often entails relatively long frontages. Based on the mission and the frontage assigned, the commander positions units that mutually support each other throughout the length and depth of the defense. Mutual support exists when positions and units support each other by direct and indirect fires to prevent the enemy from attacking one position without being subject to fire from one or more adjacent positions. Mutual support increases the strength of all defensive positions, prevents *defeat in detail*—concentrating overwhelming combat power against separate parts of a force rather than defeating the entire force at once (ADRP 3-90), and helps prevent infiltration between positions.

3-10. The commander may position all rifle companies forward to enable mutual support between companies and platoons, positioning a small reserve, such as a platoon, to block the most likely avenue of enemy approach. The commander may position security forces forward, or have rifle companies responsible for their own security and establish listening posts or outposts directly to their front. The scout platoon may be deployed forward or to the flank to provide early warning. Battalion mortars may operate as split sections occupying overwatch positions to the rear of the maneuver companies to provide depth. When employed forward in the security area, the weapons company conducts shaping operations to establish the necessary conditions for the decisive operation in the main battle area through attriting, disrupting, and delaying the enemy.

3-11. When the battalion defends a narrow frontage, the commander may deploy two rifle companies forward while retaining a rifle company in-depth or as a reserve positioned to block enemy penetrations. The weapons company is positioned in-depth. The scout platoon may be deployed forward or may secure a flank. Battalion mortars may be deployed as a platoon or in split sections to provide support fires throughout the defense. When defending in a narrow frontage or a long frontage, the IBCT commander can position a second battalion behind the battalion deployed along the obstacle allowing the forward battalion to deploy the maximum combat power forward to ensure the integrity of the obstacle.

3-12. The deployment of the weapons company and its assault platoons depends on terrain. When operating in close terrain with limited fields of fires, the weapons company may keep all or most of its assault platoons and be assigned overwatch positions. In open terrain with good fields of fire, assault platoons may be assigned to the rifle companies with the rest of the weapons company positioned as the reserve, especially if the frontage is long and the flanks are vulnerable. Battalion mortars may operate as split sections with the weapons company occupying overwatch positions to the rear of the rifle companies to provide depth.

3-13. The commander employs fires in-depth to destroy the enemy and project combat power to the enemy side of the obstacle. Forward observers may be positioned forward in observation posts and with reconnaissance forces and may be assigned to observe specific named areas of interest or target areas of interest. Forward observers within the main battle area with rifle companies and platoons also may be assigned to observe specific named areas of interest or target areas of interest. Army aviation attack and reconnaissance unit attacks may be conducted in close proximity or in direct support of the battalion in the main battle area or against enemy forces not in direct contact with the maneuver forces of the battalion. Artillery and mortar fire support plans are integrated into the direct fire plans of the forward rifle companies.

Perimeter Defense

3-14. A perimeter defense is a defense oriented in all directions. A perimeter defense by design has a secure inner area with most of the combat power located on the perimeter (figure 3-2). Perimeters vary in shape depending on the terrain and situation with the perimeter shape conforming to the terrain features that best use friendly observation and fields of fire. The commander in a perimeter defense designates the trace of the perimeter, battle positions, contact points, and lateral and forward boundaries. When the commander determines the most probable direction of enemy attack, that part of the perimeter covering that approach may be reinforced with additional resources. The commander employs patrols, raids, ambushes, air attacks, and supporting fires to harass and destroy enemy forces before they make contact with the perimeter. The commander increases the effectiveness of the perimeter by tying it into a natural obstacle, such as a river, which allows the defending unit to concentrate its combat power in more threatened areas. Normally, the reserve centrally locates to react to a penetration of the perimeter at any point.

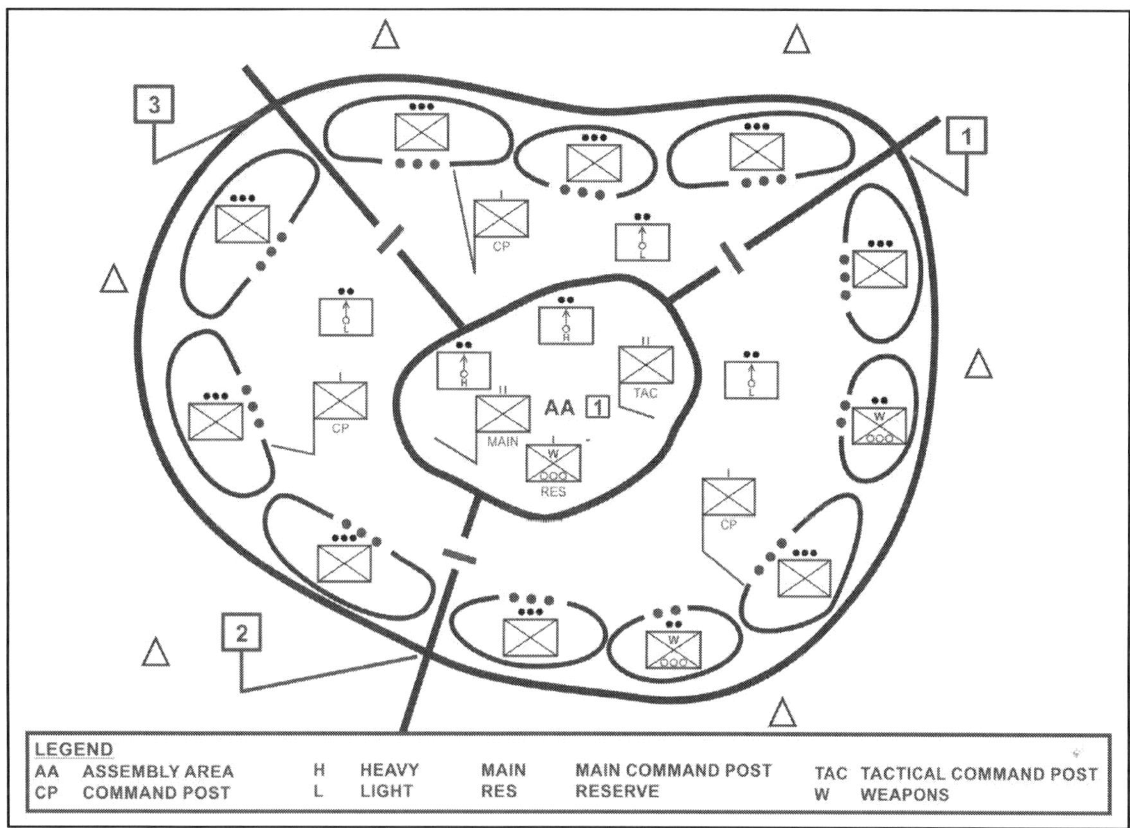

Figure 3-2. Perimeter defense (rifle companies positioned forward)

3-15. The battalion commander can employ the perimeter defense as an option when conducting an area or mobile defense. The commander establishes a perimeter defense when the battalion must hold critical terrain, such as a strong point, or when it must defend a noncontiguous area of operation where the battalion perimeter defense is not tied in with adjacent units. The battalion may also form a perimeter when conducting airborne and air assault operations, when it is bypassed and isolated by the enemy and must defend in place, or when it is securing an isolated objective, such as a bridge, mountain pass, or airfield. Unless executing a strongpoint, the perimeter defense usually does not take considerable time and resources to execute. The prerequisites for a successful perimeter defense are aggressive patrols (combat and reconnaissance) and security operations outside the perimeter. The commander designates checkpoints, contact points, passage points, and passage routes for elements operating outside the boundary of the perimeter. Forces within the perimeter can perform these activities, or another force external to the battalion can perform them.

3-16. The commander may employ all defending forces forward along the perimeter, or establish a defense in-depth within the perimeter. Figure 3-2 illustrates one of many ways the battalion can organize forces for a perimeter defense. In this illustration, the commander positions companies forward along the perimeter. The commander divided the perimeter into rifle company area of operations with boundaries and contact points. Two rifle companies, one oriented to the southeast and the other oriented to the southwest, reinforced with one assault platoon each from the weapons company establish defensive firing positions on the perimeter to cover the most likely mounted avenues of approach. The battalion reserve is centrally located to counter any penetration of the perimeter from any direction. This organization of forces reduces the possibility of fratricide and friendly fire incidents within the perimeter and maximizes combat power on the perimeter.

3-17. Figure 3-3 illustrates an organization of forces where the battalion commander constructs an outer and inner perimeter to create depth in the defense. Companies position two rifle platoons along the outer perimeter and one rifle platoon in reserve along the inner perimeter. As in the above illustration, two rifle companies, one oriented to the southeast and the other oriented to the southwest, position attached assault platoons to cover the most likely mounted avenues of approach. This organization of forces provides depth to each company's area of operations and enables each company commander to position company mortars near the reserve platoon to enhance control and security. Attacks against outer and inner perimeter elements range from long-range sniper, mortar, or rocket fire, attacks by suicide demolition squads, and attacks by major enemy ground and air forces. Within the perimeter, the battalion commander positions forces to decrease the possibility of an enemy simultaneously suppressing the inner and outer perimeter forces with the same fires.

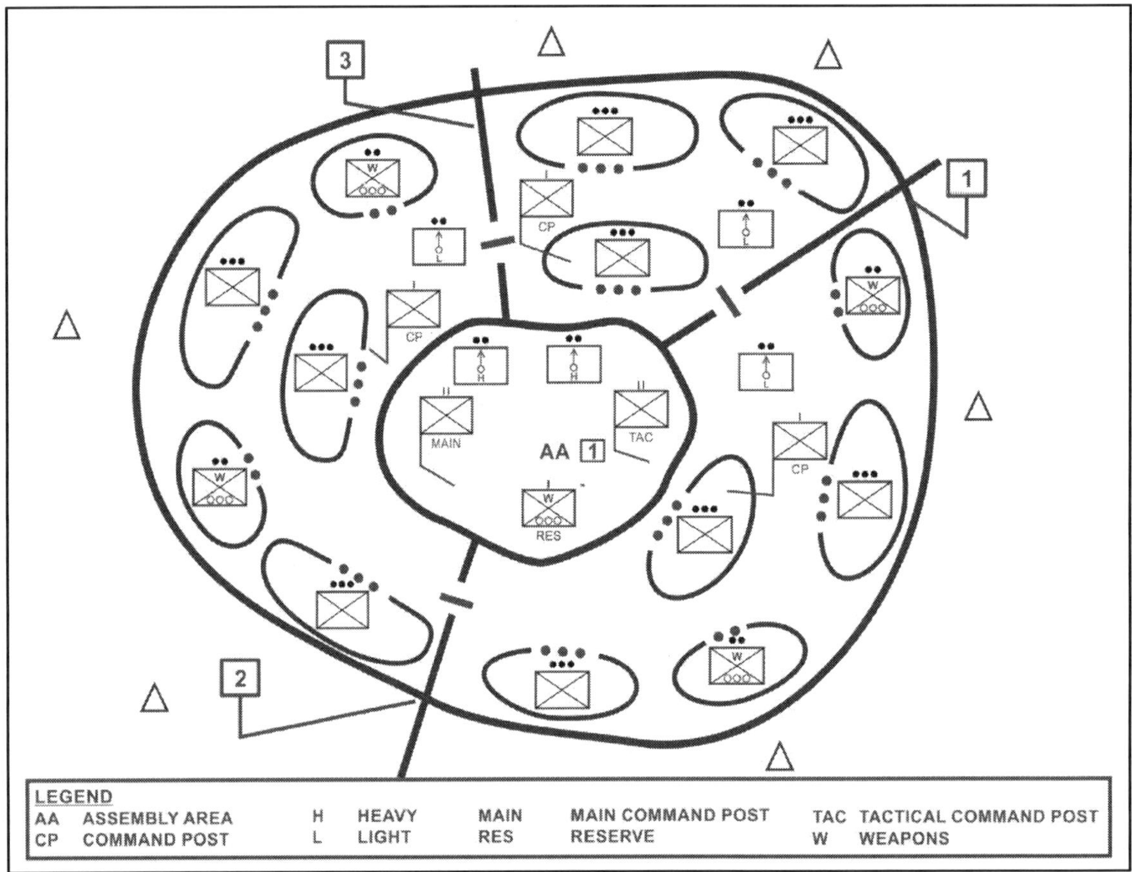

Figure 3-3. Perimeter defense (rifle platoons positioned in-depth)

3-18. Tactical positions on the outer perimeter achieve the maximum degree of mutual support when they are located to observe or monitor the ground between them, or conduct patrols to prevent enemy infiltration. In open terrain, the commander covers gaps on the outer perimeter between units with fires. The commander does not allow gaps between defensive fighting positions when the unit is in restrictive terrain. At night or during periods of limited visibility, the commander may position tactical units closer together to retain the advantages of mutual support. Defending during periods of limited visibility or nighttime conditions, subordinate unit leaders must coordinate the nature and extent of their mutual support. The ability of the attacker to create conditions of smoke—including thermal neutralizing smoke—and the smoke and dust associated with a battle also means that the defending commander must be able to rapidly modify the defense to one effective during limited visibility. In fact, the commander should assume limited visibility rather than full visibility during defensive planning. During limited visibility, the commander may—

- Move forces forward to more closely observe an obstacle.
- Use unmanned aircraft system resources to detect enemy movement.
- Emplace stay behind forces to report enemy activity.
- Plan white light and infrared illumination.

3-19. Normally, the commander employs the scout platoon and sniper teams outside the perimeter for early warning. Early warnings of pending enemy actions ensure the commander time to react to any threat. The commander may augment perimeter security with squad-sized or smaller observation posts and aerial surveillance forward of the perimeter, provided and controlled by units on the perimeter. Security forces are positioned to observe avenues of approach. Patrols cover areas that cannot be observed by stationary forces. Any security force operating outside the perimeter coordinates their passage of lines into and out of the perimeter with the appropriate perimeter units. The distance from the perimeter at which security elements operate is determined primarily by their mobility, availability of fire support, communications, and the location and mobility of the enemy.

3-20. Employment of organic and attached weapons in a perimeter defense are generally as prescribed for other defense operations. The commander uses engagement areas, target reference points, final protective fires, and principal direction of fire control measures. Figure 3-4, page 3-8, illustrates two rifle companies along the outer perimeter and one rifle company along the inner perimeter. In this example, engagement areas are established where the commander intends to contain and destroy an enemy force with the massed effects of all available weapons and supporting systems. The commander designates engagement areas to cover each enemy avenue of approach into the perimeter. The commander determines the size and shape of the engagement area by the relatively unobstructed line-of-sight from the weapon systems in their firing positions and the maximum range of those weapons.

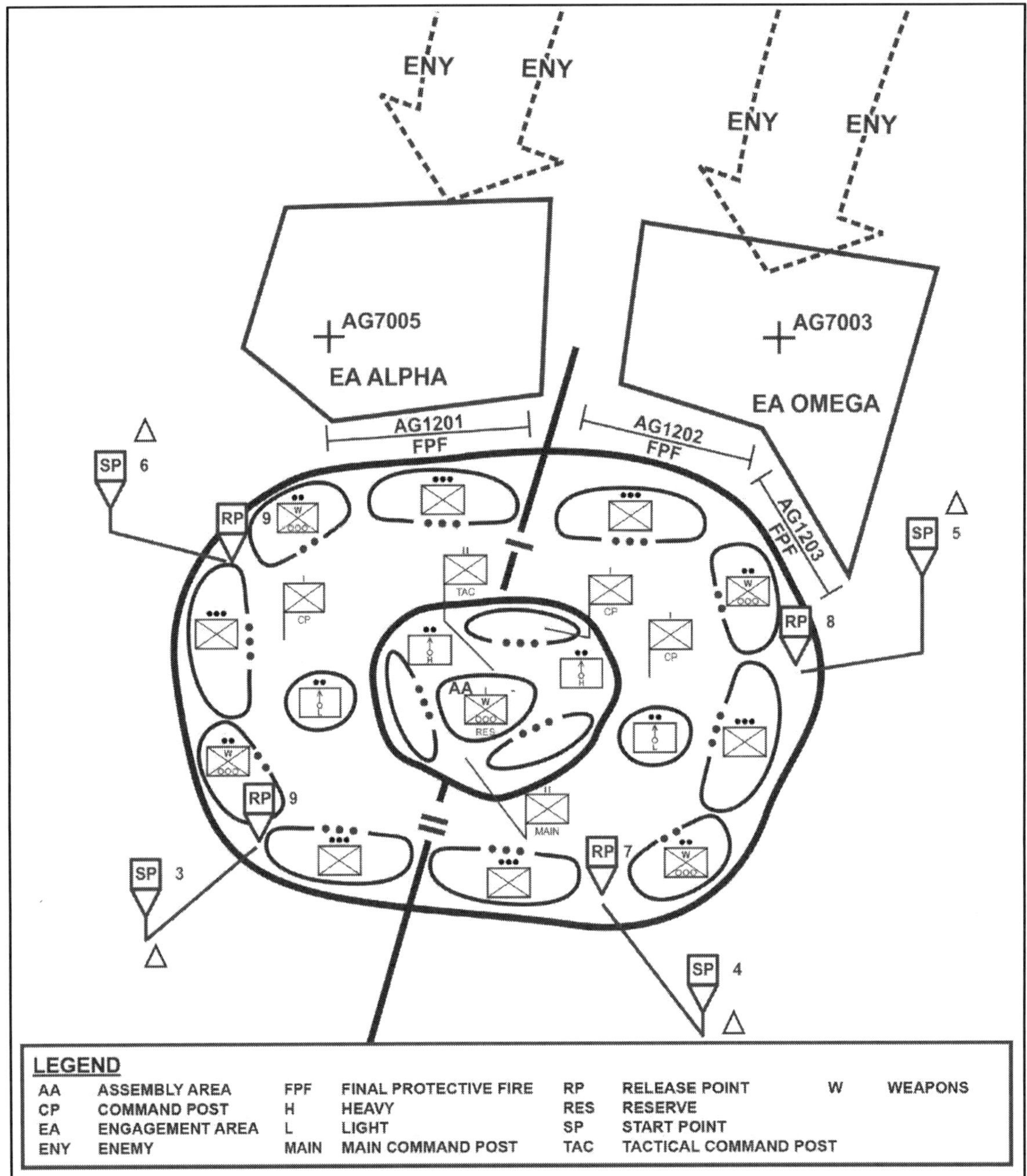

Figure 3-4. Perimeter defense (engagement area control measures)

3-21. Once engagement areas are determined, the commander arrays available forces and weapon systems in positions to concentrate overwhelming effects into these areas. The commander routinely subdivides an engagement area into smaller engagement areas for subordinates using one or more target reference points or by prominent terrain features. The commander assigns a *sector of fire*, that area assigned to a unit, crew-served weapon, or an individual weapon within which it will engage targets as they appear in accordance with established engagement priorities (FM 3-90-1), to subordinates within each engagement area. Indirect fires engage the enemy as far forward of the perimeter as possible and may support the battalion from within or outside the perimeter. Available fires from outside the perimeter are coordinated and integrated into the overall defensive plan. Using fire support from outside the perimeter conserves ammunition from within the perimeter.

3-22. The commander positions the reserve to block the most dangerous avenue of approach and assigns supplementary positions to the reserve on other avenues that may become critical. Ideally, the reserve has the mobility necessary to react to enemy action in any portion of the perimeter. Assault platoons and attached tanks, initially occupying firing positions on the perimeter, may be tasked to reinforce the reserve. If the perimeter is penetrated, the reserve blocks the penetration or counterattacks to reduce the penetration and restore the perimeter. After committing the reserve, the commander must reconstitute the reserve to meet other possible threats. This reconstitution force normally comes from an unengaged unit in another portion of the perimeter. When an unengaged force is used to constitute a new reserve, sufficient forces must be retained in the vacated area to defend that portion of the perimeter.

3-23. Sustainment elements may support from within the perimeter or from another location, depending upon the mission and status of the battalion, type of transport assets available, weather, and terrain. When resupplied by air, availability of landing and drop zones protected from the enemy's observation and fire is a critical consideration in selecting and organizing the perimeter. Since aerial supply is vulnerable to weather and enemy fires, emphasis is placed on supply economy and protection of available stocks. Sustainment assets within the perimeter should locate in protected locations from which they can provide continuous support. These assets should come under the control of the battalion's unit trains. Unit trains location and operations should not restrict or be restricted by maneuver forces, fire support assets, and the reserve. When the battalion forms the perimeter because of isolation, other units may seek the battalion's protection. These units receive missions based on their support capabilities.

Reverse Slope Defense

3-24. The commander organizes a reverse slope defense on the portion of a terrain feature or slope with a topographical crest that masks the battalion's main defensive positions from enemy observation and direct fire. The proper organization provides observation across the entire front and security to the battalion's main battle area. Success is based on denying the enemy the topographical crest through occupation or by fire. The goal of this technique is to make the enemy commit forces against the forward slope of the defending battalion, causing enemy forces to attack in an uncoordinated fashion across the exposed topographical crest. Firing from covered and concealed positions throughout the battle area, the defending battalion maintains a distinct advantage over the exposed enemy forces and canalizes the enemy through unfamiliar terrain into planned engagement areas (figure 3-5, page 3-10).

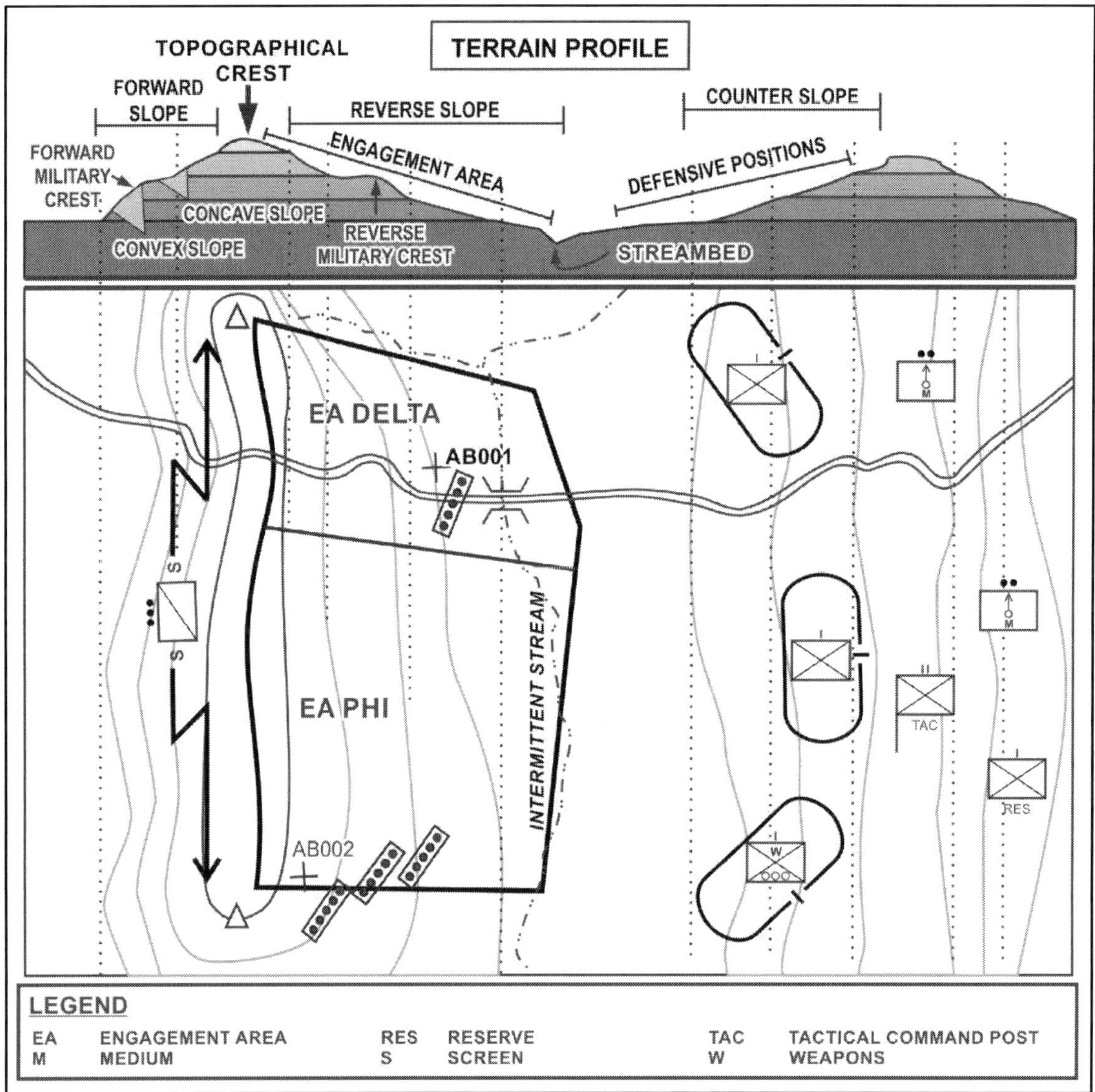

Figure 3-5. Reverse slope defense (slope terminology)

3-25. All or part of the defending battalion may employ a reverse-slope defense. This technique allows subordinate units of the battalion to concentrate direct fires into a relatively small area while being protected from the enemy's direct observation and supporting fires. The battalion destroys the enemy's isolated forward units through surprise and concentrated fires. The control of the forward slope is essential for success. Gaining control of the forward slope can be accomplished by using dominating terrain behind the defender or by using stay behind forces, such as scout and sniper teams, that can observe and call for fire on the attacker. A reverse-slope defense is generally conducted at battalion level and below within an area of operation that is conducive to the use of a reverse-slope defense. The commander may choose to conduct a reverse slope defense when—

- Enemy fire makes the forward slope untenable.
- The lack of cover and concealment on the forward slope makes it untenable.
- The forward slope has been lost or has not yet been gained.
- The forward slope is exposed to enemy direct fire weapons fired from beyond the effective range of the defender's weapons.
- Moving to the reverse slope removes the attacker's standoff advantage.
- The terrain on the reverse slope affords better fields of fire than the forward slope.

- The defender wants to avoid creating a dangerous salient (an outwardly projecting part of a line of defense).
- The commander is forced to assume a hasty defense while in contact with or near the enemy.

3-26. As this technique positions the bulk of the battalion's defense on the far side of any elevated terrain (a hill, ridge, or mountain) on the side opposite from the attacking force. The reverse slope's principal military significance lies in the fact that its position places it outside the reach of the enemy's direct observation and direct-fire weapons. In addition to hindering both the attacker's ability to observe the defender's positions as well as reducing the effectiveness of the attacker's long-range weapons such as tanks and artillery. This technique may even succeed in deceiving the enemy as to the true location and organization of the battalion's main defensive positions. Ideally, as the attacker advances and passes over the top of the hilltop or ridgeline, the attacker is engaged (ambushed) by defensive short-range fire from the battalion. Engagement areas and obstacles are generally positioned on the reverse slope with the topographical crest normally marking the far edge of the engagement area. The battalion dominates the topographical crest by fires to prevent the enemy from successfully engaging it. The forward edge of the position should be within small--arms range of the crest. The forward edge should be far enough from the crest that fields of fire allow the battalion time to place well-aimed fire on the enemy before he reaches friendly positions.

3-27. To succeed, the battalion commander prevents the attacker from conducting a detailed observation of the reverse-slope defense through the deployment of aggressive patrols (combat and reconnaissance) and security operations forward of the topographical crest. The battalion establishes observation posts and combat outposts on, or forward of the topographical crest. This allows long-range observation over the entire front, and indirect fire coverage of forward obstacles. Observation posts or combat outposts may vary in size from a few Soldiers to a reinforced platoon and should include forward observers. During limited visibility, the numbers may increase to improve security. Depending upon specific missions, security forces forward of the topographical crest, conduct patrols [both combat and reconnaissance (to include surveillance)], and security operations to engage enemy reconnaissance and main body forces. Security forces position forward to stop or delay the enemy, disorganize the attack, and deceive the enemy as to the location of battalion's main defensive positions. As security forces withdraw, the battalion may position stay behind forces or use unmanned aircraft systems to maintain observation, fire support, and security to the front. Once withdrawn, security force may either reinforce the reserve or move into a battle position for the main engagement. During security force operations, the Infantry weapons company can provide mobility, optics, and long-range direct fire capabilities to the battalion.

3-28. When defending in a valley or depression, the commander places over-watching elements forward of the topographic crest and on the flanks to protect the main defensive positions of the battalion. This is especially desirable when over-watching elements can observe and place fires on the crest and forward slope. Over-watching elements maintain observation and fires over the entire forward slope as long as possible to destroy enemy forces, thus preventing the enemy from massing for a final assault. In the battalion's main battle area on a counter slope (also known as the reverse forward slope), fires must cover the area immediately in front of the reverse slope positions to the topographical crest. The commander organizes defensive positions to permit fires on enemy approaches around and over the crest and on the forward slopes of adjacent terrain features, if applicable. The battalion's fire support plan, to destroy, disrupt, and attrit enemy forces on the forward slope, prevents the enemy's occupation and use of the topographical crest. Key factors affecting the organization of these areas are mutually supporting covered and concealed positions, numerous existing and reinforcing obstacles, the ability to bring devastating fires from all available weapons onto the crest, and a counterattack force. The counterattack plan specifies measures necessary to clear the crest or regain it from the enemy control. Depending on the terrain, the most desirable location for the reserve may be on the counterslope or the reverse military crest of the counterslope.

3-29. Another variation available to the commander, when line of sight restrictions exist to a unit's direct front, is to organize a system of reverse slope defenses firing to the oblique defilade, each covering the other (figure 3-6, page 3-12). In this example, battalion main battle area positions were unable to engage targets directly to their front, but could cover each other using oblique defilade. Line of sight restrictions can be obstacles, terrain, and vegetation driven. This system of reverse slope defenses protects defenders from enemy frontal and flanking fires and from fires coming from above the main defensive area.

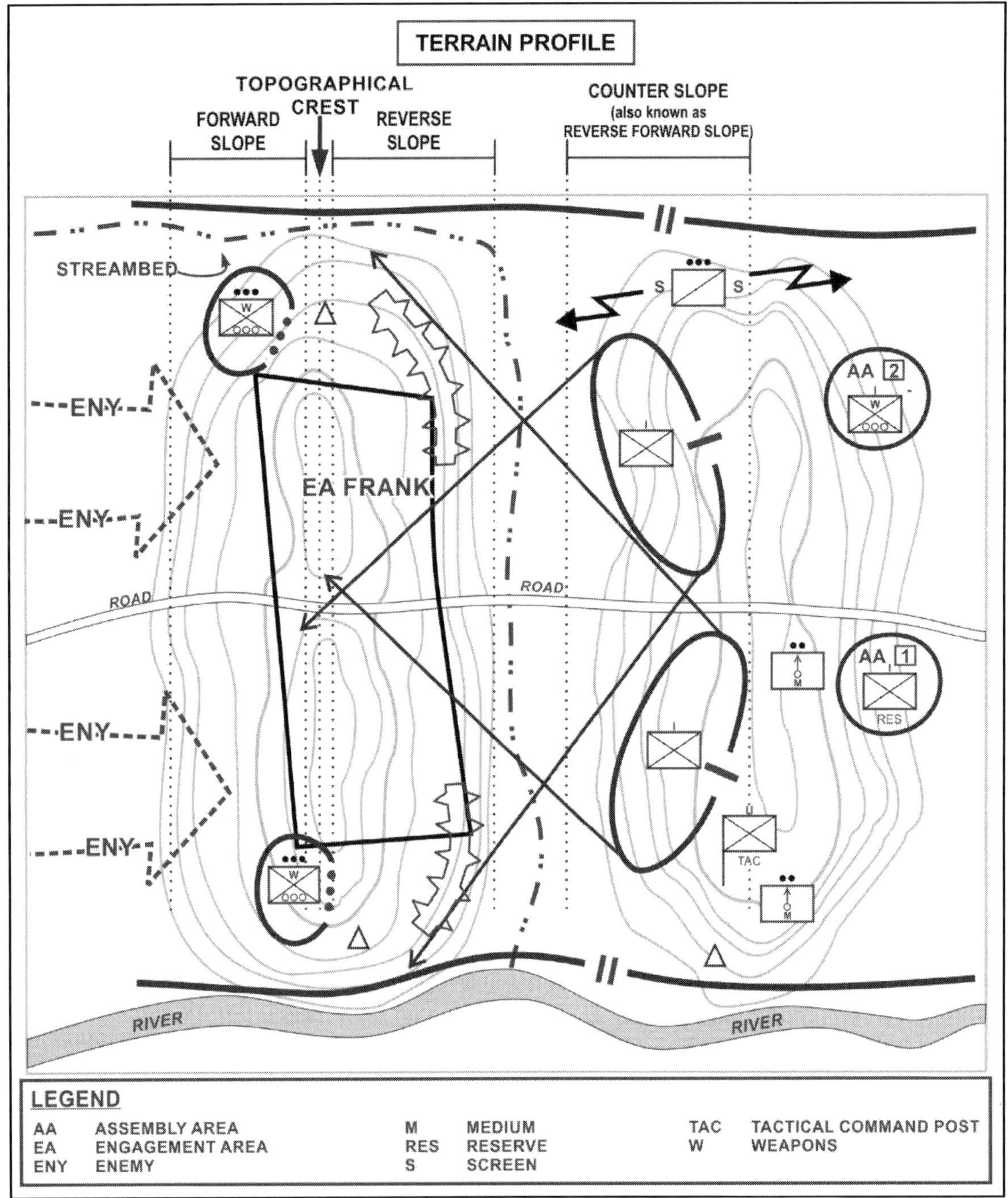

Figure 3-6. Reverse slope defense (oblique defilade)

3-30. The defending commander's major advantage over the attacker, as in any defense, is the ability to select the ground on which the battle takes place. The commander positions subordinate forces in mutually supporting positions in-depth to absorb enemy penetrations or canalize them into prepared engagement areas, defeating the enemy's attack by concentrating the effects of overwhelming combat power. Defending forces have the advantage of preparing the terrain by reinforcing natural obstacles, fortifying positions, and rehearsing operations. First preparing the ground to force the piecemeal commitment of enemy forces and their subsequent defeat in detail. Then preparing the ground to force the enemy to fight where the enemy does not want to fight, such as in open areas dominated by terrain that offers adequate cover and concealment for the occupying defending force. Advantages specific to a reverse slope defense include the following:

- Masked enemy ground observation of the main battle area.
- Degraded observation capabilities of most surveillance devices and radar.
- Ineffective enemy direct fire on main defensive positions without coming within range of the defender.
- Enemy exposure when the enemy masses the effects of direct fire weapons.
- Enemy's inability to identify obstacles on the reverse slope.
- Enemy's inability to determine the strength and location of main defensive positions.
- Enemy's inability to observe fires in main battle area.
- Tactical surprise over the enemy gained by the defender.
- Freedom of movement within the battle area due to the lack of enemy ground observation.
- Cover and concealment (topographical crest) to enable preparation of the defense.

3-31. As the reverse slope defense pursues offensive opportunities through surprise and deceptive actions. Once a reverse slope defense is employed successfully to halt an enemy attack, it may have limited further value because the effect of surprise will be difficult to attain. Disadvantages to a reverse slope defense include the following:

- Observation of the enemy may be limited, and the defender may be unable to cover obstacles on the forward slope by direct fire.
- The topographical crest may limit the range of important direct-fire weapons. These weapons may have to locate separately from the battalion's main battle area to exploit their range.
- The enemy holds the high ground in an attack and attacks downhill; any counterattack by the defending force is uphill.
- Because the reverse military crest must be controlled, the effectiveness of the reverse slope defense is reduced during limited visibility.

SITUATIONS UNIQUE TO THE CONDUCT OF DEFENSIVE TASKS

3-32. During the conduct of defensive tasks, situations requiring denial operations, defending encircled, stay-behind operations, and relief in place have their own unique planning, preparation, and execution considerations. In the defense, denial operations conducted to deprive the enemy of some or all of the short-term benefits of capturing an area may be required. In other defensive situations, the Infantry battalion may become encircled or be directed to conduct operations as a stay behind force or to conduct a relief in place. These actions may be planned or forced by the enemy. In addition to the actions addressed above, this section discusses linkup between friendly ground forces specific to these actions and selected forms of maneuver specific to the offense.

DENIAL OPERATIONS

3-33. *Denial operations* are actions to hinder or deny the enemy the use of space, personnel, supplies, or facilities (FM 3-90-1). This may include destroying, removing, and contaminating those supplies and facilities or erecting obstacles. The commander designs denial operations to deprive the enemy of some or all of the short-term benefits of capturing an area. Denial operations differ from countermobility operations that use or enhance the effects of natural and man-made obstacles to deny the enemy freedom of movement and maneuver. The principles of denial include the following:

- The commander denies the enemy the use of military equipment and supplies.
- Steps taken to deny equipment and supplies to the enemy do not if possible preclude their later use by friendly forces.
- The commander orders the destruction of military equipment and supplies only when friendly forces cannot prevent them from falling into enemy hands.
- The user is responsible for denying the enemy the use of the user's military equipment and supplies by means of its destruction, removal, or contamination.
- Deliberately destroying medical equipment and supplies and making food and water unfit for consumption is unlawful.

3-34. In denial operations, the definition of a unit's military equipment and supplies could expand to include military installations and any civilian equipment and supplies used by the friendly force. Under the law of war, the destruction of civilian property is only permitted where required by immediate military necessity. The determination of whether there is sufficient necessity to justify destruction is a complex decision that requires consideration of moral, political, and legal considerations. Additionally, civil instability increased by the destruction of civilian property, material, and equipment could have adverse effects on the outcomes of the different elements of decisive action. (Refer to FM 3-90-1 for additional information.)

Note. As stated in chapter 2, encirclement operations can be offensive or defensive in nature. Offensive encirclement operations are designed to isolate an enemy force. (See paragraph 2-7.) Defensive encirclement operations, referenced below, are the result of a unit's isolation by the actions of an enemy force. Isolation can occur due to the chaotic, intense, and highly destructive nature of combat operations. Isolation can also occur when operations extend across large areas containing relatively few units to maneuver against one another or when units operate in restrictive terrain. (Refer to FM 3-90-2 for additional information.)

DEFENDING ENCIRCLED

3-35. The Infantry battalion when encircled can continue to defend, conduct a breakout from encirclement, exfiltrate toward other friendly forces, or attack deeper into enemy-controlled territory. When defending encircled, the battalion normally establishes a perimeter on restrictive terrain, ideally controlling a choke point or other key terrain. The battalion's form of maneuver once becoming encircled depends on the commander's intent and the mission variables of METT-TC, including the—

- Availability of defensible terrain.
- Relative combat power of friendly and enemy forces.
- Sustainment status and the ability to resupply the encircled force.
- Ability to treat and evacuate wounded Soldiers.
- Morale and fighting capacity of the Soldiers.

3-36. The Infantry battalion or subordinate unit may find itself encircled as a result of offensive actions, as a detachment left in contact (commonly referred to as a DLIC), when defending a strong point, when occupying a combat outpost, or when defending an isolated defensive position. The commander and subordinate leaders anticipate becoming encircled when assigned a stay-behind force mission, or when occupying either a strong point or a combat outpost. When defending encircled the enemy will normally attempt to split the defense of the encircled force and defeat it in detail. When encircled, defensive positions within the battalion are made as strong as possible given time and resource constraints.

3-37. During encirclement, the battalion commander (or the senior commander within the encircled force) assumes command over all encircled forces and takes immediate action to protect them. The commander reestablishes unity of command and reorganizes any fragmented units and places Soldiers separated from their parent units under the control of other units. When the commander determines the unit is about to be encircled, the commander must decide quickly what assets stay and what assets leave. The commander immediately informs higher headquarters (when capable) of the situation and directs the accomplishment of the following tasks:

- Establish security.
- Reestablish a chain of command.
- Establish a viable defense.
- Maintain morale.

3-38. In most cases, the Infantry battalion establishes a perimeter defense when faced with encirclement by an enemy force. The commander maximizes the capabilities of available forces establishing mutually supporting positions forward or in-depth, depending on the terrain, within and around the perimeter along principal enemy avenues of approach. Units occupy the best available defensible terrain though it may be necessary to seize key or decisive terrain to incorporate into the perimeter defense. Once the commander assigns an area of operation and battle position to subordinates, basic preparations specific to priorities of work are similar to any perimeter defense though situation dependent to the actual assets available to the encircled force. (Section I of this chapter discusses the conduct of a perimeter defense, reference figures 3-2, 3-3, and 3-4, on pages 3-5, 3-6, and 3-8, respectively.)

Breakout from an Encirclement

3-39. A *breakout* is an operation conducted by an encircled force to regain freedom of movement or contact with friendly units. It differs from other attacks only in that a simultaneous defense in other areas of the perimeter must be maintained (ADRP 3-90). A breakout is both an offensive and a defensive mission.

General Considerations for a Breakout

3-40. An encircled force normally attempts to conduct breakout operations when one of the following four conditions exist:

- The commander directs the breakout or the breakout falls within the intent of a higher commander.
- The encircled force does not have sufficient relative combat power to defend itself against enemy forces attempting to reduce the encirclement.
- The encircled force does not have adequate terrain available to conduct its defense.
- The encircled force cannot sustain itself long enough to be relieved by forces outside the encirclement.

3-41. Prior to encirclement, when possible, the commander reorganizes the encircled force for breakout based on available resources. The commander then initiates a breakout attack as quickly as possible prior to or after the enemy encircles the force. While detailed combat information about the enemy's disposition is probably not available, the enemy is normally disorganized during and right after the enemy encircles the force and least likely to respond in a coordinated manner. As a minimum, a commander uses boundaries; a line of departure or line of contact; time of the attack; phase lines; axis of advance or direction of attack; objectives; and a limit of advance to control and synchronize the breakout. Forces located outside the encirclement assist the breakout when available by conducting shaping operations. Above all, the encircled force maintains the momentum of the breakout attack; otherwise, it is more vulnerable to destruction than it was before the breakout attempt.

Organization of Forces

3-42. Once the commander determines the scheme of maneuver for the breakout attack, the battalion organizes to give each force enough combat power to accomplish its mission. The commander typically organizes the encircled force to conduct rupture, follow-and-assume, main body, and rear guard missions. When sufficient forces exist in the encirclement, the commander may organize a reserve and a separate diversionary force. A breakout at battalion level will generally not have sufficient combat power to resource each of these forces. Normally, the commander's first priority is to resource the force with the rupture mission then prioritizes which forces to resource next based on commander's mission analysis.

Planning a Breakout

3-43. A rapid, simple, and well-executed plan is usually best. The Infantry battalion may not have a second chance. The commander's plan takes advantage of limited visibility, difficult terrain, and surprise and normally does not attack in the obvious direction. When the battalion conducts a breakout, sooner is usually better than later. The battalion's combat power will never be stronger and the enemy's combat power will not be weaker as the enemy retains the ability to resupply and reinforce. The encircled force is unable to resupply, replace casualties, or evacuate wounded, and faces the prospect of lowered morale. Regardless of whether the enemy plans to contain the force or annihilate the encircled force, it is imperative that the battalion take action as quickly as possible. Even when the enemy's intent is a complete encirclement, redistribution and positioning enemy forces will take time. It is during this time, before a cohesive encirclement can be established, that the breakout should be made.

3-44. Once the battalion is cut off or its ability to maneuver is severely limited, the battalion normally establishes a hasty perimeter defense. As the enemy establishes the encirclement, information collection continues not only on the encircling enemy forces, but also on all enemy elements between the encircled force and other friendly units. Enemy strength can be expected on direct routes to friendly forces and on good avenues out of the encirclement. The enemy cannot, however, be strong everywhere. The enemy's strength and mobility at any one point on the encirclement are the critical elements to be determined by the commander's information collection effort. As the commander uses deception to reinforce the enemy's perception of what the battalion will do. The battalion commander can then attack from an unexpected direction and location.

3-45. The sequence for a breakout has to remain flexible because the commander usually has limited information and limited information resources. For example if there is an unexpectedly weak response from the enemy, the commander may direct that the rupture force continue as the decisive attack instead of passing it over to the follow-and-assume force. The following organization of forces for a breakout attack is provided for decision purposes:

- The diversionary force, when organized, attacks to deceive the enemy as to the friendly commander's intention and direction of attack.
- The rupture force penetrates the enemy's defensive line and holds the shoulders of the penetration.
- The follow-and-assume force follows the rupture attack and is committed to maintain the momentum of the attack and seize objectives past the rupture.
- The main body consists of the main command post, the bulk of the unit's sustainment and support forces, and the unit's casualties and follows behind the follow-and-assume force.
- A reserve, when organized, moves to the rear of the main body and is committed as required.
- The rear guard protects the perimeter while the decisive and shaping operations are conducted, then conducts a withdrawal and follows the main body.
- Linkup operations with friendly forces completes the mission.

Note. The battalion commander positions where best to exercise command during the execution of the breakout without losing the ability to respond to changing situations. The commander carefully considers where to position, balancing the need to inspire Soldiers during the breakout with maintaining an overall perspective of the entire operation.

Preparing for a Breakout

3-46. Preparations by the encircled battalion have to be rapid. The establishment of a chain of command is essential to ensure orders are obeyed and required tasks prior to the attack are accomplished, to include troop movement within the encirclement. An essential aspect of the encirclement for the Infantry battalion is what the battalion can carry out. The commander objectively considers options through an analysis of the mission variables of METT-TC. Difficult decisions may have to be made to enable the force to be saved. Two of these decisions are casualty evacuation and mortuary affairs. The commander evacuates wounded from the encirclement whenever possible. This also reduces the logistic burden of providing long-term medical care to wounded Soldiers. All wounded capable of fighting are assigned duties consistent with their wounds. Those who are incapacitated are evacuated through any available means.

3-47. The breakout plan outlines the commander's destruction criteria during preparation for equipment or supplies left behind. All vehicles, critical munitions, supplies, and equipment (except medical supplies) that cannot be moved are destroyed. Class III and V supplies must be inventoried and reallocated. No equipment should be taken that cannot be manned or maintained. The destruction of excess equipment will alert the enemy to the battalion's intention if done too soon. Equipment is destroy by demolition and burning as the commander initiates breakout. If possible, all ammunition that cannot be carried should be expended through direct or indirect fire missions utilized during deception operations and the rupture to penetrate the enemy defensive positions at the point(s) of the breakout.

Executing a Breakout

3-48. During execution, the commander exploits darkness and limited visibility, taking all possible precautions to deceive the enemy about the location of the rupture attack and the positioning of subordinate forces. Through necessity, the commander assigns multiple missions to subordinate forces because there are not typically enough forces within the encirclement to have separate forces for each required mission. For example, the follow-and-assume force could receive a be-prepared mission to help extract the rear guard, a mission generally given to the reserve. Forces located outside the encirclement, when available, assist the breakout by conducting shaping operations. Above all, the encircled force maintains the momentum of the breakout attack; otherwise, it is more vulnerable to destruction than it was before the breakout attempt.

Note. The following section (paragraphs 3-49 through 3-57) illustrates a fictional scenario as a discussion vehicle illustrating one of many ways the battalion can conduct a breakout attack. The battalion defends against encirclement from a motorized Infantry force. The enemy force approaches from the northeast.

Breakout—Sequence of Events, Example

3-49. A breakout from encirclement will follow a sequence of events similar to the example sequence addressed below. This sequence of events is used for discussion purposes and is not the only way of conducting a breakout encirclement. The battalion commander uses this or a similar sequence of events to plan actions necessary to accomplish the mission. The commander understands events will vary depending on the mission variables of METT-TC and to some degree, events will overlap. Each example sequence is unique and requires variations in organization and conduct prior to and during the breakout from an encirclement.

Protect the Force

3-50. The battalion commander normally establishes a defensive perimeter to protect the force prior to or after encirclement (figure 3-7, page 3-18). The commander reorganizes any fragmented units and places Soldiers separated from their parent units under the control of other units. The commander establishes a clear chain of command throughout the encircled force, reestablishes communications with units outside the encirclement, and adjusts support relationships to reflect the new organization. The commander positions and conducts security operations as far forward as possible to reestablish contact with the enemy and provide early warning. Vigorous patrolling begins immediately. Each unit clears its position to ensure there are no enemy forces in the perimeter. The battalion conducts reconnaissance and surveillance (R&S) missions to determine enemy locations, dispositions, and intent.

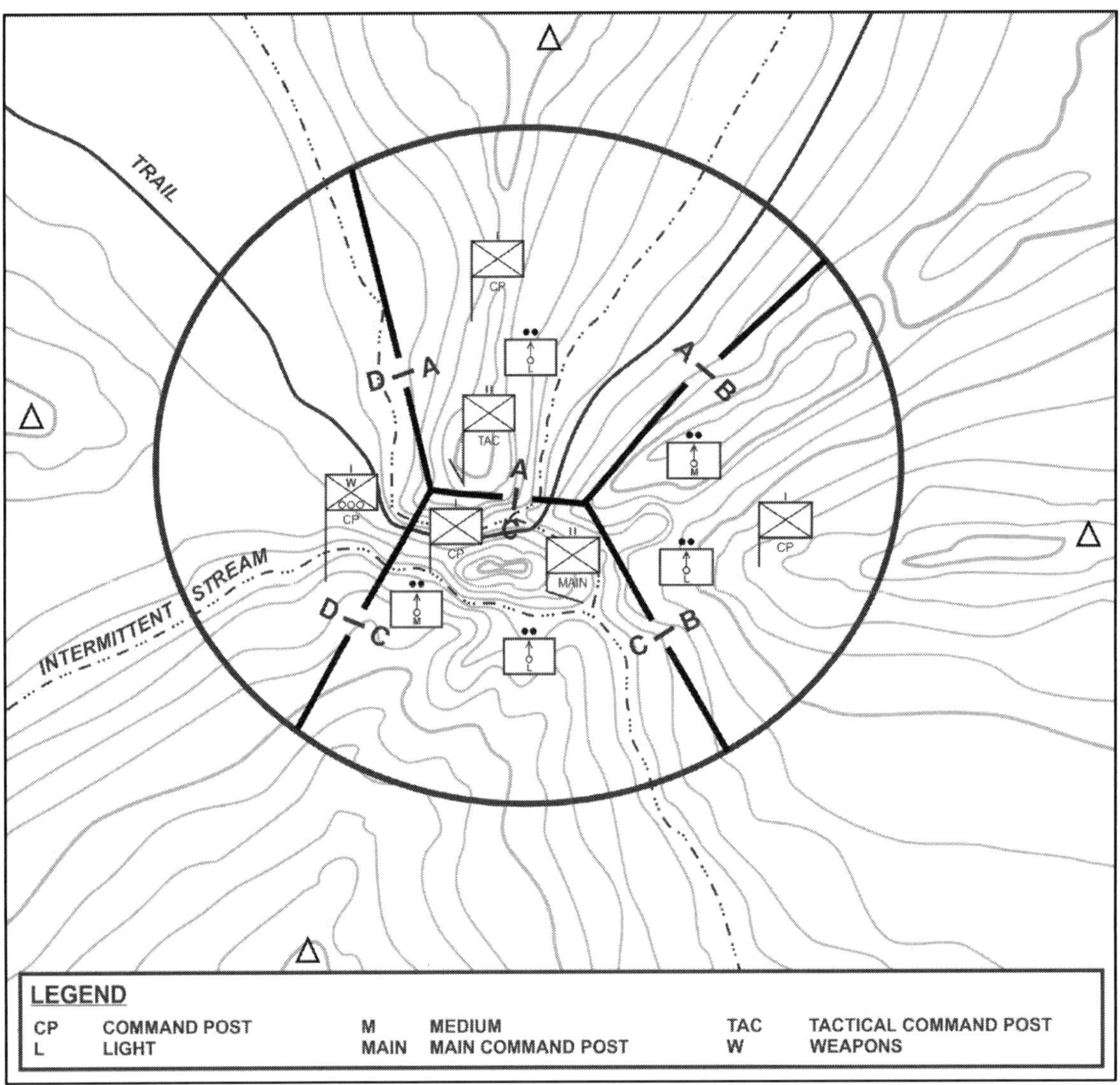

Figure 3-7. Protect the force (establish defensive perimeter), example

Organization of Forces

3-51. The commander's information collection effort (security operations and R&S missions) enables task organization within the encirclement and the movement of unit positions to allow key elements to reposition and begin preparations for the breakout (figure 3-8). The battalion typically task organizes to conduct rupture, follow-and-assume, main body, and rear guard missions. A reserve and a separate diversionary force may be established when sufficient forces exist within the encirclement.

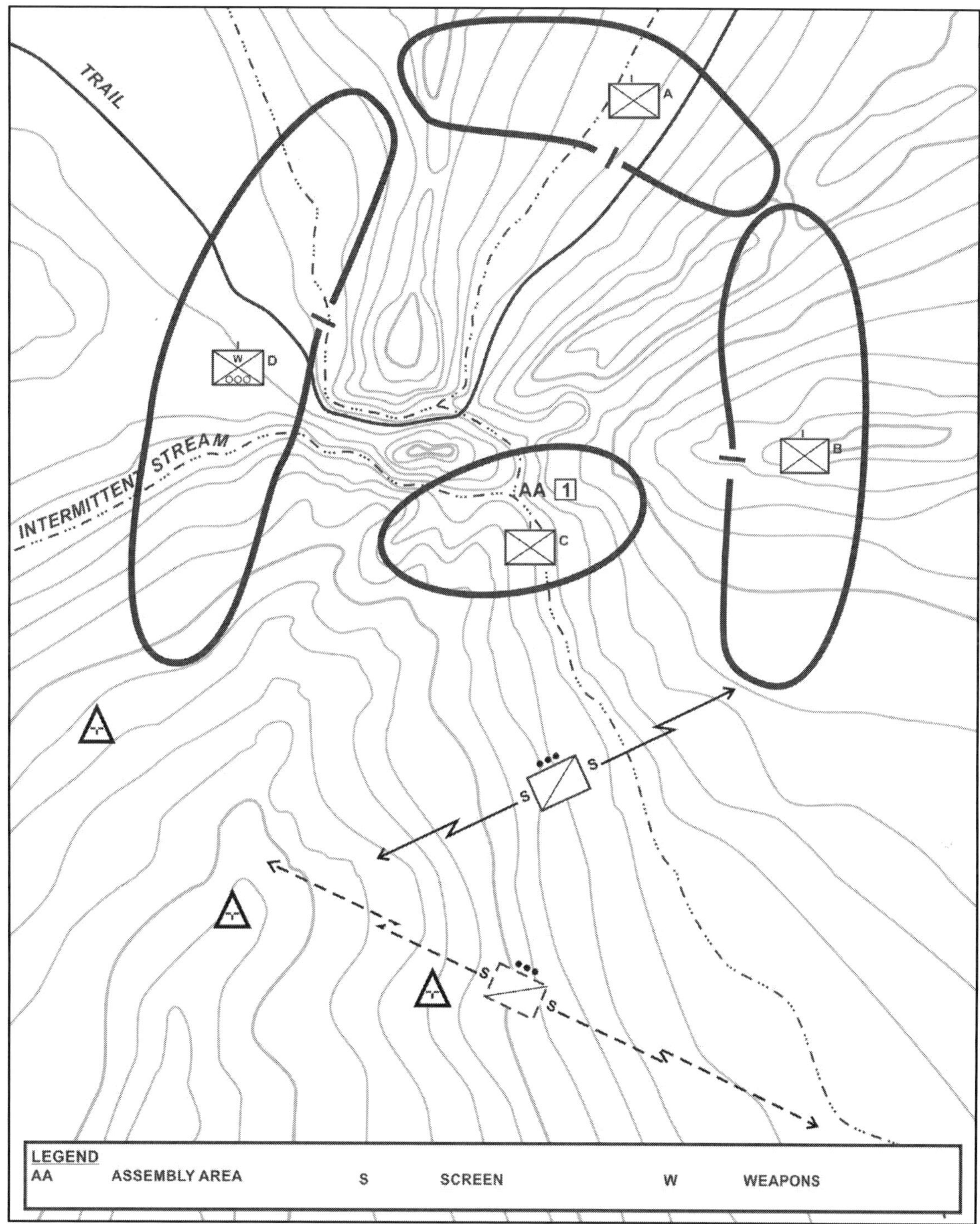

Figure 3-8. Organization of forces (breakout attack), example

Diversionary Attack

3-52. A diversionary attack is made along a likely escape route to draw the enemy away from the area of the proposed rupture. The attack to divert attention from the massing and movement of the rupture attack is always considered and may be imperative if the encirclement has been completed. The attack is directed at a point where the enemy might expect a breakout and strong enough to convince the enemy that it is the rupture attack. The rupture attack is not committed until the enemy has had time to respond to the diversion either by repositioning forces or by committing their reserve. Once the diversion is complete, the diversionary force can follow the main body and may assume the duties of the reserve. The possibility that the diversionary attack may rupture the encirclement should not be overlooked. If this occurs, the battalion must be prepared to exploit its success. When sufficient forces are not available for a diversionary attack (figure 3-9), which is generally the case for the Infantry battalion, other means of diversion to deceive the enemy such as movement of vehicles, false radio traffic, the unit improving defensive positions as if preparing to fight in place, should be planned.

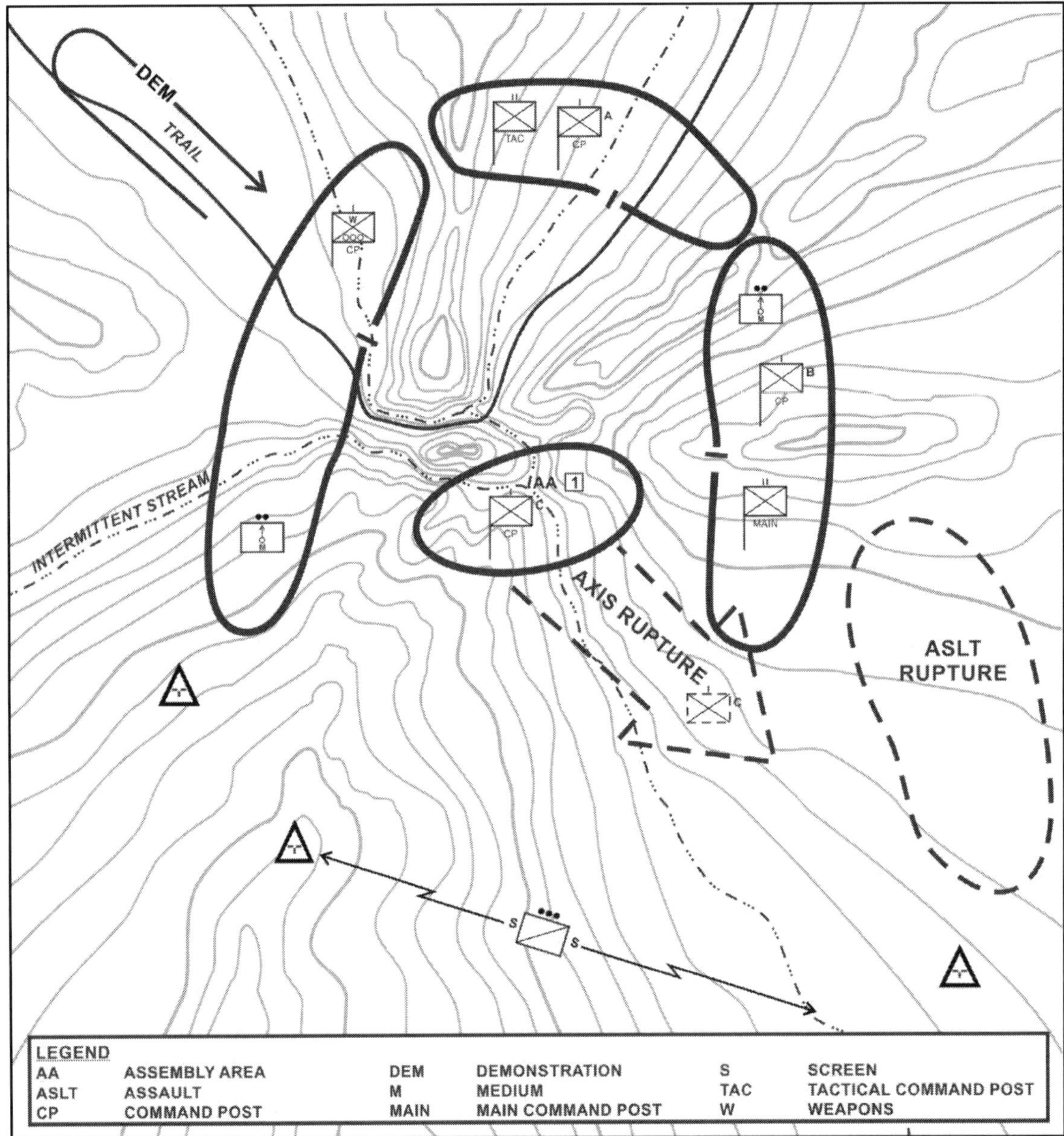

Figure 3-9. Diversionary attack (demonstration to deceive the enemy), example

Note. Forces conducting a demonstration do not seek contact with the enemy. See paragraphs 2-433 to 2-435 for additional information.

Rupture Attack

3-53. The commander takes all possible precautions to deceive the enemy concerning the location and time of the rupture attack. The commander may use one or more shaping operations (feints and demonstrations) to assist the rupture force in penetrating enemy positions to create the rupture and expand the shoulders. When the enemy reacts by attempting to close on the diversionary attack, the rupture attack initiates. Once the rupture attack force is committed, the diversionary attack breaks contact then moves as planned in the operation. When the time and place for the decisive attack is determined, sufficient combat power must be massed to ensure a rupture. The rupture force minimizes occupation of assault positions before starting the breakout. Rupture objectives are chosen to maintain the resulting gap in the encirclement and to provide defensive positions against a possible counterattack. The rupture attack force may be assigned the mission of securing these objectives while other units pass through the gap to continue the advance (figure 3-10).

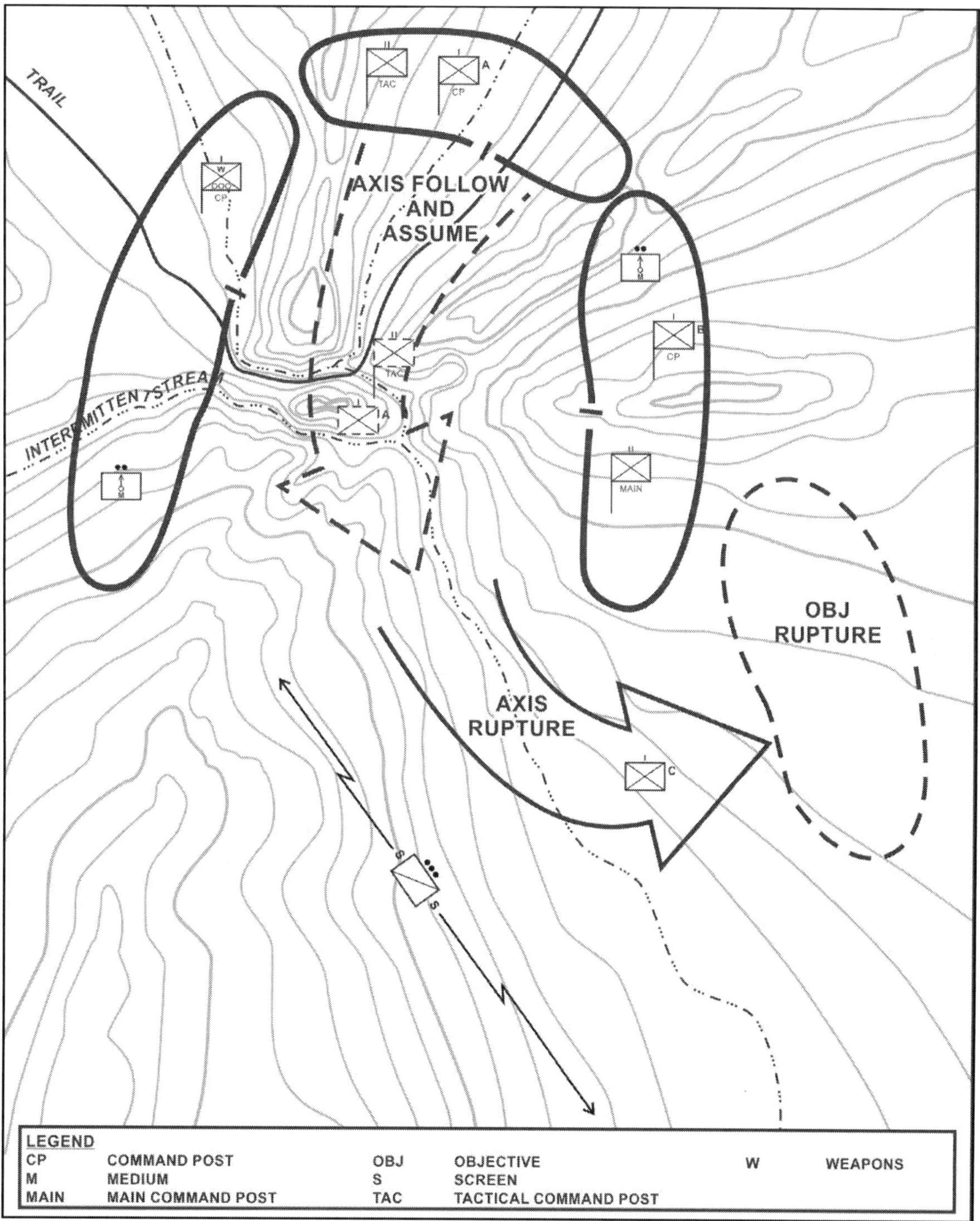

Figure 3-10. Attack to rupture the enemy's cordon (decisive attack), example

Follow-and-Assume

3-54. The follow-and-assume force follows the rupture attack and is committed to maintain the momentum of the attack or to seize objectives past the rupture. If the rupture force secures the gap, the follow-and-assume force normally becomes the battalion's lead element. If the follow and assume force secures the gap, it performs rear security once all battalion elements have passed through. When the follow-and-assume force is not committed, it continues through the rupture and normally conducts the decisive operation until completing linkup operations with another friendly force. The commander with a follow-and-assume mission in a breakout coordinates closely with the rupture force commander regarding the location of the gap, the enemy situation at the rupture point, and the enemy situation, if known, along the direction of attack past the rupture point. The battalion commander does not assign the follow-and-assume force a supporting or shaping tasks, such as clear routes and fix bypassed enemy forces when those tasks dissipate its available combat power. When these tasks are required to ensure the success of the breakout and resources permit, the commander designates a separate follow and support force to perform these tasks (figure 3-11).

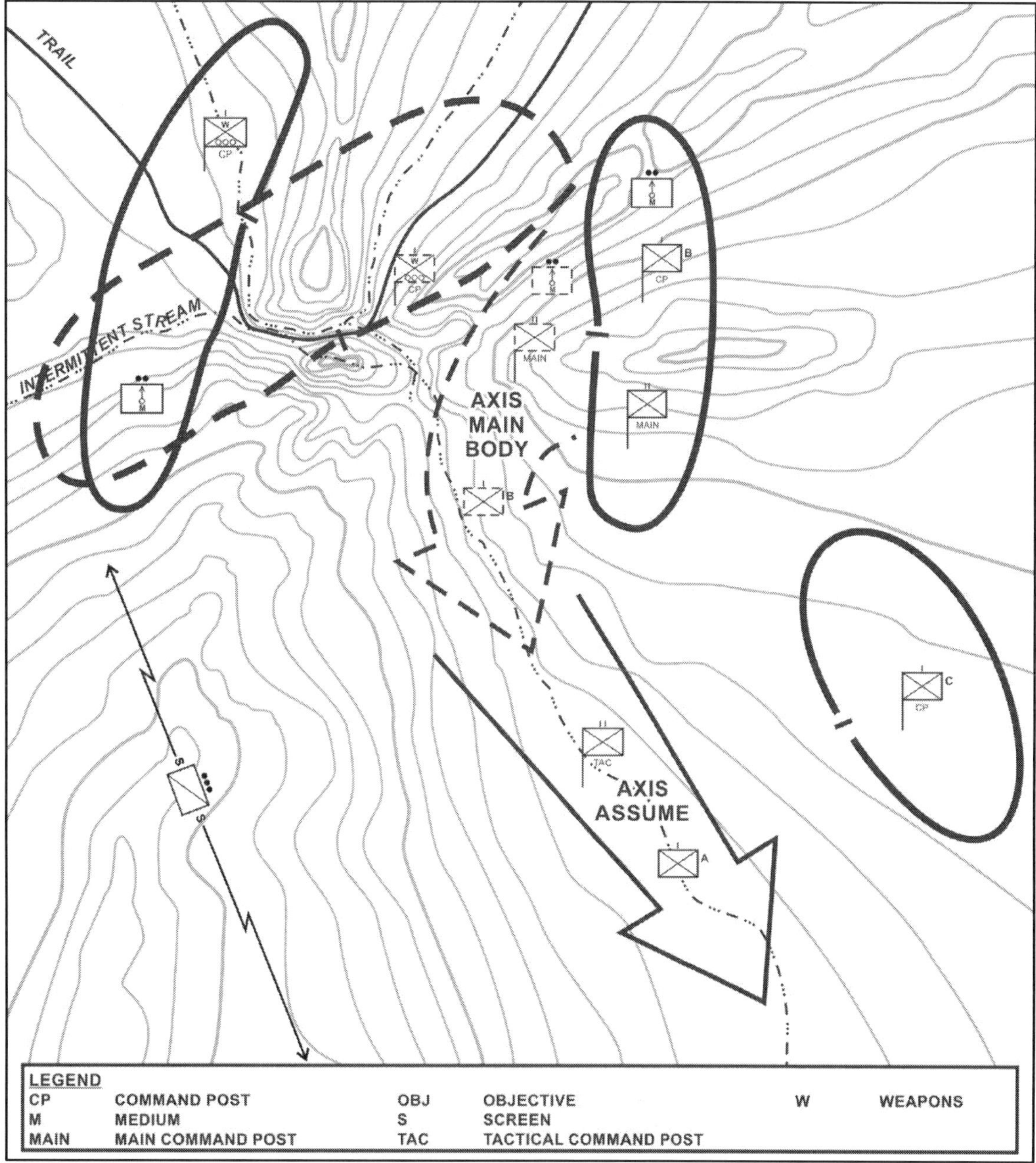

Figure 3-11. Follow-and-assume (committed to maintain the momentum), example

Main Body

3-55. The main body consists of the main command post, the bulk of encircled sustainment and supporting assets, the unit's casualties, and combat forces not required for other missions. Typically, the main body has sufficient combat power to protect itself and to establish flank security forces that deploy once the main body passes through the POP. The main body moves through the rupture and follows the follow-and-assume force (figure 3-12, page 3-24). The diversionary force follows the main body and may assume the duties of the reserve. The rear guard continues to withdraw through the now empty perimeter.

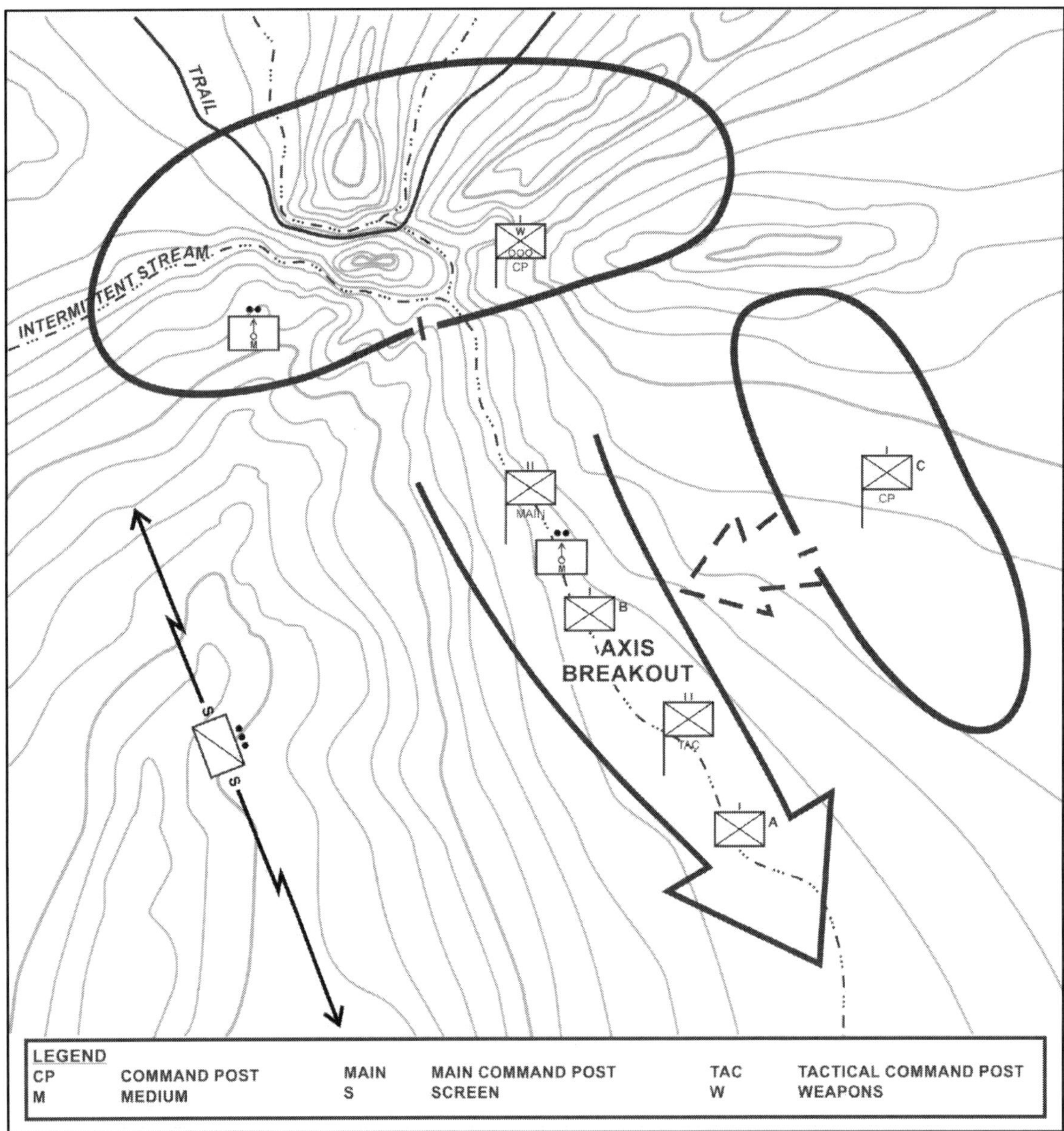

Figure 3-12. Main body (movement through gap), example

Rear Guard

3-56. To allow for the massing of the rupture attack, the follow-and-assume force, and the diversionary attack (if planned), and to protect the flanks or rear of these forces, part of the battalion must remain on the perimeter. As these forces move through the area of penetration, the rear guard commander must spread forces over an extended area. The rear guard requires flexibility and mobility to maintain the perimeter against enemy pressure. The size of the rear guard is critical. The rear guard must be sufficient to maintain the integrity of the defense, but must not draw so much combat power from the rupture force that the breakout attempt will be inadequate. Once the breakout commences, the rear guard withdraws toward the rupture. Rear guard forces left in contact must conduct a vigorous delaying operation (see paragraph 3-319 for a discussion of delaying operations) on the perimeter, so that no portion of the rear guard gets cut off. Under a single commander, the rear guard protects the main body from attack, while it moves from the area. In addition to providing security, the rear guard deceives the enemy about the intentions of the encircled force, simulating its activities until the main body clears the gap in the enemy's encirclement (figure 3-13).

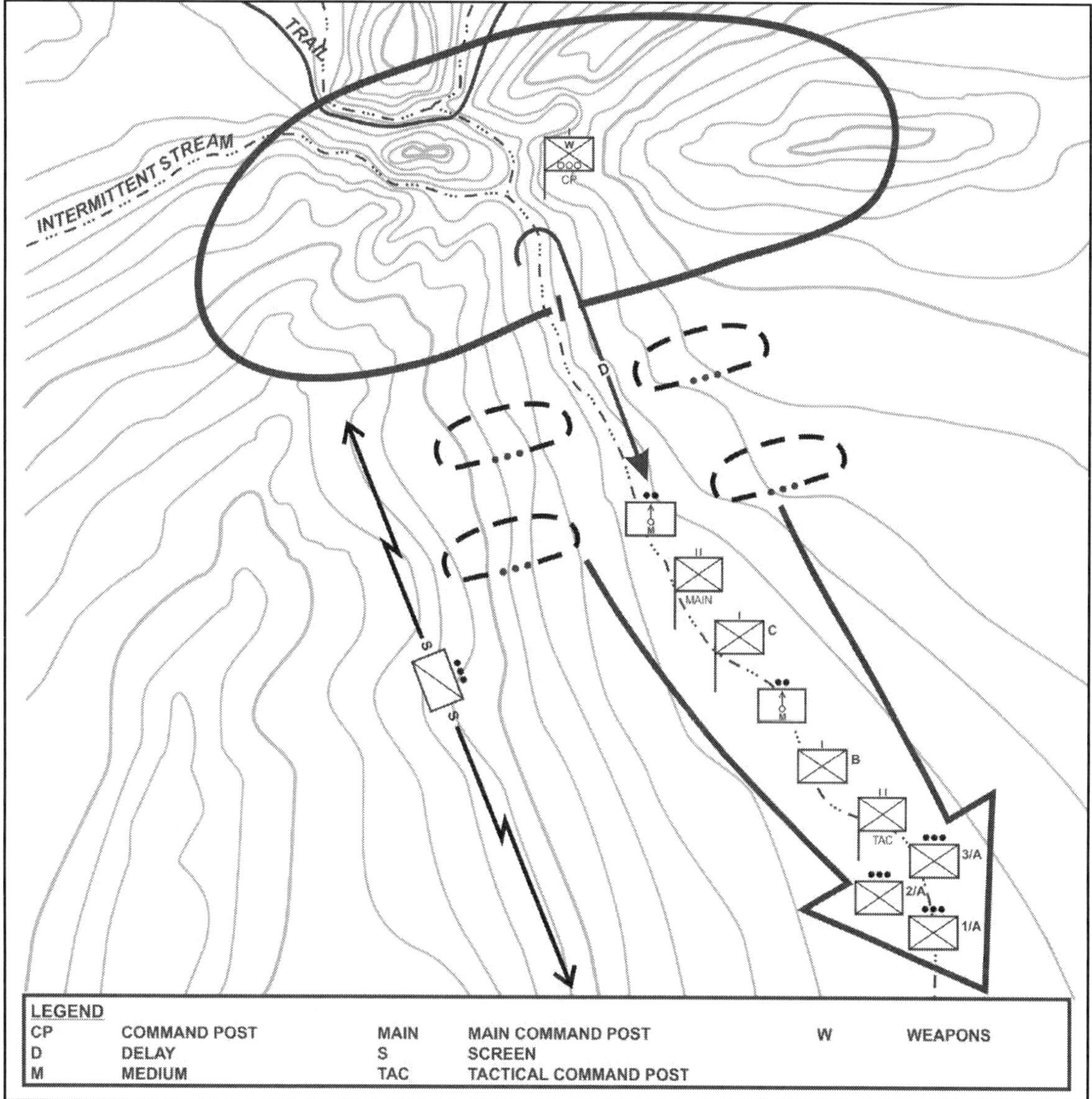

Figure 3-13. Rear guard (delaying operations on the perimeter), example

Note. The primary purpose of a reserve is to retain flexibility through offensive action. The commander makes every attempt to keep a small portion of the encircled force uncommitted, so it can be employed at the decisive moment to ensure the breakout's success. The commander may be unable to establish a separate reserve force because of the need to resource either the rupture force, the follow-and-assume force, or the rear guard. In this event, the commander assigns and prioritizes various be-prepared missions to the follow-and-assume force.

Movement after Breakout

3-57. After the breakout is complete and the battalion has passed through the gap created in the encirclement, speed and control continue to be critical. The battalion normally moves on a column axis, organized in the order that units pass through the gap. Every attempt is made to bypass enemy resistance prior to linkup with friendly forces. As rear guard forces contract the perimeter under contact with the enemy, the force conducts withdrawal (see paragraph 3-374 for a discussion of withdrawal operations) to break contact with the enemy forces. When the enemy closely pursues the breakout force, the efforts of the rear guard may become the decisive operation for the encircled force. The commander should position the reserve, when established or the follow-and-assume force where it can support the rear guard.

Exfiltrate Toward Other Friendly Forces

3-58. When the possibility of massing sufficient combat power to create a rupture during breakout seems remote or if another force cannot relieve the Infantry battalion, an *exfiltration*—the removal of personnel or units from areas under enemy control by stealth, deception, surprise, or clandestine means. See also special operations; unconventional warfare (JP 3-50)—during periods of reduced visibility and in close terrain may offer the greatest probability of success. Ideally, the Infantry battalion conducts exfiltration through rough or difficult terrain in areas likely to reduce enemy observation and fires. These conditions often allow undetected movement of small elements when movement of the entire force presents more risk. When it is unlikely that the entire force will be able to exfiltrate, part of the force may stay behind to create a diversion or perform a detachment left in contact mission.

3-59. Depending on the size of the force, the terrain, and the enemy situation the commander may plan single or multiple exfiltration routes to conduct linkup with friendly forces. The commander plans fires to cover the exfiltration movement and maneuver and to break contact with the enemy along these routes. Key control measures for the conduct of exfiltration include the establishment of exfiltration routes and lanes, restrictive fire areas, and no-fire areas. Based on reconnaissance, surveillance, and available intelligence, the exfiltrating force passes through or around enemy defensive positions and when detected tries to bypass the enemy (figure 3-14).

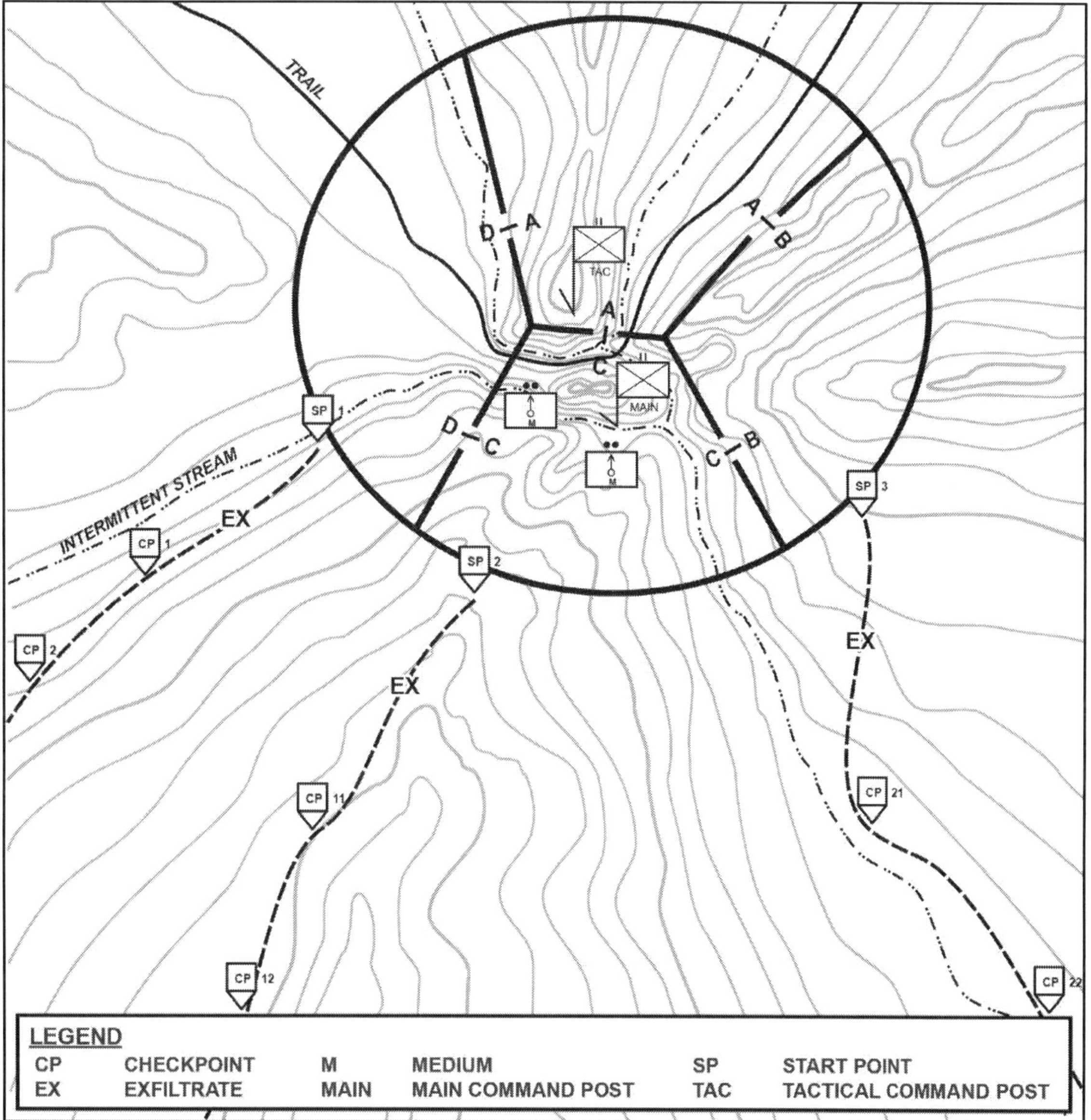

Figure 3-14. Exfiltration to other friendly forces

3-60. As with a breakout attack, the force conducts an exfiltration only after destroying or incapacitating all equipment, except medical, that is left behind. Combat and tactical vehicles, because the noise they make and the limitations they impose on exfiltration routes, make detection more likely.

Note. *Exfiltrate* is a tactical mission task where a commander removes Soldiers or units from areas under enemy control by stealth, deception, surprise, or clandestine means (FM 3-90-1). Friendly forces exfiltrate when they have been encircled by enemy forces and cannot conduct a breakout or be relieved by other friendly forces. In addition to being encircled by enemy forces, units returning from a raid, an infiltration, or a patrol behind enemy lines can also conduct an exfiltration. (Refer to FM 3-90-1 for additional information.)

Attack Deeper into Enemy-Controlled Territory

3-61. When the possibility of conducting a breakout or exfiltration from encirclement seems remote, attacking deeper into enemy territory may be a course of action that the enemy is not likely to expect. The Infantry battalion may attack deeper to seize key terrain, disrupt the enemy's offensive action, locate to more favorable defensive terrain, or provide an opportunity for linkup from another direction or extraction point. Attacking deeper is only feasible if the battalion can sustain itself while isolated, or when that sustainment can come from aerial resupply and enemy supply stocks. Though this course of action may involve greater risk, it may offer the only feasible course of action under some circumstances. The commander's form of maneuver to attack deeper into enemy territory depends on the higher commander's intent and the mission variables of METT-TC. (See paragraphs 2-5 through 2-22 for information on the six basic forms of maneuver during an attack.)

STAY-BEHIND OPERATIONS

3-62. A *stay-behind operation* is an operation in which the commander leaves a unit in position to conduct a specified mission while the remainder of the forces withdraw or retire from an area (FM 3-90-1). A stay-behind force may also result from enemy actions that bypass friendly forces. The main purpose of a stay-behind force is to destroy, disrupt, and deceive the enemy. A stay-behind force is a high-risk mission because of the danger that it will be located, encircled, and destroyed by the enemy. Resupply and casualty evacuation are extremely difficult. For these reasons, the stay-behind force should consist of enough combat power to protect and sustain its fighting capability for the duration of the mission. The commander considers assigning this mission only after a thorough analysis of the mission variables of METT-TC.

3-63. When it is unlikely that an encircled battalion will be able to breakout or exfiltrate the entire force, part of the force may stay behind to create a diversion or perform a detachment left in contact mission (figure 3-15). Infantry forces are especially suited to conduct these missions in complex terrain. Infantry forces take advantage of such terrain, reinforced by the use of situational obstacles to enhance the effects of natural obstacles to deny enemy freedom of movement and maneuver. Restricted and severely restricted terrain offers cover for the movement of the friendly force and favors using ambushes against the enemy. Infantry forces are best suited to exfiltrate by means of dismounted march or air movement once the stay-behind operation concludes.

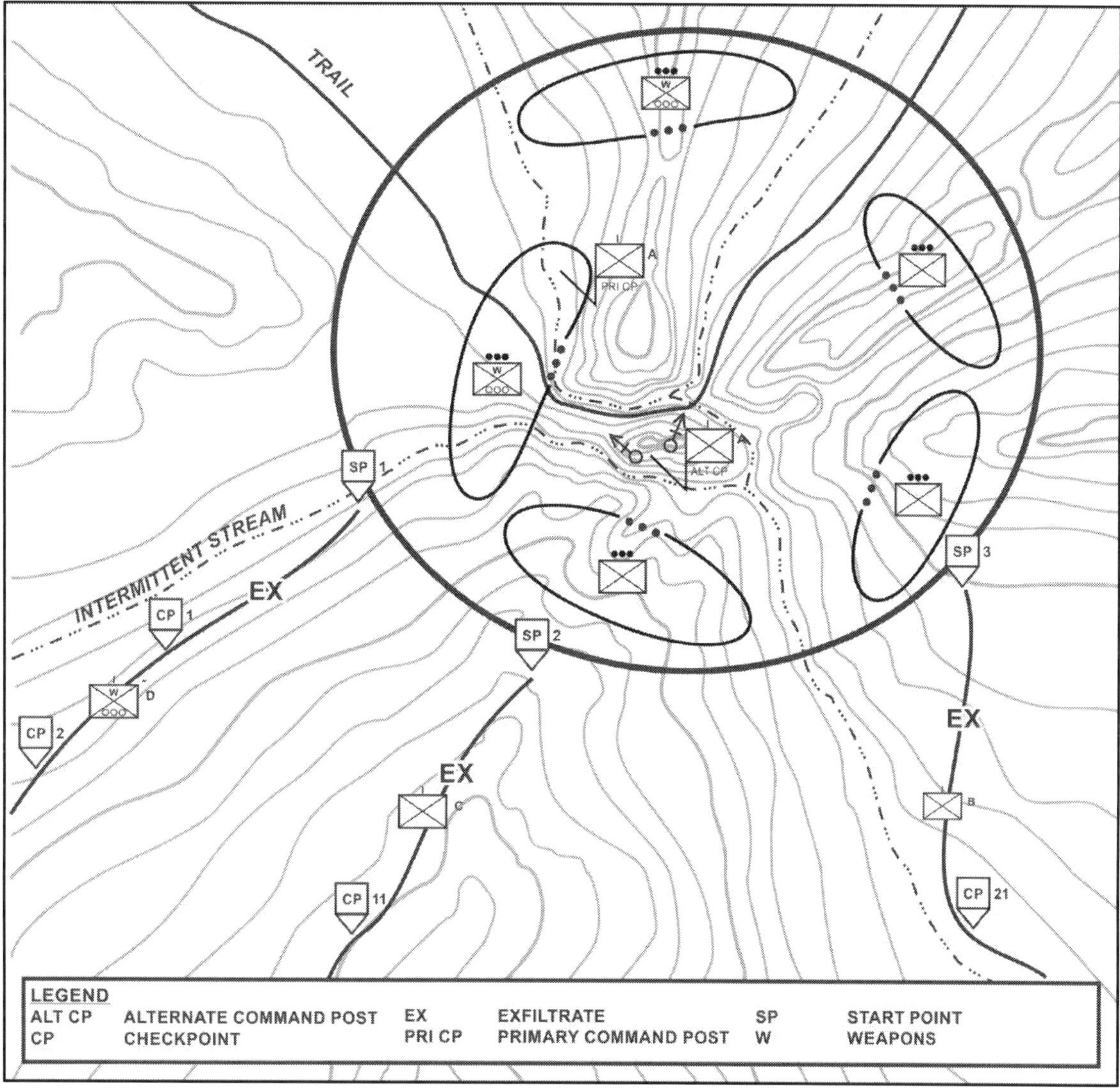

Figure 3-15. Stay-behind force (detachment left in contact mission)

3-64. As part of a defense, the Infantry battalion or subordinate units conducting a stay-behind operation may occupy hide positions (company or platoon perimeter defenses) well forward of the forward edge of the battle area. As the enemy passes, the Infantry force attacks the enemy through a series of series of raids and ambushes. The intent of these attacks may be to attrit the enemy or to cause enemy follow-on forces to be more cautious and to slow down to clear possible attack and ambush sites. An Infantry force can be inserted via infiltration, air assault, or parachute; it can also be a bypassed force. Sustaining the force(s) can be in the form of caches or aerial resupply.

3-65. An encircled stay-behind force can continue to defend encircled, conduct a breakout from encirclement, exfiltrate toward other friendly force, or attack deeper into enemy-controlled territory. After stay-behind forces accomplish their missions, the commander may require them to conduct linkup operations. Stay-behind operations eventually require the force to reenter friendly lines or linkup with other elements, often in more than one location. The commander carefully coordinates this reentry to prevent friendly fire incidents. Return routes for the stay-behind force are the best-covered and concealed routes available. The commander places guarded gaps or lanes near obstacles along these routes that cannot be bypassed.

RELIEF IN PLACE

3-66. A *relief in place* is an operation in which, by direction of higher authority, all or part of a unit is replaced in an area by the incoming unit and the responsibilities of the replaced elements for the mission and the assigned zone of operations (Army uses the term, area of operations) are transferred to the incoming unit (JP 3-07.3). The Infantry battalion normally conducts a relief in place as part of a larger operation, primarily to maintain the combat effectiveness of committed forces. The IBCT or higher headquarters directs when and where to conduct the relief and establishes the appropriate control measures. Normally, during the conduct of combat operations, the unit relieved is defending. However, a relief in place may set the stage for resuming offensive operations or serve to free the relieved unit for other tasks.

General Considerations for a Relief in Place

3-67. Upon receipt of the order to conduct the relief, relief commanders and their staffs establish liaison personnel to exchange information pertinent to the relief operation. Commanders emphasize communications, intelligence handover, and transfer of command. If possible, the incoming battalion's command post collocates with the main command post of the battalion in positon to facilitate continuous information exchanges relative to the occupation plan, fire support plan, and intelligence updates that include past, present, and probable enemy activities.

3-68. Face-to-face coordination reduces any potential misunderstandings related to relief preparation or the forthcoming operations. The relieving battalion can establish advance parties to conduct detailed face-to-face coordination and preparations for the operation, down to the company level and possibly to the platoon level. Depending on the situation, advance parties infiltrate forward to avoid detection. A battalion advance party normally includes the echelon's tactical command post, which co-locates with the main command post of the unit being relieved. The relieving battalion commander may attach additional liaison personnel as the operation develops.

3-69. Depending on the amount of planning and preparations associated with the relief in place, a relief is characterized as either a deliberate operation or hasty operation. The major differences are the depth and detail of planning and, potentially, the execution time. Detailed planning generally facilitates shorter execution time by determining exactly what the commander believes needs to be done and the resources needed to accomplish the mission. Deliberate planning allows the commander and staff to identify, develop, and coordinate solutions to most potential problems before they occur, and to ensure the availability of resources when, and where needed.

3-70. When either force gains direct fire contact with an enemy force, the force immediately notifies the other unit and the higher headquarters. When responsibility for the area of operation has not passed, the relieving unit becomes under the operational control of the force not yet relieved. The relieved unit becomes operational control to the relieving unit when the responsibility for the area of operation has passed to the relieving unit.

3-71. Responsibility for the area transfers, as directed by the IBCT commander or senior common commander, normally when the incoming unit has a majority of the fighting force in place and all mission command systems are operating. Units involved in a relief in place should be of similar type—such as mounted or dismounted—and task organized to help maintain operations security. The relieving unit usually assumes as closely as possible the same task organization as the unit being relieved, and assigns responsibilities and deploys in a configuration similar to the relieved unit. As support elements of the unit being relieved displace, they leave the relieving unit supply stocks according to previously coordinated arrangements.

Planning a Relief in Place

3-72. Once ordered to conduct a relief in place, the commander of the relieving battalion contacts the commander of the battalion to be relieved. The co-location of battalion and subordinate command posts, when conditions allow, helps achieve the level of coordination required. As a minimum, the relieving battalion establishes communications and liaison with the battalion being relieved. The WARNORD, from the IBCT or common higher headquarters, designates the time of relief, relieving and relieved units, and sequence of events. The WARNORD specifies the future missions of the relieved force, route priorities, any restrictions on advance parties, any extraordinary security measures, and the time and place for issuing the complete order.

Increased Burden on Mission Command

3-73. The meeting of friendly ground forces, inherent in a relief, places an increased burden on mission command. The consequences of mutual interference between relief forces and the complexity associated with such areas as traffic control, fire support coordination, and obstacle plans, and the relief in place of mission command systems require close coordination between all headquarters involved.

3-74. In a deliberate relief, units exchange plans and liaison personnel, conduct briefings, perform detailed reconnaissance, and publish orders with detailed instructions. In a hasty relief, the commander abbreviates the planning process and controls the execution using oral and fragmentary orders. The relieving unit receives current intelligence, operations, and sustainment information from the battalion being relieved, as well as from IBCT or common higher headquarters, adjacent units, and subordinate elements.

3-75. Commanders establish early liaison between the stationary and the relieving subordinate forces and identify measures to control the relief. Control measures associated with a relief in place are generally restrictive to prevent fratricide. As a minimum, these control measures include the area of operations with its associated boundaries, battle positions, contact points, start points, routes, release points, assembly areas, fire support coordination measures, and direct fire control measures, such as target reference points and engagement areas.

3-76. Once received, the relieving unit verifies the obstacle records of the unit being relieved. Handover of obstacles is a complex procedure. Initially, the engineer priority is on mobility to get the relieving unit into the area of operation. Mobility focuses on those routes and lanes leading into the area of operation. Once the relief occurs, priority of the mobility and survivability effort transitions to support the relieving unit's continuing mission. The commander may require supporting engineers to assist with survivability tasks to support the relieving force.

Tactical Vulnerability

3-77. A relief in place, by nature, is a tactically vulnerable operation. During initial planning care must be taken, and plans must address, security operations (screen and guard missions) that provide sufficient security to warn friendly forces in case the enemy tries to take advantage of the relief. (See paragraph 3-66.) The battalions involved must concentrate on security while preparing for and executing the operation. The intent of the operation is to complete the relief without discovery by the enemy. Consequently, commanders involved in the relief typically plan reliefs for execution during periods of reduced visibility. To enhance security, commanders' emphasis face-to-face coordination, impose light and noise discipline and electromagnetic emission control measures, such as radio silence or radio-listening silence.

Mask the Relief

3-78. While the battalions involved plan, prepare, and execute the relief in place, the IBCT or common higher headquarters and other units continue actions to mask the relief. These include using deception (demonstrations and feints), concealment, obscuration, radio silence, and harassing and interdiction fires. The IBCT or common higher headquarters executes operations to attack and disrupt the enemy's uncommitted and reserve forces during the relief. The intent is to fix or distract the enemy, so that the enemy does not detect or interfere with the relief.

3-79. Deception, in regards to the relief to achieve surprise, leads the enemy into inaccurate perceptions. Deception of the enemy requires a detailed knowledge of friendly vulnerabilities. A counterintelligence assessment of enemy collective capabilities directed against the friendly forces involved in the relief can provide that detailed knowledge. The commander secures success through deceptive techniques and procedures and cyberspace electromagnetic activities to inaccurately portray friendly forces involved in the relief, mislead enemy commanders, and deny those same enemy commanders the ability to use cyberspace and the electromagnetic spectrum against the relief.

3-80. Concealment from the enemy is a primary planning concern for relief commanders to hide necessary friendly force concentrations. Dispersion (laterally and in-depth) and multiple routes at company and subordinate levels decrease the possibility of enemy units massing on friendly elements as they organize and position prior to executing the relief. The enemy should perceive only one command structure during the relief operation—that of the battalion being relieved—until the operation is complete.

Relief Techniques

3-81. The commander directing the relief, in coordination with relief commanders and staffs during planning, determine the most appropriate method for executing the relief by using one of three techniques. The three techniques for conducting a relief in place are sequential, simultaneous, or staggered.

Sequential

3-82. A sequential relief occurs when each element in the relieved unit is relieved in succession, from right to left or left to right, depending on how it is deployed. This technique is the most deliberate and time-consuming; however, it minimizes confusion and maintains the best mission command and readiness posture. A sequential relief involves sequentially relieving maneuver companies of the battalion one at a time. Separate routes to the rear of the relieved companies' locations are planned for each maneuver company and placed on the operations overlay. Routes are labeled sequentially and correspond to the order in which the company executes them during the relief. When the lead company reaches its release point, platoons are guided into the positions they are occupying. Crews exchange range card and fire support information. Once the relief occurs, relieved units move to the rear to occupy their next location. When the lead company is in position, the next company moves along its designated route(s) to relieve its counterpart: thereby repeating the relief process. This process repeats until each company has been relieved. When transfer of supplies from the relieved unit is directed, battalion S-4s coordinate a transfer point to execute the exchange.

Simultaneous

3-83. A simultaneous relief occurs when all elements are relieved at the same time. Simultaneous relief takes the least time to execute, but is more difficult to control and more easily detected by the enemy. A simultaneous relief involves simultaneously relieving maneuver companies of the battalion at the same time. Separate routes to the rear of the relieved companies' locations are planned and labeled for each maneuver company and placed on the operations overlay. When relieving companies reach their release points, platoons are guided to the positions they are occupying. Crews exchange range card and fire support information, and the relieved unit then moves to the rear to its next location. When transfer of supplies from the relieved unit is directed, battalion S-4s coordinate a transfer point to execute the exchange.

Staggered

3-84. A staggered relief occurs when the commander relieves each element in a sequence determined by the tactical situation, not its geographical orientation. As with a sequential relief, staggered reliefs can occur over a significant amount of time. Separate routes to the rear of the relieved companies' locations are planned and labeled for each maneuver company and placed on the operations overlay. When relieving companies reach their release points, platoons are guided to the positions they are occupying. Information exchanges and transfers of supplies are the same as in the other two techniques. Once the relief occurs, relieved units move to the rear to occupy their next location. When the first company to move is in position, the next company is identified to move along its designated route(s) to relieve its counterpart: thereby repeating the relief process. This process repeats until each company has been relieved.

Preparing for a Relief in Place

3-85. The complexity of a relief in place requires extensive liaison and reconnaissance during preparation. Exchanging information about the enemy and civilian situations, friendly dispositions, terrain analysis, and fire support and obstacle plans, coupled with reconnaissance, helps relief commanders to prepare for the mission. Liaison by the IBCT commander, or common higher headquarters commander, enable leaders' reconnaissance of the area of operation they will assume. Leaders' reconnaissance includes the lowest-echelon leader allowed by the tactical situation and required by the situation. The reconnaissance focuses on the route into the positions the unit is to occupy, the positions themselves, the current disposition of the unit being relieved, and any obstacles that could affect the movement.

3-86. The commander conceals the relief from the enemy for as long as possible. At the first indication that relief is necessary, which is usually the WARNORD for the relieving unit, both the relieved unit and the relieving unit review their operations security plans and procedures. Commanders may use military deception measures during preparation activities to maintain secrecy. To maintain security during preparation, the relieving unit makes maximum use of the relieved unit's radio nets and operators. Both units involved in the relief operate on the command frequencies and encryption variables of the combat net radios of the relieved unit at all levels. The relieved unit's signal officer is in charge of communications throughout the relief operation.

3-87. Relief units conduct rehearsals to discover any weaknesses in the plan and familiarize all elements of both forces with the plan. Finding time for rehearsals requires commanders and staffs to focus on time management. R&S efforts before troop movement allow commanders to make necessary changes to the plan prior to rehearsals. During troop movement, reconnaissance elements of the relieving unit precede its movement with a route reconnaissance to intermediate assemble areas. Reconnaissance elements then conduct reconnaissance of the routes leading from the assembly areas to the positions of the unit being relieved.

3-88. Relief units exchange as much information as possible during rehearsals or at least prior to troop movement when combined relief unit rehearsals are not possible. Units involved share information including communications security (COMSEC) procedures and graphic overlays consisting of—

- Primary and alternate linkup points.
- Checkpoint and waypoint information.
- Unit disposition and activity (friendly and enemy).
- Locations and types of obstacles.
- Fire control measures including restrictive fire lines and no-fire areas.

3-89. On occasion, when units involved in a relief in place are not of similar type—such as an Infantry company relieved by a mechanized Infantry company. The relieving unit still assumes as closely as possible the same task organization as the unit being relieved, and assigns responsibilities and deploys in a configuration similar to the relieved unit. Under these conditions, commanders allocate time to construct individual vehicle fighting positions.

Executing a Relief in Place

3-90. When executing a relief in place, the linkup method chosen prior to the intermediate assembly area (when used) usually involves conducting linkup (see paragraph 3-96 for additional information on linkup) at a predetermined contact point(s). During movement to the contact point(s), the commander and subordinate leaders monitor the progress and execution of the linkup to ensure that established positive control measures are followed or adjusted as required. Contact points must be readily recognizable and posted on overlays. When possible, the moving force should halt short of the contact point and send a smaller force (patrol) forward to pinpoint the contact point. After the patrol makes contact with the force in position at the contact point, the patrol may leave a portion of patrol at the contact point then move back with the remaining members of the patrol to guide the subordinate unit back to conduct linkup. Following linkup the relieving force is guided to an assemble area, or depending on the situation, guided directly into relief positon(s) (figure 3-16, page 3-34).

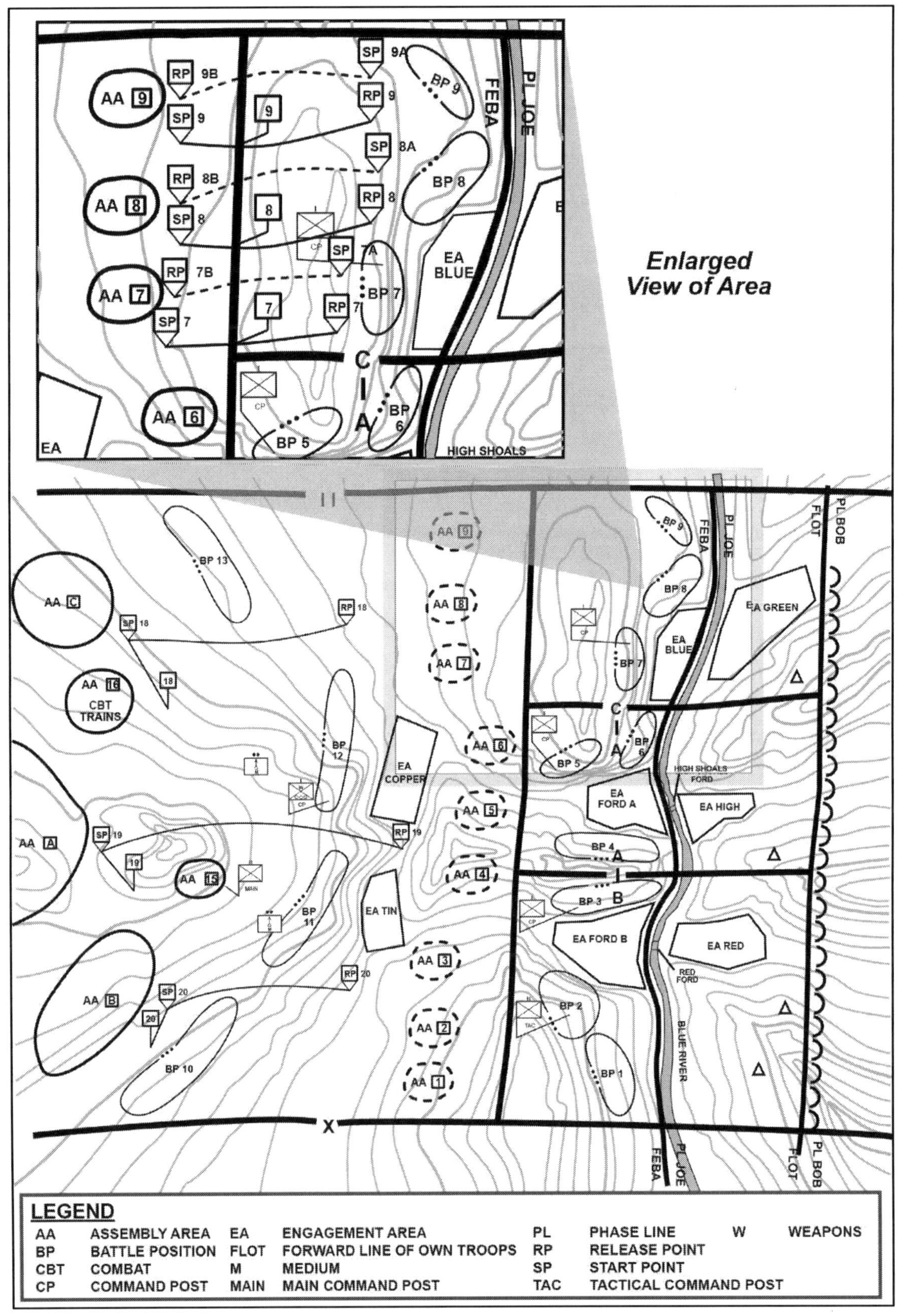

Figure 3-16. Overlay of a battalion relief in place

3-91. In situations where the commander desires to conceal the relief from the enemy, such as during a sequential or staggered relief, the relieving unit may occupy the same positions as the unit it relieves. Alternatively, the relieving unit may establish more favorable positions in the vicinity of the relieved unit's location. Occupying different positions makes early discovery by the enemy more likely. Any increase in activity in forward positions can reveal the relief to the enemy. During the relief, counterreconnaissance actives attempt to counter detection by the enemy while information collection systems attempt to detect if the enemy can discover the relief before its completion.

3-92. In a simultaneous relief, the relieving unit begins moving from its current location to assembly area (when used) in the area of operation of the force being relieved. Once the relief begins, all elements involved execute the relief as quickly as possible. Both forces are vulnerable to enemy attack because of the concentration, movement, and intermingling of forces in a simultaneous relief. Any unnecessary delay during execution provides the enemy additional time to acquire and engage the forces involved. Relief forces in the area of operation come under the operational control of the relieving unit commander at the time or triggering event previously established by the plan, for the operation.

3-93. As the first relieving element arrives from the assembly area (when used) to assume the position, it establishes a screen forward of the relieved unit's positions as the tactical situation permits. The remainder of the relieving unit moves forward to positions behind the unit being relieved. The relieving unit may use the relieved unit's alternate and supplementary defensive positions to take advantage of any previous defensive preparations. At the previously established time or event, passage of command takes place. At that point, if possible, the commander of the relieving unit informs all units involved in the relief of the passage of command.

3-94. As the relieved unit continues to defend, the relieving unit's advance parties coordinate procedures for the rearward passage of the relieved unit. On order, the relieved unit begins withdrawing through the relieving unit and moves to assembly areas. Crew-served weapons of subordinate companies are usually the last elements relieved after exchanging range cards. The relieving unit replaces them on a one-for-one basis to the maximum extent possible to maintain the illusion of routine activity. The relieved unit's supporting and sustainment assets assist both the relieved unit and the relieving unit during this period.

3-95. A relief does not normally require the battalion's fire support assets to relieve weapon system for weapon system, unless the terrain limits the number of firing positions available. Generally, the relieved battalion's fire support assets remain in place until all other relieved elements displace and are available to reinforce the fires of the relieving battalion in case the enemy tries to interfere. If the purpose of the relief is to continue the attack, the fire support assets of both forces generally remains in place to support the subsequent operation.

> ***Note.*** Relief in place during the conduct of operations centering on the stability element of decisive action involve many of the planning, preparation, and execution considerations mentioned above. Time is not normally such an important factor and most reliefs in place or transfers of authority in these types of operations are deliberate and may occur over an extensive time-period.

LINKUP

3-96. A *linkup* is a meeting of friendly ground forces, which occurs in a variety of circumstances (ADRP 3-90). Linkup operations happen when an encircled battalion or subordinate element breaks out (breakout attack) to rejoin friendly forces or exfiltrates towards friendly forces, or when a force comes to the relief of an encircled element. During the conduct of defensive and offensive tasks, linkups happen when forces conduct forward or rearward passage of lines or battle handover. During the conduct of offensive tasks linkup operations result from the linkup of two encircling arms conducting a double envelopment, when converging maneuver forces meet to complete the encirclement of an enemy force, or when an advancing force reaches an objective area previously seized by an airborne or air assault. During infiltration (see paragraph 2-6 for information on infiltration), through or into an area occupied by enemy forces, attacking forces conduct linkup when two infiltrating elements in the same or different infiltration lanes or in an area of operation are scheduled to meet to consolidate before proceeding on with their missions.

General Considerations for a Linkup

3-97. The key to a successful linkup is the commander's integration and synchronization of one of two linkup methods. The preferred method is when the moving force has an assigned limit of advance near the other force and conducts the linkup at a predetermined contact point or points to coordinate further operations. The commander uses the other method during highly fluid operations when the enemy force escapes from a potential encirclement, or when one of the linkup forces is at risk and requires immediate reinforcement. In this method, the moving force continues to move and conduct long-range recognition via radio or other measures, stopping only when it makes physical contact with the other force.

3-98. When the IBCT commander directs a linkup operation, the commander establishes minimum control measures. The commander assigns each battalion or subordinate force conducting the linkup an area of operation defined by lateral boundaries and a restrictive fire line that also acts as a limit of advance to ensure positive control and reduce the risk of fratricide. The IBCT transmits these and other restrictive fire support coordination measures to linkup forces. These restrictive fire support coordination measures are subsequently adjusted as require and overlays are updated as one or both forces moves to linkup. This process continues until a single restricted fire line is established between the forces. Usually, this is the point on the ground where the two forces plan to establish contact. Both forces may be moving toward each other, or one may be stationary. Whenever possible, joining forces exchange as much information as possible before starting an operation. The headquarters ordering the linkup establishes—

- A common operational picture using available mission command systems, such as blue force tracker.
- Command relationship and responsibilities of each force before, during, and after linkup.
- Coordination of fire support before, during, and after linkup, including control measures.
- Linkup method.
- Recognition signals and communication procedures, including pyrotechnics, armbands, vehicle markings, gun-tube orientation, panels, colored smoke, lights, and challenge and passwords.
- Operations to conduct following linkup.

3-99. In addition to the minimum control measures addressed in paragraph 3-98 above, the headquarters ordering the linkup can establish no-fire areas around one or both forces and establish a coordinated fire line beyond the area where the forces linkup. The commander establishes a no-fire area to ensure that uncleared air-delivered munitions or indirect fires do not cross either the restrictive fire line or a boundary and impact friendly forces. The establishment of a coordinated fire line allows available joint fires to expeditiously attack enemy targets approaching the area where the linkup is to occur.

Note. A restrictive fire support coordination measure prevents fires into or beyond the control measure without detailed coordination. The primary purpose of restrictive measures is to provide safeguards for friendly forces. Restrictive fire support coordination measures include an airspace coordination area, a no-fire area, a restrictive fire area, and a restrictive fire line. Establishing a restrictive measure imposes certain requirements for specific coordination before the engagement of those targets affected by the measure. (See FM 3-52 for a description of an airspace coordination area.)

3-100. When attacking forces infiltrate undetected through or into an area occupied by enemy forces, the higher commander may establish a linkup point or points to consolidate forces before proceeding on with the mission. (See paragraph 2-16 for information on infiltration.) As with the selection of infiltration routes or lanes, the commander selects linkup points that avoid the enemy, provide cover and concealment, and facilitate navigation. The commander designates alternate linkup points, as enemy action may interfere with the primary linkup point(s). These and other control measures are adjusted during the operation to provide for freedom of action prior to and during the linkup. These same general actions and control measures are required when forces exfiltrate undetected through enemy territory to linkup with other friendly forces. Linkup point or points where two exfiltrating elements are scheduled to meet should be easily identifiable point on the ground, large enough for all exfiltrating elements to assemble, and offer cover and concealment. (See paragraph 3-96 for information on exfiltration.)

Planning a Linkup

3-101. The commander directing the linkup establishes command relationships and responsibilities for the forces involved. Positive control during linkup is necessary to prevent inadvertent fratricidal as forces move to linkup. The higher headquarters plan prescribes the primary and alternate day and night identification and recognition procedures, vehicle systems, and manmade materials used to identify friend from enemy.

3-102. Whenever possible, forces conducting linkup establish liaison during planning and continue it throughout the linkup operation. Liaison parties require the capability to communicate with parent unit(s). As the distance closes between linkup forces, the necessity to track movement and maintain close liaison increases. Use of Army manned and unmanned aircraft systems can improve and expedite this process.

3-103. When linkup operations require a passage of lines (see paragraph 2-217 for information on passage of lines). Once through friendly lines, and to affect the linkup, the battalion moves out as in an exploitation; speed, aggression, and boldness characterize this action. If possible, the linkup force avoids enemy interference with its mission and concentrates its efforts on completing the linkup. When an enemy force threatens the successful accomplishment of the mission, the enemy is either destroyed or bypassed and reported.

3-104. The communication plan includes all essential frequencies and secure variables to maintain communication between the two forces. Linkup forces use voice systems as required to share combat information and to identify friend from enemy.

Preparing for a Linkup

3-105. Whenever possible, linkup forces exchange as much information as possible before starting an operation. Forces involved share information including COMSEC procedures and graphic overlays consisting of—

- Primary and alternate linkup points.
- Checkpoint and waypoint information.
- Unit disposition and activity (friendly and enemy).
- Locations and types of obstacles.
- Fire control measures including restrictive fire lines and no-fire areas.

3-106. When time is available, the directing headquarters and subordinate forces involved in the linkup, conduct rehearsals. When time is not available, the directing commander walks the linkup commanders (when available) through the operation. The commander stresses the linkup and coordination required to reduce the potential for fratricidal engagements between linkup forces. In addition, the commander ensures that each linkup commander is prepared to respond to an enemy meeting engagement or attack before the linkup.

Executing a Linkup

3-107. When conducting a linkup, both forces may be moving or one may be stationary. The linkup method chosen may involve conducting the linkup at a predetermined contact point(s) or when involved in a highly fluid situation, stopping only when one force makes physical contact with the other force. Using either method, the commander monitors the progress and execution of the linkup to ensure that established positive control measures are followed or adjusted as required.

Linkup of a Moving Force and a Stationary Force

3-108. When one of the linkup forces involved is stationary, contact points are selected near the restricted fire line or limit of advance where the axis of advance of the linkup force intersects the security elements of the stationary force. Contact points must be readily recognizable to both forces and should be posted on overlays. When possible, the moving force should halt short of the contact point and send a smaller force (patrol) forward to pinpoint the contact point. After the patrol makes contact with the stationary force, the patrol may leave a portion of patrol at the contact point then move back with the remaining members of the patrol to guide the battalion or subordinate units back to conduct linkup (figure 3-17, page 3-38).

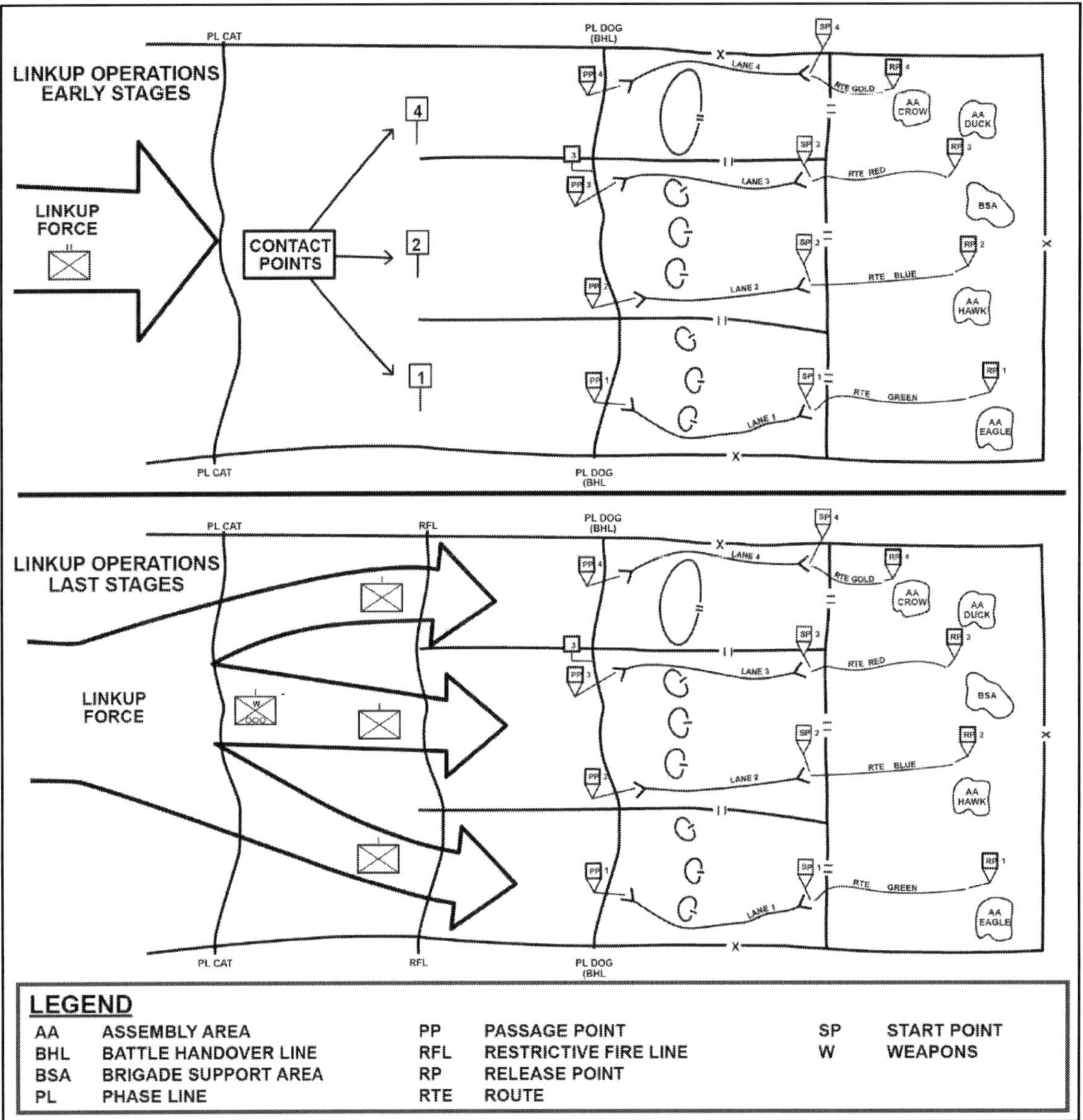

Figure 3-17. Linkup of moving force with stationary force (contiguous area of operation)

3-109. Alternate points are chosen so the units are prepared in case enemy activities cause linkup at places other than those planned. The number of contact or passage points selected depends on the terrain and number of routes used by the linkup force. The commander also may consider using multiple contact points and corresponding passage points to more rapidly move linkup forces through the friendly unit.

3-110. To facilitate a rapid passage of lines and to avoid inadvertent engagement of friendly forces, personnel in the linkup force must be thoroughly familiar with recognition signals and plans. As required, stationary forces assist in the linkup by opening lanes in minefields, breaching or removing selected obstacles, furnishing guides, providing routes with checkpoints, and designating assembly areas.

3-111. When the moving force conducts linkup with an encircled force, contact points are located and coordinated in the same manner as stated above when the stationary force is not encircled. After the moving force makes contact with the stationary force at the contact point, guides form the stationary force lead the battalion or subordinate units to linkup points entering the perimeter (figure 3-18). During this linkup operation, the battalion may carry additional supplies and materiel to reinforce the encircled force.

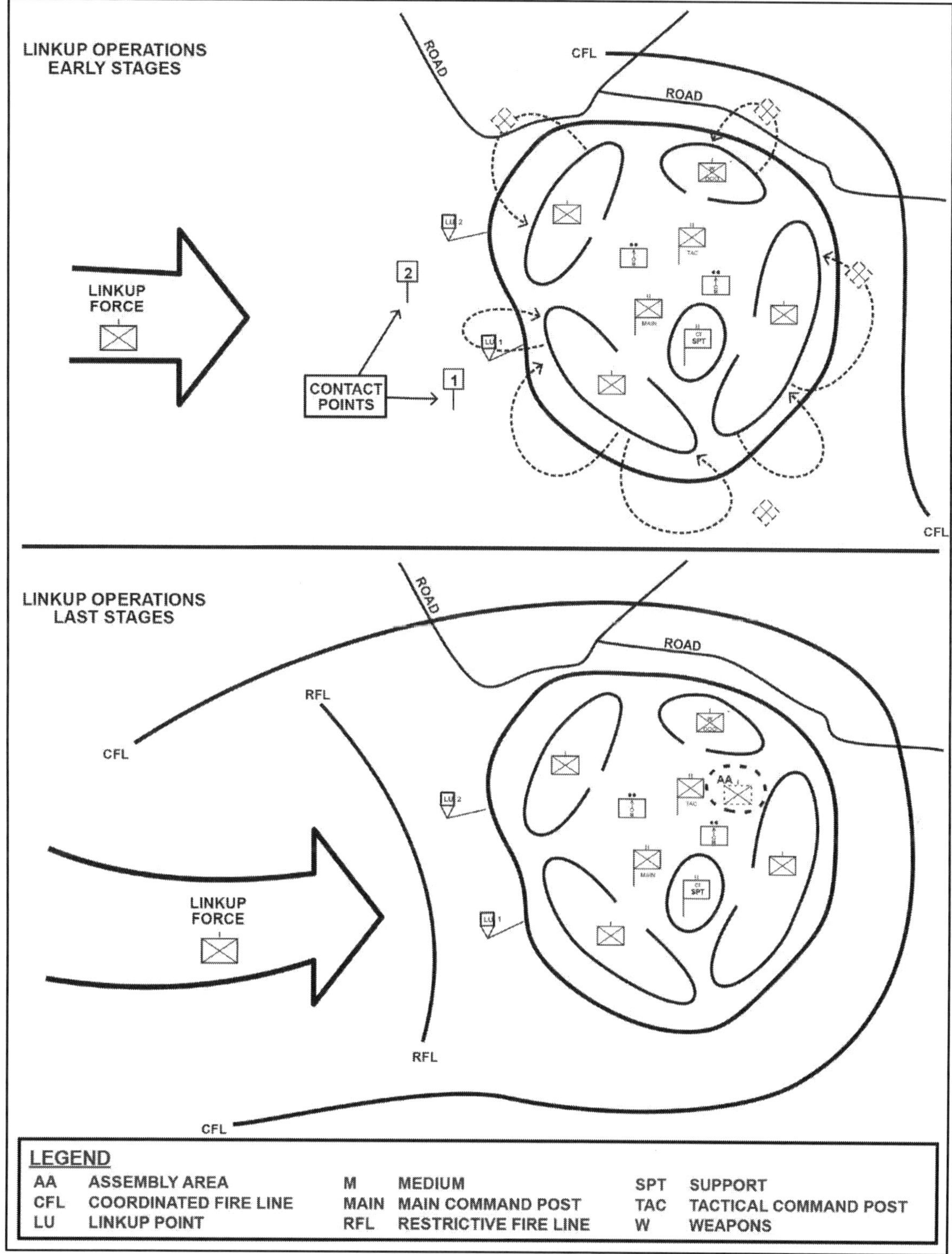

Figure 3-18. Linkup of moving force with stationary force (noncontiguous area of operation)

3-112. When the linkup is complete, the linkup force may join the stationary force, pass through the stationary force, go around the stationary force, or continue the attack. When the linkup force continues operations with the stationary force, a single commander for the overall force is designated. Objectives for the linkup provide for dispersion in relation to the stationary force. The linkup force may immediately pass through the perimeter of the stationary force, be assigned objectives within the perimeter, or be assigned objectives outside the perimeter, depending on the mission.

Linkup of Two Moving Forces

3-113. Linkup between two moving forces is one of the most difficult operations and is normally conducted to complete the encirclement of an enemy force or as the result from the linkup of two encircling arms conducting a double envelopment. (See paragraph 2-5 for information on these two forms of envelopment.) When it is tactically feasible, one of the linkup forces involved should become stationary. When this is not possible, primary and alternate linkup points for the two moving forces are established on the boundary where the two forces are expected to converge. As converging maneuver forces, move closer, positive control is coordinated to ensure they avoid firing on one another and to ensure the enemy does not escape (figure 3-19).

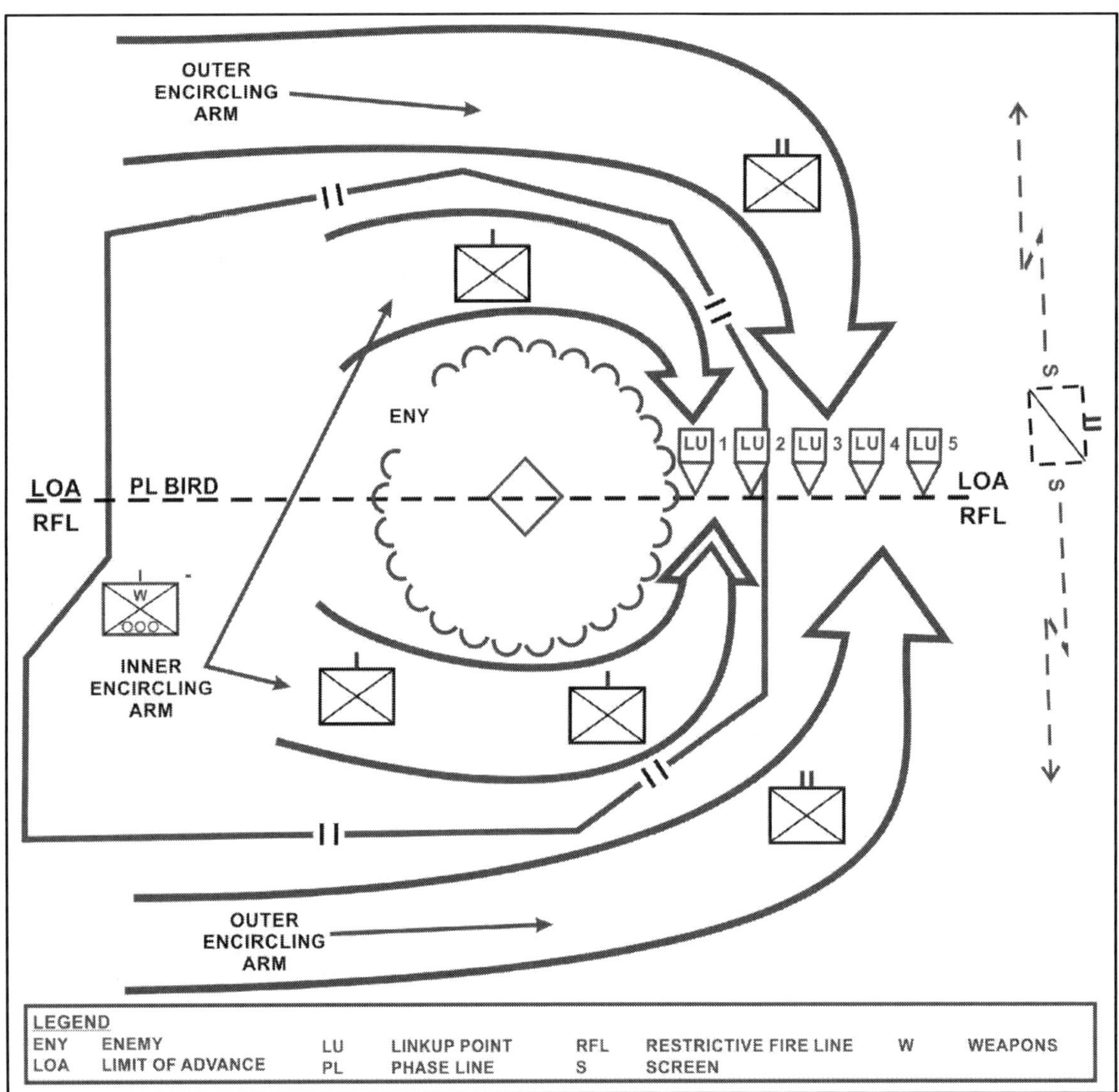

Figure 3-19. Linkup of two moving forces (encirclement operation)

3-114. During linkup, maneuver forces establish and maintain communications throughout operations. As each force maneuvers to linkup, progress is tracked and adjustments to the linkup plan are made as conditions dictate. For example, if two forces are involved in the operation and one is unable to travel at a speed commensurate with the plan, the linkup location may require adjustment. As linkup forces near each other, a restrictive fire line is established that also acts as a limit of advance at the point on the ground where linkup forces are anticipated to make contact. The tempo of the operation may be slowed to maintain positive control and to prevent fratricide.

During the timeframe right before linkup, commanders must be vigilant to ensure enemy forces do not slip between the two closing forces.

3-115. The fire support team changes or activates fire support coordination measures established for the operation based on the progress of the forces and the enemy situation. All changes are provided to subordinate fire support teams of the two maneuver forces involved in the linkup. As maneuver forces draw closer to each another, coordinated fire lines are canceled, and a restricted fire line is placed into effect to prevent fratricide between the converging forces. Once the linkup has occurred, fire support is organized per the higher headquarters plan for future operations.

3-116. The directing commander positions where best to observe the progress of the operation and maintains voice communications with the S-3. The S-3 is positioned based on the operational concerns expressed by the commander. For example, if a certain flank is of concern to the commander during the operation, or a shaping operation is required to penetrate the enemy's lines, then the S-3 locates where he can best observe the battalion's secondary action.

KEY DOCTRINAL TERMS AND DEFINITIONS

3-117. The following key doctrinal terms and definitions are used throughout this and other chapters and appendixes. Refer to referenced publications for additional information.

- *Block* is also an engineer obstacle effect that integrates fire planning and obstacle efforts to stop an attacker along a specific avenue of approach or to prevent the attacking force from passing through an engagement area (FM 3-90-1). ***Note.*** See paragraph 2-244 for the definition of the term, block, when used as a tactical mission task.
- *Counterreconnaissance* is a tactical mission task that encompasses all measures taken by a commander to counter enemy reconnaissance and surveillance efforts. Counterreconnaissance is not a distinct mission, but a component of all forms of security operations (FM 3-90-1).
- *Disengage* is a tactical mission task where a commander has the unit break contact with the enemy to allow the conduct of another mission or to avoid decisive engagement (FM 3-90-1).
- *Final protective fire* is an immediately available prearranged barrier of fire designed to impede enemy movement across defensive lines or areas. Also called FPF (JP 3-09.3).
- *Fix* is also an obstacle effect that focuses fire planning and obstacle effort to slow an attacker's movement within a specified area, normally an engagement area (FM 3-90-1). ***Note.*** See paragraph 2-245 for the definition of the term, fix, when used as a tactical mission task.
- *High-value target* is a target the enemy commander requires for the successful completion of the mission (JP 3-60).
- *Target area of interest* is the geographical area where high-value targets can be acquired and engaged by friendly forces. Also called TAI (JP 2-01.3).
- *Terrain management* is the process of allocating terrain by establishing areas of operation, designating assembly areas, and specifying locations for units and activities to deconflict activities that might interfere with each other (ADRP 5-0).

SECTION II – AREA DEFENSE

3-118. The *area defense* is a defensive task that concentrates on denying enemy forces access to designated terrain for a specific time rather than destroying the enemy outright (ADRP 3-90). The focus of the area defense is retaining terrain where the bulk of the defending force positions itself in mutually supporting, prepared positions. Units maintain their positions and control the terrain between these positions. The defeat mechanism is normally the massing of combat power to destroy the enemy in engagement areas. Area defenses are conducted when—

- Directed to defend or retain specified terrain.
- Forces available have less mobility than the enemy does.
- The terrain affords natural lines of resistance.
- The terrain limits the enemy to a few well-defined avenues of approach.

- There is time to organize the position.
- Conditions require the preservation of forces.

ORGANIZATION OF FORCES

3-119. The commander organizes an area defense around the static framework of the defensive positions seeking to destroy enemy forces by interlocking fire or local counterattacks. The commander has the option of defending in-depth or defending forward. The depth of the force positioning depends on the threat, task organization of the battalion, and nature of the terrain. When the commander defends forward within an area of operations, the force is organized so that most of the available combat power is committed early in the defensive effort. In an area defense, the commander organizes the defending force to accomplish information collection, security, main battle area, reserve, and sustainment missions. (Refer to FM 3-96 for additional information.)

INFORMATION COLLECTION (RECONNAISSANCE AND SURVEILLANCE)

3-120. The commander [assisted by the operations staff officer (S-3) and intelligence staff officer (S-2)] coordinates, integrates, and supervises the execution of information collection plans and operations to determine the locations, strengths, and probable intentions of the attacking enemy force. The commander places a high priority on early identification of the enemy's main effort and likely avenue of approach. The S-3 provides the staff coordination and integration of information collection and the allocation of resources through plans and operations to support the commander's visualization of the operation. The S-2 helps the commander focus and integrate assets and resources to satisfy information requirements of the battalion and higher headquarters.

3-121. The initial information collection plan is crucial to begin or adjust the information collection effort to help answer information requirements necessary in developing effective plans. The initial information collection plan sets reconnaissance, surveillance, and intelligence operations in motion, to include the insertion and extraction methods for reconnaissance, security, surveillance, and intelligence collection assets. The scheme of information collection will describe how the commander intends to use reconnaissance missions and surveillance tasks to support the concept of operations. This will include the primary *reconnaissance objective*—a terrain feature, geographic area, enemy force, adversary, or other mission or operational variable, such as specific civil considerations, about which the commander wants to obtain additional information (ADRP 3-90).

3-122. The commander ensures that the mission of reconnaissance forces and surveillance assets are coordinated with those of higher headquarters. In the defense, more so than in the offense, R&S operations overlap the unit's planning and preparing phases. Battalion subordinate commanders and leaders performing R&S tasks understand that these tasks often start before the commander fully develops the plan. Commanders and leaders have to be responsive to changes in orientation and mission. The commander ensures that the staff plans, prepares, and assesses the execution of the information collection portion of the overall plan.

Note. At the tactical level, reconnaissance, surveillance, security operations, and intelligence operations are the primary means by which a commander conducts information collection to answer the CCIRs and to support operations. Conduct intelligence operations, IBCT level (within the military intelligence company) and above, is conducted by military intelligence units and Soldiers to obtain information to satisfy formal information requirements. Intelligence operations collect information about the intent, activities, and capabilities of threats and other relevant aspects of the operational environment to inform the commander's decisions. (Refer to FM 2-0 for additional information.)

SECURITY

3-123. When the Infantry battalion is part of a larger unit's area defense, for example an IBCT. The IBCT commander establishes a forward security force to provide early warning, reaction time, and initial resistance to the enemy. The commander balances the need to create a strong security force to shape the battle with the resulting diversion of combat power from the main body's decisive operation. On a battlefield where forces are contiguous with one another, the location of security forces is usually in front of the main defensive positions. On a noncontiguous battlefield, security forces are located on avenues of approach between the protected force and known or suspected enemy locations. As security missions are usually time- or event-driven, the commander clearly identifies the mission of the security force by time, space, and amount of destruction to the enemy force, or by the type of enemy forces to destroy. (See FM 3-96.)

3-124. The *security area* is that area that begins at the forward area of the battlefield and extends as far to the front and flanks as security forces are deployed. Forces in the security area furnish information on the enemy and delay, deceive, and disrupt the enemy and conduct counterreconnaissance (ADRP 3-90). The IBCT commander normally identifies the security area, the battle handover line from the forward security force to the forward main battle area units, the exact trace of the forward edge of the battle area, and where the commander envisions the main fight will occur. From this, the battalion commander can determine how to structure the security area and the array of forces to employ.

3-125. Depending on the IBCT commander's guidance and plan, the Infantry battalion has several possible security force missions and options. The battalion conducting an area defense-contiguous area of operations (figure 3-20) may—

- Establish security area layered behind the IBCT security area to add depth to the operation.
- Secure its own flanks and rear while the IBCT assets conduct the primary security area operation forward of the forward edge of the battle area.
- Conduct its own security area operation, generally a task-organized combined arms team, in the absence of a higher echelon security force (figure 3-21, page 3-44).

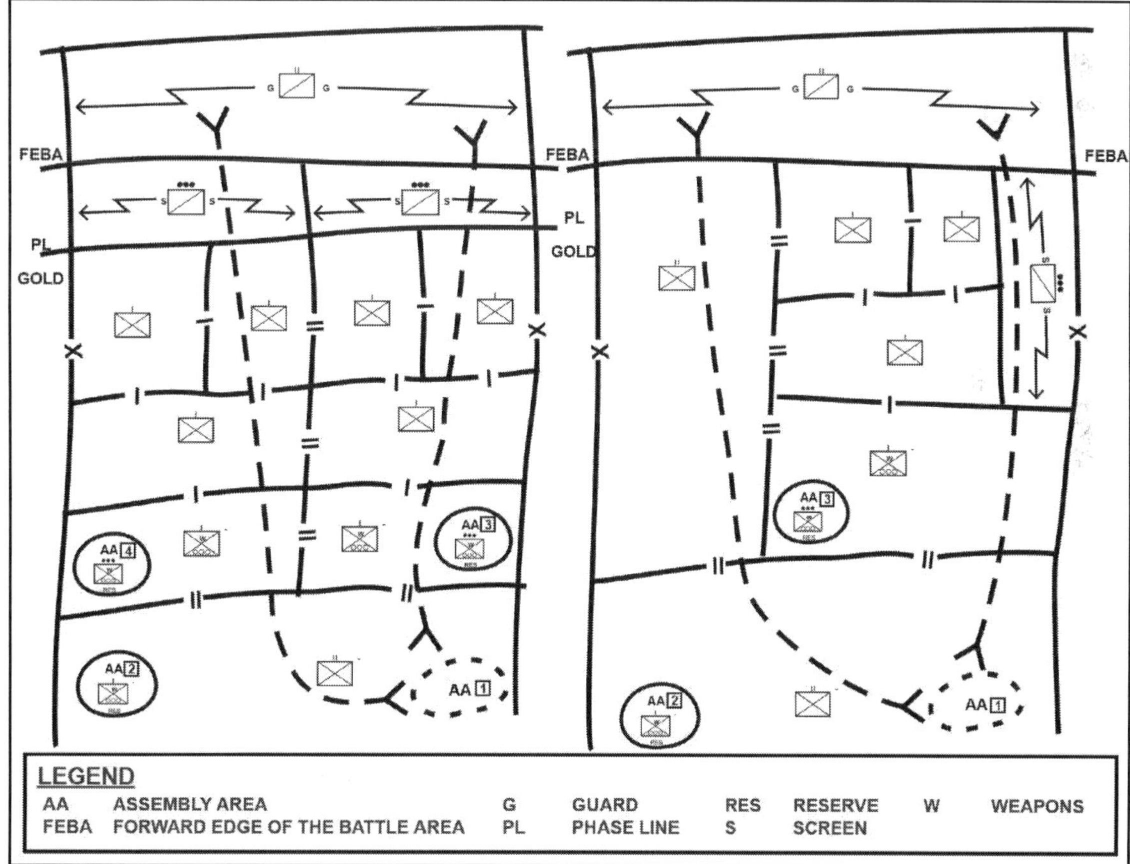

Figure 3-20. Security area–organization of forces (contiguous area of operations)

3-126. When the battalion commander organizes the battalion with its own security force forward of the battalion's main battle area, for example in an area defense-noncontiguous area of operations (figure 3-21, page 3-44). The commander may choose one or a combination of the following basic options:

- Use the scout platoon forward as a screening force.
- Use the scout platoon forward in conjunction with other maneuver elements in a screen mission.

- Use a rifle company, or in combination with the scout platoon, snipers, and assault platoon(s) forward in a guard mission.
- Use the weapons company (may be augmented with mortars, snipers, or scouts) forward, moving or stationary, to screen.

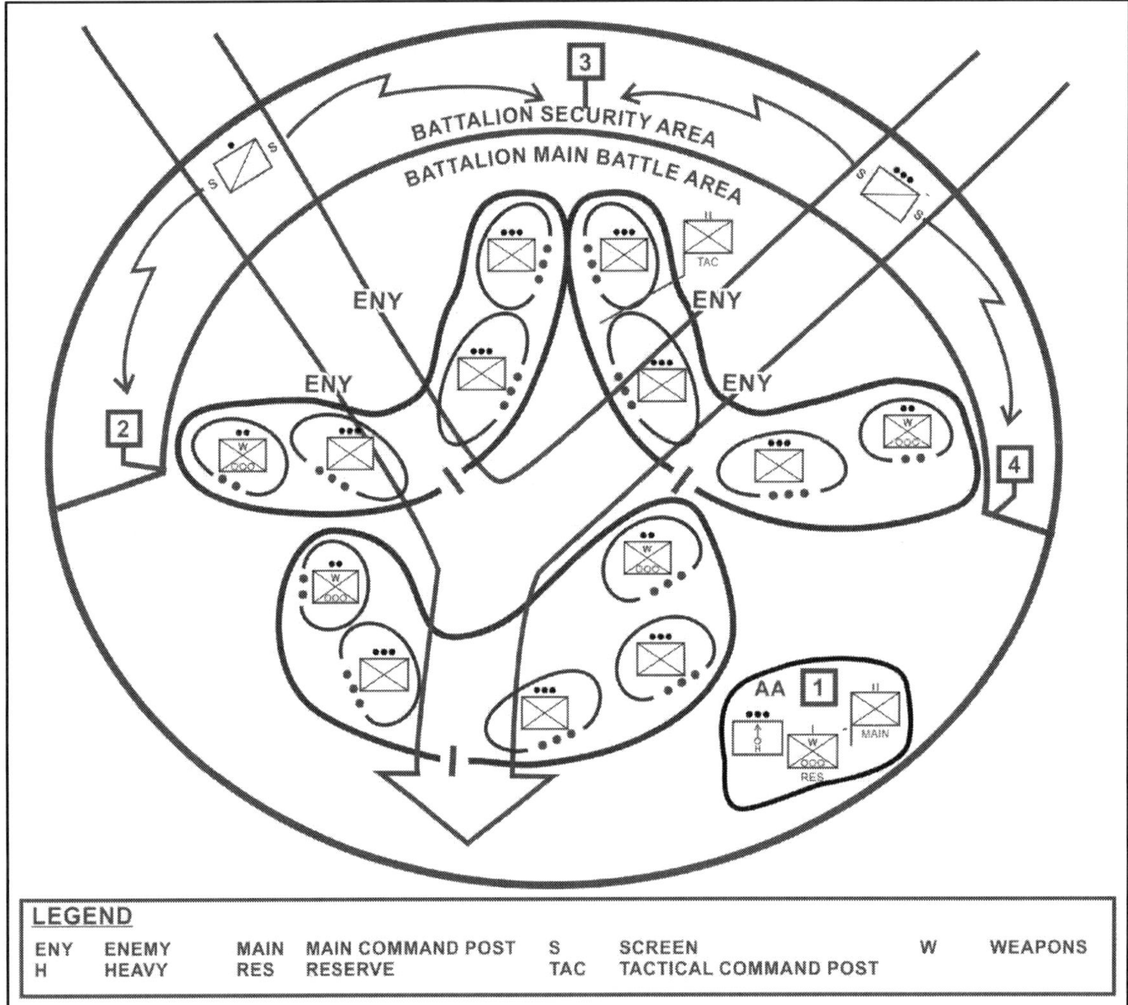

Figure 3-21. Security area–organization of forces (noncontiguous area of operations)

3-127. Although the commander usually assigns mobile units to the security area. The commander may insert Infantry rifle units and scouts into the security area through dismounted infiltration or air insertion by helicopters. These units may remain as a stay behind force or, on order, conduct dismounted exfiltration. Specific guidance and tasks provided by the commander to security area forces may include—

- Duration of the mission.
- Results to be achieved against the enemy to include specific guidance against the enemy reconnaissance forces.
- Specific CCIRs with associated named area of interest and target areas of interest and windows of observation with indicators.
- Avenues of approach to be monitored.
- Information requirements and last time information is of value.
- Mission command, fire support, sustainment, and protection.
- Engagement, disengagement, and withdrawal criteria.
- Rearward passage coordinating instructions.
- Follow-on tasks or missions.

Main Battle Area

3-128. The *main battle area* is the area where the commander intends to deploy the bulk of the unit's combat power and conduct decisive operations to defeat an attacking enemy (ADRP 3-90). The battalion's main battle area mission extends from the forward edge of the battle area to the unit's rear boundary. The forward edge of the battle area marks the foremost limit of the areas in which the majority of ground combat units deploy, excluding the areas in which security forces are operating. The commander selects the main battle area based on the higher commander's concept of operations, the IPB, the results of initial R&S efforts, and the commander's individual assessment of the situation. The commander's decisive operation in an area defense focuses on retaining terrain by using fires from mutually supporting, prepared positions supplemented by a counterattack(s) and the repositioning of forces from one location to another. The commander builds the battalion's decisive operation around identified decisive points, such as key terrain or high-payoff targets. The decisive operation normally involves close combat since an area defense emphasizes terrain retention.

3-129. Within the main battle area, the commander positions the main body of the battalion to halt, defeat, and ultimately destroy attacking enemy forces. The majority of the main body deploys into prepared defensive positions within the main battle area. Mobile elements of the force are positioned ready to deploy where and when needed. The commander delegates responsibilities within the main battle area by assigning areas of operations and establishing boundaries to and for subordinate units. The commander locates subordinate unit boundaries along identifiable terrain features and extends them beyond the forward line of own troops by establishing forward boundaries. Unit boundaries should not split avenues of approach or key terrain. When the battalion commander does not assign area of operations to subordinate units within the battalion. The commander is responsible for terrain management, security, clearance of fires, and coordination of maneuver among other doctrinal responsibilities within the entire area of operations.

3-130. As narrow frontages and deep areas of operations increase the elasticity of an area defense by increasing the commander's maneuver options. When the battalion's main body defends on a broad front the commander is forced to accept gaps and conduct noncontiguous operations. In this case the forward line of own troops will not be contiguous. Noncontiguous areas of operations place a premium on initiative, effective information operations, decentralized security operations, and innovative logistics measures. Noncontiguous operations within the battalion's main battle area complicate or hinder mutual support of combat, sustaining operations elements because of extended distances, and security risks associated with movement between subordinate units and elements. Defending shallow areas of operations reduces flexibility and requires the commander to fight well forward. As in the security area, the battalion commander may consider using the forward part of the main battle area to deploy mobile forces to destroy and delay the enemy. This includes mobile organic units as well as any attached armored, mechanized, or wheeled elements.

Reserve

3-131. The *reserve* is that portion of a body of troops, which is withheld from action at the beginning of an engagement, in order to be available for a decisive movement (ADRP 3-90). The reserve mission provides the commander with the flexibility to exploit success or deal with a tactical setback and the flexibility to respond in situations where there is a great deal of uncertainty about the enemy. The commander designates and positions the reserve in a location where it can effectively execute several contingency plans. The commander considers terrain, potential engagement areas, and probable points of enemy penetrations, commitment criteria, and routes. The commander may have a single reserve under battalion control or, if the terrain dictates, the companies may designate their own reserves. The reserve should be positioned outside the enemy's direct fire range in a covered and concealed position. Information concerning the reserve must be considered essential elements of friendly information and protected from enemy reconnaissance. The commander may choose to position the reserve forward initially to deceive the enemy or to move the reserve occasionally in order to prevent it from being targeted by enemy indirect fires.

3-132. The size of the reserve depends upon the size of the area covered in the defense, potential missions, and the clarity of expected enemy action. The battalion may need to defend an area of operations so large that only local reserves are feasible due to reaction time and the number of potential enemy course of actions. The size of the reserve is also relative to the commander's uncertainty about the enemy's capabilities and intentions. The more uncertainty that exists, the larger the reserve. When the commander knows the enemy's size, dispositions, capabilities, and intentions, a comparatively small reserve may be required. Battalion security forces and security forces of the IBCT may be able to clarify enemy intentions by collecting information on the massing of forces, electronic signals, and troop movement. The collection of information via joint, interorganizational, and multinational teams intelligence, surveillance, and reconnaissance plans and IBCT and battalion information collection plans allow the commander to gain an understanding of the situation and to better focus efforts toward the size and task organization of the reserve.

3-133. In an area defense, the most likely mission of the battalion reserve is to conduct a counterattack in accordance with previously prepared plans. The reserve conducts local counterattacks to restore the integrity of the battalion's defense or to exploit opportunities. As the reserve is not a committed force, the commander can assign it a wide variety of tasks on its commitment. In certain situations, it may become necessary to commit the reserve to block enemy penetration or to reinforce fires into an engagement area. Once committed, the reserve's operation usually becomes the battalion's decisive operation. However, the commander can commit the reserve to shaping operations to allow the ongoing decisive operation to achieve success. On its commitment, the commander should designate another uncommitted force as the reserve. If the commander does not have that flexibility, the commander holds the reserve for commitment at a decisive moment and accepts the associated risk.

3-134. The battalion commander and staff determine where and under what conditions the reserve force is likely to be employed in order to position it effectively. The commander provides specific planning guidance to the reserve to include priority for planning. The reserve force commander analyzes assigned planning priorities, conducts the coordination with units that will be affected by maneuver and commitment, and provides information to the commander and staff on routes and employment times to designated critical points on the battlefield. The reserve commander should also expect to receive specific decision points and triggers for employment on each contingency. This guidance allows the reserve commander to conduct quality rehearsals and to anticipate commitment as he monitors the fight. The commander develops a plan to reconstitute another reserve force once the original reserve force is committed, most often accomplished with a unit out of contact. Tasks issued as planning priorities may include one or more of the following:

- Counterattack locally.
- Defeat enemy air assaults.
- Block enemy penetrations.
- Reinforce a committed company.
- Protect rear area operations.
- Secure high-value assets.

3-135. When the commander is not able to resource a separate reserve, the commander may (though not the preferred option) constitute all or a portion of the reserve from the security force after it conducts a rearward passage of lines through main battle area units. In cases where the security force is the reserve for an area defense or employed in other tasks such as security operations, rear area security, or obstacle emplacement. The commander must withdraw the force to have sufficient time to occupy its reserve position, perform the necessary degree of reconstitution, and prepare plans for its reserve role. The commander must balance these uses with the needs to protect the reserve, and with the reserve commander's requirement to conduct troop-leading procedures, coordination, and reconnaissance. Before battle handover, the battalion commander must state the acceptable risk to the security force or the disengagement criteria in quantifiable terms, such as friendly strength levels, time, or event. In these cases, after completing the rearward passage, the security force moves to an assembly area to prepare for its subsequent reserve operations. The assembly area for the reserve should be free from enemy interference and clear of main battle area units, main supply routes, and the movements of other portions of the reserve.

Note. Defending commanders usually have difficulties establishing and resourcing reserve forces because they are normally facing an enemy with superior combat power. Nevertheless, commanders at each echelon down to the battalion retain reserves as a means of ensuring mission accomplishment and for exploiting opportunities through offensive action. Company commanders may retain a reserve based on the mission variables of METT-TC. Commanders do not place artillery and other fire support systems in reserve, systems committed to echelon support operations are not in reserve.

Sustainment

3-136. The sustainment mission in an area defense requires a balance between establishing forward supply stocks of ammunition, barrier material, and other supplies in sufficient amounts, and having the ability to move the supplies in conjunction with enemy advances. Planners must balance their employment of maintenance and medical support forward and the need for mobility.

3-137. Echeloned trains (combat and field) provide the commander with the flexibility to execute the sustainment mission in support of the battalion's area defense. The combat trains, when established, normally locate in the battalion's support area and is position dependent upon the mission variables of METT-TC. Sustainment missions from the combat trains are conducted throughout the battalion's area of operations as well as the brigade's support area.

Note. In contiguous areas of operations, a support area is an area for any commander that extends from its rear boundary forward to the rear boundary of the next lower level of command.

3-138. As the majority of higher-level sustainment missions in support of the battalion originate at the IBCT level, sustainment units above the battalion locate in the brigade support area (BSA), division support area, or other support area. The commander considers terrain and probable points of enemy penetrations as well as routes and the ability to support the battalion when positioning support areas. (Refer to appendix H and FM 3-96 for additional information.)

Note. A majority of the BSB is located in the BSA, but the combat sustainment support battalion and other echelon above brigade sustainment can be in a BSA, division support area, or other support area. Dependent on METT-TC the battalion's field trains may located in the BSA. (Refer to FM 3-96 and appendix H for additional information on positioning the battalion's field trains.)

DOCTRINAL BASIS FOR AN AREA DEFENSE

3-139. An area defense capitalizes on the strength inherent in a closely integrated defensive organization on the ground. The defending force limits the enemy's freedom of maneuver and channels the enemy into designated engagement areas. Shaping operations, designed to regain the initiative, limit the attacker's options and disrupt the enemy's plan. Shaping operations, coupled with sustaining operations, combine with the decisive operations of the main battle area force to defeat the enemy.

Planning

3-140. Planning an area defense is a complex effort requiring detailed planning and extensive coordination. In the defense, synchronizing the effects of battalion warfighting functions with information and leadership allows the battalion commander to apply overwhelming combat power against selected advancing enemy forces to unhinge the enemy commander's plan and destroy the enemy's combined-arms team. An area defense is a mix of static and dynamic actions. As an operation evolves, the commander may shift decisive and shaping operations to disrupt and maintain pressure on the enemy and to deny the enemy freedom of maneuver and the initiative. The commander's defensive plans must address how the preparations for, and the conduct of, the area defense impact the civilian population of the area of operations. (Refer to FM 3-96 and FM 3-90-1 for additional information.)

Mission Command

3-141. The battalion commander considers the mission variables of METT-TC to determine how best to concentrate efforts and economize forces. A detailed terrain analysis might be the most important process that the commander and staff complete. A successful defense relies on a complete understanding of terrain in order to determine likely enemy course of action and the best positioning of the battalion assets to counter them. Initially, integrated with the staff's IPB, the commander visualizes the enemy's anticipated actions. The battalion commander and staff refine the IBCT's IPB to focus on the details of the operation in the battalion's area of operation. The IBCT commander normally defines where and how the IBCT will destroy or defeat the enemy and how he envisions the battalion executing its portion of the IBCT fight.

3-142. The battalion commander and staff base their determination of how and where to defeat the enemy, on where they believe the enemy will go, the terrain, the forces available, and the IBCT commander's intent. The battalion commander may define a defeat mechanism that includes the use of single or multiple counterattacks to achieve success. The battalion commander and staff analyze their unit's role in the IBCT fight, and determine how to accomplish the IBCT commander's intent. In an area defense, the battalion usually achieves success by massing the cumulative effects of obstacles and fires to defeat the enemy forward of a designated area, often in conjunction with an IBCT counterattack. When follow by a delaying operation, the battalion achieves success by combining maneuver, fires, and obstacles, and by avoiding decisive engagement until conditions are right to achieve the desired effect of gaining time or shaping the battlefield for a higher echelon counterattack.

3-143. The commander and staff analyze the forces and assets available, paying particular attention to the obstacle assets and fire support allocated by the IBCT. The staff defines the engineer and fire support allocation in terms of capability, resources, and priority. For example, it should define engineer capability in terms of the number of obstacles of a specific effect engineers can emplace in the time available. Fire support analysis should include the number of targets that can be engaged with an expected result at what point in the battle. Control measures within a commander's area of operation include designating the security area, the battle handover line, the main battle area with its associated forward edge of the battle area, and the echelon support area. The commander uses battle positions, direct fire control, and fire support coordination measures, in addition, to those control measures referenced in chapter 2 of this publication to further synchronize the employment of combat power.

3-144. With a definitive understanding of the assets available, the commander and staff determine what effects combat forces, fires, and obstacles must achieve on enemy formations by avenue of approach and how these effects will support both the IBCT's and the battalion's defeat mechanism. The battalion commander, supported by the staff, assigns a mission for each subordinate unit, establishes priorities for protection and sustainment, and develops obstacle and fire support plans concurrently with the defensive force array, defining a task and purpose for each obstacle and target in keeping with the commander's stated fire support tasks and intended obstacle effects. The desired end state is a plan, which defines how the commander intends to mass the effects of direct and indirect fires with obstacles and use of terrain to shape the battlefield and to destroy or defeat the enemy. For example, the commander may plan to *canalize*, a tactical mission task in which the commander restricts enemy movement to a narrow zone by exploiting terrain coupled with the use of obstacles, fires, or friendly maneuver (FM 3-90-1), the enemy's movement into a predetermined position where the enemy is vulnerable to piecemeal destruction.

3-145. In planning the transition to countermobility and survivability work in detail, the commander ensures adequate time for subordinate troop-leading procedures. Prior to transition, the commander—

- Sites situational obstacles early.
- Plans multiple locations to support depth and flexibility in the defense.
- Ensures adequate time, resources, and security for obstacle emplacement systems.
- Integrates triggers for execution of situational and reserved obstacles in the decision support template.
- Focuses the countermobility effort to shape the enemy's maneuver into positions of vulnerability.
- Ensures adequate mobility for withdrawing security forces, the reserve, and repositioning of main battle area forces.
- Plans appropriately for Class IV and Class V (mines) download sites as near to the emplacement location as is practical.
- Establishes early on the priority of effort and the priority of support.

3-146. The commander considers the entire area of operations, the enemy, and information collection activities (intelligence operations, reconnaissance, security operations, and surveillance) necessary to shape an operational environment and civil conditions. This includes the conduct of noncombatant evacuation operations for U.S. civilians and other authorized groups. The commander's legal obligations to that civilian population must be met. Ideally, the host nation government will have the capability to conduct the six primary stability tasks. (See chapter 4.) To the extent that a host nation government is unable to conduct the immediate subordinate stability tasks, the defending unit will have to attempt to make up the shortfall.

Movement and Maneuver

3-147. Maneuver allows the battalion commander to take full advantage of the area of operations and to mass and concentrate when desirable. Maneuver, through movement in combination with fire, allows the battalion to achieve a position of advantage over the enemy to accomplish the mission. The commander studies the ground and selects positions that allow the massing of fires on likely approaches. The commander concentrates force within the main battle area and positions security, intelligence, and reconnaissance force and surveillance assets throughout the area of operations.

3-148. When conducting an area defense the commander combines static and mobile actions to accomplish the mission. Static actions usually consist of fires from prepared positions. Mobile actions include using the fires provided by units in prepared positions as a base for counterattacks and repositioning units between defensive positions. The commander can use the reserve and uncommitted forces to conduct counterattacks and spoiling attacks to desynchronize the enemy forces or prevent them from massing.

Area of Operations

3-149. A defense in an area of operation (area defense) provides the greatest degree of freedom of maneuver (movement in combination with fires) to achieve a position of advantage in respect to the enemy. The battalion most often uses this method when it has an adequate amount of depth and width to the battlefield. For the battalion defense to be cohesive, phase lines, engagement areas, battle positions, and obstacle belts help to coordinate subordinate maneuver forces and achieve synchronized action. Assignment of area of operations allow flexibility and prevents the enemy from concentrating overwhelming firepower on the bulk of the defending force. Forces defending against an enemy with superior mobility and firepower must use the depth of their positions to defeat the enemy. The depth of the defense must come from the initial positioning of units throughout the area of operation, not from maneuvering. A properly positioned and viable reserve and counterattack force enhances depth.

3-150. In *contiguous area of operations*—where all of a commander's subordinate forces' areas of operations share one or more common boundaries (FM 3-90-1)—the IBCT and subordinate battalions deploy the bulk of their combat power in the main battle area (see figure 3-7, page 3-18). The IBCT's main battle area extends from the forward edge of the battle area to the forward battalions' rear boundaries. Battalion main battle areas are subdivisions of the IBCT's main battle area. The forward edge of the battle area marks the foremost limit of the areas in which the preponderance of ground combat units deploy, excluding the areas in which security forces are operating. The IBCT commander assigns the battalion main battle area by establishing unit boundaries. The IBCT and battalion commanders establish area of operations and battle positions (primary, alternate, supplementary, subsequent, and strongpoints) to implement their concept of operation. As in all operations, the commander promotes freedom of action by using the least restrictive control measures necessary to implement tactical concepts. On the battlefield where forces are contiguous with one another, the location of security forces is usually in front of the main defensive positions. The commander uses a reserve force to reinforce fires, add depth, block penetrations, restore positions, or counterattack to destroy enemy forces and seize the initiative.

3-151. In *noncontiguous area of operations*—where one or more of the commander's subordinate force's areas of operation do not share a common boundary (FM 3-90-1)—the battalion often must defend either on a broad front or in an area of operation so large that employing units in mutually supporting positions is unrealistic. Noncontiguous area of operations complicate or hinder mutual support of combat, sustaining operations elements because of extended distances, and security risks associated with movement between subordinate units and elements. In noncontiguous operations when there is no cohesive battalion main battle area, company battle areas extend from the unit's location to the battalion's area of influence. Noncontiguous operations require a judicious effort by the commander and staff in determining the positioning of maneuver forces. During the terrain analysis, the commander and staff must look closely for choke points, intervisibility lines, and reverse slope opportunities in order to take full advantage of the battalion's capabilities to mass and concentrate the effects of overwhelming combat power while providing protection for the units.

3-152. When defending in-depth on a noncontiguous battlefield, the commander may position defending subordinate units in successive layers of battle positions along likely enemy avenues of approach (see figure 3-8, page 3-19). Battle position perimeters vary in shape depending on the terrain and situation. Perimeter shapes conform to the terrain features that best use friendly observation and fields of fire. The commander can increase the effectiveness of unit perimeters by tying it into a natural obstacle, such as a river, which allows the defending unit to concentrate its combat power in more threatened areas. As the commander determines the most probable direction of enemy attack, that part of the perimeter covering that approach may be reinforced with additional resources. The commander positions the reserve to block the most dangerous avenue of approach and assigns on-order positions on other critical avenues. Security forces locate on avenues of approach between the protected force and known or suspected enemy locations. Noncontiguous operations place a premium on initiative, decentralized security operations, and innovative logistics measures.

Position Selection

3-153. During an area defense, the scout platoon may position forward of the Infantry battalion's main battle area. The commander gives the scout platoon specific priority information requirements to allow for its efficient deployment within the battalion's security area and to position itself for the preparation and execution of the area defense. The battalion scout platoon, often with augmentation, attempts to discern enemy intentions by collecting information on the massing of forces and troop movement. On a noncontiguous battlefield, the scout platoon is positioned between main body forces and known or suspected enemy locations. Significant consideration must be given when planning the communications package, sustainment, mobility assets, engagement, disengagement, bypass, and withdrawal criteria; and the indirect fire support coverage for all elements participating within the security area. The staff must consider redundant information collection means in the event one of the elements operating in the security area is compromised or incapacitated.

3-154. Once the battalion commander has assigned area of operations to maneuver companies, the commander determines any potential area between higher headquarters, adjacent, and subordinate units that is unassigned. Any area within the battalion area of operation that is not assigned to a subordinate unit remains the responsibility of the battalion. The battalion may plan to cover this area with reconnaissance and surveillance assets, to include ground sensors and unmanned aircraft systems along with higher echelon information collection assets. However depend upon mission analysis, the commander may accept risk by placing no assets to monitor or react to this unassigned area. The battalion plans local counterattacks to isolate and destroy any enemy that manages to penetrate through a gap in the area of operations, including unassigned areas.

3-155. When conditions favors a defense that takes advantage of the mobility of the weapons company and scout platoon, combined with the defensive abilities but relatively immobility of its rifle companies. The battalion can increase the depth of the security area or main battle area, emphasizing the mobility and firepower of the weapons company while the rifle companies develop strong defensive positions. Once completing their tasks, these mobile elements then conduct a passage of lines and revert to battalion control or are attached to rifle companies. The rifle companies then pick up the fight and defend in place (see figure 3-10, page 3-21). The need for flexibility for the companies requires graphic control measures to assist in control during local counterattacks and repositioning of forces. Specified routes, phase lines, attack-by-fire positions, support areas, engagement areas, target reference points, and other fire control measures are required for the effective synchronization of maneuver.

3-156. When deciding where to place the reserve, the commander decides whether to orient the reserve on its most likely mission or its most important mission. The commander and staff expend significant effort during the planning process to ensure the commander can effectively use the reserve when needed. The commander may locate the reserve within the area of operation where it can employ the road network to rapidly displace throughout the area of operation in response to a number of opportunities or contingencies. In restrictive terrain that lacks routes for movement, the commander can task organize the reserve into small elements and position them where they can react quickly to local combat developments. This dispersion provides increased protection but reduces the ability of the reserve to mass fires.

3-157. Battalion sustainment operations (see appendix H for information on echeloned trains) are positioned within or outside the battalion's area of operations. During contiguous operations the IBCT will direct, coordinate, and monitor most support area operations for the battalion from the BSA. During noncontiguous operations, positioning is critical to the responsiveness and survivability of support areas. The battalion commander protects support forces and critical assets by conducting area security operations. The commander uses area security operations to protect the rapid movement of combat and field trains or protect forward positioned stocks and cached commodities. The commander clearly defines responsibilities for the security of units within echelon support areas and coordinates to mitigate the effects of security operations on the primary functions of units located within echelon support areas.

Battle Positions

3-158. A *battle position* is a defensive location oriented on a likely enemy avenue of approach. (ADRP 3-90). The battle position is an intent graphic that depicts the location and general orientation of the majority of the defending forces. The commander's use of a battle position does not direct the position of the subordinate's entire force within its bounds since it is not an area of operation. The subordinate commander can move elements freely within the assigned battle position. To comply with the commander's intent, a force can maneuver outside the battle position to adjust fires or to seize opportunities for offensive action. Battalion security, fire support, protection, and sustainment assets are often positioned outside the battle position with approval from the headquarters assigning the battle position. Repositioning of units between battle positions must be carefully coordinated to prevent fratricide. (Refer to FM 3-96 and FM 3-90-1 for additional information.)

3-159. The IBCT commander assigns a battle position to a battalion to control the battalion's fires, maneuver, and positioning (figure 3-22, page 3-52). The commander assigns boundaries to provide space for the battalion security, support, and sustainment elements that operate outside a battle position. When the IBCT commander does not establish unit boundaries, the IBCT is responsible for fires, security, terrain management, and maneuver between positions of adjacent battalions. The battle position prescribes a primary direction of fire by the orientation of the position. The IBCT commander defines when and under what conditions the battalion can displace from the battle position or maneuver outside it. Planning considerations for a battle position, although not inclusive, may include—

- Establishment of outposts and observation posts.
- Development of integrated fires plans that include final protective fires.
- Priorities of work.
- Counterattack plans.
- Stockage of supplies.
- Integration and support of subordinate forces outside the strong point.
- Actions of adjacent units.

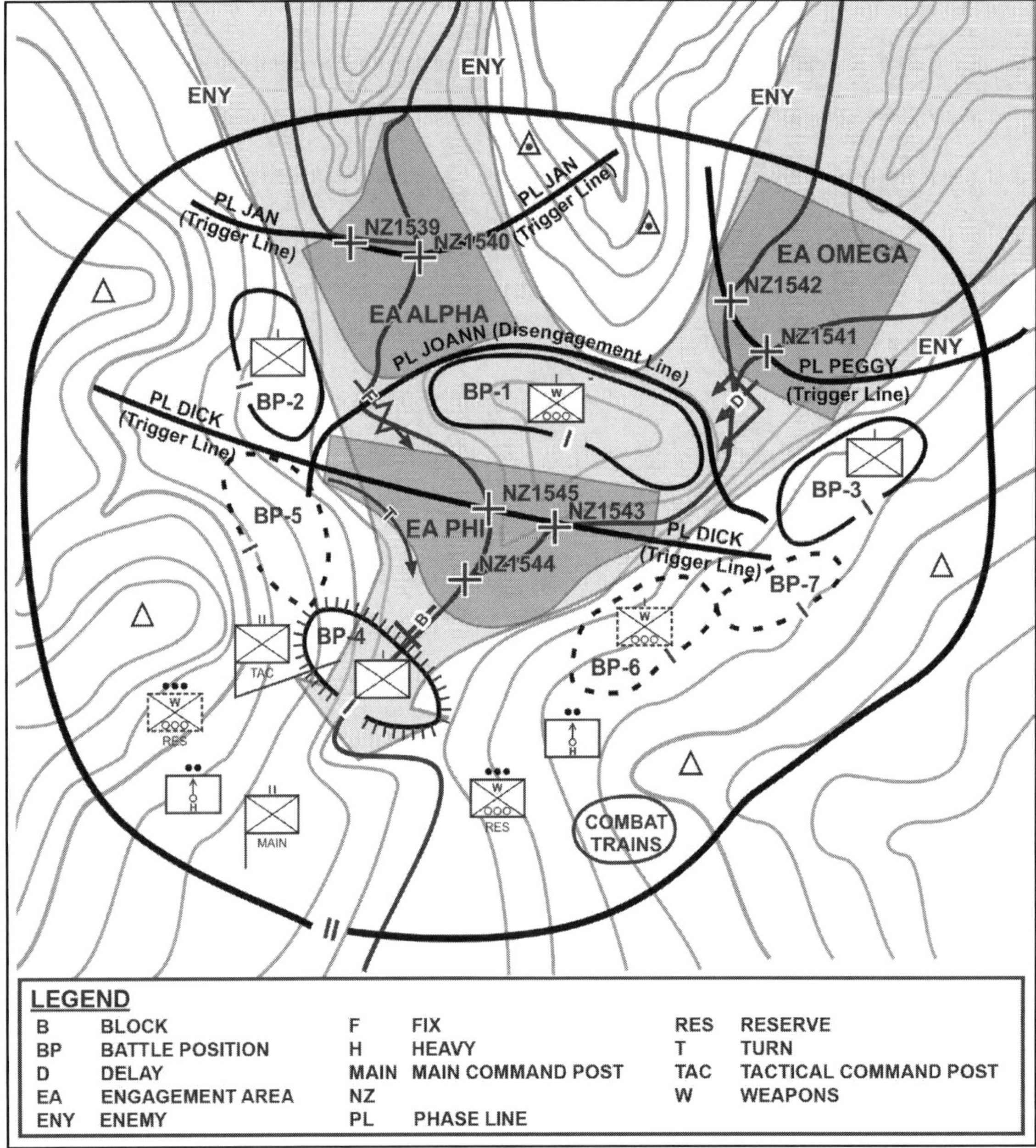

Figure 3-22. Battalion battle position (disposition of forces), example

3-160. The three levels of preparation for a battle position are occupy, prepare, and plan. Occupy is the complete preparation of the position where subordinate units will initially defend. Units fully plan, prepare, and occupy positions before the "defend no later than" time specified in the battalion operations order. Companies rehearse the occupation, and company commanders establishes a trigger for occupation of the position. Units occupying the battle position, despite time constraints, dig in survivability positions, construct fighting positions, designate target reference points, develop direct and indirect fire plans, emplace obstacles, clear fields of fire, and preposition ammunition.

3-161. The use of on-order battle positions with the associated tasks prepare or plan adds flexibility and depth to the defensive plan. Prepare missions, normally critical to the defense maintain security on the position and on the routes to the position. Prepare per se means that the unit fully reconnoiters the position and the corresponding engagement area, marking positions in the battle position and fire control measures in the engagement area. From the battle position, units accomplish all actions to enable the execution of the mission immediately on occupation.

Planning, coordination, and rehearsals are required for the unit to displace to the battle position and accomplish the mission. Plan means that the unit fully reconnoiters the engagement area and battle position. The unit specifically plans tentative unit positions in the battle position, and establishes fire control measures in the engagement area. The unit also coordinates and plans for defense from the position. Leaders reconnoiter, select, and mark positions, routes, and locations for security elements. Then coordinate movement and other actions, such as preparing obstacles and occupation plans, with other elements of the battalion.

3-162. The commander allocates space to elements within the battle position area based on the space available, terrain, and mission task. The battalion commander thinks two levels down or in terms of platoon battle positions when he selects a battle position for a subordinate company. When practical, the commander should allow enough space on each battle position for dispersed primary, alternate, and supplementary positions for key weapons. The battalion commander can vary the number of maneuver elements in the battle position by allocating larger company battle positions. Battle positions can also reflect positions in-depth. They may take a shape other than the standard oblong shape, which suggests a linear defense within the battle position. Large positions also increase dispersion to counter enemy fires. The commander can combine company area of operations and battle positions in the battalion area of operation to suit the tactical situation.

3-163. The five types of battle positions are primary, alternate, supplementary, subsequent, and strong point. The commander always designates the primary battle position. A *primary position* is the position that covers the enemy's most likely avenue of approach into the area of operations (ADRP 3-90). A primary position is the best position to accomplish the assigned mission. Routes between positions should be well known and rehearsed (optimally under the same conditions expected during execution). The commander designates and prepares alternate, supplementary, and subsequent positions as required.

3-164. An *alternate position* is a defensive position that the commander assigns to a unit or weapon for occupation when the primary position becomes untenable or unsuitable for carrying out the assigned task (ADRP 3-90). The alternate position covers the same area as the primary position. These positions allow the defender to carry out original task, such as covering an avenue of approach or engagement area, using the original direct fire plan. Alternate positions increase the defender's survivability by allowing engagement of the enemy from multiple positions and movement to other positions in case of suppressive or obscuring fires.

3-165. A *supplementary position* is a defensive position located within a unit's assigned area of operations that provides the best sectors of fire and defensive terrain along an avenue of approach that is not the primary avenue where the enemy is expected to attack (ADRP 3-90). A supplementary position is assigned when more than one avenue of approach into a unit's area of operations exist.

3-166. A *subsequent position* is a position that a unit expects to move to during the course of battle (ADRP 3-90). The defending unit may have a series of subsequent positions (particularly in delay operations), each with associated primary, alternate, and supplementary positions.

3-167. A *strong point* is a heavily fortified battle position tied to a natural or reinforcing obstacle to create an anchor for the defense or to deny the enemy decisive or key terrain (ADRP 3-90). A strong point implies retention of terrain to control key terrain and blocking, fixing, or canalizing enemy forces. Before assigning a strong point mission, the commander ensures that the strong point force has sufficient time and resources to construct the position, which requires significant engineer support. Defending units require permission from the higher headquarters to withdraw from a strong point. Once the strong point is occupied, all units and equipment not essential to the defense are displaced from the strong point. All combat, maneuver enhancement, and sustainment assets within the strong point require fortified positions. Extensive protective and tactical obstacles are required to provide an all-around defense.

Engagement Areas

3-168. An *engagement area* is an area where the commander intends to contain and destroy an enemy force with the massed effects of all available weapons and supporting systems (FM 3-90-1). The success of any engagement depends on how effectively the battalion integrates the direct fire plan, the indirect fire plan, the obstacle plan, Army aviation fires, close air support, and the terrain within the engagement area to achieve the battalion's tactical purpose.

3-169. Effective use of terrain reduces the effects of enemy fires, increases the effects of friendly fires, and facilitates surprise. Terrain appreciation—the ability to predict its impact on operations—is an important skill for every leader. For tactical operations, commanders analyze terrain using the five military aspects of terrain, expressed in the Army memory aid OAKOC (obstacles, avenues of approach, key terrain, observation and fields of fire, and cover and concealment). See ATP 2-01.3 and ATP 3-34.80 for information on analyzing the military aspects of terrain.

3-170. Figure 3-22, on page 3-52 depicts three company engagement areas used within the context of a battalion area defense. Within the defense, the battalion commander positioned defending companies (defense in-depth) in successive layers of battle positions (primary, alternate, and strong point) along the enemy's most likely avenue of approach.

Engagement Area Development

3-171. The commander develops engagement areas, to include engagement criteria and priority, to cover each enemy avenue of approach. Within the battalion's main battle area, the commander determines the size and shape of the engagement area(s) by the relatively unobstructed line-of-sight from the weapon systems firing positions and the maximum range of those weapon systems. Once the commander and staff select engagement areas, the commander arrays available forces and weapon systems in positions to concentrate overwhelming effects into these areas. The commander routinely subdivides engagement areas into smaller engagement areas for subordinates using one or more target reference points or by key terrain or prominent terrain feature. The commander assigns sector of fires to subordinates to ensure complete coverage of engagement areas and to prevent fratricide and friendly fire incidents. Responsibility for an avenue of approach or key terrain is never split.

3-172. Security area forces, to include field artillery fire support teams and observers, employ fires to support operations forward of the battalion's main battle area using precision and other munitions to destroy enemy reconnaissance and security forces and identified high-payoff targets, and to attrit enemy forces as they approach the battalion's main battle area. The employment of fires within the security area also help to deceive the enemy about the location of the battalion's main battle area. The battalion fire support officer plans the delivery of fires at appropriate times and places throughout the area of operations to slow and canalize the enemy force as the enemy approaches. The employment of fires allows security area forces to engage the enemy without becoming decisively engaged. To prevent fratricide, the commander designates no-fire areas where security area forces are positioned. The commander uses fires to support the withdrawal of security forces once shaping operations are completed within the security area and the defending unit is prepared to conduct main battle area operations.

3-173. *Engagement criteria* are protocols that specify those circumstances for initiating engagement with an enemy force (FM 3-90-1). Engagement criteria may be restrictive or permissive in nature. For example, the battalion commander may instruct a subordinate company commander not to engage an approaching enemy unit until the enemy commits to an avenue of approach. The commander establishes engagement criteria in the direct fire plan in conjunction with engagement priorities and other direct fire control measures to mass fires and control fire distribution.

3-174. *Engagement priority* is the order in which the unit engages enemy systems or functions (FM 3-90-1). The commander assigns engagement priorities based on the type or level of threat at different ranges to match organic weapon systems capabilities against enemy vulnerabilities. Engagement priorities are situationally dependent and used to distribute fires rapidly and effectively. Subordinate elements can have different engagement priorities but will normally engage the most dangerous targets first, followed by targets in-depth or specialized systems, such as engineer vehicles.

3-175. A *target reference point is* a predetermined point of reference, normally a permanent structure or terrain feature that can be used when describing a target location. Also called TRP (JP 3-09.3). The battalion and subordinate units may designate target reference points to define unit or individual sectors of fire and observation, usually within the engagement area. Target reference points, along with trigger lines, designate the center of an area where the commander plans to distribute or converge the fires of all weapons rapidly to further delineate sectors of fire within an engagement area. Once designated, target reference points may also constitute indirect fire targets.

3-176. A ***trigger line*** is a phase line located on identifiable terrain that crosses the engagement area—used to initiate and mass fires into an engagement area at a predetermined range for all or like weapon systems. The commander can designate one trigger line for all weapon systems or separate trigger lines for each weapon or type of weapon system. The commander specifies the engagement criteria for a specific situation. The criteria may be either time-or event-driven, such as a certain number or certain types of vehicles to cross the trigger line before initiating engagement. The commander can use a time-based fires delivery methodology or a geography based fires delivery.

Note. The example below addresses the general steps to engagement area development for the area defense illustrated on page 3-285. In this example, a battalion task force conducts an area defense (defense in-depth) against a motorized Infantry and armor threat. The fictional scenario within this example, used for discussion purposes, is not the only way to develop an engagement area. For clarity, many graphical control measures, such as phase lines, are not shown.

Engagement Area Development (Motorized Infantry/Armor Threat), Example

3-177. Although often identified as a method to defeat enemy armor, engagement areas are an effective method to defeat any enemy attack whether the attack is primarily an armor, Infantry, or a mixed armor and Infantry force. The key is the identification of the likely enemy avenues of approach and actions, and the placement of adequate friendly forces, obstacles, and fires to defeat the enemy. The following seven-step engagement area development process, used for discussion purposes, represents one way an Infantry battalion builds an engagement area. The commander and staff (specifically the S-3) integrate these steps within the military decision-making process and the IPB. Steps (asterisks denote steps that occur simultaneously) include the following:

- Identify likely enemy avenues of approach.
- Identify most likely enemy course of action.
- Determine where to kill the enemy.
- Position subordinate forces and weapons systems.*
- Plan and integrate obstacles.*
- Plan and integrate fires.*
- Rehearse the execution of operations within the engagement area.

Note. Within the scenario, the IBCT commander focuses the IPB effort on the characteristics of the operational environment that can influence enemy and friendly operations and how the operational environment influences friendly and enemy courses of actions. The IBCT staff (specifically the S-2 and S-3) identified three likely enemy avenues of approach to and through the IBCT's area of operation. Two enemy avenues of approach were identified within the Blue River Valley, Avenue of Approach 1 and Avenue of Approach 2 (area of operations assigned to Infantry Battalion 2). A third enemy avenue of approach was identified north of the Blue River Valley (area of operations assigned to Infantry Battalion 1, not illustrated). Success, against these likely enemy avenues of approach, results in allowing the commander to quickly choose and exploit terrain, weather, and civil considerations to best support the mission. (See ATP 2-01.3)

3-178. Step 1. Identify likely enemy avenues of approach. The brigade and battalion staffs identified significant characteristics of the operational environment to determine the effects of the terrain, weather, and civil considerations on enemy and friendly operations. The primary analytic tools used to aid in determining this effect, specific to terrain, are the modified combined obstacle overlay, (commonly referred to as MCOO) and the terrain effects matrix. Figure 3-23, page 3-56, identifies the two enemy avenues of approach and the terrain within the Infantry Battalion 2's area of operation that impedes friendly and enemy movement (severely restricted and restricted areas) and the terrain where enemy and friendly forces can move unimpeded (unrestricted areas). Key terrain forward of the Green River is critical to the battalion defense because occupying these position will allow the engagement of enemy forces forward of the river, preventing the establishment of an enemy force on the east bank and the use of crossing sites to support movement into less restrictive terrain west of the Green River.

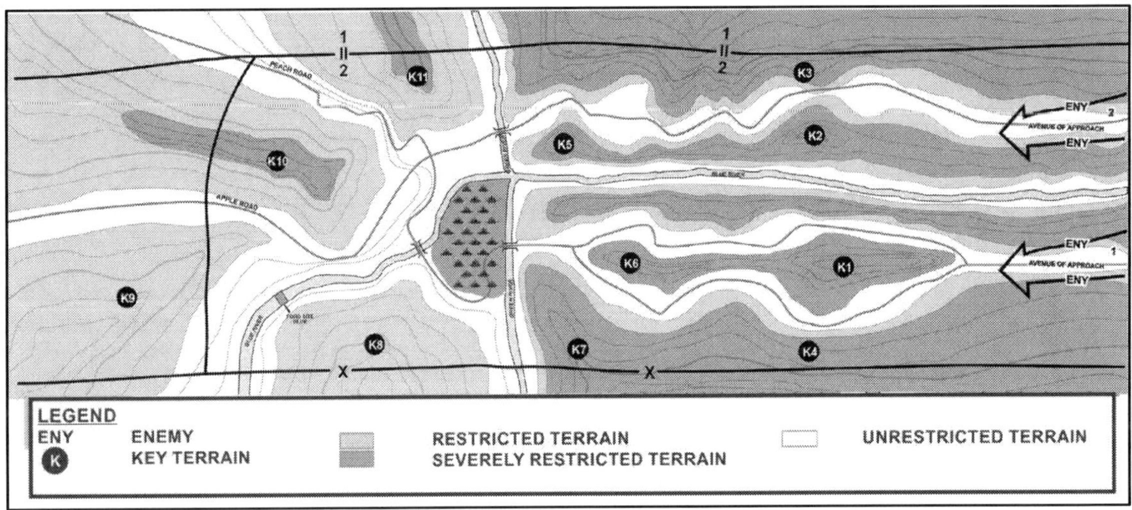

Figure 3-23. Identify likely enemy avenues of approach, example

3-179. Enemy Avenue of Approach 1 and Avenue of Approach 2, within the scenario, support mounted movement though the terrain forward (west) of the Green River is restricted, thus hindering enemy movement to some degree. The terrain typically consists of moderate-to-steep slopes or moderate-to-densely spaced obstacles, such as trees or rocks. Enemy forces within this restricted area will have difficulty maintaining preferred speeds, moving in combat formations, and transitioning from one formation to another. Enemy movement will require zigzagging or frequent detours. A poorly developed road system will hinder the enemy's ability to sustain its attack along both enemy avenues of approach. The unrestricted terrain further west from Green River will allow the enemy to move unimpeded along enemy Avenue of Approach 1 and Avenue of Approach 2 once clear of the river.

3-180. The terrain along enemy Avenue of Approach 3, identified to the North of the Blue River valley, is severely restricted (not illustrated). Steep slopes and large or densely spaced obstacles with little or no supporting roads characterize the terrain. Though suitable for dismounted movement, the terrain within Avenue of Approach 3 impedes motorized Infantry and armor movement. Swamps and the rugged terrain within this area are examples of restricted areas for dismounted infantry forces. The road system utilized to sustain the enemy's attack is very limited along Avenue of Approach 3. (Refer to ATP 2-01.3 for additional information on determining terrain characteristics and the terrain's effect on operations.)

> ***Note.*** The IBCT commander and staff, during step 4 of the IPB, identify and develop possible enemy course of actions that can affect the IBCT's mission. Enemy course of action development requires identifying and understanding the significant characteristics related to enemy, terrain, weather, and civil considerations of the operational environment and how these characteristics affect friendly and enemy operations, steps 1 and 2 of the IPB, respectively. The purpose of evaluating the enemy, step 3 of the IPB, is to understand how an enemy can affect friendly operations. The commander, in order to plan for all possible contingencies, understands all course of actions an enemy commander can use to accomplish the enemy objective(s). To aid in this understanding, the staff determines all valid enemy course of actions and prioritizes them from most to least likely. The staff also determines which enemy course of action is the most dangerous to friendly forces. To be valid, enemy course of actions should be feasible, acceptable, suitable, distinguishable, and complete—the same criteria used to validate friendly course of actions. (See ATP 2-01.3.)

3-181. Step 2. Identify most likely enemy course of action. The commander and staff (specifically the S-2 and S-3) determine the enemy's most likely course of action, within the scenario, is to attack with two battalions (motorized Infantry battalion task forces) abreast, one along Avenue of Approach 1 and one along Avenue of Approach 2 (figure 3-23). The enemy's approach, compartmentalized forward of the Green River, restricts movement and prevents the attacking enemy force from fully exploiting its combat superiority. The terrain forward of the Green River allows for the massing of friendly fires with the enemy piecemeal commitment into friendly engagement areas. The terrain requires the enemy to zigzag and commit to frequent detours (due to the compartmentalization during movement), exposing portions of the enemy force for destruction without giving up the advantage of friendly forces fighting from protected positions (figure 3-23).

3-182. The enemy's main effort, predicted to move along Avenue of Approach 2, requires crossing one river, the Green River. A secondary effort of the enemy, predicted to move along enemy Avenue of Approach 1, requires crossing both the Green and Blue Rivers. The least likely enemy avenue of approach, Avenue of Approach 3 to the north, though the largest area of operation to defend requires the enemy to move through severely restricted terrain to the east and west of the Green River. The enemy, predicted to establish multiple infiltration lanes (company and platoon size elements) along this approach, infiltrates forces to the rear to disrupt friendly operations. Enemy follow-on forces, anticipated armor battalion task force, will attempt to exploit enemy successes along enemy Avenues of Approach 1 and 2.

Note. The desired end state of step 4, determine threat course of actions, of the IPB process is the development of graphic overlays (enemy situation templates) and narratives (enemy course of action statements) for each possible enemy course of action identified. Generally, there will not be enough time during the military decision-making process to develop enemy situation overlays for all course of actions. A good technique is to develop alternate or secondary course of actions, write a course of action statement, and produce a list of high-value targets to use during the mission analysis briefing and course of action development during the military decision-making process. Once these tools and products are complete, the staff constructs overlays depicting the enemy's most likely and most dangerous course of action to use during the friendly course of action development and friendly course of action analysis steps of the military decision-making process. (Refer to ATP 2-01.3 and FM 6-0 for additional information.)

3-183. Step 3. Determine where to kill the enemy. Whether planning deliberately or rapidly when determining where to kill enemy, the IBCT commander, subordinate commanders, and staffs maintain a shared understanding of the steps within the IPB and the military decision-making process. Within the scenario, the IBCT commander focuses this effort to determine the effects of the terrain, weather, and civil considerations on the enemy avenues of approach identified within and north of the Blue River Valley. During step 4 of the IPB, the IBCT and battalion staffs (specifically the S-2 and S-3) identify and develop possible enemy course of actions that can affect the IBCT mission. Based on the results of this analysis, the commander concentrates efforts and economizes forces to kill the enemy east of the Green River along three enemy avenues of approach to best utilize the restricted and severely restricted areas forward in the IBCT's area of operations.

3-184. Step 4. Position subordinate forces and weapons systems. Within the scenario, the battalion commander's concept for the area defense (Infantry Battalion 2) required the positioning of subordinate forces and weapon systems to accomplish their mission independently and in combination by means of fires, the employment of obstacles, and absorbing the strength of the attack within defensive battle positions. The commander assigned subordinate maneuver companies an area of operation, based on the mission variables of METT-TC, to maximize decentralized execution empowering subordinate commanders to position battle positions within their assigned area of operations. At the same time, each subordinate commander addressed security requirements for the flanks of assigned area of operations by assigning responsibility to a subordinate element or organizing a security force or observation post(s) to accomplish that mission. The battalion commander retained a reserve (tank platoon) to contain enemy penetrations between positions, to reinforce fires into an engagement area, or to help a portion of the security force or main body disengage from the enemy if required.

3-185. Step 5. Plan and integrate obstacles. During the conduct of the area defense, countermobility planning (see paragraph 3-207) is the primary concern of the battalion's engineer staff NCO, in coordination with the battalion S-3, S-2, and fire support officer. (External to the battalion, the engineer planner would coordinate with the brigade engineer battalion and assistant brigade engineer). The plan addresses how security area and main battle area forces reinforce the natural defensive characteristics of the terrain with the employment of obstacles to block, disrupt, fix, and turn attacking enemy forces into planned engagement areas. Countermobility planning also includes the positioning of protective obstacles to prevent the enemy from closing with defensive battle positons within the battalion area defense (figure 3-24, page 3-58).

3-186. Within the scenario, the battalion commander's concept for the employment of obstacles within the area defense (Infantry Battalion 2) forces the enemy to enter established engagement areas positioned where the commander intends to kill the enemy. To succeed, the battalion through the employment of obstacles and the static positioning of company and platoon battle positions control, stop, or canalize attacking enemy forces to counteract the enemy's initiative. The commander, through dynamic actions of the battalion reserve covers gaps between positions and takes advantage of available offensive opportunities such as a local attack or counterattack that do not risk the integrity of the defense (figure 3-24).

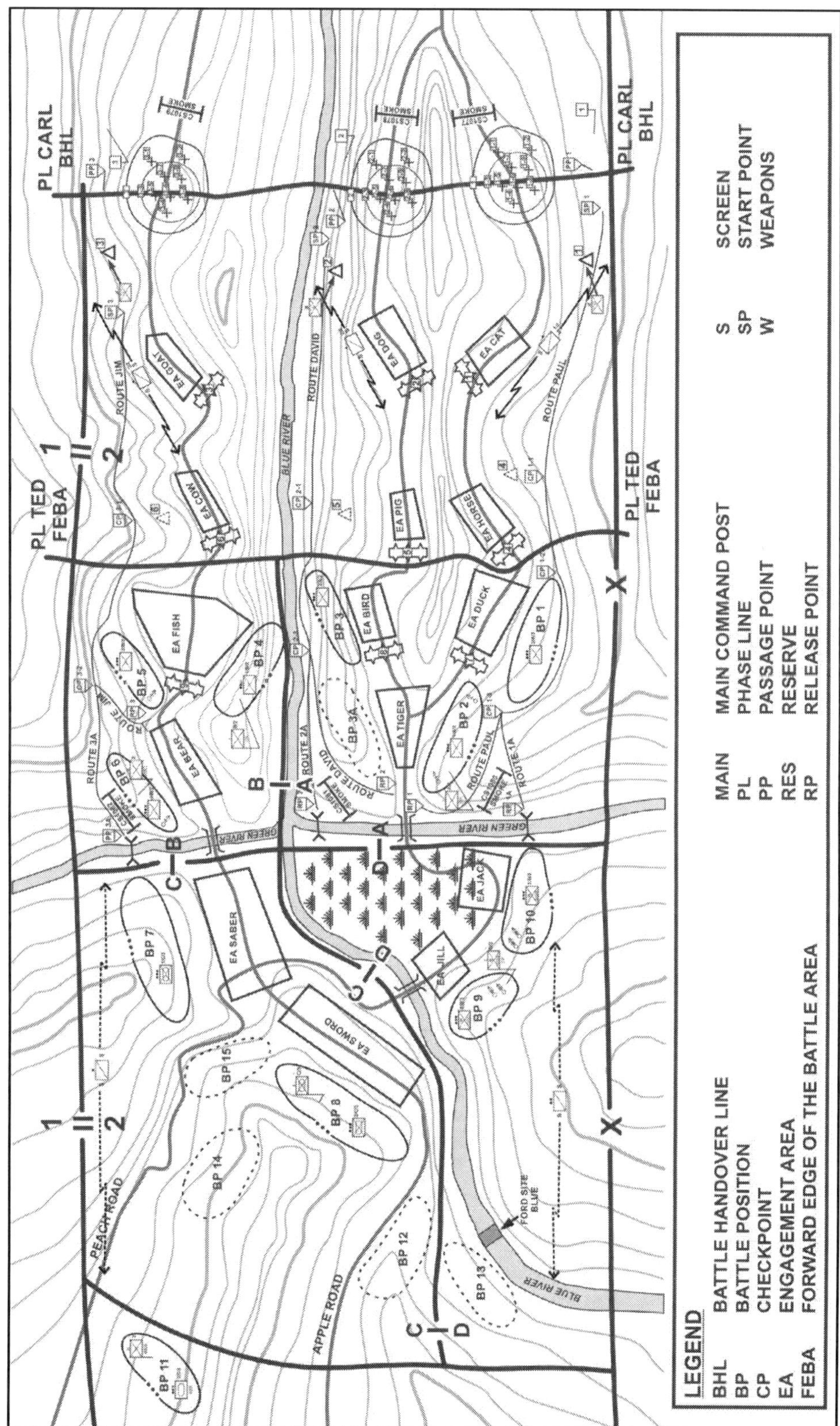

Figure 3-24. Engagement areas (motorized Infantry/Armor threat), example

3-187. Step 6. Plan and integrate fires. Within the scenario, the IBCT and division higher area of interest and area of operations extends far enough beyond the Infantry battalion's forward line of own troops that the IBCT commander has the time and resources to identify approaching enemy forces, assess options, and recommend targets for attack to enable the battalion's mission. Fires conducted by joint fire assets or through the provision of mission orders to the division's combat aviation brigade and field artillery units at the IBCT and division echelons further enable the battalion's ability to seize the initiative before the advancing enemy makes contact with battalion forward defensive positions. The division joint air-ground integration center, in coordination with the IBCT fire support cell, plans and coordinates joint fires, suppression of enemy air defenses, airspace coordination areas, ingress and egress routes, and other airspace requirements to deliver aerial and surface-delivered fires simultaneously into a given engagement area or target area.

3-188. Before the enemy closes into direct fire engagement areas, in either the security area or the main battle area, the Infantry battalion directs the initiation of fires. The commander and staff plan to provide the most effective fires resources and mitigate the risk of fratricide as the attacking enemy nears the designated engagement area while supporting air conducts army aviation and close air support attacks. During engagement area development, fire control measures, such as target reference points, trigger lines, and final protective fires enable observed fires and the obstacle plan to force the enemy to use avenues of approach covered by friendly engagement areas. These shaping operations typically focus on enemy high-payoff targets, such as command and control nodes, engineer, fire support, and air defense assets and follow-on forces for destruction or disruption (figure 3-24, page 3-59).

3-189. Step 7. Rehearse the execution of operations within the engagement area. The battalion coordinates and rehearses engagement area actions on the ground, gaining intimate familiarity with the terrain. The battalion commander, the S-2, the S-3, engineer planner, and the fire support officer, at a minimum, rehearse the sequence of events with the company commanders and separate element leaders for each engagement area.

3-190. During rehearsals, the commander confirms designated target reference points, trigger lines, final protective fires, engagement areas, and other direct- and indirect-fire control measures in each engagement area within the battalion's area of operations. Once in position, the commander may modify subordinate unit positions and preplanned control measures during rehearsals to improve defensive capabilities as required. The commander ensures the integration of fires by adjusting the planned positions of weapon systems to obtain maximum effectiveness against targets in the planned engagement area. The commander coordinates all fires, including those of supporting Army aviation and close air support, used to isolate the targeted enemy force in the planned engagement area while preventing the target's escape or reinforcement. The battalion rehearses the confliction of fires to ensure maximum damage before the enemy can respond. The commander rehearses the actions of the reserve to reinforce fires, add depth, or block, to restore a position by counterattack, or to reinforce the destruction of enemy forces within planned engagement areas.

3-191. Company commanders rehearse their planned actions within the company engagement area(s). Platoon leaders reconnoiter and identify positions and identify movement or withdrawal routes and revises them as required. Company commanders rehearse assigned weapon system primary sectors of fire and secondary sectors of fire to increase the capability of concentrating fire in certain areas in accordance with established criteria and priorities for engagement. Secondary sectors of fire, when there are no targets in the primary sector or when the commander needs to cover the movement of another friendly element, correspond to another element's primary sector of fire to obtain mutual support. Secondary sectors of fire are rehearsed and confirmed depending on the availability of time before execution. Subordinate commanders may impose and rehearse additional fire control measures as required and as time permits. (See FM 3-21.10 for additional information.)

Forms of Defensive Maneuver

3-192. The battalion commander may choose between two defensive maneuver forms when planning an area defense: a defense in-depth or a forward defense. The commander usually selects the form of defensive maneuver, but the higher headquarters' commander may define the general defensive scheme for the battalion. These two employment choices are not exclusionary. Part of a defending unit can conduct a forward defense, while the other part conducts a defense in-depth (figure 3-26, page 3-63). The specific mission may also impose constraints such as time, security, and retention of certain areas, which are significant factors in determining how the battalion defends.

Defense In-Depth

3-193. A defense in-depth is the preferred option when tactical conditions allow. Defense in-depth reduces the risk of the attacking enemy penetrating the defense and affords some initial protection from enemy indirect fires.

A defense in-depth limits the enemy's ability to exploit a penetration through additional defensive positions employed in-depth. The defense in-depth provides more space and time to exploit information collection efforts and fire support to reduce the enemy's options, weaken the enemy force, and set the conditions for the enemy's destruction, disintegration, or dislocation. The defense provides the commander more time to gain information about the enemy's intentions and likely future actions before decisively committing to a plan. (See paragraph 3-285 for an example illustrating a battalion task force defending in-depth against a motorized Infantry and armor threat.)

3-194. When the commander has the option of conducting a defense in-depth, he uses security forces and forward main battle area elements to identify, define, and control the depth of the enemy's main effort while holding off secondary thrusts. Doing so allows the commander to conserve combat power, strengthen the reserve, and better resource the counterattack. Even if the enemy is initially successful, the defense in-depth allows the battalion to execute decisive maneuver by effectively repositioning subordinate units to conduct counterattacks or to prevent penetrations. Dependent on the mission variables of METT-TC, it may require forces with at least the same mobility as the enemy to maneuver to alternate, supplementary, and subsequent positions. The mobility of the enemy force can determine the disengagement criteria of the defending forces as they seek to maintain depth. The battalion commander considers using a defense in-depth when—

- The mission allows the battalion to fight throughout the depth of the area of operations.
- The terrain does not favor a forward defense and there is better defensible terrain deeper in the area of operations.
- Sufficient depth is available in the area of operations.
- Cover and concealment forward in the area of operations is limited.
- Chemical, biological, radiological, and nuclear weapons may be used.
- The terrain is restrictive and limits the enemy's maneuver and size of attack.

Forward Defense

3-195. When the commander defends forward in an area defense the battalion employs the majority of its combat forces near the forward edge of the battle area. The scout platoon may establish a relatively narrow security area forward of the main battle to limit the terrain over which the enemy can gain influence or control. The battalion fights to retain these forward positions, conducts counterattacks against enemy penetrations, or destroys enemy penetration in forward engagement areas. To accomplish this, the commander deploys forces or plans counterattacks well forward in the main battle area or even beyond the main battle area. Due to its inherent lack of depth, the forward defense is the least preferred option. While the battalion may lack depth, companies and platoons array forces as able in-depth.

3-196. During a forward defense, the commander uses R&S forces and security forces forward to find the enemy in vulnerable situations and exploit the opportunity to conduct a spoiling attack (see paragraph 2-445) to weaken the enemy's main attacking force and disrupt the enemy's operations. The battalion commander uses a forward defense when a higher commander directs the commander to retain forward terrain for political, military, economic, and other reasons. Alternatively, a commander may choose to conduct a forward defense when the terrain in that part of the area of operations—including natural obstacles—favors the defending force because—

- Terrain forward in the area of operations favors the defense.
- Strong, existing natural or manmade obstacles, such as a river or a canal, are located forward in the area of operations.
- Assigned area of operations lacks depth due to the location of the protected area.
- Natural engagement areas occur near the forward edge of the battle area.
- Cover and concealment in the rear portion of the area of operations is limited.
- Directed by higher headquarters to retain or initially control forward terrain.

Forward Defense Against an Infantry Threat (Static Actions Oriented on Terrain Retention), Example

3-197. In this example, the Infantry battalion conducts an area defense, retains terrain west of the Blue River. The enemy can attack within the battalion's area of operations with an estimated Infantry regiment and can reinforce with tanks if a bridgehead is secured. The terrain is broken and hilly with alternating wooded and cleared areas. Steep slopes and the moderate to heavy wooded areas restrict vehicular movement. The Blue River is not fordable to vehicles and cannot be crossed by Infantry without some delay or special equipment except at High Shoals Ford and Red Ford (figure 3-25).

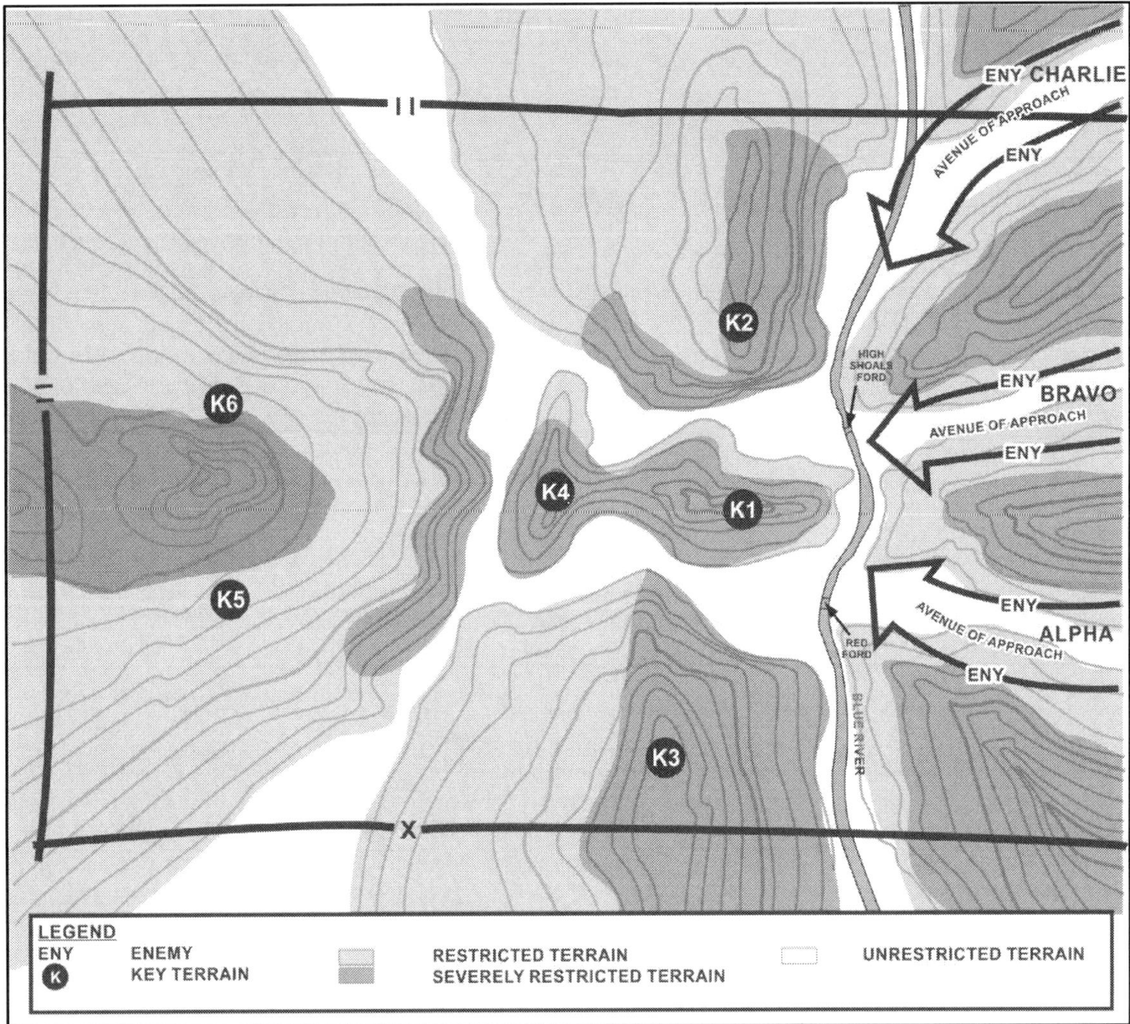

Figure 3-25. Key terrain and avenues of approach, example

3-198. The battalion commander and staff identify *key terrain*—any locality, or area, the seizure or retention of which affords a marked advantage to either combatant (JP 2-01.3)—as shown in figure 3-25. Key Terrain 1, 2, and 3 are critical to the defense because occupying them will allow the engagement of enemy forces forward of the river, preventing the establishment of an enemy force on the east bank and the use of crossing sites to support continuous movement into and through the battalion's area of operations. The commander designates Key Terrain 1 as decisive terrain in the concept of operations to communicate its importance to battalion staff and subordinate commanders. *Decisive terrain*—when present is key terrain whose seizure and retention is mandatory for successful mission accomplishment (FM 3-90-1). From Key Terrain 4, elements can slow or stop enemy forces, which may seize Key Terrain 1, 2 and 3 and attempt to push through the battalion area of operations toward Key Terrain 5 and Key Terrain 6. Elements positioned on Key Terrain 5 and 6 can counterattack to regain control of area vicinity Key Terrain 1, 2, 3, and 4.

Note. In operations over mountainous terrain, the analysis of key and decisive terrain is based on the identification of these features at each of the three operational terrain levels. Understanding there are few truly impassable areas in the mountains. The commander recognizes that what may be key terrain to one force may be an obstacle to another force. The commander also recognizes that properly trained forces can use high obstructing terrain as a means to achieve decisive victories with comparatively small-sized combat elements. (Refer to ATP 2-01.3 and ATTP 3-21.50 for additional information.)

3-199. The commander and staff identified two *avenues of approach*—the air or ground route leading to an objective (or to key terrain in its path) that an attacking force can use (ADRP 3-90)—into and through the battalion's area of operations. Although the enemy can attack on both, Avenue of Approach Alpha is the largest and most dangerous to the defense (figure 3-25, page 3-62). The seizure of Key Terrain 1 by the enemy would give the enemy a vehicle-crossing site (Red Ford) and a protected area from which the enemy could continue the attack on less restrictive terrain. The battalion commander decides to establish a forward defense, which allows detection and engagement of the enemy at any point the enemy may choose to attack. The battalion commander determines that by positioning Infantry elements along the river, the commander could exploit the linear obstacle to gain good observation and fires on enemy forces attempting to seize crossing sites.

3-200. The commander visualizes the allocation of battalion forces based on the location and size of the enemy avenues of approach, the terrain, and the width of the battalion's area of operations. The commander positions rifle companies forward, allowing each company to gain mutual support between platoons and with adjacent units. The depth allotted to a forward company allows enough space to deploy available combat elements and provides suitable terrain for alternate and supplementary positions, the command post, company mortars, and company trains. In deciding on the depth of forward company area of operations, the commander considers the location of the battalion reserve. The areas provided to forward companies does not include the terrain required for the battalion reserve (figure 3-26, page 3-64).

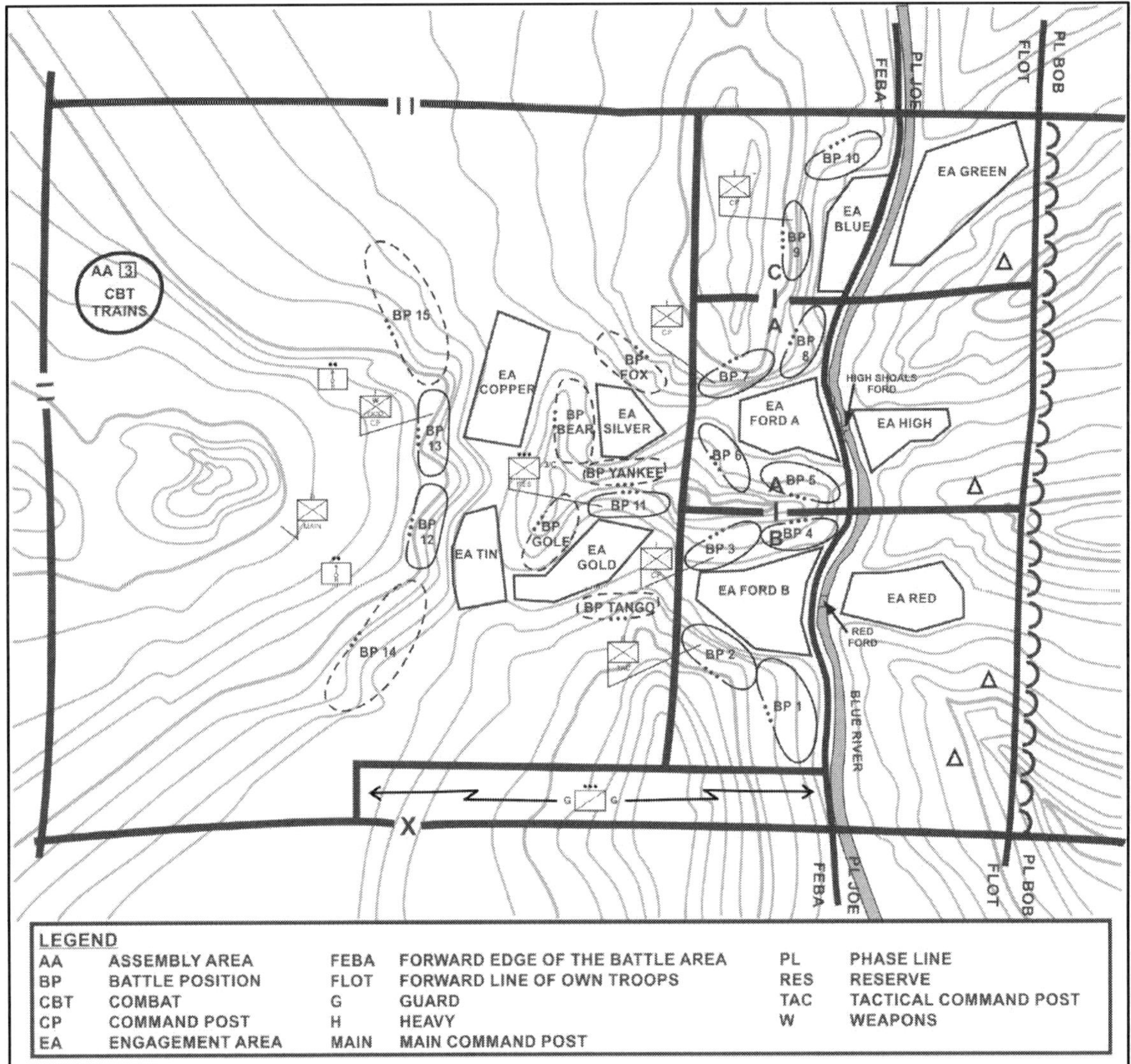

Figure 3-26. Company and platoon employment, example

3-201. The commander determines that three rifle platoons and one assault platoon (battalion main effort) is required to defend Avenue of Approach Alpha, three rifle platoons and one assault platoon to defend Avenues of Approach Bravo, and two rifle platoons to defend Avenue of Approach Charlie (north of High Shoals Ford). The commander weights the battalion's main effort with priority of fires to block enemy penetration on Avenue of Approach Alpha. The battalion defends with two companies forward, with platoons in-depth, along the two most dangerous avenues of approach, Alpha and Bravo. Company C defends with two platoons forward on Avenue of Approach Charlie (least likely avenue of approach). Company D (weapons company-minus) establishes overwatch positions from primary and alternate battle positions (Battle Positions 12, 13, 14, and 15) to the rear of forward rifle companies. On order, establish supplementary (overwatch) positions (Battle Positions Tango, Golf, Bear, and/or Fox) to the rear of Company B, along Avenue of Approach Alpha and Company A, along Avenue of Approach Bravo. The commander designates one rifle platoon from Company C (vicinity Battle positon 11) as the battalion reserve with a be-prepared mission to block the most dangerous avenue of approach, Alpha. The commander also assigns an on-order position to the reserve, Yankee, to block Avenue of Approach Bravo.

3-202. Due to the limited number of firing positions along the river, the battalion commander retains two of the four assault platoons under the Company D commander's control. Assault platoons, controlled by the Company D commander, initially establish battle positions 12 and 13 to assist in-depth to limit the enemy advance. Should the enemy establish a crossing site across the river, on order, Company D conducts counterattacks (see figure 3-28, page 3-78) to establish attack by fire positions (tentative battle positions are Tango, Golf, Bear, or Fox) to reinforce forward companies A and B. The commander positions the scout platoon initially to conduct R&S forward of the battalion (not illustrated), on order, conduct guard mission to the right flank of Company B to defeat, cause to withdraw, or fix enemy dismounted movement south of Battle Position 1and along the southern boundary of the battalion. Battalion mortars operate split section, initially positioned forward vicinity Battle Positions Golf and Tango (not illustrated) with priority of fires to security area actions. On order, reposition by section to firing positions vicinity Battle Positions 12 and 13 with priority of fires to main battle area actions initially in Engagement Area Red, then Engagement Area Ford B (see figure 3-26, page 3-64).

Army Aviation Support

3-203. In the defense, the speed and mobility of aviation can assist the Infantry in the concentration of forces and tactical flexibility. Army aviation attack and reconnaissance units support the Infantry battalion with aerial R&S, observed fires (in contact), and deep operations (out of contact) independent of the battalion's ground maneuver. Army Aviation attacks against enemy forces in or out of contact can be the decisive or shaping operation at the tactical or operational level and enable the combined arms team to maintain the tempo of operations while presenting multiple dilemmas to the enemy at the ground maneuver commander's time and place of choice.

3-204. During deep operations, Army aviation attack and reconnaissance units conduct operations as a maneuver force with manned and unmanned systems maneuvering interdependently. *Manned-unmanned teaming* is the integrated maneuver of Army Aviation rotary wing and unmanned aircraft systems to conduct movement to contact, attack, reconnaissance, and security tasks (FM 3-04). Manned-unmanned teaming (expressed in the acronym MUM-T) enables increased depth and breadth of aviation reconnaissance and maneuver, longer persistence over the reconnaissance objective, increased ability to gain and maintain enemy contact, greater survivability and more options to develop the situation with enhanced maneuver, fires, and mission command. Army aviation attack and reconnaissance units can attack deep high-payoff targets, enemy concentrations and moving columns, and disrupt the enemy at the decisive point. Aviation forces employ in-depth to attack exploitation forces or follow-on echelons before they can move forward to the close battle. (Refer to FM 3-04 and ATP 3-04.64 for additional information.)

3-205. When Army aviation attack and reconnaissance units maneuver independent of ground maneuver forces, attacks can be hasty or deliberate. The Army aviation maneuver commander controls Army aviation maneuver and fires within an area of operations assigned by a higher headquarters but the attack is still synchronized and integrated with the overall higher ground scheme of maneuver. Based on the complexity of the targeted enemy force and operational environment. Army Aviation attacks against enemy forces out of friendly contact frequently are higher risk operations that require detailed planning by the supported ground maneuver headquarters for the proper allocation, synchronization and integration of joint fires (to include suppression of enemy air defenses), collection assets and other enabling capabilities. (Refer to ATP 3-09.32 and ATP 3-01.4 for additional information.)

3-206. Army aviation attack and reconnaissance units, in close coordination with the Infantry battalion, attack to destroy, defeat, disrupt, divert, or delay enemy forces to enable the combined arms team to seize, retain, or exploit the initiative. These attacks can also be either hasty or deliberate. In either case, the ground maneuver battalion commander (in contact) is responsible for the integration and synchronization of Army aviation in the ground scheme of maneuver and controls the distribution and de-confliction of Army aviation maneuver and fires during maneuver. Synchronization of aviation assets into the defensive plan is important to ensure aviation assets are capable of massing fires and to prevent fratricide. Detailed air-ground integration and coordination is necessary to ensure efficient use of aviation assets.

3-207. In support of the battalion's area defense, Army aviation forces support forward security area operations and mass fires during the battalion's main battle area fight. When assigned aviation assets the battalion commander gives careful consideration to engagement area development and direct fire planning (see paragraph 3-169). Supporting aviation units, through aviation liaison officers, are involved throughout the battalion's planning and preparation for the area defense. Aviation liaison officers, the battalion S-3 and fire support officer conducts Army airspace control coordination and deconfliction with indirect fires and close air support as required with the appropriate airspace control authority. (Refer to ATP 3-91.1 for additional information.)

Countermobility Operations

3-208. *Countermobility operations* are those combined arms activities that use or enhance the effects of natural and man-made obstacles to deny enemy freedom of movement and maneuver (ATP 3-90.8). Primary purposes of countermobility operations are to shape enemy movement and maneuver and to prevent the enemy from gaining a position of advantage. In support of the battalion's area defense, countermobility operations are conducted to disrupt enemy attack formations and assist the battalion in defeating the enemy in detail. Countermobility operations channel attacking enemy forces into engagement areas throughout the depth of the defense and protect the flanks of friendly counterattack forces. Countermobility operations shape engagements, maximize the effects of fires, and provide close-in protection around defensive positions to defeat the final assault of the enemy and to prevent and warn of intrusion into critical support area sites and fixed sites such as bases. (Refer to ATP 3-90.8 for additional information.)

Countermobility Planning

3-209. The commander and staff develop the countermobility plan concurrently with the fire support plan and defensive scheme of maneuver, guided by the commander's intent. The conduct of countermobility operations typically involves engineers and includes proper obstacle integration with the maneuver plan, adherence to obstacle emplacement authority, and positive obstacle control. Combined arms obstacle integration synchronizes countermobility operations into the concept of operations. Because most obstacles have the potential to deny the freedom of movement and maneuver to friendly forces and enemy forces, it is critical that the commander properly weighs the risk and evaluates the trade-off of employing various types of obstacles.

3-210. Obstacle control is essential in supporting the commander's plan. Responsibilities for executing tasks within countermobility operations can be broadly divided into two entities: emplacing unit and owning unit. This framework helps the commander plan for and assign responsibilities for obstacle execution to subordinate units. The responsibilities of each may vary based on the type of obstacle and the situation. The commander's concept of operations will include the following tasks:

- Site obstacles.
- Construct, emplace, or detonate obstacles.
- Mark, report, and record obstacles.
- Maintain obstacle integration.

Terrain Reinforcement

3-211. Countermobility operations typically reinforce the terrain to block, fix, turn, or disrupt the enemy's ability to move or maneuver, giving the commander opportunities to exploit enemy vulnerabilities or react effectively to enemy actions. The commander reinforces the terrain to prevent the enemy from gaining a position of advantage taking full advantage of the natural restrictiveness of the existing terrain to minimize the time, effort, and materiel needed to achieve the desired obstacle effects. Reinforcing the terrain focuses on existing and reinforcing obstacles. Existing obstacles are inherent aspects of the terrain that impede movement and maneuver. Existing obstacles may be natural (rivers, mountains, wooded areas) or man-made (enemy explosive and nonexplosive obstacles and structures, including bridges, canals, railroads, and embankments associated with them).

3-212. Reinforcing obstacles are those man-made obstacles that strengthen existing terrain to achieve a desired effect. Reinforcing obstacles must be planned and emplaced to support the maneuver commander's plan, while not hindering friendly-force mobility. Obstacle plans are developed based on a thorough understanding of the commander's intent and concept of operations, enemy mobility capabilities, and the effects of the natural terrain and existing obstacles. Only then can the true value of integrating obstacles, observation, fires, and maneuver be realized. The basic employment principles for reinforcing obstacles are—

- Support the maneuver commander's plan.
- Integrate with observation and fires.
- Integrate with other obstacles.
- Employ in-depth.
- Employ for surprise.

3-213. Reinforcing obstacles on land consist of land mines, networked munitions, and demolition and constructed obstacles. A *land mine* is a munition on or near the ground or other surface area that is designed to be exploded by the presence, proximity, or contact of a person or vehicle (ATP 3-90.8). Land mines can be further defined as antivehicle or antipersonnel. They can be air-, artillery-, or ground-delivered. Land mines can be employed in quantities within a specific area to form a minefield, or they can be used individually to reinforce nonexplosive obstacles. Land mines fall into the two general categories—persistent and nonpersistent. Persistent land mines are not capable of self-destructing or self-deactivating. Nonpersistent land mines are capable of self-destructing or self-deactivating.

Note. As of 1 January 2011, U.S. forces are no longer authorized to employ persistent (those that are not self-destructing or self-deactivating) or nondetectable land mines. See JP 3-15 for more information on the laws, agreements, and policies that are most significant to the employment of obstacles.

3-214. *Networked munitions* is a remotely controlled, interconnected, weapons system designed to provide rapidly emplaced ground-based countermobility and protection capability through scalable application of lethal and nonlethal means (JP 3-15). Demolition obstacles are created using explosives. Examples include bridge or other structure demolition (rubble) and road craters. (See ATP 3-90.8, appendix B for more information on demolition obstacles.) Constructed obstacles are created without the direct use of explosives. Examples include wire obstacles, antivehicle ditches, or similar construction that typically involves the use of heavy equipment. (See ATP 3-90.8, appendix C for more information on construction obstacles.)

3-215. Reinforcing obstacles, categorized as tactical and protective, are employed as part of the movement and maneuver and protection (see paragraph 3-207) warfighting functions. Tactical obstacles help shape enemy maneuver and prevent the enemy from gaining a position of advantage, while protective obstacles protect people, equipment, supplies, and facilities against threats. The primary purposes of tactical obstacles are to shape enemy maneuver and to maximize the effects of fires. Tactical obstacles directly attack the ability of a force to move, mass, and reinforce; therefore, they affect the tempo of operations. Commanders integrate obstacles into the scheme of movement and maneuver to enhance the effects of fires. Preexisting obstacles that a unit reinforces and integrates with observation and fires may become tactical obstacles. The types of tactical obstacles are clearly distinguished by the differences in execution criteria. The three types are—

- *Directed obstacle*, an obstacle directed by a higher commander as a specified task to a subordinate unit (ATP 3-90.8).
- Situational obstacle, an obstacle that a unit plans and possibly prepares prior to starting an operation, but does not execute unless specific criteria are met.
- *Reserved obstacle*, an obstacle of any type, for which the commander restricts execution authority (ATP 3-90.8).

Obstacle Intent

3-216. An *obstacle* is any natural or man-made obstruction designed or employed to disrupt, fix, turn, or block the movement of an opposing force, and to impose additional losses in personnel, time, and equipment on the opposing force (JP 3-15). Obstacle intent describes how obstacles support the commander's concept of operations. Obstacle intent consists of the— target, effect, relative location. The target is the enemy force that the commander wants to affect with tactical obstacles. The commander usually identifies the target in terms of the enemy size and type, the echelon, the avenue of approach, or in combination. Tactical obstacles and fires—direct and indirect—manipulate the enemy in a way that supports the commander's intent and scheme of movement and maneuver.

3-217. Obstacle effect describes the effect that the commander wants the obstacle(s), combined with fires, to have on the enemy. The obstacle effect—drives integration, focuses subordinate fires, focuses obstacle effort, and multiplies firepower effects. Important to remember, obstacle effects occur because of the combined effects of fires and obstacles, rather than from obstacles alone. Tactical obstacles produce one of the following effects: disrupt, turn, fix, and block (figure 3-27, page 3-68). Obstacle effect symbols are used as control measures for obstacle groups and as elements of the control measures for obstacle zones and belts. During course-of-action development, obstacle effect symbols are also used in developing and showing the initial obstacle plan that supports each course of action.

APPLICATION	DESCRIPTION	PURPOSE	FIRES AND OBSTACLES MUST:	OBSTACLE CHARACTERISTICS
DISRUPT 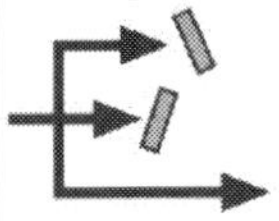	The arrows indicate the direction of enemy advance. The length of the arrows indicate where the enemy is slowed or allowed to bypass.	• Breakup enemy formations. • Interrupt the enemy's timetable and C2. • Cause premature commitment of breach assets. • Cause the enemy to piecemeal his attack.	• Cause the enemy to deploy early. • Slow part of his formation while allowing part to advance unimpeded.	• Do not require extensive resources. • Difficult to detect at long range.
TURN 	The heel of the arrow is the anchor point. The direction of the arrow indicates the desired direction of the turn.	• Force the enemy to move in the direction desired by the the friendly commander.	• Prevent the enemy from bypassing or breaching the obstacle belt. • Maintain pressure on the enemy force throughout the turn. • Mass direct and indirect forces at the anchor point of the turn.	• Tie into impassable terrain at the anchor point. • Consist of obstacles in depth. • Provide a subtle orientation relative to the enemy's approach.
FIX 	The arrow indicates the direction of enemy advance. The irregular part of the arrow indicates where enemy advance is slowed by obstacles.	• Slow an attacker within an area so he can be destroyed. • Generate the time necessary for the friendly force to disengage.	• Cause the enemy to deploy into attack formation before encountering the obstacles. • Allow the enemy to advance slowly in an EA or AO. • Make the enemy fight in multiple directions once he is in the EA or AO.	• Arrayed in depth. • Span the entire width of the avenue of approach. • Must not make the terrain appear impenetrable.
BLOCK 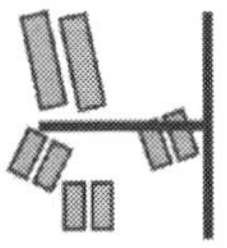	The vertical line indicates the limit of enemy advance and where the obstacle ties into severely restricted terrain. The horizontal line shows the depth of the obstacle effort.	• Stop an attacker along a specific avenue of approach. • Prevent an attacker from passing through an AO or EA. • Stop the enemy from using an avenue of approach and force him to use another avenue of approach.	• Prevent the enemy from bypassing or penetrating through the belt. • Stop the enemy's advance. • Destroy all enemy breach efforts.	• Must tie into impassable terrain. • Consistent of complex obstacles. • Defeat the enemy's mounted and dismounted breaching effort.
Direction of enemy attack →				

LEGEND
AO AREA OF OPERATIONS C2 COMMAND & CONTROL EA ENGAGEMENT AREA

Figure 3-27. Tactical obstacle effects

3-218. Relative location refers to the location of a tactical or protective obstacle in relation to maneuver or fire control measures such as assembly areas, battle positions, or engagement areas. Engineers and other countermobility planners describe planned obstacle locations in relation to maneuver or fire control measures to help maneuver commander visualize linkages between obstacles, fires, and maneuver and to ensure obstacle integration. (Refer to ATP 3-90.8 for additional information.)

Obstacle Control Measures

3-219. *Obstacle control measures* are specific measures that simplify the granting of obstacle-emplacing authority while providing obstacle control (FM 3-90-1). The commander establishes obstacle control by— delegating or withholding emplacement authority; marking, reporting, recording, and tracking obstacles; and restricting types or locations of obstacles through obstacle control measures. The commander uses obstacle control measures and other specific guidance or orders to grant or withhold obstacle emplacement authority to subordinate commanders and provide obstacle control. Obstacle control measures consist of—obstacle zones, obstacle belts, obstacle groups, obstacle restrictions.

3-220. An *obstacle zone* is a division-level command and control measure, normally done graphically, to designate specific land areas where lower echelons are allowed to employ tactical obstacles (JP 3-15). Obstacle zones are permissive, allowing an IBCT to place reinforcing obstacles to support IBCT's scheme of maneuver without interfering with future operations. Obstacle zones are assigned to a single subordinate unit to ensure unity of effort, keeping tactical obstacle responsibility along the same lines as control of direct and indirect fires.

Normally assign an obstacle effect (block, fix, turn, or disrupt) is not assigned to an obstacle zone, allowing subordinate commanders flexibility in using obstacles.

3-221. An *obstacle belt* is a brigade-level command and control measure, normally given graphically, to show where within an obstacle zone the ground tactical commander plans to limit friendly obstacle employment and focus the defense (JP 3-15). An obstacle belt assigns an intent to the obstacle plan and provides the necessary guidance on the overall effect of obstacles within a belt. The commander plans obstacle belts within assigned obstacle zones to grant obstacle-emplacement authority to subordinate units. Obstacle belts focus obstacles to support the IBCT scheme of maneuver and ensure obstacles do not interfere with the maneuver of any higher headquarters.

3-222. *Obstacle groups* are one or more individual obstacles grouped to provide a specific obstacle effect (FM 3-90-1). The Infantry battalion uses obstacle groups to ensure subordinate units emplace individual obstacles that support the battalion's scheme of maneuver. Individual obstacle may include antitank ditches, abatis, booby traps, mines and minefields, roadblocks, craters, and wire obstacles. Subordinate units integrate obstacle groups with direct- and indirect-fire plans. The battalion commander can plan the placement of obstacle groups anywhere in an obstacle zone or belt, respectively. Unlike obstacle zones or belts, obstacle groups are not areas but relative locations for actual obstacles. Obstacle groups are displayed using the obstacle-effect graphics. When detailed planning is possible (to include detailed on-the-ground reconnaissance), the commander may show obstacle groups using individual obstacle graphics.

Note. In rare cases, brigades, divisions, or even corps may use obstacle groups for specific tactical obstacles.

3-223. The commander may use obstacle restrictions to provide additional obstacle control and to limit the specific types of obstacles used, such as restricting the use of buried mines. Obstacle restrictions ensure that subordinates do not use obstacles with characteristics that impair future operations. These restrictions also allow the commander to focus the use of limited resources for the decisive operation by restricting their use elsewhere. An *obstacle restricted area* is a command and control measure used to limit the type or number of obstacles within an area (JP 3-15). The commander with emplacement authority uses obstacle restricted areas to restrict obstacle placement. The obstacle restricted area graphic depicts the impacted area, the unit imposing the restriction, and the restrictions in effect. (Refer to ATP 3-90.8 for additional information.)

Mobility

3-224. When planning an area defense, the commander and staff identify the battalion's mobility requirements by analyzing the scheme of maneuver, counterattack options, reserve planning priorities, fire support, protection, and sustainment movement requirements, and adjacent and higher unit mission, movement and maneuver. The staff (specifically the engineer staff planner) integrates this analysis into the battalion obstacle plan while avoiding the impediment of friendly maneuver when possible. Because the bulk of the engineer force is committed to countermobility and survivability during preparation, the commander uses clear obstacle restrictions on specific areas within the battalion's area of operation to maintain mobility. Mobility support linkup and coordination is factored into the overall defensive preparation timeline.

3-225. When obstacles must be constructed along a mobility corridor that primarily supports friendly movement. A lane or gap and associated closure procedures are planned and rehearsed. Lanes or gaps may be closed with situational or reserved obstacles. Beyond preparing and marking lanes and gaps through obstacles, engineers normally perform mobility tasks once defensive preparations are complete. Reduction assets (see appendix F) may then be positioned to counter templated enemy situational obstacles, or be task organized to the reserve, counterattack force, or any other unit that must maneuver or move subsequent to the execution of the defense. To do this effectively, mobility asset and supported maneuver units integrate, prepare, and rehearse.

3-226. Although not specifically designed or intended as an obstacle, structures may pose as an obstacle based on existing characteristics or altered characteristics that result from combat operations or a catastrophic event. Structures such as bridges and overpasses present an inherent impediment to mobility based on weight and clearance restrictions. Existing obstacles are shown on the combined obstacle overlay developed as part of the IPB. As described in ATP 3-34.80, geospatial engineering is critical in accurately predicting the effects that existing obstacles will have on enemy and friendly movement and maneuver.

3-227. On occasion when the battalion requires significant mobility support during defensive preparation. For example, route reduction or clearance, road repair or maintenance, and landing zone and pickup zone clearance. Engineers have resources to perform these tasks, but cannot perform them and simultaneously prepare the defense. Engineer augmentation above the IBCT, when available can perform general engineering tasks, leaving IBCT engineer assets from the brigade engineer battalion to assist in the construction of the battalion defense.

Civil Consideration

3-228. The commander considers how to minimize civilian interference with the battalion's combat operations while protecting civilians from future hostile actions according to the law of war. The commander considers the type and size of the area of responsibility, line-of-communication security, and the threat and plan for detainee operations and dislocated civilians to determine how their presence may affect movement and maneuver. The commander must also consider the threat civilians pose to defending force and its operations if enemy agents or saboteurs are part of the civilian population.

Intelligence

3-229. IPB is a critical part of defensive planning. (See ATP 2-01.3.) IPB helps the commander determine where to concentrate combat power, where to accept risk, and where to plan the potential decisive operation. The staff integrates intelligence from the higher echelon's collection efforts and from units operating forward of the battalion's area of operations. Information collection includes collection from spot reports, tactical unmanned aircraft systems, and other higher-level collection assets. Early warning of enemy air attack, airborne or helicopter assault or insertion, and dismounted infiltration are vitally important to provide adequate reaction time to counter these threats as far forward as possible. To aid in the development of a flexible defensive plan, the intelligence preparation of the battlefield presents all feasible enemy course of actions. The essential areas of focus are terrain analysis, determination of enemy force size and likely course of actions with associated decision points, and determination of enemy vulnerabilities.

3-230. In the defense, key terrain is usually within and behind the defensive area, such as terrain that gives good observation over avenues of approach to and through the defensive position; terrain that permits the defender to cover an obstacle by fire; or areas along an lines of communications that affect the use of reserves or sustainment operations. Key terrain may include portions of the population, such as political, tribal, and religious groups or leaders; a localized population; infrastructure; or governmental organizations. Weather conditions can affect visibility as well. Temperature can affect the use of thermal sights. Cloud cover can negate illumination provided by the moon. Additionally, precipitation and other obscurants can have varying effects as well. Low visibility is beneficial to offensive and retrograde operations because it conceals concentration of maneuver forces, thus enhancing the possibility of surprise. Low visibility hinders the defense because cohesion and control become difficult to maintain, reconnaissance operations are impeded, and target acquisition is degraded.

3-231. Avenues of approach are air or ground routes used by an attacking force leading to its objective or to key terrain in its path. The identification of avenues of approach is important because all course of actions that involve maneuver depend on available avenues of approach. During offensive tasks, the evaluation of avenues of approach leads to a recommendation on the best avenues of approach to a command's objective and identification of avenues available to the enemy for counterattack, withdrawal, or the movement of reinforcements or reserves. In a defense operation, it is important to identify avenues of approach that support enemy offensive capabilities and avenues that support the movement and commitment of friendly reserves. Avenues of approach are developed by identifying, categorizing, and grouping mobility corridors and evaluating avenues of approach.

3-232. Depending on the size and capabilities of the attacking force, functions can be allocated to disruption, the enemy's main defense, support, and reserve. Reconnaissance and other security measures are a constant activity to observe Avenues of approach near a complex battle position in order to provide early warning to the enemy. Staff analysis throughout the IPB focuses on determining the enemy's—

- Main, supporting, and reinforcing efforts.
- Use of reserves.
- Use of special munitions.
- Use of air support.

Fires

3-233. Supporting the commander's concept of operations during the defense involves attacking and engaging targets throughout the battalion's area of operations with massed or precision fires. Fire support planners, internal and external to the battalion, make maximum use of any preparation time available to plan and coordinate supporting fires. Planners ensure fires complement and support all security forces forward of the main battle area as these fires play a key role in disrupting the attacker's tempo and synchronization during the defense. Fire support planning and execution must address flexibility through operations in-depth and support to defensive maneuver. The commander promotes freedom of action within the main battle area by using the least restrictive control measures necessary to implement the scheme of maneuver. When required, specifically at the IBCT level and above, massing overwhelming close supporting fires and interdiction at critical places and times gains maximum efficiency and effectiveness in suppressing direct and indirect fire systems and repelling and slowing the tempo of the enemy assault, respectively.

3-234. The battalion may utilize unmanned aircraft systems, remote sensors, and reconnaissance and security forces to call for fire on the enemy throughout the area of operations. Quick, violent, and simultaneous action throughout the depth of the defender's area of operations can degrade, confuse, and paralyze an enemy force just as that enemy force is most exposed and vulnerable. Though the battalion may receive priority of fires for a specific mission or phase of the defense, the commander must not overly rely on indirect fire assets available from the IBCT. Battalion and company mortars may be the primary indirect fire assets for the battalion. (See FM 3-09.) Additional fire support considerations for supporting the commander's concept of operations include—

- Allocating initial priority of fires to the forward security force.
- Planning targets along enemy reconnaissance mounted and dismounted avenues of approach.
- Engagement of approaching enemy formations at vulnerable points along their route of march.
- Planning the transition of fires from the security area to through the main battle area fight.
- Planning the echelonment of fires (see appendix C).
- Incorporating existing fire support coordination measures and detailed triggers to adjust them.
- Developing clear triggers to initiate fires and adjust priority of fires.
- Ensuring integration of fires in support of obstacle effects.
- Ensuring the integration of fires with the battalion's counterattack plan and repositioning contingency plans.
- Identifying and targeting high priority targets.
- Airspace deconfliction.
- Integration and positioning of organic mortars.

Sustainment

3-235. Because sustainment operations in support of the defense requires more centralized control, the battalion S-4 ensures sustainment planning is coordinated fully with the rest of the staff and subordinate units. The S-4 coordinates with the battalion S-3 to ensure that supply routes do not interfere with maneuver and obstacle plans, but still support the full depth of the defense. The battalion's sustainment concept of support considers prepositioning Class IV, Class V, and Class III (bulk) far forward initially to support the security area during the counterreconnaissance fight, followed by the main battle area so that the battalion can rapidly transition from defense to offense. When transitioning initially from the offense to the defense, the S 4 considers cross leveling classes of supply and sustainment assets. (See http://www.army.mil/article/105838/CASCOMreleases_OPLOG_PlannerVersion_8_0/.)

3-236. As sustainment considerations within the Infantry battalion are characterized by constrained organic assets. Planning for sustainment operations throughout the security area is critical to sustaining reconnaissance and security operations to prevent enemy forces from determining friendly force disposition. Forces operating within the security area are configured prior to departure of the main battle area with a minimum of 72-hour logistics package (LOGPAC) of Class I (subsistence), Class III (petroleum, oil, and lubricants), and Class V. Preconfigured combat loads are positioned in the combat trains to expedite resupply operations. (See appendix H.) Classes of supply (Classes IV and V) are pre-positioned, as required within the defense. Sustainment support to the security area must include planning for both ground and aerial resupply and medical evacuation of long duration observation points.

3-237. Enemy actions and the maneuver of combat forces complicate medical operations, as will the depth and dispersion of the defense. Defensive operations must include health service support to medical personnel who have much less time to reach a patient, complete vital emergency medical treatment, and remove the patient from the battle site. With the enemy's initial attack and the battalion's counterattack producing the heaviest patient workload, they are also the most likely times for the enemy's use of artillery and CBRN weapons. These enemy attacks can disrupt ground and air routes and delay evacuation of patients to and from treatment elements.

Protection

3-238. Because the battalion defends to conserve combat power for use elsewhere or later, the commander must secure the force. The commander enables security, by means of providing information about the activities and resources of the enemy, through the employment reconnaissance forces and surveillance assets within the battalion's assigned area of operation. The commander may employ counterreconnaissance forces, and establish combat outposts and observation posts to counter enemy R&S efforts. The commander integrates battalion R&S efforts, and security operations with those of the IBCT and intelligence, surveillance and reconnaissance assets above the IBCT.

3-239. As discussed in chapter 2, personnel and physical assets have inherent *survivability*—a quality or capability of military forces which permits them to avoid or withstand hostile actions or environmental conditions while retaining the ability to fulfill their primary mission (ATP 3-37.34), which can be enhanced through various means and methods. One way the battalion can enhance survivability when existing terrain features offer insufficient cover, which is the protection from the effects of fires (FM 3-96) and concealment, the protection from observation or surveillance (FM 3-96) is to alter the physical environment to provide or improve cover and concealment. Similarly, natural or artificial materials may be used as camouflage to confuse, mislead, or evade the enemy. Together, these are called *survivability operations*—those military activities that alter the physical environment to provide or improve cover, concealment, and camouflage (ATP 3-37.34). Although survivability encompasses capabilities of military forces both while on the move and when stationary, survivability operations focus more on stationary capabilities—constructing fighting and protective positions and hardening facilities.

3-240. Within the battalion's area defense, protective obstacles—employed to protect people, equipment, supplies, and facilities against threats—are key enablers to survivability operations. Protective obstacles provide local, close-in protection to prevent the enemy from delivering a surprise assault from areas close to defending positions. Protective obstacles protect the emplaced position by warning, mitigating, and preventing hostile actions and effects. (See ATP 3-90.8.) The commander uses protective obstacle effects (warning, mitigation, and prevention) to convey intent and facilitate protective obstacle planning and design. Protective obstacles employed in support of the battalion's area defense must be capable of being rapidly emplaced, and recovered or destroyed. Final protective fires are integrated within the protective obstacle plan to defeat the final assault of the enemy.

3-241. As subordinate units within the battalion conduct survivability operations within the limits of their capabilities. Engineer and CBRN assets provide additional capabilities to support survivability operations in support of the battalion. Engineer support to survivability operations is a major portion of the enhance protection line of engineer support and the integration of survivability priorities for critical systems and units within and supporting the battalion. (See FM 3-34). CBRN support to survivability operations includes the employment of immediate and operational decontamination techniques to enhance survivability of forces in CBRN environments. CBRN R&S assets determine likely locations for enemy employment of CBRN weapons. CBRN defense consists of measures to minimize or negate the vulnerabilities and effects of a CBRN incident. (See appendix G.)

Preparation

3-242. Preparation activities help the commander, staff, and subordinate units of the Infantry battalion (including attachments) understand the situation and their roles in the overall operation. The commander takes every opportunity to improve situational understanding prior to execution, through the integration of intelligence and operations. The commander and staff continuously plan, task, and employ aggressive and continuous information collection assets and forces throughout the preparation of the defense. The commander may conduct spoiling attacks during preparation to disrupt the enemy's offensive preparations. The following paragraphs discuss key activities (although not inclusive) that the battalion commander and staff, and subordinate units conduct to ensure the battalion is protected and prepared for execution.

Rehearsals

3-243. Rehearsals allow the commander and staff to assess subordinate preparations and identify areas that may require more supervision. The commander considers time, preparation activities, and operations security when selecting a rehearsal type. (See appendix B.) Rarely will the battalion be able to conduct a full-force rehearsal given the tempo of operations and the potentially large size of the area of operations. When possible the commander considers conducting key leader map and terrain board rehearsals at night to focus attention during periods of increased visibility on inspecting preparations and working with subordinate leaders. Battalion rehearsals should cover—

- R&S missions.
- Security operations.
- Battle handover and passage of lines.
- Security area and main battle area engagement.
- Engagement, disengagement, reposition, and withdrawal criteria.
- Reserve employment options and commitment criteria.
- Actions to deal with enemy penetrations, major enemy efforts along areas of risk or flank avenues of approach, and enemy actions in the rear area.
- Sustainment, particularly casualty evacuation, emergency resupply operations, and reorganization.
- Execution of routes for repositioning, movement of the reserve, withdrawal, and movement to casualty collection points.
- Execution of follow-on missions to exploit defensive success.
- Integration of aviation assets, when applicable.

Survivability and Countermobility

3-244. Much of the strength of a defense rests on the integration and construction of reinforcing obstacles, exploitation of existing obstacles, and actions to enhance the survivability of the force through construction of fighting positions and fortifications. The commander's intent focuses survivability and countermobility preparation through the articulation of obstacle intent (target, effect, and relative location), and priorities and establishment of priorities for survivability and countermobility. Guided by that intent, the battalion engineer staff planner develops a scheme of engineer operations that includes engineer task organization, priorities of effort and support, subordinate engineer unit missions, and survivability and countermobility instructions for all units.

Monitoring of Preparation Activities

3-245. During preparation, the commander and staff monitor preparatory actions and track the higher and adjacent unit situations and the enemy situation. The staff updates and refines plans based on additional reconnaissance and updated intelligence information. The staff continues to disseminate these modifications through fragmentary orders. The commander and staff conduct much of the preparation phase simultaneously with operations conducted in the security area, continuing even as forward-deployed forces gain contact with the enemy. Throughout the preparation, the battalion commander, company commanders, and key staff members physically inspect preparatory activities. Weapons positioning, sitting of obstacles, direct and indirect fire plans and associated triggers, sustainment operations, and Soldier knowledge of their missions are all critical checks.

3-246. As subordinate units position elements and execute defensive preparations, the battalion staff coordinates their activities within the overall situation. The S-2 monitors the enemy situation through information collection efforts at the battalion and IBCT level. The information collection matrix focuses battalion efforts on indicators that reveal the enemy's likely time and direction of attack. The staff continually analyzes this assessment to determine the effects on preparation time available and any changes to the course of action. Priority intelligence requirements are updated as the situation changes and as the information collection effort answers information requirements.

3-247. The S-3 monitors the status of rehearsals and updates the plan as needed based on continuously updated intelligence and the status of preparations. The S-4 analyzes the status of logistics and maintenance of equipment within the battalion to determine any required adjustments to the plan or task organization. The engineer planner for the battalion monitors the progress of all engineer efforts within the area of operation and continually projects the end state of this effort based on the current and projected work rates. The engineer planner identifies potential shortfalls early and determines how to shift assets to make up for the shortfalls or recommend where to accept risk. As the enemy closes on the battalion's area of operations, the battalion begins final preparations that typically include—

- Final coordinating of battle handover and passage of lines.
- Positioning of situational obstacle employment systems.
- Verifying communications status.
- Evacuating unused Class IV and V to prevent capture or loss to enemy action.
- Withdrawing engineer forces from forward areas.
- Linking up fire support, protection, and sustainment assets with reserve or other supported combat forces (if not previously accomplished).
- Reviewing the R&S plan to ensure it still meets the commander's priority intelligence requirements.
- Final positioning or repositioning of R&S assets, security forces, and observers.
- Positioning of teams to close lanes in obstacles or execute reserved obstacles.
- Executing directed, reserve, or situational obstacles.
- Periodic situation updates and issuing of final guidance to subordinates.
- Registering indirect fire targets with mortars, if not already done.
- Conducting a final radio or even map rehearsal with key leaders.
- Conducting targeting meetings to update targets, resources, and priorities.
- Cover gaps between defensive positions, reinforce those positions as necessary, and counterattack to seal penetrations or block enemy attempts at flanking movements.

EXECUTION

3-248. A defending force within the main battle area uses a variety of tactics, techniques, and procedures to accomplish the mission. At one end of the defensive continuum is a static defense oriented on terrain retention. At the other end is a dynamic defense focused on the enemy. Within the Infantry battalion, the commander combines static actions (see figure 3-26, page 3-63) to control, stop, or canalize the attacking enemy forces and dynamic actions to cover gaps between defensive positions, reinforce those positions as necessary, and counterattack to seal penetrations (see figure 3-28, page 3-78) or to block enemy attempts at flanking movements.

3-249. Throughout the area defense, the commander conducts shaping operations designed to regain the initiative by limiting the attacker's options and disrupting the enemy's plan. Shaping operations prevent enemy forces from massing and create windows of opportunity for the conduct of decisive maneuver, allowing the defending force to defeat the attacking enemy in detail. The mission variables of METT-TC determine how closely the commander synchronizes shaping operations (or supporting efforts) with the decisive operation (or main effort). R&S missions and security operations are normally components of the Infantry battalion's shaping operations.

Gain and Maintain Enemy Contact (Security Area Actions)

3-250. Once security area forces have moved into the security area, actions in the security area predominantly focus on reconnaissance, counter reconnaissance, target acquisition, reporting, destruction, delay of the enemy main body, and battle handover. Battalion security area forces integrate these actions with friendly forces forward of them, maintaining information flow and security. Battalion security area forces may execute battle handover with forward elements then assist them in executing a rearward passage. Throughout security area operations, security forces coordinate and crosstalk with units to their rear. When battalion security forces execute rearward passage of lines and battle handover they may then move to the flanks of the main battle area or occupy an assembly area to the rear to plan for future operations. On approaches that the enemy does not use, the commander may desire to leave elements of the security force forward to preserve observation and access to enemy flanks.

3-251. Information collection within the security area provides the commander with information to support their decision making, to provide early warning and reaction time, and to support targeting. Guided by the CCIRs, the four primary tasks conducted as part of information collection (reconnaissance, security operations, surveillance, and intelligence operations) help provide the following information—

- Location, movement, and destruction of enemy reconnaissance and security forces and surveillance assets.
- Speed, direction, composition, and strength of enemy formations.
- Locations of high-payoff targets (for example-indirect fire, bridging, and command and control assets).
- Enemy actions at decision points.
- Enemy flanking actions, breaching operations, and force concentrations.
- Battle damage assessment.
- Movement of follow-on forces.

3-252. As the enemy's attack begins, reconnaissance and security forces identify committed enemy unit positions and capabilities, determines the enemy's intent and direction of attack, and gains time to react. The commander uses the information available, in conjunction with military judgment, to determine the point at which the enemy commits to a course of action. The staff integrates the information provided by reconnaissance and security forces and surveillance assets with information received from higher and adjacent units, subordinate units, unified action partners, or special operations forces operating within the area of operation. The commander ensures the distribution of a common operational picture throughout the force during the battle as a basis for subordinate commander actions. In an area defense, critical decisions for the commander normally include—

- Initiation and employment of direct and indirect fires against enemy formations.
- Modifications or adjustments to the defensive plan.
- Execution of situational and reserved obstacles.
- Withdrawal of forward security forces.
- Commitment of the reserve, counterattack, or both.

Disrupt and Fix the Enemy (Security Area Engagement)

3-253. Engagements in the battalion security area normally are limited. Counterreconnaissance forces focus on locating and destroying enemy reconnaissance elements. As the enemy closes into the area, observers initiate indirect fires and the execution of reserved obstacles. The focal points are normally early warning and identification of the enemy's decisive and shaping operations, strength, and composition of threat forces, and direction of attack in order for the commander to make decisions and position forces. In the event enemy reconnaissance and security forces and surveillance assets penetrate the security area, battalion forces operating in the security area must be prepared to conduct target handover with the battalion's main battle area forces.

3-254. After making contact with the enemy, the commander seeks to disrupt the enemy's plan, the enemy's ability to control forces, and the enemy combined arms team. Ideally, the results of the commander's shaping operations should force a disorganized enemy, whose ability to synchronize its elements has been degraded, to conduct a movement to contact against prepared defenses. Once the process of disrupting the attacking enemy begins, it continues throughout the defense of the security area. The commander may use assault platoons to engage enemy formations at longer distances under the control of the battalion, the weapons company, or other security force commander. Ensuring though that these elements are not decisively engaged, and that they retain their ability to maneuver. These forward units also call for close air support, Army attack aviation, and precision guided munitions from artillery and mortars. Security area engagements can provide the following advantages:

- Depth to the area of operation.
- More time to prepare in the main battle area.
- A weakened enemy.
- Confusion to the location of the friendly defensive positions.
- Forces the enemy to deploy and more clearly indicate their main attack or intentions.

3-255. Within the security area, the commander does everything possible to limit the options available to the enemy. To limit the enemy's options, in addition to disrupting the enemy, the commander conducts shaping operations to constrain the enemy into a specific course of action, control enemy movements, or fix the enemy in a given location. While executing these operations, the commander continues to find, delay, or attrit enemy follow-on and reserve forces to keep them from entering the main battle area. The commander has several options to help fix an attacking enemy force. The commander can design shaping operations—such as securing the flanks and point of a penetration—to fix the enemy and allow friendly forces to execute decisive maneuver elsewhere. Combat outposts and strong points can also deny enemy movement to or through a given location. The commander uses obstacles covered by fire to fix, turn, block, or disrupt to limit the enemy's available options. Properly executed obstacles (situational and reserved) are a result of the synthesis of top-down and bottom-up obstacle planning and emplacement. Blocking forces can also affect enemy movement. A blocking force may achieve its mission from a variety of positions depending on the mission variables of METT-TC.

3-256. The commander coordinates the battle handover between security forces and main battle area forces as quickly and efficiently as possible to minimize their vulnerability to enemy fire. When the battle handover is a transfer of responsibility for the battle from the IBCT's or a higher unit's security area force to the Infantry battalion. The higher commander who established the security force prescribes criteria for the handover and designates the location where the security forces will pass through, routes, contact points, and the battle handover line. The battle handover line is normally forward of the forward edge of the battle area where the direct fires of the forward combat elements of the battalion can effectively overwatch the elements of the passing unit. The IBCT commander or other higher headquarters commander coordinates the battle handover with the battalion commander. This coordination overlaps with the coordination for the passage of lines, and the two should be conducted simultaneously. Normally, coordination includes—

- Establishing communications, this includes ensuring linkage on tactical radios and tactical radio networks (see ATP 6-02.53) and effective information overlap.
- Providing updates on both friendly and enemy situations and the addition of appropriate command posts and leaders to the message groups on situation reports and updates.
- Coordinating passage, which includes identifying passage points and lanes, and recognition signals and exchanging or disseminating graphics of these and obstacle overlays.
- Collocating command posts.
- Dispatching representatives to contact points and establishing liaisons.
- Coordinating recognition signals.
- Reporting status of obstacles and routes, including overlays.
- Coordinating fire support, protection, and sustainment requirements, with particular attention given to casualty and equipment evacuation requirements.
- Coordinating actions to assist the security force with breaking enemy contact.
- Coordinating and exchanging maneuver, obstacle, and fire plans.
- Coordinating location, communications plans, and fire support coordination measures (specifically no fire areas) to any stay-behind forces.

3-257. Within the Infantry battalion, the battle handover between battalion security forces and the forward Infantry companies in the main battle area are less complicated, but equally as critical and must be planned in detail. Security forces and forward companies identify rearward passage points and lanes, coordinate movement with the individual units or units covering them and through which they are moving. Frequently, the first elements to displace are the maneuver forces that were executing counterreconnaissance, moving to initial defensive positions in the main battle area, or acting as the battalion or IBCT reserve. The battalion scout platoon normally displaces to vantage points on the flanks, moves to establish surveillance on other avenues of approach, or infiltrates back to the rear portion of the battalion's area of operation. When battle handover occurs within the battalion, companies within the main battle area—

- Assist passage of lines and disengagement.
- Gain and maintain contact with enemy forces as battle handover occurs.
- Maintain security.
- Execute on order, reserved obstacles and prepared to emplace situational obstacles in the security area as the passing force withdraws.

Maneuver (Main Battle Area Engagement)

3-258. In an area defense, the battalion decisive operation is decided in the main battle area. The battalion commander shapes and decides the engagement by massing the effects of combat power. Effects are synchronized in time and space and should be rapid and unexpected so that they break the enemy's offensive tempo and disrupt the enemy's attack. Synchronized prior planning and preparation bolster the effects of combat power, increasing the effectiveness of the defense.

Scheme of Maneuver

3-259. Depending on the defensive scheme of maneuver, the battalion may fight primarily from a single series of positions or it may conduct delay operations capitalizing on movement and repeated attacks to defeat the enemy in-depth. Forward positioned forces, obstacles, and fires are used to break the enemy's momentum, force the enemy to deploy earlier than desired, reduce the enemy's numerical advantage, disrupt enemy formations and tempo, and force the enemy into positions of vulnerability. The battalion masses fires (see paragraph 3-231) and integrates obstacles (see paragraph 3-206) to disrupt, turn, fix, block, canalize and then destroy attacking enemy forces in engagement areas throughout the battalion's area of operations.

3-260. As the operation evolves, the commander knows that there will probably be a requirement to shift the decisive operation and shaping operation(s) or the main effort and supporting effort(s) to press the engagement and keep the enemy off balance. The commander integrates information collection tasks to shift the effects of fires and maneuver forces so that they are repeatedly focused, and refocused to achieve decisive, destructive, and disruptive effects upon the enemy's attack. IPB enables information collection to determine likely enemy actions, while security area forces and main battle area forces confirm or deny those actions.

3-261. Throughout the area defense, the battalion must maintain a cohesive defense if it is to defeat the enemy. This does not mean, however, that the forces must be massed close together or that companies must have mutually supporting fires. With forces dispersed, companies can maintain cohesion by maintaining the common operational picture, crosstalk among subordinates, and the continual tracking and reporting of the enemy. The commander and staff, and subordinate units continually assess the enemy's options and movement while identifying means to defeat them. With forces widely dispersed, continual assessment of time and distance variables are essential. To maintain defensive cohesion, subordinate companies keep their movement, positioning, and fires consistent with the battalion commander's intent, the defensive scheme of maneuver, and the obstacle plan.

3-262. Unless the IBCT or other higher headquarters plan makes other provisions (for example-a higher echelon reserve or counterattack force is responsible), the Infantry battalion is responsible for controlling enemy advances within its area of operations. When a penetration threatens the battalion, the commander may take several actions to counter the situation. In order of priority, the commander may do any or all of the following:

- Allocate priority of all available fires, to include artillery and mortar fires, Army aviation attack, and close air support, to the threatened unit. (This is the most rapid and responsive means of increasing the combat power of the threatened unit.)
- Direct or reposition adjacent units to engage enemy forces that are attacking the threatened unit. (This may not be possible if adjacent units are decisively engaged.)
- Commit the reserve to reinforce the threatened unit.
- Commit the reserve to block, contain, or destroy the penetrating enemy force.
- Accept penetration of insignificant enemy forces and maintain contact with them as they move deeper into the main battle area.
- Move forces to alternate, supplementary, or subsequent positions or to withdraw forces.
- Commit attached engineers or other element to assist in containing the penetration or to constitute a new reserve.

3-263. When a penetration occurs, units within the main battle area continue to fight, refuse their flanks, and engage the enemy's flanks and rear. The penetrated force must try to minimize the penetration to prevent the area of penetration from widening and to protect adjacent unit flanks. Adjacent units take immediate action to secure their exposed flanks, which may include security missions or the establishment of a blocking position(s). Adjacent units also may need to reposition forces or direction of fire, readjust subordinate area of operations and tasks, or commit their reserve. Forces within the main battle area try to reestablish contact across the area of penetration when possible.

3-264. During combat operations, sustainment operations are tailored in response to changes in tactical requirements. The battalion S-4, in coordination with the forward support company commander, fully integrates sustainment operations with the battalion battle rhythm through planning and oversight of on-going operations. Logistical synchronization matrices and logistics reports are used to initiate and maintain synchronization between operations and sustainment functions.

3-265. Protection of sustainment operations and locations ensures continuity of logistics operations. Because committing combat forces to sustaining operations and locations such as the combat or field trains diverts combat power from the main battle area, the commander carefully weighs the need for such diversions against the possible consequences to the overall operation. Generally, support elements in the battalion area of operations rely on positioning, movement, and self-protection for survival. They—

- Establish sustainment operations in covered and concealed areas away from likely enemy avenues of approach.
- Establish and maintain perimeter security and early warning observation posts, integrating weapons and crews that are in the rear for repair operations.
- Keep sustainment units ready to move on short notice as the security battle begins.
- Maintain internal security for any movement while executing sustaining operations.

3-266. Early warning to sustainment elements to the rear of the battalion's area of operations is critical to survival in the event of a penetration of the main battle area or an enemy attack from an unexpected area. Sustainment plans and rehearsals address actions to be taken in the event of attacks on sustaining operations, including defensive measures, displacement criteria, casualty evacuation, routes, rally points, and subsequent position. (See appendix H for additional information.)

Note. The example below is a continuation of the scenario from paragraph 3-195, specifically figures 3-25 on page 3-61 and 3-26 on page 3-63. As this scenario continues, the enemy commits its main attack along Avenue of Approach Alpha. Once the enemy intent is confirmed, the commander moves mobile elements of the battalion to reinforce Company B, the battalion's main effort.

Example - Forward Defense against an Infantry Threat (Mobile Actions Focused on the Enemy)

3-267. In this example, forward observers along the battalion's forward line of own troops confirmed the enemy's commitment to Avenue of Approach Alpha. Simultaneously, the commander allocates additional fire support to the area threatened and repositions mobile forces to reinforce the battalion's main effort along Avenue of Approach Alpha. The commander alerts the IBCT commander to the threat and that the battalion confirms the enemy's commitment to Avenue of Approach Alpha. The battalion commander's goal is to prevent the enemy's further advance by using a combination of indirect fires, direct fires from prepared positions, obstacles, and mobile elements of the battalion (figure 3-28).

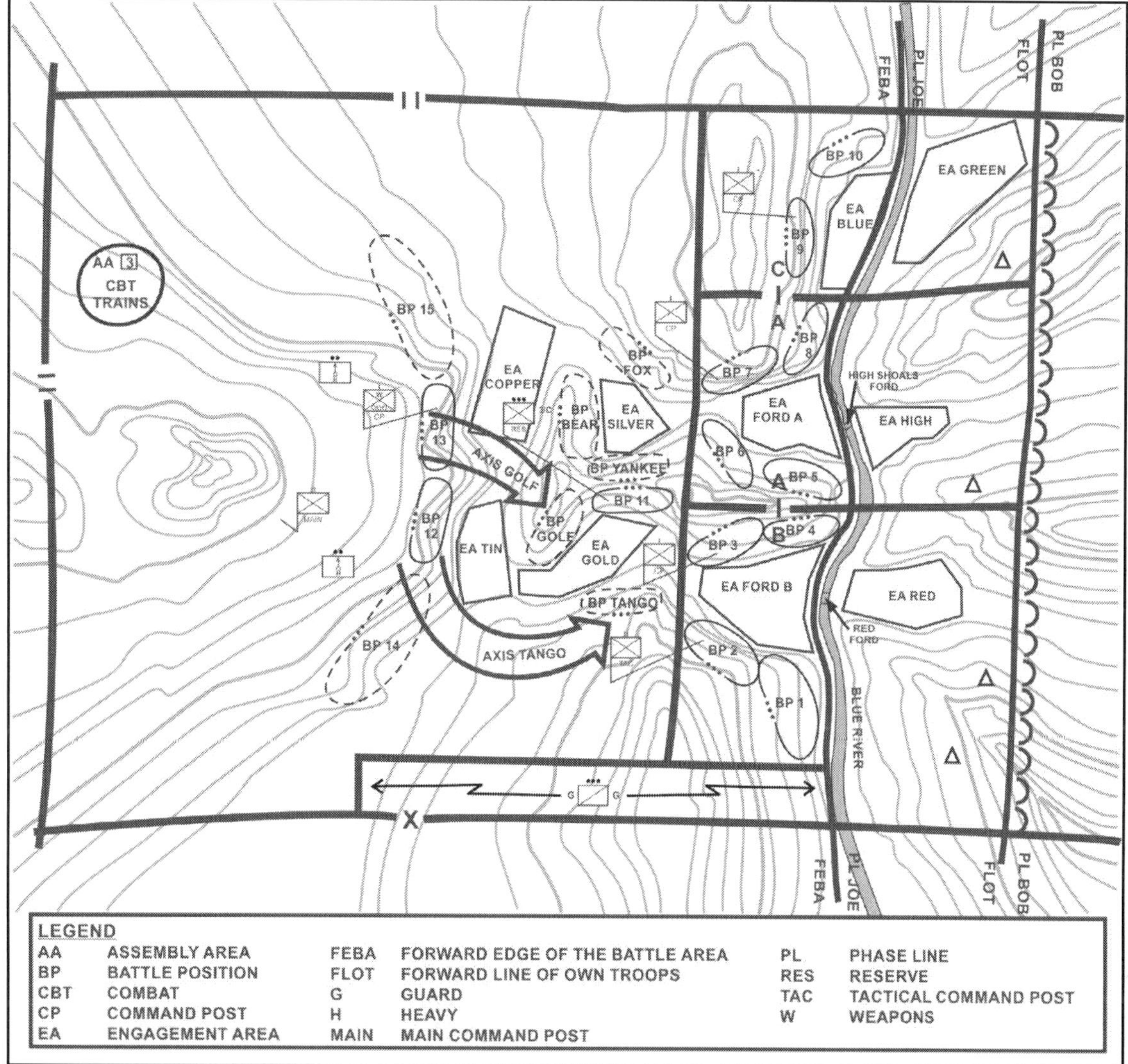

Figure 3-28. Mobile actions focused on the enemy, example

3-268. As in the earlier example (see paragraph 3-265), due to the limited number of firing positions along the river, the battalion commander retained two of the four assault platoons under Company D commander's control. Company D initially establish battle positions 12 and 13 to assist in limiting the enemy advance along Avenues of Approach Alpha and Bravo. Battalion mortars operating split section, initially positioned forward vicinity Battle Positions Golf and Tango with priority of fires to security area actions (not illustrated). On order, the mortar platoon repositions by section to firing positions vicinity Battle Positions 12 and 13 with priority of fires to Company B main battle area actions initially in Engagement Area Red, then Engagement Area Ford B. On order, Company D occupies attack by fire positions located vicinity Battle Positions Tango and Golf to reinforce the battalion's main effort, Company B along Avenue of Approach Alpha.

> ***Note.*** As the battle develops, one of the commander's most critical decisions is when to commit the reserve (see paragraph 3-283).

Follow Through

3-269. The Infantry battalion may conduct local counterattacks to restore or preserve defensive integrity. Unless the conduct of the defense has left the battalion largely unscathed, the battalion usually lacks the ability to conduct a significant counterattack by itself. If the battalion has the ability to organize a counterattack force, this force must have mobility or be prepositioned in a position of advantage to attack the enemy from an unexpected flank. Within the context of the IBCT's operations, a defending battalion may execute a counterattack in support of the IBCT's defensive posture, as part of a larger force seeking to complete the destruction of the enemy's attack, or as part of a transition to offensive tasks.

AREA DEFENSE—SEQUENCE OF EVENTS, EXAMPLE

3-270. The Infantry battalion may assume a defensive mission following an attack of its own or in anticipation of an enemy attack. Most defenses will following a sequence of events similar to the example sequence addressed below. This sequence of events is used for discussion purposes and is not the only way to sequence an area defense. With any sequence, the commander understands events will vary depending on the mission variables of METT-TC and to some degree, events will overlap.

LEADER'S RECONNAISSANCE

3-271. Before occupying any position, to include those in the forward security area, leaders at all echelons conduct some type of reconnaissance of their area of operation and position(s). The reconnaissance effort is as detailed as possible in regards to the mission variables of METT-TC. Reconnaissance can consist of a simple map reconnaissance, or a more detailed leader's reconnaissance and initial layout of the new position. When feasible, the commander and subordinate leaders conduct a leader's reconnaissance of the complete area of operation to develop plans based on their view of the actual terrain. When available, the commander may use aviation assets to conduct the leader's reconnaissance.

3-272. The success of the leader's reconnaissance is critical to conduct occupation without hesitation and to begin priorities of work prior to and immediately upon occupation. Participants in the reconnaissance include the commander, subordinate commanders, and selective key staff members and element leaders. The goals are, but not limited to, identification of enemy avenues of approach, engagement areas, sector of fires, the tentative obstacle plan, indirect fire plan, observation post locations, and command post locations. The commander and staff develop a plan for the leader's reconnaissance that includes the following:

- Provisions for security.
- Areas to reconnoiter.
- Priorities and time allocated for the reconnaissance.
- Considerations for fire support, communications, and casualty and medical evacuation plans.
- Contingency plan in the event that the reconnaissance is compromised.

ESTABLISH SECURITY

3-273. When the Infantry battalion is part of a larger units area defense, for example the IBCT. The IBCT commander establishes a forward security area before the Infantry battalion moves to defend. Even with this forward security area established, the battalion must still provide for its own security, especially over large geographical areas or in complex terrain and noncontiguous area of operations. In order to prevent the enemy from observing and interrupting defensive preparations and identifying unit positions, the battalion establishes the security area well forward of the planned main battle area for the battalion, but within indirect fire and communications range. When the battalion commander is unable to push the security area forward to achieve this objective, the battalion may have to hold its positions initially, as it transitions and then withdraws units to the defensive main battle area, establishing a forward security force in the process.

3-274. When security forces of the IBCT are forward of the Infantry battalion, the battalion commander integrates the battalion security force actions with that of the IBCT and adjacent battalions. In contiguous area of operations, the battalion commander normally organizes and defines the security area forward of the forward edge of the battle area and assigns company areas of operation to prevent gaps in the battalion area defense. In noncontiguous area of operations, the battalion commander normally organizes and defines the security area forward of the main battle area, or along likely avenues of approach. In noncontiguous area of operations, individual companies will have more responsibility for independent security areas actions. The battalion commander ensures the independent security area actions at company level align with the security area plan for the Infantry battalion and if applicable the IBCT.

3-275. As the commander arrays security forces forward of the battalion's main battle area. The battalion commander and subordinate commanders also plan security operations within the main battle area to prevent enemy reconnaissance, reduction of obstacles, targeting of friendly positions, and other disruptive actions. Subordinate units secure obstacles, battle positions, command posts, and sustainment sites (for example when established, battalion combat trains and battalion field trains) throughout the battalion's area of operation. The threat force and battlefield organization will dictate the commander's decision whether elements in the battalion conducting sustaining operations are allocated security forces, or if they provide their own security. With extended lines of communication, the battalion may also secure logistical elements moving forward from the field trains, generally location in the BSA, to support the battalion.

OCCUPATION AND PRIORITIES OF WORK

3-276. When the battalion establishes a security area independent of the IBCT or other higher echelon. Battalion security forces deploy forward as remaining forces occupy and prepare positions in the main battle area. Security of the main battle area is critical during occupation to ensure subordinate units avoid detection and maintain combat power for the actual defense. The plan of occupation for the battalion must be thoroughly understand to maximize the time available for occupation and preparation of the defense. When establishing the security area, the battalion commander may lead with the scout platoon to conduct reconnaissance and establish observation post along the forward edge of the security area. From these observation posts, the scout platoon uses long-range fires to hinder the enemy's preparations, to reduce the force of the enemy's initial blows, and to start the process of wresting the initiative from the enemy. The commander can reinforced the scout platoon, based on the mission variables of METT-TC, understanding though this reduces the available combat power for the main battle area engagement. Throughout the security area, security forces position and reposition to—

- Prevent enemy observation of defensive preparations.
- Defeat infiltrating enemy reconnaissance forces.
- Prevent the enemy from delivering direct fires or observed indirect fires into the battalion area defense.
- Provide early warning of the enemy's approach.

3-277. Within the main battle area, and depending on the situation, the commander may send a subordinate force to initially secure positions prior to the main body's arrival. The mission of this force is to continue to conduct reconnaissance of key terrain and obstacles, guide and provide local security as the battalion main body occupies the defense and initiates priorities of work. As all elements within the main battle area establish local security, priorities of work continues to include refining the plan, positioning of forces, preparing positions, constructing obstacles, planning and synchronizing fires, positioning logistics, and conducting inspections and rehearsals. To aid in operations security and to reduce vulnerability, the staff balances the benefits of dispersion against the requirements and resources for the security area. Usually, the greater the dispersion between companies, the larger the security area.

3-278. Throughout occupation and priorities of work in the main battle area, security area actions continue without interruption. Security forces may be assigned screen, or area security missions. The scout platoon or other security force may position to screen and provide early warning along the most likely enemy avenues of approach, reinforced in-depth with sections or platoons from Infantry or weapons companies. When applicable, the commander integrates reconnaissance and ground maneuver units in the security forces. This provides the forces required for the hunter-killer technique where reconnaissance forces primarily locate the enemy element(s) and attack them with indirect fires, not to engage in direct fire attack except in self-defense. The reconnaissance force then guides the maneuver force to destroy, neutralize, or repel threat forces with direct fires.

SECURITY AREA ENGAGEMENT

3-279. When the Infantry battalion is part of the IBCT or other higher echelon's area defense, the battalion commander integrates engagements within the security area of the battalion with that of the higher echelon. When the battalion establishes a security area independent of the IBCT or other higher echelon, the battalion's planned indirect fires usually include forward observers and fire support teams executing indirect fire targets on a primary enemy avenue of approach. This can be in support of the higher headquarters' scheme of fires using IBCT or higher echelon artillery, or in support of the battalion scheme with the use of organic mortars and allocated artillery fires.

3-280. The scheme of fires within the security area combined with the use of situational obstacles serve to disrupt the enemy and canalize the enemy in the engagement areas, and to force the enemy to commit enemy engineer assets prior to the main battle area engagement. As situational obstacles are planned and triggered relative to specific enemy attack options, they are related to accomplishing a specific essential countermobility task (see ATP 3-90.8) and fire support task (see FM 3-09) allowing for more effective engagements within the security area. Forward security forces employed forward may cover these situational obstacles with direct fires prior to their withdrawal to positions within the main battle area.

3-281. When the battalion supports a higher echelon's or its own scheme of maneuver by fighting a delay through the depth of the security area and into the main battle area. The purpose may be to take advantage of restrictive avenues of approach, to set the conditions for a counterattack, or to avoid a decisive engagement until favorable conditions are set. As security forces complete the rearward passage of lines, battalion main battle area forces assumes control of the battle at the battle handover line. Battle handover from forward security forces to forward main battle area forces requires firm, clear arrangements—

- For assuming command of the action.
- For coordinating direct and indirect fires.
- For the security force's rearward passage of lines.
- For closing lanes in obstacles.
- For movement of the security force with minimal interference to main battle area actions.

3-282. As security area engagements transition into the main battle area, security area forces withdraw to battle positions within the main battle area and counterattack or reserve positions. Security area forces may move to a flank or to the rear of the main battle area to provide security.

Main Battle Area Engagement

3-283. The battalion commander seeks to defeat, disrupt, or neutralize the enemy's attack forward of or within the main battle area. The commander integrates direct and indirect fires with the obstacle plan, local counterattacks and reserve forces to destroy the enemy in designated engagement areas or to force the enemy transition to a retrograde or hasty defense. The commander focuses fires in an effort to attack the enemy throughout the depth of the battalion's area of operation. However, fire support to the battalion may be limited to critical points and times. Control measures allow the commander to rapidly concentrate the use of combat power at the decisive point, provide flexibility to respond to changes, and allocate responsibility of terrain and obstacles to synchronize the employment of combat power.

3-284. As attacking forces reach the forward edge of the battle area, the enemy will try to find weak points in the battalion defense and attempt to force a passage, possibly by a series of probing attacks. Battalion forward elements engage the enemy's lead forces as the enemy attack develops along identified enemy avenues of approach. The commander arrays forces and establishes engagement areas using obstacles and fires to canalize enemy forces. When shaping operations allow for the canalization the enemy. The enemy advance slows and the increased density of forces present good targets for defensive fires within battalion engagement areas. The maximum effects of these simultaneous and sequential fires, brought to bear at this stage of the battle, enable the battalion's destruction of the attacking enemy force.

3-285. When the battalion is unable to bring sufficient combat to shape the enemy's advance prior to entering the main battle area. The commander masses direct and indirect fires and conducts movement to gain or regain a positional advantage over the advancing enemy force. The commander reassigns priorities of support and reposition forces to meet the enemy where the enemy actually is rather than where the commander's defensive plan projected that the enemy would be. The commander controls the commitment of local counterattack and reserve forces to engage enemy penetrations. Within this situation, the IBCT or other higher echelon engages enemy follow-on forces with long-range rockets and air support to reinforce the battalion. The slowing or delay of enemy follow-on forces into the battalion's main battle area enables the defeat of the enemy's attack by the battalion in detail, one echelon at a time. Defeat of these forces in the main battle area can disrupt the enemy's timetable and lead to the creation of exploitable gaps between committed and subsequent echelons to create conditions for a counteroffensive.

CONSOLIDATION, REORGANIZATION, AND FOLLOW THROUGH

3-286. Following a successful defense, the Infantry battalion can exploit by counterattack based on branches and sequels to the plan before the enemy can secure gains or organize a defense. The higher commander's concept of operations and the mission variables of METT-TC dictate the battalion's follow-on mission, for example a counteroffensive that allows the battalion to regain the initiative. When the situation prevents offensive action, the battalion continues to defend. As in the initial establishment of the defense, gaining depth in the security area is critical. A local counterattack can provide space for establishment or reestablishment of the security area and time for the rest of the battalion to consolidate and reorganize. Any attack option must pay particular attention not only to the terrain and enemy, but also to friendly obstacles (and their destruction times, if applicable) and areas where dual-purpose improved conventional munitions or bomblets have been used. If the battalion cannot counterattack to gain adequate space and time, then the battalion may have to direct one company to maintain contact with the enemy and guard the area of operation while others move to reestablish the defense farther to the rear. The battalion must consolidate and reorganize whether it continues to defend or transition to offensive actions.

Note. The following section illustrates a fictional scenario as a discussion vehicle illustrating one of many ways an Infantry battalion can conduct an area defense. In this scenario the IBCT defends against three enemy avenues of approach (see figure 3-23, page 3-56). Infantry Battalion 1 defends against enemy Avenue of Approach 3 the least likely enemy approach. Infantry Battalion 2 defends against enemy Avenue of Approach 1 and Avenue of Approach 2, the most likely avenues of approach. This illustration will mainly focus on Infantry battalion 2. Engagement area development for this illustration is found in paragraphs 3-175 through 3-189, pages 3-54 to 3-59. The third Infantry battalion of the IBCT is under division control as part of the counterattack force (air assault-not illustrated) for the corps.

ILLUSTRATION OF AN AREA DEFENSE

3-287. In this illustration, an IBCT conducts an area defense against a motorized Infantry and armor threat. Army aviation attack and reconnaissance units, initially control by division and higher echelon, conducted attacks against enemy forces not in direct contact with the ground maneuver forces of the IBCT. As the enemy advanced, aviation attacks continued in close proximity or in direct support of IBCT and battalion security forces and main battle area forces. Artillery and mortar fire support plans were integrated into forward security areas actions and the direct fire plans of maneuver companies in the main battle area. Engineer priorities of work were initially to countermobility, then to survivability. As the IBCT prepared for the defense, the brigade support battalion established the BSA just forward of the division support area to support the IBCT area defense.

INFANTRY BRIGADE COMBAT TEAM – TASK ORGANIZATION AND SCHEME OF MANEUVER

3-288. Within this scenario, the IBCT tasked-organized with two Infantry battalions, a combined arms battalion, a cavalry squadron, a field artillery battalion, a brigade engineer battalion, and a brigade support battalion. Infantry Battalion 2, task-organized with two Infantry rifle companies, a mechanized Infantry company team (two mechanized Infantry platoons and one tank platoon), and a weapons company. Company C, the third Infantry rifle company from Infantry Battalion 2, was placed under IBCT control as the reserve for the IBCT. Infantry Battalion 1 task-organized with its three Infantry rifle companies and weapons company. The combined arms Battalion tasked-organized with two armor company teams, each with two tank platoons and one mechanized Infantry platoon. The commander weighted the main effort by attaching the mechanized Infantry company team from the combined arms battalion to Infantry Battalion 2, as stated above. No change in task organization for the IBCT cavalry squadron, field artillery battalion, and brigade engineer battalion. The brigade support battalion tasked-organized with the logistical elements required to support the combined arms battalion. Subordinate unit task organization and scheme of maneuver are as follows:

- Infantry Battalion 2, main effort, conducts an area defense. The battalion defends in-depth with two Infantry rifle companies, with one assault platoon each attached, forward, and a mechanized Infantry company team (two mechanized Infantry platoons) and weapons company (two assault platoons) back. The tank platoon from the mechanized Infantry company team is the battalion reserve. (Illustrated.)

- Infantry Battalion 1, supporting effort, conducts an area defense to the north of Infantry Battalion 2. The battalion defends in-depth with two Infantry rifle companies forward, and one Infantry rifle company back. The weapons company (two assault platoons) is the battalion reserve. (Not illustrated.)
- The combined arms battalion, two armor company teams, is the counterattack force for the IBCT. (Not illustrated.)
- Infantry rifle company C from Infantry Battalion 2 is the reserve for the IBCT. Company C is mounted and has two attached assault platoons from Infantry Battalion 1. (Not illustrated.)
- The cavalry squadron establishes the security area forward of the IBCT main battle area. (Not illustrated.)
- The brigade engineer battalion priorities of work, countermobility, survivability, and then mobility. Priority of engineer effort initially to Infantry Battalion 2, then to the mobility of the IBCT counterattack force.
- The field artillery battalion provides priority of fires initially to security area forces, then to Infantry Battalion 2, on order to the IBCT counterattack force. (Not illustrated.)
- The brigade support battalion establishes the BSA just forward of the division support area. Priority of support initially to security area forces, then to Infantry Battalion 2, finally to the IBCT counterattack force. (Not illustrated.)

Infantry Battalion 2—Task Organization and Scheme of Maneuver

3-289. Infantry Battalion 2, tasked-organized as a battalion task force, conducts an area defense against a motorized Infantry and armor threat. The battalion commander organized the defense in-depth around a static framework of defensive positions seeking to destroy enemy forces and dynamic local counterattacks designed to defeat or destroy an enemy force, restore an original position, or block enemy penetration (figure 3-29).

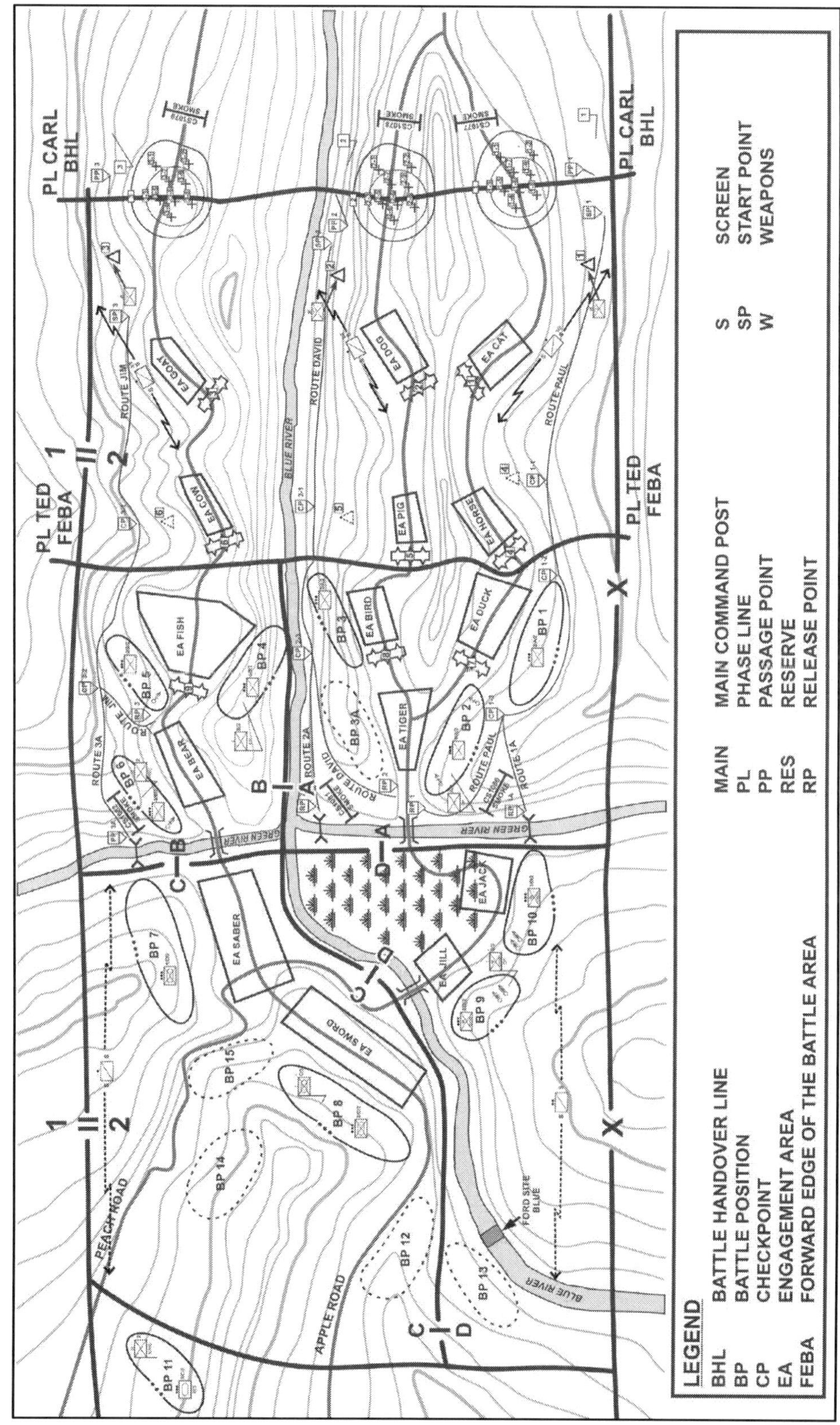

Figure 3-29. Battalion area defense (defense in-depth), example

3-290. The commander weighted the battalion main effort (Company B) with priority of fires, priority of engineer effort, and priority of support. The commander organized the task force with a forward security force, a main battle area force, and a reserve force.

3-291. Within the battalion security area, established just forward of the battalion's main battle area and to the rear of the IBCT security area, the commander assigned security force missions, established observation posts, and specified named areas of interest and target areas of interest to observe. In the main battle area, the commander organized the battalion task force to defend in-depth with two Infantry rifle companies forward and a mechanized Infantry company team and weapons company back. The tank platoon from the mechanized Infantry company team, placed under battalion control, established battle position 11 as the battalion reserve just forward of the BSA.

3-292. The commander, through the targeting process (see ATP 3-60), determined target sets and fire support priorities and assigned priority of fires initially to forces within the security area. In the main battle area, the priority of engineer effort went to countermobility, then survivability. As the battalion task force establish the defense, battalion unit trains (includes sustainment support for mechanized company team) organize to establish echelon trains—battalion field trains and battalion combat trains—to support the battalion's defense in-depth. The following subordinate unit task organization and scheme of maneuver is as follows:—

- The scout platoon screens forward of the battalion's main battle area. The platoon establishes contact with IBCT security area forces and prepares to assist in their rearward passage of lines. On order, the scout platoon conducts a rearward passage of lines and conducts screens west of the Green River.
- Sniper squad, with attached forward observers, establishes observation posts to observe named areas of interest and target areas of interest within the battalion security area along enemy Avenues of Approach 1 and 2. On order, the sniper squad with attachments conducts a rearward passage of lines to occupy subsequent observation posts within the main battle area or may stay in place to provide combat information as the enemy advances.
- Company B, battalion main effort, with attached assault platoon occupies Battle Position 4 and Battle Position 5, destroys enemy targets in Engagement Area Fish; occupies Battle Position 6, destroys enemy targets in Engagement Area Bear. Battalion tactical command post initially collocates vicinity Battle Position 6.
- Company A, with attached assault platoon occupies Battle Position 1, destroys enemy targets in Engagement Area Cat; occupies Battle Position 2, destroys enemy targets in Engagement Area Dog; and occupies Battle Position 3, destroys enemy targets in Engagement Area Bird.
- The mechanized Infantry company occupies Battle Position 7, destroys enemy targets in Engagement Area Saber and occupies Battle Position 8, destroys enemy targets in Engagement Area Sword.
- Company D occupies Battle Position 9, destroys enemy targets in Engagement Area Jill and occupies Battle Position 10, destroys enemy targets in Engagement Area Jack.
- Tank platoon, battalion reserve establishes battle position 11, receives be prepared mission to conduct local counterattacks to defeat, destroy, or block enemy penetration or reinforce position and gaps as necessary. Battalion main command post initially collocates vicinity Battle Position 11.
- Battalion mortar platoon operates by split section vicinity Battle Positons 9 and 10. On order, displaces by section to subsequent firing positions to provide continuous fire support, priority of fires initially to forward security forces, then Company B the battalion main effort.
- Battalion combat trains establish support area to the rear of Battle Position 11in the battalion's main battle area. This echelon of support includes maintenance support and treatment team alpha (clinically staffed with the battalion surgeon), and carries mostly Class I, V, and limit Class III.
- Battalion field trains establish support area to the rear of the battalion's main battle area in the BSA. This echelon of support includes all battalion sustainment not located with the combat trains, includes treatment team bravo, clinically staffed with the physician assistant. (Not illustrated.)

Gain and Maintain Enemy Contact

3-293. Gaining and maintaining enemy contact without becoming decisively engaged, in the face of the enemy's determined efforts to destroy friendly reconnaissance, surveillance, and security actions is vital to the success of defensive actions. Within the scenario, security area forces (scout platoon, sniper squad, and attached forward observers) performed R&S tasks and security operations to gain information, using every opportunity through limited offensive action to attrit, delay, and harass the enemy prior to engagement in the main battle area. As security area forces conduct R&S to determine the enemy has chosen course of action, these same forces seek to defeat enemy reconnaissance forces, and hide the Infantry battalion's dispositions, capabilities, and intent. The battalion commander uses collected information and other combat information provided by means belonging to higher echelons, in conjunction with military judgment, to determine the point at which the enemy commits to a course of action.

3-294. Within the battalion security area (figure 3-30, page 3-88), the scout platoon conducts two squad screens south of the Blue River and one squad screen to the north of the river with the reconnaissance objective to determine the enemy course of action. Secondary tasks include providing early warning of enemy dismounted movement along enemy avenues of approach, and establishing contact with IBCT security forces forward of the battle handover line, Phase Line Carl. Although primary passage points for the IBCT Cavalry squadron are through the supporting effort (Infantry Battalion 1, not illustrated) to the north. The scout platoon prepares to assist elements of the Cavalry squadron as required through passage points 1, 2, and 3, vicinity the battle handover line. The scout platoon may, or in some combination, assist in the passage of lines manning contact points 1, 2, and 3; establish Observation Posts 4, 5, and 6 to facilitate future offensive actions; or conduct a rearward passage of lines itself. The sniper squad, with attached forward observers, establishes Observation Posts 1, 2, and 3 along the battle handover line, two teams south of the Blue River and one team to the north of the river. The mission of the sniper teams is to observe named areas of interest and target areas of interest within the IBCT and Infantry battalion security areas along enemy Avenues of Approach 1 and 2.

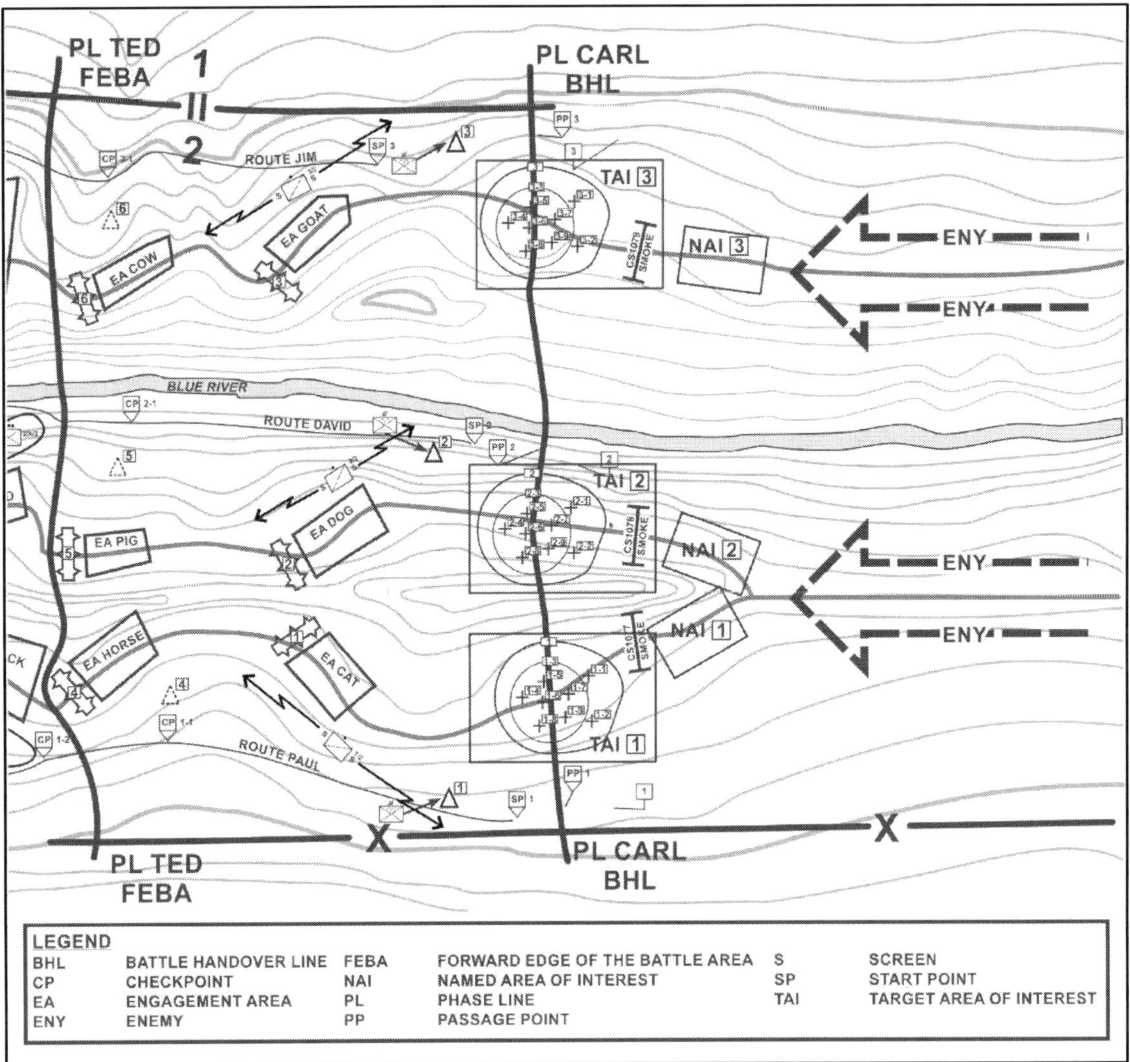

Figure 3-30. Battalion area defense (security area engagement), example

3-295. As the enemy forces approach the main battle area, security area forces when their positioning allows may continue limited offensive action to attrit the enemy, or displace to alternate positions to facilitate future offensive actions. When security area forces withdraw to the main battle area, they conduct a rearward passage of lines as quickly as possible using multiple passage points and lanes along the forward edge of the battle area. Throughout this process, security area forces take advantage of previous liaison and plans, making any required last-minute coordination with main battle area forces at contact points to ensure a rapid rearward passage of lines. (See paragraphs 2-117 through 2-153 on pages 2-35 to 2-49 for the doctrinal basis and illustration of a passage of lines and battle handover.)

Disrupt and Fix the Enemy

3-296. When security area forces make contact with the enemy, the battalion commander seeks to disrupt the enemy's plan, the enemy's ability to control forces, and the enemy combined arms team by countering the enemy's initiative and preventing the enemy from massing overwhelming combat power. The commander initiates actions to force the enemy into engagement areas forward of the main battle area to destroy the enemy's cohesion and disrupt the tempo of the enemy's approach to the main battle area. In addition to disrupting the enemy, the commander conducts limited offensive actions to constrain the enemy into a specific course of action, control enemy movements, or fix the enemy in a given location. Throughout the conduct of the defense, long-range fires, unexpected defensive positions and obstacles, local counterattacks, and attacks delivered by reserve forces combine to disrupt the enemy's attack and break the enemy's will to continue offensive operations.

3-297. During the defense, the battalion commander concentrates the engineer effort on countering the enemy's mobility and canalizing the enemy in engagement areas. The scheme of engineer operations when synchronized with the static and dynamic actions of the battalion takes advantage of the enemy force's forward orientation by fixing the enemy and then delivering a blow to the enemy's flank or rear. When the enemy's attacking force assumes a defensive posture, the commander rapidly coordinates and concentrates all defending fires against unprepared and unsupported segments of the attacking enemy force. The Infantry battalion delivers these fires simultaneously or sequentially.

Note. In this scenario, the commander establishes priorities among countermobility and survivability efforts and synchronizes these efforts with static and dynamic actions. Throughout the depth of the battalion defense, the priority of engineer effort for the Infantry battalion is countermobility, then survivability. Countermobility priority of support is first to the security area then to the main battle area. Survivability priority of support is to the mechanized Infantry company team, specifically Battle Position 7 and Battle Position 8. Mobility priority of support focuses at the IBCT echelon, specifically the counterattack force.

Maneuver

3-298. During engagement within the battalion security area, Infantry battalion 1 (not illustrated) and 2 assume responsibility for the battle at the battle handover line, Phase Line Carl. As forward security forces moved towards and through the battle handover line, the battalion commander if required increases the intensity of fires in target groups 1, 2, and 3 and along withdrawal routes forward of the battle handover line from battalion level and above fire support assets to enable security forces to break contact. Both static and dynamic actions initially from battalion security area forces, then from main battle area forces and higher echelon enablers, provide support to cover the withdrawal of the IBCT security force, and if required to close passage points. The battalion commander planned obscurants (linear smoke targets CS1078, CS1079, and CS1080) to assist security forces with breaking contact with the enemy forward of the battle handover line. Sniper teams 1, 2, and 3 established observation posts 1, 2, and 3 respectively, within the battalion's forward security area to collect information on the enemy and to provided overwatch during battle handover.

Note. As stated early, the primary reward passage of lines for the IBCT cavalry squadron is through Infantry Battalion 1 (not illustrated) to the north of the IBCT main effort (Infantry Battalion 2). Infantry Battalion 2 still plans and prepares for a rearward passage of lines to assist subordinate elements of the cavalry squadron. For discussion purposes, the cavalry squadron employed one troop (Cavalry troop A) south of the Blue River requiring a rearward passage of lines through Infantry Battalion 2.

3-299. As Cavalry troop A began movement to withdraw through Passage Points 1 and 2, Scout squad 1 and Scout squad 2 receive a change of mission from screening south of Engagement Area Cat and north of Engagement Area Dog, respectively (figure 3-31, page 3-90). Scout squad 1, (change of mission) conduct movement to and establish Passage Point 1, then reconnoiter Start Point 1 to Checkpoint 1-1 along Route Paul. Scout squad 2, (change of mission) conduct movement to and establish Passage Point 2, then reconnoiter Start Point 2 to Checkpoint 2-1 along Route David. Scout squad 3 continues to screen north of Engagement Area Goat to provide early warning of enemy dismounted movement north of Engagement Area Goat. Through previous liaison and coordination with IBCT security forces (Cavalry troop A), the scout platoon successfully initiated the rearward passage Cavalry troop A through Passage Points 1 and 2.

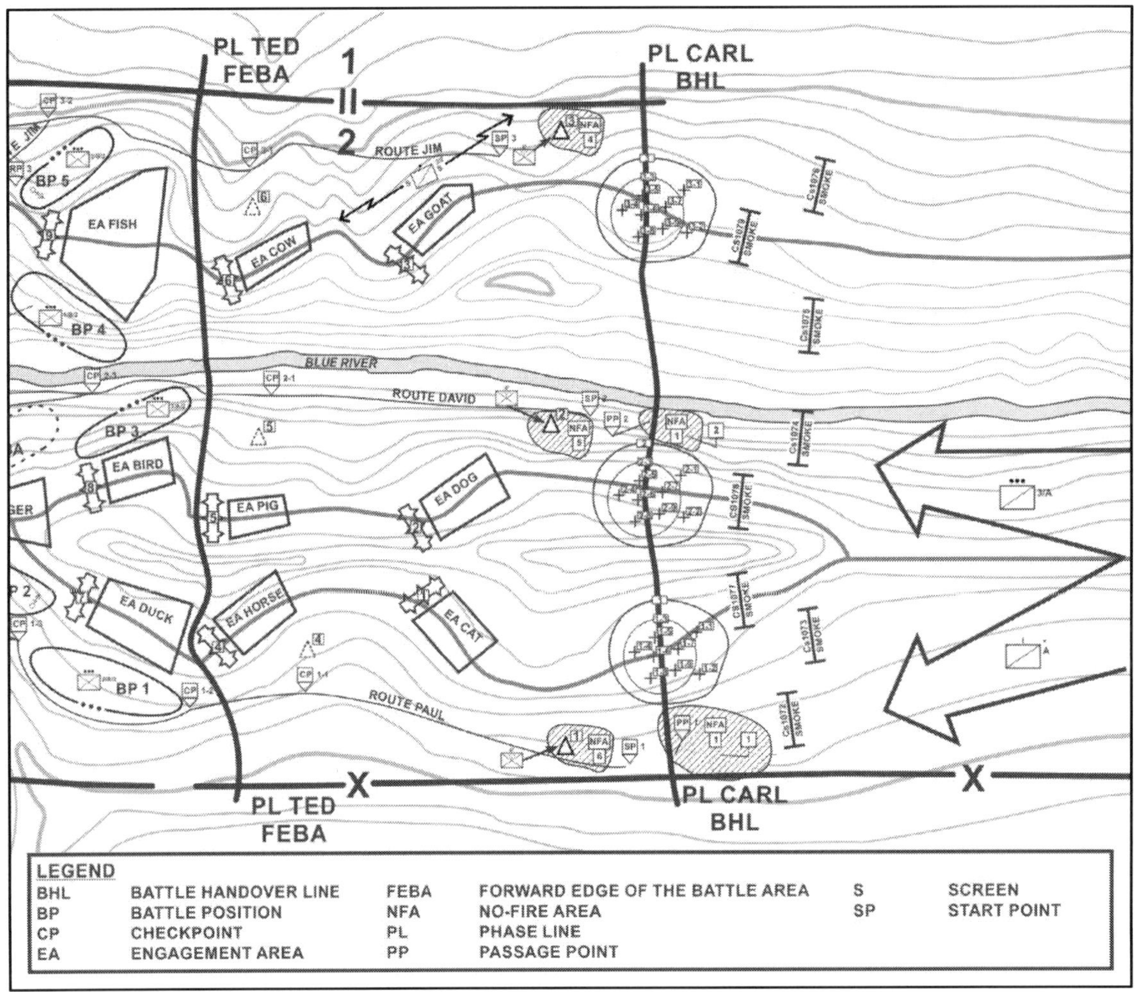

Figure 3-31. Battalion area defense (battle handover), example

3-300. After Cavalry troop A successfully moved alone Route Paul and Route David through checkpoints 1-1 and 2-1, Scout squads 1 and 2 moved to establish Observation Posts 4 and 5, respectively, to gain information on the enemy and to conduct limited offensive action to attrit the enemy and facilitate future offensive actions. Cavalry Troop A continued rearward movement along Route Paul and Route David to Release Point 1 and Release Point 2, then continued movement to the brigade rear for follow-on mission. Scout squad 3 (change of mission) conduct movement to and establish Observation Post 6 to gain information on the enemy and to conduct limited offensive action to attrit the enemy and facilitate future offensive actions.

3-301. During engagement within the security area and main battle area, the battalion commander does not allow the attacking enemy to consolidate, unless it fits the battalion scheme of maneuver (figure 3-32). As attacking enemy forces reach the main battle area of Infantry Battalion 2 (battalion task force), the enemy may try to find weak points in the defense and attempt to force a passage, possibly by a series of probing attacks. As the enemy attack develops, battalion forward elements engage the enemy's lead forces along identified enemy Avenues of Approach 1 and 2. The battalion commander arrays forces, based on the mission variables of METT-TC, and establishes engagement areas using obstacles and fires to canalize enemy forces. (See engagement area development, paragraphs 3-168 through 3-189 on pages 3-53 to 3-59.) The battalion takes actions to increase the kill probabilities of various weapon systems at different ranges, to include establishing range markers for direct fire weapons, confirming the zero on weapons, or clearing obstacles that might snag the cables over which travel the commands of wire-guided munitions.

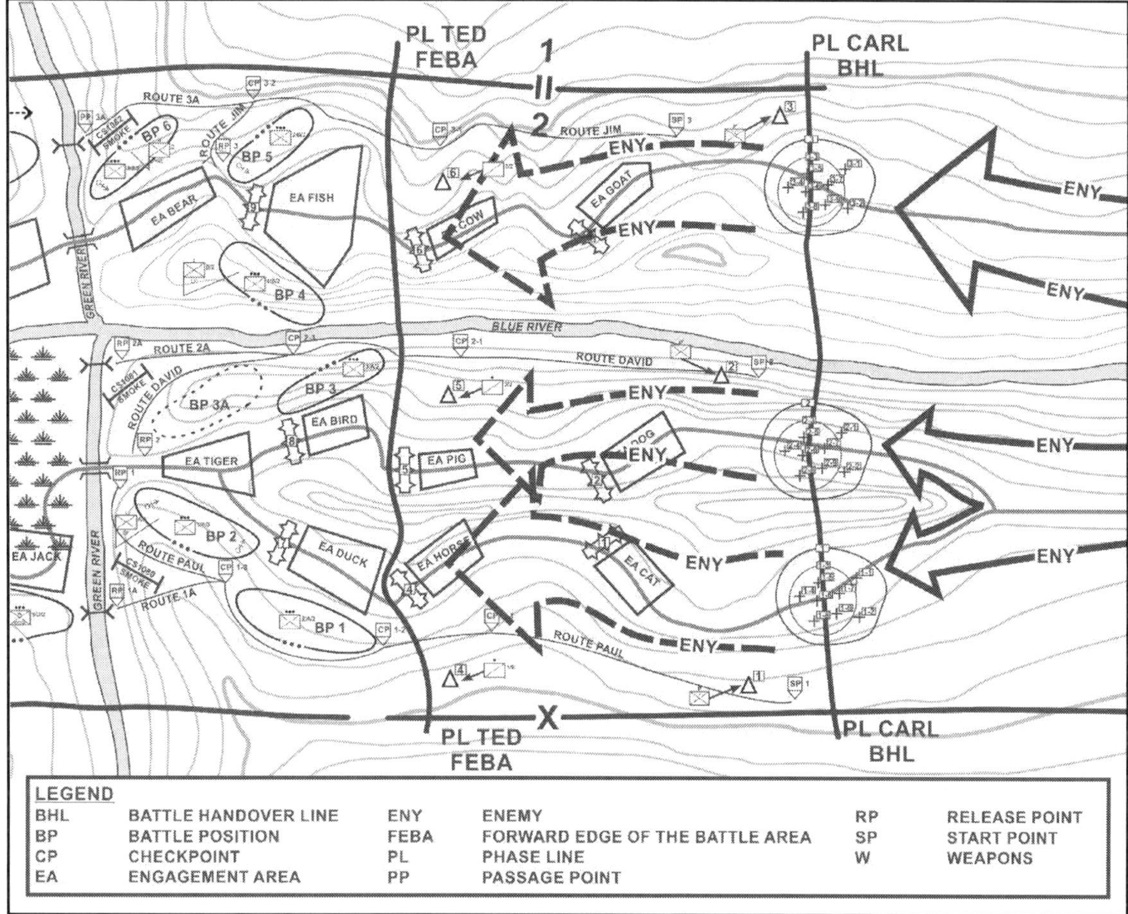

Figure 3-32. Battalion area defense (engagement), example

3-302. Subordinate companies within the task force, maneuver using massed direct and indirect fires and movement to gain positional advantage over the assaulting enemy force. The battalion commander directs the engineer obstacle and sustainment effort by the assignment of priorities. During the battle, the commander repositions forces to meet the enemy where the enemy actually is rather than where the commander projected that the enemy would be. The enemy advance may slow because of canalization and the increased density of forces resulting from limited maneuver space, presenting good targets for defensive direct and indirect fires and Army aviation fires and close air support. The maximum effects of simultaneous and sequential fires are brought to bear at this stage of the battle.

3-303. Maneuver company commanders keep gaps between platoon defensive battle positions under surveillance, covered by fire or, where possible, blocked by barriers or repositioned friendly forces. Commanders leverage the use of choke points and obstacles to prevent enemy penetration. Company defensive schemes of maneuver clearly define the responsibility for dealing with any enemy penetration. If the enemy succeeds in penetrating forward employed forces in the main battle area, the commander blocks the penetration immediately and destroys this enemy force as soon as possible, using a local counterattack force within the company area of operation. The lowest possible echelon conducts this local counterattack; however, the commander must be aware of the problem of piecemeal commitment.

Note. Subordinate units do not abandon a position unless it fits within the higher commander's intent, or that higher commander grants permission to do so. In this scenario, forward defending forces (Company A and Company B) plan contingency missions for withdrawal (dismounted) to subsequent battle positions during the course of battle on the west side of the Green River. Elements of Company A would move from Checkpoint 1-3 along Route 1a to raft site (Release Point 1a) and from Checkpoint 2-3 along Route 2a to raft site (Release Point 2a). Elements of Company B move from Checkpoint 3-3 along Route 3a to raft site (Release Point 3a). In addition to contingency planning for the withdrawal of subordinate units, the battalion commander plans for and organizes a detachment left in contact and stay-behind forces when the scheme of maneuver requires them. (Refer to Section IV later in this chapter for more information.)

3-304. The battalion commander does not counterattack as an automatic reaction to an enemy penetration, nor does the commander commit the reserve solely because the enemy has reached a certain phase line or other location. The battalion commander may extend actions within the depth of the area of operation to counter enemy penetrations that cannot be stopped farther forward within the main battle area. The commander may employ fire support assets and local counterattacks by forces already defending to destroy, disrupt, or attrit enemy penetrations, thus avoiding the need to commit the counterattack force or the reserve. In this example (figure 3-33), enemy penetrations on mounted avenues of approach reach the Green River. The IBCT commander makes the decision not to destroy the two bridges within the area of operation of Infantry Battalion 2. Enemy forces that move west of the river, where the terrain opens up more to long range direct fire systems, are engaged from Battle Positions 7, 8, 9, and 10.

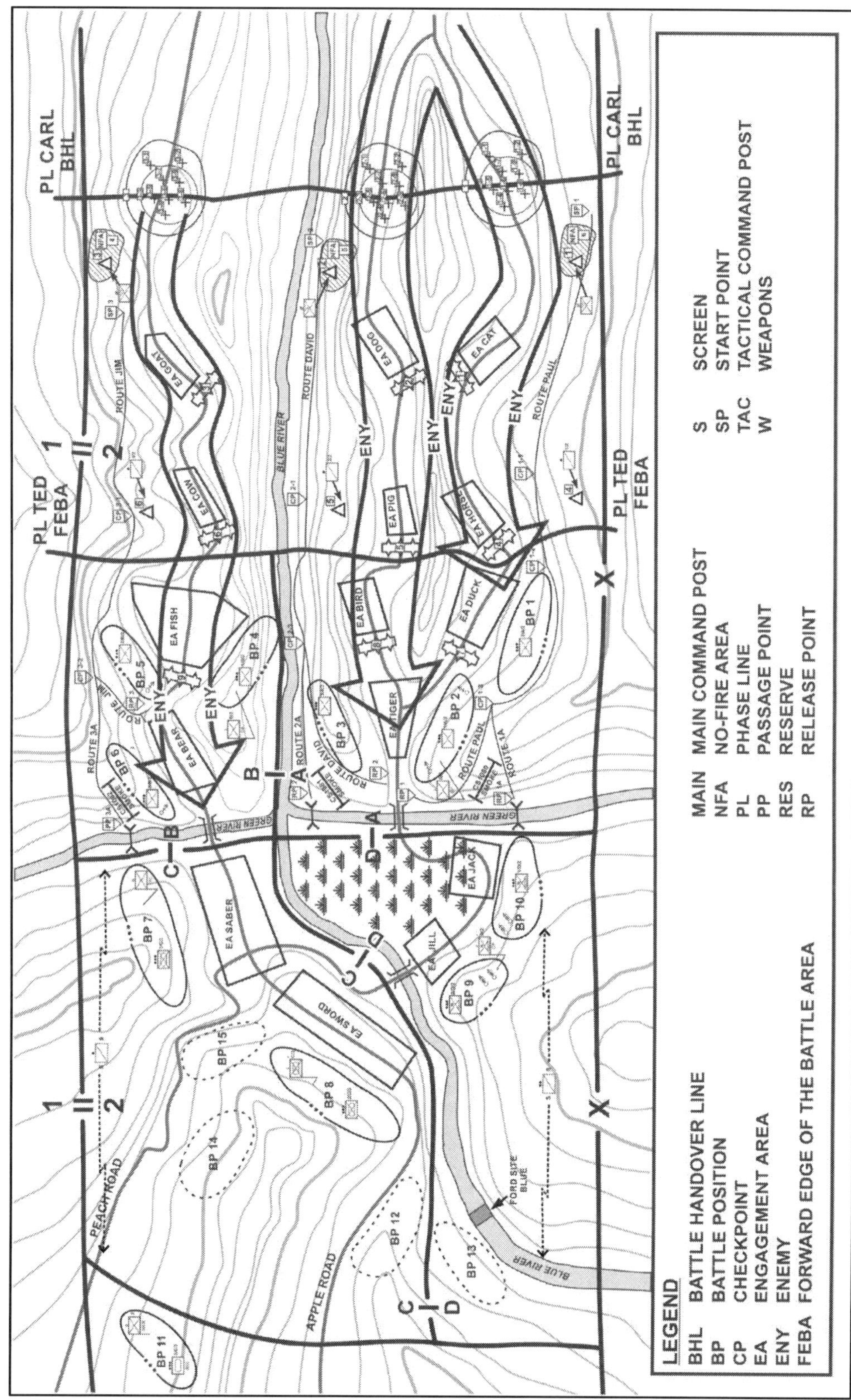

Figure 3-33. Battalion area defense (extend actions in-depth), example

3-305. As a battle develops, one of the commander's most critical decisions is when to commit the reserve. The commander controls commitment of the reserve with all available local resources to prevent the enemy from consolidating gains deep within the battalion's main battle area. When possible, the commander launches the reserve's counterattack when the enemy presents a flank or rear, overextends, or the enemy's momentum dissipates. Once the flanks of the enemy's main effort are identified, the commander can target counterattacks to isolate and destroy enemy forces within the main battle area. Sometimes the commander may determine that the reserve is unable to conduct a successful counterattack. In this case, the commander uses available resources to block, contain, or delay the enemy to gain time to employ IBCT or higher-echelon counterattack forces or reserve. In the scenario, the reserve is committed to block (Battle Position 15) the enemy's advance along Avenue of Approach 2 (enemy attack along Avenue of Approach 1 was defeated). Within this situation, the commander and staff plan how to integrate reinforced company and platoon battle positions into the defensive scheme, and adjust or establish new boundaries or positions (figure 3-34).

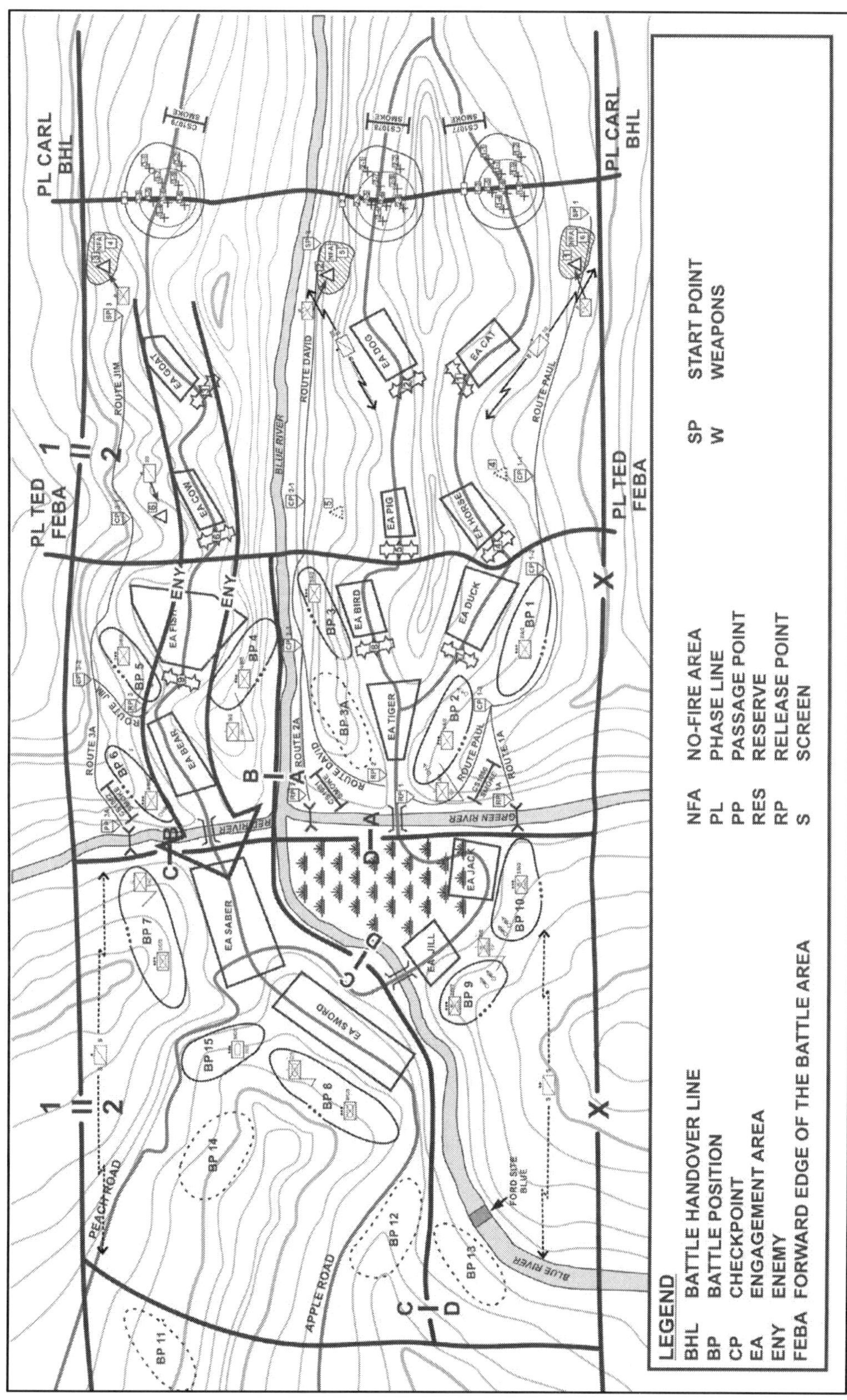

Figure 3-34. Battalion area defense (commitment of the reserve), example

Follow Through

3-306. The purpose of the defensive action illustrated within this scenario was to retain terrain and create conditions for a counteroffensive to regain the initiative. In the scenario the IBCT tasked-organized with two Infantry battalions, a combined arms battalion, a cavalry squadron, a field artillery battalion, a brigade engineer battalion, and a brigade support battalion to defend against three enemy avenues of approach. The third Infantry battalion of the IBCT, under division control, participated as part of the corps counterattack force (air assault-not illustrated within the scenario).

3-307. Initially the IBCT cavalry squadron established a security area forward of IBCT's main battle area. As enemy forces advanced, the squadron attrited enemy motorized Infantry and armor forces within its capability. The cavalry squadron conducted battle handover then moved to the rear of the IBCT's main battle area to conduct follow-on missions. Infantry Battalions 1 and 2 defended in-depth, within the IBCT's main battle area, causing the enemy to sustain unacceptable losses short of any decisive objectives. The IBCT commander counterattacked to regain the initiative as the division transition to offensive actions. The combined arms battalion, two armor company teams, was the counterattack force for the IBCT (IBCT counterattack-not illustrated within the Infantry battalion scenario).

SECTION III – MOBILE DEFENSE

3-308. The *mobile defense* is a defensive task that concentrates on the destruction or defeat of the enemy through a decisive attack by a striking force (ADRP 3-90). A mobile defense orients on the destruction of the attacking enemy force, as opposed to retaining terrain, by permitting the enemy to advance into a position that exposes the enemy to a decisive counterattack. The commander may yield ground in some areas to allow the enemy commander to think the attack has been successful or to entice the enemy force to move toward an engagement area where the enemy is vulnerable to the striking force's attack. (Refer to FM 3-90-1 for additional information.)

Note. A division is the smallest unit that can conduct, versus participate in, a mobile defense. This is because of its ability to fight multiple engagements throughout the width, depth, and height of the division area of operations, while simultaneously resourcing fixing, striking, and reserve forces.

3-309. When the division plans to conduct a mobile defense, subordinate BCTs may shape the penetration of the enemy attack as part of the division fixing force. The division commander allocates only an absolute minimum amount of combat power to the fixing force within the division area of operations. For a typical division with four to six attached BCTs this may be as small as a single BCT or two BCTs at the most. The division commander allocates the maximum available combat power at the time of attack to the striking force. The mobile striking force should possess greater combat power than that of the enemy force it seeks to defeat or destroy and be capable of equal or greater mobility. At the division level, this translates into two or more BCTs supported by the field artillery brigade and combat aviation brigade, and joint fires. The reserve of a division conducting a mobile defense may consist of a single Stryker or combined arms battalion task force or may be a reinforced armored or Stryker BCT. (Refer to ATP 3-91 for additional information.)

3-310. In a mobile defense, an IBCT attached to the division is normally part of the fixing force. The fixing force conducts either an area defense or a delay structured to establish the conditions necessary for the successful conduct of the striking force's attack. The division commander takes advantage of the fixing force fighting a mix of static (defensive positions) and dynamic (local counterattacks) actions. Within the mobile defense, fixing forces reposition as necessary and conduct local counterattacks to control the depth and breadth of an enemy penetration and ensure the retention of ground from which the striking force can launch the decisive counterattack. When facing large enemy penetrating forces, division shaping operations or supporting efforts repeatedly isolate portions of the enemy force that are then attacked by the striking force and defeat the enemy in detail. An IBCT or a subordinate infantry battalion task force can garrison a strongpoint to shape the enemy's penetration. This and other graphic control measures help the division commander direct the division's BCTs and supporting brigades throughout the execution portion of a mobile defense. (Refer to sections II and IV of this chapter for additional information on the conduct of an area defense or delay, respectively.)

3-311. The attack by the *striking force,* a dedicated counterattack force in a mobile defense constituted with the bulk of available combat power (ADRP 3-90), in the engagement area isolates the targeted penetrating enemy force and defeats or destroys that enemy force, if possible. When shaping the commitment of the striking force, the commander may use Infantry to isolate targeted enemy forces through vertical envelopment. This form of maneuver requires local air superiority and the suppression of most enemy air defense systems during the time Infantry units move along air movement corridors to their respective landing zones. Once on the ground, air assault forces require direct fire and indirect fire support capable of defeating, when the threat exist, counterattacking enemy armor systems. Fires include a situationally appropriate mixture of dismounted anti-armor systems, Army attack aviation, close air support, and precision guided munitions delivered by cannon and rocket. (Refer FM 3-99 for additional information on conducting an air assault.)

SECTION IV – RETROGRADE

3-312. A *retrograde* is a defensive task that involves organized movement away from the enemy (ADRP 3-90). The enemy may force these operations, or a commander may execute them voluntarily. In either case, the higher commander of the force executing the operation must approve the retrograde. Retrograde operations are transitional operations; they are not considered in isolation. In a retrograde the battalion is usually part of a larger scheme of maneuver designed to regain the initiative and defeat the enemy. (Refer to FM 3-90-1 for additional information.)

GENERAL CONSIDERATIONS FOR THE RETROGRADE

3-313. Retrograde movements may be classified as delaying, withdrawal, or retirement actions. Delaying actions trade space for time, preserve friendly combat power, and inflict maximum damage on the enemy. Withdrawal actions involve a planned voluntary disengagement from the enemy conducted with or without enemy pressure. Retirement involves an organized movement to the rear by a force that is not in contact with the enemy. In each action, a force moves to the rear, using combinations of combat formations and marches. (Chapter 2 discusses combat formations and troop movement.) The commander may use all three actions singularly or in combination with other offensive or defensive tasks.

3-314. The commander executes retrogrades to—

- Disengage from operations.
- Gain time without fighting a decisive engagement.
- Resist, exhaust, and damage an enemy in situations that do not favor a defense.
- Draw the enemy into an unfavorable situation or extend the enemy's lines of communications.
- Preserve the force or avoid combat under undesirable conditions, such as continuing an operation that no longer promises success.
- Reposition forces to more favorable locations or conform to movements of other friendly troops.
- Position the force for use elsewhere in other missions.
- Simplify sustainment of the force by shortening lines of communications.
- Position the force where it can safely conduct reconstitution.
- Adjust the defensive scheme to secure more favorable terrain.
- Deceive the enemy.

Note. A retrograde can negatively affect the participating Soldiers' attitudes more than any other type of operation because they may view the retrograde as a defeat. The commander must maintain the unit's aggressiveness and not allow retrograde operations to reduce or destroy unit morale. The commander can counter any negative effects of the operation on unit morale by planning and efficiently executing the retrograde and ensuring that Soldiers understand the purpose and duration of the operation.

3-315. A critical consideration for the battalion commander is the relative mobility of the enemy to the Infantry battalion. Although the elements of the headquarters and headquarters company, weapons company, and forward support company are vehicle-mounted, rifle companies are primarily foot-mobile. Ideally, the commander is able to place the rifle companies on restrictive or severely restrictive terrain that reduces any mobility advantage that the enemy may have. Elements of the weapons company can fight in more open terrain and take advantage of their relative long-range fires. The weapons company also can rapidly reinforce the rifle companies with heavy direct fire weapons.

3-316. Relative mobility is most critical during delaying operations. Additional assets or the assignment of advantageous terrain is required to enable the Infantry battalion to conduct an effective delaying action against an enemy with greater mobility. Restrictive and severely restrictive terrain enhances the strengths and reduces the limitations of the Infantry battalion during the conduct of a delaying operation. Against a mobile enemy, to the degree possible, the Infantry battalion avoids open terrain with high-speed avenues of approach.

3-317. Enemy vehicles, especially mechanized Infantry vehicles, can rapidly cross seemingly substantial obstacles. Much of the strength of an obstacle rests on the integration and construction of reinforcing obstacles, exploitation of existing obstacles, and actions to enhance the survivability of the force conducting a retrograde. Combined arms countermobility (see paragraph 3-206) operations used to enhance the effects of natural and man-made obstacles deny enemy freedom of movement and maneuver forcing the enemy to conduct a breach or gap crossing.

3-318. Military police forces assisting the retrograde are involved primarily in security and mobility support operations to support and preserve the retrograding unit's freedom of movement. Military police support to the retrograde may include, although not inclusive, establishing traffic control posts, and route and convoy security. (Refer to 3-39.30 for additional information.)

3-319. The commander anticipates the effects on sustainment elements during the retrograde to ensure adequate support for the operation and the prompt evacuation of casualties. Retrograde movements generally result in increased distances between sustainment and combat units, which makes providing support more difficult. This in turn increases the need for movement management and prepositioned services and supplies and demands on transportation if the Infantry battalion's movement is to be mounted. The commander assigns transportation priorities for the movement of combat troops and their supplies, the movement of obstacle materials to impede the enemy, and the evacuation of casualties and repairable equipment. The commander prevents unnecessary supplies from accumulating in areas that will be abandoned, only allowing essential medical and logistics support to be located in the area involved in the retrograde.

3-320. The commander establishes maintenance, recovery, and evacuation priorities and destruction criteria for inoperable equipment in the concept of support for the retrograde. Sustainment units place as much maintenance, recovery, and evacuation assets forward as possible to augment or relieve combat elements of the burden of repairing unserviceable equipment. The recovery and evacuation plan positions assets at critical locations to keep disabled vehicles from blocking movement routes. These assets evacuate systems that cannot be repaired within established timelines. Sustainment units use all available means to accomplish this, including equipment transporters and armored or other protective vehicles with inoperative weapon systems. When recovery and evacuation are impossible, retrograde units destroy inoperable equipment to prevent capture. When possible, these units destroy the same vital components in each type of system to prevent the enemy from rapidly exploiting captured friendly systems through battlefield cannibalization.

3-321. Assignment of medical evacuation precedence provides the supporting medical unit and controlling headquarters with information that is used in determining priorities for committing evacuation assets. For this reason, correct assignment of precedence cannot be overemphasized; over classification remains a continuing problem. Patients are evacuated as quickly as possible, consistent with available resources and pending missions. Medical elements supporting the retrograding force must provide rapid evacuation of casualties to medical facilities. Medical evacuation requirements are especially demanding in the large area of operations common to the retrograde. Commanders may augment the ground ambulance capabilities of supporting forward medical units.

Note. During a retrograde (mainly when conducting a delaying operation or withdrawal operation) the Infantry battalion may become encircled or be directed to conduct operations as a stay-behind force. This may be planned or forced by enemy action. When encircled or acting as a stay behind force the battalion normally establishes a perimeter on restrictive terrain, ideally controlling a choke point or other key terrain. When adequately sustained, often by Army aviation or Air Force assets, the battalion can remain in position indefinitely and attack the enemy in its rear or against more vulnerable support units. If the battalion cannot be adequately sustained, then it must conduct a breakout as soon as possible. See paragraphs 3-35 and 3-62 for information on defending encircled and conducting a stay-behind operation, respectively.

DELAY

3-322. A *delaying operation* is an operation in which a force under pressure trades space for time by slowing down the enemy's momentum and inflicting maximum damage on the enemy without, in principle, becoming decisively engaged (JP 3-04). When conducting a delay, the Infantry battalion yields ground to gain time while retaining flexibility and freedom of action. The battalion may execute a delay when it has insufficient combat power to attack or defend or when the higher unit's plan calls for drawing the enemy into an area for a counterattack, as in a mobile defense.

ORGANIZATION OF FORCES

3-323. The battalion's organization of forces depends on how the IBCT has structured its forces unless the battalion operates independently. The IBCT normally organizes into a security force, main body, and reserve, though operations extended across large areas may preclude the use of an IBCT-controlled security force and reserve. In this case, the IBCT may direct the battalion to organize its own security, main body, and reserve forces; the same as if the battalion was operating independently. The IBCT commander can designate a battalion as the security or reserve force for the IBCT.

3-324. When the battalion operates independently or establishes its own security force within the IBCT's area of operation, the battalion normally uses the scout platoon and the sniper squad as a screening force. These elements position to observe the most likely enemy avenues of approach and can initiate fires to slow and weaken the enemy. These elements may be reinforced with other elements, for example, assault platoons from the weapons company, forward observers, and fire support teams executing direct and indirect fire targets on a primary enemy avenue of approach.

3-325. The battalion's main body, which contains the majority of the force's combat power, may use alternate or subsequent positions to conduct the delay. The commander usually deploys the main body as a complete unit into a forward position when conducting a delay from subsequent positions. The commander divides the main body into two parts, roughly equal in combat power, to occupy each set of positions when conducting a delay from alternate positions. The commander retains a reserve, normally a company or company minus, to defeat enemy penetrations, reinforce positions, or assist units with breaking contact. Reserve missions require the force tasked to be the reserve to have the mobility and strength to strike with such force that an enemy has no option but to react to it.

DOCTRINAL BASIS FOR A DELAY

3-326. The delay is one of the most demanding of all ground combat operations. A delay wears down the enemy so that friendly forces can regain the initiative through offensive action, buy time to establish an effective defense, or determine enemy intentions as part of a security operation. The purpose of the delay is to control the enemy's tempo by forcing the enemy to deploy multiple times and repeatedly concentrate its combat power to defeat the delaying force. Although the battalion must establish and maintain contact, it should avoid becoming decisively engaged, except when directed to prevent enemy penetration of a phase line for a specific duration. It is critical that the commander's intent defines what is more important to the mission: gaining time, inflicting casualties on the enemy, or protecting the force. Normally in a delay, inflicting casualties on the enemy is secondary to gaining time. The commander establishes risks for each delay but ordinarily maintaining freedom of action and avoiding decisive engagement is of ultimate importance.

Planning

3-327. Conducting a delay requires the close coordination of forces and a clear understanding by subordinates of the commander's intent, the scheme of maneuver, and detailed mission graphics. The potential for the loss of control is high in delay operations, making cross talk and coordination between commanders and subordinate leaders extremely important. Subordinate initiative is critical, but it must be in the context of close coordination with others. Plans must be flexible, with control measures throughout the area of operation allowing forces to maneuver to address all possible enemy options. Planning considerations for the area defense address in section II of this chapter apply to delaying operations. The six-warfighting functions below are the framework for discussing planning considerations that apply to delaying operations for the battalion.

Mission Command

3-328. Centralized planning and decentralized execution characterize a delaying operation. Critical to the success of the delay is the commander and staff's shared understanding of the operational environment, the operation's purpose, problems, and approaches to solving them. The battalion commander's intent provides a clear and concise expression of the purpose of the delaying operation and the desired military end state for the delay. The commander's intent becomes the basis on which staffs and subordinate leaders develop plans and orders that transform thought into action. The higher commander's intent provides the basis for unity of effort in delaying operations.

Parameters of the Delay

3-329. The battalion commander clearly articulates the parameters of the delay in the order, specifically subordinate missions in terms of space, time, and friendly strength. Through these parameters, normally stated in paragraph 3 of the delay order- tasks to subordinate units, the commander provides direction for actions during the delaying operation as planned and when subordinate commanders are unable to meet the initial terms of the delay mission.

3-330. First within these parameters, the commander directs one of two alternatives within the order: delay within the area of operation (figure 3-37, page 3-111) or delay forward of a specified line or terrain feature for a specified time (figure 3-38, page 3-112). Time during the conduct of a delay is usually based on another unit completing its activities, such as establishing rearward defensive positions. A mission of delay within an area of operation implies that force integrity is a prime consideration. In this case, the battalion delays the enemy as long as possible while avoiding decisive engagement. Generally, this force displaces once predetermined criteria have been met, such as when the enemy force reaches a *disengagement line*—a phase line located on identifiable terrain that, when crossed by the enemy, signals to defending elements that it is time to displace to their next positions (ADRP 3-90).

3-331. The second parameter the order must specify is what is considered acceptable risk. Acceptable risk ranges from accepting decisive engagement by holding terrain for a given period to avoiding decisive engagement in order to maintain the delaying force's integrity. The depth available for the delay, the time needed by the higher headquarters and subsequent missions for the delaying force determine the amount of acceptable risk.

3-332. Third, the order must specify whether the delaying force may use the entire area of operation or whether it must delay from specific battle positions. A delay using the entire area of operation is preferable, but a delay from specific positions may be required to coordinate two or more units in the delay.

Command Posts Operations

3-333. The battalion main command post is normally the first command post within an echelon to displace during a delay, leaving the tactical command post to control the delay, until it can be reestablished in a secure location. The main command post may displace by echelon, leaving a residual mission command capability in the original location. Communications are essential to the success of this type of operation, and the commander builds redundancy into the communications architecture. Digital information systems help ensure that redundancy by providing a common operational picture and a distributed database allowing one command post to assume the duties of another command post during movement or when one is destroyed. The communications architecture also provides for operating through the degradation of friendly communications systems by the enemy or critical failure points within the friendly system. (See ATP 6-02.53.) As a contingence, battalion trains, (combat or field) can establish an alternate command post though with limited capabilities. (Refer to appendix A for additional information.)

Control Measures

3-334. Control measures are the same for both alternatives addressed in paragraph 3-324 above, except that during a delay forward of a specified line for a specified time, the commander annotates the phase line with the specified time. If the delaying force is ordered to hold the enemy forward of a given phase line (delay line) for a specified time, mission accomplishment outweighs preservation of the force's integrity. Such a mission may require the force to defend a given position until ordered to displace. Control measures established by the battalion commander, for example battle positions, engagement areas, and attack-by-fire positions, allow subordinate commanders to direct the fight more closely giving subordinates a clearer picture of how the battalion commander envisions fighting the delay.

3-335. The battalion commander may dictate specific events to control the delay, for example, the enemy penetration of a phase line can trigger the initial repositioning of subordinate forces to subsequent positions during the course of the battle. The commander may also use phase lines to control the timing and movement of delaying units though assigning time minimums to delays by phase line can limit subordinate commanders to delaying on or forward of those lines, at least until the specified times. A *delay line* is a phase line where the date and time before which the enemy is not allowed to cross the phase line is depicted as part of the graphic control measure (FM 3-90-1). Contact points, coordination points, restrictive fire lines, coordinated fire lines, trigger lines, target reference points, checkpoints, and other control measures are established to avoid fratricide and support subordinate unit coordination.

***Note*.** Designating delay lines is a command decision that imposes a high degree of risk on the delaying unit. The delaying unit must do everything in its power—including accepting decisive engagement—to prevent the enemy from crossing that line before the time indicated. A delay line may also be event driven. For example, a commander can order a delaying unit to prevent penetration of the delay line until supporting engineer assets complete construction of a rearward obstacle belt. (Refer to FM 3-90-1 for additional information.)

Deception

3-336. The delay must include the integration of direct and indirect fires and situational obstacles to make the enemy doubt the nature of the friendly mission and leave no choice but to deploy and maneuver. Engagement at maximum ranges for all weapons systems causes the enemy to take time-consuming measures to deploy, develop the situation, and maneuver to drive the delaying force from its position. An aggressive enemy commander will not deploy if friendly forces are determined to be delaying; the enemy commander will use mass and momentum to develop sufficient pressure to cause subordinate units of the battalion to fall back, or to become decisively engaged.

Movement and Maneuver

3-337. The delay order addresses the conduct of movement prior to the execution phase of the operation, the scheme of maneuver and priorities for the delay and defines how much freedom subordinate leaders have in maneuvering their forces. The battalion commander specifies constraints on maneuver and requirements for coordination. The commander defines the criteria for disengagement and maneuver to alternate and subsequent positions or area of operations, and identifies the series of battle positions or phase lines from which or forward of which the companies must fight.

Scheme of Maneuver

3-338. The scheme of maneuver must allow the battalion to dictate the pace of the delay and maintain the initiative. The commander selects positions that allow the battalion to inflict maximum damage on the enemy, support disengagement, and enable delay actions to alternate or subsequent positions. The commander may choose to delay from alternate or subsequent positions depending on the strength of the companies and the size of the area of operation.

3-339. During delaying operations, the battalion normally assigns deep and parallel area of operations to delaying companies. This provides enough terrain for companies to operate in-depth, and maximizes the ability for battalion assets to support simultaneously multiple units throughout the operation. Generally, each enemy avenue of approach is assigned to only one subordinate unit. The commander and staff make provisions for coordinated action along enemy avenues of approach that diverge and pass from one subordinate area of operation to another. When determining the scheme of maneuver, positions should incorporate as many of the following characteristics as possible:

- Good observation and long-range fields of fire.
- Covered or concealed routes of movement to the rear.
- A road network or areas providing good cross-country trafficability.
- Existing or reinforcing obstacles to the front and flanks.
- Maximize use of highly defensible terrain.

3-340. As the battalion staff plans the delay the staff considers maneuver actions, fires, obstacles, and the employment of other supporting assets necessary to degrade the enemy's mobility and support friendly forces' disengagement to alternate or subsequent positions. This is especially critical at locations and times when companies or the entire battalion may become decisively engaged with the enemy. As the staff develops and refines the plan, it develops decision points for key actions. This includes triggers for the employment of fires and situational or reserve obstacles; displacement of subordinate units to alternate or subsequent positions; and movement of indirect fire assets, mission command systems and facilities, and sustainment assets. The staff also selects routes for reinforcements, artillery, command posts, and sustainment elements to use and synchronizes their movements with the delaying actions of forward units.

Engineer Tasks and Support

3-341. Engineer priorities during a delaying operation are normally countermobility first, then mobility. However, restrictive terrain that impedes friendly movement may require the commander to reverse priorities. Close coordination is necessary so that engineer obstacles are covered by fire and do not impede the planned withdrawal routes of delaying forces or the commitment of a counterattacking and reserve force. In addition to engineering assets, when possible, the delaying force should have a greater-than-normal allocation of fire support systems to include Service and joint aviation to enable the coverage of friendly obstacles and to assist the delay force to break contact if necessary. Engineer-focused considerations include the following:

- When operating within a large area of operation, task-organize countermobility assets to companies and decentralize their control and execution.
- Task-organize reduction assets (see appendix F) to companies to support rearward breaching and reducing requirements.
- Develop the obstacle plan to support disengagement of delaying forces and to shape the enemy's maneuver to meet the commander's intent. Consider countermobility requirements for all delaying positions throughout the depth of the area of operation. Integrate scattterable mines at delay positions to support disengagement and movement to subsequent positions.
- Consider the impact of the obstacle effort on the movement of friendly forces and future operations. Develop obstacle restrictions, establish lanes and guides, and employ situational or reserved obstacles to support mobility requirements. Plan for closing lanes behind friendly forces with scatterable or hand-emplaced munitions. Develop clear criteria for execution of situational and reserved obstacles. Integrate decisions for their execution in the decision support template.

Avoiding Decisive Engagement

3-342. A key to avoiding decisive engagement is to maintain a mobility advantage over the attacking enemy. The commander seeks to increase the battalion's mobility while degrading the enemy's ability to move. The battalion improves its mobility by—

- Maintaining contact with the enemy, maintaining reconnaissance and security on flanks, and coordinating with adjacent units to prevent forces from being isolated.
- Prioritizing and task-organizing mobility assets to maximize the ability of the battalion to perform the delay.
- Reconnoitering routes and battle positions.

- Improving routes, combat trails, bridges, and fording sites between delay positions, as time and resources permit.
- Using indirect fires and obstacles to support disengagement and to cover movement between positions.
- Task-organizing and positioning breaching assets within subordinate formations to breach enemy scatterable mines rapidly.
- Using multiple routes.
- Controlling traffic flow and restricting refugee movements to unused routes.
- Keeping logistical assets uploaded and mobile.
- Caching ammunition on rearward routes and ensuring that units know the locations of these supply points. If possible, supply points are guarded and prepared for destruction when not used by delaying forces.
- Task-organizing additional medical and equipment evacuation assets to subordinate companies to increase the ability to disengage and displace rapidly.
- Positioning air defense assets to protect bridges and chokepoints on rearward routes.

3-343. As the commander seeks to increase the battalion's mobility. The commander degrades the enemy's ability to move through planning and an understanding of the area of operation. The battalion degrades the mobility of the enemy by—

- Maintaining continuous pressure on the enemy throughout the area of operation.
- Attacking logistics as well as maneuver and fire support assets.
- Securing and controlling chokepoints and key terrain that dominates high-speed avenues of approach.
- Destroying enemy reconnaissance and security forces to cause the enemy to move more deliberately.
- Employing a combination of directed, situational, and reserved obstacles.
- Employing indirect fires, obscurants, manned-unmanned teaming, and close air support, if available.
- Using deception techniques such as dummy positions.

Delay Techniques

3-344. When conducting a delay the battalion commander normally assigns subordinate units contiguous areas of operations that are deeper than they are wide. The commander synchronizes the employment of these combined arms teams throughout the depth of each assigned area of operation for the delay. When the commander expects to delay for only a short time or the area of operation lacks depth, the delaying unit may be forced to fight from a single set of positions. When the commander expects the delay to last for a longer period, or if sufficient depth is available, the delaying unit may delay from either alternate or subsequent positions. In both techniques, delaying units normally reconnoiter delay positions before occupying them and, if possible, post guides on one or two positions. In executing both methods of delay, it is critical that the delaying units maintain contact with the enemy between delay positions. Table 3-1 summarizes the comparison of two delay techniques.

Table 3-1. Comparison of two delay techniques

Method of Delay	*Use When*	*Advantages*	*Disadvantages*
Delay from alternate positions.	• Area of operation is narrow. • Forces are adequate to split between different positions (in-depth).	• Allows positioning in-depth. • Harder for enemy to isolate units. • Increases flexibility. •Allows more time for maintenance.	• Requires continuous coordination. • Requires passage of line, increasing vulnerability and fratricide potential. •Engages only part of the force at one time.
Delay from subsequent positions.	• Area of operation is wide. • Forces available are not adequate to position in-depth.	• Reduced fratricide risk. • Ease of mission command. • Repeated rearward passages not required.	• Limited depth to the delay positions. • Easier to penetrate or isolate units. • Less time is available to prepare each position. • Less flexibility.

3-345. In a delay from alternate positions (figure 3-35), two or more units in a single area of operation occupy delaying positions in-depth. As the first unit engages the enemy, the second occupies the next position in-depth and prepares to assume responsibility for the operation. The first force disengages and passes around (preferred method) or through the second force. The force then moves to the next position and prepares to reengage the enemy while the second force takes up the fight. Both the IBCT and Infantry battalion can use this method of delay. If the area of operation is narrow, the battalion employs companies in-depth occupying alternate positions. This enables the battalion to develop a strong delay, with forces available to counterattack or assist in the disengagement of the company in contact. At the battalion level, using alternate positions helps maintain pressure on the enemy and helps prevent platoons or companies from being decisively engaged. A delay from alternate positions is particularly useful on the most dangerous avenues of approach because it offers greater security and depth than a delay from subsequent positions. However, it also poses the highest potential for fratricide and vulnerability as units pass near or through each other.

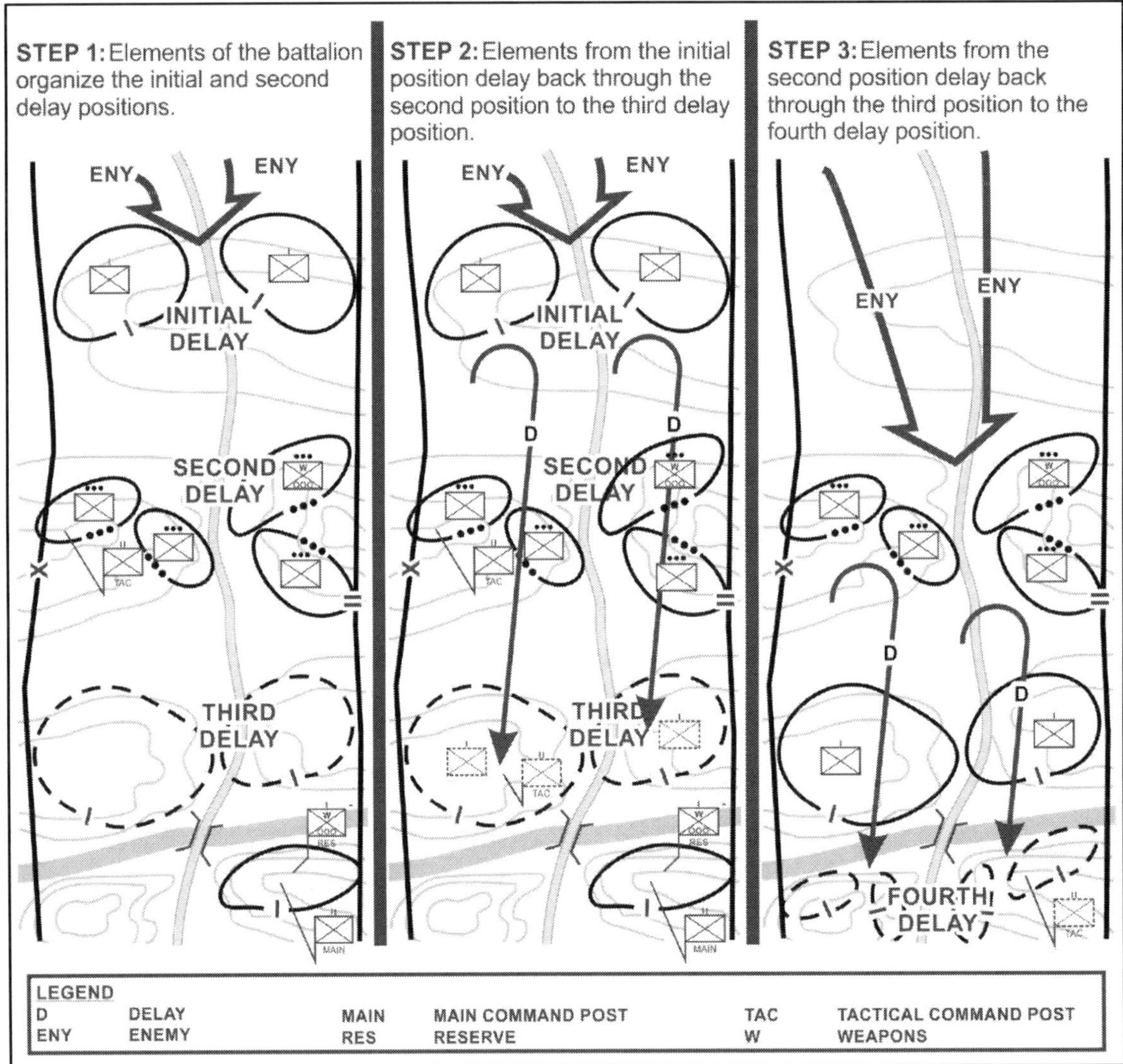

Figure 3-35. Delay from alternate positions

3-346. The battalion uses a delay from subsequent positions when the assigned area of operation is so wide that available forces cannot occupy more than a single layer of positions. In a delay from subsequent positions, the majority of forces are arrayed along the same phase line or series of battle positions. The forward forces delay the enemy from one phase line to the next within their assigned area of operations. At battalion level, this is the least preferred method of delaying since there is a much higher probability of forces becoming isolated or decisively engaged, particularly if the delay must be maintained over more than one or two subsequent positions. The battalion also has limited ability to maintain pressure on the enemy as it disengages and moves to subsequent positions unless it has been allocated additional and adequate indirect fire support. The commander may have to use the Infantry weapons company, with its greater mobility, as a reserve to ensure that the rifle companies can delay at the same rate (figure 3-36).

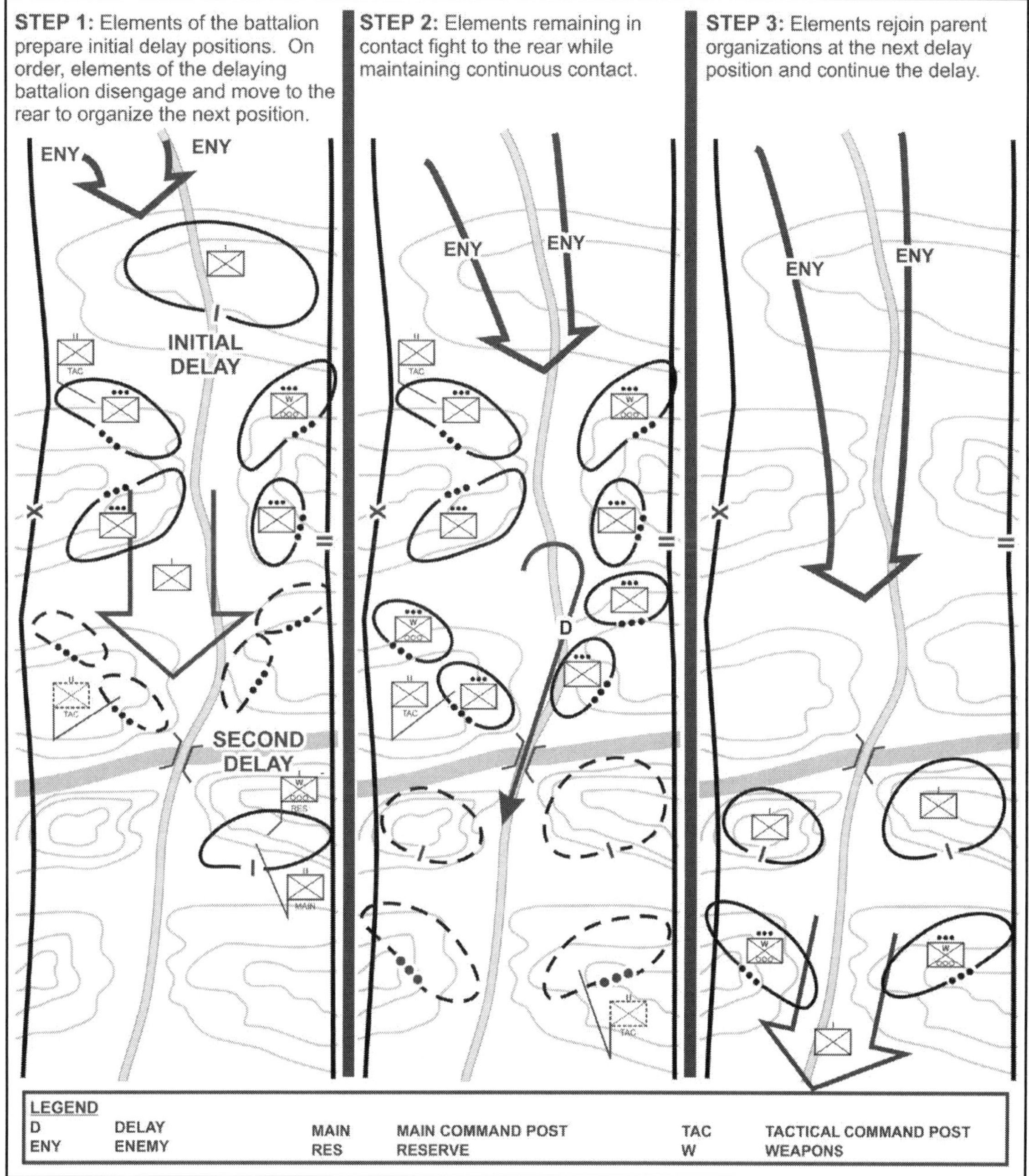

Figure 3-36. Delay from subsequent positions

Intelligence

3-347. During the battalion's IPB, the staff, led by the S-2 and S-3, analyzes the effects of terrain and the anticipated enemy situation to identify positions that offer the best opportunity to engage, delay, and inflict damage on the enemy force. As the staff develops delay, alternate, and subsequence positions and control measures, the staff calculates enemy closure rates and compares them to friendly displacement rates between positions. The staff calculates time and space factors for each enemy avenue of approach to dictate the amount of time subordinate units have to engage the enemy and move, before becoming decisively engaged. From this analysis, the staff develops triggers and lines for displacement to alternate and subsequence positions in-depth. (Refer to ATP 2-01.3 for additional information.)

Fires

3-348. The extended frontages and ranges common to delaying operations, and in general to retrograde operations, make the provision of fire support difficult and limit the commander's ability to mass fires. For that reason, delaying forces often have more than the normal allocation of fire support assets. The commander's risk of losing supporting artillery systems and their ammunition also increases when conducting delaying operations. The commander balances the decision to commit fire support systems forward against anticipated requirements in subsequent battle stages. In particular, the commander takes precautions to protect towed artillery systems from being overrun by a mobile enemy. The commander also uses available rotary- and fixed-wing aircraft to augment fires or replace artillery systems and to reduce risk through the ability to conduct air movement of artillery systems (rotary-wing only). The following are key considerations for the fire support plan:

- Attack the enemy throughout and forward (IBCT and above) of the battalion's area of operations.
- Engage the enemy with fires to inflict casualties and disrupt their approach before they get to friendly delaying positions.
- Plan final protective fires for each series of delaying positions to support disengagement.
- Mass fires on high-payoff targets and canalizing terrain to limit the momentum of the enemy's attack.
- Plan and designate priority targets along routes from one delaying position to the next.
- Mass all available fires to support disengagements.
- Use obscuration and screening fires to conceal and screen friendly positions and movements.
- Plan appropriate mortar positions, for example split section, to provide support throughout the mission.
- Establish clear priorities and detailed triggers to adjust them.
- Develop detailed triggers to initiate and lift fires for each target.

Sustainment

3-349. The requirement to maintain continuous support during the delay requires sustainment organizations to echelon (see appendix H) their assets throughout the area where the retrograde will take place. This echeloning, coupled with the wide dispersion of combat forces that is inherent in a delay, complicates the conduct of the delay. Communication within the sustainment system, accurately tracking the battle, and anticipating support requirements are especially important. The following are key sustainment planning considerations:

- Keep the sustainment assets mobile and supplies uploaded.
- Task-organize ambulances and recovery vehicles to the companies.
- Emphasize maintenance support forward for the weapons company with short evacuation times.
- Synchronize refueling and resupply operations with the scheme of maneuver and the anticipated enemy situation to ensure continuity of support. Increase class III and V stocks and position forward.
- Do not coordinate for throughput too far forward, which might cause assets to be caught in the fight or add to route congestion. This may not apply during the initial preparations for the delay.
- Plan routes for sustainment assets that do not conflict with maneuver elements or refugee movement.

Protection

3-350. Because of the importance of countermobility and mobility tasks, a battalion conducting a delay probably has few engineer assets to devote to the survivability operations. Battalion subordinate units construct survivability positions, within the limits of their capabilities, in-depth, as required, to support repositioning forces. The battalion maximizes the use of camouflage, concealment, and cover when constructing primary, alternate, and subsequence fighting and protective positions. In the case of camouflage and concealment, survivability operations include both stationary and on-the-move capabilities. Military deception, part of the mission command warfighting function, can be enabled by the use of survivability operations intended to help mislead enemy decision makers. This may include the use of dummy or decoy positions or devices. (Refer to ATP 3-37.34 for additional information.)

3-351. Obscuration fires on or near enemy positions decrease an enemy's capability to visually sight friendly forces. The commander employs obscuration, when and where weather conditions allow, to provide concealment for movement and assemblies. Obscuration curtains, blankets, and haze (see ATP 3-11.50) can protect friendly withdrawing columns, critical points, positons, and routes however; the commander takes precautions to ensure that the obscuration does not provide a screen for the enemy's advance. The commander may employ obscuration to assist with breaking contact with the enemy or to deceive the enemy of the battalion's actual intentions. Terrain that hinders the mobility and surveillance capabilities of enemy combat systems and supporting tactical vehicles can offers concealment and cover for the movement of friendly forces.

Note. Smoke generators, in general, are very low-density items. If smoke generators are available, the battalion may employ them for deception, obscuring movement and positions, or obscuring portions of the battlefield to reduce enemy visibility and ease of movement.

3-352. Ground-based air defense artillery units execute most Army air and missile defense operations, though air and missile defense support to the IBCT is generally limited. To ensure adequate air defense coverage of forces during movements from one delaying position to another, the Infantry battalion should expect to use its organic weapons systems for self-defense against enemy air threats. When available air and missile defense units prevent the enemy from interdicting delaying forces of the battalion, while freeing the commander to synchronize movement and firepower. Air defense of a delaying force has three main considerations—the protection of the force while it is in position, the protection of any forces left in contact, and the protection of the force as it moves to alternate or subsequent positions or to the rear. Priority should be toward maintaining the mobility of the force in most cases.

Note. Air defense assets remain mobile yet able to engage aerial targets with little advance warning. During the delay, these assets should work in teams and move to the rear in alternating bounds. This ensures that dedicated air defense assets will always be in position, with the flexibility needed to keep pace with the operations. Selected firing points should not be obvious positions that an enemy would target as part of preparatory or supporting fires. Coverage includes movement and maneuver routes, chokepoints, assembly areas, battle and firing positions, and bridges that delaying forces intend to use. Early warning of enemy air attack is provided over combat net radios using the command net at the brigade echelon and below. (Refer to ATP 3-01.7 for information on air defense operations at the tactical level.)

3-353. When planning CBRN operations in the delay, the battalions may coordinate for CBRN reconnaissance assets if available from the IBCT. Decontamination operations in the delay focus on individual and crew operational decontamination procedures until the conclusion of the operation, when thorough decontamination can be accomplished.

Preparation

3-354. Defensive preparations for the conduct of an area defense discussed in Section II of this chapter also apply during the conduct of a delay. Resources—including the time available to prepare (specifically in regards to not becoming decisively engaged)—determine the extent of preparations. Throughout preparation, the commander assigns a high priority to R&S missions, and security operations. Additionally, the preparation of alternate, supplementary, and subsequent positions receives a higher priority than in either a mobile or an area defense. Understanding that it is not always possible to complete all preparations before starting the delaying operation, delaying units continue to prepare and adapt plans as the situation develops.

Organization of a Battle Position

3-355. In the delay, the battalion prepares battle positons in a manner similar to the area defense. However, when organizing battle positions, the commander places more emphasis on width than depth, as well as reconnaissance and preparing routes for displacing. Within each battle position most of the available firepower is oriented toward the expected enemy avenue of approach. Flank and rear security units are normally manned with forces internal to the delaying force. The commander plans and reconnoiters withdrawal routes from primary positions to alternate, supplementary, and subsequent positions in accordance with the plan. In preparing a battle position, the commander places less emphasis on installing protective obstacles, final protective fires, and ammunition stockpiling than would occur in either an area or a mobile defense. Battle positions within a delaying operation are sometimes referred to as delay positions and alternate positions and subsequent positions during the conduct of a delay.

Rehearsals

3-356. When conducting a rehearsal, key leaders, as a minimum rehearse the operation against all feasible enemy course of actions to promote flexibility during decision-making. The commander examines each subordinate unit commander's plan as they fight the delay during the rehearsal, paying close attention to the following:

- Direct and indirect fire instructions.
- Timing of movements (to include in limited visibility).
- Delaying actions from one position to the next, to include disengagement criteria and triggers.
- Means and methods of disengaging from the enemy.
- Maintaining contact with the enemy as the force moves to alternate and subsequent positions.
- Execution of situational and reserved obstacles to include closure of lanes.
- Movement times, routes, and positioning of fire support, engineer, protection, and sustainment assets.

3-357. The commander also rehearses contingences to deal with enemy penetrations and decisive engagement, and the opportunity to resume the offense. Battalion rehearsals serve to synchronize the movement of maneuver forces, fire support, protection, and sustainment units. During rehearsal, it is especially important to portray movement times and required routes realistically to identify potential conflicts.

Inspections and Preoperations Checks

3-358. Preparations throughout the battalion include inspections and pre-operations checks, especially at subordinate echelons. These inspections and checks ensure subordinate units, Soldiers, and systems are as fully capable and ready to execute the mission as time and resources permit and to ensure delaying forces have the resources necessary to accomplish the mission. Within the Infantry battalion, the loads that Soldier's carry is of particular importance during delaying operations. How much Soldiers carry, how far, and in what configuration are critical mission considerations requiring command emphasis. The commander and subordinate leaders of the battalion inspect and check subordinate units to ensure—

- Movement, maneuver, fire support, and obstacle plans are consistent with the commander's intent and concept of operations.
- Delaying units coordinate to maintain cohesion and mutual support during the delay.
- Subordinate unit engagement areas enable the battalion and higher echelon scheme of maneuver.
- Engagement area development includes disengagement criteria, routes, and triggers that support the battalion's maneuver within its area of operation.

Execution

3-359. When the Infantry battalion is part of a larger scheme of maneuver designed to regain the initiative and defeat the enemy, or when the battalion operates independently, the complex nature of a delay require maneuver elements within the delaying operation to execute different, yet complementary, actions. In a single delaying operation, attacks, area defenses, mobile defenses, and other actions may occur in any sequence or simultaneously. When conducting a delay, as in an area defense (see section II of this chapter), the Infantry battalion defends using a variety of tactics, techniques, and procedures to accomplish the mission. The commander deploys security forces well forward of the initial delay positions of the main body to buy time, to establish an effective delay, and to give early warning of any enemy approach. Forward security forces detect and report as enemy forces approach to confirm the enemy's probable course of action.

Gain and Maintain Enemy Contact

3-360. When subordinate to the IBCT or other higher headquarters the battalion may position the scout platoon, or a task organized company in a screen behind the cavalry squadron or higher headquarters force to establish contact with friendly forward elements. Though not the preferred option, delaying forces of the Infantry battalion must be prepared to occupy the forward line of own troops without a forward security force to their front from a higher echelon (see paragraph 3-123). Once battalion security forces (or forward positioned forces) make contact with the enemy, they maintain contact. Security forces use covered, concealed, and coordinated routes to avoid enemy and friendly fires.

Disrupt and Fix the Enemy

3-361. Security forces fix, defeat, and destroy the enemy's reconnaissance and security elements without risking decisive engagement. These forces direct fires at the approaching enemy force as far forward of the delay positions as possible to disrupt and fix the enemy. Engaging a moving enemy at long ranges tends to inflict far more casualties on an attacking enemy than the enemy can inflict on the delaying force; it also slows the enemy force's tempo of operations. The more a delaying force can blind an enemy force through the elimination of that force's reconnaissance assets, the more likely the enemy force is to hesitate and move with caution.

3-362. As the enemy closes with security forces, IBCT and battalion forces move back through or around the initial positions of the main body to subsequent positions that allow them to observe the main body area and assist in the disengagement and movement of forces to their next positions. This also prevents the enemy from finding gaps between delaying units and attacking the exposed flanks of delaying units. When the Infantry battalion occupies the forward line of own troops, engagements forward of the battalion's initial delaying positions are normally limited to observed fires to continue the disruption and attrition of the attacking enemy.

Maneuver

3-363. The battalion maneuvers to force the enemy to deploy multiple times and repeatedly concentrate its combat power to defeat the delaying forces of the battalion. When operating independently, or when keeping the entire battalion synchronized with the remainder of the IBCT, the commander makes decisions about disposition, displacement, timing, and engagement in the context of the higher commander's intent and priority for the delay. For example, when time is more important than force preservation, or vice versa? In many instances, the battalion or its elements must accept decisive engagement to execute the mission in conjunction with the actions of another force.

Disposition

3-364. As delaying operations evolve, the battalion commander closely controls the disposition of security, main body, and reserve forces in order to maintain the cohesion of the battalion. When conducting a delay the Infantry battalion masses effects and concentrates actions quickly for a short period to inflict the maximum damage on the enemy at the maximum range. To avoid decisive engagement, the Infantry battalion disengages before the enemy can breach obstacles or mass effective fire on delaying forces.

3-365. In determining the disposition of the delaying force, the commander takes advantage of the terrain by selecting terrain that favors friendly actions and hampers enemy actions. The terrain dictates where elements of the battalion can orient on a moving enemy force and ambush it. During a delay, compartmentalized terrain facilitates shorter displacements initiated at closer range to the enemy. Subordinate commanders conducting operations in compartmentalized terrain select locations that restrict the enemy's movement and prevent the enemy force from fully exploiting its combat superiority. In restricted terrain, positions may be close together, except when conducting a delay using air assault or air movement techniques.

3-366. When the Infantry battalion delays in flat or more open terrain, delay positions are often far apart. In selecting positions, subordinate commanders consider natural and artificial obstacles, particularly when the enemy has numerous armored combat systems. Earlier displacements at greater distances with good, long-range fields of fire are generally required to stay in front of the advancing enemy. Under these conditions (flat and open terrain), delaying forces of the Infantry battalion are usually augmented with motorized transportation assets and increased indirect fire support and Army aviation and Air Force assets.

Displacement

3-367. As delaying forces of the battalion displace, they move to the flanks of delay positions and do not move through friendly engagement areas or target reference points, unless the tactical situation makes such movement necessary. Delaying forces ensure their routes do not reveal the locations of other friendly elements to include stay-behind forces and forward observers. Delaying forces may move by bounds within the battalion or company to maintain direct fires on the enemy and cover movement. Short, intense engagements at near maximum range with sustained fires and covering obscurants, are key to forcing the enemy into deploying early and often for a decisive engagement. Observers position to the flanks in-depth to observe and shift fires as forces delay to alternate and/or subsequent positions.

3-368. Once a delay starts, subordinate units of the battalion displace rapidly between positions using obstacles and defensive positions in-depth to slow and canalize the enemy. The battalion commander exploits the mobility of the weapons company's combat systems to confuse and defeat the enemy. Whenever possible, the commander grasps any fleeting opportunity to seize the initiative, even if only temporarily. By aggressively contesting the enemy's initiative through offensive action, the delaying force avoids passive patterns that favor the attacking enemy. The battalion commander may conduct a counterattack (though still seeking to avoid a decisive engagement) from an unexpected direction, temporarily confusing the enemy commander. Attacks of this nature throw the enemy off stride, disorganizes the enemy force, confuses the enemy commander's picture of the fight, and helps prolong the delay. In turn, this confusion may affect the enemy's tempo and momentum. These attacks also affects the movement of the enemy's reserve and other follow-on forces.

Timing and Engagement

3-369. As the advancing enemy force approaches, the enemy crosses one or more trigger lines and moves into engagement areas within the range of the delaying force's anti-armor missiles and heavy and small arms direct fires. The commander holds the delaying force's direct fire, until the enemy is positioned where the fire plan and scheme of maneuver require their use. The commander controls fires from the delaying force in the same manner as in any defense. The more damage the delaying force can inflict on the enemy, the longer the force can stay in position.

3-370. As the enemy force presses its attack and maneuvers against the delaying force, the commander constantly assesses the action to guide the engagements of delaying units. Throughout the delay, the battalion relies heavily on fires external to the battalion (field artillery, and Army and Air Force aviation assets) to suppress the enemy, so subordinate units of the battalion can disengage, move, and occupy new positions. When a subordinate unit is unable to maintain separation from the enemy, the commander can shift additional combat multipliers and other resources to that particular area of operation to counter the enemy's success. As one subordinate unit displaces, the commander may order other subordinate units to change their orientation to cover the move.

3-371. When subordinate units of the battalion pass between friendly positions, each unit travels along a designated route, using demolitions as required and requesting additional fire support if the enemy maintains contact with the delaying unit. When a delaying unit's route passes through a friendly linear obstacle, that unit becomes vulnerable to enemy attack because of the danger of congestion within the lane and on the far side of the obstacle. The commander takes into consideration the increased time required for a passing unit to transit through an obstacle area and plans contingencies to prevent the enemy from engaging a passing unit until it can redeploy into a tactical formation.

3-372. The commander retains the reserve for the decisive moment. Typically, the commander commits the reserve to help a subordinate unit disengage and regain its ability to maneuver or to prevent the enemy from exploiting an advantage. Due to the inherent mobility of the Infantry weapons company versus other units organic to the Infantry battalion. Normally, the weapons company is tasked with the reserve mission establishing support by fire positions for this task. When the commander commits the reserve, the commander's ability to influence the battle is greatly reduced unless the commander can reconstitute a new reserve. The reserve is not committed early in the delay unless the integrity of the delay is threatened.

3-373. Sustainment elements position to the rear of the battalion's delay as far as possible but close enough to provide adequate support. Ammunition stocks must be capable of sustaining the quantity of fire support required in the delay. Maintenance operations supporting the battalion focus on evacuating rather than returning damaged vehicles and equipment to combat. Vehicles and equipment are fixed quickly in position, evacuated to the echelon support area, or destroyed to prevent enemy capture.

Parameters of the Delay Mission (Area of Operation and Specific Line or Terrain Feature for a Specific Period)

3-374. The commander must specify certain parameters in an order for a delay mission. First, the commander must direct one of two alternatives: delay within the area of operation (AO) or delay forward of a specified line or terrain feature for a specified time. A mission of delay within the AO implies that force integrity is a prime consideration. Usually, a mission of delay for time is based on another unit completing its activities, such as establishing rearward defensive positions.

3-375. When the Infantry battalion is assigned a mission to delay within an area of operation, the intent of the operation is to slow and control the enemy tempo and to defeat as much of the enemy as possible without sacrificing the integrity of the delaying force. The higher commander provides guidance regarding intent and desired effect on the enemy, though restrictions regarding terrain, time, and coordination with adjacent forces are minimized. Normally, a delay within an area of operation is assigned when force preservation is of higher priority and there is considerable depth within the assigned area of operation (figure 3-37, page 3-112).

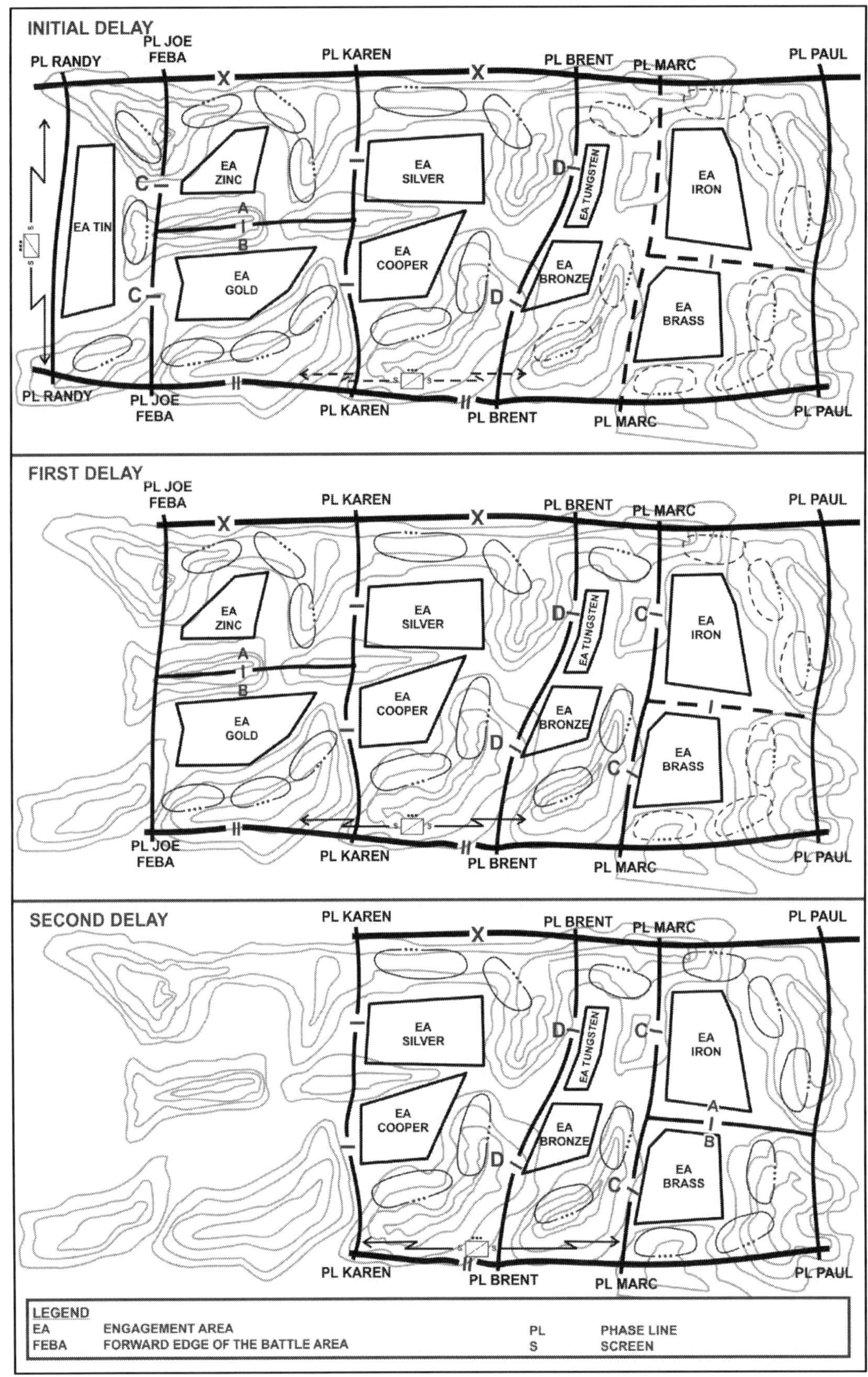

Figure 3-37. Delay within an area of operations

3-376. When the mission is to delay forward of a specific line or terrain feature for a specific period. The Infantry battalion must control the enemy's attack and retain specified line or terrain to achieve some purpose relative to another unit that can include setting the conditions for a counterattack, for completion of defensive preparations, or for the movement of other forces or civilians. Normally in a delay, inflicting casualties on the enemy is secondary to gaining time. This parameter carries a much higher risk for the battalion, with the likelihood that part of, or the entire battalion becoming decisively engaged. The timing of the operation is most often controlled graphically by a series of phase lines with associated dates and times to define the desired delay-until period (figure 3-38).

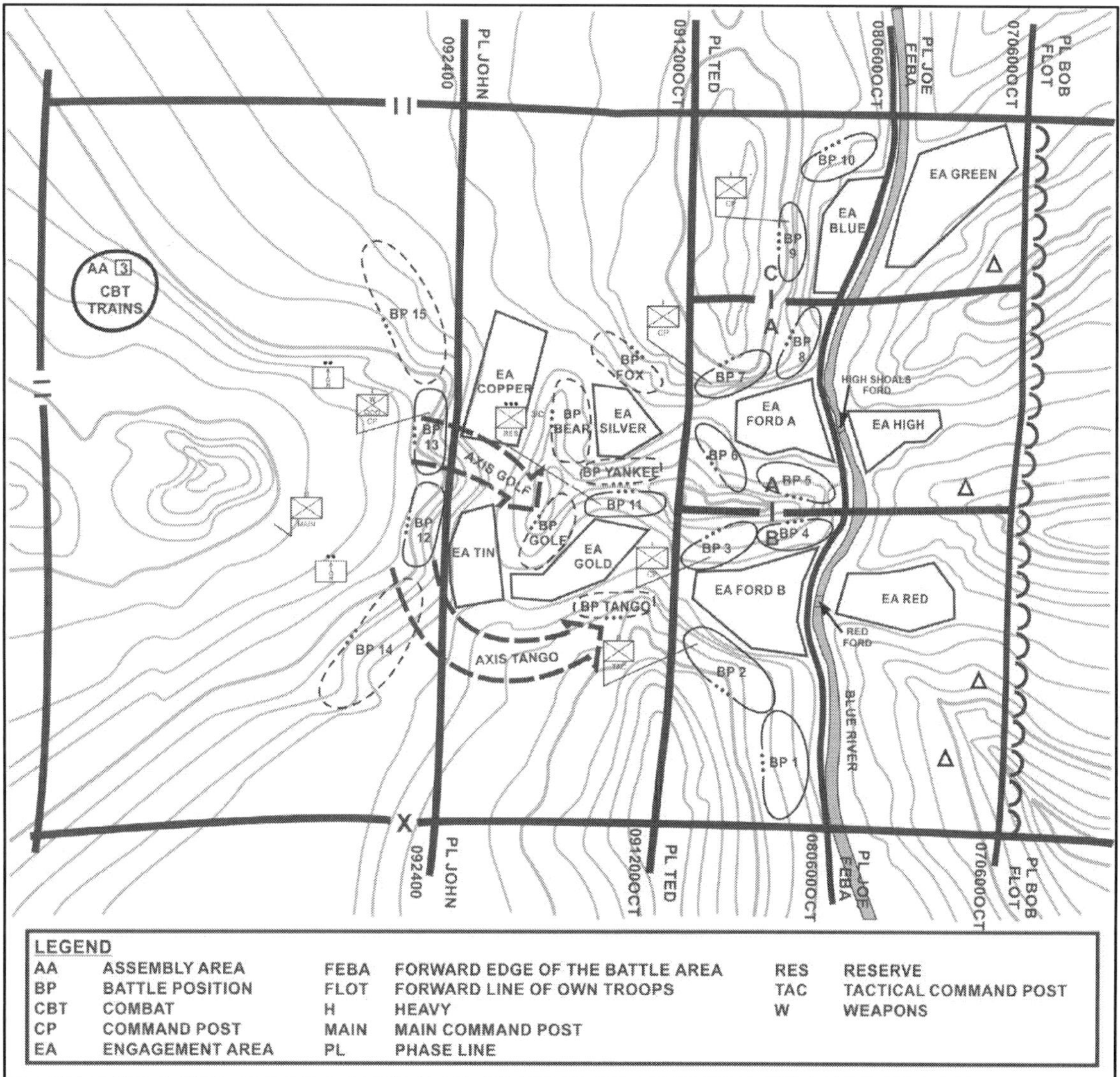

Figure 3-38. Delay forward of a terrain feature for a specified time

Follow Through

3-377. A delaying operation terminates when the delaying force conducts a rearward passage of lines through a defending force, the delaying force reaches defensible terrain and transitions to the defense, the advancing enemy force reaches a culminating point, or the delaying force goes on the offense once reconstituted. Ideally, a battalion that has been delaying conducts a rearward passage of lines through the established defense of another friendly force. When the advancing enemy force reaches a culminating point, the delaying force may maintain contact in its current position, withdraw to perform another mission, or transition to the offense. In all cases, the commander plans for the expected outcome of the delay and actions taken after termination of the delay based on the situation and the higher commander's plan.

WITHDRAWAL

3-378. A *withdrawal operation* is a planned retrograde operation in which a force in contact disengages from an enemy force and moves in a direction away from the enemy (JP 3-17). Withdrawing units, whether all or part of a committed force, voluntarily disengage from the enemy to preserve the force or release it for a new mission. Based on the higher headquarter order and the enemy situation, the battalion's withdrawal may be assisted or unassisted and may take place with or without enemy pressure.

ORGANIZATION OF FORCES

3-379. As in the delay, the battalion's organization of forces depends on how the IBCT has structured its forces unless the battalion operates independently. The IBCT normally organizes into a security force, main body, and reserve, though operations extended across large areas may preclude the use of an IBCT-controlled security force and reserve. In this case, the IBCT may direct the battalion to organize its own security, main body, and reserve forces; the same as if the battalion was operating independently. The IBCT commander or the Infantry battalion commander organizes a *detachment left in contact*—an element left in contact as part of the previously designated (usually rear) security force while the main body conducts its withdrawal (FM 3-90-1)—and/or a stay-behind force (see paragraphs 3-62 to 3-65, pages 3-28 and 3-29) if the scheme of maneuver requires them. The IBCT commander can designate a battalion as the security or reserve force for the IBCT.

3-380. When the battalion operates independently or establishes its own security force within the IBCT's area of operation, the battalion normally uses the scout platoon and/or the sniper squad as a screening force when the withdrawal is not under pressure. These forces position to observe the most likely enemy avenues of approach and can initiate fires to slow and weaken the enemy. When the withdrawal is under enemy pressure these elements can be reinforced with other elements, for example, assault platoons from the weapons company, forward observers, and/or fire support teams executing direct and indirect fire targets on a primary enemy avenue of approach. When withdrawing under pressure the commander may make provision to resource a detachment left in contact, normally established with an Infantry rifle company or company combined arms team, to cover the remaining elements of the battalion (main body minus the reserve) as they withdraw.

3-381. The battalion's main body consists of all elements remaining after the commander resources a security force and the reserve. The battalion commander retains a reserve, normally a company or company minus, to counter penetrations between positions, reinforce threatened areas, and protect withdrawal routes. When the complete formation withdraws under pressure, the reserve may take limited offensive action, such as spoiling attacks, to disorganize, disrupt, and delay the enemy. Reserves may also extricate encircled or heavily engaged forces. The force tasked with the reserve mission requires the mobility and combat power to accomplish assigned tasks.

DOCTRINAL BASIS FOR A WITHDRAWAL

3-382. Withdrawals are inherently dangerous because they involve moving units to the rear and away from what is usually a stronger enemy force. The heavier the previous fighting and the closer the contact with the enemy, the more difficult the withdrawal. Ideally, the battalion commander avoids withdrawing from action under enemy pressure, though this is not always possible.

Planning

3-383. The commander plans and coordinates a withdrawal in the same manner as a delay, though some mission variables of METT-TC apply differently because of the differences between a delay and a withdrawal. A withdrawal may precede a retirement operation or follow a delaying operation. Control measures used in the withdrawal are the same as those in a delay or a defense).

General Considerations for the Withdrawal

3-384. Because a withdrawing force is most vulnerable if the enemy attacks, the commander normally plans for a withdrawal under enemy pressure. The commander then develop contingencies for a withdrawal without pressure. The commander's main considerations include the following:

- Plan for the next mission following the withdrawal.
- Disengagement criteria (time, friendly situation, enemy situation).
- Plan for a deliberate break in contact from the enemy.

- Plan for deception to conceal the withdrawal for as long as possible.
- Rapid displacement of the main body, safeguarded from enemy interference.
- Selection and protection of primary withdrawal routes and alternate withdrawal routes.
- Sitting of obstacles behind the detachment left in contact to complicate the enemy's pursuit.
- Ensuring fire support and sustainment assets remain within distance to support withdrawing units, security forces, and/or detachments left in contact.

3-385. Planning for a withdrawal normally begins with the preparation of the plan for the next mission. Initial planning includes the development of disengagement criteria, route selection, and displacement timing based on the friendly and enemy situation. The follow-on mission for the battalion drives the end state of the withdrawal in order to best position units to accomplish the next mission. The desired end state can include withdrawing to an assembly area for follow-on missions or the establishment of a new defensive position. Alternatively, subordinate units of the battalion can withdraw indirectly to either area through one or more intermediate positions. When preparing the new defensive position, the commander balances the need for security with the need to get an early start on the defensive effort.

3-386. The commander's plan for the withdrawal clearly defines how to deceive the enemy as to the execution of the withdrawal; how the battalion is to disengage from the enemy; and the end state of the operation in terms of time, location, and disposition of friendly and enemy forces. The commander usually confines the battalion's rearward movement to times and conditions when the advancing enemy force cannot observe the activity and easily detect the withdrawal operation. To help preserve secrecy and freedom of action, for example, the commander must consider visibility conditions and times when the enemy's R&S effort can observe friendly movements.

3-387. When planning for the deliberate break from the enemy the commander has essentially two options: break contact using deception and stealth or break contact quickly and violently under the cover of supporting fires reinforced by obstacles to delay enemy's pursuit. In either option, the commander may employ obscuration to assist with breaking contact with the enemy or to deceive the enemy of the battalion's actual intentions. Terrain that hinders the mobility and surveillance capabilities of enemy combat systems and supporting tactical vehicles can offers concealment and cover for the movement of friendly forces.

Assisted and Unassisted Withdrawal

3-388. When the withdrawal is assisted, the assisting force(s) occupies positions to the rear of withdrawing forces and prepares to accept control of the situation. The assisting force can also assist withdrawing forces with route reconnaissance, route maintenance, fire support, protection, and sustainment. Both forces closely coordinate the withdrawal. After coordination, the withdrawing force delays to a battle handover line, conducts a passage of lines, and moves to its final destination. Generally in an assisted withdrawal, the withdrawing force coordinates the following with the assisting force:

- Rearward passage of lines.
- Reconnaissance of withdrawal routes.
- Forces to secure choke points or key terrain along the withdrawal routes.
- Forces to assist in movement control such as traffic control posts.
- Required combat, fire support, protection, and sustainment to assist the withdrawing battalion in disengaging from the enemy.

3-389. In an unassisted withdrawal, the battalion establishes its own security and disengagement from the enemy. Subordinate units of the battalion reconnoiter and secure routes used in its rearward movement while fire support and sustainment echelons within the battalion support the withdrawal. The battalion establishes a security force as the rear guard while the main body withdraws. The commander designates a flank guard or screen as the situation requires. Sustainment and other support forces normally withdraw first, followed by combat forces not tasked with the security or reserve mission. However, sustainment and other support forces as they move to the rear must continue to maintain the ability to support the battalion. To deceive the enemy as to the battalion's movement and if withdrawing under enemy pressure, the commander establishes a detachment left in contact. As subordinates of the battalion withdraw, the detachment left in contact disengages from the enemy and follows the main body to its final destination.

Withdrawal Under and Without Enemy Pressure

3-390. When withdrawing under enemy pressure, all subordinate units withdraw simultaneously when available routes allow, using delaying tactics (see paragraph 3-334) to fight their way to the rear. In the usual case, when simultaneous withdrawal of all forces is not practical, the commander decides the order of withdrawal. The commander then makes three interrelated key decisions: when to start the movement of selected sustainment and main body elements, when forward elements should start thinning out, and when the security force should start its disengagement operations. The commander avoids premature actions that lead the enemy to believe a withdrawal is being contemplated. The commander anticipates the enemy's means of interference and plans the employment of security forces, field artillery, and Army and Air Force aviation assets to counter this interference. Additional factors influencing this decision may include—

- Subsequent missions.
- Availability of transportation assets and routes.
- Disposition of friendly and enemy forces.
- Level and nature of enemy pressure.
- Degree of urgency associated with the withdrawal.

3-391. When withdrawing without enemy pressure the commander plans when to begin the withdrawal and has the option of taking prudent risks to increase the displacement capabilities of the withdrawing force. For example, the main body may be ordered to conduct a tactical road march instead of moving in tactical formations. The commander can plan for stay-behind forces (see paragraph 3-62, page 3-28) as part of the operation.

Detachment Left in Contact

3-392. A detachment left in contact, when used, remains behind to deceive the enemy into believing the battalion is still in position while most of the unit withdraws. The detachment simulates—as nearly as possible—the continued presence of the main body until it is too late for the enemy to react to the main body's withdrawal. The battalion commander develops specific instructions about what the detachment is to do when the enemy attacks and when and under what circumstances the detachment continues to delay or conduct withdrawal. When the detachment left in contact disengages from the enemy, the detachment uses the same techniques as in the delay. When required, and if available, the battalion commander provides the detachment with additional recovery, evacuation, and transportation assets to use after disengagement to speed its rearward movement.

3-393. The commander uses two methods to resource the detachment left in contact. The first is for each forward subordinate maneuver element (generally the Infantry rifle company) of the battalion to leave a sub-element in place (figure 3-39). For example, each forward rifle company leaves a task- organized platoon or detachment in contact. This is the least desirable option since it complicates mission command and task organization. The battalion commonly uses this option when the subordinate companies have lost significant portions of their mission command systems. Typically, these elements fall under a detachment commander designated by the battalion commander.

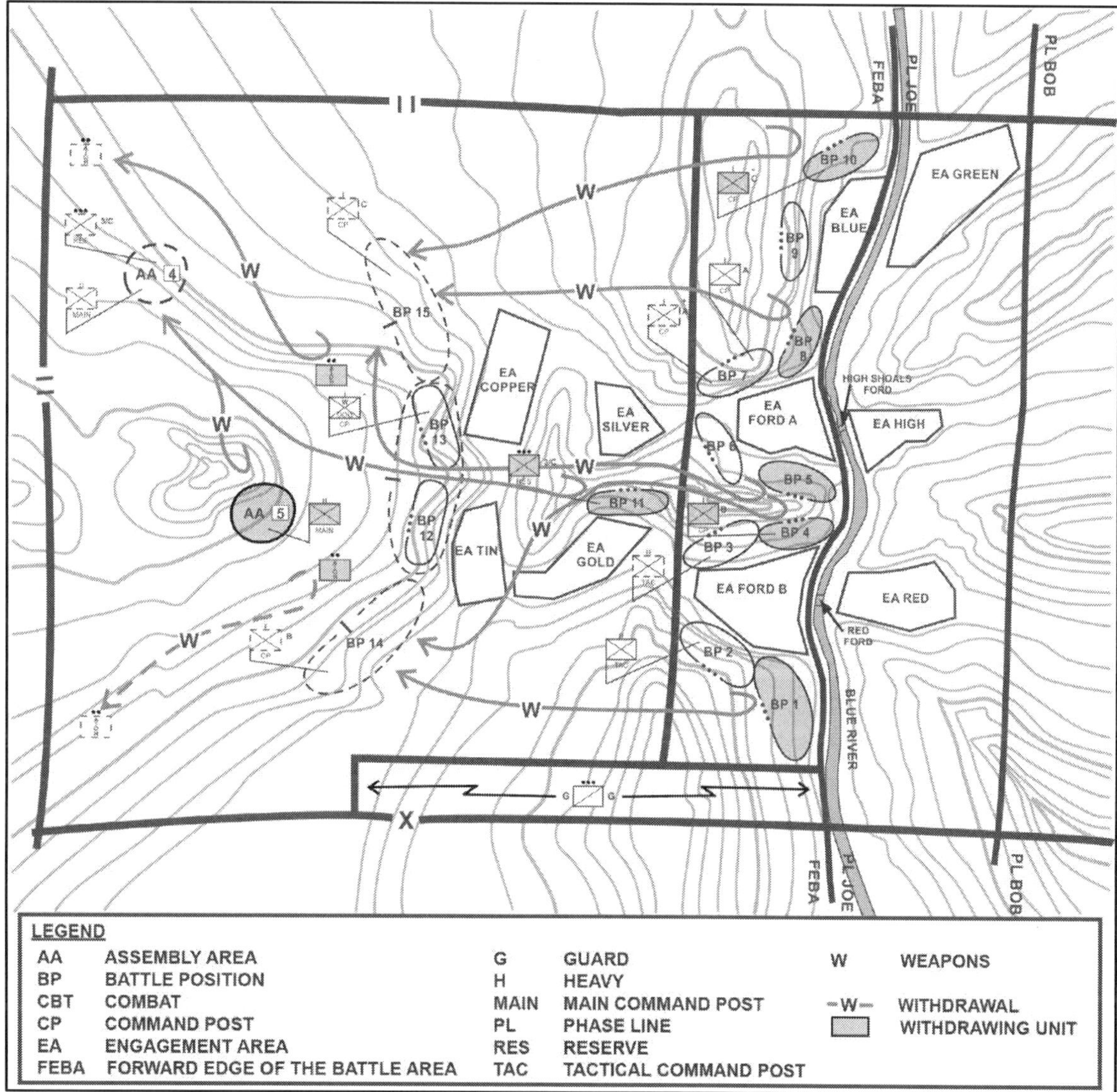

Figure 3-39. Detachment left in contact (multiple subelements left in place)

> ***Note.*** When subordinate units of the battalion are widely dispersed or the battalion's withdrawal area of operation is in an area with multiple corridors, the commander may have subordinate units control separate detachments left in contact. Each forward subordinate maneuver element (generally the Infantry rifle company) of the battalion establishes and controls its detachment, allowing for effective dispersion of forces while maintaining mission command.

3-394. The second method involves one forward subordinate maneuver element (generally a subordinate Infantry rifle company) of the battalion staying behind as the detachment left in contact. For example, a battalion with three or two maneuver companies positioned forward leaves one of the forward positioned companies as the detachment left in contact (figure 3-40, page 3-118). The detachment left in contact normally repositions its forces (expanding its security responsibilities) to cover the width of the battalion's area of operation.

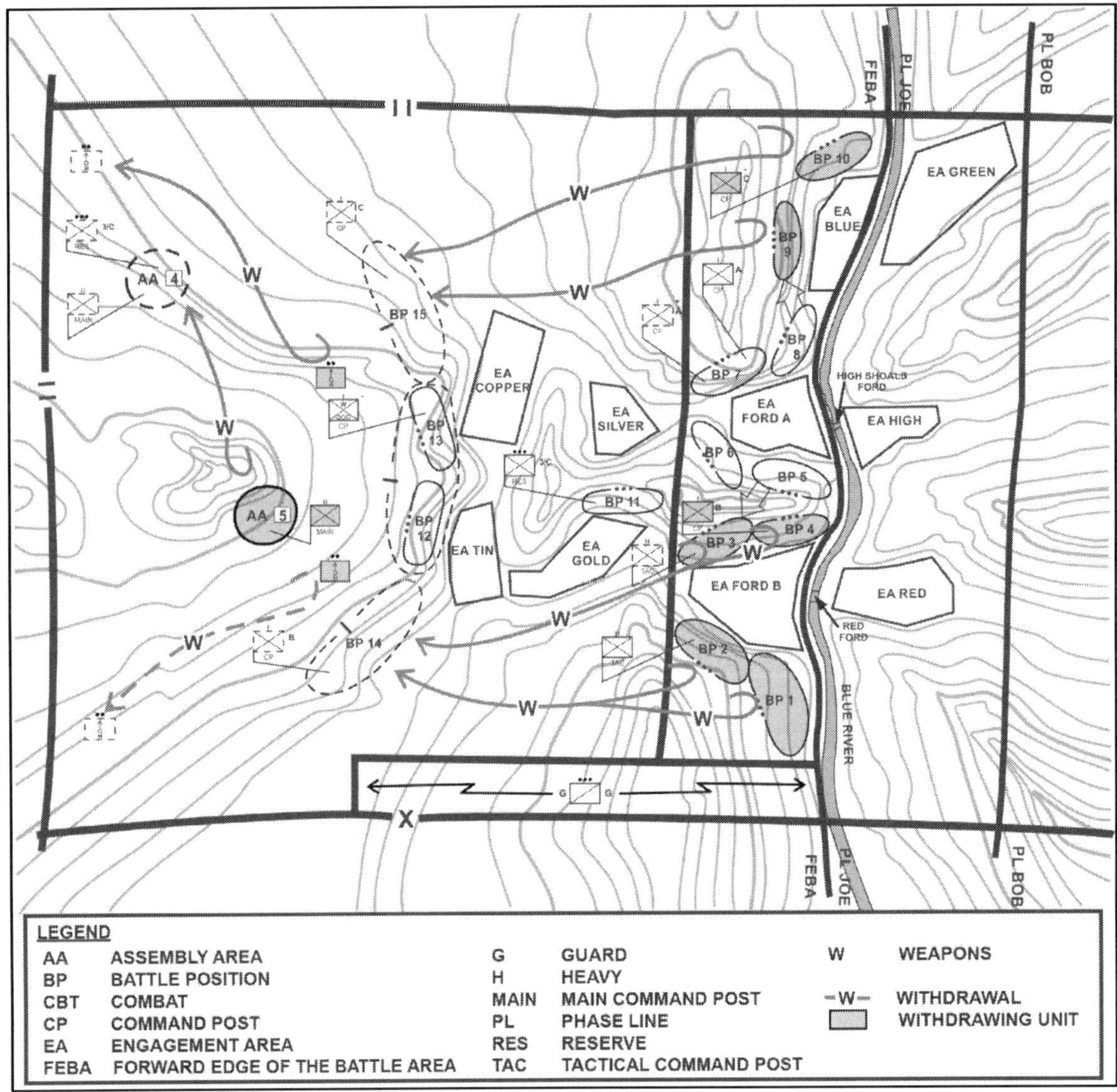

Figure 3-40. Detachment left in contact (one task-organized unit left in place)

3-395. An additional security force behind the existing main defensive positions of the detachment left in contact can be established to assist in the withdrawal process. The security force can be created from withdrawing units or from an assisting unit. The detachment left in contact can delay to the security force to its rear and join it, or delay back, conduct battle handover, and then conduct a rearward passage of lines. In either case, the additional security force becomes the rear guard.

Preparation

3-396. The battalion prepares for a withdrawal in the same manner as a delay. Preparation activities ensure subordinate commanders and units, the battalion staff, and Soldiers have a clear understanding of the withdrawal plan and the current enemy situation. To the extent possible, subordinate commanders and leaders conduct inspections and rehearse key portions of the plan to ensure maneuver units and Soldiers understand their portion of the plan or role and that supporting elements and equipment are positioned and ready to execute the withdrawal.

3-397. When preparing for an assisted withdrawal, the commander ensures adequate coordination for battle handover and passage of lines. The focus of the rehearsal for the withdrawal is on actions to maintain security, disengagement from the enemy (when under enemy pressure), and the movement of forces. When possible, key leaders or liaisons from the assisting force attend rehearsals. During rehearsals, control measures are confirmed to include fire support coordination measures. Leaders rehearse the plan against the full range of possible enemy actions. The commander rehearses contingencies for reverting to a delay, commitment of the reserve, and enemy interdiction of movement routes.

3-398. In an unassisted withdrawal, the battalion establishes its own security force and reserve and coordinates those actions with the battalion's main body. The battalion reconnoiters and secures routes to the rear and the support areas it will use during movement to the rear. In both unassisted and assisted withdrawals, the battalion rehearses the plan to disengage from the enemy. Because the force is most vulnerable if the enemy attacks, the commander always plans for a withdrawal under pressure, then develops contingencies for a withdrawal without pressure.

3-399. When support positions are located along the movement route, they are normally secured and concealed. In addition to simplifying support requirements during movement to the rear these support areas reduce the enemy's ability to interfere with logistics operations. They also allow sustainment units to withdraw earlier (prior to execution) than they otherwise could. When advising the commander, sustainment planners carefully considers whether to place supplies in caches along the route(s) understanding that once cached, supplies are difficult to recover if the operation does not go as planned. During preparation, the battalion evacuates or destroys all supplies (other than medical supplies) that the unit is unable to evacuate to prevent capture. The commander establishes destruction criteria, which is time- or event-driven, for each class of supply.

Note. Before withdrawing to the rear, the main body may dispatch quartering parties (paragraphs 2-51 and 2-80, pages 2-22 through 2-29 details the responsibilities of a quartering party) to help in occupying the new position.

Execution

3-400. As the Infantry battalion executes the withdrawal, security forces counter the enemy's try to disrupt the withdrawal or pursue the battalion. The battalion reserve remains well forward to assist the security force and other units by employing supporting direct and indirect fires and counterattacks. If the security force and the reserve cannot prevent the enemy from closing on the main body, the commander commits some or all of the main body to prevent the enemy from interfering further with the withdrawal. The main body delays, attacks, or defends as required by the situation. In this event, the withdrawal resumes at the earliest possible time. If the enemy blocks movement to the rear, the battalion must adjust its order of withdrawal march to ensure sustainment and supporting elements are not the primary fighting force to eliminate the threat. Friendly forces shift to alternate routes and bypass the interdicted area. Alternatively, they may attack through the enemy.

Gain and Maintain Enemy Contact

3-401. Typically, when under enemy pressure, the security force maintains contact with the enemy until ordered to disengage or until another force takes over the task. When performing the role of a detachment left in contact, the security fore simulates the continued presence of the main body, which normally requires the additional allocation of combat multipliers beyond that normally allocated to a force of its size. The security force, or when established a detachment left in contact, provides a way to break contact from the enemy sequentially. To conceal the security force's withdrawal, the movement is generally conducted during times of limited visibility or under obscuration to screen friendly movement and to reduce both the accuracy of enemy direct-fire systems and the enemy's ability to observe friendly movement. During withdrawal, the security force uses alternate and successive positions until the entire force breaks contact with the enemy.

Disrupt and Fix the Enemy

3-402. With the most probable threat to a withdrawing force being a pursuing enemy, the commander organizes the majority of available combat power to the security force as a rear guard or a detachment left in contact. When an enemy security zone exists between friendly and enemy forces, the existing security force can transition on order to a rear guard mission. When the withdrawing force is in close contact with the enemy, this security zone does not normally exist. Withdrawal under these conditions require that security forces, performing a rear guard mission, adopt different techniques. A detachment left in contact provides a way to sequentially break contact with the enemy.

Note. When conducting the withdrawal without enemy pressure, the security force acts as a rear guard.

3-403. When the enemy can infiltrate or insert forces ahead of the withdrawing main body force, the commander may establish an advance guard to clear the route or area of operation as the main body withdraws. The commander may designate a company or the scout platoon and/or sniper squad reinforced with Infantry and mortars as the advance guard. The commander task-organizes the advance guard, in addition to rear guard security forces or the detachment left in contact with engineers when available, with mobility assets going to the advance guard and countermobility assets, and to a lesser extent mobility assets going to the rear guard or the detachment left in contact.

Maneuver

3-404. With security forces positioned forward, the main body moves as rapidly as possible rearward on multiple routes to reconnoitered intermediate or final positions. Usually support assets and sustainment units, along with their convoy escorts, move first and precede combat units in the movement to the rear. After the main body withdraws a safe distance, the commander orders the security force to begin its rearward movement. When not pursued by the enemy, the security force may move in a march column. Once the security force begins moving, it assumes the duties of a rear guard. Security elements balance security and deception with speed as it disengages. Security forces maintain tactical movement and security techniques until they break contact and are clear of the enemy; it then withdraws as rapidly as possible. The main body moves rapidly on multiple routes to designated positions and may occupy a series of intermediate positions before completing the withdrawal. Despite confusion and enemy pressure, subordinate units follow specified routes and movement times.

Follow-Through

3-405. Once the Infantry battalion successfully disengages from the enemy, the command has two options. The battalion can rejoin the overall defense under more favorable conditions or transition into a retirement and continue to move away from the enemy and towards its next mission.

RETIREMENT

3-406. A *retirement* is a form of retrograde in which a force out of contact moves away from the enemy (ADRP 3-90). Retirements are conducted to reposition forces for future operations or to accommodate the current concept of operation. The Infantry battalion normally conducts retirement as a tactical road march (see paragraphs 2-39 through 2-79, page 2-19 through 2-29) where security and speed are the most important considerations. When moving to an assembly area the retiring force's ability to defend from the assembly area and protect itself during movement are major factors in positioning the assembly area(s) and identifying the retirement route(s). Though interference from enemy ground forces is not anticipated, mobile enemy forces, unconventional forces, air strikes, air assault operations, or long-range fires may attempt to interdict the retiring force. Typically, within this type retrograde another unit's security force covers the movement of the retiring force.

3-407. The Infantry battalion's organization of forces for a retirement depend on how the IBCT has structured its forces unless the battalion operates independently. The IBCT normally designates security elements and a main body in a retirement, though operations extended across large areas may preclude the use of IBCT-controlled security elements. In this case, the IBCT may direct the battalion to organize its own security elements for the movement; the same as if the battalion was operating independently. The formation employed during the battalion retirement depends on the number of available routes and the potential for enemy interference. Limited road nets or a flank threat may require echelonment (march units and serials) of the movement in terms of time and ground locations.

3-408. A battalion size retirement march column(s) normally requires an advance guard (may include a quartering party to quarter the new assembly area) and flank and rear security. The advance guard is made up of reconnaissance and security forces and when available engineers focused on mobility. The commander assigns a flank security element to prevent potential enemy interference with the retiring force's extended columns or the commander may designate flank security responsibilities to subordinate march units. The size and composition of these security forces is dependent on the strength and imminence of the enemy threat.

3-409. The terrain and the enemy threat dictate whether the retiring force commander establishes a single rear security force, usually a rear guard, or whether each column forms a separate rear security force. Rear guard security forces, and when available engineers focused on countermobility, protect the rearward moving column(s) from surprise, harassment, and attack by any pursuing enemy force. Rear security element(s) generally remain in march columns, unless there is a potential for enemy interference. If the enemy establishes contact, the rear security element(s) conducts a delay.

3-410. In the main body, maneuver support and sustainment units precede the movement of combat forces. When necessary, elements of the main body can reinforce the rear guard or any other security element. Because fire support elements moving within the main body can respond most rapidly, they are usually the first elements tasked for this mission.

3-411. When the Infantry battalion receives the mission to retire form an area the commander plans and prepares for actions on contact and for the protection the force prior to and during the movement. In the initial phase of the movement, the force retires in multiple small columns. As the distance from the enemy increases, smaller columns can consolidate into larger ones for ease of movement control. Road nets and the potential for hostile interference influence how and when this consolidation occurs.

3-412. Control measures used in a retirement are the same as those in a delay and a withdrawal. As in a withdrawal, thorough planning and strict adherence to routes and movement times facilitate an orderly retirement. Typically, the commander controls movement using movement times, routes, and checkpoints.

3-413. When a battalion withdrawal from action precedes a retirement the actual retirement begins after the unit breaks contact. At the designated time, retiring forces of the battalion execute a withdrawal from action and form into march formation. Prior to, or after the battalion breaking contact with the enemy, the battalion moves nonessential sustainment units and supplies, when located forward, to the rear. The battalion may move into an assembly area, if necessary after the battalion withdrawal breaks contact with the enemy, before moving into a march formation to reestablish mission command or resupply. Once subordinate forces form their march formations, the battalion is prepared to initiate the retirement.

Note. While a force withdrawing without enemy pressure can also use march columns, the difference between the two situations is the probability of enemy interference.

This page intentionally left blank.

Chapter 4
Stability

Operations focused on stability ultimately aim to establish conditions the local populace regards as legitimate, acceptable, and predictable. Stabilization is the process to identify and mitigate underlying sources of instability to establish the conditions for long-term stability. Stability tasks focus on identifying and targeting the root causes of instability and building the capacity of local institutions.

Army forces accomplish stability missions and perform tasks across the range of military operations in coordination with other instruments of national power. Stability missions and tasks are part of broader efforts to establish and maintain the conditions for stability in an unstable area before or during hostilities, or to reestablish enduring peace and stability after open hostilities cease.

The first two sections of this chapter discuss the doctrinal foundation and organization of forces for operations focused on the stability element of decisive action. Sections III and IV introduce scenarios, as discussion vehicles, illustrating the methods and ways an Infantry battalion conducts operations in support of the stability tasks. Scenarios focus on the challenges confronting the commander and staff and subordinate commanders and leaders in accomplishing stability-focused missions or tasks. These scenarios are not intended to be prescriptive of how the Infantry battalion performs any particular operation.

SECTION I – FOUNDATION FOR OPERATIONS FOCUSED ON STABILITY

4-1. Ensuring a state's long-term stability depends on applying combat power to those tasks that are, in fact, essential. Essential stability tasks lay the foundation for success in sustaining the burdens of governance, rule of law, and economic development that represent the future viability of a state. Establishing this foundation depends on applying combat power to the essential stability tasks identified during the initial assessment of the situation and the framing of the basic problem. This section provides the foundation for the conduct of military operations focused on stability.

STABILITY PRINCIPLES

4-2. Based on the four principles (conflict transformation, unity of effort, legitimacy and host-nation ownership, and building partner capacity) that lay the foundation for long-term stability, Army units conduct operations focused on stabilizing the environment and transforming conditions of the environment and the state toward normalization. Units at different echelons balance these principles to mitigate fragile state characteristics prevalent at the national, regional, and local levels. Long-term stabilization efforts within an operational environment transform the drivers of conflict while maintaining *unity of effort, which is* coordination and cooperation toward common objectives, even if the participants are not necessarily part of the same command or organization, which is the product of successful unified action (JP 1), among diverse actors. Fundamental to long-term stability and critical to the host-nation's legitimacy is its involvement and ownership to build trust and confidence among the states populace. Building partner capacity addresses potentially the most important effort to support and enable partners so they can perform their roles effectively. (Refer to ADRP 3-07 and FM 3-96 for additional information.)

UNDERSTANDING THE OPERATIONAL ENVIRONMENT

4-3. Operations focused on stability require the commander to demonstrate cultural understanding and a clear appreciation of the myriad stability tasks to determine which are fundamentally essential to mission success. The commander and staff must understand the potential for conflicts among individuals and agencies with differing cultural backgrounds. For example, interagency conflict may arise because of perceived differences in organizational goals or attitudes about the appropriateness of military involvement. Anticipating counterproductive confrontations and taking steps to resolve individual and organizational conflicts constructively is paramount to successful collaboration. Additional preemptive strategies for managing conflict include ensuring all are stakeholders identified and included in making decisions.

Interorganizational Cooperation

4-4. During stability, the commander ensures the key players support interagency partnership, established ground rules, and collaborative interagency strategies to accomplish the mission. The commander must understand the importance of ensuring that interagency partners explore various alternatives, and that all partners participate. The commander adopts consensus-building leadership behavior, to include open discourse, friendly debate, and discussion with opinion sharing and feedback from participants.

4-5. Understanding an operational environment includes understanding organizational goals or attitudes for all stability partners. Within operations focused on stability, the commander must act cooperatively rather than competitively, building relationships to achieve coordinated goals. Organizations can increase collaboration by providing their representatives with a clear understanding of their organization's functions and authority within the larger civil-military partnership. Regular interaction with interagency partners also contributes to an increased understanding of roles and mission requirements. Success in operations focused on stability requires an awareness of trends that influence views of the actors and an understanding of factors that shape or constrain options and capabilities for partner organizations.

Military and Civilian Organizational Cultures

4-6. Military and civilian organizational cultures differ in significant ways comprising factors such as shared values, norms, expectations, and practices. An organizational culture influences how individuals approach work and what they regard as mission accomplishment. When team members with different organizational cultures interact with one another, differences become evident and can create tension in the group. The commander and staff can minimize difficulties by educating themselves on these organizational differences—in mission objectives, size, and resource capabilities, and neutrality among others—and challenges in information sharing and improving their understanding of and attitudes toward partners. The commander and staff develop the information needed to understand partner organizations, their component teams, and their place in the stability activities and goals to achieve the desired end state conditions. This understanding forms the backdrop for assessing the effect of military actions, plans, and decisions on partner organizations. A poor understanding of the partners must be avoided because it can hamper trust and impair integration of military team members in interagency decision-making.

STABILITY IN OPERATIONS

4-7. Operations focused on stability, range across all military operations and offer perhaps the most diverse set of circumstances the Infantry battalion faces. The objective of operations focused on stability is to create conditions that the local populace regards as acceptable in terms of violence; the functioning of governmental, economic, and societal institutions; and that adhere to local laws, rules, and norms of behavior. During decisive action, the battalion seeks to create and maintain the conditions necessary to seize, retain, and exploit the initiative; and to *consolidate gains,* which is the activities to make permanent any temporary operational success and set the conditions for a sustainable stable environment allowing for a transition of control to legitimate civil authorities (ADRP 3-0)—through partnership with associated diverse enabling organizations. The battalion, in coordination with these partner organizations, provide the means to secure and stabilize the operational environment and to conduct operations to establish and maintain stability or to reestablish stability. The commander keeps in mind how these operations transition in a comprehensive approach to avoid considering them in isolation. (Refer to FM 3-07 and FM 3-96 for additional information.)

Secure and Stabilize the Operational Environment

4-8. During operations focused on stability, the Infantry battalion provides the means to secure and stabilize its operational environment enough so the host nation can begin to resolve the root causes of conflict and state failure. Battalion operations focused on stability establish conditions that support the transition to legitimate host-nation governance, a functioning civil society, and a viable market economy. These operations establish the foundation for a safe and secure environment that facilitates reconciliation among local or regional adversaries. The battalion commander shapes the operational environment through action, influences the population and its leaders, and consolidates gains to seize, retain, and exploit the initiative and set the conditions for a stable environment. The commander and subordinate leaders identify and mitigate sources of instability, understand and nest operations within political objectives, and achieve unity of effort and a shared vision across the operational environment. (Refer to FM 3-07 for additional information.)

Identify and Mitigate Sources of Instabiltiy

4-9. To identify and mitigate sources of instability, the battalion in coordination with higher headquarters, conducts information collection (reconnaissance, security operations, surveillance, and intelligence operations) to gain a detailed understanding of the sources of instability and the capabilities and intentions of key actors. Sources of instability are actors, actions, or conditions that exceed the legitimate authority's capacity to exercise effective governance, maintain civil control, and ensure economic development. Enemy forces leverage sources of instability to create conflict, exacerbate existing conditions, or threaten to collapse failing or recovering states.

District Stability Framework

4-10. The United States Agency for International Development (known as USAID) developed the district stability framework to increase the effectiveness of stability missions. The district stability framework was designed to guide and support stabilization efforts by helping civilians and military organizations identify the causes of instability, develop activities to diminish or mitigate them, and evaluate the effectiveness of the activities in fostering stability at the tactical or operational level. The district stability framework supports unity of effort by providing partners with a common framework to—

- Understand an operational environment from a stability-focused perspective.
- Maintain focus on the local population and its perceptions.
- Identify the sources of instability in a specific local area.
- Design activities that specifically address the identified sources of instability.
- Monitor and evaluate activity outputs and impacts, as well as changes in overall stability.

4-11. The district stability framework helps overcome many of the challenges to successful operations focused on stability. The framework helps to—

- Keep military formations focused on the center of gravity for operations focused on stability—the population and its perceptions.
- Provide a common operational picture for all interagency teams in an area of operations. By focusing on sources of instability, partner organizations can focus their varied resources and expertise on shared priorities.
- Prioritize activities based on their importance to the local populace and their relevance to the overarching mission of stabilizing the area.
- Enhance continuity between military formations. Units can easily pass district stability framework data along from one unit to the next, establishing a clear baseline that identifies sources of instability and the steps taken to mitigate them.
- Empower tactical-level formations by giving them hard data useful for decision making at their level and for influencing decisions at higher levels.
- Identify MOP and MOE for unit activities rather than simply tracking MOP.
- Track indicators of overall stability by assessing whether an area is becoming more stable.
- Identify issues that matter most to the population; the district stability framework helps identify information themes that resonate with the population.

District Stability Framework Process

4-12. The district stability framework process has four steps. Ideally, all interagency partners in the area participate in the process, organized through the creation of an interagency stability working group. This work group is generally establish at the IBCT level. The four basic steps are situational awareness, analysis, design, and monitoring and evaluation.

Situation Awareness

4-13. The district stability framework process uses four lenses to achieve a population-centric, stability oriented, situational awareness of an area of operations. The stability working group examines the area of operations from four perspectives: an operational environment; cultural aspects; stability and instability dynamics; and local perceptions. This examination helps military and civilian leaders achieve a situational awareness of stability conditions and underlying factors.

Analysis

4-14. After gaining this initial situational awareness, the district stability framework process provides tools to analyze and identify potential sources of instability, their causes, the desired objectives, and the indicators that measure progress in addressing each source of instability. The analysis consists of four tasks: identify potential sources of instability, vet each source against instability criteria, determine if the source meets two of the three instability criteria, and prioritize the sources of instability. The instability criteria are—

- Decreased support for the government or legitimate governance institution.
- Increased support for adversaries or enemies.
- Undermining of the normal functioning of society.

4-15. After identifying and prioritizing the sources of instability, a tactical stability matrix is filled out for each source of instability. Examples of sources of instability include, but are not limited to—

- Insurgents forming a shadow governmental structure.
- Religious, ethnic, economic, and political friction between different groups within the local population.
- Natural disasters or resource scarcity.
- Super-empowered individuals disrupting legitimate governance.
- Severely degrading infrastructure or environment.
- Immature, undeveloped, or atrophied government, social, or economic systems.
- Ineffective or corrupt host-nation security forces.

Design

4-16. In the design step, working group members (includes interagency partners) design, prioritize, and synchronize stabilization activities. The stability working group develops activities to diminish the sources of instability identified during the analysis step. The process begins by brainstorming potential stabilization activities and continues by filtering and refining the proposed activities against a set of stability criteria, design principles, and resource availability. The design step is integrated with and similar to the Army design methodology's activities, develop an operational approach and develop the plan, described in ADRP 5-0.

Monitoring and Evaluation

4-17. The final step in the district stability framework process takes place during and after the implementation of stability activities. Monitoring and evaluation are conducted in three ways:

- Measures of performance—track implementation of an activity.
- Measures of effectiveness—measure the effect that an activity achieved.
- Overall stability—assess the overall stabilizing effect of all the activities conducted over a longer period, as well as the influence of external factors.

4-18. As work group members monitor and evaluate, they identify lessons about what worked, what did not work, and what partners can do to improve their stability efforts as they repeat the district stability framework process in the future. Effective monitoring and evaluating supports the commander's decision making throughout the operations process, focusing on the perceptions of the population and to inform a common operational picture for the IBCT, subordinate battalions and their interagency partners. Monitoring and evaluating informs and influences audiences by identifying themes that resonate with the population.

Nest Operations within Political Objectives

4-19. Operations focused on stability nest within political objectives. Proper nesting requires an ability to achieve unity of effort through comprehensive engagement and a thorough depth of cultural astuteness. A detailed analysis, based on careful consideration of operational and mission variables, helps the commander to understand and visualize the civil component and shapes future stabilization activities to secure and stabilize the operational environment within the area of operation. To ensure the most comprehensive analysis, the staff's analysis of operational and mission variables includes all relevant information retained by each military and nonmilitary actor. Relevant information includes the results of past assessments and related analyses to understanding threats to civilians and determining ways to shape the environment to enhance the legitimacy and host-nation ownership and building partner capacity.

Shared Vision Across a Stable Environment

4-20. Shared vision among the participating military forces, civilian and governmental agencies (both U.S. and international), and host-nation organizations, institutions, and forces reflects a comprehensive approach and the norms and collective experience of a diverse group of actors. In operations emphasizing stability tasks; time, space, purpose, and resources affect the environment and end state conditions and objectives. Military and nonmilitary actors integrate military and nonmilitary means to achieve shared objectives while understanding that many of the nonmilitary considerations are most important to ensure long-term stabilization of the operational environment. The commander and staff engage and establish links to all partners to facilitate and integrate plans and operations with those of other partners through comprehensive engagement. Examples of comprehensive engagement includes engaging with key leaders and the population, conducting multinational operations with international and host-nation police and military partners, building the capacity of (or enabling) other partners, developing effective civil-military operations centers, and enabling humanitarian assistance.

Establish, Maintain, or Reestablish Stability

4-21. Operations to establish, maintain, or reestablish stability involve numerous military and civilian organizations and are often protracted. In addition to the fundamentals of the operations process described in ADP 5-0 and ADRP 5-0, when planning these operations the commander and staff recognize complexity; balance resources, capabilities, and activities; recognize planning horizons; and avoid planning pitfalls.

Recognize Complexity

4-22. Given the inherent complex and uncertain nature of operations dominated by stability and that, the multifaceted drivers of instability are difficult to identify, the commander and staff use the Army design methodology (see ADRP 5-0) to help understand the root cause of instability and approaches to solve problems. The Army design methodology is an iterative process of understanding and problem framing that uses elements of operational art (see ADRP 3-0) to conceive and construct an *Operational approach*—a description of the broad actions the force must take to transform current conditions into those desired at end state (JP 5-0)—to solve identified problems. The Army design methodology results in an improved understanding of an operational environment. Based on improved understanding, the commander issues planning guidance, to include an operational approach, to guide more detailed planning using the military decision-making process. The understanding developed through Army design methodology continues throughout the operations process in the form of continuous assessment. Assessment, for example updated running estimates, helps the commander measure the overall effectiveness of employing forces and capabilities.

Operational Approach

4-23. An operational approach provides a unifying purpose and focus to all operations and provides the framework that relates tactical tasks to the desired end state. The operational approach conceptualizes the commander's vision for establishing the conditions that define the desired end state. When developing an operational approach, the commander considers how to employ a combination of defeat and stability mechanisms. Defeat mechanisms are dominated by offensive and defensive tasks, while stability mechanisms are dominant in stability tasks that establish and maintain security and facilitate consolidating gains in an area of operation.

4-24. A *defeat mechanism* is a method through which friendly forces accomplish their mission against enemy opposition (ADRP 3-0). The battalion uses a combination of four defeat mechanisms: destroy, dislocate, disintegrate, and isolate. Applying focused combinations produces complementary and reinforcing effects not attainable with a single mechanism. Used individually, a defeat mechanism achieves results proportional to the effort expended. Used in combination, the effects are likely to be both synergistic and lasting. Defeat mechanisms are not tactical missions; rather, they describe broad tactical effects. The commander translates these effects into tactical tasks.

4-25. A *stability mechanism* is the primary method through which friendly forces affect civilians to attain conditions that support establishing a lasting, stable peace (ADRP 3-0). As with defeat mechanisms, combinations of stability mechanisms produce complementary and reinforcing effects that accomplish the mission more effectively and efficiently than single mechanisms do alone. The four stability mechanisms are compel, control, influence, and support. Compel means to use, or threaten to use, lethal force to establish control and dominance, effect behavioral change, or enforce compliance with mandates, agreements, or civil authority. Control involves imposing civil order. Influence means to alter the opinions, attitudes, and ultimately behavior of foreign friendly, neutral, adversary, and enemy populations through information operations, presence, and actions. Support is to establish, reinforce, or set the conditions necessary for the instruments of national power to function effectively.

Primary Stability Tasks

4-26. *Stability tasks* are tasks conducted as part of operations outside the United States in coordination with other instruments of national power to maintain or reestablish a safe and secure environment, and provide essential governmental services, emergency infrastructure reconstruction, and humanitarian relief (ADP 3-07). Stability tasks are part of every operation. However, the proportion of stability tasks, in relation to offensive and defensive tasks, may change. Whether an operation is a peace operation preventing conflict or a large-scale combat operation, forces will always integrate offensive, defensive, and stability tasks. For example, in a peace operation, the Infantry battalion may still perform offensive tasks such as a raid (see appendix D), cordon and search, and search and attack (see chapter 2) during the conduct of counterinsurgency operations. Conversely, in large-scale combat operations, the battalion performs stability tasks to control captured areas or to provide emergency essential services. The Army's six primary stability tasks are:

- Establish civil security.
- Establish civil control.
- Restore essential services.
- Support governance.
- Support economic and infrastructure development.
- Conduct security cooperation. (See paragraph 4-31, and FM 3-22 for additional information.)

4-27. The combination of stability tasks conducted during operations depends on the situation. In some operations, the host nation can meet most or all of the population's requirements. In those cases, Army forces work with and through host-nation authorities. Commanders use civil affairs operations to mitigate how the military presence affects the population and vice versa. Conversely, Army forces operating in a failed state may need to support the well-being of the local population. That situation requires Army forces to work with civilian organizations to restore basic capabilities. Civil affairs operations prove essential in establishing the trust between Army forces and civilian organizations required for effective, working relationships.

Note. Section III of this chapter illustrates two scenarios where the Infantry battalion conducts area security missions during transition after open hostilities.

4-28. Six Army primary stability tasks (figure 4-1) correspond directly to the six stability sectors, used by the Department of State, Office of the Coordinator for Reconstruction and Stabilization, and directly support the broader efforts within the stability sectors. Together these six primary stability tasks and the Department of State stability sectors provide a mechanism for interagency tactical integration, linking the execution of discreet tasks among the instruments of national power required to establish end state conditions that define success. Tasks performed in one sector inevitably create related effects in another sector; planned and performed appropriately, carefully sequenced activities complement and reinforce these effects. The subordinate tasks performed by the battalion under the primary stability tasks directly support broader efforts within stability executed as part of unified action. *Unified action* is the synchronization, coordination, and/or integration of the activities of governmental and nongovernmental entities with military operations to achieve unity of effort (JP 1). Refer to ADRP 3-07 and FM 3-07 for additional information.

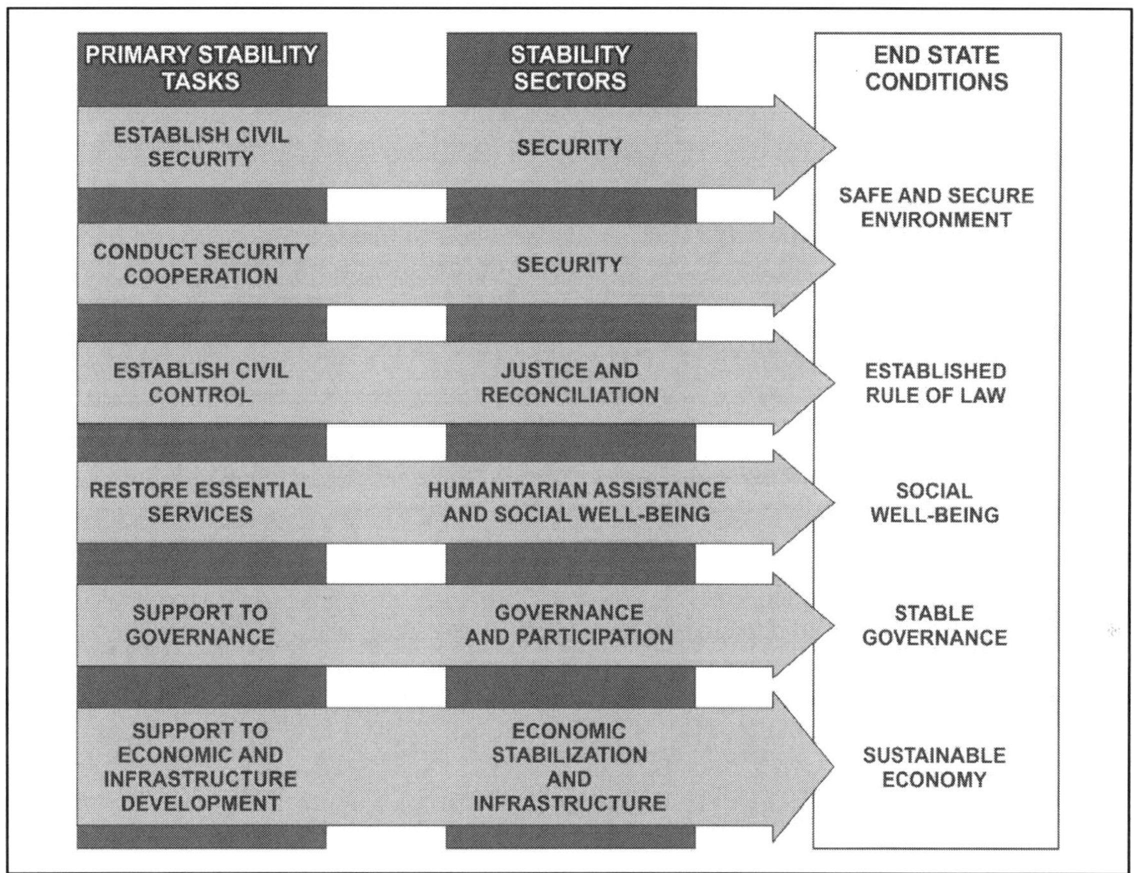

Figure 4-1. Mechanism for interagency tactical integration

4-29. Over time, to ensure safety and security is sustained, unified action partners perform numerous tasks across all stability sectors. As part of a joint team working with unified action partners, achieving a specific objective or setting certain conditions often requires the battalion to perform a number of related tasks among the six primary stability tasks. An example of this is the effort required to provide a safe, secure environment for the local populace. Rather than the outcome of a single task focused solely on the local populace, safety and security are broad effects. The battalion can help achieve safety and security by performing a number of related tasks to assist in ending hostilities, isolating belligerents and criminal elements, demobilizing armed groups, eliminating explosives and other hazards, and providing public order and safety. (Refer to FM 3-96 for additional information.)

Related Activities and Missions

4-30. Operations characterized by stability tasks often combine with certain activities and missions common to Army operations. These activities and missions cut across all stability tasks regardless of the focus and require increased emphasis and attention by the commander. Some activities—such as security sector reform and disarmament, demobilization, and reintegration—cut across the primary stability tasks and may be the centerpiece of an operation. For example, *security sector reform* is a comprehensive set of programs and activities undertaken by a host nation to improve the way it provides safety, security, and justice. Also called SSR (JP 3-07). Within the security sector reform, battalion transformation tasks (see paragraphs 4-33) may focus on these programs and activities undertaken to improve the way a host nation provides safety, security, and justice to help enable conditions for enduring stability and peace.

4-31. The primary stability task of establishing security cooperation (SC) (see FM 3-22) may include the Infantry battalion, depending on the missions assigned, conducting security force assistance (see section IV of this chapter) as a subset of SC. Security force assistance offers a means of support for SC activities in support of building capacity of a foreign security force. As soon as the foreign security force can perform this task, the battalion transitions this task within civil security to the host nation. Related activities and missions (see FM 3-07 for additional information) include—

- Information operations.
- Protection of civilians.
- Mass atrocity response operations.
- Security sector reform.
- Disarmament, demobilization, and reintegration.
- Destruction, monitoring, and redirection of weapons of mass destruction (WMD) and mitigation of CBRN hazards.
- Security cooperation.
- Peace operations.
- Foreign humanitarian assistance.
- Counterinsurgency.
- Foreign internal defense.

Note. *Security cooperation* involves all Department of Defense interactions with foreign defense establishments to build defense relationships that promote specific U.S. security interests, develop allied and friendly military capabilities for self-defense and multinational operations, and provide U.S. forces with peacetime and contingency access to a host nation. Also called SC (JP 3-22).

Stability Framework

4-32. The stability framework refers to the range of failed, failing, and recovering states. The distinction among them is rarely clear, as fragile states do not travel a predictable path to failure or recovery. This framework encompasses the stability tasks performed by military and civilian actors across the range of military operations. .When applied, the stability framework helps the commander identify the types and ranges of tasks performed in the phases and identify lead responsibilities and priorities. Stability tasks occur in three phases: initial response, transformation, and fostering sustainability phases. (Refer to FM 3-96 for additional information.)

Initial Response Phase

4-33. The initial response phase generally reflects tasks executed to stabilize an operational environment in a crisis state. During this phase, the battalion conducts these tasks during or directly after a conflict or disaster where the security situation hinders the introduction of civilian personnel. The battalion conducts operations that safeguard the local population and prevent factions or actors contributing to sources of instability. Identifying actors and their intentions during this phase through information collection allows the battalion to seize the initiative during this phase.

Transformation Phase

4-34. The transformation phase represents the broad range of post-conflict reconstruction, stabilization, and capacity building tasks where host-nation security forces and, potentially, intergovernmental organization peacekeepers begin to contribute. During the transformation phase, the battalion can conduct security force assistance (see section IV of this chapter) as a subset of SC to develop the ability of the host nation to defend against internal and external threats, contribute to multinational operations, and assist other partner nations to provide for their security. SC activities of these types transition in a relatively secure environments, free from most wide-scale violence, often to support broader civilian efforts. Throughout transformation, the battalion continues in partnership with unified action partners according to the legitimate government binding agreements.

Fostering Sustainability Phase

4-35. Fostering sustainability encompasses long-term efforts that capitalize on capacity-building and reconstruction activities to establish conditions that enable sustainable development. The battalion transitions to a steady state posture focused on advisory duties and continued SC to enable the host nation to sustain development. The battalion usually performs fostering sustainability phase tasks only when the security environment is stable enough to support tasks.

> ***Note.*** SC supports the implementation of national and theater strategies, and is a key element of global and theater shaping activities supporting stabilization, building security relationships, building partner capacity, and providing access. SC activities typically include security assistance programs, security force assistance activities, joint combined exercise training, and other Service training opportunities with partner nations and other friendly foreign security forces. SC is a primary focus for shaping all geographic combatant commanders' theater campaign plans, and can be conducted across the range of military operations and during all phases of a joint operation or campaign. Service components posture forces to conduct SC activities and to execute theater campaigns and operations, as directed. These activities help shape the operational environment and produce the conditions necessary for a joint force commander, when directed, to seize the initiative, dominate, and establish stability within the operational area during a joint operation/campaign. Conducting sustained SC activities in an area of responsibility typically requires a combination of assigned and attached forces, composed of conventional forces and special operations forces. Those forces may include Department of Defense civilian personnel and contractors. (Refer to FM 3-22 for additional information.)

End State Conditions

4-36. To achieve conditions that ensure a stable and lasting peace, stability tasks in operations capitalize on coordination, cooperation, integration, and synchronization among the indigenous population and institutions, unified action partners, other civil entities, and the interagency. These complementary civil-military efforts aim to strengthen legitimate governance, restore or maintain rule of law, support economic and infrastructure development, and foster a sense of national unity. These complementary efforts also seek to reform institutions to achieve sustainable peace and security and create the conditions that enable the host-nation government to assume responsibility for civil administration.

Balance Resources, Capabilities, and Activities

4-37. When planning operations dominated by stability the commander, weights limited resources and capabilities according to priority of effort. When requirements for operations dominated by stability outpace available resources and capabilities necessary to reestablish conditions of peace and stability, planning involves focusing efforts toward accomplishing the mission while carefully rebalancing resources, capabilities, and activities across multiple lines of effort. While the commander typically focus resources on the decisive operation, the commander must also provide sufficient resources to capitalize on unforeseen opportunities and to provide impetus for other efforts. The numerous stability tasks involved in an operation require specific capabilities that are often just as limited in availability. An effective plan judiciously applies these capabilities where and when most needed. The commander synchronizes the activities in time and space to create the greatest effect, one that achieves broad success in one line of effort while reinforcing progress in the others. (Refer to ADRP 3-07 for additional information.)

Recognize Planning Horizons

4-38. Based on imperfect knowledge and involving assumptions about future operations dominated by stability that are fundamentally uncertain in nature affect the commander's planning horizon and ability to plan. A *planning horizon* is a point in time commanders use to focus the organization's planning efforts to shape future events (ADRP 5-0). Uncertainty increases with the length of the planning horizon and the rate of change in the environment. A fundamental tension in planning for operations emphasizing stability tasks is the tension between short-term needs and long-term objectives. Immediate security or humanitarian concerns can create a need for short-term solutions with negative impacts for longer-term objectives, such as in establishing stable governance or a sustainable economy. (Refer to ADRP 5-0 for additional information.)

Avoid Planning Pitfalls

4-39. The challenges for operations dominated by stability involving various systems, cultures, and personalities can create significant pitfalls to developing a coherent, integrated plan. Collaborative planning for stability in an operation, especially among the many diverse participants, presents unique challenges and opportunities.

4-40. The first pitfall consists of attempting to forecast and dictate events too far into the future based on the assumption that events will progress on a logical, linear path to the future. Plans often underestimate the scope of changes in direction that may occur, especially in operations that occur among populations, where predictability is elusive at best. Effective planning includes sufficient branches and sequels to account for the nonlinear nature of events.

4-41. The second pitfall consists of trying to plan in too much detail. While sound plans must include detail, planning in more detail than needed only consumes limited time and resources.

4-42. The third pitfall consists of using planning as a scripting process that tries to prescribe the course of events with precision. When planners fail to recognize the limits of foresight and control, the plan tends to become a coercive and overly regulatory mechanism that restricts initiative and flexibility.

4-43. The fourth pitfall is the danger of institutionalizing rigid planning methods that lead to inflexible or overly structured thinking. This tends to make planning rigidly process-focused and produce plans that overly emphasize detailed procedures.

> ***Note.*** Planning provides a disciplined framework for approaching and solving complex problems. Familiarity with the requisite processes and steps typically speeds the planning effort, and repetition only serves to imbue it with an inherent efficiency. The danger is in taking that discipline to the extreme. This especially proves dangerous in the collaborative environments typical of operations with a dominant stability component, where the mix of different planning cultures and processes can stymie progress. Stakeholders may want to follow a rigid, institutionalized planning methodology or, in some situations, not use any planning methodology whatsoever. In a collaborative environment, the commander streamlines the planning effort, providing economy of effort and coordination among team members working on the same problem.

Considerations for Transitions

4-44. In the context of stabilization activities, transitions occur across each of the six primary stability tasks to achieve the end state. All partners must understand transition, the transfer of responsibility, authority, power, and accountability incrementally on several levels and by numerous partners. Stabilization activities within the stability framework are designed within three transition phases: repair and (re)establish systems, normalize systems, and transfer and exits. The MOP or MOE indicating the appropriate time to complete tactical transitions must be, to the extent possible, clear and specific.

Transition Planning

4-45. A well-structured transition plan nests the battalion's short-term stabilization actions with long-term end states. The numerous layers of tasks conducted by the battalion, its higher headquarters and other partners, compound the challenges to the transition plan. Numerous lines of effort crossing different transition phases potentially add complexity to partner organizations within the area of operation. To ensure success, the battalion, its higher headquarters, and unified action partners clearly plan and express the steps for transition and their withdrawal so local actors have time to adapt to their responsibilities and structures. The phases of the stability framework (see paragraphs 4-29 to 4-32) nests with the three stability transition phases, with initial response tasks generally occurring in transition phase 1, transformation tasks in transition phase 2, and fostering sustainability tasks in transition phase 3 (table 4-1).

Table 4-1. Phases of the stability framework and stability transition phases

Phases of the stability framework:	*Initial response*	*Transformation*	*Fostering sustainability*
Stability transition phases:	1—Repair and (re)establish systems	2—Normalize systems	3—Transfer and exit

Phase 1—Repair and (Re)Establish Systems

4-46. Transition phase 1, based on the security situation, initiates the repair or replacement of systems that were once active or in control of the host nation. The commander initiates the transition phase with a comprehensive approach towards mentoring and advising host-nation military leaders to ensure the host nation becomes willing to and capable of assuming its legitimate responsibilities. The Army rules of allocation provide a civil affairs company to each BCT to support civil-military operations. A civil military affairs team is then normally attached to the battalion and is organized to assist in planning and execution of subsequent transition phases and coordinate horizontally and vertically within the civil component of the operational environment to ensure that military efforts support host-nation agencies and organizations. The commander and staff cooperate and coordinate with all partner agencies and organizations recognizing that military efforts are just one part of successfully building host-nation autonomy.

Phase 2—Normalize Systems

4-47. Transition phase 2, in this context, refers broadly to host-nation political, economic, social, and military structures that ensures—a safe and secure environment, the rule of law, social well-being, stable governance, and a sustainable economy. Phase 2 begins as the host-nation government becomes operational, and its management of civil security, civil control, essential services, governance, economic development, and infrastructure tasks becomes routine practice. This phase takes time. Host-nation personnel take the lead, while partners continue to mentor and coach. The host nation owns and operates the systems and processes. Mentors seek to understand the culture, norms, and customs of the host nation to help determine when to terminate education, training, mentoring, and coaching support.

Phase 3—Transfer and Exit

4-48. As systems normalize during phase 2 with the host nation, the battalion, its IBCT and unified action partners begin to remove themselves from the local bureaucracy within the area of operation. With clear transfer plans developed in transition phase 2, the process of transferring responsibilities may begin to occur. The commander expects to transfer different aspects of an operation at different times based on setting conditions. Additionally, the commander considers that delays in one sector may delay the transfer in another and plans accordingly to anticipate and account for potential delays.

Standard Measures

4-49. MOPs and MOEs are standard measures used to analyze progress when moving from one transition phase to another. Reports on what is being done to achieve success and whether or not what is being done is successful in each transition phase is essential to determining progress. Standard measures must be clear and achievable to provide a true picture of progress. The commander and staff integrate assessments with host-nation, interagency, and interorganizational partners in determining the criteria for MOPs and MOEs and determining what standard to use. Integrating MOEs and MOPs help ensure all partners work to the same purpose and the same goal. (Refer to ADRP 5-0 for additional information on the assessment process.)

SECTION II – ORGANIZATION OF FORCES

4-50. When conducting operations focus on stability tasks, just as when conducting primarily offensive and defensive tasks, the organization of forces for a battalion is based on the mission variables of METT-TC. As the overall plan develops and the required task organization for the battalion is determined, the request for forces process gives the commander the ability to request capabilities not normally available at the battalion or company echelon. In all cases, the commander alters the command and support relationships of subordinate elements attached to the battalion or under the battalion's operational control according to the mission variables of METT-TC.

4-51. While each stability-focused operation takes place in its own specific operational circumstances, there are common organizational considerations that apply. Operations focused on the conduct of stability tasks typically involve the battalion with more multinational partners than do operations centric to the conduct of offensive and defensive tasks. Throughout the operations process, the structure of the supported host nation government or transitional military authority impacts the organizations or individuals interacting with the battalion commander and staff. During the operations process, the commander may designate additional staff officers to work nontraditional actions with the host nation government or transitional military authority at different levels in addition to the liaison the battalion will have to conduct during planning, preparation, execution, and assessment.

4-52. As the commander determines the required organization of forces, the staff integrates capabilities internal to the battalion with attached capabilities to accomplish assigned missions and to meet changing situations. When considering which capabilities to allocate to subordinate units, and which to maintain at the battalion headquarters, the commander understands that task organization and support arrangements will change frequently due to the relative nature and duration of operations in support of stability tasks. The unique aspects of stability tasks requires individual augmenters and augmentation cells to support unique force-tailoring requirements and personnel shortfalls. Augmentation, driven by METT-TC considerations, enables coordination with the media, government agencies, nongovernmental organizations, international organizations, other multinational forces, and U.S. Army civil affairs forces.

4-53. Specific requirements, driven by a stability task(s), can result in the augmentation of the battalion with additional functional capabilities. Civil affairs, military police, military working dogs, public affairs teams, special forces teams, joint units, host nation security forces (when operational control or tactical control), multinational units, military information support units, and additional engineer and sustainment assets are tailored to enable these requirements. (***Note.*** Special forces teams and joint units usually are a coordinating relationship, not a command relationship.) When situations destabilize within the local population, the commander determines which capability, through augmentation or liaison team (through either a command relationship or coordinating relationship), is best to monitor and report on the situation. Depending on the situation, unit ministry, engineers, psychological operations, civil affairs, counterintelligence, linguistics, and logistics personnel may be candidates for such teams.

4-54. Most operations focused on the conduct of stability tasks requires the Infantry battalion to coordinate, integrate, and synchronize operations with a variety of outside organizations. These include other U.S. Armed Services or government agencies as well as international organizations, including nongovernmental organizations, coalition, and United Nation military forces or agencies. During stability, coordination and integration of civilian and military activities must take place at every level within the battalion to ensure synchronization efforts. Normally, the battalion plans operations to complement those of participating government and private agencies. The commander ensures battalion civil-military objectives and concept of operations are understood by the indigenous population and institutions government and private agencies. If the battalion has an attached civil affairs team, that team in conjunction with the S-9 and coordinating centers such as the civil-military operations center, normally established at IBCT level are designed to accomplish this task. The outcome of this civil coordination helps to nest the battalion civil-military operations plan with the BCT civil-military objectives resulting in unity of effort.

4-55. Deployments associated with the conduct of stability-focused operation often require the battalion to plan and prepare for the potential that it will provide one or more mobile training teams (see ATP 3-05.2) to host national security forces. When a separate U.S. training organization does not exist, for example, an Infantry battalion should be able to provide host nation security forces with training, advice, and assistance in infantry tactics and the use of Infantry weapons up through battalion level. Training and assistance includes counterinsurgency tactics and techniques (see FM 3-24.2). When provided by the battalion, mobile training teams may be placed under the operational control of a U.S. security assistance organization and should be able to provide limited advice and assistance on the conduct of other stability tasks. At a minimum, small units within each company will be expected to mentor/advise those host nation counterparts with which they work. (Refer to FM 3-22 for additional information.)

4-56. Early on in the planning process the battalion personnel staff section, in coordination with brigade and higher echelon manpower and personnel staff sections, screen personnel files to review the records of identified Soldiers that might have specific skill sets useful to the battalion or higher echelon during the conduct of stability focused operations. The review by brigade and battalion manpower and personnel staff sections might not be completed before deployment, especially as it pertains to newly assigned Soldiers. These skill sets include individuals with professional certification or work experience in those non-military fields that might have utility during operations focused on the conduct of stability tasks. Individuals that have the necessary degree of cultural understanding and foreign language skills enable the effective augmentation of information operations and public affairs requirements.

Note. A database, usually maintained at division level, is created to track the college degrees, civilian work experience, cultural knowledge, and language proficiencies of Soldiers identified during the personnel files screening. (Refer to ATP 3-91 for additional information.)

4-57. During prolonged operations when the battalion commander does not employ the tactical command post, the staff assigned to it reinforces the main command post. This allows economy of scale to the skill sets appropriate to the conduct of stability tasks with those of the conduct of offensive and defensive tasks in the battalion's area of operations. Battalion standard operating procedures address the specifics for this, including procedures that quickly detach the tactical command post from the main command post. When participating in operations focused on the conduct of stability tasks as part of short-term military engagement or limited contingency operations, the battalion commander may choose to deploy only the tactical command post into the area of operation specific to the engagement or operation.

SECTION III – OPERATIONAL AREA SECURITY

4-58. Population-centric operational area security is common across the range of military operations, but is almost a fixture during the conduct of stability-focused operations. Population-centric operational area security typically combines aspects of the area defense and offensive tasks (for example search and attack, cordon and search, raid, and ambush) to eliminate the efficacy of internal defense threats. During the conduct of area security operations, the commander must understand the relationship with host nation authorities and the civilian population. A clear understanding of the commander's authority is essential in exercising that degree of control necessary to ensure security and safety to all military forces and the civilian population located within the Infantry battalion's area of operation.

OPERATIONAL OVERVIEW

4-59. *Operational area security* is a form of security operations conducted to protect friendly forces, installations, routes, and actions within an area of operations (ADRP 3-37). During the conduct of stability-focused operations, area security operations establish and maintain the conditions for stability in an unstable area before or during hostilities, or enduring peace and stability after open hostilities cease. Operational area security is often an effective method of providing civil security and civil control during operations focused on stability. Security objectives, regardless of which element of decisive action (offense, defense, or stability) currently dominants, ensure freedom of action over a prolonged period in consonance with the battalion commander's concepts of operations and intent.

CIVIL CONSIDERATIONS

4-60. Civil considerations reflect the influence of manmade infrastructure, civilian institutions, and attitudes and activities of the civilian leaders, populations, and organizations within the operational environment on the conduct of military operations. Commanders and staffs analyze civil considerations in terms of the categories expressed in the memory aid known as ASCOPE (areas, structures, capabilities, organizations, people, and events). See ATP 2-01.3 for additional information on these categories.

4-61. Since civilians are normally present in operations with a dominant stability component, the battalion normally restrains its use of force when conducting area security operations. However, the commander remains responsible for protecting the force and considers this responsibility when considering rules of engagement. Restrictions on conducting operations and using force must be clearly explained and understood by everyone. Subordinate leaders and Soldiers must understand that their actions, no matter how minor, may have far-reaching positive or negative effects. Leaders and Soldiers must realize that media (either hostile or neutral) and adversaries can quickly exploit their actions, especially the way they treat the civilian population.

Area Security

4-62. *Area security,* a security task conducted to protect friendly forces, installations, routes, and actions within a specific area (ADRP 3-90), takes advantage of the local security measures performed by all units, regardless of their location in the area of operations. Local security includes any local measure taken by units against enemy actions. Local security, dependent upon the situation, may involve avoiding enemy detection or deceiving the enemy about friendly positions and intentions. Local security may include finding any enemy forces in the immediate vicinity and knowing as much about their positions and intentions as possible. Local security prevents a unit from being surprised, and is an important part of maintaining the initiative during area security.

4-63. The requirement for maintaining local security is an inherent part of any area security mission. Units use both passive and active measures to provide local security. Passive local security measures include using camouflage, movement control, noise and light discipline, operations security, and proper communications procedures. Measures also include employing available sensors, night-vision devices, and daylight sights to maintain surveillance over the area immediately around the unit. Active measures, dependent upon the situation, may include—

- Using observation posts, combat outposts, combat patrols, and reconnaissance patrols.
- Establishing specific levels of alert based on the mission variables of METT-TC.
- Establishing stand-to times. (Unit standard operating procedures (SOPs) detail activities during the conduct of stand-to.)

Economy-of-Force Missions

4-64. The Infantry battalion, charged with execution, conducts an area security operation as an economy-of-force mission. Area security missions are numerous, complex, and generally never ending. For this reason, the commander and staff synchronize and integrate security efforts, focusing on protected forces, installations, routes, and actions within the battalion's assigned area of operation. Protected forces within the battalion range from subordinate units and elements, echeloned command posts, and sustainment elements within the battalion's support area (when established). Protected installations can be part of the sustainment base, or they can constitute part of the area's civilian infrastructure. Protected ground lines of communication include the route network to support the numbers, sizes, and weights of tactical and support area movement within the battalion's area of operations. Actions range from securing key points (bridges and defiles) and terrain features (ridgelines and hills) to large civilian population centers and their adjacent areas.

Offensive and Defensive Activities

4-65. During the conduct of stability-focused tasks, area security missions are a mixture of offensive and defensive activities involving not only subordinate companies and platoons, but also those host nation security forces over which the battalion has a command relationship such as operational control, or can otherwise influence. Offensive area security activities include subordinate tasks of movement to contact [search and attack (see paragraph 2-259) or cordon and search (see paragraph 2-219) missions] and combat patrols (see ATP 3-21.8), when required, designed to ambush detected enemy forces and/or to conduct raids (see appendix D) within the battalion's area of operation. Defensive area security activities include the establishment of base perimeter security (See appendix I), combat outposts, observation posts, surveillance, moving and stationary screen and guard missions, and reconnaissance and counterreconnaissance missions.

> ***Note.*** During the conduct of offensive or defensive-focused tasks, battalion forces engaged in area security operations can saturate an area or position on key terrain to provide protection through early warning, reconnaissance, surveillance, and/or security operation (screen or guard missions), and to guard against unexpected enemy or adversary attack with an active response. Early warning may come from ground base forces and/or ground- and space-based sensors. Forces engaged in area security operations are typically organized in a manner that emphasizes their mobility, lethality, and communications capabilities. (Refer to chapter 2 and chapter 3 for additional information.)

4-66. During offensive or defensive-focused tasks, area security operations are often designed to ensure the continued conduct of sustainment operations to support decisive and shaping operations by generating and maintaining combat power. Area security operations may be the predominant method of protecting support areas that are necessary to facilitate the positioning, employment, and protection of resources required to sustain, enable, and control forces. (Refer to appendix H for additional information.)

Readiness

4-67. During area security operations, forces must retain readiness over longer periods without contact with the enemy. This occurs most often when the enemy commander knows that enemy forces or insurgents are seriously overmatched in available combat power. In this situation, the enemy commander normally tries to avoid engaging friendly forces unless it is on terms favorable to the enemy. Favorable terms include the use of mines and booby traps. Area security forces must not develop a false sense of security, even if the enemy appears to have ceased operations in the secured area. The commander must assume that the enemy is observing friendly operations and is seeking routines, weak points, and lax security for the opportunity to strike with minimum risk. This requires the commander to influence subordinate small-unit leaders to maintain the vigilance and discipline of their Soldiers to preclude this opportunity from developing.

PLANNING CONSIDERATIONS

4-68. During operational area security planning, the commander apportions combat power and dedicates assets to protection tasks and systems based on an analysis of the operational environment, the likelihood of threat action, and the relative value of friendly resources and populations. Based on an initial assessment of the operational environment, the commander task organizes subordinate units and elements and assigns security areas within the battalion's area of operation. Although all resources have value, the mission variables of METT-TC make some resources, assets, or locations more significant to successful mission accomplishment from enemy or adversary and friendly perspectives. Throughout the operations process the commander relies on the risk management process and other specific assessment methods to facilitate decision making, issue guidance, and allocate resources (see appendix B). Criticality, vulnerability, and recoverability are some of the most significant considerations in determining protection priorities that become the subject of the commander's guidance and the focus of area security operations.

Mission Command

4-69. During area security operations, the battalion commander devotes considerable time and energy to the problems of coordination and cooperation due to the joint, interagency, and multinational nature of stability-focused tasks. The battalion plans and conducts area security operations in concert with partner participants towards a unified effort, often as a supporting organization rather than the lead organization. The battalion commander uses liaisons to enable unity of effort between partner elements and the coordination centers established by the IBCT or higher commander.

Interagency and Multinational Organizations

4-70. One factor that distinguishes the conduct of stability-focused tasks from the conduct of offensive- and defensive-focused tasks is the requirement for interagency coordination at battalion level and below. During area security operations with interagency partners, the commander has inherent responsibilities. These responsibilities include the requirement to clarify the mission; to determine the controlling legal and policy authorities; and to task, organize, direct, sustain, and care for the organizations and individuals for whom the battalion provides the interagency effort. The commander also ensure seamless termination of the mission under conditions that ensure the identified objectives are met and can be sustained after the operation.

4-71. When operating inside or with multinational organizations, the battalion commander and subordinate commanders and leaders should expect to integrate foreign units down to the company level. Security force assistance activities within an area security mission require carefully selected and properly trained and experienced personnel (as trainers or advisors) who are not only subject matter experts, but also have the sociocultural understanding, language skills, and seasoned maturity to more effectively relate to and train foreign security forces. Additionally, company commanders and subordinate leaders within the battalion train with the fact that they will routinely interact with multinational partners during other area security missions. Battalion standard operating procedures will require modification to incorporate multinational small units that do not have compatible communications and information systems.

Desired End State

4-72. The battalion commander's definition of the desired end state is a required input to area security operations. While end state is normally described as a stable, safe, and secure environment during stability-focused operations, this description is not sufficient. Initial MOE and performance quantifying that environment are determined during the planning process. (Refer to appendix B and section I of this chapter for additional information.) Measures of effectiveness and performance are important in stability-focused area security operations since traditional combat measures, such as territory gained, enemy personnel killed or captured, and enemy combat vehicles destroyed or captured do not apply. The commander also ensures the desired end state reflects the prolonged time-period associated with many stability-focused area security operations.

4-73. Achieving the desired end state requires a knowledge of operational design (see section I), the ability to achieve unity of effort, and a thorough depth of cultural awareness (see section I) relating to the battalion's area of operations. Through economy of forces, the commander identifies a finite amount of available combat power to apply against the essential tasks associated with a given area security operation. Identifying essential tasks lays the foundation for the success of security area operations that represent the future stability of a state. Decisions about use of combat power are more than a factor of the size of the force deployed, its relative composition, and the anticipated nature and duration of the mission. Assuring the long-term stability depends on applying unity of effort to the tasks that are, in fact, essential.

Human Component

4-74. No other military activity has as significant a human component as operations that occur among the people. Human beings capture information and form perceptions based on inputs received through all the senses. Humans see actions and hear words. Humans compare gestures and expressions with the spoken word. Humans weigh the messages presented to them by the battalion and other sources with the conditions that surround them. When the local and national news media are unavailable or unreliable, people turn to alternative sources, such as the internet—where information flows freely at unimaginable speeds—or rumor and gossip. Perception equals truth to people lacking objective sources of information. Altering perceptions requires shaping information according to how people absorb and interpret information, molding the message for broad appeal and acceptance.

Information Operations

4-75. The final success or failure of the battalion's area security operation rests with the perceptions of the inhabitants within and external to the battalion's area of operations and goes beyond defeating the enemy. Securing the trust and confidence of the civilian population is the chief aim of information operations, which integrates and synchronizes information-related capabilities to generate effects in the information environment necessary to influence enemy, adversary, neutral, and friendly audiences.

4-76. Information operations synchronization of information-related capabilities promotes the legitimacy of the mission and reduces bias, ignorance, and confusion by persuading, educating, coordinating, or influencing targeted audiences. Further, it promotes–through Soldier and leader engagement, civil affairs operations, and military information support operations, among other information-related capabilities–interaction at all echelons with these audiences so these target audiences understand the objectives and motives of battalion and that of the IBCT, and the scope and duration of area security actions. Combined with broad efforts to build partner capacity, for example, security force assistance (see Section IV). Information operations are essential to achieving decisive results: a stable host nation government and peaceful civilian population.

4-77. When an information operations officer or NCO is attached or established within the battalion (one technique is the use the electric warfare NCO), an information operations officer planner coordinates with the IBCT information operations officer to synchronize information-related capabilities into the battalion's information operations planning. Synchronization requires the battalion information operations staff planner to participate in battalion targeting within the fire support cell as well as the various working groups and meetings chaired by the current and future operations and other integrating cells within the battalion. Participation allows for the development of a holistic understanding of the information environment within the problem sets facing the battalion staff. A staff-wide understanding helps synchronize the information-operations related planning and targeting and allows for shifts in priorities. This synchronization, in coordination with the information operations and civil-military operations managed at the IBCT enables united action partners to be incorporated into planning.

Note. When attached or established within the battalion, an information operations officer or NCO planner is responsible for synchronizing and deconflicting information-related capabilities employed in support of battalion operations. In coordination with the information operations officer at the IBCT, the information operations officer or NCO staff planner synchronizes capabilities within the battalion staff that communicate information to audiences and affect information content and flow of enemy or adversary decision-making while protecting friendly information flow. The information operations planner prepares Appendix 15 and a portion of Appendices 12, 13, and 14 to Annex C (Operations) to the operation order when established. (Refer to FM 3-13 and FM 3-96 for additional information.)

4-78. Within the security environment, enemies, adversaries, and other organizations use propaganda and disinformation against the commander's efforts to influence various civilian populations within and external (area of interest) to the battalion's area of operations. The battalion's public affairs staff NCO, in coordination with the IBCT public affairs officer works closely with the intelligence staff officer to be proactive, rather than reactive, to such attacks. A coordinated information operations plan informs and counters the effects of propaganda and misinformation. The plan (generally developed at the IBCT level in coordination with the division information operations officer and public affairs officer) establishes mechanisms, such as a media center and/or editorial board, to educate and inform local and international media, which in turn, informs the public, with accurate and timely information. Additionally, civil affairs operations and military information support operations are integrated into counterpropaganda efforts at the IBCT level through the information operations working group.

4-79. When needed, the battalion chaplain can play an important role in bridging gaps with religious leaders that set conditions for future successful key leader engagements and civil affairs operations. During planning, the chaplain advises the commander concerning matters of religion, culture, and religious key leaders in the area of operation and areas of interest. The chaplain and unit ministry team provide important, up-to-date perspectives concerning local, provincial, and national atmospherics not often included or clear in other sources. Their efforts should always be coordinated with the battalion information operations NCO and IBCT information operations working group.

4-80. Without a detailed Soldier and leader engagement plan, different units and staff elements meet with and engage local leadership with different desired end states thereby undermining the ability of any or all forces to build capacity and work towards transition to host nation lead. Coordination between staff elements or units within the battalion, when working with the same host nation individual or office, enables unity of effort and the desired end state for the battalion's area security operation. The creation of a detailed engagement plan includes identifying differences between provinces or localities within the province and sets out the objectives to reach the desired end state. Host nation leaders in a city, district or province have face-to-face meetings with these leaders to advance the creation and building of host nation capacities.

4-81. Operations security is as important during the conduct of stability-focused operations as it is during the conduct of offensive- and defensive-focused operations. Operations security contributes to the battalion's ability to achieve surprise during area security missions, thus enabling its chances for success. Within the battalion area of operation, human adversaries/enemies monitor the battalion's normal activities to detect variations in activity patterns that forecast future operations. They monitor the conversations of Soldiers both on duty and off duty to gain information and intelligence. Adversaries/enemies monitor commercial internet activity and phone calls from battalion operational and recreation facilities. They will look at trash created by battalion activities. The absence of operations security about battalion activities contributes to excessive friendly casualties and possible mission failure in area security operations just like it does in combat operations. The battalion's information superiority hinges in no small part on effective operations security; therefore, measures to protect essential elements of friendly information cannot be an afterthought. (Refer to FM 3-96 for additional information.)

Notes. The need to maintain transparency of the battalion's intentions during area security operations is a factor when balancing operations security with information release. Release authority for information–to include foreign disclosure rules–must be fully understood by commanders and staffs within the IBCT. The public affairs and information operations officers (see FM 3-61 and FM 3-13, respectively) lead the coordination and synchronization processes within the IBCT. Release authority for information rests with the commander at the appropriate level.

4-82. Multinational staffs result in additional security problems. Each nation has different access to U.S. information systems. Maintaining operations security with multinational staff members is difficult and sometimes the security rules restrict the ability of multinational partner staff officers to contribute. The chief of staff and foreign disclosure officer at division level develop workarounds when required. One such workaround is to provide the multinational staff officer a U.S. assistant to get on a U.S. secured information system to ensure the multinational staff officer has the information needed to contribute. The division assistant chief of staff, signal (known as the G-6) establishes and maintains two separate sets of different information systems when this occurs. (Refer to ATP 3-91 and ATP 6-02.75 for additional information.)

MOVEMENT AND MANEUVER

4-83. During the conduct of stability-focused operations, the battalion plans its operational area security movement and maneuver simultaneously with offensive and defensive movement and maneuver, though with an extensive emphasis on security and engagement skills (negotiation, rapport building, cultural awareness, and critical language phrases). Movement and maneuver within the battalion area of operation is normally decentralized to the company, platoon, and squad level. Through economy of force, the battalion commander determines the right mix of forces to quickly transition between operations as the situation requires. During area security operations, the commander plans for future movement within the battalion's area of operation and, as required, in adjacent areas of operation. The battalion's lethal capabilities make the execution of area security operations possible even if the probability of combat is remote. When new requirements develop the battalion commander plans for the shifting of priorities when the need arises.

Fire and Movement

4-84. The application of fire and movement lends itself to several offensive and defensive tasks [for example search and attack (see paragraph 2-259), cordon and search (see paragraph 2-219), and area defense (see paragraph 3-139)] within the civil security and civil control primary stability tasks. Across the range of military operations, the Infantry battalion and its subordinate units play a major role in ensuring the outcome of these primary stability tasks. The battalion and its subordinate units are useful in the conduct of other primary stability tasks because of their deterrence value and the flexibility and labor the battalion provides to the IBCT commander.

Mobility and Countermobility

4-85. Mobility (see paragraph 3-222) and countermobility (see paragraph 3-206) operations are key enablers to operational area security. In stability focused area security operations, mobility operations allow civilian traffic and commerce to continue or resume. Resuming normal civilian activities in the battalion's area of operation is an important objective within stability focused area security operations. Countermobility operations indirectly support stability focused area security operations in regards to offensive and defensive tasks.

4-86. Mobility operations focus on keeping ground lines of communications open for both civilian and military activities and on reducing the threat of mines and other unexploded ordnance to the same. During operational area security, the battalion commander and staff develop the countermobility plan concurrently with the fire support plan and defensive scheme of maneuver, guided by the commander's intent. When combat engineer support falls under the mobility and countermobility tasks, it can include—

- Constructing combat roads and trails.
- Breaching existing obstacles (including minefields).
- Marking minefields, including minefield fence maintenance.
- Clearing mines and debris from roads.
- Conducting route reconnaissance to support the main supply routes and civilian lines of communications.
- Creating obstacles between opposing factions to prevent easy movement between their positions.

4-87. The battalion employs roadblocks not only to restrict traffic for security purposes, but also to control the movement of critical cargo. Cargo could be generators designed to restore electric power in a large area or items that support the population and resources within the battalion's area of operation.

Occupy an Area

4-88. Planning for the battalion's occupation of an area or relief in place (see paragraph 3-66) begins before the battalion deploys or when being relieved, redeploys. Planning includes not only battalion forces and their activities, but also other governmental agencies, multinational partners, host nation agencies, and potential international organizations. The mission variables of METT-TC determine the occupation or relief in place that occurs. Sometimes occupation, much like occupying an initial area of operation is appropriate. This can take place when the battalion's stability-focused area security operation occurs within limited intervention or peace operations. A relief in place may be appropriate during the conduct of an area defense (figure 3-16, page 3-34). However, a stability-focused area security transition by function may be more effective if the relief in place is to takes place with host nation military forces and civil authorities within the range of military operations of irregular warfare. Some of these functions include medical and engineer services, local security, communications, and sustainment. Battalion plans do not remove a provided capability from the area of operation until the replacement capability is operating.

Surveillance, Reconnaissance, and Security Operations

4-89. In restrictive and unrestrictive terrain, the commander relies on manned and unmanned surveillance assets and reconnaissance and security forces to collect information within the battalion's area of operation. In restrictive terrain, reconnaissance and security forces within the battalion's area of operation focus on key terrain such as potential choke points to assist in this collection. Using a *combat outpost*, a reinforced observation post capable of conducting limited combat operations (FM 3-90-2), is a technique for employing reconnaissance and security forces in restrictive terrain that precludes mounted reconnaissance and security forces from covering the assigned area. While the mission variables of METT-TC determine the size, location, and number of combat outposts a unit establishes, a reinforced platoon typically occupies a combat outpost. (Refer to FM 3-21.10 for additional information.) A combat outpost must have sufficient resources to accomplish its designated missions, such as conducting aggressive combat patrolling and reconnaissance patrolling. Combat outposts are established when observation posts (see chapter 2 and chapter 3 for information on observation post activities) are threaten by insurgency or in danger of being attacked by enemy forces infiltrating into and through the battalion's assign area of operation.

Note. During the conduct of defensive-focused operations, the battalion commander uses a combat outpost to extend the depth of the security area, to keep friendly forward observation posts in place until they can observe the enemy's main body, or to secure friendly forward observation posts that will be encircled by enemy forces. Mounted and dismounted forces can employ combat outposts. (Refer to chapter 3 for additional information.)

Army Aviation

4-90. Army aviation attack and reconnaissance units with manned and unmanned systems—when deployed early with initial response forces—can be a significant deterrent on the indigenous combatants, particularly if factions or insurgence are not yet organized during the initial response phase (see paragraph 4-32). Attack and reconnaissance helicopters may be employed to act as a response force against enemy threats. Along with unmanned aircraft systems, attack and reconnaissance helicopters may conduct reconnaissance, surveillance, and/or security over wide areas and provide the battalion a means for visual route reconnaissance and early warning. Utility helicopters provide an excellent mission command capability to support stability focused area security operations and to transport patrols or security elements throughout the battalion's area of operation. Cargo helicopters provide the capable to move large numbers of military and civilian security force personnel and to conduct resupply when surface transportation is unavailable or routes become impassable.

Note. Battalion plans include measures for the effective use of all resources, to include, exploiting airpower for transportation and resupply over extended distances and, where appropriate, tightly controlled close air support.

Reserve and Response Force Operations

4-91. Maintaining a reserve during any operations is difficult. Often, the commander finds that the battalion has more tasks than units do, and stability focused area security operations are no exception. Nonetheless, contingencies or missions may arise that require establishing a reserve. Maintaining a reserve allows the establishing commander to plan for worst-case scenarios and to exploit opportunities, provide flexibility, and conserve the force during long-term operations.

4-92. The response force, see paragraph I-40, differs from a reserve in that it is not in support of a particular engagement. A response force is a dedicated force on a base with adequate tactical mobility and fire support designated to defeat Level I and Level II threats and shape Level III threats until a tactical combat force (see paragraph I-32) can defeat them or other available response forces. The response force answers to the establishing headquarters. (See ATP 3-37.10.) Considerations when establishing a response force include—

- Threats.
- Communication equipment and procedures.
- Alert procedures.
- Transportation.
- Training priorities.

4-93. To counter an indirect fire threat, the commander employs counterfire radars throughout and area of operation to locate hostile indirect fire systems. The use of quick reactionary forces, attack helicopter, or local friendly forces are ideal for response to counterfire radar acquisitions as clearance of fire procedures are often time consuming and not necessarily reliable when determining locations for host nation forces. Additionally, indiscriminate use of indirect fire on counterfire radar acquisitions can lead to unwanted collateral damage.

INTELLIGENCE

4-94. The conduct of stability focused tasks demands greater attention to civil considerations—the political, social, economic, and cultural factors in an assigned area of operation—than does the conduct of conventional offensive and defensive focused tasks. During operational area security the commander expands the IPB process beyond geographical and force capability considerations. (See ATP 2-01.3 for additional information on IPB for stability missions.) Information collection, specifically plan requirements and assess collection, enables relevant, predictive, and tailored intelligence within an area of operation. (See ATP 2-01 for additional information on the specific functions for stability missions.) Intelligence cells and knowledge management elements within the IBCT and battalion headquarters develop procedures to share collected intelligence data and products, both vertically and horizontally, throughout the force. (See ATP 2-19.4 for additional information on intelligence techniques for stability missions.)

Understanding

4-95. Operational area security requires the integration of the IBCT and battalion's information collection effort to develop a clear understanding of all potential threats and the populace. Success in the stability environment requires a cultural understanding to gauge the reaction of the civilian population within and external to the battalion's area of operations to a particular course of action conducted, to avoid misunderstandings, and to improve the effectiveness of the execution of that course of action by the battalion and/or IBCT. Changes in the behavior of the populace may suggest needed change in tactics, techniques, and/or procedures or even strategy. Biographic information, leadership analysis, and methods of operation within the existing cultural matrix are keys to understanding the attitudes and ability of positional and reference civilian leaders to favorably or unfavorably influence the outcome of battalion area security operations.

Indicators of Change

4-96. During area security operations, the commander and staff tie priority intelligence requirements to identifiable indicators of change within the operational environment, to include, civil inhabitants and their cultures, politics, crime, religion, economics, and related factors and any variances within affected groups of people. The commander often focuses on named areas of interest in an effort to answer critical information requirements to aid in tactical decision making and to confirm or deny threat intentions regardless of which element of decisive action currently dominants. During area security operations, priority intelligence requirements related to identifying enemy and adversary activities are tracked where appropriate.

Commander's Critical Information Requirements

4-97. Due to the increased reliance on human intelligence (HUMINT), when conducting area security operations, the commander emphasizes the importance of commander's critical information requirements (CCIRs) to all personnel within the battalion. CCIRs are information requirements identified by the commander as being critical to facilitating timely decision making, and answers to CCIRs can come from staff at all levels. All personnel must be given appropriate guidance to improve information-gathering capabilities throughout the battalion. Interpreters, military source operations, speaking to local civilian personnel, security operations, and patrolling (combat and reconnaissance) are primary sources for assessing the economic and health needs, military capability, and political intent of those receiving assistance who or are otherwise a party to the area security operation. (Refer to ADRP 5-0 and ATP 3-55.4 for additional information.)

4-98. Planners at the IBCT and battalion ensure that any HUMINT assets assigned from outside the battalion are employed effectively, which is typically accomplished by integrating HUMINT collectors at the lowest level possible. The gaining unit accounts for HUMINT asset security and establishes tasking priorities and command relationships for temporary and long-term commitments. (Refer to FM 2-22.3 for additional information.)

Note. Medical personnel must know the Geneva Convention restrictions against medical personnel collecting information of intelligence value except that observed incidentally while accomplishing their humanitarian duties.

Employment and Control of Human Intelligence Collection Teams

4-99. Commanders consider security when planning for the employment of HUMINT collection teams. (See FM 2-0.) Generally, three security conditions exist: permissive, uncertain, and hostile.

Permissive Environment

4-100. In a permissive environment, HUMINT collection teams normally travel throughout the AO without escorts or a security element. HUMINT collectors may frequently make direct contact with overt sources, view the activity, or visit the area that is the subject of the information collection effort. They normally use debriefing and elicitation as their primary collection techniques to obtain firsthand information from local civilians and officials.

Uncertain Environment

4-101. In an uncertain environment, security considerations increase, but risk to the collector is weighed against the potential intelligence gain. An uncertain environment limits use of controlled sources and requires additional resources. HUMINT collection teams should still be used throughout the AO but normally are integrated into other ground reconnaissance or other missions. For example, a HUMINT collector may accompany a patrol visiting a village. Security for the team and their sources is a prime consideration. HUMINT collection teams are careful not to establish a fixed pattern of activity or arrange contacts in a manner that could compromise the source or the collector. Debriefing and elicitation are still the primary collection techniques. Teams are frequently deployed to conduct collection at checkpoints, refugee collection points, and detainee collection points. They may conduct interrogations of detainees within the limits of applicable laws and policies.

Hostile Environment

4-102. In a hostile environment, the three concerns for HUMINT collection are access to the sources of information, timeliness of reporting, and security for the HUMINT collectors. A hostile environment requires significant resource commitments to conduct controlled source operations. Prior to the entry of a force into a hostile area, HUMINT collectors may be used to debrief civilians, particularly refugees, and to interrogate other detainees who have been in the area. HUMINT collection teams are normally located with the friendly units to facilitate timely collection and reporting. HUMINT collectors accompany the BCT lead elements or ground reconnaissance forces during operations. They interrogate detainees and debrief refugees, displaced persons, and friendly force patrols.

Security Missions

4-103. Due to the possibility of tying forces to fixed installations or sites, security missions may become defensive in nature. When this occurs the battalion commander carefully balances with the need for offensive action. Early warning of enemy activity through information collection is paramount in the conduct of area security missions to provide the commander with time to react to any threat or other type change identified within the stability environment. The battalion's IPB identifies the factors effecting security missions within the assigned area of operation. Factors, although not inclusive, include—

- The natural defensive characteristics of the terrain.
- The existing roads and waterways for military lines of communication and civilian commerce.
- The control of land and water areas and avenues of approach surrounding the area security.
- The control of airspace.
- The proximity to critical sites such as airfields, power generation plants, and civic buildings.

FIRES

4-104. The conduct of fires in support of stability-focused tasks is essentially the same as for offensive- and defensive-focused tasks. However, constraint is vital in the conduct of fires during stability-focused tasks. Such constraint typically concerns the munitions employed and the targets engaged to obtain desired effects. Constraint increases the legitimacy of the organization that uses it while potentially damaging the legitimacy of an opponent.

Employment of Fires

4-105. Employment of fires provide continuous deterrents to hostile action and are a destructive force multiplier for the commander, regardless of which element of decisive action currently dominants. Within stability-focused tasks, the planning and delivering of fires precludes fires on protected targets, unwanted collateral damage, and the political ramifications of perceived excessive fire. In addition to lethal effects, the targeting functions of the battalion fire support cell includes nonlethal effects input to the information collection plan and the targeting work groups at the IBCT and battalion headquarters (see appendix B for targeting functions within the battalion fire support cell).

4-106. During the employment of fires, the commander having the ability to employ a weapon does not mean it should be employed. In addition to collateral damage considerations, the employment of fires could have second and third order negative effects. Collateral damage could adversely affect efforts to gain or maintain legitimacy and impede the attainment of both short- and long-term goals. For example, excessive force can antagonize those friendly and neutral parties involved. The use of nonlethal capabilities should be considered to fill the gap between verbal warnings and deadly force to avoid unnecessarily raising the level of conflict. Key considerations for employment of fires in support of stability-focused tasks include—

- Stability-focused tasks conducted in noncontiguous areas of operation complicate the use of fire support coordination measures, the ability to mass and shift fires, and clearance of fires procedures.
- Key terrain may be based more on political, cultural and/or social considerations than physical features of the landscape; fires may be used more frequently to defend key sites than to seize them.
- Rules of engagement are often more restrictive than in combat operations; commander's guidance for fires requires careful consideration during development and wide dissemination to all levels.
- Precision-guided munitions and/or employment of nonlethal capabilities may be necessary to limit collateral damage.
- Fires that may be used to demonstrate capabilities, as a demonstration (see paragraph 2-433), or during a denial operation (see paragraph 3-33).

Note. Mortars at the IBCT and below, due to their smaller bursting radius, reduce collateral damage. Mortars are generally more responsive to the small unit operations common to area security missions. In addition to lethal fires, mortars may provide illumination to demonstrate deterrent capability, observe contested areas, or support area security missions [including patrolling (reconnaissance and combat)].

Application of Lethal and Nonlethal Capabilities

4-107. Though highly effective for their intended purpose, lethal capabilities may not always be suitable. For example, during stability-focused tasks, the application of lethal fires is normally greatly restricted, making the use of nonlethal capabilities the dominant feasible option. The considerations for use of nonlethal capabilities in targeting should not pertain to only specific phases or missions, but should be integrated throughout the area of operation. Escalation of force measures can be established in order to identify hostile intent and deter potential threats at checkpoints, entry control points and in convoys. Such measures remain distinct from other use of force guidance such as fire support coordination measures and are intended to protect the force, minimize the use of force against civilians while not interfering with self-defense if attacked by adversaries. One of the primary mechanisms for employing non-lethal capabilities and generating non-lethal effects is information operations. Participating in the targeting process, information operations synchronizes a range of non-lethal capabilities to produce non-lethal effects that advance the desired end state. Thus, information operations participates in the targeting process.

Fire Support Coordination Measures

4-108. As during offensive- and defensive-focused tasks, fire support coordination measures are established for stability-focused tasks to facilitate the attack of high-payoff targets throughout the area of operations. Restrictive fire support coordination measures are those that provide safeguards for friendly forces and noncombatants, facilities, or terrain. For example, no-fire areas and restrictive fire areas may be used not only to protect forces, but also to protect populations, critical infrastructure, and sites of religious or cultural significance. Regardless of which element of decisive action currently dominants, coordination measures are required to coordinate ongoing activities to create desired effects and avoid undesired effects.

> ***Note.*** Fire support coordination, planning, and clearance demands special arrangements with joint and multinational forces and local authorities. These arrangements include communications and language requirements, liaison personnel, and procedures focused on interoperability. The North Atlantic Treaty Organization standardization agreements (commonly called STANAGs) provide excellent examples of coordinated fire support arrangements. These arrangements provide participants with common terminology and procedures.

PROTECTION

4-109. Battalion activities associated with executing operational area security (ADRP 3-37), physical security (see ATP 3-39.32), operations security (ADRP 3-37), and antiterrorism (ATP 3-37.2) tasks enhance the security of the command within an area of operation. In large part, the measures within these four tasks are the same or complementary. Stability-focused operations closely resemble battalion activities for these tasks during the conduct of offensive- and defensive-focused operations though the battalion generally works closer with civilian inhabitants. (Refer to FM 3-96 for additional information.)

Assessments to Support Protection Prioritization

4-110. Initial planning by the commander and staff requires various assessments to support protection prioritization; namely threat, hazard, vulnerability, criticality, and capability within the battalion's area of operation. The commander uses a vulnerability assessment methodology that includes the review of site-specific characteristics, mission, threat analysis, security plans and procedures, and any specific command concerns. Assessment determines which assets can be protected given no constraints (critical assets) and which assets are protected with available resources (defended assets). The commander makes decisions on acceptable risks and provides guidance to the staff so that they can employ protection capabilities based on the critical asset list and defended asset list. These lists are coordinated with the battalion's higher headquarters and subordinate units. All forms of protection are used and employed during preparation and continue through execution to reduce friendly vulnerability.

> ***Note.*** Deploying battalions and higher echelons should have a trained Level II antiterrorism officer assigned. An assigned antiterrorism officer works to ensure that security considerations are integrated in base designs and unit operations. These individuals guide their units in conducting threat assessment, criticality assessments, and vulnerability analysis to determine each unit's vulnerability to terrorism. (Refer to ATP 3-91 and ATP 3-37.2 for additional information.)

Protection Template

4-111. The protection template lists and integrates all protection tasks in an appropriate way for use by subordinate units, and any base and base cluster operations envisioned to be established during the battalion's area security operation. The protection cell when established within the battalion S-3 section, augments the staff with a small protection planning cell that maintains and publishes the template in coordination with the IBCT protection cell. The template is use as a reference prior to or during employment. Battalion and base/base cluster situational modifications to this template, and their regular rehearsal of all parts of their protection plans are inspected periodically by the IBCT protection working group. During inspections, the protection working group identifies weak areas in subordinate protection plans, ensures that area of operation protection best practices are incorporated into the plans of the battalion, and provides protection-related observations, insights, and lessons learned to subordinate units, and any unit relieving the battalion or subordinate unit within its area of operation.

> ***Note.*** When a protection cell officer and/or noncommissioned officer is not designated within or attached to the battalion, protection cell functions and tasks are the responsibility of the battalion operations officer and/or noncommission officer. Key protection tasks conducted within the IBCT's and battalion's area security operation include area security, CBRN operations, coordinating air and missile defense, personnel recovery, explosive ordnance disposal, and detainee operations. (Refer to FM 3-96 for additional information on integrating and synchronizing protection tasks.)

Protective Services

4-112. The commander may determine that it is necessary (or be required) to provide protective services from within the battalion to protect high value host nation civil and military authorities or other selected individual(s). This requirement usually occurs when host nation security forces have been so extensively penetrated by hostile elements that they cannot be trusted to provide protective services or when host nation security forces lack the technical skills and capabilities to provide the desired degree of protection. The element(s) tasked to perform protective services for designated personnel receives as much training and specialized equipment as is possible prior to the mission. (Refer to ATP 3-39.35 for additional information.)

Allocation of Combat Power

4-113. Protection of installations or areas of operation (including route and convoy security) by the battalion requires significant allocation of combat power when a threat beyond organized crime exists. Conducting resupply from one base to another is treated as a tactical action and tracked in the battalion main command post current operations cell. When the battalion establishes a response force(s), care is taken so that the response force does not establish patterns when responding to incidents. Establishment of patterns—same route, same combat formation, configuration and order of vehicles, and same response force responding from the same base—allows an enemy to ambush the response force at a point of its choosing.

> ***Note.*** Dependent on the situation, host nation security forces are involved as much as possible in the performance of the above protection tasks. Host-nation support is important in the variety of services and facilities that can support security and protection of assets within the battalion's area of operation. Services provided by the host nation relieve the battalion of the need to provide equivalent capabilities thereby reducing the battalion's sustainment and protection footprint.

Threat Levels

4-114. Threats within the battalion's security area operation are categorized by the three levels of defense required to counter them. Any or all threat levels may exist simultaneously in the battalion's area of operation. Emphasis on base defense and security measures may depend on the anticipated threat level. Within the battalion's area of operation all elements protect themselves from Level I threats. This includes medical elements although they have reduced defensive capabilities since they can only use their non-medical personnel to provide their own local security. Locating medical elements on bases with other units mitigate this factor.

4-115. The battalion commander positions response forces to respond to a level II threat (enemy force or activities that can be defeated when augmented by a response force) in appreciation of time-distance factors so that no element is left outside supporting distance from a response force. The commander integrates fire support assets into the composition of the response because of the speed at which these assets can react over the extensive distances involved in area security operations. Where possible, host nation security assets constitute part of the response to smooth the interactions of these forces with the civilian population.

Note. A Level III threat is an enemy force or activities beyond the defensive capability of any local reserve or response force. The response to a Level III threat is a tactical combat force, generally established no lower than division level due to the inability to resource at lower echelons. (See appendix I for additional information on threat levels.)

Survivability

4-116. Precautions should be taken to protect positions, headquarters, support facilities, and accommodations including the construction of obstacles, protective bunkers, fighting positions, and shelters. Battalion subordinate units practice alert procedures and develop drills to occupy positions. Engineer forces enable, when available, survivability needs. Units maintain proper camouflage and concealment based on the mission variables of METT-TC. Area security forces are vulnerable to personnel security risks from local employees and other personnel subject to bribes, threats, or compromise. The threat from local criminal elements is a constant threat and protection consideration. The most proactive measure for survivability is individual awareness by Soldiers in all circumstances. Soldiers look for things out of place and patterns preceding aggression. Commanders and subordinate leaders ensure Soldiers remain alert, do not establish routines, and maintain appearance and bearing. (Refer to chapter 3 and ATP 3-37.34 for additional information.)

Notes.

In stability-focused operations, the enemy sniper poses a significant threat to dismounted (or mounted) movement and marches. Counter-sniper drills should include rehearsed responses, reconnaissance and surveillance (R&S), and cover and concealment. The battalion's rules of engagement provide instructions on how to react to sniper fire, including restrictions on weapons used depending on the circumstances. For example, rules of engagement may allow units to use weapon systems, such as a sniper rifle team, to eliminate a positively identified sniper even in a crowded urban setting because of the reduce possibility for collateral damage. (Refer to Appendix E and ATP 3-21.18 for additional information.)

An enemy improvised explosive device (IED) attack is another major threat to dismounted (or mounted) movement and marches. Prior to the conduct of any area security mission, commanders and subordinate leaders brief personnel on the latest IED threat types, usage, and previous emplacements within an area of operation or along mounted and dismounted movement or march routes. All Soldiers maintain situational awareness by looking for IEDs and IED hiding places. Units vary routes and times, enter overpasses on one side of the road and exit out the other, train weapons on overpasses as the movement passes under, and avoids chokepoints to reduce risk. Units should expect an IED attack at any time during movements and expect an ambush immediately after an IED detonation. Early mornings and periods of reduced visibility are especially dangerous since the enemy has better opportunities to emplace IEDs without detection. (Refer to ATP 3-21.18 and ATP 3-21.8 for additional information.)

Air and Missile Defense

4-117. Offensive and defensive air defense planning considerations continue to apply when the battalion conducts stability-focused operations. However, the air threat trends toward Group 1 and 2 unmanned aircraft systems (see ATP 3-04.64) employed by enemy forces opposing the battalion's effort to provide a stable, safe, and secure environment. Air and missile defense sensors and mission command elements external to the battalion provide early warning against aerial attack, and populate the battalion's common operational picture. Soldiers train in aircraft recognition and on rules of engagement due to multiple factions using the same or similar aircraft, to include international and private organizations employing their own or charter civilian aircraft. (Refer to ATP 3-01.8 for additional information.)

> *Note*. See ATP 3-01.15 for information on the tactics, techniques, and procedures for an integrated air defense system. See ATP 3-01.50 for information on the operations of the air defense and airspace management cell established within the IBCT fire support cell.

4-118. Counterrocket, artillery, and mortar batteries may be located in or near the battalion's area of operation to support its area security mission. Battery sensors detect incoming rockets, artillery, and mortar shells and may be used to detect Group 1 and 2 unmanned aircraft systems. The battery's fire control system predicts the flight path of incoming rockets and shells, prioritizes targets, and activates the supported area of operation's warning system according to established rules of engagement. Exposed elements within the area of operation then can take cover and provide cueing data that allows the battery's weapon system to defeat the target before the target can impact the area. The commander clearly defines command and support relationships between counter-rocket, artillery, and mortar elements and the battalion during planning. (Refer to ATP 3-01.60 for additional information.)

4-119. The battalion commander and subordinate leaders ensure all passive and active air defense measures (see chapter 2) are well planned and implemented. Passive measures include use of concealed routes and assembly areas, movement on secure routes, marches at night, increased intervals between elements of the columns, and dispersion. Active measures include use of organic and attached weapons according to the operations order (OPORD) and unit SOP. Air guard duties assigned to specific Soldiers during dismounted (or mounted) movements and marches give each a specific search area. For movements and marches, seeing the enemy first gives the unit time to react. Leaders understand that scanning for long periods decreases the Soldier's ability to identify enemy aircraft. During extended or long movements and marches, Soldiers are assigned air guard duties in shifts. (Refer to ATP 3-21.18 and ATP 3-21.8 for additional information.)

Force Health Protection

4-120. The nature of area security in support of stability-focused tasks requires the battalion surgeon to stress planning for the provision of preventive medicine, veterinary services, and combat and operational stress control over that inherent in supporting offensive- and defensive-focused tasks. The battalion area security focused within the conduct of stability-focused tasks interacts with the civilian population of its area of operations to a far greater degree. Under these conditions, the probability of Soldiers exposure to zoonotic diseases, toxic industrial chemicals and other pollutants, and bad food and water increases. The prolonged tours of duty typically associated with these operations and the enemy's use of unconventional weapons, such as mines and suicide bombers, tends to increase psychiatric casualties. The battalion surgeon coordinates the employment of combat stress teams with the chaplain to best meet the needs of battalion Soldiers for stress control. (Refer to FM 3-96 and ATP 4-02.8 for additional information.)

Chemical, Biological, Radiological, and Nuclear Operations

4-121. CBRN operations, measures taken to minimize or negate the vulnerabilities and effects of a CBRN incident, involve a combination of active and passive defense measures to reduces the effectiveness or success of CBRN weapon employment. An effective CBRN defense by the battalion counters enemy threats and attacks and the presence of toxic industrial materials in its area of operations by minimizing vulnerabilities, protecting friendly forces, and maintaining an operational tempo that complicates enemy or terrorist targeting.

4-122. The Infantry battalion employs key CBRN passive defense activities organized within two overarching CBRN principles (protection and contamination mitigation) to survive and sustain area security operations in a CBRN environment. The commander and staff, in coordination with the IBCT, integrates these principles regardless of the mission type. (See appendix G for additional information.)

4-123. The commanders considers the requirement for CBRN support if evidence exists that enemy forces or terrorists have employed CBRN agents or have the potential for doing so. A mix of different CBRN units—such as decontamination, hazard response, reconnaissance, and surveillance—are necessary to balance capabilities. The CBRN staff officer at the battalion participates in the intelligence process to advise the commander of commercial and toxic industrial materials in the local area. (Refer to FM 3-96 and FM 3-11 for additional information.)

Convoy Security

4-124. Convoy security is a specialized kind of area security operations conducted to protect convoys. Units conduct convoy security operations anytime there are insufficient friendly forces to secure routes continuously in an area of operations and there is a significant danger of enemy or adversary ground action directed against the convoy. The battalion may conduct convoy security operations in conjunction with route security operations within its area of operation. Planning includes designating units for convoy security; providing guidance on tactics, techniques, and procedures for units to provide for their own security during convoys; or establishing protection and security requirements for convoys carrying critical assets. Local or theater policy typically dictates when or which convoys receive security and protection. (Refer to ATP 4-01.45 for additional information.)

SUSTAINMENT

4-125. The battalion commander's responsibilities during area security includes support areas and extends to self-protection of battalion assets operating outside of the battalion echelon support area(s. Forces engaged in area security operations protect the force, installation, route, area, or asset. Area security operation are often designed to ensure the continued conduct of sustainment operations to support decisive and shaping operations by generating and maintaining combat power. Area security operations may be the predominant method of protecting echelon support areas that are necessary to facilitate the positioning, employment, and protection of resources required to sustain, enable, and control forces.

4-126. The battalion commander and staff must plan for and coordinate protection for subordinate units and detachments located within and away from the battalion support area. While the battalion S-3 is responsible overall for developing the support area(s) security plan, the battalion S-2 assists by developing the information collection plan to support intelligence operations, reconnaissance, surveillance, and security operations within the battalion support area. The battalion commander uses the intelligence preparation of the battlefield to analyze the mission variables of enemy, terrain, weather, and civil considerations to determine their effect on sustainment operations.

4-127. The enemy may avoid maneuver forces, preferring to attack targets commonly found in sustainment areas. Sustainment elements must organize and prepare to defend themselves against ground or air attacks. The security of the trains at each echelon is the responsibility of the individual in charge of the echeloned trains. All elements in, or transiting the support area, assist with forming and defending the area. Based on mission analyses, the battalion S-3 subdivides the area, and assigns subordinate and tenant units to those subdivided areas. When a subordinate or tenant unit receives a change of mission or can no longer occupy an assigned area, area adjustments are made to the support area by the battalion S-3. When a particular supply point is sufficiently large, it may be assigned its own area for defense, and a security force may be attached to provide security. (Refer to ADRP 3-37 for additional information.) Additional activities to enable echelon support area(s security include—

- Select sites that use available cover, concealment, and camouflage.
- Use movement and positioning discipline, as well as noise and light discipline, to prevent detection.
- Establish area defenses.
- Establish observation posts and conduct patrols.
- Position weapons (small arms, machine guns, and antitank weapons) for self-defense.
- Plan mutually supporting positions to dominate likely avenues of approach.
- Prepare a fire support plan.
- Make area of operations sketches and identify sectors of fires.
- Emplace target reference points to control fires.
- Integrate available combat vehicles within the trains into the plan and adjust the plan when vehicles depart.
- Conduct rehearsals.
- Establish rest plans.
- Identify an alarm or warning system to enable rapid execution of the defense plan.

- Designate a response force (see ATP 3-37.10) with appropriate fire support.
- Ensure the reaction force is equipped to perform its mission.
- Ensure the reaction force is well-rehearsed or briefed on—
  - Unit assembly.
  - Friendly and threat force recognition.
  - Actions on contact.

4-128. The Infantry battalion S-4, in coordination with the battalion S-3, select battalion supply routes within the area of operation. A supply route is selected based on the terrain, friendly disposition, enemy situation, and scheme of maneuver. Alternate supply routes are planned in the event that a main supply route is interdicted by the enemy or becomes too congested. In the event of CBRN contamination, either the primary or the alternate supply route(s) can be designated as the dirty main supply route to handle contaminated traffic. Alternate supply routes should meet the same criteria as the main supply route. Military police may assist with regulating traffic and the security of routes and convoys on those routes, and engineer units, if available, can maintain routes. (Refer to ADRP 3-37 for additional information.) Battalion supply route(s) considerations include—

- Location and planned scheme of maneuver for subordinate units.
- Location and planned movements of other units moving through the battalion's area of operation and within the IBCT's area of operations.
- Route classification, width, obstructions, steep slopes, sharp curves, and roadway surface.
- Two-way, all-weather trafficability.
- Classification of bridges and culverts. Location and planned scheme of maneuver for subordinate units.
- Requirements for traffic control such as choke points, congested areas, confusing intersections, or through built-up areas.
- Location and number of crossover routes from the main supply route to alternate supply routes.
- Requirements for repair, upgrade, or maintenance of the route, fording sites, and bridges.
- Route vulnerabilities that must be protected, such as bridges, fords, built-up areas, and choke points.
- Enemy threats such as air attack, mines, ambushes, and CBRN attacks.
- Known or likely locations of enemy penetrations, attacks, CBRN attacks, or obstacles.
- Known or potential civilian and refugee movements that must be controlled or monitored.

4-129. Security of supply routes may require the battalion commander or IBCT commander to commit combat units. The security and protection of supply routes along with lines of communications are critical to military operations since most support traffic moves along these routes. (Refer to appendix H for additional information.)

PREPARATION

4-130. During preparation activities, the Infantry battalion continues to plan, train, organize, and equip for area security missions within its area of operation. The conduct of preparation activities in support of stability-focused tasks is essentially the same as for offensive- and defensive-focused tasks. (See ADRP 5-0 for a complete discussion.) However, factors that distinguish stability-focused tasks are the increased requirement for interagency coordination at battalion level and below and the demands on the battalion staff to perform tasks or functions outside their traditional scope of duties. The commander's realignment of organizations and functions during operational area security reflect carefully weighing and acceptance of risk (for example—economy of force) to reflect the demands of the battalion's area security mission.

MISSION COMMAND

4-131. Stability-focused tasks within area security operations tend to be more complex and involve to a greater extent unified action partners. Battalion preparatory activities stress authoritative relationships established between the battalion and the other military service components or agencies that operate in assigned or projected areas of operation. Though difficult, the battalion commander strives to achieve unity of command, spending a great deal of effort during preparations to clarify the roles and functions of the various, often completing agencies. The battalion commander, when required, modifies standard command and support relationships to meet the requirements of the situation.

Inherent Responsibilities

4-132. The battalion commander has inherent responsibilities—including the requirements to clarify the mission; to determine the controlling legal and policy authorities; and to organize, direct, sustain, and care for the organizations and individuals for whom they provide the effort in interagency and multinational operations. The commander serves as the unit's chief engager, responsible for informing and influencing audiences inside and outside the organization. For example, the commander often integrates host nation security forces and interagency activities with subordinate companies and platoons and down to the individual Soldier level for support units. With this in mind, obtaining the necessary numbers of scalable communications packages and linguist to support the battalion's planned operations and training are important preparatory activities.

Continue to Coordinate and Conduct Liaison

4-133. Coordinating and conducting liaison ensures that subordinate commanders and leaders internal and external to the battalion understand their unit's role in upcoming operations, and that they are prepared to perform that role. In addition to military forces, many civilian organizations may operate in the same area of operation. Their presence can both affect and be affected by battalion operations. Continuous coordination and liaison between the command and unified action partners helps to build unity of effort, especially with civilian organizations because of the variety of external organizations and the inherent coordination challenges.

4-134. Available resources and the need for direct contact between sending and receiving headquarters determines when to establish liaison. Establishing and maintaining liaison enables direct communications between the sending and receiving units or headquarters beginning with planning and continue through preparing and executing, or it may start as late as execution. Commander and staff coordinate with higher, lower, adjacent, supporting, and supported units and civilian organizations. Coordination includes, but is not limited to the following:

- Sending and receiving liaison teams.
- Establishing communication links that ensure continuous contact during execution.
- Exchanging standard operating procedures.
- Synchronizing security operations with R&S plans to prevent breaks in coverage.
- Facilitating civil-military coordination among those involved.

Continue to Build Partnerships and Teams

4-135. As part of the battalion's coordination efforts, the commander may establish or utilize (from higher echelon) special negotiation elements that move wherever they are needed to build partnerships or teams and diffuse or negotiate confrontations within the battalion area of operation. Echeloned elements partner with linguist support and personnel with the authority to negotiate on behalf of the appropriate level chain of command. As the battalion and these elements conduct preparatory activities, subordinate units of the battalion rehearse activities supporting these operations and when required ensures that these elements have access to required transportation and communications assets.

Initiate the Information Network

4-136. During preparation, the information network is tailored and engineered to meet the specific needs of each operation and partnered participant. This includes not only communications, but also how the commander expects information to move between and be available for subordinate commanders and leaders and their units within an area of operation. During preparation, the staff and subordinate units prepare and rehearse the information network supporting the plan. Network considerations include the following:

- Management of available bandwidth.

- Availability and location of data and information.
- Positioning and structure of network assets.
- Tracking status of key network systems.
- Arraying sensors, weapons, and the information network to support the concept of the operation.

Note. Defining the ground rules for sharing unclassified information between the battalion, other military forces and foreign governments, nongovernmental organizations and international agencies according to higher commander policy is an important function of the division and IBCT knowledge management and foreign disclosure officers. The division G-6 and IBCT signal staff officer (S-6) staff sections are responsible for disseminating and implementing those ground rules to the Infantry battalion.

Movement and Maneuver

4-137. Success in operational area security hinges on protecting the battalion forces within the area of operation and their ability to act in support stability-focused tasks. The positioning and repositioning of forces address the early detection and defeat of enemy forces attempting to operate within the battalion's area of operation. Enemy attacks within the battalion's area security range from individual saboteurs and terrorist acts to enemy insurgent operations.

Assign and Define Responsibility

4-138. During preparation activities, the commander assigns and defines responsibilities for the security of units within the battalion's area of operation and/or respective base or base cluster. Subordinate areas of operation and/or base and base cluster commanders are responsible for the local security of their respective area and/or base and base cluster. Individual area of operation and base commanders designate protection standards and defensive readiness conditions (in coordination with the battalion's security plan) for tenant units and units transiting through their area or base. Commanders coordinate with the battalion main command post to mitigate the effects of security operations on the primary functions of units located within the area of operation.

Degree of Risk

4-139. The degree of risk the battalion commander accepts within an area security operation, regarding the enemy threat, invariably passes to the subordinate unit commander assigned the area security mission. For example, the subordinate unit commander moves security forces to decrease the threat's impact on logistics and medical units to support the battalion's continued operations at the anticipated level. When available and to not divert any battalion assets from their primary area security missions, military police (see ATP 3-39.30) or other available security force (possibly host nation) screen or guard friendly command post facilities and critical sites from enemy observation or attack. Subordinate unit security plans, to protect command posts, critical sites, base, base clusters, and security corridors, are rehearsed and inspected by the commander. These plans address support unit, site, and base and convoy defense against Level I threats. Plans also address response force operations directed against Level II and Level III threats (see paragraphs 4-105 and 4-106 and appendix I for additional information on threat levels).

Terrain Management

4-140. Terrain management is the process of allocating terrain by establishing areas of operation. The commander designates assembly areas and specifies locations for units and activities to deconflict movements and repositioning of units, and other activities that might interfere with each other. Subordinate commanders assigned an area security mission manage terrain within their boundaries and identify and locate key terrain in the area. The battalion operations officer, with support from others in the staff, deconflict operations, control movements, and deter fratricide as units move to execute planned area security missions. The commander and staff also track and monitor unified action partners and their activities in the battalion's area of operation.

Terrain Preparation

4-141. Terrain preparation starts with the situational understanding of the terrain through proper terrain analysis. Terrain preparation involves shaping the terrain to gain an advantage, such as improving cover, concealment and observation, fields of fire, new obstacle effects through reinforcing obstacles, or mobility operations for initial positioning of forces. Terrain preparation can make the difference between the area security operation's success and failure. Commanders must understand the terrain and the infrastructure of their area of operations as early as possible to identify potential for improvement and establish priorities of work, and to begin preparing the area.

INTELLIGENCE

4-142. As the battalion prepares, the commander takes every opportunity to improve situational understanding prior to and during operations specific to aggressive and continuous collection. The commander executes collection, focused on requirements tied to the execution of tactical missions [normally intelligence operations (undertaken by military intelligence units and Soldiers), reconnaissance, surveillance, and security operations], early in planning and continues it through preparation and execution.

Note. *Intelligence operations* are tasks undertaken by military intelligence units and Soldiers to obtain information to satisfy validated requirements (ADRP 2-0).

Information Collection

4-143. Through information collection, the commander and staff continuously plan, task, and employ collection forces and assets to collect timely and accurate information. Collection helps to satisfy the CCIRs, in addition to other information requirements. Collection efforts within the battalion worked through the battalion intelligence cell (specifically the intelligence staff officer) to the IBCT intelligence cell. Intelligence cells, in coordination with the IBCT provost marshal, work to develop a readily searchable database—including biometric data if possible—of potential insurgents, terrorists, and criminals within the battalion's area of operation. This information is use by patrols to identify individuals, according to applicable guidance, when encountered during civil reconnaissance patrols (see ATP 3-21.8) and other operations. (Refer to appendix B for additional information.)

Analysis and Dissemination of Information and Intelligence

4-144. *Intelligence analysis* is the process by which collected information is evaluated and integrated with existing information to facilitate intelligence production (ADRP 2-0). The commander and staff refine security requirements and plans (including counterterrorism and counterinsurgency) as answers to various requests for information become available. Timely, relevant, accurate, predictive, and tailored intelligence analysis; reporting; and products enable the commander to determine the best locations to place area security measures and to conduct area security missions in support of stability-focused tasks. Rehearsal of area security measures and missions enable subordinate units to understand how these measures and missions fit into the battalion's area security operation, and that of the host nation when applicable. (Refer to ATP 2-19.4 for additional information.)

PROTECTION

4-145. As preparation activities continue, the commander's situational understanding may change over the course of the area security operation, enemy actions may require revision of the security plan, or unforeseen opportunities may arise. Protection assessments made during planning may be proven true or false. Intelligence analysis from R&S may confirm or deny enemy actions or show changed security conditions in the area of operations because of shaping operations. The status of friendly forces may change as the situation changes. In any of these cases, the commander identifies the changed conditions and assesses how the changes might affect upcoming area security missions. Significant new information requires commanders to make one of three assessments listed below regarding the area security plans:

- The new information validates the plan with no further changes.
- The new information requires adjustments to the plan.
- The new information invalidates the plan, requiring the commander to reframe and develop a new plan.

4-146. Protecting information during preparation activities is a key factor in protecting battalion subordinate units and the overall IBCT area security operation. The secure and uninterrupted flow of data and information allows the battalion to multiply its combat power and synchronize IBCT and other unified action partner capabilities and activities. The need to be candid and responsive to requests for information balance the need to protect operational information, such as troop movements, security plans, and vulnerabilities identified during preparation (inspections and rehearsals). Working closely with all partners develop the essential elements of friendly information to preclude inadvertent public disclosure of critical or sensitive information. Information protection includes cybersecurity, computer network defense, and electronic protection. All three are interrelated. (Refer to ADRP 3-37 for additional information.)

Sustainment

4-147. Resupplying, maintaining, and the issuing of supplies or equipment occur during temporary and long term area security commitments. Repositioning of sustainment assets also occur. During preparation, sustainment planners take action to optimize means (force structure and resources) for supporting the commander's area security plan. These actions include, but are not limited to, identifying and preparing bases, host-nation infrastructure and capabilities, contract support requirements, and lines of communications. They also include forecasting and building operational stocks as well as identifying endemic health and environmental factors. Integrating environmental considerations will sustain vital resources and help reduce the logistics footprint. Planners focus on identifying the resources currently available and ensuring access to them. During preparation, sustainment planning continues to support operational planning (branch and sequel development) and the targeting (lethal and nonlethal) process.

4-148. Dependent on the mission variables of METT-TC, sustainment elements may support the battalion from within its area of operation or from echelon support areas located outside the area of operation. The threat within the assigned area of operation is generally the major consideration in determining the size and composition of forces (support and operational) arrayed during an area security operation. Support elements (and any other force) within the battalion's area of operation must be able to defend themselves against a level I threat, a small enemy force that can be defeated by those units normally operating in the echelon support area or by the perimeter defenses established by friendly bases and base clusters. The battalion commander uses a response force to response to a level II threat (see paragraph 4-108). Host nation security forces, when feasible, may be an effective means of reinforcing the security of sustainment elements supporting from within and external to the battalion's area of operation because of their knowledge of the area, its language, and customs. (Refer to appendix H for additional information.)

Notes.

Base and base cluster defense is the cornerstone of successful operational area security and support area efforts. The commander achieves the application of effective area security for base and base clusters and their tenant and transient units by developing a comprehensive plan linked to site selection, layout, and facility design. Appendix I outlines the organization of forces, control measures, and considerations pertaining to planning, preparing, and executing base and base cluster operations.

The commander and staff assess the need for providing protection to contractors operating within the battalion's area of operation and designate forces to provide security to them when appropriate. The mission of, threat to, and location of each contractor determines the degree of protection needed. Protecting contractors involves not only active protection to provide escort or perimeter security, but also training and equipping of contractor personnel in self-protection (protective equipment and weapons). Under certain conditions, contract security forces may be another means of reinforcing the security of sustainment elements supporting from within and external to the battalion's area of operation, and base and cluster defenses.

EXECUTION

4-149. Though close combat dominance remains the principal means to influence enemy actions, the conditions and standards of performance are modified by the mission variables of METT-TC and the more restrictive rules of engagement required during the conduct of stability-focused tasks. The general scope of battalion missions supporting stability-focused tasks include security operations, patrols and patrolling (reconnaissance and combat), intelligence operations (for example human intelligence assets from outside the battalion), surveillance (ground forces and aerial assets), convoy security, and Soldier and leader engagements. Additionally, missions often require the establishment of static security posts, base and base clusters, searches, roadblocks, checkpoints, observation posts, and combat outposts supports the conduct of stability-focused tasks. The condition set surrounding each mission differs and requires detailed analysis and planning.

APPORTIONMENT OF COMBAT POWER AND DEDICATED ASSETS

4-150. The battalion commander, during area security operations, apportions combat power and dedicates assets to protection tasks based on an analysis of the operational environment, the likelihood of enemy action, and the relative value of friendly resources and populations. Although all resources have value, the mission variables of METT-TC make some resources, assets, or locations more significant from enemy or adversary and friendly perspectives. The commander relies on risk management (see appendix B) and other assessment methods to facilitate decision-making, issue guidance, and allocate resources. Criticality, vulnerability, and recoverability are some of the most significant considerations in determining protection priorities that become the subject of the commander's guidance and the focus of battalion's area security efforts.

Note. In the illustrations below, the IBCT conducts area security operations to establish stability after open hostilities cease. With the complex and dynamic nature of an area security operation, it is important to remember that area security tasks and activities change from day to day, based upon the mission variables of METT-TC.

ILLUSTRATION OF AN INFANTRY BRIGADE COMBAT TEAM AREA SECURITY OPERATION (NONLINEAR AND NONCONTIGUOUS AREAS OF OPERATION)

4-151. In this illustration, the IBCT commander emphasizes the conduct of area security by the IBCT when subordinate battalions do not share a boundary (noncontiguous areas of operation). The IBCT headquarters, usually the current operations integrating cell within the main command post, retains responsibility for the area not assigned to subordinate battalions. Key responsibilities of the IBCT include area security; information collection, integration, and synchronization; and clearance of fires within those portions of the IBCT's area of operation not assigned to subordinate battalions. [This is in addition to the other six doctrinal responsibilities of area owning unit commanders that includes terrain management, civil affairs operations, movement control, personnel recovery, airspace control of assigned airspace, and minimum-essential stability tasks (see ADRP 3-90 for a discussion of each responsibility)].

4-152. The following scenario, used for discussion purposes, represents one way the IBCT may employ forces during the conduct of an area security operation in nonlinear and noncontiguous areas of operation. In this scenario, the IBCT's area security operation spans an area of operation approximately 95 kilometers wide and 70 kilometers in length. The mountain range separating Infantry Battalion 1 and 2 spans an area approximately 20 kilometers wide. Example IBCT area security task organization and concept of operation follows:

Infantry Brigade Combat Team

4-153. The IBCT conducts area security, across two provinces separated by a mountain range (unassigned area), to support host-nation operations (figure 4-2). The IBCT main command post positions in the eastern province within *Forward Operating Base (FOB)*—an airfield used to support tactical operations without establishing full support facilities (JP 3-09.3)—Talon. Infantry Battalion 2, the Cavalry squadron, the brigade engineer battalion, and the brigade support battalion headquarters collocate with the IBCT headquarters elements in FOB Talon.

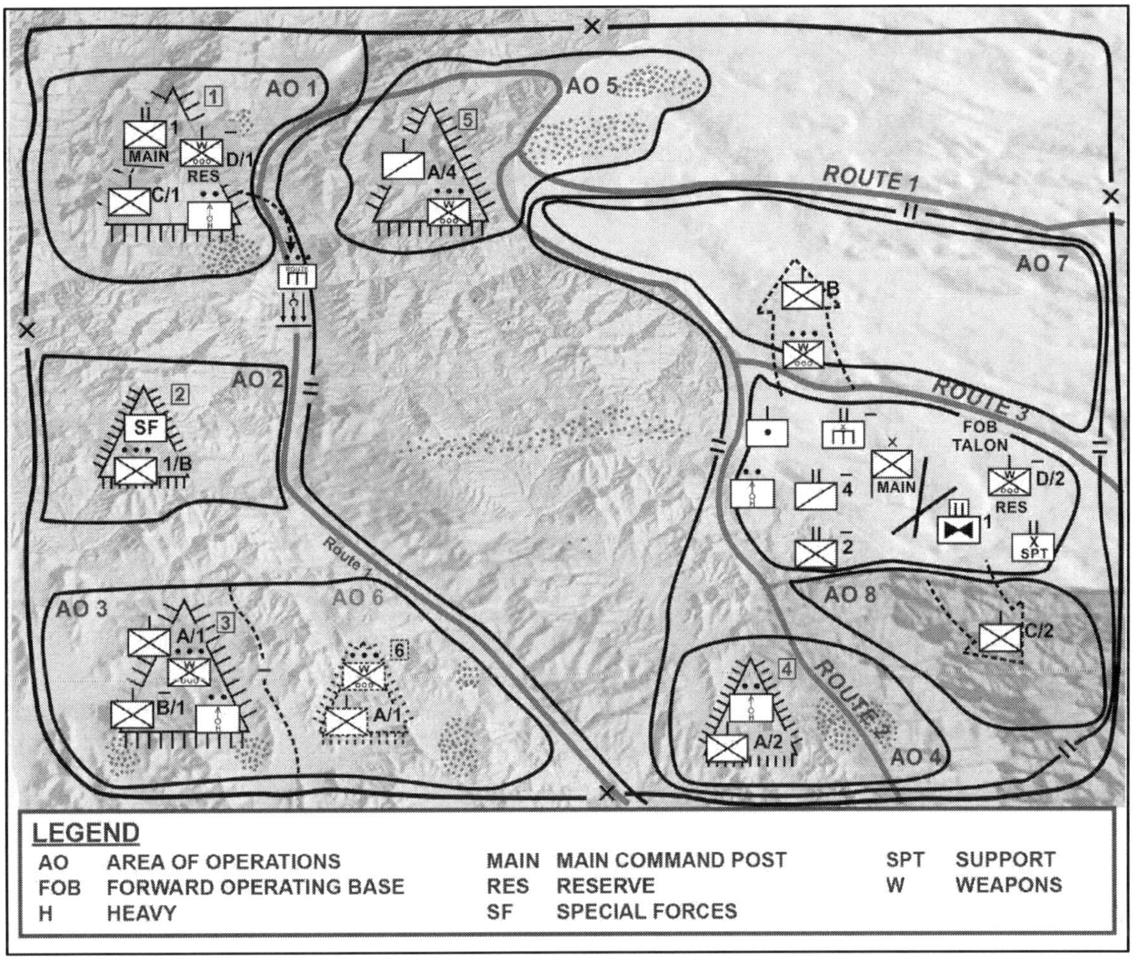

LEGEND
AO AREA OF OPERATIONS
FOB FORWARD OPERATING BASE
H HEAVY
MAIN MAIN COMMAND POST
RES RESERVE
SF SPECIAL FORCES
SPT SUPPORT
W WEAPONS

Figure 4-2. IBCT area security operation (nonlinear and noncontiguous areas of operation), example

4-154. The IBCT is task organized with one field artillery battery (M777) positioned within FOB Talon. An explosive ordnance disposal company, positioned in FOB Talon, supports operations across the IBCT's area of operation. A Special Forces Detachment locates in the western province to support IBCT operations. An Infantry rifle platoon augments the detachment to assist in base defense and day-to-day operations. Company D (from Infantry Battalion 2), with two assault platoons, establishes IBCT reserve in the eastern province. Company D's, IBCT reserve, be prepared mission is to respond to activities within unassigned areas of the IBCT and Infantry Battalion 2.

Infantry Battalion 1

4-155. Infantry Battalion 1, responsible for assigned area of operation in the western province, conducts area security and maintains lines of communication running north to south along route 1. Battalion main command post locates in Combat Outpost 1. Battalion mortars operate in split section with one mortar section located Combat Outpost 1 and one mortar section located Combat Outpost 3. The battalion, augmented with a route clearance package, prevents the enemy from influencing operations along route 1. Rifle companies, illustrated in the example, are responsible for area security within their area of operation in support of host nation operations. Company A, with attached assault platoon, initially occupies Combat Outpost 3 then moves to establish Combat Outpost 6 in Area of Operation 6. Company B initially collocates with Company A in Combat Outpost 3. Company C occupies Combat Outpost 1. Company D, with two assault platoons, establishes battalion reserve in the western province. Company D's, battalion reserve, be prepared mission to respond to activities within the unassigned areas of the battalion.

Infantry Battalion 2

4-156. Infantry Battalion 2, responsible for assigned area of operation in the eastern province, conducts area security and maintains lines of communication running north to south along Route 2. Battalion main command post locates FOB Talon. Battalion mortars operate in split section with one mortar section located FOB Talon and one mortar section located Combat Outpost 4. Rifle companies, illustrated in the example, are responsible for area security within their areas of operation in support of host nation operations. Company A occupies Combat Outpost 4. Company B, with attached assault platoon, operates out of FOB Talon to conduct operations in Area of Operation 7. Company C operates out of FOB Talon to conduct operations in Area of Operation 8. Company D, with two assault platoons, establishes IBCT reserve in the eastern province. Company B and Company C based out of FOB Talon supports, as required, the brigade engineer battalion's base defense of FOB Talon.

Infantry Battalion 3

4-157. Infantry Battalion 3, under division control, is the tactical combat force (see paragraph I-32) for level III threats to divisional assets (not illustrated).

Cavalry Squadron

4-158. Cavalry squadron commander and staff, located FOB Talon, conduct mission command for reconnaissance, surveillance, and security operations within the unassigned areas (specifically the mountain range separating the eastern and western provinces) of the IBCT as required. Troop A occupies Combat Outpost 5 with attached assault platoon, conducts area security within Area of Operation 5. Troop B, response force, positions at the airfield on FOB Talon. Troop C supports route clearance operations along route 2, on order, conducts reconnaissance and security operations within the unassigned areas of the IBCT.

Aviation Task Force 1

4-159. Aviation task force 1, direct support (attack and assault aircraft) to the IBCT, conducts air operations in support of the IBCT's area security operation. The IBCT commander, assisted by the combat aviation brigade commander, brigade aviation officer, air liaison officer, air defense and airspace management officer, and aviation task force commander, applies the appropriate level of combat power necessary to achieve mission success across the IBCT's area of operation. When the aviation task force cannot resource mission requirements, the brigade aviation element coordinates with the division joint air ground integration center (see ATP 3-91.1) for the additional aviation assets. (Refer to FM 3-96 and FM 3-04 for additional information.)

> ***Note.*** The ability of the Army aviation commander, in coordination with the ground commander, to exercise mission command is essential to the execution of air-ground operations (see appendix C). The optimal establishment of command posts, integration of the air and ground staffs, and utilization of mission command systems are integral to both commanders' ability to understand, visualize, describe, direct, lead, and assess operations. Air-ground operations are complicated more when host nation and multinational partners participate in, or are in support of area security operations. Army aviation facilitates mission command with airborne command and control systems and communication relay packages, a key enabler in area security operations. (Refer to ATP 3-04.1 for additional information.)

Special Forces (Operational Detachment Alpha)

4-160. Operational Detachment Alpha, with a coordinating relationship to the IBCT, establishes Combat Outpost 2 and is responsible for Area of Operation 2 within the western province. Company B attaches one Infantry rifle platoon to the detachment. Specific tasks, although not inclusive, in support of host nation operations include special reconnaissance, high value target extraction, and security force assistance to host nation special forces within the IBCT's area of operation.

> ***Note.*** *Special reconnaissance* is reconnaissance and surveillance actions conducted as a special operation in hostile, denied, or politically sensitive environments to collect or verify information of strategic or operational significance, employing military capabilities not normally found in conventional forces. Also called SR (JP 3-05).

Brigade Engineer Battalion

4-161. The brigade engineer battalion headquarters, brigade signal company, and military intelligence company collocate in FOB Talon. A route clearance package from the brigade engineer battalion locates Combat Outpost 1, transitions to Combat Outposts 2 and 3 based upon clearance efforts for a given mission. Capabilities provided by the brigade engineer battalion, although not inclusive, include a CBRN reconnaissance platoon, an analysis and integration platoon (human intelligence), a multisensor ground platoon, and a tactical unmanned aerial system platoon. (Refer to ATP 3-34.22 for additional information.) As illustrated in this example, the brigade engineer battalion is responsible for FOB Talon base defense mission. IBCT headquarters and headquarters company is task organized within the brigade engineer battalion and assumes the duties and responsibilities (what is commonly referred to as the mayor cell) to coordinate for basic life support within FOB Talon.

> ***Note.*** The route clearance platoon from the brigade engineer battalion provides for its own security during clearance operations. As part of a larger effort with an increased threat, an Infantry rifle company or platoon conducts route clearance operations with an attached route clearance package.

Brigade Support Battalion

4-162. Brigade support battalion, located FOB Talon, provides direct support to subordinate units within the IBCT area of operation. Company A, distribution company, conducts replenishment operations. Company B, field maintenance company provides field maintenance support, to include recovery. Company C, brigade support medical company provides Role 1, unit level medical care, and Role 2, basic primary care, Army Health System support. The brigade support battalion provides mortuary affairs for the IBCT.

Illustration of a Battalion Area Security Operation (Nonlinear and Contiguous Areas of Operation)

4-163. In this illustration, battalion areas of operation within the IBCT's area security mission do not share a boundary. The IBCT retains responsibility for the area not assigned to subordinate battalions. (See paragraph 4-146 for a discussion of noncontiguous areas of operation and unassigned areas at the IBCT level.) The IBCT commander assigned each battalion an area of operation (noncontiguous areas of operation) in the central province to conduct area security and maintain lines of communication running within the central province. Infantry Battalion 1 (figure 4-3, page 4-38) positioned in the center; Infantry Battalions 2 and 3 (not illustrated) positioned in the eastern and western portions, respectively; and the cavalry squadron positioned in the northern portion of the central province execute area security activities within the IBCT's area of operation. The IBCT main command post locates, Base (see appendix I) Gecko, within Infantry Battalion 1 area of operation. IBCT headquarters and headquarters company assumed the duties and responsibilities (commonly referred to as the mayor cell) to coordinate for basic life support within Base Gecko. The IBCT enables subordinate battalion operations from Base Gecko with the necessary enablers to augment combat power within battalion areas of operation. Critical area security enablers, although not inclusive, include the following:

- Route clearance packages.
- Intelligence operations, for example human intelligence teams.
- Provincial reconstruction teams.

- Close air support.
- Army aviation rotary-wing and unmanned aircraft systems (attack and reconnaissance)
- One system remote viewing terminal, real-time unmanned aircraft system or manned aircraft video.

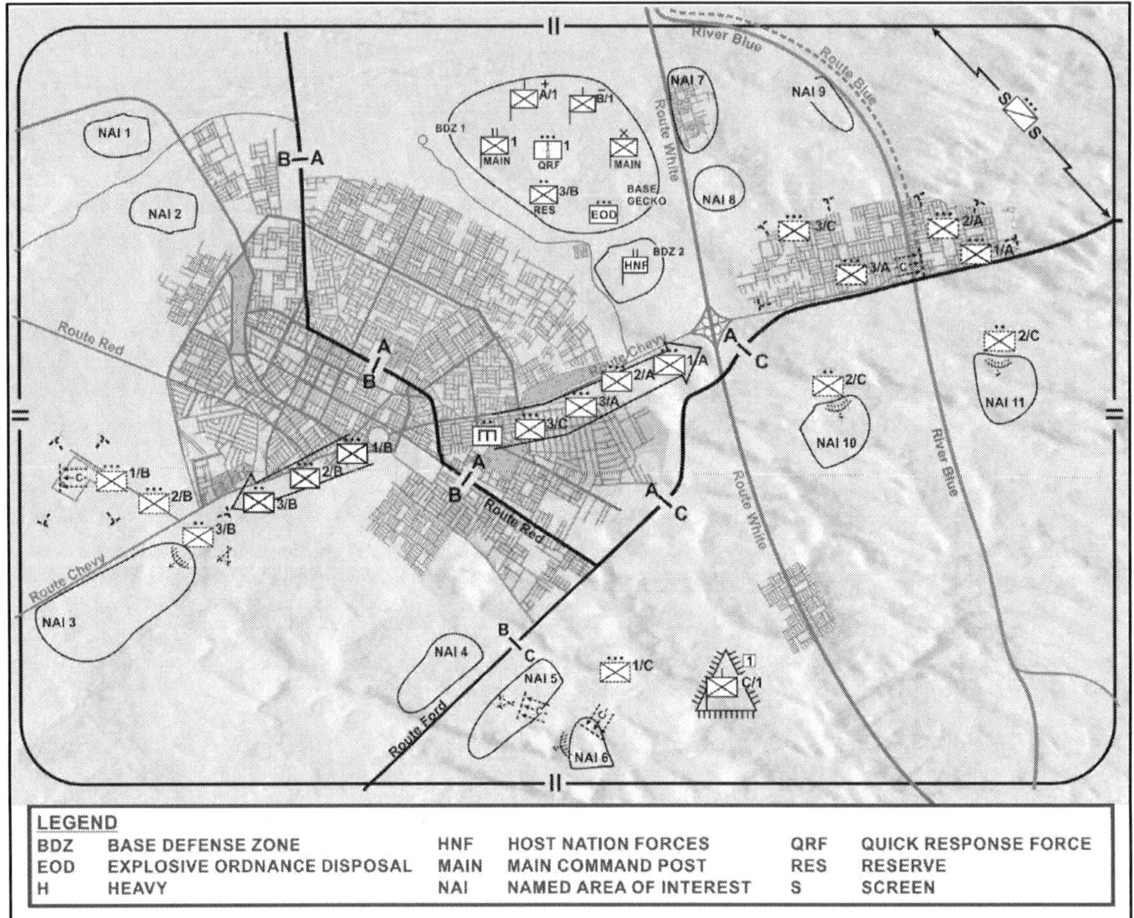

Figure 4-3. Battalion area security operation (nonlinear and contiguous areas of operation), example

> ***Note.*** The following scenario, used for discussion purposes, represents one way the Infantry battalion may employ forces during the conduct of an area security operation. In this scenario, the Infantry battalion's area security operation spans an area of operation approximately 45 kilometers wide and 35 kilometers in length. Example battalion employment follows:

Infantry Battalion 1 - Support to Host-Nation Operations

4-164. Infantry Battalion 1 conducts area security in support of host nation operations. The Battalion, minus Company C, stages out of Base Gecko adjacent to the host nation military base for the central province. The battalion commander task organizes, and assigns areas of operation and essential tasks to subordinate units. Depending on the mission, subordinate units may collocated with the higher headquarters, while other subordinate units (typically company and platoon) establish individual combat outposts or observation post. As operations progress, the commander may re-task units, or change the array of forces to meet changes in the threat, assist host nation security forces, or provide persistent surveillance to an area for a specific timeframe. Battalion reserve (normally a platoon size force) rotates between companies, dependent upon METT-TC. Battalion scout and sniper elements conduct R&S missions and security operations tasks in support of battalion and company stability-focus tasks. Battalion mortars and other allocated fire support assets position where they are able to most effectively influence the area of operation.

4-165. In this example, the battalion conducts security force assistance to help organize, train, equip, rebuild, and advise host nation forces. Select forces within the battalion, tasked organize to align with the appropriate counterpart, plan, prepare, and execute security force assistance. When conducting unilateral and partnered operations and training it is important for subordinate commanders and leaders to assess the potential for an insider attack. Commanders and leaders take the appropriate precautions to prevent insider threats by identifying personnel to pull security (covertly) at each echelon, and having all participants to remain vigilant in identifying insider threat behavior. (See section IV for additional information on security force assistance.)

4-166. During mission analysis, the commander and staff identify specific targets and areas likely to benefit ongoing operations through lethal and nonlethal means. (Lethal [for example, mortars and artillery, Army attack aviation-manned and unmanned, and close air support] and nonlethal [for example, electronic warfare, see appendix B of this publication] effects are planned for and allocated to companies in support of operations.) The company commander identifies priority of effort to subordinate platoons based upon METT-TC. Companies resource details from subordinate elements to secure the battalion commander, command sergeants major, or other headquarters personnel necessary to be on a mission outside a secured perimeter, typically referred to as a personal security detachment.

4-167. The battalion commander and subordinate commanders ensure appropriate measures are taken to account for all Soldiers at all times. *Personnel recovery* is the sum of military, diplomatic, and civil efforts to prepare for and execute the recovery and reintegration of isolated personnel (JP 3-50). Commanders conduct contingency planning and coordinate actions to be taken for the potential of missing personnel (commonly called duty status whereabouts unknown [DUSTWUN]) to expedite personnel recovery in the event it happens. Contingency planning and coordination covers immediate actions to recover missing personnel. Examples of these actions may include securing avenues an enemy may use to flee with kidnapped friendly personnel, clearing operations that clear an area of known enemy and facilitate locating personnel gone missing, and coordination/communication outside of the unit to expedite recovery. (Refer to FM 3-50 for additional information.)

Note. Regardless of level of command, guidance must be communicated to organizations or individuals on expectations in an isolating event. Effective personnel recovery planning guidance accounts for the operational environment and the execution of operations. Personnel recovery guidance broadly describes how the commander intends to employ combat power to accomplish personnel recovery tasks within the higher commander's intent. This guidance is developed based on three interrelated categories, which are personnel recovery guidance, isolated Soldier guidance, and evasion plan of action. Commanders develop and include personnel recovery guidance in execution documents. Personnel recovery appears in Appendix 13, *Personnel Recovery,* of Annex E, *Protection,* of the operations order (see FM 6-0). Commanders translate the personnel recovery guidance and develop isolated Soldier guidance. Isolated Soldier guidance provides Soldiers with guidance concerning isolating incidents. Isolated Soldier guidance focuses on awareness, accountability, reporting of isolation incidents, and actions to take when isolated. There is no set format; isolated Soldier guidance is intended to be flexible for the mission, area, and threat, and adjusted by the tactical commander as required. Evasion plan of action are specific instructions and are developed when the risk of isolation is elevated. Evasion plan of actions are developed by units for specific missions or when conditions change. All echelon battalions have trained personnel recovery specialists. The personnel recovery specialist and the personnel recovery officer responsibilities fall into four broad categories: advisor to the commander, point of contact for personnel recovery efforts to the staff and others, coordinator of PR activities across the command, and trainer.

4-168. The battalion commander, with the assistance of the battalion S-3, ensures subordinate unit movement and maneuver is coordinated to prevent bottlenecks, and allow friendly freedom of movement. As operations continue the commander arrays forces as needed to meet mission requirements for a given day. As the commander assigns each company an area of operation, within each company the company commander assigns platoon missions based on the battalion concept of operation and upon changes within their area of operation.

Company and Platoon—Area Security Tasks and Activites

4-169. The following paragraphs, acts as a vehicle to explain how companies and platoons conduct tasks and activities within the battalion on any given day during the course of area security operations. [When published, MCoE Army techniques publication FM 3-21.10 will discuss in detail the tasks and activities conducted by the Infantry rifle company within this scenario.] Subordinate company and platoon tasks and activities within this example include:

Company A

4-170. Company A, assigned the northeastern area of operation within the battalion, conducts cordon and knock due to the permissive nature of the threat. (See ATTP 3-06.11 for additional information on the cordon and knock technique.) Company A is the main effort for the battalion on this day. The purpose of Company A's mission is threefold: gather information from the local populace regarding possible enemy in the area, identify suspected weapons staging/cache sites used to enable smuggling into the eastside of the battalion's area of operation, and provide humanitarian aid to the local village by handing out blankets and having healthcare personnel screen children for possible illnesses.

Company B

4-171. Company B, assigned the northwestern area of operation within the battalion, conducts cordon and kick due to the non-permissive nature of the threat. (See ATTP 3-06.11 for additional information on the cordon and kick technique.) Company B is a shaping operation for the battalion on this day. The purpose of Company B's mission is to kill or capture members of a known insurgent cell and weapons smuggling ring in an area identified during the targeting process. The company collects information on enemy activities in the area and criminal facilitators during the operation. The company searches potential weapons cache sites suspected of use prior to the movement of these weapons to the eastern side of the battalion's area of operation.

Company C

4-172. Company C, assigned the southeastern area of operation within the battalion, establishes Combat Outpost 1 and executes operations south of Route Ford. Company C conducts clearance of NAI 5 and NAI 6 and overwatches NAI 10 and NAI 11 to deny enemy lines of communication, and access to cache and enemy indirect fire locations. The establishment of small kill teams/ambushes, strictly adhering to rules of engagement practices, engage enemy who attempt to influence friendly operations. Third platoon, located Base Gecko, conducts battalion reserve mission.

Company D

4-173. Company D (not illustrated), under IBCT control, establishes the IBCT response force within Infantry Battalion 2's area of operation in the eastern portion of the IBCT's area of operation. Infantry Battalion 2 has an area of operation four times the size of Infantry Battalion 1's area of operation.

Battalion Scout Platoon

4-174. The battalion scout platoon, northeast of Company A, conducts a screen to provide early warning. Scouts overwatch Company A to identify enemy indirect fire teams attempting to influence the cordon and search, humanitarian assistance operations, and fires from the battalion mortars.

Battalion Mortar Platoon

4-175. The battalion mortar platoon, response force, operates in split section to allow for two simultaneous employments within the battalions area of operation, if necessary. As the battalion's response force, the mortar platoon provides support across the battalion in a wide range of roles. For example securing explosive ordinance detachment elements, retrieving detainees, providing security for casualty evacuation and medical evacuation, and augmenting combat power through fire support and/or maneuver. The mortar platoon leader and/or platoon sergeant conduct time distance analysis to ensure they are able to arrive on scene in an appropriate amount of time.

SECTION IV – SECURITY FORCE ASSISTANCE

4-176. Security force assistance (SFA) contributes to unified action by the United States Government to support the development of the capacity and capability of foreign security forces (FSF) and their supporting institutions, whether of a partner nation or an intergovernmental organization (regional security organization). The development of capacity and capability is integral to successful stability missions and extends to all organizations and personnel under partner nation control that have a mission of securing its population and protecting its sovereignty from internal and external threats. FSF are considered to be duly constituted foreign military, paramilitary, police, and constabulary forces such as border police, coast guard, and customs organizations, as well as prison guards and correctional personnel; and their supporting institutions.

OPERATIONAL OVERVIEW

4-177. SFA activities are conducted to organize, train, equip, rebuild (or build), and advise foreign security forces from the ministerial/department level down through the tactical units. The Department of Defense maintains capabilities for SFA through conventional forces, special operations forces, the civilian expeditionary workforce, and when necessary contractor personnel in both joint operational area and a non-joint operational area environment. SFA activities require carefully selected and properly trained and experienced personnel (as trainers or advisors) who are not only subject matter experts, but also have the sociocultural understanding, language skills, and seasoned maturity to more effectively relate to and train FSF. Ideally, SFA activities help build the FSF capacity to train their own forces independent of sustained United States Government efforts.

Note. Department of Defense Instruction 5000.68, Security Force Assistance (SFA), dated 27 October 2010. With, through, and by. Describes the process of interaction with FSF that initially involves training and assisting (interacting "with" the forces). The next step in the process is advising, which may include advising in combat situations (acting "through" the forces). The final phase is achieved when foreign security forces operate independently (act "by" themselves).

SUPPORT TO SERVICE AND JOINT OPERATIONS/MISSIONS

4-178. SFA activities are conducted across the range of military operations and across the conflict continuum (from peace through war) supporting Service and joint operations and missions (figure 4-4, page 4-42.) Significant SC and military engagements are routinely conducted worldwide for peacetime theater and global shaping through the geographic combatant commanders' theater campaign plans. Some of those SC activities are likely to include SFA activity efforts in the lower range of the conflict continuum. Timely and effective execution of relevant SFA activities as part of SC for shaping in the theater campaign may contribute to stabilization and perhaps a measure of deterrence to prevent the requirement for U.S. forces having to conduct a contingency operation. Joint forces must have the ability to conduct SFA activities throughout all phases of an operation/campaign to effectively partner with FSF supporting U.S. or U.S.-led multinational requirements. (Refer to FM 3-22 for additional information.)

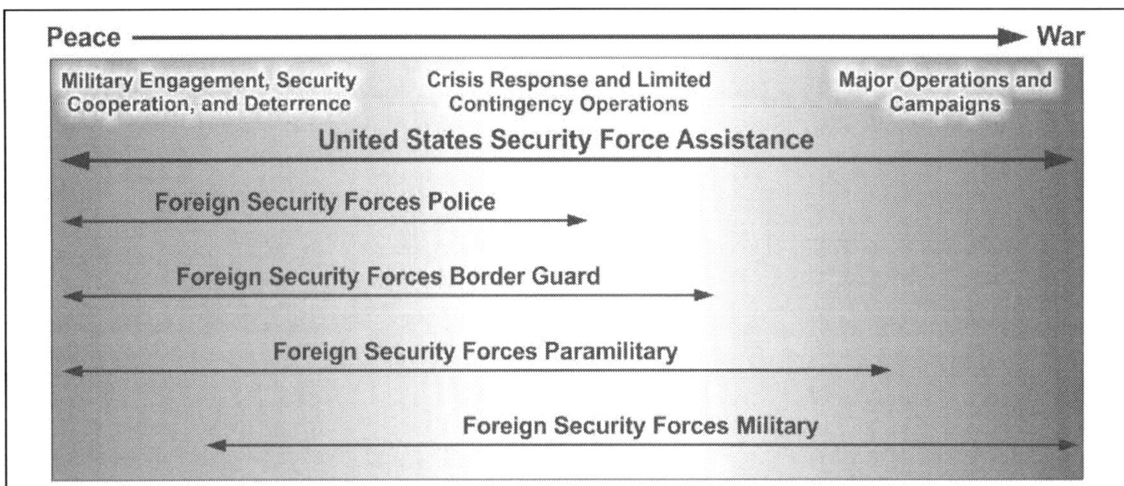

Figure 4-4. Security force assistance in the conflict continuum

> ***Note.*** For the purpose of the following discussion, SFA is addressed within the initial response, transformation, and fostering sustainability phases of the stability framework.

Phasing for Security Force Assistance

4-179. Phasing for SFA, initial response, transformation, and fostering sustainability, mirror the stability framework described in section I of the chapter and are based on the operational environment. SFA can start in any phase or may even move to a previous phase due to changes in the conditions of the operational environment. Differences within and between phases may not change on the surface, but relationships with FSF can change drastically. For example, the latter stage of the transformation phase can differ greatly from the initial stages of the transformation phase. Span of control and the area of operation for SFA can expand within a phase and as operations continue within the stability framework. As the three phases are based on the operational environment, they provide a baseline for augmentation. Potential augmentation may require military police, legal, public affairs, civil affairs, psychological operations, engineering, sociocultural experts, sustainment, and SFA team personnel.

> ***Note.*** A provincial reconstruction team embedded at BCT level is a key element during the conduct of security force assistance. The BCT leads the effort to establish civil security, establish civil control (when approved by Congress), and to develop and enable foreign security forces. The embedded provincial reconstruction team has the lead for support to economic and infrastructure development, restore essential services, and support to governance. Together the BCT and an embedded provincial reconstruction team are able to effectively support the FSF and execute all six primary stability tasks.

Initial Response Phase

4-180. The initial response phase occurs during or immediately after a conflict where the operational environment prevents civilian personnel from operating effectively. The operational environment is typified as non-permissive. Thus, the objective of this phase is to improve the security situation, reducing the threat to the populace and creating the conditions that allow civilian personnel to safely operate.

4-181. SFA in the initial response phase is normally required when FSF lack the capability or capacity to provide the required level of security. This phase often requires SFA efforts to help generate and train or assist new and existing FSF. This phase may require a combination of the types of SFA and considerable support, sustainment, and medical resources. IBCT activities during the initial response may have to be conducted with multinational major combat operations, to include providing a safe, secure environment for the local populace. SFA efforts during this phase focus on improving the FSF capability and capacity so all security forces—U.S., other, and FSF—provide a secure environment and reduce the threat. As security conditions improve, transition to the transformation phase begins.

Transformation Phase

4-182. In the transformation phase, SFA activities seeks to assist FSF to stabilize the operational environment in a crisis or vulnerable state. The operational environment in this phase is more permissive than the initial response phase; however, military forces will often be required to provide security to some actors. Activities in this phase normally include a broad range of post-conflict reconstruction, stabilization, and capacity-building efforts, which the embedded provincial reconstruction team is essential for long-term success. Objectives in this phase include continuing efforts to improve the security situation, reducing the threat to the populace, building host-nation capacity across the stability sectors, and facilitating the comprehensive approach to assist FSF.

4-183. The transformation phase represents a broad range of SFA activities to support FSF. The initial response phase differs from the transformation phase in the FSF capability to provide for a safe and secure environment. More specifically, FSFs may have a level of proficiency to no longer need a permanent U.S. and FSF relationship for tactical operations. However, they may still need full-time advisors and support, sustainment, and medical assistance. Embedded provincial reconstruction team members will continue to play a vital role in assisting governance and development efforts throughout this phase. SFA end state for this phase seeks to establish conditions so the host nation's security sector can provide a secure environment with its own security forces.

Fostering Sustainability Phase

4-184. In this phase, the focus of SFA continues to shift toward assisting institutions required to sustain FSFs. This phase encompasses long-term efforts to assist FSF. FSF conduct independent operations and can provide a safe, secure internal environment. While SFA activities may be initially required during this phase, activities reduced as FSF become more capable and viable. The determination for the battalion to receive a change of mission from SFA is based on the policy and conditions of the operational environment. Provincial reconstruction teams and other forces may remain to support a theater SC plan.

Transitions

4-185. Transitions during SFA are dependent upon the conditions within the operational environment. Transitions are initially identified during planning using a comprehensive approach (see paragraphs 4-174). Transitions can occur simultaneously or sequentially in different levels or war and in separate echelons, to include having potentially at the tactical level, transitions for different units within the battalion's area of operations. Major transitions can include the Infantry battalion in the beginning of an initial response phase being the supported unit with the FSF transitioning to the supported unit later on in the phase. At this point in the transformation phase, the area in which the battalion conducts SFA will expand. This expansion can occur multiple times during the transformation phase, which is based on conditions, especially the capability and capacity of FSF. The commander, to facilitate flexibility, visualizes and incorporates branches and sequels into the overall plan to enable transitions. Unless planned, prepared for, and executed efficiently, transitions can reduce the tempo of the operation, slow its momentum, and surrender the initiative.

PLANNING CONSIDERATIONS

4-186. Planning for SFA, like any other operation, begins either with the anticipation of a new mission or the receipt of mission as part of the military decision-making process. The Army design methodology is particularly useful as an aid to conceptual planning when integrated with the detailed planning typically associated with the military decision-making process to develop the capacity and capability of FSF and their supporting institutions. Planning helps the commander create and communicate a common vision between the staff, subordinate commanders, and unified action partners. Planning results in a plan and orders that synchronize the action of participating partners in time, space, and purpose to achieve objectives and accomplish missions.

Comprehensive Approach

4-187. SFA planning requires a comprehensive approach, as well as an in-depth understanding of the operational environment (see paragraph 4-175). Planning must be nested within policy, internal defense and development strategy, the campaign plan, and any other higher-echelon plans. Continuous and open to change, planning for SFA includes identifying how to best assist the FSF and developing a sequence of actions to change the situation. Planning involves anticipating consequences of actions and developing ways to mitigate them.

Note. Internal defense and development (known as IDAD) focuses on building viable institutions (political, economic, social, and military) that respond to the needs of society. Ideally, internal defense and development is a preemptive strategy. However, if an insurgency or other threat develops, it becomes an active strategy to combat that threat. To support the host nation effectively, U.S. forces, especially planners, consider the host-nation's internal defense and development strategy. (Refer to FM 3-22 for additional information.)

4-188. Considering the elements of operational art provides the IBCT commander and staff with a combination of conventional forces while leveraging the unique capabilities of special operations forces, to assist in achieving SFA objectives. The planning for and selection of the appropriate mix of military forces, civilian expeditionary workforce, and/or civilian personnel and contractors should be a deliberate decision based on thorough mission analysis and a pairing of available capabilities to requirements. Important factors to consider in these decisions include the nature of the host-nation force, the nature of the skills or competencies required by the host-nation force, and the nature of the situation and environment into which U.S. forces will deploy.

Understanding

4-189. Understanding the operational environment is fundamental to all operations, and essential to SFA activities. An in-depth understanding of the operational environment includes the size, organization, capabilities, disposition, roles, functions, and mission focus of host-nation forces, opposing threats, regional players, transnational actors, joint operational area, or non-joint operational area of responsibility, especially the sociocultural factors of the indigenous and other relevant populations. Identifying all actors influencing the environment and their motivations will help planners and practitioners define the goals and methods for developing host-nation security forces and their institutions.

4-190. The plan, which includes the commander's intent, provides understanding to U.S. and FSF on the actions to take. (SFA planning may involve the development of non-military security forces and their supporting institutions.) Plans and orders provide decision points and branches that anticipate options that enable the force to adapt as the operation unfolds. This is especially important for SFA, as these operations tend to be prolonged efforts. Units conducting SFA often rotate before achieving all objectives. As a result, planning should establish objectives and milestones that can be achieved during the battalion's mission. These objectives and milestones must support higher echelon plans, including the campaign plan and internal defense and development strategy.

Note. SFA planning may involve the development of nonmilitary security forces and their supporting institutions.

Legitimacy

4-191. Legitimacy of the forces providing SFA may be tenuous during some phases of a complex operation, but it is an essential consideration for achieving long-term objectives. SFA should aim to ensure that all FSF operate within the bounds of domestic and international laws, respect human rights, and that they support wide-ranging efforts to enforce and promote the rule of law, thus supporting legitimacy and transparency. Legitimacy fosters transparency and confidence among host-nation government, FSF, host-nation population, and United States Government agencies. Another aspect of legitimacy is supporting host-nation ownership in the SFA effort, because it facilitates a sense of sustainability for building a capacity or security reform through acceptance by the host-nation population.

4-192. Throughout planning, the commander and staff consider how each SFA activity affect popular perceptions, and focus on the activities that enable the legitimacy of the host-nation government and FSF, not just make them technically competent. Commander and staff must ensure an appropriate information management plan is developed for SFA in coordination with interagency partners and the IBCT or other higher headquarters. SFA advisors/trainers must work with the FSF to give a positive context and narrative to the FSF professionalization efforts and capacity to secure the population. Coordination of the information themes and messages among the U.S., FSF, and the host-nation government, and the presentation or availability of information to the indigenous population can limit or mitigate the propaganda efforts of insurgents or hostile forces. This may serve to mitigate the potential for destabilizing influences of hostile forces or criminal elements to propagandize SFA efforts and damage the host-nation government's credibility and legitimacy.

SECURITY FORCE FUNCTIONS

4-193. Security forces perform three generic functions: executive, generating, and operating. The executive function includes strategic and operational direction that provides oversight, policy, and resources for the FSF generating and operating functions. The generating function develops and sustains the capabilities of the operating forces. In the U.S., the generating function is primarily performed by the Services. For the U.S., this function is performed by its military schools, training centers, and arsenals. FSF generating forces refer to the capability and capacity of the FSF to organize, train, equip, and build operating force units. FSF operating forces form operating capabilities through the use of concepts similar to warfighting functions to achieve FSF security objectives.

Note. Employing operational forces to fill SFA capabilities associated with developing the FSF generating function (FSF tasks such as "develop FSF doctrine" or "stand up a staff officer's college"), and possibly in the FSF executive function (ministries) would likely be beyond the inherent capability of the operating force and would likely require special training or augmentation by subject matter experts drawn from U.S. generating organizations.

4-194. U.S. operating forces are typically better suited to develop FSF operating force capabilities than they are to developing FSF generating forces of generating capabilities. Typically, the battalion is tasked to train and/or advise FSF operating forces. The operating function employs military capabilities through application of warfighting functions of mission command, movement and maneuver, intelligence, fires, protection, and sustainment during actual operations. Operating, as it applies to police security forces, may include training and actual operations with the integration of patrolling, forensics, apprehension, intelligence, investigations, incarceration, communications, and sustainment. Operating forces are responsible for collective training and performing missions assigned to the unit.

SECURITY FORCE ASSISTANCE TASKS

4-195. SFA activities normally use the following general developmental tasks of organize, train, equip, rebuild and build, advise and assist, and assess (OTERA-A). These functional tasks, serving as SFA capability areas, are used to develop the capabilities required by the FSF. OTERA-A tasks are a tool to develop, change, or improve the capability and capacity of FSF. Through a baseline assessment of the FSF, and considering U.S. interests and objectives, the commander and staff planners can determine which OTERA-A tasks will be required to build the proper capability and capacity levels within the various units of the FSF. Assessments of the FSF against a desired set of capabilities will assist in developing an OTERA-A based plan to improve FSF. (Refer to FM 3-22 for additional information.) The following are basic descriptions of the OTERA-A tasks:

- Organize. All activities taken to create, improve, and integrate doctrinal principles, organizational structures, capability constructs, and personnel management. This may include doctrine development, unit or organization design, command and staff processes, and recruiting and manning functions.

- Train. All activities taken to create, improve, and integrate training, leader development, and education at the individual, leader, collective, and staff levels. This may include task analysis, the development and execution of programs of instruction, implementation of training events, and leader development activities.
- Equip. All activities to design, improve, and integrate materiel and equipment, procurement, fielding, accountability, and maintenance through life cycle management. This may also include fielding of new equipment, operational readiness processes, repair, and recapitalization.
- Rebuild or Build. All activities to create, improve, and integrate facilities. This may include physical infrastructures such as bases and stations, lines of communication, ranges and training complexes, and administrative structures.
- Advise/Assist. All activities to provide subject matter expertise, guidance, advice, and counsel to FSF while carrying out the missions assigned to the unit or organization. Advising may occur under combat or administrative conditions, at tactical through strategic levels, and in support of individuals or groups.
- Assess. All activities for determining progress toward accomplishing a task, creating an effect, or achieving an objective using MOE and MOP to evaluate foreign security force capability. Once an objective is achieved, the focus shifts to sustaining it.

DECISIONS TO REDUCE OR OFFSET RISK

4-196. Risk management (see appendix B is the Army's process for helping organizations and individuals make informed decisions to reduce or offset risk. Risk management measures identified in SFA planning add to the plan's flexibility during execution. A flexible plan can mitigate risk by partially compensating for a lack of information. SFA planning requires a thorough, comprehensive approach to analyzing and agreeing upon risk reduction measures. Each SFA activity is distinct based on context and changes over time. There is a risk of focusing SFA efforts in one area or type of relationship at the expense of others based on short-term goals. To mitigate this risk, SFA activities should be regarded as the providing means and ways to achieve meaningful mid- to long-term objectives with partners as well as the end states.

4-197. Reducing or offsetting risk does not only rely on the SFA force and supporting agencies but also on the FSF elements in question. Conditions determine when to use an element of the FSF. The battalion commander and staff use assessments to determine objectives and requirements for reducing or offsetting risk. Risk applies to how well the FSF, the Infantry battalion, and other host-nation and partner organizations agencies can tolerate changes in the operational environment, as well as the challenges and conditions inherent to the operation.

4-198. The Infantry battalion commander and FSF commander assess the risk associated with the employment forces and mitigate that risk as much as possible. For example, advisors from the Infantry battalion play a significant role in SC mission such as the SFA. They live, work, and sometimes are required to fight with their partner FSF. The relationship between advisors and FSF is vital. Advisors are not merely liaison officers. Though they do not command foreign security force units, they are a necessary element to understanding the human dimension, specifically managing relationships and mitigating risk between the SFA forces and FSF, across the range of military operations.

SUSTAINING ACTIVITIES

4-199. Sustaining SFA activities consists of two major components: the ability of the United States and other partners to sustain the SFA activities successfully and the ability of host-nation security forces to sustain their capabilities independently over the long term. The first component may be predicated on the host-nation maintaining legitimacy while the second component should be considered holistically when working with the host nation to build their security forces. It is important to consider the sociocultural factors, infrastructure, and education levels of prospective FSF when fielding weapons systems and maintaining organizations. Though this is not a battalion, BCT, or division decision, a strong recommendation through the SFA chain should be made in regards to this consideration.

INTELLIGENCE

4-200. Intelligence provides an assessment of host nation and potential adversaries' capabilities, capacities, and shortfalls. It involves understanding sociocultural factors, information and intelligence sharing, and intelligence training. Information sharing between the battalion and FSF must be an early consideration for planners. A continuous intelligence effort will gauge the reaction of the local populace and determine the effects on the infrastructure of SFA efforts as well as evaluate strengths, weaknesses, and disposition of opposition groups in the area. Ultimately, intelligence supports the SFA and FSF leaders' decision-making processes, and supports the protection of friendly forces and assets.

Note. Train personnel two deep in every staff section or advisory subunit on foreign disclosure before deployment. Interaction with host nation and FSF, even North Atlantic Treaty Organization or other coalition allies requires foreign disclosure officer approval. This will become a huge bottleneck if not trained for and decentralized.

PROTECTION

4-201. Protection is incumbent upon the commander to fully understand the threat environment within the battalion's area of operation. By having access to fused intelligence from local, regional and national resources, the commander can accurately assess threats and employ measures to safeguard SFA personnel and facilities. Protection planning considerations should address additional support requirements for the response force, emergency procedures, personnel recovery, or the requirement to integrate SFA personnel into the host-nation protection plan.

4-202. Nontraditional threats, such as the insider threat, can undermine SFA activities as well as the cohesion of U.S. forces and FSF. Tactically, the breakdown of trust, communication, and cooperation between host nation and U.S. forces can affect military capability. Adversaries may view attacks against U.S. forces as a particularly effective tactic, especially when using co-opted host-nation forces to conduct these attacks. While these types of insider or green on blue" attacks have been context-specific to a particular theater, the commander should ensure that protection plans take into account the potential for these types of attacks and plan appropriate countermeasures.

Note. More stringent protection controls and measures that are overtly heavy handed must be well balanced yet culturally sensitive enough to not send the wrong message to the very people and organizations the United States is trying to assist.

LOGISTICS

4-203. Logistics planners at the battalion level must understand the IBCT's concept of support and sustainment estimates that outline the responsibilities and requirements for maintaining logistics support for deployed forces within the IBCT's area of operation. Logistics support might include support of SFA augmentees and FSF within the battalion's area of operation to conduct operational missions (supporting host-nation civilians or military forces with medical, construction, power generation, maintenance and supply, or transportation capabilities).

PREPARATION

4-204. Preparation for SFA creates conditions that improve the Infantry battalion's opportunities for success. The degree to which the battalion is tasked within SFA operations depends on preparation in terms of cultural knowledge, language, functional skills, and the ability to apply these skills within the operational environment. Preparation includes, but is not limited to, initiate security and information collection, continue to coordinate and conduct liaison, refine the plan, complete task organization, conduct pre-mission training, conduct rehearsals and inspections; continue to build partnerships and teams, and initial movement. Preparation facilitates and sustains plans-to-operations transitions, including those to branches and sequels, which are of vital importance for the often-dynamic operational environment for SFA.

Premission Preparations

4-205. After receiving a mission, the battalion continues detailed preparation activities, prepares for and rehearse classes given in country, and conduct extensive briefings on the area of operation. Key staff and subordinate unit actions particular to SFA include the following:

Current Operations

4-206. The battalion operations staff officer (S-3) ensures pre-deployment training for Soldiers, to include preparation for training FSF and rehearsals for movement. The S-3 reviews the program of instruction for training FSF, to include getting approval from the commander, and higher headquarters if necessary. The S-3, in coordination with IBCT S-9 (civil affairs officer), ensures the operation plan minimizes how operations affect the civilian population and addresses ways to mitigate the civilian impact on military operations. The civil-military operations plan is coordinated with the indigenous population and institutions, unified action partners, other civil entities, and interagency as necessary. This coordination might include civil affairs battalions or brigades, provincial reconstruction teams, or United States Agency for International Development project officers in the area of operation.

Note. The primary staff officers of the current operations cell may be called upon to be the primary advisors to the host-nation forces staff sections and cells.

4-207. The battalion intelligence staff officer (S-2) supervises the dissemination of intelligence and other operationally pertinent information within the unit and, as applicable, to higher, lower, or adjacent units or agencies. The S-2 monitors the implementation of the intelligence collection plans to include updating the commander's priority intelligence requirements, conducting area assessment, and coordinating for additional intelligence support. The S-2 establishes liaison with FSF intelligence and security agencies (within the guidelines provided by applicable higher authority). The S-2 assesses the intelligence threat and resulting security requirements, including coordination with the S-3 on specific security and operations security measures.

4-208. The battalion personnel staff officer (S-1) supervises the battalion personnel staff section, in coordination with brigade and higher echelon manpower and personnel staff sections, screen personnel files to review the records of identified Soldiers that might have specific skill sets useful to the battalion or higher echelon during the conduct of stability focused operations. Skill sets include individuals with professional certification or work experience in those non-military fields that might have utility during operations focused on the conduct of stability tasks. (See paragraph 4-55 for additional information.)

4-209. The battalion logistics staff officer (S-4) supervises, as required, the logistics support of SFA augmentees and FSF within the battalion area of operation to conduct operational missions (supporting host-nation civilians or military forces with medical, construction, power generation, maintenance, supply, or transportation capabilities). (See paragraph 4-89 for additional information.)

4-210. The battalion signal staff officer (S-6), in coordination with the IBCT S-6, ensures depth in communication and synchronization between organizations both horizontally and vertically within the battalion's proposed area of operation. In coordination with the IBCT S-9 and S-6, establishes communications as early as possible upon arrival with the civil-military operations center (normally established at IBCT level), civil liaison teams, civil information management architecture, and supporting networks to facilitate communication and coordination with the nonmilitary agencies.

Predeployment Training

4-211. During predeployment training, Soldiers receive training, materials, and briefings on the operational area. This training can cover the history, culture, religion, language, tribal affiliations, local politics, and cultural sensitivities as well as any significant nongovernmental organizations operating in the operational area. Advisors focus their premission training on the specific requirements of developing FSF. The training emphasizes the host-nation culture and language and provides cultural tips for developing a good rapport with foreign personnel. (Refer to FM 7-0 for premission training for SFA.)

4-212. Based on the battalion commander's, or higher commander's training guidance, subordinate unit commanders assign missions and approve the draft mission-essential task list that supports SFA. The staff plans, conducts, and evaluates training to support this guidance and the approved mission-essential task list for SFA missions. Subordinate commanders prioritize tasks that need training. Since there is never enough time to train in every area, commanders focus on tasks essential for mission accomplishment.

4-213. Once commanders select tasks for training, the staff builds the training schedule and plans on these tasks. The staff provides the training requirements to the commander. After approving the list of tasks to be trained, the commander includes the tasks in the unit training schedule. The staff then coordinates the support and resource requirements with the S-3 and S-4. Finally, the commander ensures standards are enforced during training.

Evaluation

4-214. Evaluations can be either internal or external. Internal evaluations occur at all levels, and they must be inherent in all training. External evaluations are usually more formal and conducted by a headquarters one or two levels above the unit being evaluated. This subject must be carefully planned and discussed with FSF leaders to account for cultural sensitivities and current capabilities. A critical weakness in training is the failure to evaluate each task every time it is executed. Every training exercise provides potential for evaluation feedback. Every evaluation is also a training session. Leaders and trainers must continually evaluate to optimize training. Evaluation must occur as the training takes place. Emphasis is on direct, on-the-spot evaluations. However, leaders allow Soldiers to complete the task first. Leaders plan after action reviews at frequent, logical intervals during exercises. This technique allows the correction of shortcomings while they are still fresh in everyone's mind. The after action review eliminates reinforcing bad habits.

Specified Training

4-215. Augmentation elements require area orientation, refresher combat training, field-training exercises, and the like. Unit training objectives are for developing capabilities to conduct internal and external defense activities for tactical operations, intelligence operations, psychological operations, populace and resources control operations, and civil affairs and advisory assistance operations in the host nation language. Units identified for SFA begin intensified training immediately upon deployment notification.

4-216. After deployment to the host nation and before commitment to operations, the unit may receive in-country training at host-nation training centers or at designated training locations. This training helps personnel become psychologically and physically acclimated to the host nation. This training also allows commanders and staff some time to coordinate and plan within their own command and with civilian and military joint and multinational organizations. After commitment, training continues and is stressed between operations, using needed improvements identified in operations as the basis for training.

4-217. Insider attacks are a threat in any area of operation. The battalion commander ensures that military forces, civilian expeditionary workforce, and/or civilian personnel and contractors are trained to identify behavioral indicators of possible insider threats and the means to apply prevention tools to mitigate this threat. Cultural awareness yields situational awareness and leads to increased force protection for SFA personnel. Eliminating and/or minimizing the insider threat, especially by proper preparation and training of forces, is critical to mission success.

> ***Note.*** To reduce the potential for insider attacks, FSF should be further vetted to identify individuals whose motivations toward the host nation and United States Government are in question.

Build Partnerships and Teams

4-218. The Infantry battalion of the IBCT, augmented for SFA, will have subordinate units whose sole focus is working with FSF. Advisor teams may be formed from BCT or battalion organic resources, external augmentation, or a combination. These teams optimally are embedded with the counterpart unit(s), or they may reside on a U.S. camp and commute to FSF they support. The method depends on policy, direction from higher headquarters, the conditions of the operational environment, and capacity of the FSF camps to accommodate the U.S. forces.

Security Force Assistance Brigade

4-219. Security force assistance brigades (when established) will provide SFA to host-nation FSF. The security force assistance brigade provides organic forces to form the basis for the security force assistance brigade mission to support FSF. The company team is the foundation for the security force assistance brigade's mission and is augmented with additional personnel and assets to accomplish the mission.

Support to Security Force Assistance Activities

4-220. The Infantry battalion of the IBCT may support security force assistance activities, including potentially supporting multiple FSF organizations in its area of operation. Additionally, these FSF organizations may each report through different host-nation government channels and even to different ministries. To synchronize efforts in this case, U.S. forces must achieve unity of effort. Similarly, each of the FSF organizational commanders should synchronize their efforts with the host-nation government representatives, as appropriate.

4-221. Figure 4-5 depicts an example task organization, used for discussion purposes, for a battalion supporting multiple SFA activities. Within the task organization, one company team acts as a response force with adequate tactical mobility and designated fire support to defeat Level I and Level II threats. The response force shapes Level III threats until a tactical combat force or other capable response force can defeat it. (See paragraphs I-32 and I-30, respectively.) Additionally, the task organization depicts how a company team may form the foundation for host nation military and border support. Finally, the task organization depicts how a company team may provide police support. Support is in the form of two platoons supporting police assistance teams and a third platoon in a combined (multinational) security station providing support to a police assistance team.

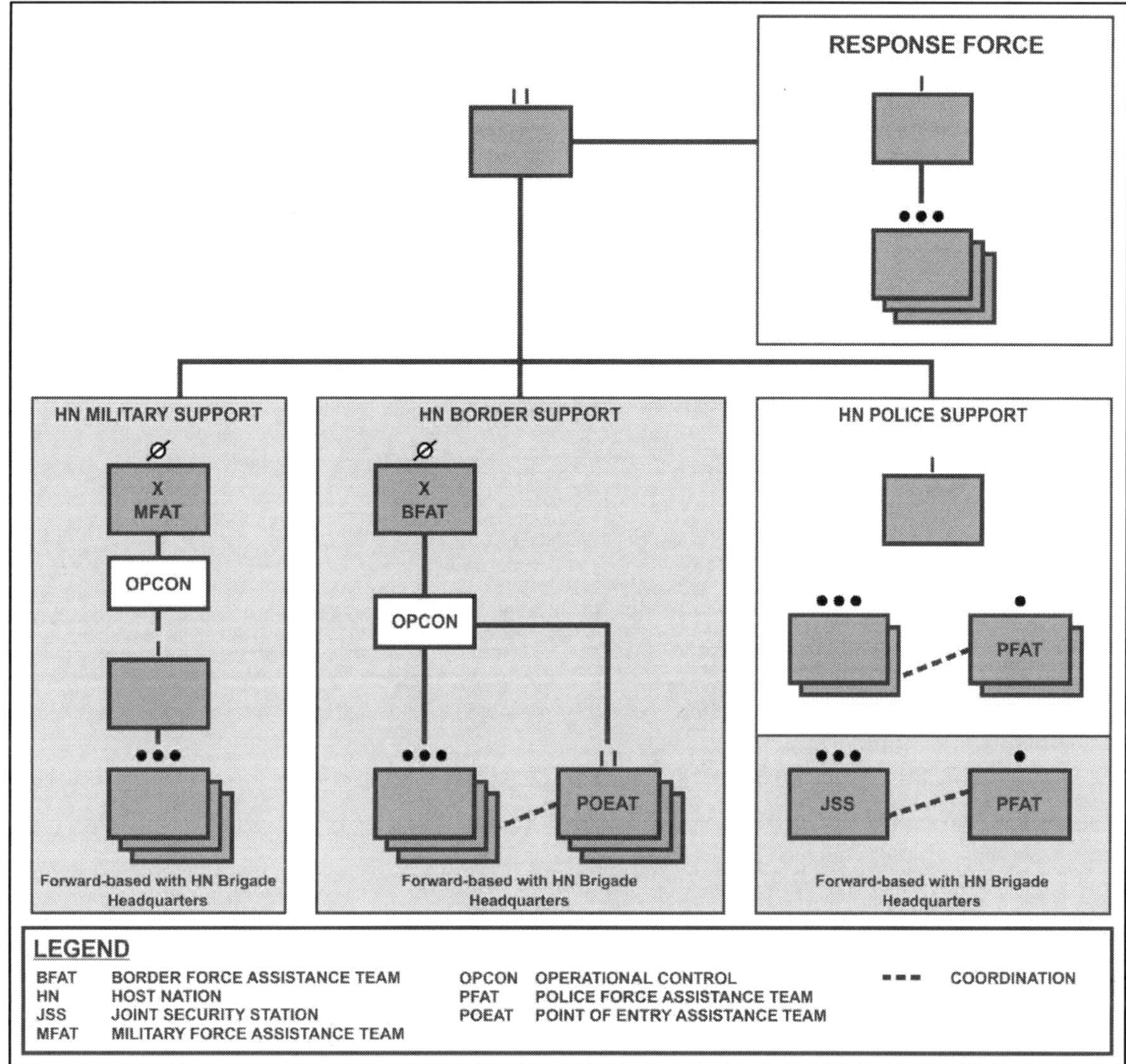

Figure 4-5. Battalion support to security force assistance, example

Note. The designation of force assistance teams used in figure 4-5 are not to be prescriptive of how the Infantry battalion might support a particular SFA activity. Designations are intended to be used as a guide illustrating one way the battalion may task organize to support multiple SFA activities within an area of operation.

4-222. Subordinate units of the IBCT conducting SFA are best located inside the base of the FSF to be trained. Collocation facilitates integration with the FSF and allows the two forces to form mutual understanding and trust. Collocation and the close cooperation often facilitates and improves the population's perception of the legitimacy of U.S. and foreign security forces, which can be an essential condition of the overall mission's end state.

4-223. When protection conditions require, a U.S. area may be established in the FSF base, although this is not optimal. Key considerations for collocation may include the threat, FSF acceptance, physical space inside the FSF base, sustainment capabilities, medical facilities, and availability of response forces.

4-224. When U.S. forces are operating out of smaller outposts in an urban environment, the local populace sees the integration and presence of the U.S. and foreign security forces working together. This integration not only enhances overall operational effectiveness and trust, living and working together builds legitimacy of the two forces as well as FSF; it reinforces trust between the FSF and the people they are tasked to protect.

Deployment

4-225. SFA is often conducted in operational environments in which U.S. forces are guests of the host nation or partner organization. When not already in country, SFA units move into the operational area by following their deployment OPORD and standing operating procedures. (See FM 3-22 for information on deployment activities.) When located within the operational area, units conduct troop movement (see paragraphs 2-34 to 2-36) to their assigned area of operation.

Note. See FM 3-22 for information on redeployment and post-deployment activities.

In-Country Preparations

4-226. Upon arrival, the commander and S-3 brief the higher headquarters on the planned execution of the mission and reconfirm the required command relationship. Local conditions may require the unit to confirm or establish its in-country and external mission command systems and sustainment functions relationships from outside its operational area upon arrival. The SFA unit establishes direct working relationships with the next higher in- or out-of-country supporting element to—

- Determine the limits of the available support and expected reaction time between the initiation of the support request and fulfillment.
- Confirm or establish communications procedures between the supporting element and the SFA unit, to include alternative and emergency procedures for mission command, all support operations, and medical evacuation.

4-227. The SFA unit establishes procedures to promote interagency cooperation and synchronization. The unit—

- Identifies the location of the concerned host nation, U.S., or other agencies.
- Contacts the concerned agency to establish initial coordination.
- Exchanges information or intelligence.
- Confirms or establishes other coordination protocols as necessary.
- Incorporates the newly established or changed procedures into the plans for mission execution.

4-228. The unit immediately establishes operations security procedures to support its mission execution and identifies rally points incorporated into its defensive, evasion, and personnel recovery plans.

4-229. After receiving a detailed briefing and further guidance from the advance party, unit personnel continue to develop effective rapport with the FSF commander and counterparts. They also assess their working, storage, and living areas for security and verify the location of the training site, communications center, dispensary area, and FSF troop area. With the FSF commander, the unit commander—

- Establishes rapport.
- Conducts introductions in a businesslike, congenial manner.
- Briefs on the unit's mission, its capabilities, and the restrictions and limits imposed on the detachment by the higher U.S. commander.
- Ensures all unit personnel fully support FSF and firmly believes a joint U.S.-FSF effort will succeed. Requests counterpart linkup be made under the mutual supervision of the FSF commander and the unit commander.
- Ensures all current unit plans are tentative and that assistance is needed to finalize them.
- Deduces or solicits the actual estimate of unit capabilities and perceived advisory assistance and material requirements.
- Recommends the most desirable courses of action while emphasizing how they satisfy present conditions, achieve the desired training, and meet advisory assistance goals.
- Informs the higher in-country U.S. commander of any significant changes in the unit's plan to assist FSF.

4-230. Through the S-2, the commander's priority intelligence requirements are based on the latest information available and requirements for additional priority intelligence requirements that arise from modified estimates and plans. The S-2 also—

- Analyzes the foreign unit's status to finalize unit plans for advisory assistance. These plans can include task organization of unit with counterparts, staff functions for planning SFA, and advisory assistance for executing SFA.
- Explains analysis to counterparts and encourages them to help with—and participate in—analyzing, preparing, and briefing the analysis to the foreign unit commander.
- Prepares and briefs the plans for training and advisory assistance.
- Helps the foreign unit inspect the available facilities to identify deficiencies. If the unit finds deficiencies, the S-2 prepares estimates of courses of action for the FSF commander to correct them.
- Supervises the preparation of the facilities with their counterparts and informs unit and FSF commanders on the status of the facilities.

4-231. The unit ensures its security is based on the present or anticipated threat. Some recommended actions the unit may take include—

- Hardening its positions based on available means and requirements to maintain low visibility.
- Maintaining unit internal guard system with at least one Soldier who is awake and knows the locations of all other unit personnel. The guard reacts to an emergency by following an internal alert plan and starting defensive actions.
- Maintaining communications with all subordinate unit personnel deployed outside the immediate area controlled by the main body.
- Establishing plans for immediate defensive actions in the event of an attack or a loss of rapport with hostile reaction.
- Discussing visible security measures with foreign counterparts to ensure understanding and to maintain effective rapport. Unit personnel do not divulge sensitive information for the sake of possible rapport benefits.
- Encouraging the foreign unit, through counterparts, to adopt additional security measures identified when analyzing the foreign unit's status and inspecting its facilities.
- Coordinating defensive measures with the foreign unit to develop a mutual defensive plan. Unit personnel obtain from the unit's present reaction and defensive plans for attack. They encourage the foreign unit to conduct mutual full-force rehearsals of defensive plans; if unsuccessful, the unit conducts internal rehearsals of the plans.

EXECUTION

4-232. In execution, the battalion commander, staff, and subordinate commanders focus efforts on translating decisions, made during planning and preparation, into actions supporting the SFA mission. Once the Infantry battalion arrives in-country, it begins the employment of forces to support the development of FSF capabilities and capacities. Employment of the battalion occurs generally with the establishment of advising, assisting, and training teams and key individuals. These teams and key individuals partner with foreign counterparts during FSF planning [preparing the FSF for the mission(s) itself] to increase the capability and capacity of FSF planning processes, as well as to increase the probability of success.

Note. SFA activities normally use the general developmental tasks (known as FSF development tasks) of organize, train, equipment, rebuild and build, advise and assist, and assess (OTERA-A) to develop the functional capabilities required by the FSF. See paragraph 4-85 for information on organize, equip, and rebuild and build developmental tasks.

FOREIGN SECURITY FORCE DEVELOPMENT TASKS—ADVISE, ASSIST, AND TRAIN

4-233. The Infantry battalion conducting SFA missions normally task organizes into smaller rotational teams, and identifies key individuals, for execution. These teams and key individuals focus on advising, assisting, or training a specific partner individual, unit, or activity. These teams and key individuals include, but are not limited to, Infantry battalion training, advising, or advisory teams and individuals. Specialized teams and individuals may also be required for partner sustainment, engineer, or police units.

Advise and Assist Foreign Security Forces

4-234. The Infantry battalion, with possible additional augmentation teams, advises and assists FSF to improve their capability and capacity. Advising establishes a personal and a professional relationship where trust and confidence define how well the advisor will be able to influence the foreign security force. Assisting is providing the required supporting or sustaining capabilities so FSF can meet objectives and the end state. The level of advice and assistance is based on conditions and continues until FSF can establish required systems or until conditions no longer require it. Advising and assisting teams from the battalion do not permit the FSF to fail critically at a point that would undermine the overall effort.

Battalion Security Force Assistance – Advise and Assist Scenario, Example

4-235. The following example, advise and assist scenario, used for discussion purposes, represents different ways to advise and assist FSF operations. Key Infantry battalion advisors include the—

Battalion Commander

4-236. Before the mission, the Infantry battalion commander advises and assists the FSF commander. The FSF commander then issues planning guidance for planning the execution of the mission and clarifies commander's intent. The battalion commander advises and assists the FSF commander throughout the operations process for the tactical operation(s). By accompanying the FSF commander when the mission is received from higher headquarters, the battalion commander assists any subsequent missions. The battalion commander monitors how FSF subordinate units understand the commander's intent and all specified and implied tasks.

4-237. During the execution of the mission, the battalion commander helps the FSF commander provides mission command (also called, command and control) during operations. While monitoring the tactical situation, the battalion commander recommends changes to the chosen course of action(s) to exploit the situation. After monitoring the flow of information, the battalion commander recommends improvements to the use of intelligence collection assets and the processes used by subordinates to report required information.

Battalion Executive Officer

4-238. The battalion executive officer performs the organizational analysis of the FSF coordinating staff sections to ensure efficiency during the planning process according to the FSF commander's initial planning guidance. With the foreign counterpart, the executive officer advises and assists the counterpart in directing foreign staff sections as they develop estimates, plans, and orders. The executive officer monitors the liaison and coordination with FSF higher headquarters, recommending changes to improve efficiencies.

Battalion Staff

4-239. Before the mission, members of the battalion staff advise and assist foreign counterparts in preparing staff estimates and courses of actions for essential tasks. The battalion staff helps write tentative plans and/or orders based on the FSF commander planning guidance and FSF standing operating procedures. Plans, depending on the situation, may include primary, alternate, contingency, and emergency plans.

4-240. During execution, the battalion staff helps foreign counterparts coordinate the execution of FSF tasks. The staff assists in the dissemination of FSF plans and/or orders to senior and adjacent staff sections and supporting elements. The battalion staff helps notify higher, lower, or adjacent staff sections of modified estimates and plans. The staff—led by the S-3 and S-2 and the S-3 and S-2 counterparts—helps update the CCIRs with the latest information and future requirements.

Personnel Staff Officer, S-1

4-241. The personnel staff officer provides advice, assists, and makes recommendations to the foreign counterpart for all matters concerning human resources support. This includes monitoring the maintenance of foreign unit strength, pay, accountability of casualties, and unit morale. The S-1 must emphasize to subordinates the need to assist counterparts in paying troops and accounting for funds. Close observation of disbursement and unobtrusively polling FSF troops about their pay is a vital, but an unfamiliar, skill set amongst U.S. troops.

Note. U.S. forces' automated pay systems are nothing like the cash-only transactions in FSF. Graft, corruption (ghost soldiers/policemen), and extortion are rife in these circumstances.

Intelligence Staff Officer, S-2

4-242. The intelligence staff officer advises and assists the monitoring of FSF operations security to protect classified and sensitive material and operations and recommends improvements. By helping the foreign counterpart update the situation map, the intelligence staff officer helps to keep both commands up to date on the current situation. The intelligence officer recommends improvements to the standing operating procedures of the main command post (when established the tactical operations command post) communications framework so the intelligence section receives situation reports. The intelligence officer helps the counterpart monitor the collection, evaluation, interpretation, and the dissemination of information. The intelligence officer assists in the examination of captured insurgent documents and material. The intelligence officer helps gather and disseminate intelligence reports from available sources to ensure the exploitation of all unit operations assets. The intelligence officer helps the counterpart to brief and debrief patrols operating as a part of R&S activities. The intelligence officer works with the advisor operations officer to develop R&S plans with the FSF partner.

Note. As above, train personnel two deep in every staff section or advisory subunit on foreign disclosure before deployment. Interaction with host nation and FSF, even NATO or other coalition allies requires foreign disclosure officer approval. This will become a huge bottleneck if not trained for and decentralized.

Operations Staff Officer, S-3

4-243. The operations Staff officer helps the foreign counterpart to prepare tactical plans and/or orders using estimates, predictions, and information. The operations officer monitors command and communications nets, assists in preparing all plans and orders, and helps to supervise the training and preparation for operations. The operations officer monitors the planning process and makes recommendations for consistency with FSF partner objectives and goals.

Logistics Staff Officer, S-4

4-244. The logistics staff officer advises and assists the foreign counterpart in maintaining equipment readiness; monitoring the support provided to the foreign unit, its subunits, and attachments; and in recommending improvements. The logistics officer helps to supervise the use of transportation assets.

Signal Staff Officer, S-6

4-245. The signal staff officer advises and assists the foreign counterpart for all matters concerning network operations, network transport, information services, and spectrum management operations within the battalion's SFA and FSF area of operation. The signal officer monitors communications security throughout planning, preparation, and execution of SFA and FSF activities. The signal officer ensures SFA personnel are trained in the protection of sensitive communications equipment and cryptographic materials during the execution of FSF operations. The signal officer identifies SFA and FSF communications requirements, obtains communications resources for austere locations, and ensures redundant and backup systems are available and tested.

4-246. The Signal officer, in coordination with the IBCT, continuously assess and assist interorganizational information management coordination, normally required among participating interagency partners and the affected partner nation organizations. The signal officer uses assessments as part of the SFA and FSF communications synchronization plan. The signal officer uses foreign disclosure procedures and a tailored and responsive information- sharing process as part of the SFA and FSF assessment plan for dissemination with interagency partners and/or multinational audience.

Civil Affairs Team

4-247. A civil Affairs team is generally assigned to each battalion to conduct civil affairs operations in support of the civil-military operations plan. The civil affairs team is the basic civil affairs tactical support element provided to a supported commander. The civil affairs team executes civil affairs operations and is capable of conducting civil reconnaissance and civil engagement along with assessments of the civil component of the operational environment. The success of the overarching civil affairs operations plan is predicated on the actions of the civil affairs team at the lowest tactical levels. The civil affairs team, due to its limited capabilities, relies on its ability to leverage other civil affairs assets and capabilities through reachback to the civil affairs company civil-military operations center in order to shape operations. The civil-military operations center is a standing capability formed by civil affairs units and is tailored to the specific tasks associated with the mission and normally augmented by other enablers such as engineer, medical, and transportation resources available to the supported commander. The civil affairs team attached to the battalion will interface with the S-9, civil-military operations center and civil affairs company at the IBCT level to ensure all battalion civil-military operations are nested with the IBCT commander's civil-military operations plan.

Civil-Military Teams

4-248. Upon deployment, civil-military teams advise the SFA and FSF commanders and staffs on civil-military considerations and coordinate efforts of any civil affairs units supporting the FSF operation. Civil-military teams mentor counterpart teams and the supported foreign element staff on civil-military operations and the importance of respecting human rights. Civil-military teams may introduce counterparts to relevant nongovernmental organizations, United States Agency for International Development project officers, and provincial reconstruction team staff.

Note. The judge advocate (judge advocate general corps) mentors (provide legal mentorship) and/or coordinates the legal and moral obligations of military commanders to civilian populations under their control. (See AR 27-1 and FM 1-04.)

Company-Level and Below Advisors

4-249. Company-level and below advisors assist foreign counterparts to analyze the FSF mission and commander's intent from higher headquarters. Company advisors assist FSF company commanders and subordinate leaders restate the mission, conduct an initial risk assessment, identify a tentative decisive point, and define their own intent. Company advisors assist their foreign counterparts to analyze the mission and operational variables. From these variables, advisors help their foreign counterparts to develop a course of action that meets the higher headquarters concept of operations and commander's intent. Company advisors assist in the conduct of operations and the flow of information to the FSF higher commander. (Refer to FM 3-21.10 for additional information.)

Train Foreign Security Forces

4-250. Battalion trainers (or advisors) consistently provide and instill leadership at all levels of the FSF organization. Depending on the circumstances, the battalion may execute an SFA training missions unilaterally, or as part of a multinational force. In any case, leadership is especially important in the inherently dynamic and complex environment associated with SFA. SFA activities require the personal interaction of battalion trainers (or advisors) and FSF trainees, and other military and civilians organizations/agencies. A high premium is placed on effective leadership from junior, to the most senior noncommissioned and commissioned officers. This leadership must fully comprehend the operational environment and be prepared, fully involved, and supportive for FSF training to succeed. An effective FSF requires leadership from both the provider and the recipient sides throughout training to help build the FSF capacity to train their own forces.

Battalion Security Force Assistance —Training Scenario, Example

4-251. Battalion trainers work with the FSF to give a positive context and narrative to the FSF professionalization efforts and capacity to secure the population. Coordination of the information themes and messages among the Infantry battalion, FSF, and the host-nation government, and the presentation or availability of information to the indigenous population can limit or mitigate the propaganda efforts of insurgents or hostile forces. This may serve to mitigate the potential for destabilizing influences of hostile forces or criminal elements to propagandize the battalion's training effort and damage the FSF credibility and legitimacy. (Refer to FM 7-0 for additional information.) The following example - training scenario, used for discussion purposes, represents different ways to support FSF training.

Training Assessment

4-252. Prior to training the FSF, the battalion commander begins with a training assessment, in coordination with the FSF commander, of the training plans designed prior to the battalion's employment. This assessment is important to evaluate the FSF and to exercise the working relationship between the Infantry battalion and the FSF. The training assessment covers all aspects of leadership, training, sustainment, and professionalization. To support an assessment, the battalion commander analyzes the following specific foreign unit considerations:

- The unit's mission and mission-essential task list and capability to execute them.
- Staff capabilities.
- Personnel and equipment authorization.
- Physical condition.
- Any past or present foreign influence on training and combat operations.
- Operational deficiencies identified during recent operations or exercises with U.S. personnel.
- Sustainment capabilities, to include training programs.
- Internal training programs and personnel.
- Training facilities.

4-253. The battalion commander assesses the level of professionalism of FSF, both units and individuals. Adhering to established rules of engagement, ethics that meet the established laws and regulations of the commanding authority, laws for land warfare, and human rights are key areas that require assessment. The FSF support of civilian leaders and political goals also fall within this assessment.

4-254. Battalion subordinate leaders, working with FSF leaders, evaluate current members of the FSF for past military skills and positions. Often military reorganizations arbitrarily shift personnel to fill vacancies outside their knowledge and experience.

Analysis of the Prepared Training Plan

4-255. After completing the training assessment, the battalion commander and subordinate commanders analyze the prepared training plans and determine if changes are necessary. Training plans stress the deficiencies identified in the training assessment. The training plan identifies those in the host nation able to help train FSF to strengthen the legitimacy of the process. Using a comprehensive approach within the battalion's area of operations can provide support and expertise that enhance the training and operations process, and the FSF eventual self-sustainment. As the FSF gains sufficient capacity and capabilities to perform independently, trainers/advisors transition from a leading role to a mentoring role.

Program of Instruction

4-256. In coordination with the FSF staff and subordinate units, the staff and subordinate units of the Infantry battalion develop programs of instruction. These programs incorporate all training objectives that satisfy the training requirements identified during assessment. Training programs support these requirements.

The FSF commander approves these programs of instruction prior to execution by the battalion. When executing programs of instruction, trainers/advisors adhere to training schedules consistent with changes in the mission variables. Trainers/advisors ensure through their counterparts and the FSF commander that all personnel receive training. Foreign counterpart trainers rehearse all classes approved on the programs of instruction.

Presentation of Instruction

4-257. Presenting the training material properly, trainers follow lesson outlines approved in the programs of instruction. All training clearly states the task, conditions, and standards desired during each lesson, ensuring the FSF understand them. Trainers/advisors state all warning and safety instructions (through interpreters when required) to the FSF. The training to reinforce the concepts includes demonstrations of the execution of each task, stressing the execution as a step-by-step process. Trainers monitor FSF progress during instruction and practical exercises, correcting mistakes as they are made.

Training Methodology (Crawl-Walk-Run)

4-258. An effective method of training used is the crawl-walk-run training methodology to assist trainers in teaching individual tasks, battle drills, and collective tasks, and when conducting field exercises. This methodology is employed to develop well-trained leaders and units. Crawl-walk-run methodology is based on the three following characteristics in lane training:

- Crawl (explain and demonstrate). The trainer describes the task step-by-step, indicating what each individual does.
- Walk (practice). The trainer directs the unit to execute the task at a slow, step-by-step pace.
- Run (perform). The trainer requires the unit to perform the task at full speed, as if in an operation, under realistic conditions.

4-259. During all phases, the training must include the mission of the unit in the context of the higher unit's mission to assist with the practical application of the training. Identifying the higher commander's mission and intent, as well as the tasks and purposes of other units in the area, adds context to the training. This method is expanded to include the role of other actors.

4-260. Trainers continue individual training to improve and sustain individual task proficiency while units train on collective tasks. Collective training requires interaction among individuals or organizations to perform tasks, actions, and activities that contribute to achieving mission-essential task proficiency. Collective training includes performing collective, individual, and leader tasks associated with each training objective, action, or activity. (Refer to FM 7-0 for additional information.)

Collective Training

4-261. Collective training starts at squad level. Squad battle drills provide key building blocks to support FSF operations. Battalion trainers link battle drills and collective tasks through a logical, tactical scenario in situational training exercises. Although this exercise is mission-oriented, it results in more than mission proficiency. Battle drills and collective tasks support situational training exercises, while these exercises support operations. Battalion trainers/advisors must understand the operational environment when training FSF; training incorporates how internal and external threats and civilians affect the environment.

4-262. Flexibility in using Army doctrine in training enhances efforts to make training realistic. Battalion trainers/advisors modify Army doctrine to fit the FSF level of expertise, mission command systems, the tactical situation, and sustainment base. Often the structure and capabilities of FSF differ from that required by Army doctrine. When FSF counter an insurgency, these exercises emphasize interplay among psychological and tactical, populace and resources control, intelligence, and civil affairs operations. (Refer to FM 7-0 for additional information.)

Individual Training

4-263. Individual training within the FSF by the battalion emphasizes physical and mental conditioning, tactical training, basic rifle marksmanship, first aid, combatives, and the operational environment. Individual training includes general tactics and techniques of security operations and the motivation, operations, and objectives of internal and external threats. Tough and realistic training conditions troops to mentally and physically withstand the strain of continuous operations. The battalion cross-trains the FSF on all types of weapons, communications and other equipment, and skills particular to their unit. Personnel losses must never cause weapons, communications equipment, or essential skills to be lost due to a lack of fully trained replacement personnel.

Small-Unit Leader Training

4-264. SFA activities frequently entail rapidly changing circumstances; thus, FSF small-unit leaders must be able to plan and execute operations with little guidance. Battalion trainers/advisors stress small-unit leadership training concurrently with individual training. Tools the trainer uses to train leaders are manuals, previously established training, tactical exercises without troops, and unit missions. Small-unit leader training by the battalion develops aggressiveness, tactical proficiency, and initiative. Small-unit leader training should include combined arms technical training procedures for forward observer and close air support. Leadership training includes land navigation in difficult terrain and under conditions of limited visibility. Mission readiness and the health and welfare of subordinates are continuous parts of training.

FOREIGN SECURITY FORCE DEVELOPMENT TASKS —ASSESS

4-265. The functional tasks of OTERA-A (see paragraph 4-85) serve as SFA capability areas used by the SFA battalion to develop, change, or improve the capability and capacity of the FSF. By conducting an assessment of the FSF, the SFA battalion can determine which area or areas within the OTERA-A construct to use to improve the FSF to the desired capability and capacity. In essence, the SFA battalion conducts an assessment of the FSF against desired capabilities and then develops an OTERA-A plan to help the FSF build capability and capacity.

Example, Battalion Security Force Assistance, Assessment Scenario

4-266. The assessment developmental task, not limited to planning, preparing, or executing, by the Infantry battalion is ongoing throughout the operations process. Assessment involves continuously monitoring and evaluating the operational environment to determine what changes might affect the conduct of training and operations. The following example – assessment scenario, used for discussion purposes, represents different ways to assess training and operations.

Security Force Assistance Activity Assessment

4-267. During SFA assessments to evaluate the status of FSF capabilities and capacity, assessments by the Infantry battalion, establish a measurement (see paragraphs 4-43 and 4-48) at a particular time and can be compared to other assessments to observe differences and progress attributable to SFA activities. Activity assessment by the battalion involves deliberately comparing forecasted outcomes with actual events to determine the overall effectiveness of the battalion's employment. More specifically, assessment helps the battalion commander determine progress toward attaining the desired end state, achieving objectives, and performing tasks.

Foreign Security Force Training and Evaluation

4-268. In training, the after action review provides the critical link between training and evaluation. The review is a professional discussion that includes the training participants and focuses directly on the training goals. An after action review occurs after all collective FSF training. Effective after action reviews review training goals with the responsible FSF commander. During the review, SFA trainers/advisors ask leading questions, surface important tactical lessons, explore alternative course of actions, assist the retention teaching points, and keep the after action review positive.

Comprehensive Review

4-269. The battalion commander encourages the FSF commander to conduct a comprehensive review of collective training events with the entire unit, or at a minimum, with key subordinate leaders. If possible, the review occurs during the field portion of the training when the unit assembles at logical stopping points.

During the review, the battalion commander and subordinate trainer/advisors avoid criticizing or embarrassing the FSF commander or subordinates. After action reviews provide feedback to increase and reinforce learning, providing a database for key points. During reviews within subordinate echelons, evaluators draw information from FSF subordinate leaders to form possible alternative course of actions for future activities.

Note. It is important to conduct comprehensive after- action reviews and reports, focusing on the specifics of the SFA activities, to gather information as soon as possible after execution.

Short-, Mid-, and Long-term Success

4-270. During SFA activities, including FSF operations, success is defined within the context of three periods: short-, mid-, and long-term. In the short-term period, FSF make steady progress in fighting threats, meeting political milestones, building democratic institutions, and standing up security forces. In the mid-term period, FSF lead fighting threats and provide security, have a functioning government, and work towards achieving economic potential. In the long-term period, FSF are peaceful, united, stable, and secure; integrated into the international community; and a full partner in international security concerns.

Monitor the Current Situation

4-271. The Infantry battalion commander and subordinate advisors help foreign counterparts monitor the current situation for unanticipated successes, failures, or enemy actions. As the battalion commander assesses the progress of FSF operations, the commander looks for opportunities, threats, and acceptable progress. The commander considers, as part of the military decision-making process, the second- and third-order effects of the FSF operation. The battalion commander and subordinate advisors develop a cultural awareness and use this awareness so that operations and relationships achieve the desired end state.

Operational Success

4-272. Throughout the operation, the battalion commander assists the FSF commander in addressing changes to the operation and the feeding the assessments of the progress or regression back into the planning process. The closer SFA and FSF commanders work with trainer/advisor teams and the more they interact with local political and cultural leaders, the better the overall chances of mission success. Keys to operational success within the SFA and FSF area of operation, although not all inclusive, include the following:

- Establish MOE to provide benchmarks against which the commander assesses progress toward accomplishing the mission.
- Establish MOP to determine whether a task or action was performed to standard.
- Establish close and continuing relationships with all advisor teams, other actors operating in its area of operation, and foreign area officers with local or regional expertise.
- Establish close and continuing relationships with all foreign units (military, police, and others) operating in the area of operation.
- Establish close and continuing relationships with all political entities and actors within the area of operation.
- Establish redundant communications within the area of operation, especially when the battalion shares its area of operation with other entities that have cultural differences and lack of or degraded communications.

Appendix A

Command Post Operations and Organizations

A *command post* (CP) is a unit headquarters where the commander and their staff perform their activities. Also called CP (FM 6-0). The commander balances the need to create a capable command post organization(s) to support the capacity to plan, prepare, execute, and continuously assess operations with the resulting diversion of capabilities to fight the enemy due to the size of the CP itself. Larger CPs ease face-to-face coordination; however, they are vulnerable to multiple acquisitions and means of attack. Smaller CPs can be hidden and protected more easily, but they may not exercise the degree of mission command necessary to control all battalion subordinate units. Striking the right balance provides a responsive yet agile organization.

Appendix A addresses how the Infantry battalion commander organizes the headquarters into CPs and organizes the staff and personnel within CPs into functional and integrating cells, staff sections, and meetings during the conduct of operations. This appendix provides guidelines for CP operations to include the importance of establishing standard operating procedures (SOPs) and a battle rhythm. This appendix considers various factors that degrade the efficiency of mission command systems within organizations and considerations for digital and analog mission command systems techniques. (Refer to FM 3-96, FM 6-0, and ATP 6-0.5 for additional information.)

COMMAND POST ORGANIZATION

A-1. The battalion commander organizes the staff within each CP to perform essential staff functions to aid with planning and controlling operations. The commander positions CPs within areas of operation to maintain flexibility, redundancy, survivability, and mobility. Activities common in all CPs include, but are not limited to—

- Maintaining running estimates.
- Controlling operations.
- Assessing operations.
- Developing and disseminating orders.
- Coordinating with higher, lower, and adjacent units.
- Conducting knowledge management and information management.
- Conducting network operations.
- Providing a facility for the commander to control operations, issue orders, and conduct rehearsals.
- Maintaining the common operational picture (COP).
- Performing CP administration (examples include sleep plans, security, and feeding schedules).
- Supporting the commander's decision-making process.

A-2. CPs provide staff expertise, communications, and information systems that work in concert to aid the commander in planning and controlling operations. The headquarters design, combined with robust communications, gives the commander a flexible mission command structure consisting of a main CP and a tactical CP, and as required, the battalion may establish a combat trains CP and/or a field trains CP. *Information system* is the equipment that collects, processes, stores, displays, and disseminates information. This includes computers—hardware and software—and communications, as well as policies and procedures for their use. (ADP 6-0)

A-3. The commander mans, equips, and organizes CPs to control operations for extended periods. CP personnel, information systems, and equipment must be able to support 24-hour operations while in continuous communication with all subordinate units and higher and adjacent units. The commander arranges CP personnel and equipment to facilitate internal coordination, information sharing, and rapid decision making. CP personnel ensure they have procedures to execute the operations process within the headquarters to enhance how they exercise mission command. The commander uses the following to assist effective CP operations: SOPs, battle rhythm, meetings, which are described below:

- SOPs serve two purposes. Internal SOPs standardize each CP's internal operations and administration. External SOPs developed for the entire force standardize interactions among CPs and between subordinate units and CPs. For SOPs to be effective, all Soldiers must know their jobs requirements and train to their standards.
- *Battle rhythm* is a deliberate cycle of command, staff, and unit activities intended to synchronize current and future operations (FM 6-0). A headquarters' battle rhythm consists of a series of meetings, briefings, and other activities synchronized by time, purpose, and prioritization.
- Meetings are gatherings to present and exchange information, solve problems, coordinate action, and make decisions. Meetings may involve the staff; the commander and staff; or the commander, subordinate commanders, staff, and other partners. Who attends depends on the issue(s) and requirement(s) for the meeting.

Main Command Post

A-4. The *main command post* is a facility containing the majority of the staff designed to control current operations, conduct detailed analysis, and plan future operations (FM 6-0). A main CP is the battalion commander's principal CP. The main CP includes representatives of all staff sections and a full suite of information systems to plan, prepare, execute, and assess operations. It is larger, has more staff members, and is less mobile than the tactical CP. The main CP operates from a stationary position and moves as required to maintain control of the operation. CP locations are METT-TC dependent. Techniques include—in contiguous AOs, the main CP may locate behind company CPs and the battalion tactical CP, and out of enemy medium artillery range, if practical. In noncontiguous AOs, the main CP usually locates within a subordinate company's AO. The battalion XO supervises all staff activities and functions within the main CP.

A-5. The main CP serves the following functions:

- Synchronizes combat and sustainment activities in support of an overall operation.
- Provides a focal point for the development of intelligence.
- Supports information understanding for the battalion commander and subordinates by monitoring, analyzing, and disseminating information.
- Allows the commander the flexibility to control the operation from a forward location.
- Receives, processes, produces, and submits required reports.
- Creates and disseminates orders.
- Supports continuous operations.
- Monitors and anticipates the commander's decision points.
- Plans future operations.
- Coordinates with higher headquarters and adjacent units.
- Serves as net control station for the operations and intelligence radio net and backup net control station for a command radio net.
- Plans and controls battalion reconnaissance and surveillance, and security operations.

A-6. The jump CP (when utilized) is part of the main CP that is used to maintain mission command while the main CP displaces. The jump CP may be led by the battalion assistant S-3 and consist of sufficient staff and communications to establish communications and control the battalion while the main CP displaces. When acting as the advance party for the main CP, the jump CP marks locations for the rest of the main CP. It is austere and temporary. (See paragraph A-17 for additional information displacement.)

TACTICAL COMMAND POST

A-7. The *tactical command post* is a facility containing a tailored portion of a unit headquarters designed to control portions of the operation for a limited time (FM 6-0). The commander employs the tactical CP as an extension of the main CP to help control execution of the operation or a specific task. The tactical CP is fully mobile. As a rule, in addition to the commander, it includes leaders, Soldiers, and equipment essential to the tasks assigned. The tactical CP relies on the main CP for planning, detailed analysis, and coordination.

A-8. The tactical CP typically performs the following functions:

- Monitors and controls current operations.
- Maintains and updates the COP.
- Allows the commander to be at the decisive point.
- Monitors and assesses the progress of operations.
- Monitors and assesses the progress of higher and adjacent units.
- Performs targeting for current operations.
- Provides a facility for the commander to control operations and issue orders.
- Provides input to targeting and future operations planning.
- Provides a facility for the commander to conduct rehearsals.

A-9. When the commander does not employ the tactical CP, the staff assigned to it reinforces the main CP. Unit SOPs should address the specifics for this, including procedures to detach the tactical CP from the main CP.

COMBAT TRAINS COMMAND POST

A-10. When established, the combat trains CP plans and coordinates sustainment operations in support of the tactical operations. The combat trains CP serves as the focal point for all administrative and logistical functions for the battalion. The combat trains CP may serve as an alternate CP for the battalion main CP. The battalion S-4 usually serves as the combat trains CP officer in charge, and the maintenance control officer usually serves as the maintenance collection point officer in charge. The headquarters and headquarters company commander usually exercises mission command for the combat trains CP. The combat trains CP serves the following functions:

- Tracks the current battle.
- Controls sustainment support to the current operation.
- Provides sustainment representation to the main CP for planning and integration.
- Monitors supply routes and controls the sustainment flow of materiel and personnel.
- Coordinates evacuation of casualties, equipment, and detainees.

FIELD TRAINS COMMAND POST

A-11. When established, the field trains CP serves as the battalion commander's primary direct coordination element with the supporting BSB generally located in the brigade support area. The field trains CP usually consists of the headquarters and headquarters company executive officer and first sergeant, a battalion S-4 and S-1 representative(s), and supply sergeant or representative(s). The field trains CP can be controlled by the headquarters or headquarters company executive officer or designated representative. The field trains CP serves the following functions:

- Synchronizes and integrates the Infantry battalion's concept of support.
- Coordinates logistics requirements with the BSB support operations officer.
- Configures logistical packages tailored to support requirements.
- Coordinates with the IBCT for personnel services and replacement operations.
- Forecasts and coordinates future sustainment requirements.
- Coordinates retrograde of equipment and personnel (casualty evacuation, personnel movement, and human remains).

FACILITY CONSIDERATIONS

A-12. CP facilities consist of the vehicles and locations where the commander, assisted by the staff directs the battle and sustains the force. CPs must have the ability to track the battle and assume control of the current fight. CP survivability depends mostly on concealment and mobility. The best way to protect a CP is to prevent the enemy from detecting it. Good camouflage and proper noise, light, and signal discipline enhance the security provided by a good location.

LOCATION

A-13. The commander, dependent upon an analysis of the mission variables of METT-TC, determines CP locations within an AO. Built-up areas can be good locations for CPs because they provide cover and concealment, access to electricity and other services, and good access and regress routes. However, they also can put indigenous populations at risk and can provide enemy units with covered and concealed positions to monitor and attack the CP. Locating a CP in built-up areas for longer periods tend to degrade its ability to displace quickly. A CP not located in a built-up area should be located on a reverse slope, when possible, with cover and concealment. Positioning should avoid key terrain features such as hilltops and crossroads. Locate CPs on ground that is trafficable, even in poor weather. Other actions when positioning CPs include—

- Ensuring line-of-sight frequency modulation communications with higher, lower, and adjacent units.
- Using terrain to mask communications signals from the enemy.
- Using terrain for passive security, that is, for cover and concealment.
- Collocating with tactical units for mutual support and local security.
- Avoiding possible enemy targeting from enemy observed artillery and aircraft attacks.
- Locating the CP near an existing road network, out of sight from possible enemy observation.

A-14. CPs should be near, but not next to, a high-speed avenue of approach with no more than one or two routes leading into the CP. These routes should provide cover, concealment, and access to other routes of communication. When possible, locate near a helicopter-landing zone. The area selected must be large enough to accommodate CP elements including liaison teams and attachments from other units, communications support, and eating, sleeping, latrine, and maintenance areas. Sufficient space must be available for positioning security elements and vehicle dismount points and parking. Dryness and light are vital when working with maps and producing orders and overlays. CPs should be sheltered from weather conditions and should have lights for night work while exercising proper light discipline. Buildings are the best choice, if not available, CPs can operate from their organic vehicles, tents, or any field expedient means.

OPERATIONS SECURITY

A-15. Operations security considerations for positioning CPs include the following:

- Avoid posting signs advertising CP locations.
- Disperse CP vehicles and thoroughly camouflage all vehicles and equipment.
- Maintain noise and light discipline.
- Position CP assets off major enemy avenues of approach to reduce the probability of detection.
- Use an observation post or combat outpost to maintain surveillance and to secure any remote antennas located outside the perimeter.
- Sound proof and dig in generators, if possible.
- Provide all subordinate units and elements of the CP with near and far recognition signals. Ensure the CP uses these signals, challenges, and passwords to control access into the perimeter.
- Designate a rally point(s) and an alternate CP, ensuring all members of the unit know their locations for use during enemy ground, artillery, or air attacks.

SECURITY FORCE

A-16. As threat levels increase, and when possible, the commander can establish a security force within a CP facility or a response force (see appendix I) for several facilities within an AO. Basic considerations for CP security force operations include the following:

- Establish security force positions with a 360-degree perimeter far enough out to prevent enemy direct fire and observed indirect fires on the CP.
- Establish communication procedures between the security force and CP operations personnel.
- Ensure the security force has adequate available weapons, based on the potential enemy threat.
- Rehearse the execution of the perimeter defense.

DISPLACEMENT

A-17. The main CP may displace as a whole or, more often, by echelon. Displacement as a whole normally is reserved for short movements with communications maintained by alternate means and minimal risk of degrading CP operations. A portion of the main CP, called a jump CP, moves to the new location, sets up operations, and takes over operational control from the main CP. The remaining portion of the main CP then moves to rejoin the jump CP. The jump CP consists of the necessary vehicles, personnel, and equipment to assume CP operations while the remainder moves. The battalion XO or S-3 selects a general location for the jump CP site. The jump CP may include a quartering party, which can consist of a reconnaissance and security elements, and equipment and personnel for quartering the remainder of the CP. The signal officer, who is usually part of the quartering party, ensures communications on all nets are possible from the new site. When the jump CP becomes operational, it also becomes the net control station for the unit. The remainder of the CP then moves to rejoin the jump CP. Another technique is to hand off control to the tactical CP or the combat trains CP or field trains CP and move the main CP as a whole. The tactical CP also can split, with the commander moving with the decisive operations (or main effort) and the S-3 moving with a shaping operation (or supporting effort) respectively.

ECHELONED COMMAND POSTS

A-18. Echeloned CPs control with varying levels of staff participation at each unit level. The two primary CPs within the battalion are the tactical CP (graphically depicted as the TAC) and the main CP (graphically depicted as the MAIN). The commander, in coordination with the staff, determines the composition, nature, and tasks of each CP. For example, the battalion commander can operate forward, accompanied by selected personnel in the tactical CP (see paragraph A-7) to control fire and movement during the battle. This CP normally includes the battalion S-3, fire support officer, and air liaison officer (ALO) but there is no requirement for these individuals to collocate. The commander may collocated with one element or in one area while the S-3 is collocated with another element or in another area. The main CP (see paragraph A-4), normally led by the XO, positions with the majority of the staff designed to control current operations, conduct detailed analysis, and plan future operations. The combat trains CP (graphically depicted as CTCP) see paragraph A-10, and field trains CP (graphically depicted as FTCP) see paragraph A-11, when established, include the logistics and services personnel that sustain the battalion.

TACTICAL STANDARD OPERATING PROCEDURES

A-19. CPs are organized to permit continuous and rapid execution of operations. SOPs for each CP should be established, known to all, and rehearsed. These SOPs should include at a minimum the following:

- Organization and setup.
- Plans for teardown and displacement.
- Eating and sleeping plans.
- Shift manning, shift changes and operation guidelines.
- Physical security plans.
- Priorities of work.
- Loading plans and checklists.
- Orders production.

- Techniques for monitoring enemy and friendly situations.
- Posting of map boards.
- Maintenance of journals and logs.

Information

A-20. CPs within the battalion monitor communications nets, receive reports, and process information to satisfy commander needs or critical information requirements. This information is maintained, in addition to digital systems, on maps, charts, and logs. Each section or cell maintains daily journals to log messages and radio traffic. CPs maintain information as easily understood map graphics and charts. Status charts can be combined with situation maps to give the commander and staff friendly and enemy situation snapshots for the planning process. This information is updated continuously. For simplicity, all map boards should be the same size and scale, and overlay mounting holes should be standard on all map boards. This allows easy transfer of overlays from one board to another. The following procedures for posting friendly and enemy information on the map aid the commander and staff in following the flow of battle:

- Friendly and enemy unit symbols are displayed on clear acetate placed on the operations overlay. These symbols can be marked with regular stick cellophane tape or with marking pen.
- Units normally keep track of subordinate units, two levels down. This may be difficult during offensive tasks. It may be necessary to track locations of immediate subordinate units instead.

Battle Captain

A-21. Battle captain positions are habitually filled, and found in most, if not all CPs. They coordinate the day-to-day staff activities, in effect acting as an assistant XO, and provide continuity for the staff's actions. The battle captain's informal role is to plan, coordinate, supervise, and maintain communication flow throughout the CP to ensure the successful accomplishment of all assigned missions. The battle captain assists the commander, XO, and S-3 by being the focal point in the CP for communications, coordination, and knowledge and information management. The battle captain is also the CP officer in charge in the absence of the commander, XO, and S-3. To function effectively, the battle captain must have a working knowledge of all elements in the CP, understand unit SOPs, and ensure CP personnel use them. The battle captain must know the current plan and task organization of the unit and understand the commander's intent.

A-22. Battle captains integrate into the decision-making process and know why certain key decisions were made. Battle captains must know the technical aspects of the battle plan and understand the time-space relationship to execute any specific support task. Battle captains must understand and enforce the battle rhythm, the standard events or actions that happen during a normal 24-hour period and ensure that the CP staff is effective throughout the period. Understanding their assigned authorities, battle captains use judgment to adjust activities and events to accomplish the mission across different shifts, varying tactical circumstances, and changes in the CP location. Battle captains have the overall responsibility for the smooth functioning of the facility and its staff elements. This range of responsibility includes—

- Maintaining continuous operations (while static and mobile).
- Tracking the current situation.
- Ensuring communications are maintained and all messages and reports are logged.
- Assisting the XO in ensuring a smooth and continuous information flow.
- Processing essential data to ensure tactical and logistical information is gathered and provided to staff members on a regular basis.
- Tracking commander's critical information requirements and providing recommendations.
- Approving fabrication and propagation of manual unit icons.
- Sending reports to higher and ensuring relevant information passes to subordinate units.
- Monitoring security within and around the CP.
- Organizing the CP to displace rapidly.
- Conducting battle drills and enforcing the SOP.

A-23. The battle captain ensures all staff elements in the CP understand their actions in accordance with the SOP and operations order, and provides coordination for message flow, staff briefings, updates to charts, and other coordinated staff actions. As the focal point in the CP, the battle captain processes essential information from incoming data, assesses it, ensures dissemination, and makes recommendations to the commander, XO, and S-3. The battle captain ensures the consistency, accuracy, and timeliness of information leaving the CP, including preparing and issuing fragmentary orders and WARNORDs. The battle captain monitors and enforces the updating of charts and status boards necessary for battle management and ensures this posted information is timely, accurate, and accessible.

Digital and Analog Mission Command Systems and Techniques

A-24. Digital mission command systems within a CP bring a dramatic increase in the level of informational dominance units may achieve. Techniques for digital procedures and for integrating analog and digital units contribute to battlefield lethality and tempo, and the ability to maintain information dominance. These techniques can significantly speed the process of creating and disseminating orders, allow for extensive collection of information, and increase the speed and fidelity of coordination and synchronization of battlefield activities. At the same time, achieving the potential of these systems requires extensive training, a high level of technical proficiency by both operators and supervisors, and the disciplined use of detailed SOPs. Communications planning and execution to support the digital systems is significantly more demanding and arduous than is required for units primarily relying on frequency modulation communication and joint network node and the CP node communications. (See ATP 6-02.53 for additional information on CP tactical radio operations.)

A-25. Whether to use frequency modulation or digital means for communication is a function of the situation and SOPs. Some general considerations can help guide the understanding of when to use which mechanism at what time. Frequency modulation communication is normally the initial method of communications when elements are in contact. Before and following an engagement, the staff and commanders use digital systems for disseminating orders and graphics and conducting routine reporting. During operations, however, the staff uses a combination of systems to report and coordinate with higher and adjacent units.

A-26. The Infantry battalion staff must remain sensitive to the difficulty and danger of using digital systems when moving or in contact. The staff should not expect digital reports from subordinate units under such conditions. Other general guidelines include the following:

- Initial contact at any echelon within the battalion should be reported on frequency modulation voice; digital enemy spot reports should follow as soon as possible to generate the enemy COP.
- Elements moving about the battlefield (not in CPs) use frequency modulation voice unless they can stop and generate a digital message or report.
- Emergency logistical requests, especially casualty evacuation requests, should be initiated on frequency modulation voice with a follow-up digital report, if possible.
- Combat elements moving or in contact should transmit enemy spot reports on frequency modulation voice; their higher headquarters should convert frequency modulation reports into digital spot reports to generate the COP. At company level, the XO, the first sergeant, or the company CP converts the reports.
- Calls for fire on targets of opportunity should be sent on frequency modulation voice; fire support teams submit digitally to advanced field artillery tactical data system.
- Plan calls for fire digitally and execute them by voice with digital back-up.
- Routine logistical reports and requests are sent digitally.
- Routine reports from subordinates to battalion before and following combat are sent digitally.
- Orders, plans, and graphics should be done face-to-face, if possible. If these products are digitally transmitted, they should be followed by frequency modulation voice call to alert recipients that critical information is being sent. The transmitting element should request a verbal acknowledgement of both receipt and understanding of the transmitted information by an appropriate Soldier, who usually is not the computer operator.
- Obstacle and CBRN-1 reports should be sent initially by voice followed by digital reports to generate a geo-referenced message portraying the obstacle or contaminated area across the network.

Friendly Common Operational Picture

A-27. The creation of the friendly COP is extensively automated, requiring minimal manipulation by CPs or platform operators. Each platform creates and transmits its own position location and receives the friendly locations, displayed as icons, of all the friendly elements in that platform's wide area network. This does not necessarily mean that all friendly units in the general vicinity of that platform are displayed, because some elements may not be in that platform's network. For example, a combat vehicle in an Infantry battalion probably will not have information on an above IBCT level artillery unit operating nearby because the two are in different networks. The COP generated from individual Joint Capabilities Release platforms is transmitted to CPs through the battalion's transport network. The battalion S-6 ensures the proper alignment and interoperability between the mission command applications and the battalion's transport network. (See ATP 6-02.53 for additional information on tactical radio networks.)

A-28. Commanders must recognize limitations in the creation of the friendly COP which results from vehicles or units that are not equipped with the Joint Capabilities Release. The following are two aspects to consider:

- Not all units will be equipped with all mission command system components, particularly multination partners and organizations. It is likely analog units or organizations will enter the IBCT and battalion AOs.
- Most dismounted Soldiers will not be equipped with a digital device that transmits information.

A-29. The following are ways to overcome these limitations:

- A digitally-equipped element tracks the location of specified dismounts and manually generates and maintains an associated friendly icon.
- The main CP tracks analog units operating within the area and generates associated friendly icons. The main CP must keep the analog equipped unit informed of other friendly units' locations and activities.
- A digitally equipped platform acts as a liaison or escort for analog units moving or operating in the same area. Battalion and higher elements must be informed of the association of the liaison officer icon with the analog unit.
- Do not use friendly positional information to clear fires because not all elements will be visible. Friendly positional information can be used to deny fires and can aid in the clearance process, but it cannot be the sole source for clearance of fires. This holds true for all Army mission command systems.

Enemy Common Operational Picture

A-30. The most difficult and critical aspect of creating the COP is creating the picture of the enemy. The enemy COP is the result of multiple inputs (for example, frequency modulation spot reports, unmanned aircraft system and joint surveillance target attack radar system reports, reports from Joint Capabilities Release-equipped platforms in subordinate units, electronic or signal intelligence feeds) and inputs from battalion information collection efforts through the battalion intelligence staff officer (S-2). Enemy information generation is a complex process requiring automated intelligence all source inputs and detailed analysis from within the battalion.

A-31. Generation of the enemy COP occurs at all echelons. At battalion level and below, the primary mechanism for generating information is the joint capabilities release. When an observer acquires an enemy element, they create and transmit a spot report, which automatically generates an enemy icon that appears throughout the network. Only those in the address group to whom the report was sent receive the text of the report, but all platforms in the network can see the icon. As the enemy moves or its strength changes, the observer must update this icon. If the observer must move, the observer ideally passes responsibility for the icon to another observer. If multiple observers see the same enemy element and create multiple reports, the S-2 or some other element that has the capability must eliminate the redundant icons.

A-32. Unit SOPs must clearly establish who has the ability, authority, and responsibility to create and input enemy icons. Without the establishment of these procedures, it is highly probable that the enemy COP will not be accurate.

A-33. Joint capabilities release spot reports must include the battalion S-2 in the address group for the data to be routed through the CP server into the All Source Analysis System (ASAS) to feed the larger intelligence picture. Frequency modulation reports received at a CP can be inputted manually into the ASAS database by the S-2 section. Joint capabilities release and frequency modulation voice reports are the primary source of enemy information in the battalion's AO.

A-34. The IBCT S-2 section and the supporting analysis control team support the Infantry battalion by receiving ASAS intelligence feeds from higher and adjacent units along with feeds from Joint Surveillance Target Attack Radar System, unmanned aircraft system, and the common ground station. They enter enemy information from these sources into the ASAS database and send this information through Joint Capabilities Release to the battalion S-2. These feeds, along with frequency modulation voice and Joint Capabilities Release reports, are the primary sources of enemy COP inputs for the Infantry battalion.

A-35. Fusion of all the intelligence feeds normally occurs at IBCT and higher levels. The IBCT S-2 routinely (every 30 minutes to every hour) sends the updated enemy picture to subordinate units down to platform level. Since the fused ASAS database focuses on the deep areas of the battlefield, its timeliness may vary. Subordinate element of the battalion normally use only the Joint Capabilities Release-generated COP. Companies should stay focused entirely on the Joint Capabilities Release-generated COP. Battalion leaders and staffs refer occasionally to the Joint Capabilities Release-generated intelligence picture to keep track of enemy forces they might encounter in the near future, but that are not yet in the battalion's close area.

A-36. Automation and displays contribute enormously to the ability to disseminate information and display it in a manner that aids comprehension. However, information generation must be rapid for it to be useful. Information also must be accompanied by analysis; pictures alone cannot convey all that is required nor will they be interpreted the same by all viewers. The battalion S-2 and section must be particularly careful about spending too much time operating an ASAS terminal while neglecting the analysis of activities for the battalion and subordinate commanders and other staff members. (Refer to ATP 6-02.53 for additional information.) The success of the battalion's intelligence effort depends primarily on the ability of staffs to—

- Analyze enemy activities effectively.
- Develop and continuously refine effective intelligence preparation of the battlefield.
- Create effective collection requirements management.
- Execute effective collection operations management.

Graphics and Orders

A-37. The advent of digitization does not mean that acetate and maps have no use and will disappear, at least not in the near future. Maps remain the best tools when maneuvering and fighting on the battlefield, and for controlling and tracking operations over a large area. The combination of a map with digital information and terrain database is ideal; both are required and extensively used.

A-38. Army mission command system components support the creation and transmission of operations orders. Staff sections normally develop their portions of orders and send them to the S-3 where they are merged into a single document. The S-3 deconflicts, integrates, and synchronizes all elements of the order. Once the order is complete, it is transmitted to subordinate, higher, and adjacent units. The tactical internet does not possess high transmission rates and therefore orders and graphics should be concise to reduce transmission times. Orders transmitted directly to Joint Capabilities Release-equipped systems within the battalion must meet the size constraints of the order formats in the Joint Capabilities Release. Graphics and overlays are constructed with the same considerations for clarity and size.

A-39. Digital graphics must interface and be transmittable. The interface and commonality of graphics will continue to evolve technologically and will require further software corrections. The following guidelines apply when creating graphics:

- Create control measures based on readily identifiable terrain, especially if analog units are part of the task organization.
- Boundaries are important, especially when multiple units must operate in close proximity or when it becomes necessary to coordinate fires or movement of other units.

- Intent graphics that lack the specificity of detailed control measures are an excellent tool for use with warning and fragmentary orders and when doing parallel planning. Follow them with appropriately detailed graphics, as required.
- Use standardized colors to differentiate units. This is articulated in the tactical SOP. For example, IBCT graphics may be in black, Battalion A in purple, Battalion B in magenta, and Battalion C in brown. This adds considerable clarity for the viewer. Subordinate company and team colors are then specified.
- Use traditional doctrinal colors for other graphics (green for obstacles, yellow for contaminated areas, and so forth).

A-40. In order to accelerate transmission times when creating overlays, use multiple smaller overlays instead of a single large overlay. System operators can open the overlays they need, displaying them simultaneously. The technique also helps operators in reducing screen clutter. The S-3 should create the initial graphic control measures on a single overlay and distribute it to the staff. The overlay is labeled as the operations overlay with the appropriate order number. Staff elements should construct their appropriate graphic overlays using the operations overlay as a background but without duplicating the operations overlay. This avoids unnecessary duplication and increase in file size and maintains standardization and accuracy. Each staff section labels its overlay appropriately with the type of overlay and order number. Before overlays are transmitted to subordinate, higher, and adjacent units, the senior battle captain or the XO checks them for accuracy and labeling. Hard copy (traditional acetate) overlays are required for the CPs and any analog units. Transmit graphics for on-order missions or branch options to the plan before the operation as time permits. If time is short, transmit them with WARNORDs.

Digital Standard Operating Procedures

A-41. The battalion SOP should contain standards for digital operations, in addition to analog operations. Most of digital operating procedures are established at the IBCT level with the battalion SOP complying and adding detail when required. One of the critical requirements when task organized to another unit is to receive and disseminate that units SOP.

A-42. To create a common picture, Joint Capabilities Release platforms must have the same information filter settings. This is particularly important for the enemy COP so that as icons go stale, they purge at the same time on all platforms. Standard filter settings should be established in unit SOPs and be the same throughout the battalion. For enemy offensive operations, the filter setting times should be short; for enemy defensive operations, the setting times should be longer, reflecting the more static nature of the enemy picture.

A-43. The standardization of friendly and enemy situational filter settings is of great importance in maintaining a COP. Joint capabilities release provides three methods for updating individual vehicle locations: time, distance, and manual. When the system is operational, it automatically updates friendly icons using time, distance traveled, or both, based on the platform's friendly situational filter settings. The unit should standardize filter settings across the force based on both the mission and the function of the platform or vehicle. Use shorter refresh rates for combat vehicles and vehicles that frequently move and longer refresh rates for static vehicles such as CPs. Tailoring the frequency of these automatic updates reduces the load on the tactical internet, freeing more capacity for other types of traffic.

A-44. The IBCT node is probably the most effective place to standardize the situational filter settings using the IBCT tactical SOP. There are no set rules for what these settings should be. The commander must establish them based on the unit's experience using Joint Capabilities Release and the capacity of the tactical internet. The Infantry battalion should use the capability to update a vehicle's position manually only when a platform's system is not fully functional and it has lost the ability to maintain its position automatically.

Reporting and Tracking of Battles

A-45. Having all platforms and units on the battlefield send spot reports digitally may result in mass confusion. However, in order to eliminate confusion, there should be one designated individual within the unit authorized to initiate digital spot reports. While the designated individual will be somewhat removed from the fight, that individual can assist those who execute the direct firefight by filtering multiple reports of the same event.

A-46. Another technique used to eliminate duplicate reporting problems is to limit the creation of enemy icons through digital spot reports to reconnaissance elements and the company leadership (commander, XO, or 1SG) or other designated individual. Others report to their higher headquarters, which creates and manages the icon. At company level, the XO, 1SG, or CP personnel become the primary digital reporters. These assignments cannot be completely restrictive. Unit SOPs and command guidance must allow for and encourage Soldiers who observe the enemy and know they are the sole observer (because there is no corresponding enemy icon displayed in the situational COP) to create a digital spot report. IBCT and battalion SOPs should define the schedule for report submissions, the message group for the reports, and the medium (digital system or verbal) used.

A-47. Battle tracking is the process of monitoring designated elements of the COP tied to the commander's criteria for success. Battle tracking requires special attention from all staff officers, and normally done digitally and manually with situation maps and boards. The XO and S-3 must continue to monitor the progress of the operation and recommend changes as required.

A-48. Establish a routine schedule of system updates. For example, the S-2 section should continuously update the ASAS database and should transmit the latest COP to the network every 30 minutes during operations if the battalion commander, S-3, or reconnaissance elements need it. Staff sections should print critical displays on an established schedule. These printed snapshots of the COP are used for continuity of battle tracking in the event of system failures, and can contribute to after action reviews and unit historical records.

A-49. SOPs define the technical process for creating, collating, and transmitting orders and overlays, both analog and digital and in degraded environments. For interoperability and clarity, IBCT SOPs should define the naming convention and filing system for all reports, orders, and message traffic. This significantly reduces time and frustration associated with lost files or changes in system operators or the environment. Information systems will inevitably migrate to a web-based capability. This allows information into a database to be accessed by users as needed or when they are able to retrieve it. For example, the S-2 may transmit an intelligence summary to all subordinates. Inevitably, some will lose the file or not receive it. The S-2 can simultaneously post that same summary to an established homepage so users can access it as required. If this technique is used, the following are a few things to consider—

- Posting a document to a homepage does not constitute communications. The right people are alerted when the document is available.
- Keep documents concise and simple. Elaborate PowerPoint slide briefings take longer to transmit, causing delays in the tactical internet.
- The amount of information entered in a database and personnel who have access is carefully controlled, both to maintain security and to keep from overloading the tactical internet.
- Assign responsibility to personnel who are authorized to input and delete both friendly and enemy unit icon information.

A-50. Procedures for integrating digital and analog units and operations within degraded environments are essential and should consider the following:

- Frequency modulation and joint network node/command post node are the primary communications mediums with the analog unit.
- Hard copy orders and graphics are required.
- Graphical control measures require a level of detail necessary to support operations of a unit without situational information. This requires more control measures tied to identifiable terrain, especially during operations within degraded environments.
- Liaison teams are critical in both digital and analog situations through direct liaison with the partner unit(s).
- The staff must recognize that integrating an analog unit into a digital unit requires retention of most of the analog control techniques. In essence, both digital and analog control systems must be in operation, with particular attention paid to keeping the analog unit apprised of all pertinent information that flows digitally.

- The staff establishes redundant communication, especially when the battalion shares its area of operation with other entities that have cultural differences and lack of or degraded communications.

Note. The techniques and procedures addressed above apply equally to conditions with degraded networks. (See appendix B for information on electronic protection.)

Considerations Concerning the Degradation of Battalion Mission Command Systems

A-51. As the staff supports the commander in the exercise of mission command, assists subordinate units, and informs units and organizations outside the battalion a broad array of actors and activities challenge the battalion's freedom of action in cyberspace and space. Enemies and adversaries utilize cyberspace and space to degrade the battalion's capability to communicate and operate mission command systems. The ability for adversaries and enemies to operate in the cyberspace and space domains increases the need for the battalion to maintain the capability to conduct offensive and defensive cyberspace operations to affect the operating environment and to protect friendly mission command systems. For example, enemy global positioning satellite jamming capabilities could render precision fires and blue force tracker inaccurate.

A-52. During the operations process, the battalion prepares for degraded mission command systems and reduced access to cyberspace and space capabilities. Considerations concerning the battalion in denying and degrading adversary and enemy use of cyberspace and the electromagnetic spectrum (see appendix B) and other effects that degrade friendly mission command systems include:

- Enemy capabilities (cyberspace, space, electronic warfare [EW]) to degrade, planned and targeted against mission command systems. Enemy efforts include jamming, spoofing, intercepting, hacking, and direction finding (leads to targeting).
- Friendly effects that degrade include:
- Lack of familiarity with Army mission command systems.
- Lack of protection and/or countermeasures at battalion and below echelons.
- Lack of understanding of threat capabilities and doctrine for employment.
- Terrain and weather, and other environmental variables.

A-53. Key indicators that battalion mission command systems are being degraded include:

- Reliable voice communications are degraded.
- Increased latency for data transmissions.
- Frequent and accurate targeting by threat lethal and nonlethal effects.
- Increased pings/network intrusions.
- Inconsistent digital COP, for example spoofing.
- Inaccurate Global Positioning System (commonly known as GPS) data/no satellite lock and inconsistency between inertial navigation aids and GPS-enabled systems.

A-54. Battalion efforts the counter the effects of degraded mission command systems include—

- Train to recognize indicators that it is happening.
- Develop contingency plans and rehearse implementation during the planning process and preparations.
- Maintain analog COP at all echelons.
- Train to operate from the commander's intent, and analog graphics and synchronization matrixes.
- Keep plans as simple as possible that are less susceptible to friction.

A-55. Battalion efforts to prevent degraded mission command systems include—

- Minimize length of frequency modulation transmissions.
- Use terrain to mask transmission signatures.
- Employment of directional antennas.
- Require physical presence of leaders at briefings, for example distribute information via analog means in person.

- Use of camouflage and deception in all domains.
- Use of communications windows to reduce transmissions.
- Employment of encryption/cypher techniques.

Appendix B
Planning and Preparation

Planning and preparation for operations leads the battalion commander to make decisions during execution. At its core, decision making is knowing if to decide, then when and what to decide. Decision making includes understanding the consequences of decisions. The military decisionmaking process (MDMP) is an established and proven analytical process. The tool helps the commander and staff to develop estimates and a plan. The MDMP drives preparation.

Since time is a factor in all operations, the commander and staff conduct a time analysis early in the planning process. The analysis helps them determine what actions they need to take and when to begin those actions to ensure forces are ready and in position before execution. The plan may require the commander to direct subordinates to start necessary movements; conduct task-organization changes; begin reconnaissance and surveillance (R&S) missions, and security operations; and execute other preparation activities before completing the plan. (Refer to FM 3-96 and ADRP 5-0 for additional information.)

Appendix B addresses commander driven planning and preparation considerations within the battalion. This appendix discusses various factors that degrade operating efficiencies and ways to counter these inefficiencies through planning and preparation techniques and procedures. Appendix B addresses electronic warfare planning considerations and electronic protection techniques and procedures used to prevent enemy jamming and intrusion into friendly mission command systems.

PLANNING

B-1. Planning is the art and science of understanding a situation, envisioning a desired future, and laying out effective ways of bringing that future about. Planning consists of two separate but interrelated components, a conceptual component and a detailed component. Successful planning requires the integration of both these components. Commanders and subordinate leaders within Infantry battalion employ three methodologies for planning: the Army design methodology (see paragraph 1-4, the MDMP (battalion echelon), and troop-leading procedures (company echelon and below, see FM 3-21.10). The battalion commander determines how much of each methodology to use based on the scope of the problem, familiarity with it, and the time available. Planning helps the battalion commander create and communicate a common vision between the staff, subordinate commanders and leaders, and unified action partners. Planning results in an order that synchronizes the action of forces in time, space, and purpose to achieve objectives and accomplish missions.

Note. The Army design methodology assist the battalion commander and staff in understanding ill-structured problems and developing operational approaches to manage or solve those problems (see ATP 5-0.1). Troop leading procedures provide small-unit leaders with a framework for planning and preparing for operations. (See FM 3-21.10.)

PARALLEL, COLLABORATIVE, AND DISTRIBUTED PLANNING

B-2. Whether planning deliberately or rapidly, all planning requires the skillful use of available time to optimize planning and preparation throughout the battalion. Taking more time to plan often results in greater synchronization; however, any delay in execution risks yielding the initiative—with more time to prepare and act—to the enemy. When allocating planning time to subordinate unit commanders, the battalion commander must ensure subordinates have enough time to plan and prepare their own actions prior to execution. Both parallel, collaborative, and distributed planning help optimize available planning time. Parallel planning allows each echelon to make maximum use of time available. Collaborative planning is the real-time interaction of commanders and staffs. Distributed planning allows the commander and staff members to execute planning from different locations.

Parallel Planning

B-3. *Parallel planning* is two or more echelons planning for the same operation sharing information sequentially through WARNORDs from the higher headquarters prior to the higher headquarters publishing their operation plan or operation order (ADRP 5-0). Parallel planning requires significant interaction between echelons. Parallel planning can happen only when higher headquarters produces timely WARNORDs and shares information with subordinate headquarters as it becomes available (figure B-1).

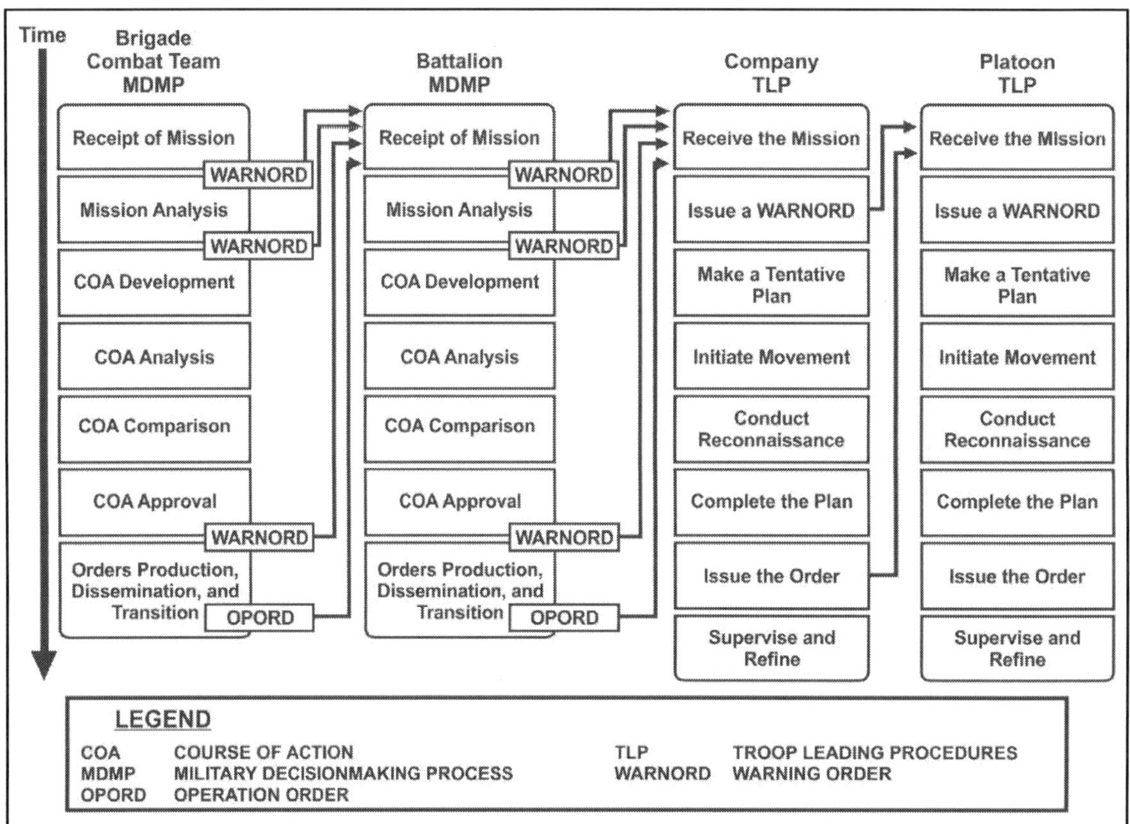

Figure B-1. Parallel planning

Collaborative Planning

B-4. *Collaborative planning* is commanders, subordinate commanders, staffs, and other partners sharing information, knowledge, perceptions, ideas, and concepts regardless of physical location throughout the planning process (ADRP 5-0). Collaborative planning is the real-time interaction among commanders and staffs at two or more echelons developing plans for a single operation. It must be used judiciously.

B-5. Collaborative planning is most appropriate when time is scarce and a limited number of options are being considered. It is particularly useful when the commander and staff can benefit from the input of subordinate commanders and staffs.

B-6. Collaborative planning is not appropriate when the staff is working a large number of course of actions or branches and sequels, many of which will be discarded. In this case, involving subordinates wastes precious time working options that are later discarded. Collaborative planning also is often not appropriate during ongoing operations in which extended planning sessions take commanders and staffs away from conducting current operations.

B-7. As a rule of thumb, if the commander is directly involved in time-sensitive planning, some level of collaborative planning probably is needed. The commander, not the staff, must make the decision to conduct collaborative planning. Only the commander can commit subordinate commanders to using their time for collaborative planning.

Distributed Planning

B-8. Digital communications and information systems enable members of the same staff to execute the MDMP without being collocated. Distributed planning saves time and increases the accuracy of available information in that it allows for the rapid transmission of voice and data information, which can be used by staffs over a wide geographical area. (See paragraph B-83 and FM 3-12 for information on appropriate electronic protection active and passive measures.)

Role of the Commander and Executive Officer

B-9. The commander is in charge of the MDMP. From start to finish, the commander's personal role is central. The commander's participation in the process provides focus and guidance to the staff; however, there are responsibilities and decisions that are the commander's alone. The amount of direct involvement is driven by the time available, personal preferences, and the experience and accessibility of the staff. The less time available, the less experienced the staff, and the less accessible the staff, the greater the commander's involvement in the MDMP.

B-10. The executive officer manages, coordinates, and disciplines the staff's work and provides quality control. The executive officer ensures the staff has the information, guidance from the commander, and facilities it needs. As the senior knowledge management officer in the battalion, the executive officer directs the activities of each staff section and subordinate unit to integrate and synchronize knowledge and information management within the battalion. The executive officer establishes the battle rhythm within the battalion, determines timelines for the staff, establishes confirmation brief times and locations, enforces the information management plan, and provides any unique instructions needed to guide the staff in completing the MDMP.

B-11. Warning orders are used to facilitate parallel planning. By issuing guidance and participating in formal and informal briefings, the commander and executive officer guide the staff through the MDMP. In a collaborative environment, the commander can extend this participation directly to the staff and subordinate commanders and leaders. Such interaction helps the staff and subordinates to resolve questions, and involves all staff and subordinates in the complete process.

Role of Reconnaissance and Surveillance

B-12. The battalion commander deploys the battalion scout platoon and other reconnaissance forces and surveillance assets early in the planning process to facilitate information collection. The commander and staff ensure R&S is continuous during planning, preparation, and execution of the mission. Information collected during R&S may result in initial plans or course of actions being modified, or even discarded. Further, when the plan changes, the commander must modify their R&S objective to support the new plan.

B-13. Commander's critical information requirements and decision points focus the staff's monitoring activities and prioritize the unit's collection efforts. Information requirements concerning the enemy, terrain and weather, and civil considerations are identified and assigned priorities through R&S. The battalion operations officer, in coordination with the battalion intelligence officer, uses friendly reports to coordinate other assessment-related information requirements. To prevent duplicated collection efforts, information requirements associated with assessing the operation are integrated into both the R&S plan and friendly force information requirements.

B-14. R&S assists significantly in developing course of actions during the planning process. Conducted early, R&S helps confirm or deny the commander's initial assessment. Information also may allow the commander to focus immediately on a specific course of action or to eliminate course of actions that R&S show to be infeasible.

B-15. When conducting R&S, the commander must determine if the benefits outweigh the risks. During the conduct of defensive tasks and operations in support of stability tasks, R&S often can be conducted with little risk. During the conduct of offensive tasks, R&S involves more risk.

B-16. When the commander deploys reconnaissance forces and surveillance assets, particularly human intelligence, planning guidance must be given to ensure the survival of the force and/or assets while still enabling mission accomplishment. At a minimum, this guidance will include—

- Mission statement to include eyes-on-target time and anticipated length of mission.
- Priority intelligence requirement.
- Enemy situation in the AO.
- Commander's intent for intelligence, which can be stated by the intelligence staff officer (S-2) or operations staff officer (S-3).
- Method of deployment and insertion with abort criteria. Coordination time and place are included, if applicable.
- Fire support plan to include assets available and restrictive fire support coordination measures.
- Communication plan (primary and backup).
- Casualty evacuation plan.
- Exfiltration plan.
- Resupply plan (ground and aerial).

INTEGRATING PROCESSES AND CONTINUING ACTIVITIES

B-17. Throughout the operations process, the battalion commander and staff integrate warfighting functions to synchronize the force in accordance with the commander's intent and concept of operations. Integrating processes and continuing activities, used throughout the process, assist in synchronizing the battalion's operation.

Integrating Processes

B-18. In addition to the major activities of the operations process, the commander and staff use several integrating processes to synchronize specific functions throughout the operations process. The integrating processes are—

- Intelligence preparation of the battlefield.
- Targeting.
- Risk management.

Intelligence Preparation of the Battlefield

B-19. *Intelligence preparation of the battlefield* (IPB) is the systematic process of analyzing the mission variables of enemy, terrain, weather, and civil considerations in an area of interest to determine their effect on operations (ATP 2-01.3). Led by the battalion S-2, the entire staff participates in the IPB to develop and sustain an understanding of the enemy, terrain and weather, and civil considerations. IPB helps identify options available to friendly and threat forces.

B-20. IPB supports all activities of the operations process. IPB identifies gaps in current intelligence. IPB products help the commander and staff, and subordinate commanders and leaders understand the threat, physical environment, and civil considerations throughout the operations process. (Refer to ATP 2-01.3 for additional information.)

Note. See paragraph B-54 for information on the relationship between the IPB and the MDMP.

Targeting

B-21. *Targeting* is the process of selecting and prioritizing targets and matching the appropriate response to them, considering operational requirements and capabilities (JP 3-0). Targeting personnel within the battalion identify critical target subsets that when successfully acquired and attacked significantly diminish enemy capabilities. The commander synchronizes combat power to attack and eliminate critical target(s) using the most effective system in the right time and place.

B-22. Targeting is a complex and multidiscipline effort that requires coordinated interaction among many command and staff elements within and external to the battalion. The functional and integrating cell members within the battalion necessary for effective collaboration are represented in the targeting working group (see paragraph C-8). Close coordination among all cells is crucial for a successful targeting effort. Sensors and collection capabilities under the control of external agencies must be closely coordinated and carefully integrated into the execution of attacks especially those involving rapidly moving, fleeting, or dangerous targets. In addition, the appropriate means and munitions must attack the vulnerabilities of different types of targets. (Refer to ATP 3-60 for additional information.)

Commander's Targeting Guidance

B-23. The commander's targeting guidance must be articulated clearly and simply to enhance understanding. Targeting guidance must focus on essential threat capabilities and functions that could interfere with the achievement the battalion's objectives. The commander's targeting guidance describes the desired effects to be generated by fires, physical attack, cyberspace electromagnetic activities, and other information related capabilities against threat operations. Targeting enables the commander through various lethal and nonlethal capabilities the ability to produce the desired effects. Capabilities associated with one desired effect may also contribute to other desired effects. For example, delay can result from disrupting, diverting, or destroying enemy capabilities or targets. (See ATP 3-60 for a complete listing of desired effects.) The commander can also direct a variety of nonlethal actions or effects separately or in conjunction with lethal actions or effects.

B-24. The commander can also provide restrictions as part of their targeting guidance. Targeting restrictions fall into two categories—the no-strike list and the restricted target list.

B-25. The no-strike list consists of objects or entities protected by—

- Law of war.
- International laws.
- Rules of engagement.
- Other considerations.

B-26. A restricted target list is a valid target with specific restrictions such as—

- Limit collateral damage.
- Preserve select ammo for final protective fires.
- Do not strike during daytime.
- Strike only with a certain weapon.
- Proximity to protected facilities and locations.

B-27. The targeting process supports the commander's decision making with a comprehensive, iterative, and logical methodology for employing the ways and means to create desired effects that support achievement of objectives. Once actions are taken against targets, the commander and staff assess the effectiveness of the actions. If there is no evidence that the desired effects were created, reengagement of the target may be necessary, or another method selected to create the desired effects.

Targeting Categories

B-28. The targeting process can be generally grouped into two categories: deliberate and dynamic. Deliberate targeting prosecutes planned targets. These targets are known to exist in the area of operations and have actions scheduled against them. Examples range from targets on target lists in the applicable plan or order, targets detected in sufficient time to place in the joint air tasking cycle, mission type orders, or fire support plans. *Dynamic targeting* is targeting that prosecutes targets identified too late, or not selected for action in time to be included in deliberate targeting (JP 3-60). Dynamic targeting (see ATP 3-60.1) prosecutes targets of opportunity and changes to planned targets or objectives. Targets of opportunity are targets identified too late, or not selected for action in time, to be included in deliberate targeting. Targets engaged as part of dynamic targeting are previously unanticipated, unplanned, or newly detected.

B-29. The two types of planned targets are scheduled and on-call:

- Scheduled targets exist in the area of operation and are located in sufficient time so that fires or other actions upon them are identified for engagement at a specific, planned time.
- On-call targets have actions planned, but not for a specific delivery time. The commander expects to locate these targets in sufficient time to execute planned actions.

B-30. The two types of targets of opportunity are unplanned and unanticipated:

- Unplanned targets are known to exist in the area of operations, but no action has been planned against them. The target may not have been detected or located in sufficient time to meet planning deadlines. Alternatively, the target may have been located, but not previously considered of sufficient importance to engage.
- Unanticipated targets are unknown or not expected to exist in the area of operation.

Targeting Methodology

B-31. Targeting and the decide, detect, deliver, and assess (D3A) methodology is designed to be performed by the commander's staff in planning the engagement of targets. (See engagement area development, paragraphs 3-169.) The D3A methodology organizes the efforts of the commander and staff to accomplish key targeting requirements. Targeting is an outgrowth of the commander's decisions and establishes the requirements for the development of an effective information and intelligence collection effort. The D3A methodology helps the staff and targeting working group decide which targets must be acquired and engaged. Targeting develops options used to engage targets. Options can be lethal or nonlethal, organic or supporting at all levels throughout the range of military operations as listed–maneuver, electronic attack, psychological, attack aircraft, surface-to-surface fires, air to surface, other information related capabilities, or a combination of these operations. In addition, D3A assists in the decision of who will engage the target at the prescribed time. It also assists targeting working groups determine requirements for combat assessment to assess targeting and attack effectiveness. (Refer to ATP 3-60 for additional information.) The four functions of D3A are listed below:

- Decide which targets to engage.
- Detect the targets.
- Deliver the appropriate effects (conduct the operation).
- Assess the effects of the engagement(s).

Note. See paragraph B-59 for information on the relationship between targeting and the MDMP.)

Risk Management

B-32. Risk management is the Army's process for helping organizations and individuals make informed decisions to reduce or offset risk. Using this process increases the force's operational effectiveness and the probability of mission accomplishment. This systematic approach identifies hazards, assesses them, and manages associated risks. Risk management outlines a disciplined approach to express a risk level in terms readily understood at all echelons. For example, the commander may adjust the level of body armor protection during dismounted movement balancing an increased risk level to individual Soldiers to improve the likelihood of mission accomplishment (see paragraphs B-35 through B-54 for information on Soldier load).

Principles of Risk Management

B-33. The principles of risk management (see ATP 5-19) are—

- Integrate risk management into all phases of missions and operations.
- Make risk decisions at the appropriate level.
- Accept no unnecessary risk.
- Apply risk management cyclically and continuously.

Five-Step Process

B-34. Risk management is a cyclical and continuous five-step process to identify and assess hazards; develop, choose, implement, and supervise controls; and evaluate outcomes as conditions change. Except in time-constrained situations, planners complete the process in a deliberate manner—systematically applying all the steps and recording the results. In time constrained conditions, the commander, staff, subordinate leaders, and Soldiers use judgment to apply risk management principles and steps. The five steps of risk management are—

Step 1–Identify the hazards.

Step 2–Assess the hazards.

Step 3–Develop controls and make risk decisions.

Step 4–Implement controls.

Step 5–Supervise and evaluate.

Linkage to Operations Process

B-35. The five steps of risk management follow a logical sequence that correlates with the battalion's operations process activities. Steps 1 and 2 normally have greatest emphasis in the planning activities for the operation. Step 3 normally begins in planning and continues throughout the preparing activities for the operation. The majority of step 4 normally occurs within the preparing and executing activities for the operation, with some continuing emphasis in planning. Step 5 normally occurs during executing with some continuing emphasis in planning. The assessment activity of the operations process is continuous. While table B-1 is in a bar format, both processes are cyclical, fluid, and dynamic. Activities and steps can overlap or be revisited during operations.

Table B-1. Risk management and the operations process

Risk Management Steps		*Operations Process Activities*	
1	Identify the Hazards	Planning	Assess
2	Assess the Hazards	Planning	
3	Develop Controls and Make Risk Decisions	Planning and preparing	
4	Implement Controls	Planning, preparing, and executing	
5	Supervise and Evaluate	Planning and executing	

Note. See paragraph B-59 for information on the relationship between risk management and the MDMP.

B-36. DD Form 2977, Deliberate Risk Assessment Worksheet, is the Army's standard form for deliberate risk assessment (ATP 5-19, appendix A). DD Form 2977 captures the information analyzed during the five steps of risk management and the operations process. Commander and staff use the form to track hazards and risks in a logical manner to help users in thinking through the five steps and then sharing the resulting assessment. For example, weather conditions can create specific hazards and risks during operations. Common weather hazards to assess are cold, ice, snow, rain, fog, heat, humidity, wind, dust, visibility, and illumination. (Refer to ATP 5-19 for a detailed discussion on the analysis of risk.)

Note. DD form 2977 is a living document. Pen and pencil changes on hard copies are acceptable and encouraged since changes will occur during operations. Aviation; explosive; chemical, biological, radiological, or nuclear; and other highly technical activities may require additional specialized documentation.

Continuing Activities

B-37. While units execute numerous tasks throughout the operations process, the commander and staff always plan for and coordinate continuing activities. Continuing activities include the following:

- Liaison. (See FM 6-0.)

- Information collection. (See FM 3-55.)
- Security operations. (See FM 3-90-2.)
- Protection. (See ADRP 3-37.)
- Terrain management. (See ADRP 5-0.)
- Airspace control. (See FM 3-57.)

Note. In addition to the source references listed above, continuing activities are addressed throughout the publication.

Running Estimate

B-38. A *running estimate* is the continuous assessment of the current situation used to determine if the current operation is proceeding according to the commander's intent and if planned future operations are supportable (ADP 5-0). The commander and each staff element maintain a running estimate (see FM 3-96). The commander maintains a running estimate to consolidate understanding and visualization of the operation. The commander's running estimate summarizes the problem and integrates information and knowledge of each staff's running estimate. In their running estimates, the commander and each staff element continuously consider the effects of new information and update the following:

- Facts.
- Assumptions.
- Friendly force status.
- Enemy activities and capabilities.
- Civil considerations.
- Conclusions and recommendations.

B-39. Running estimates help the staff to track and record pertinent information and provide recommendations to the commander. Running estimates represent the analysis and expert opinion of each staff element by functional area. Staffs maintain running estimates throughout the operations process to assist commanders in the exercise of mission command. Each staff element and command post functional cell maintains a running estimate focused on how its specific areas of expertise are postured to support future operations. Because an estimate may be needed at any time, running estimates must be developed, revised, updated, and maintained continuously during operations. Running estimates can be presented verbally or in writing. (Refer to FM 6-0 for additional information.)

Military Decisionmaking Process

B-40. The *military decisionmaking process* is an iterative planning methodology to understand the situation and mission, develop a course of action, and produce an operation plan or order. (ADP 5-0). The MDMP helps the commander and staff to apply thoroughness, clarity, sound judgment, logic, and professional knowledge to understand situations, develop options to solve problems, and reach decisions. The process helps the commander, staff, and others think critically and creatively while planning.

B-41. During the MDMP, the IBCT headquarters solicits input and continuously shares information concerning future operations through planning meetings, WARNORDs, and other knowledge management and information management (see FM 3-96) means. The process enables the sharing information with subordinate and adjacent units, supporting and supported units, and unified action partners.

B-42. The MDMP is used by the commander and staff to develop and thoroughly examine numerous friendly and enemy courses of action. Based on the commander's visualization, the MDMP can be adjusted to meet the current situation. The commander and staff typically conduct this examination when developing the commander's visualization and operation plans, when planning for an entirely new mission, and during extended operations. The MDMP can be performed slowly and deliberately or quickly with heavy commander involvement.

B-43. The MDMP has seven steps. Each step (process) builds on the outputs from the previous steps. Each step then produces its own output that drives subsequent steps. Errors committed early in the process, especially with a faulty mission analysis, affect later steps. Figure B-2, page B-10 and 11, provides an overview of the MDMP. (Refer to FM 6-0 for additional information).

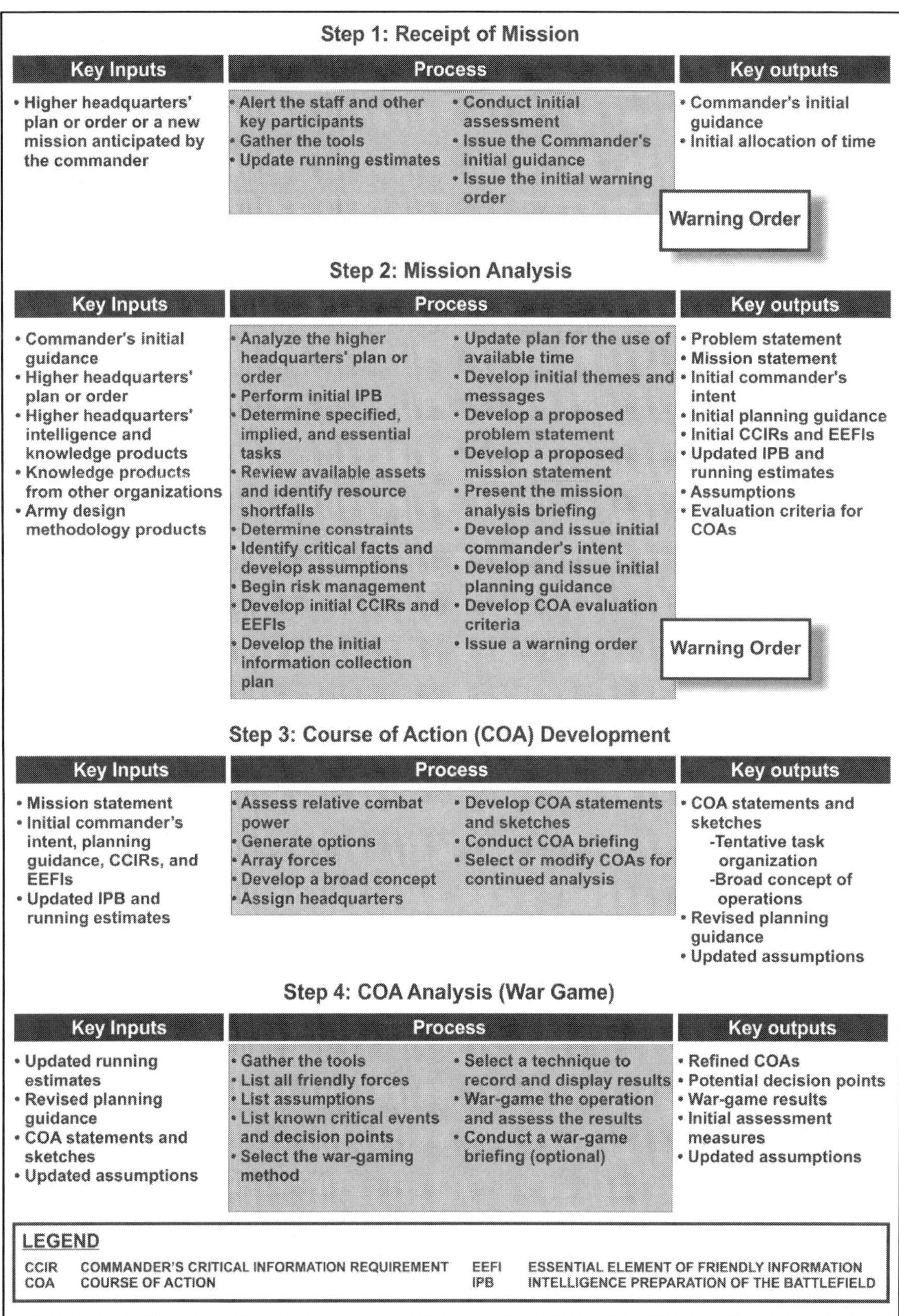

Figure B-2. Military decisionmaking process overview

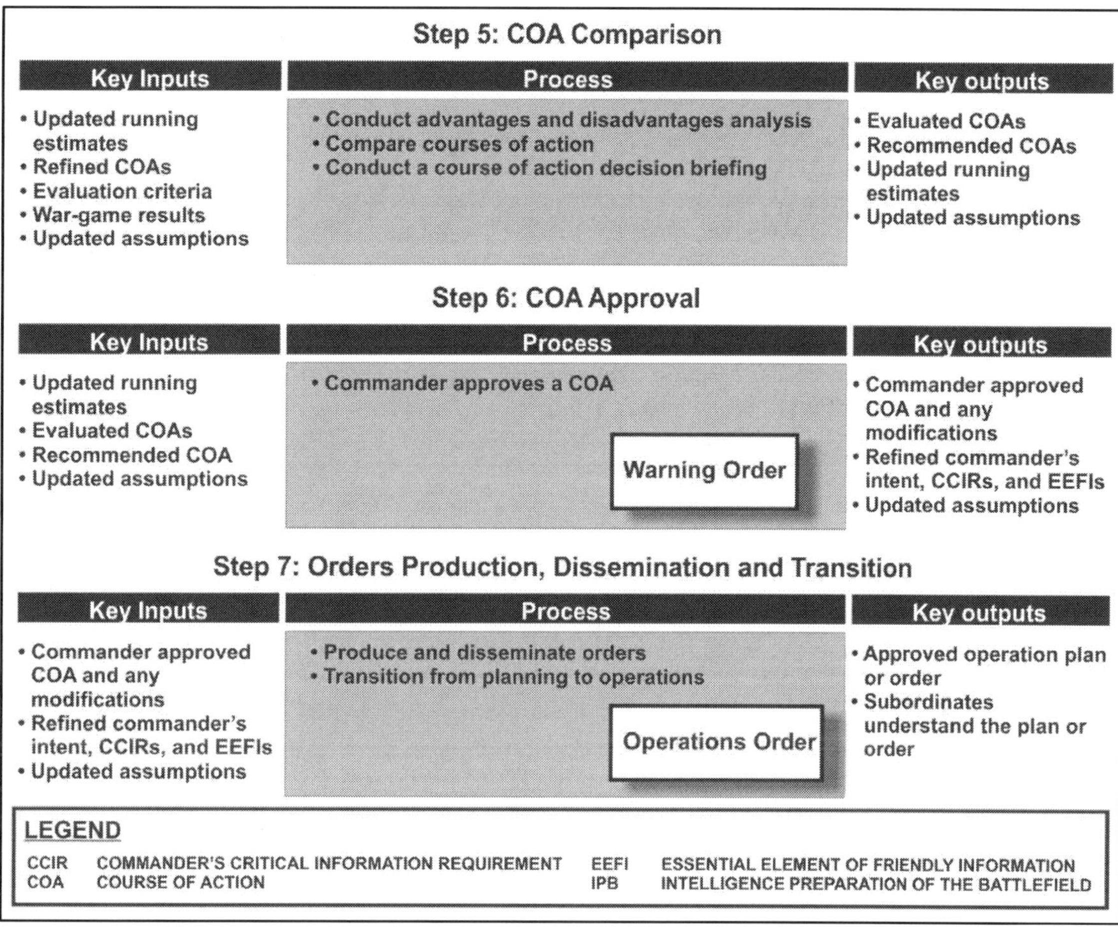

Figure B-2. Military decisionmaking process overview (continued)

Planning in a Time Constrained Environment

B-44. Any planning process aims to quickly develop a flexible, sound, and fully integrated and synchronized plan. However, any operation may "outrun" the initial plan. The most detailed estimates cannot anticipate every possible branch or sequel, enemy action, threat action, or reaction from the local population, unexpected opportunity, or change in mission directed from higher headquarters. Fleeting opportunities or unexpected enemy action may require a quick decision to implement a new or modified plan. When this occurs, the battalion staff and subordinate units often find themselves pressed for time in developing a new plan.

B-45. Before the battalion can conduct decision making in a time-constrained environment, it must train on, and master, all of the steps in the MDMP. The battalion only can abbreviate the MDMP if it fully understands the role of each step of the process and the requirements to produce the necessary products. Training on these steps must be thorough and result in a series of staff battle drills tailored to the time available. Training on the MDMP must be stressful and replicate realistic conditions and timelines. There is only one process, and omitting steps of the MDMP to meet time constraints is not the solution, though they may be done in a shortened timeframe. Anticipation, organization, and prior preparation are the keys to success in a time-constrained environment. Well-trained staffs will know where they can abbreviate actions and planning, focusing on only the most critical factors.

B-46. The commander and staff can use the time saved on any step of the MDMP to—

- Refine the plan more thoroughly.
- Conduct a more deliberate and detailed war game.
- Consider potential branches and sequels in detail.

- Focus more on rehearsing and preparing the plan.
- Allow subordinate units more planning and preparation time.

Abbreviated Military Decisionmaking Process

B-47. The battalion abbreviates the MDMP when there is too little time for a thorough and comprehensive application of the process. The most significant factor to consider is time. It is the only nonrenewable resource and often the most critical one.

Techniques

B-48. There are four primary techniques for abbreviating the MDMP:

- Increase the battalion commander's involvement, allowing time to make decisions without waiting for detailed briefings after each step.
- Limit options. When the commander is more prescriptive it saves the staff time by allowing it to focus more closely.
- Maximize parallel planning. Although parallel planning should be normal during the MDMP, maximizing its use in a time-constrained environment is critical.
- Limit the number of courses of action. If the commander conducts a personal assessment and chooses a course of action, he can direct the staff to refine only that one course of action. This technique normally saves the most time. It is highly dependent on the commander having an accurate grasp of the relevant tactical situation facing the battalion.

B-49. In a time-constrained environment, the importance of WARNORDs increases as available time decreases; a verbal WARNORD now, followed by a written order later (or posted to a database), are worth more than a written order one hour from now. The same WARNORDs used in the MDMP should be issued when abbreviating the process. In addition to WARNORDs, units must share all available information, (particularly IPB products) with subordinates as soon as possible. The information systems greatly increase this sharing of information and the commander's visualization through collaboration with their subordinates.

B-50. While the steps used in a time-constrained environment are the same, many of them may be done mentally by the battalion commander or with less staff involvement than during the MDMP. The products developed when the process is abbreviated may be the same as those developed for the MDMP; however, they may be much less detailed and some may be omitted altogether. Battalion standard operating procedures (SOPs) and mission requirements tailor this process to the commander's preference for orders in this environment.

B-51. When developing the plan, the staff initially may use the MDMP and develop branches and sequels. During execution, they may abbreviate the process. The IBCT may use the complete process to develop the plan while the subordinate Infantry battalion headquarters abbreviates the process.

Advantages

B-52. The advantages of using the abbreviated MDMP include the following:

- It maximizes the use of available time, and may be the required solution due to mission requirements.
- It may allow subordinates more planning and preparation time.
- It focuses staff efforts on the commander's specific and directive guidance.
- It facilitates adaptation to a rapidly changing situation.

Disadvantages

B-53. An abbreviated MDMP—

- Is much more directive and limits staff flexibility and initiative.
- Ignores some available options during the development of friendly courses of action.
- Can result in only an oral operations order or fragmentary order.
- Increases the risk of missing a key factor or failing to uncover a better option.

- Might decrease the coordination and synchronization of the plan.
- Requires more focus to rehearse, and could require more face time between commanders.

Intelligence Preparation of the Battlefield and the Military Decisionmaking Process

B-54. During planning, the commander focuses activities on understanding, visualizing, and describing, while directing and assessing. The IPB is one of the processes the commander uses to aid in planning (see ATP 2-01.3). The IPB supports the MDMP methodology integrating the activities of the commander, staff, subordinate units, and other partners to—

- Understand the situation and mission.
- Develop and compare course of actions.
- Decide on a course of action that best accomplishes the mission.
- Produce an operation order for execution.

Four Steps of the Intelligence Preparation of the Battlefield

B-55. The IPB consists of four steps. Each step is performed or assessed and refined to ensure that IPB products remain complete and relevant. Figure B-3, page B-14, shows the relationship between IPB and the steps of MDMP along with key inputs and outputs during the process. The four IPB steps are—

- Define the operational environment.
- Describe environmental effects on operations.
- Evaluate the threat.
- Determine threat course of actions.

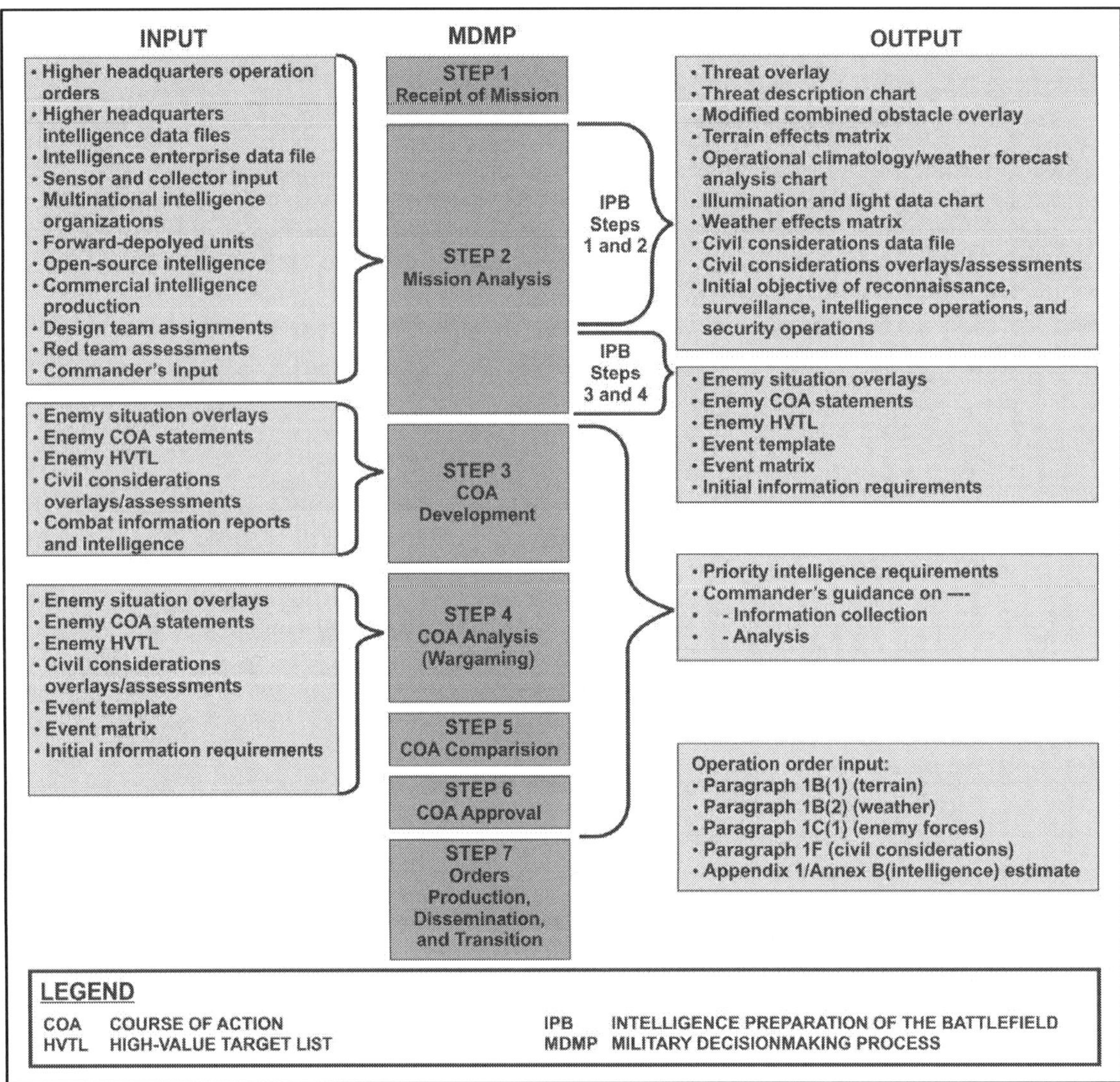

Figure B-3. IPB and the military decisionmaking process

Intelligence Support Teams from the Military Intelligence Company

B-56. The military intelligence (MI) company (K series) within the IBCT distributes intelligence support teams regardless of which element of decisive action (offense, defense, or stability) currently dominants. Dependent on the situation these teams can be employed down to maneuver company level. The intelligence support teams' mission is to provide basic analytic support, develop basic-level intelligence products, serve as a conduit for effective intelligence communications, and when resourced, manage some information collection programs. Some of those information collection programs include friendly force debriefings, document and media exploitation, and biometric and forensic collections.

B-57. The IBCT can employ anywhere from two intelligence analysts, for example to a maneuver company, or a large team of intelligence analysts as an intelligence support team to support, based on the situation, an Infantry battalion, brigade engineer battalion, fires battalion, Cavalry squadron, brigade support battalion, or to further augment the IBCT intelligence cell or brigade intelligence support element. A supported maneuver unit or element may subsequently augment the intelligence analysts with non-MI Soldiers to form a larger intelligence support team. When this occurs, it is critical that the appropriate S-2 section thoroughly train all non-MI personnel on intelligence support team activities.

B-58. The IBCT S-3 and S-2 work together with the battalion S-3s and S-2s to determine the intelligence support teams' task organization, based on the mission variables of METT-TC, using standard command and support relationships as part of the overall IBCT intelligence architecture. Planning considerations for the intelligence support team includes the supported unit's—

- Commander's guidance.
- Decisive and shaping operations and main and supporting efforts.
- Specific tasks and the requirement for quick analysis at the point of action or to help manage a unit's information collection effort.
- Ability to provide transportation and logistical support.
- Communications capacity for the intelligence support teams.
- Use of a specific intelligence support team to support or train with a specific unit.

Targeting and the Military Decisionmaking Process

B-59. Targeting methodology (see paragraph B-31) is an integral part of the MDMP. Targeting begins with the receipt of the mission and continues through operations process's execution and assessment phases. Like MDMP, targeting is a commander driven process. As the MDMP is conducted, targeting becomes more focused based on the commander's guidance and intent. Figure B-4 illustrates the relationship between the targeting methodology (decide, detect, deliver, and assess) and the MDMP along with products generated during targeting.

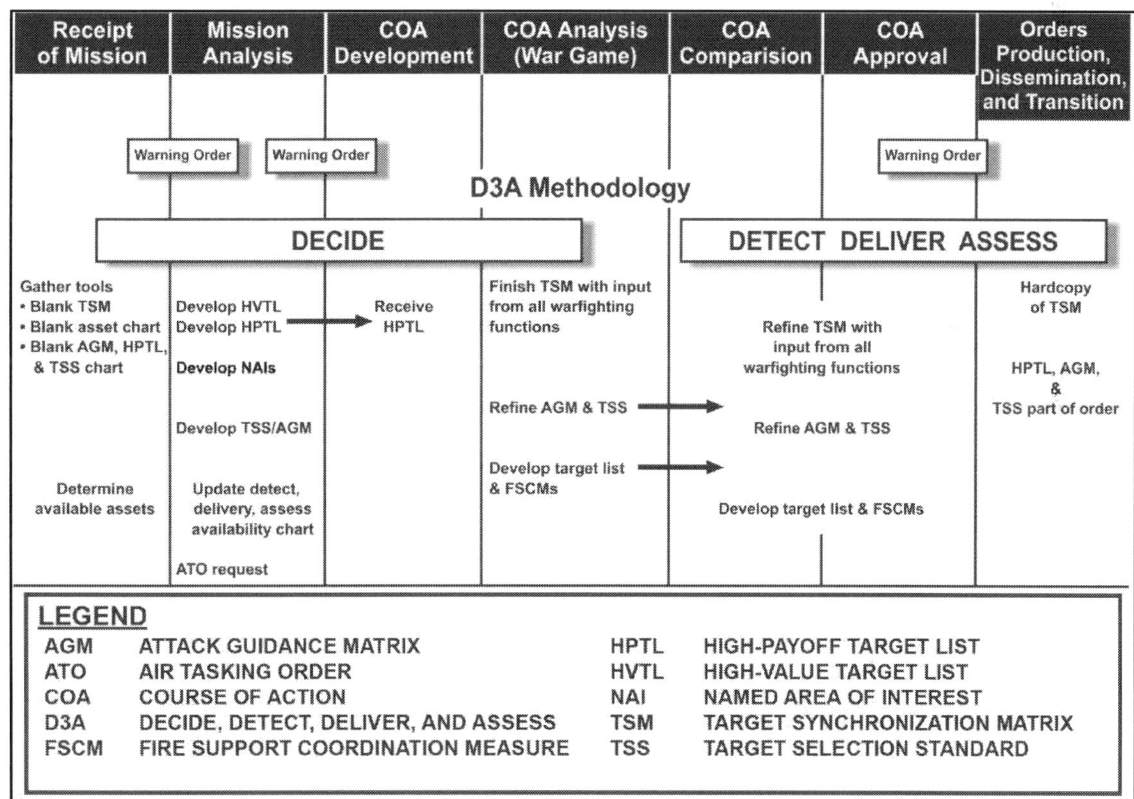

Figure B-4. Targeting methodology and the military decisionmaking process

B-60. The decide function coincides with the MDMP from the receipt of mission through the issuing of the approved plan or order. The detect function is a continuing function that starts with the commanders approval of the plan or order and is accomplished during execution of the plan or order. Once detected, targets are attacked and assessed as required. Targeting working groups are used as a vehicle to focus the targeting process within specified time.

B-61. D3A methodology functions occur simultaneously and sequentially during the operations process. Decisions are made during the planning of future operations. Current operations simultaneously detect, deliver, and assess targets based on current targeting decisions.

Risk Management and the Military Decisionmaking Process

B-62. The commander and staff use risk management to identify, assess, and control hazards, reducing their effect on operations and readiness. The five steps of risk management tend to require emphasis at different times during the MDMP (table B-2). While planning doctrine places the beginning of formal risk management in mission analysis, the commander and staff can begin identifying hazards upon receipt of the operations order. For example, when conducting unilateral and partnered operations and training it is important for the commander to assess early in the process the potential risk for an insider attack (see chapter 4 for additional information on insider attacks).

Note. The representation in table B-2 is not intended to be prescriptive. Risk management is an adaptable integrating process. The five steps are dynamic and cyclical.

Table B-2. Risk management and the military decisionmaking process

	Risk Management Steps				
Steps in the Military Decisionmaking Process	*Identify the Hazards*	*Assess the Hazards*	*Develop Controls and Make Risk Decisions*	*Implement Controls*	*Supervise and Evaluate*
RECEIPT OF MISSION	X				
MISSION ANALYSIS	X	X			
COURSE OF ACTION DEVELOPMENT	X	X	X		
COURSE OF ACTION ANALYSIS	X	X	X		
COURSE OF ACTION COMPARISON			X		
COURSE OF ACTION APPROVAL			X		
ORDERS PRODUCTION, DISSEMINATION, AND TRANSITION	X	X	X	X	X

Rapid Decision-Making and Synchronization Process

B-63. The rapid decision-making and synchronization process is a technique use during execution. While the MDMP seeks the optimal solution, the rapid decision-making and synchronization process seeks a timely and effective solution within the commander's intent, mission, and concept of operations. While identified here with a specific name and method, the commander and staff develop this capability through training and practice. When using this technique, the following considerations apply:

- Rapid is often more important than process.
- Much of it may be mental rather than written.
- It should become a battle drill for the current operations cell, and when established the plans cell.

Note. A *command post cell* is a grouping of personnel and equipment organized by warfighting function or by planning horizon to facilitate the exercise of mission command (FM 6-0). See FM 3-96 for information functional and integrating cells.

B-64. Using the rapid decision-making and synchronization process lets leaders avoid the time-consuming requirements of developing decision criteria and comparing courses of action. As operational and mission variables change during execution, this often invalidates or weakens courses of action and decision criteria before leaders can make a decision. Under the rapid decision-making and synchronization process, leaders combine their experience and intuition to quickly reach situational understanding. Based on this, they develop and refine workable courses of action.

B-65. The rapid decision-making and synchronization process facilitates continuously integrating and synchronizing the warfighting functions to address ever-changing situations. This process meets the following criteria for making effective decisions during execution:

- It is comprehensive, integrating all warfighting functions. It is not limited to any one warfighting function.
- It ensures all actions support the decisive operation by relating them to the commander's intent and concept of operations.
- It allows rapid changes to the order or mission.
- It is continuous, allowing commanders to react immediately to opportunities and threats.

B-66. The rapid decision-making and synchronization process is based on an existing order and the commander's priorities as expressed in the order. The most important of these control measures are the commander's intent, concept of operations, and commander's critical information requirements. The rapid decision-making and synchronization process includes five steps (figure B-5). The first two may be performed in any order, including concurrently. The last three are performed interactively until commanders identify an acceptable course of action. (Refer to FM 6-0 for additional information.)

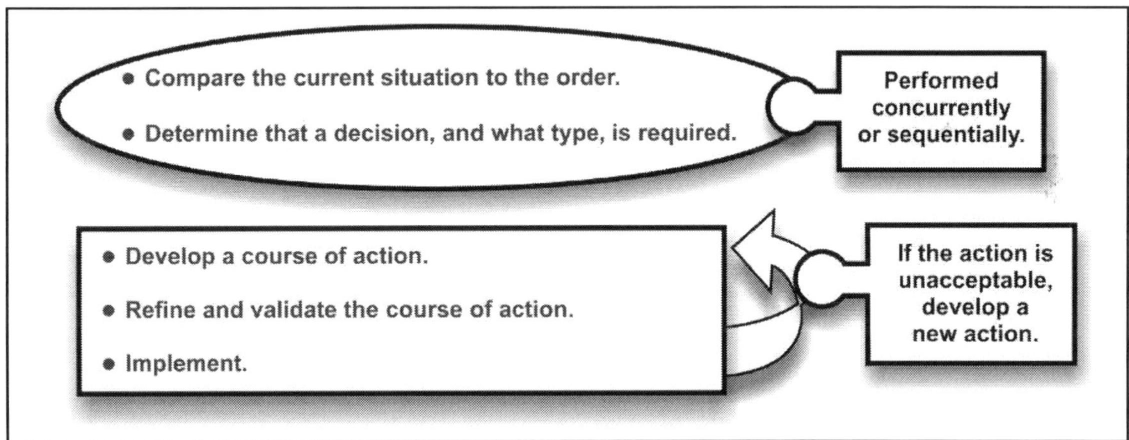

Figure B-5. Rapid decision-making and synchronization process

SOLDIER LOAD

B-67. Soldier load is an area of concern for Infantry commanders and subordinate leaders. How much is carried, how far, and in what configuration are critical mission considerations. The commander balances the risk to Soldiers from the enemy against the risk to mission accomplishment due to excessive loads and Soldier exhaustion and injury. Soldier load is limited to mission essential equipment to sustain continuous operations. The commander accepts prudent risks to reduce Soldier load based on a through mission analysis. (Refer to ATP 3-21.18 for additional information on Soldier load.)

Echeloning Loads

B-68. Maximum effort should be placed on echeloning loads The commander resists the mindset to carry everything to be prepared for every eventuality. Subordinate leaders at the lowest levels enforce load discipline to ensure that Soldiers do not voluntarily carry excess weight. The commander and staff plan for the delivery of nonessential equipment forward for subsequent operations. Echeloning loads to be secured or transported, breaks down supplies and equipment into three echelons: combat load, sustainment load, and contingency load.

Combat Load

B-69. Combat load is the minimum mission essential equipment required for a Soldier to operate (fight if required) and survive during the immediate operation. Supplies and equipment not carried in a combat load are secured and transported in sustainment loads (see paragraph B-70), and contingency loads (see paragraph B-71). Soldier's carrier loads in one of three type combat loads depending on the situation. The three types of combat loads are fighting load, approach march load, and emergency approach march load.

Fighting Load

B-70. Fighting load is the essential items a Soldier needs to maneuver on, close with, and destroy enemy forces in direct fire contact. Fighting load is the sum of everything worn or carried by the Soldier (see table B-3 for a listing of possible items carried depending on the current situation and mission). For missions requiring mobility, speed, and stealth, carrying heavy loads is a disadvantage. Cross-loading machine gun ammunition, mortar rounds, shoulder fired munitions, water, and radiotelephone operator's equipment across the unit may be necessary to balance out the loads being carried within the unit. For example, selected rifle platoon members may carry a mortar round (usually address in unit SOP) to drop at the mortar firing position once the unit reaches its area of operation to reduce loads within the mortar section.

Table B-3. Fighting load – possible items carried

Fighting Load	Pounds
Uniform complete (ACU, T-shirt, socks, boots, belt, patrol cap)	7.8
Advance combat helmet	4.2
Field load carrier (FLC)	5.8
Knee pads	0.5
M4 (no magazine)	6.4
5.56 unit basic load (UBL) (210 rounds and 7 magazines)	7.0
AN/PEQ-15 (ATPIAL)	0.5
M68 (CCO)	0.9
AN/PAS-13(V)1 (LWTS)	1.9
AN/PVS-14 (MNVD) Includes helmet mount.	1.2
Soldier Plate Carrier System (SPCS)	5.9
SAPI (front and back), SBI (sides)	16.0
1 quart canteens (2 each, with water)	4.6
MRE (1 each, stripped)	1.4
M67 fragmentary grenades (2 each)	2.0
Compass	0.5
Bayonet with scabbard	1.3
Individual first aid kit	1.0
Total	68.9

Legend:

ATPIAL	advanced target pointer illuminator aiming light	SAPI	small arms protective inserts
ACU	Army combat uniform	SBI	side ballistic insert
CCO	close combat optic	SPCS	Soldier Plate Carrier System
LWTS	light weapon thermal sight		
MRE	meal, ready to eat		
MNVD	monocular night vision device		

Note. When the improved outer tactical vest is substituted for the Soldier plate carrier system, a medium size improved outer tactical vest complete with all components [soft armor panel inserts, four ballistic plate inserts (front and back plates and two side plates), collar, and groin protectors] weighs 30 pounds, with a large improved outer tactical vest weighing about 35 pounds.

Approach March Load

B-71. An approach march load consists of the fighting load plus additional essential equipment (table B-4). The approach march load may include an assault pack or rucksack and all other items not needed in the fighting load and now required. Approach march loads, where direct contact with the enemy is intended, are dropped in an assault position, objective rally point, or when receiving effective enemy fires prior to an assault. (The key word is effective enemy fires, not contact, dropping carried loads will always delay the operation during recovery of dropped gear/equipment.) Execution of this technique must be planned for and rehearsed, and may require transportation assets, when available, to retrieve equipment later in the operation. An approach march load allows Soldiers to fight and sustain themselves until resupplied, though, approach march loads should not exceed 100 pounds.

Table B-4. Approach march load—possible items carried

Approach March Load	Pounds
Assault pack	3.1
T-Shirt	0.2
Socks (2 pair)	0.4
Wet weather top and bottom	3.0
Poncho	1.0
Poncho liner	1.5
Weapons cleaning kit	1.0
Entrenching tool with carrier	3.5
2-quart canteens (2 each)	10.0
Meals, ready to eat (3 each, stripped)	4.2
Subtotal	27.9
Add Fighting Load	68.9
Total	96.8

Note. Depending on the mission and environmental conditions, items carried normally in an approach march load may be carried as a fighting load. For example, wet weather gear top and bottom, and additional water, MREs, and ammunition.

Emergency Approach March Load

B-72. Operations requiring an emergency approach march load should only be used when absolutely necessary to accomplish a specific mission. For example, Soldiers may be required to carry heavier loads through terrain impassable by vehicles or when ground and air transportation resources are unavailable. Excessive weights associated with these loads significantly impact the unit's ability to move to the final destination without physically exhausting Soldiers. Physical exhaustion significantly limits the cognitive ability and mental focus of leaders and Soldiers, and inhibits their combat effectiveness. When carrying an emergency approach march load, commanders should seek to limit march distances and provide Soldiers time to recover before executing follow-on activities.

Sustainment Load

B-73. Sustainment loads consist of equipment required to sustain operations. Sustainment loads are generally positioned within the battalion or company support area and brought forward when needed. Sustainment loads may include rucksacks, duffel bags, and sleeping bags. Depending on the situation, personal protection items can be stored in preconfigured unit sustainment loads. Coordination is made to ensure sustainment load items are available when required.

Contingency Load

B-74. Contingency loads includes all other items not necessary for ongoing operations, such as extra clothing and personal items, or Javelin Close Combat Missile System in threat environments where the enemy lacks an armored capability. Contingency loads might be stored in duffel bags or palletized. Determining what goes in these loads and who is responsible for their storage and delivery is a critical decision for the commander.

B-75. B-bags should be palletized in unit loads and a contingency table of organizational equipment should be centralized in battalion packs. Contingency loads are generally not flown into deployment areas as part of the initial deployment. When contingency loads arrive in theater, plans should be made at or above brigade level to store unit contingency supplies and equipment. Items then can be returned to units if require by the situation. For example, if units are deployed into an area where items of contingency load are needed or if units are staged in assembly areas.

B-76. Instructions are issued to Soldiers before deployment, listing individual and organizational equipment not part of initial deployment. Contingency equipment could remain in continental United States, be stored at base areas in unit packs, or be reserve equipment issued by a higher headquarters when necessary.

B-77. Upon arrival in theater, provisions must be made for some items of equipment to be back loaded from the company, battalion, or brigade to division control. This allows units to deploy heavy for maximum flexibility. As the situation becomes clearer units can back load items not immediately needed to higher headquarters control as contingency loads for use at a later time.

Load Determination

B-78. Load determination is managed at company and platoon levels; however, standards are established at battalion level during planning to ensure Soldiers are properly equipped and physically ready for the conduct of operations. During this process, the commander and subordinate leaders determine the factors affecting Soldier load, and the capabilities and limitations of the unit. Load determination factors are address in the following paragraphs.

Load Configurations

B-79. As addressed earlier, loads are configured in three echelons: combat load (one of three configurations—fighting load, approach march, or emergency approach march), sustainment load, and contingency load. When configuring combat loads, the commander considers what ammunition, supplies, and equipment are mission essential. Effective load configuration requires the commander and staff to manage risk in a logical and control manner based on a detailed mission analysis. The commander tailors the combat load (see ATP 3-21.18 chapter 3, section II for additional information) to be carried with the unit based on this analysis and arranges for sustainment and contingency loads to be transported at a later time.

B-80. The commander's situational understanding, personal experience, and knowledge of the capabilities and limitation of the unit enables the determination of load configuration. The commander adapts to circumstances and situations encountered, makes decisions when to drop equipment, and cross loads equipment during movement. The commander maintains enough firepower and protection to defeat the enemy, when required, without burdening the unit with excessive loads.

Load Impact

B-81. The load a Soldier carries is a major concern to the commander and subordinate leaders planning a tactical movement. How much is carried, how far, and under what configuration are important mission considerations, requiring command emphasis. A Soldiers' ability to fight is directly related to the loads they carry. The commander attempts to minimize Soldier load to improve stealth, speed and survivability.

B-82. Excessive Soldier loads reduce energy and agility. Soldiers carrying an excessive load are at a disadvantage when reacting to enemy contact and during the conduct of follow-on actions at the conclusion of the tactical movement. Conversely, if the load is reduced, leaders may make decisions to leave behind mission essential or crucial equipment. Sometimes Soldiers must carry more than the recommended weights for a combat load. However, leaders must realize how this impacts the unit's overall combat power to accomplish the mission.

Weight Categories

B-83. Personal protective equipment, specifically Soldier body armor, constitutes the largest weight category of Soldier load. Body armor limits the Soldier's ability to maintain body core temperature and, to varying degrees, regulate breathing due to constriction of the torso. Depending on the mission variables of METT-TC, the commander may adjust the level of body armor protection balancing an increased risk to individual Soldiers to improve the likelihood of mission accomplishment.

B-84. Ammunition, supplies, and equipment carried by the Soldier is tailored to the requirements of the mission. For example, if the enemy threat does not include armor formations, a Soldier's combat load may not include the Javelin. In certain circumstances, it may be appropriate for units to carry additional ammunition due to sustainment constraints. In other circumstances, based on the enemy threat and historical analysis it may be necessary to carry mine detectors but not electronic countermeasure equipment. Planning and preparation processes include detailed load planning and calculation to assist the commander and subordinate leaders in organizing tactical loads to manage energy expenditure and combat effectiveness.

> ***Note.*** When exact equipment weights are required, refer to the appropriate technical manual for the item's weight.

Time Available and Terrain

B-85. The burden of load reduces the Soldier's ability to react to the enemy. Loads cause fatigue and lack of agility, placing Soldiers at disadvantages when rapid reaction to the enemy is required. Commanders consider the variables of METT-TC when determining loads. Two variables, time available and terrain, have the greatest effect on loads. Leaders must assess and balance the risk of the assigned mission against the risk to the units' ability to execute the mission given their physical condition.

B-86. Time available to move the unit, under a given load, may constrain the unit's ability to arrive in the most efficient manner with maximum available energy to accomplish assigned tasks. Time constraints may force the commander to reduce Soldier loads. When loads are not reduced, Soldier physical exertion increases requiring the commander and subordinate leaders to assess the condition of their units more often.

B-87. Difficult terrain naturally slows movement due to vegetation (thick brush compared to pine forest), grade (generally flat compared to steep hills or mountain), and composition (hard packed ground or roads compared to sand, snow/ice, or scree). Heavy loads over difficult terrain quickly exhaust Soldiers and significantly reduce their physical effectiveness and cognitive capacity to accomplish assigned tasks.

> ***Note.*** When necessary, dismounted marches can be hurried by conducting a forced march. Forced marches require speed, exertion, and more hours marched per day. This is normally accomplished by increasing marching hours for each day rather than rate of march. (See ATP 3-21.18, chapter 2, for a detail discussion on forced march.)

Electronic Warfare

B-88. *Electronic warfare* is military action involving the use of electromagnetic and directed energy to control the electromagnetic spectrum or to attack the enemy (JP 3-13.1). The commander integrates EW activities into operations through cyberspace electromagnetic activities. EW capabilities are applied from the air, land, sea, space, and cyberspace by manned, unmanned, attended, or unattended systems. *Cyberspace electromagnetic activities* is the process of planning, integrating, and synchronizing cyberspace and electronic warfare operations in support of unified land operations (ADRP 3-0). *Cyberspace operations* is the employment of cyberspace capabilities where the primary purpose is to achieve objectives in or through cyberspace (JP 3-0).

B-89. EW capabilities assist the commander in shaping the operational environment to gain an advantage. For example, electronic warfare may be used to set favorable conditions for cyberspace operations by stimulating networked sensors, denying wireless networks, or other related actions. Operations in cyberspace and the electromagnetic spectrum (EMS) depend on electronic warfare activities maintaining freedom of action in both. EW consists of three functions, electronic attack, electronic protection (EP), and electronic warfare support. This section primarily focuses on EP considerations as it relates to communications within the Infantry battalion. (See FM 3-12 for additional information on electronic warfare functions.)

Electronic Attack

B-90. Electronic attack involves the use of electromagnetic energy, directed energy, or antiradiation weapons to attack personnel, facilities, or equipment with the intent of degrading, neutralizing, or destroying enemy combat capability. Electronic attack is a form of fires. Electronic attack includes—

- Actions taken to prevent or reduce an enemy's effective use of the EMS.
- Employment of weapons that use either electromagnetic or directed energy as their primary destructive mechanism.
- Offensive and defensive activities, including countermeasures.

B-91. Examples of offensive electronic attack include—

- Jamming enemy radar or electronic command and control systems.
- Using antiradiation missiles to suppress enemy air defenses. (Antiradiation weapons use radiated energy emitted from a target, as the mechanism for guidance onto the target.)
- Using electronic deception to confuse enemy intelligence, surveillance, reconnaissance, and acquisition systems.
- Using directed-energy weapons to disable an enemy's equipment or capability.

B-92. Defensive electronic attack uses the EMS to protect personnel, facilities, capabilities, and equipment. Examples include self-protection and other protection measures such as the use of expendables (flares and active decoys), jammers, towed decoys, directed-energy infrared countermeasures, and counter radio-controlled improvised explosive device systems.

Electronic Protection

B-93. EP involves the actions taken to protect personnel, facilities, and equipment from any effects of friendly or enemy use of the electromagnetic spectrum that degrade, neutralize, or destroy friendly combat capability. For example, EP includes actions taken by the commander to ensure friendly use of the EMS, such as frequency agility in a radio or variable pulse repetition frequency in radar. The commander avoids confusing EP with self-protection. Both defensive electronic attack and EP protect personnel, facilities, capabilities, and equipment. However, EP protects from the effects of electronic attack (friendly and enemy) and electromagnetic interference, while defensive electronic attack primarily protects against lethal attacks by denying enemy use of the EMS to guide or trigger weapons.

Commander's Electronic Protection Responsibilities

B-94. EP is a command responsibility. The more emphasis the commander places on EP, the greater the benefits, in terms of casualty reduction and combat survivability, in a hostile environment or degraded information environment. The commander ensures the battalion trains on and practices sound EP techniques and procedures. The commander continually measures the effectiveness of EP techniques and procedures used within the battalion throughout the operations process. Commander EP responsibilities are—

- Review all information on jamming and/or deception reports, and assess the effectiveness of defensive EP.
- Ensure the battalion S-6 and S-2, in coordination with the EW NCO, report and properly analyze all encounters of electromagnetic interference, deception, and jamming.
- Analyze the impact of enemy efforts to disrupt or destroy friendly communications systems on friendly operation plans.
- Ensure the battalion exercises COMSEC techniques daily. Subordinate units should—
  - Change network call signs and frequencies often (in accordance with the signal operating instructions).

- Use approved encryption systems, codes, and authentication systems.
- Control emissions.
- Make EP equipment requirements known through quick reaction capabilities designed to expedite procedure for solving, research, development, procurement, testing, evaluation, installations modification, and logistics problems as they pertain to EW.
- Ensure quick repair of radios with mechanical or electrical faults; this is one way to reduce radio-distinguishing characteristics.
- Practice network discipline.

Staff Electronic Protection Responsibilities

B-95. The battalion staff assists the commander in accomplishing EP requirements. Specifically, the staff responds immediately to the commander and subordinate units. The staff—

- Keeps the commander informed.
- Reduces the time to control, integrate, and coordinate operations.
- Reduces the chance for error.

B-96. The battalion staff provides information, furnishes estimates, and provides recommendations to the commander. Specific battalion staff officer responsibilities include the—

- S-2. Advise the commander of enemy capabilities that could be used to deny the unit effective use of the electromagnetic spectrum. Keep the commander informed of the battalion's signal security posture.
- S-3. Exercise staff responsibility for EP. Include electronic warfare support and electronic attack considerations throughout the operations process and evaluate EP techniques and procedures employed. Ensure EP training is included in all unit-training programs and troop leading procedures during operations.
- S-6. Exercise staff responsibility for signal security and support EP. The S-6 in coordination with the EW NCO—
- Prepares and conducts the unit EP training program.
- Ensures alternate means of communications for those systems most vulnerable to enemy jamming.
- Ensures distribution of available COMSEC equipment to those systems most vulnerable to enemy information gathering activities.
- Ensures measures are taken to protect critical friendly frequencies from intentional and unintentional electromagnetic interference.

Signal Security

B-97. EP and signal security are closely related; they are defensive arts based on the same principle. If adversaries and enemies do not have access to the essential elements of friendly information, they are much less effective. The battalion's goal of practicing sound EP techniques is to ensure the continued effective use of the electromagnetic spectrum. The battalion's goal of signal security is to ensure the enemy cannot exploit the friendly use of the electromagnetic spectrum for communication. Signal security techniques are designed to give the commander confidence in the security of battalion transmissions. Signal security and EP is planned by the battalion based on the enemy's ability to gather intelligence and degrade friendly communications systems. (Refer to ATP 6-02.53 for additional information.)

Communications Planning Considerations

B-98. The battalion staff, specifically the S-6 in coordination with the S-2, S-3, and EW NCO, assesses threats to friendly communications during the communications planning process. Planning counters the enemy's attempts to take advantage of the vulnerabilities of friendly communications systems. Ultimately, the commander, subordinate commanders, and staff planners and radio and network operators are responsible for the security and continued operation of all mission command systems.

Communications Planning

B-99. When conducting communications planning, the S-6 uses spectrum management tools to assist in electromagnetic spectrum planning and to define and support requirements. The S-6 coordinates all frequency use before any emitter is activated to mitigate or eliminate electromagnetic interference or other negligible effects and considers the following when conducting electromagnetic spectrum management planning:

- Transmitter and receiver locations.
- Antenna technical parameters and characteristics.
- Number of frequencies desired and separation requirements.
- Nature of the operation (fixed, mobile land, mobile aeronautical, and over water or maritime).
- Physical effects of the operational environment (ground and soil type, humidity, and topology).
- All electromagnetic spectrum-dependent equipment to be employed to include emitters, sensors, and unmanned aerial sensors.
- Start and end dates for use.

Primary, Alternate, Contingency, and Emergency Plan

B-100. The primary, alternate, contingency, and emergency (PACE) plan is a communication plan that exists for a specific mission or task, not a specific unit, as the plan considers both intra- and inter-unit sharing of information. The PACE plan designates the order in which an element will move through available communications systems until contact can be established with the desired distant element. The S-6 develops a PACE plan for each phase of an operation to insure that the commander can maintain mission command of the formation. The plan reflects the training, equipment status, and true capabilities of the formation. The IBCT S-6 evaluates its communication requirements with the battalion and works with the battalion S-6 to develop an effective plan. Upon receipt of an order, the S-6 evaluates the PACE plan for two key elements as follows:

- Does the battalion have the assets to execute the plan?
- How can the battalion nest with the plan when it develops its own plan?

B-101. Accurate PACE plans are crucial to the commander's situational awareness. A subordinate unit (considerations include those for host nation and/or multinational forces) that is untrained on a particular communication system or lacks all of the subcomponents to make the system mission capable, does not ensure continuity of mission command by including the communication system in the PACE plan. The commander's ability to exercise mission command during an operation can suffer due to communication systems that are in transit or otherwise unavailable. If the battalion or a subordinate unit does not have four viable methods of communications, it is appropriate to issue a PACE plan that may only have two or three systems listed. If the unit cannot execute the full PACE plan to its higher command, it must inform the issuing headquarters with an assessment of shortfalls, gaps, and possible mitigations as part of the mission analysis process during the military decision-making process. During course of action development, the S-6 nest the subordinate unit's plan with the higher command's plan whenever practical. This aids in maintaining continuity of effort. (Refer to ATP 6-02.53 for additional information.)

Geometry

B-102. The S-6 analyzes the terrain, and determines the method(s) to make the geometry of the operations work support the commander's plan. Adhering rigidly to standard CP deployment makes it easier for the enemy to use the direction finder and aim jamming equipment. Deploying units and communications systems perpendicular to the forward line of own troops enhance the enemy's ability to intercept communication by aiming transmissions in the enemy's direction. When possible, install terrestrial line of sight communications parallel to the forward line of own troops. This supports keeping the primary strength of U.S. transmissions in friendly terrain.

B-103. Single-channel tactical satellite systems reduce friendly CP vulnerability to enemy direction efforts. Tactical satellite communication systems are relieved of this constraint because of their inherent resistance to enemy direction finder efforts. When possible, utilize terrain features to mask friendly communication from enemy positions. This may require moving headquarters elements farther forward and using more jump or tactical CPs to ensure the commander can continue to direct units effectively.

B-104. Location of CPs requires carefully planning as CP locations generally determine antenna locations. The proper installation and positioning of antennas around CPs is critical. Disperse and position antennas and emitters at the maximum remote distance and terrain dependent from the CP to ensure that not all of a unit's transmissions are coming from one central location system design. (See appendix A for additional information on CP locations.)

B-105. Establish alternate routes of communication when designing communications systems. This involves establishing sufficient communications paths to ensure that the loss of one or more routes does not seriously degrade the overall system. The commander establishes the priorities of critical communications links. Provide high priority links with the greatest number of alternate routes. Alternate routes enable friendly units to continue to communicate despite the enemy's efforts to deny them the use of their communications systems. Alternate routes can also be used to transmit false messages and orders on the route that is experiencing electromagnetic interference, while they transmit actual messages and orders through another route or means. A positive benefit of continuing to operate in a degraded system is that the problematic degraded system cause the enemy to waste assets used to impair friendly communication elsewhere. Three routing concepts, or some permutation of them, can be used in communications as follows:

- Straight-line system. Provides no alternate routes of communications.
- Circular system. Provides one alternate route of communications.
- Grid system. Provides as many alternate routes of communications as can be practically planned.

B-106. Avoid establishing a pattern of communication. Enemy intelligence analysts may be able to extract information from the pattern, and the text, of friendly transmissions. If easily identifiable patterns of friendly communication are established, the enemy can gain valuable information.

B-107. The number of friendly transmissions tends to increase or decrease according to the type of tactical operation being executed. Execute this deceptive communication traffic by using false peaks, or traffic leveling. Utilize false peaks to prevent the enemy from connecting an increase of communications with a tactical operation. Transmission increases on a random schedule create false peaks. Tactically accomplish traffic leveling by designing messages to transmit when there is a decrease in transmission traffic. Traffic leveling keeps the transmission traffic constant. Coordinate messages transmitted for traffic leveling or false peaks to avoid operational security violations, electromagnetic interference, and confusion among friendly equipment operators.

B-108. During operations, dismounted tactical unit area coverage and distance extension is a major concern to unit commanders. Communications inside buildings or over urban terrain is a challenge. For these conditions, the multiband inter/intra team radio system provides a "back-to-back" (two radios) retransmission (known as RETRANS) capability for COMSEC and plain text modes. Beside two radios, the only hardware required for retransmission is a small cable kit and some electronic filters. When configured for retransmission operations, a true digital repeater forms. Since the radios repeat the transmitted digits and since the radios do not have to have any COMSEC keys loaded in them, the radios do not degrade signal quality.

B-109. Automated Communications Engineering Software equipment and subsequent signal operating instructions development resolve many problems concerning communications patterns; they allow users to change frequencies often, and at random. This is an important aspect of confusing enemy traffic analysts. Enemy traffic analysts are confused when frequencies, network call signs, locations, and operators are often changed. Communications procedures require flexibility to avoid establishing communications patterns. (Refer to ATP 6-02.53 for additional information.)

Emission Control

B-110. The control of electromagnetic emissions is essential to successful defense against the enemy's attempts to destroy or disrupt the battalion's communications. Emission control is the selective and controlled use of electromagnetic, acoustic, or other emitters to optimize mission command capabilities while minimizing, for operations security. When operating radios, the battalion exercises emission control at all times within all echelons and only transmitters when needed to accomplish the mission. Enemy intelligence analyst look for patterns they can turn into usable information. Inactive friendly transmitters do not provide the enemy with useable intelligence. Emission control can be total; for example, the commander may direct radio silence whenever desired. *Radio silence* is the status on a radio network in which all stations are directed to continuously monitor without transmitting, except under established criteria (ATP 6-02.53).

B-111. Unit operators keep transmissions to a minimum (20 seconds absolute maximum, 15 seconds maximum preferred) and transmit only mission-critical information. Good emission control makes the use of communications equipment appear random, and is therefore consistent with good EP practices. This technique alone will not eliminate the enemy's ability to find a friendly transmitter; but when combined with other EP techniques, it makes locating a transmitter more difficult.

Replacement

B-112. Replacement involves establishing alternate routes and means of doing what the commander requires. Frequency modulation voice communications are the most critical communications used by the commander during enemy engagements and require reserving critical systems for critical operations. The enemy should not have access to information about friendly critical systems until the information is useless.

B-113. The battalion utilizes alternate means of communication before enemy engagements. This ensures the enemy cannot establish a database to destroy primary means of communication. If the primary means degrades, replace primary systems with alternate means of communication. Replacements require preplanning and careful coordination; if not, compromise of the alternate means of communication occurs and is no longer useful as the primary means of communication. Users of communications equipment require knowledge of how and when to use the primary and alternate means of communication. This planning and knowledge ensures the most efficient use of communications systems.

Concealment

B-114. The commander and subordinate leaders ensure effective employment of all communications equipment, despite the enemy's concerted efforts to degrade friendly communication to the enemy's tactical advantage. Operation plans should include provisions to conceal communications personnel, equipment, and transmissions. It is difficult to conceal most communications systems; installing antennas as low as possible on the backside of terrain features, and behind man-made obstacles, helps conceal communications equipment while still permitting communication.

Training and Procedures for Countering Enemy Electronic Attack

B-115. EP includes the application of training and procedures for countering enemy electronic attack. Once the threat and vulnerability of friendly electronic equipment to enemy electronic attack are identified, the commander takes appropriate actions to safeguard friendly combat capability from exploitation and attack. EP measures minimize the enemy's ability to conduct electronic warfare support and electronic attack operations successfully against the battalion. To protect friendly combat capabilities, units—

- Regularly brief friendly force personnel on the EW threat.
- Ensure that they safeguard electronic system capabilities during exercises, workups, and pre-deployment training.
- Coordinate and deconflict EMS usage.
- Provide training during routine home station planning and training activities on appropriate EP active and passive measures under normal conditions, conditions of threat electronic attack, or otherwise degraded networks and systems.

Electronic Warfare Support

B-116. Electronic warfare support is a division of electronic warfare involving actions tasked by, or under direct control of, an operational commander to search for, intercept, identify, and locate or localize sources of intentional and unintentional radiated electromagnetic energy for the purpose of immediate threat recognition, targeting, planning, and conduct of future operations. Electronic warfare support assist the battalion in identifying the electromagnetic vulnerability of an enemy or adversary's electronic equipment and systems. The commander take advantage of these vulnerabilities through EW operations.

B-117. Electronic warfare support systems are a source of information for immediate decisions involving electronic attack, EP, avoidance, targeting, and other tactical employment of forces. Electronic warfare support systems collect data and produce information to—

- Corroborate other sources of information or intelligence.
- Conduct or direct electronic attack operations.
- Initiate self-protection measures.
- Task weapons systems.
- Support EP efforts.
- Create or update EW databases.
- Support information related capabilities.

PREPARATION

B-118. Preparation consists of those activities performed by units and Soldiers to improve their ability to execute an operation. Preparations require commander, staff, subordinate unit, and Soldier actions to ensure the battalion is trained, equipped, and ready to execute operations. Preparations help the commander, staff, and subordinate units, and Soldiers understand the situation and their roles in upcoming operations.

B-119. Since time is a factor in all operations, the commander and staff conduct a time analysis early in the planning process to determine what preparation activities need to take place and when to begin those activities to ensure forces are ready and in position before execution. The plan may require the commander to direct subordinates to start necessary movements; conduct task-organization changes; begin reconnaissance, surveillance, and security operations; and execute other preparation activities before completing the plan.

B-120. Commander driven key preparation activities (although not inclusive) are addressed in the following paragraphs. See ADRP 5-0 for a complete listing of preparation activities.

INITIATE INFORMATION COLLECTION

B-121. *Information collection* is an activity that synchronizes and integrates the planning and employment of sensors and assets as well as the processing, exploitation, and dissemination systems in direct support of current and future operations (FM 3-55). FM 3-55 describes an information collection capability as any human or automated sensor, asset, or processing, exploitation, and dissemination system that can be directed to collect information that enables better decision making, expands understanding of the operational environment, and supports warfighting functions in decisive action.

B-122. Information collection highlights aspects that influence how the battalion operates as a ground force in close and continuous contact with the environment, including the enemy, terrain and weather, and civil considerations. Information collection involves the acquisition of information and the provision of this information to processing elements and consists of the following tasks:

- Plan requirements and assess collection.
- Task and direct collection.
- Execute collection.

Plan Requirements and Assess Collection

B-123. Plan requirements and assess collection is the task of analyzing requirements, evaluating available assets (internal and external), recommending to the operations staff taskings for information collection assets, submitting requests for information for adjacent and higher collection support, and assessing the effectiveness of the information collection plan. It is a commander-driven, coordinated staff effort led by the S-2. The continuous functions of planning requirements and assessing collection identify the best way to satisfy the requirements of the commander and staff. These functions are not necessarily sequential. (ATP 2-01 discusses the planning requirements and assessing collection functions.)

Task and Direct Collection

B-124. The S-3 (based on recommendations from the staff) tasks, directs, and, when necessary, retasks the information collection assets. Tasking and directing of limited information collection assets is vital to their control and effective use. The staff tasks information collection assets by issuing WARNORDs, fragmentary orders, and operation orders. It accomplishes directing information collection assets by continuously monitoring the operation. The staff conducts retasking to refine, update, or create new requirements. (See FM 3-55.)

Execute Collection

B-125. Executing collection focuses on requirements tied to the execution of tactical missions [normally reconnaissance, surveillance, security operations, and (intelligence operations-tasks undertaken by MI units and Soldiers)]. Typically, collection activities begin soon after receipt of mission and continue throughout preparation for and execution of the operation. Collection activities do not cease at the conclusion of the mission but continue as required. This allows the commander to focus combat power, execute current operations, and prepare for future operations simultaneously. (See FM 3-55.)

B-126. To provide effective support to execution, planning requirements and assessing collection must be linked to planned and ongoing operational activities. *Intelligence synchronization* is the "art" of integrating information collection and intelligence analysis with operations to effectively and efficiently support decision making (ADRP 2-0). Plans and orders direct and coordinate information collection by providing information collection tasks based on validated requirements essential for mission accomplishment. Plans and orders help allocate scarce information collection assets effectively and efficiently. The intelligence staff must collaborate with higher, lower, and adjacent intelligence staffs to ensure the effectiveness of planning requirements and assessing collection.

Conduct Confirmation Briefs

B-127. The confirmation brief is a key part of preparation. A *confirmation brief* is a briefing subordinate leaders give to the higher commander immediately after the operation order is given. It is the leaders' understanding of the commander's intent, their specific tasks, and the relationship between their mission and the other units in the operation (ADRP 5-0). A confirmation brief ensures the battalion commander that subordinate commanders understand—

- The commander's intent, mission, and concept of operations.
- Their unit's tasks and associated purposes.
- The relationship between their unit's mission and those of other units in the operation.

Note. Ideally, the commander conducts confirmation briefs in person with selected staff members of the higher headquarters present.

Conduct Rehearsals

B-128. A *rehearsal* is a session in which the commander and staff or unit practices expected actions to improve performance during execution (ADRP 5-0). The commander uses this tool to ensure the staff and subordinates commanders understand the concept of operations and commander's intent. Rehearsals also allow subordinate commanders and leaders to practice synchronizing operations at times and places critical to mission accomplishment. Effective rehearsals imprint a mental picture of the sequence of the operation's key actions and improve mutual understanding and coordination of subordinate commanders and supporting leaders and units. The extent of rehearsals depends on available time. In cases of short-notice requirements, detailed rehearsals may not be possible. (Refer to FM 6-0 for additional information.)

Effective and Efficient Rehearsals

B-129. Effective and efficient rehearsals ensure the staff and subordinate commanders understand the commander's intent and concept of operation. Rehearsals allow the commander and staff to identify shortcomings in unit SOPs not recognized previously. Rehearsals contribute to external and internal coordination as the staff or subordinates identify additional coordinating requirements. Subordinate commanders and leaders use rehearsals to—

- Practice essential tasks to improve performance.
- Reveal weaknesses or problems in the plan.
- Coordinate actions of subordinate elements.
- Improve understanding of concept of the operation to foster confidence.

B-130. Adequate time is essential when conducting rehearsals. Time required varies with complexity of mission, type and technique of rehearsal, and level of participation. Units conduct rehearsals at the lowest possible level, using thorough techniques, given time availability. Under time constraints, subordinate commanders and leaders conduct abbreviated rehearsals, focusing on critical events determined by reverse planning. Each unit has different critical events based on mission, unit readiness, and commander's assessment or intent.

B-131. Whenever possible, subordinate commanders and leaders base rehearsals on completed SOP. However, units may rehearse contingency plans to prepare for an anticipated mission. Rehearsals are coordinating events, not an analysis. They do not replace war-gaming. The commander and subordinates war-game during the MDMP or troop leading procedures to analyze different courses of action determining the optimal one. Rehearsals practice selected actions. The commander and subordinates avoid making major changes to SOPs during rehearsals. They make those changes essential to mission success and risk mitigation.

B-132. Subordinate units may begin rehearsals of battle drills and other SOPs before receiving the operations order. Once the order has been issued, units can rehearse mission-specific tasks. Key tasks to rehearse include—

- Actions on the objective or in the assembly area.
- Assaulting enemy positions.
- Actions on reacting to indirect fire.
- Breaching obstacles (mine and wire).
- Using special weapons or demolitions.
- Actions on unexpected enemy contact.
- Encounters with civilians.

Rehearsal Assessment

B-133. The commander establishes standards for rehearsals. Properly executed rehearsals validate each leader's role and how each unit contributes to the overall mission, what each unit does, when each unit does it relative to times and events, and where each unit does it to achieve desired effects. An effective rehearsal ensures subordinate commanders have common vision of the enemy, their own forces, terrain, and relationship between them. It identifies specific actions requiring immediate staff resolution and informs higher headquarters of critical issues or locations the commander and key individuals must oversee.

B-134. The commander and subordinate leaders assess and critique all parts of rehearsals. Critiques center on how well the unit SOP achieves the commander's intent and on coordination necessary to accomplish its goal. Usually, the commander leaves internal execution tasks within rehearsals to subordinate unit commanders' judgment and discretion.

B-135. An SOP rehearsal provides opportunities for the commander and subordinates to identify and fix unresolved problems. Subordinate leaders and key individuals ensure all participants understand changes to unit SOP and the recorder captures all coordination done at rehearsals. All changes to published SOP are in effect as soon as possible. The staff publishes these changes to the SOP and distributes to all units and personnel affected.

Types of Rehearsals

B-136. Each rehearsal type achieves different results and has specific place in the preparation timeline. The four types of rehearsals are—

- Backbrief.
- Combined arms rehearsal.
- Support rehearsal.
- Battle drill or standard operating procedures rehearsal.

Backbrief

B-137. *Backbrief* is a briefing by subordinates to the commander to review how subordinates intend to accomplish their mission (FM 6-0). Subordinates perform backbriefs throughout preparation allowing commanders to clarify the commander's intent early in subordinate planning. Backbriefs differs from the confirmation brief (a briefing subordinates give their higher commander immediately following receipt of an order) in which subordinate leaders are given time to complete their plan.

Combined Arms Rehearsal

B-138. Combined arms rehearsal is a rehearsal in which subordinate units synchronize their plans with each other. The battalion normally executes a combined arms rehearsal after subordinate units issue their operations order.

Support Rehearsal

B-139. Support rehearsal helps synchronize each warfighting function with the overall operation. Throughout preparation, units conduct support rehearsals within framework of a single or limited number of warfighting functions. These rehearsals typically involve coordination and procedure drills for aviation, fires, engineer support, or casualty evacuation.

Battle Drill or Standard Operating Procedure Rehearsal

B-140. Battle drill is a collective action rapidly executed without applying deliberate decision-making process. Battle drill or SOP rehearsal ensures all participants understand techniques or specific set of procedures. Throughout preparation, units and staffs rehearse battle drills and SOP.

Methods of Rehearsal

B-141. Methods for conducting rehearsals are limited by the commander's imagination and available resources. (Refer to FM 6-0 for more information.) Resources required for each technique range from broad to narrow and each rehearsal technique imparts different level of understanding to participants. Rehearsal techniques generally used are—

- Full-dress rehearsal.
- Key leader rehearsal.
- Terrain-model rehearsal.
- Digital terrain-model rehearsal.
- Sketch-map rehearsal.
- Map rehearsal.
- Network rehearsal.

Full-Dress Rehearsal

B-142. A full-dress rehearsal produces the most detailed understanding of the operation. It includes every participating Soldier and system. Leaders rehearse their subordinates on terrain similar to the area of operation, initially under good light conditions, and then in limited visibility if METT-TC permits. Leaders repeat small unit actions until executed to standard. Full-dress rehearsals help Soldiers clearly understand what commanders expect of them. It helps them gain confidence in their ability to accomplish the mission. Supporting elements, such as aviation crews, meet and rehearse with Soldiers to synchronize the operation.

B-143. Full-dress rehearsals consume more time than any other rehearsal type. For companies and smaller units, full-dress rehearsals effectively ensure all units in the operation understand their roles. However, the battalion commander considers how much time their subordinates need to plan and prepare when deciding whether to conduct a full-dress rehearsal.

Key Leader Rehearsal

B-144. Circumstances may prohibit a rehearsal with all members of the unit. A reduced-force rehearsal involves only key leaders of the organization and its subordinate units. It normally takes fewer resources than a full-dress rehearsal. Terrain requirements mirror those of a full-dress rehearsal, even though fewer Soldiers participate. The commander first decides the level of leader involvement. Then selected leaders rehearse the plan while traversing the actual or similar terrain. Often the commander uses this technique to rehearse fire control measures for an engagement area during defensive operations. The commander often uses a reduced-force rehearsal to prepare key leaders for a full-dress rehearsal. It may require developing a rehearsal plan mirroring the actual plan but fits the terrain of the rehearsal.

B-145. A reduced-force rehearsal normally requires less time than a full-dress rehearsal. The commander considers how much time subordinates need to plan and prepare when deciding whether to conduct a reduced-force rehearsal.

Terrain-Model Rehearsal

B-146. The terrain-model rehearsal is the most popular rehearsal technique. It takes less time and fewer resources than a full-dress or reduced-force rehearsal. (A terrain-model rehearsal takes a brigade and below unit between one to two hours to execute to standard.) An accurately constructed terrain model helps subordinate leaders visualize the commander's intent and concept of operations. When possible, the commander places the terrain model where it overlooks the actual terrain of the area of operation. However, if the situation requires more security, the commander places the terrain model on a reverse-slope within walking distance of a point overlooking the area of operation. The model's orientation coincides with the terrain. The size of the terrain model can vary from small (using markers to represent units) to large (on which the participants can walk). A large model helps reinforce the participants' perception of unit positions on the terrain.

B-147. Often, constructing the terrain model consumes the most time during this technique. Units require a clear SOP that states how to build the model so it is accurate, large, and detailed enough to conduct the rehearsal. A good SOP establishes staff responsibility for building the terrain model and a timeline for its completion.

Digital Terrain-Model Rehearsal

B-148. With today's digital capabilities, users can construct terrain models in virtual space. Units drape high-resolution imagery over elevation data thereby creating a fly-through or walk-through. Holographic imagery produces the view in three dimensions. Often, the model hot links graphics, detailed information, unmanned aircraft systems, and ground imagery to key points providing more insight into the plan. Digital terrain models reduce the operations security risk because they do not use real terrain. The battalion's geospatial engineers (when assigned) or imagery analysts can assist in digital model creation.

Sketch-Map Rehearsal

B-149. The commander can use the sketch-map technique almost anywhere, day or night. The procedures are the same as for a terrain-model rehearsal except the commander uses a sketch map in place of a terrain model. Large sketches ensure all participants can see as each participant walks through execution of the operation. Participants move markers on the sketch to represent unit locations and maneuvers. Sketch-map rehearsals take less time than terrain-model rehearsals and more time than map rehearsals.

Map Rehearsal

B-150. A map rehearsal is similar to a sketch-map rehearsal except the commander uses a map and operation overlay of the same scale used to plan the operation. The map rehearsal itself consumes the most time. Normally, a map rehearsal is the easiest technique to set up since it requires only maps and graphics for current operations. This technique requires the least terrain of all rehearsals. A good site ensures participants can easily find it yet stay concealed from the enemy. An optimal location overlooks the terrain where the unit executes the operation.

Network Rehearsal

B-151. Subordinate units conduct network rehearsals over wide-area networks or local area networks. The commander and staff practice this rehearsal by talking through critical portions of the operation over communications networks in a sequence the commander establishes. The organization rehearses only the critical parts of the operation. These rehearsals require all information systems needed to execute that portion of the operation. All participants require working information systems, the operation order, and overlays. Command posts can rehearse battle tracking during network rehearsals.

Rehearsal Area Coordination

B-152. Rehearsal area coordination is conducted with key leaders and commanders to facilitate the unit's safe, efficient and effective use of rehearsal area before its mission. Rehearsal area coordination includes—

- Identification of your unit.
- Mission.
- Terrain similar to the objective site.
- Security of the area.
- Availability of aggressors.
- Use of blanks, pyrotechnics, and ammunition.
- Available mock-ups.
- Time area is available (preferably, when light conditions approximate light conditions for the operation).
- Transportation.
- Coordination with other units using the area.

Conduct Preoperations Checks and Inspections

B-153. Subordinate unit preparation includes completing preoperations checks and inspections. These checks ensure Soldiers, units, staffs, and systems are as fully capable and ready to execute the mission as time and resources permit. Inspections ensure the force has the resources necessary to accomplish the movement. Leaders should conduct initial inspections shortly after receipt of the WARNORD. Leaders should spot check throughout the unit's preparation for combat. Key leaders should conduct final inspections prior to executing the mission. Key leaders should inspect—

- Weapons and ammunition.
- Uniforms and equipment.
- Mission essential equipment.
- Soldiers understanding of the mission and individual responsibilities.
- Communications.
- Rations and water.
- Camouflage.
- Deficiencies noted during earlier inspections.

This page intentionally left blank.

Appendix C
Fires

Fires is the use of weapons systems or other actions to create a specific lethal or nonlethal effect on a target (JP 3-09). Lethal fire support comes from IBCT organic indirect fires assets, Army artillery and aviation assets, and joint and multinational artillery and aviation assets. Nonlethal effects can come from a wide range of military and civilian, joint and multinational partners.

FIRE SUPPORT

C-1. *Fire support* is the fires that directly support land, maritime, amphibious, and special operations forces to engage enemy forces, combat formations, and facilities in pursuit of tactical and operational objectives (JP 3-09). Fire support is the collective and coordinated use of indirect fire weapons and armed aircraft in support of the commander's scheme of maneuver. Fire support planning is the process of analyzing, allocating, and scheduling fire support assets. Fire support assets include mortars, field artillery cannons and rockets, Army attack aviation, close air support (CAS), naval gunfire, and electronic attacks.

FIRE SUPPORT SYSTEM

C-2. The fire support system acquires and tracks targets, delivers timely and accurate lethal fires, provides counterfire, and plans, coordinates, and orchestrates fire support. The fire support system achieves desired effects (lethal and nonlethal means) through a combination of fire support assets. The integration of fire support assets is critical regardless of which element of decisive action currently dominants. For example in the defense, fire support systems support security forces by using both precision and area munitions to destroy enemy reconnaissance and high-payoff targets, and by delivering on-call fires at appropriate times and places. Fire support facilitates the withdrawal of security forces at the completion of their mission. Fire support systems cover barriers, gaps, and open areas within the defense. Disruptive means to temporarily deny, degrade, deceive, delay, or neutralize enemy aircraft systems can include electronic attack during an area defense.

PLANNING PRINCIPLES

C-3. The Infantry battalion's fire support officer (FSO) receives guidance from the commander in coordination with the plans developed by the battalion S-3 and from higher headquarters. The FSO plans and integrates fire support assets to achieve the desired effects to support the commander's concept of the operation. Successful fire support planning is the result of the FSO aggressively contributing to the battalion commander's planning and decision-making process. In advising the commander on the application of fire support, the FSO reviews fire support requirements against basic fire support planning principle during the development of the battalion's fire support plan. (Refer to FM 3-09 for additional information on planning principles.) Fire support principles include the following:

- Plan early and continuously.
- Ensure the continuous flow of targeting information.
- Consider the use of all capabilities.
- Use the lowest echelon capable of furnishing effective support.
- Furnish the type of support requested.
- Use the most effective fire support means.
- Avoid unnecessary duplication.
- Coordinate airspace.
- Provide adequate support.

- Provide for rapid coordination.
- Protect the force.
- Provide for flexibility.
- Use fire support coordination measures.

PLANNING AND THE INTEGRATION OF FIRES

C-4. *Fire support planning* is the continuing process of analyzing, allocating, and scheduling fires to describe how fires are used to facilitate the actions of the maneuver force (FM 3-09). Fire support planning is focused on using the timely and effective delivery of fires to enhance the actions of the maneuver force. Fire support planning involves the assignment of command or support relationships and positioning of fire support systems. It also identifies the types of targets to attack and the collection assets that acquire and track the targets, specifies the fire support assets to attack each identified target, and establishes the criteria for target defeat. The objective of fire support planning is to optimize the application of combat power. Fire support planning is performed as part of the operations process (see paragraph C-11). Fire support planning includes developing fire plans (target lists and overlays) and determining forward observer control options to ensure fire support is integrated into the commander's scheme of maneuver and can be executed in a timely manner. (Refer to ATP 3-09.42 for additional information.)

Fire Support Plan

C-5. The *fire support plan* is a plan that addresses each means of fire support available and describes how Army indirect fires, joint fires, and target acquisition are integrated with maneuver to facilitate operational success (FM 3-09). The IBCT fire support coordinator (field artillery battalion commander), FSO, and fire support cell planners develop an effective and integrated fire support plan to support IBCT operations. An effective fire support plan clearly defines fire support requirements, focuses on the tasks and their resulting effects, uses all available acquisition and attack assets, and applies the best combination of fire support assets against high-payoff targets. The fire support plan identifies critical times and places where the commander anticipates the need to maximize effects from fire support assets. Fire support planning considers existing limitations on the employment of fires, such as rules of engagement and positive identification requirements, weather effects on fires assets, the presence of special operations forces within the area of operations, desired conditions of subsequent phases, and requirements for collateral damage avoidance.

Commander's Guidance

C-6. The IBCT commander's guidance for fires provides subordinates with the general guidelines and restrictions for the employment of fires and their desired effects. The guidance emphasizes in broad terms when, where, and how the commander intends to synchronize the effects of fires with the other elements of combat power to accomplish the mission. Commander's guidance should include priorities and how the commander envisions the operation unfolding and the impact that fires will have on its success. Priority of fires is the commander's guidance to the subordinate commanders, fires planners, and supporting agencies to organize and employ fires in accordance with the relative importance of the unit's mission.

C-7. The Infantry battalion typically uses top-down fire support planning, with bottom-up refinement of the plans. The battalion commander develops guidance for fire support in terms of task, purpose, and effect. In turn, the battalion FSO, in conjunction with the S-3, determines the method used to accomplish each task. Battalion subordinate units then incorporate assigned tasks into their fire support plans. Units tasked to initiate fires refine and rehearse their assigned tasks. The commander refines the battalion's fire support plan, ensuring that designated targets achieve the intended purpose. The commander conducts rehearsals to prepare for the mission and, as specified in the plan, directs subordinate units to rehearse their assigned targets.

Role of the Battalion Commander

C-8. The key role of the commander during planning is to ensure the employment of forces in the battalion's area of operation achieves a position of advantage in respect to the enemy. The commander achieves this position of advantage through movement in combination with fires. Concurrent movement and fires planning requires the battalion commander to—

- Decide precisely what fire support tasks enable the battalion's mission.
- Take an active role in the development of the fire support tasks.
- Clearly articulate to staff, not just the FSO, the sequenced fire support tasks in terms of desired effects for each target and the purpose of each target as it relates to the scheme of maneuver.
- Ensure the FSO and subordinate commanders understand the fire support guidance.
- State the required effects for fire support, describing a targeting effect against a specific enemy formation's function or capability.
- State the fire support purpose, describing how the effect contributes to accomplishing the mission within the intent. (*Note.* The FSO, in coordination with the S-3, develops the method to achieve the desired effects and the purpose for each target.)
- Ensure that fire support missions are clearly synchronized with the scheme of maneuver and that movement of fire support assets is synchronized with this maneuver.

Role of the Battalion Fire Support Officer

C-9. The battalion FSO advises the commander and staff on fire support matters. The FSO plans, coordinates, and executes fire support for the commander's concept of the operation. The FSO makes recommendations for integrating field artillery assets and maneuver battalion mortars into the scheme of maneuver, and recommends their movement within the scheme of maneuver. Additional responsibilities include the following:

- Recommend to the commander how to best employ and control fire support teams.
- Supervise all functions of the battalion fire support cell (when established).
- Ensure all fire support personnel are trained properly.
- Participate in IBCT and battalion combined arms and fire support rehearsals.
- Prepare and disseminate the fire support execution matrix or the fire support plan (see ATP 3-09.42).
- Coordinate with the tactical air control party (TACP) on CAS missions and for terminal control personnel.
- Assist in the coordination for positioning or movement of lethal and nonlethal assets in the battalion area of operation.
- Establish and maintain communications with supporting field artillery units.
- Process requests for more fire support with the IBCT fire support teams and CAS with the TACP.
- Disseminate the approved target list and matrix to subordinate fire support teams and elements.
- Recommend appropriate changes in the target list and attack guidance when required.
- Determine, recommend, and process time-sensitive high payoff targets to the IBCT fire support team.
- Coordinate with the battalion intelligence staff officer (S-2) and operations staff officer (S-3) for target acquisition (see ATP 3-09.12) coverage and processing of battalion high payoff targets. Plan and supervise the execution of assigned and developed fire support tasks.
- Participate in the targeting meeting(s) to update targets, high payoff targets, priorities, and asset allocation.

Fire Support Task

C-10. A fire support task is a task that a fire support team must accomplish in order to support a combined-arms operation. Failure to achieve a fire support task may require the commander to alter their tactical or operational plan. A fully developed fire support task has a task, purpose, and effect. The task describes what targeting objective fires, such as delay, disrupt, limit, or destroy, must achieve on an enemy formation's function or capability. The purpose describes why the task contributes to maneuver. The effect quantifies successful accomplishment of the task.

FIRE SUPPORT PLANNING AND MILITARY DECISIONMAKING PROCESS

C-11. As a member of the battalion staff, the FSO plays a crucial role in the military decision-making process (MDMP) both as the fire support expert on the staff and as a member of the targeting team. Figure C-1, page C-4 through C-7, describes the sequence of inputs, actions, and outputs of fire support planning within the MDMP. When there is limited time to conduct the MDMP, the commander and staff use an abbreviated or accelerated decision-making process. Though omitting steps of the MDMP to meet time constraints is not the solution, steps may be done in a shortened timeframe. (See paragraphs B-63 to B-66 for information on planning in a time constrained environment or within an abbreviated MDMP.)

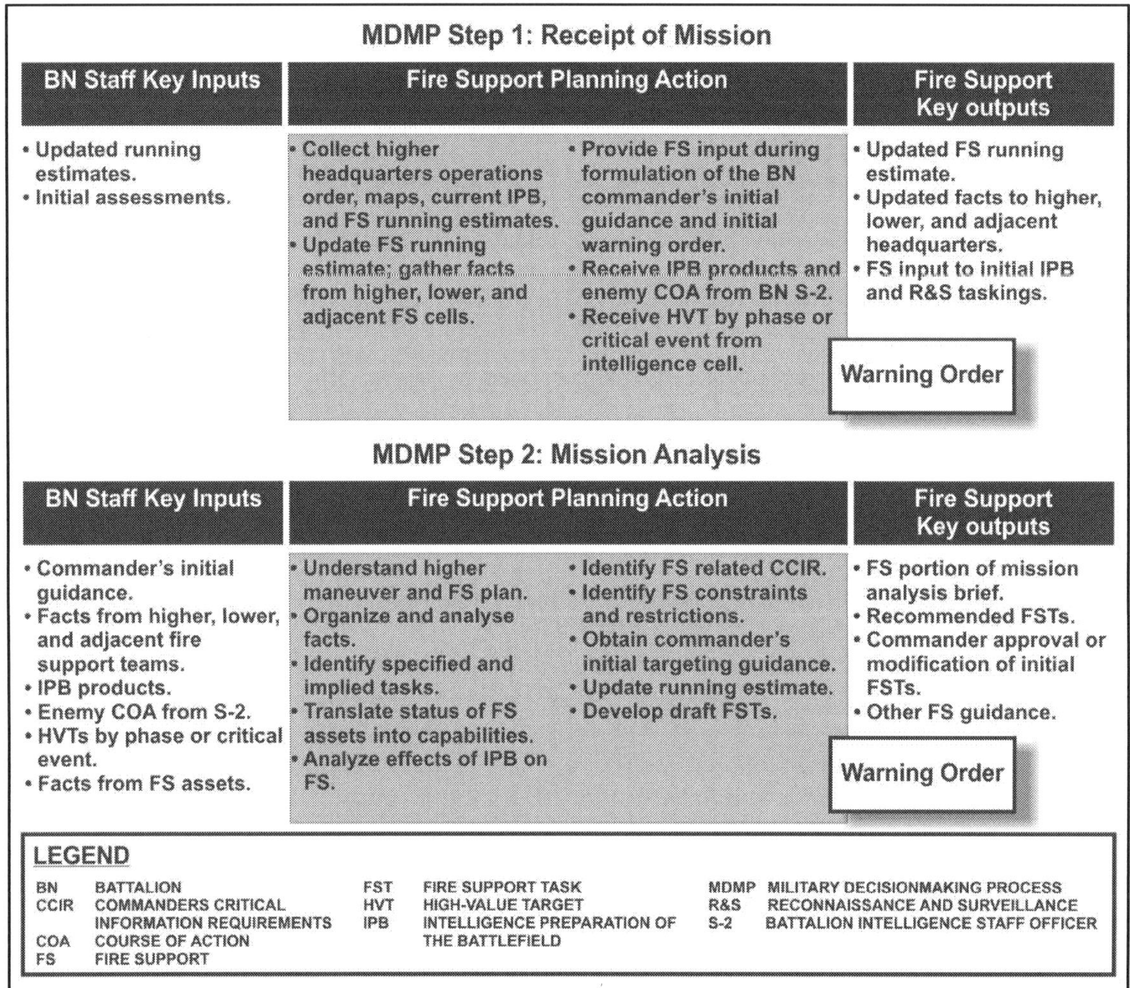

Figure C-1. Fire support planning during the military decisionmaking process

MDMP Step 3: Course of Action (COA) Development

BN Staff Key Inputs	Fire Support Planning Action		Fire Support Key outputs
• Relative combat power assessment. • Force array. • Broad concept of operation. • Develop COA statements and sketches. • Conduct a COA briefing. • Select or modify COA for continued analysis.	• Determine FSTs for each COA. • Determine where to find and attack FST formations. • Identify HPTs in those formations. • Quantify the effects for FSTs. • Plan methods for FSTs. • Allocate assets to acquire. • Allocate assets to attack. • Integrate triggers with maneuver COA.	• Assist S-2 in R&S development to support FS. • Prepare FS portion of COA/sketch. • Develop COA for each: (1) Concept of fires. (2) Draft FSEM. (3) Draft target list/overlay. (4) Draft TSM or modified TSM. (5) R&S support to FS. (6) FS running estimate.	• FS system status. • FS asset range arc depictions. • FS limitations and constrains. • FS-input to the CCIR. • Commander's approval of initial FST or modification.

MDMP Step 4: COA Anaylsis (War Game)

BN Staff Key Inputs	Fire Support Planning Action		Fire Support Key outputs
• List all friendly forces. • List assumptions. • List known critical events and DPs. • Select the war-gaming method. • Select a technique to record and display results. • War-game the operation and assess the results. • Conduct a war-game briefing (optional).	• Confirm FS assets for all COAs. • Validate FS relevant facts and assumptions. • Determine FST and the FA contribution to FST. • Develop evaluation criteria to measure the effectiveness of the FS contributions for each COA. • Develop a FSEM for each COA. • Provide likely adversary FS actions to the S-2; determine where to find and attack enemy FS capabilities. • Formulate a list of advantages and disadvantages of COA from FS perspective.	• Identify synchronization requirements including modification to FSCM and ACM. • Identify DPs, NAI, DT, and additional critical events and how these may influence positioning or posturing of FS assets. • Identify HVTs, HPTs, the FS portion of event templates, and develop a draft HPTL, target selection standards and attack guidance matrix. • Integrate IO and cyber electromagnetic activities into these targeting products.	• Refined scheme of fires. • Refined draft Annex D (FIRES) and appendices. • Refined draft FSEM. • Refined draft target list worksheet (automated or manual) and target overly. • Refined draft TSM or modified TSM (HPTL, target selection standards, attack guidance matrix). • Refined draft FSCMs. • Refined draft NAI and TAI. • Running estimate.

LEGEND

ACM	AIRSPACE COORDINATING MEASURES	FSCM	FIRE SUPPORT COORDINATION MEASURES	MDMP	MILITARY DECISIONMAKING PROCESS
BN	BATTALION	FSEM	FIRE SUPPORT EXECUTION MATRIX	NAI	NAMED AREA OF INTEREST
CCIR	COMMANDERS CRITICAL INFORMATION REQUIREMENTS	FST	FIRE SUPPORT TASK	R&S	RECONNAISSANCE AND SURVEILLANCE
COA	COURSE OF ACTION	HPT	HIGH-PAYOFF TARGET	S-2	BATTALION INTELLIGENCE STAFF OFFICER
DP	DECISIVE POINT	HPTL	HIGH-PAYOFF TARGET LIST	TAI	TARGET AREA OF INTEREST
DT	DECISIVE TERRAIN	HVT	HIGH-VALUE TARGET	TSM	TARGETING SYNCHRONIZATION MATRIX
FA	FIELD ARTILLERY	IO	INFORMATION OPERATIONS		
FS	FIRE SUPPORT				

Figure C-1. Fire support planning during the military decisionmaking process (continued)

MDMP Step 5: COA Comparision

BN Staff Key Inputs	Fire Support Planning Action		Fire Support Key outputs
• Conduct advantages and disadvantages analysis. • Compare COA. • Conduct a COA decision briefing.	• Participate with BN staff in comparing strengths, weakness, advantages, and disadvantages of each of action, emphasizing FS aspects. • Update FS and IO estimates. • Brief results of FS analysis including best COA from FS perspective and adequacy of scheme of fires and supporting assets (for example: sustainment and protection).	• Develop draft fires paragraphs and annexes to include FST, FSEM, target and overlay, TSM or modified (HPTL, target selection standards, attack guidance matrix). • Integrate IO and cyber electromagnetic input into targeting products. • Provide inputs to the information collection plan. • Update FS running estimate. • FS plan brief for each COA.	• Final drafts of the : - Scheme of fires. - Annex D (FIRES) and appendices. - FSEM. - Target list worksheet (automated or manual). - Target overlay. - Observer plan. - TSM or modified (HPTL, target selection standards and attack guidance matrix). - FSCMs. - NAI and TAI. - Running estimate.

MDMP Step 6: COA Approval

BN Staff Key Inputs	Fire Support Planning Action		Fire Support Key outputs
• Recommends a COA, usually in a decision briefing. • The BN commander decides which COA to approve. • The BN commander issues final planning guidance. • Issues warning order to subordinate units.	• Assess implications and take actions as necessary to finalize selected scheme of fires including attendant HPTL, target selection standards, and attack guidance matrix. • Integrate IO and cyber electromagnetic activities input into these targeting products. • Assist S-3 with participation in COA approval briefing. • Include scheme of fires and FST.	• Assist development of refined commander's intent and planning guidance. • Prepare FS portions of warning order including changes to CCIR, risk guidance, time sensitive reconnaissance tasks and FST requiring early initiation. • Prepare tentative FS portions of BN operation order. • Participate in required backbriefs and rehearsals. Warning Order	• For the approved course of action: - Refined scheme of fires. - Refined Annex D (FIRES) and appendices. - FSEM. - Target list worksheet (automated or manual). - Target overlay. - Observer plan. - TSM or modified (HPTL, target selection standards and attack guidance matrix). - FSCMs. - NAI and TAI. - Updated estimate.

LEGEND

BN	BATTALION	FSEM	FIRE SUPPORT EXECUTION MATRIX	NAI	NAMED AREA OF INTEREST
CCIR	COMMANDERS CRITICAL INFORMATION REQUIREMENTS	FST	FIRE SUPPORT TASK	S-3	BATTALION OPERATIONS STAFF OFFICER
COA	COURSE OF ACTION	HPTL	HIGH-PAYOFF TARGET LIST	TAI	TARGET AREA OF INTEREST
FS	FIRE SUPPORT	IO	INFORMATION OPERATIONS	TSM	TARGETING SYNCHRONIZATION MATRIX
FSCM	FIRE SUPPORT COORDINATION MEASURES	MDMP	MILITARY DECISIONMAKING PROCESS		

Figure C-1. Fire support planning during the military decisionmaking process (continued)

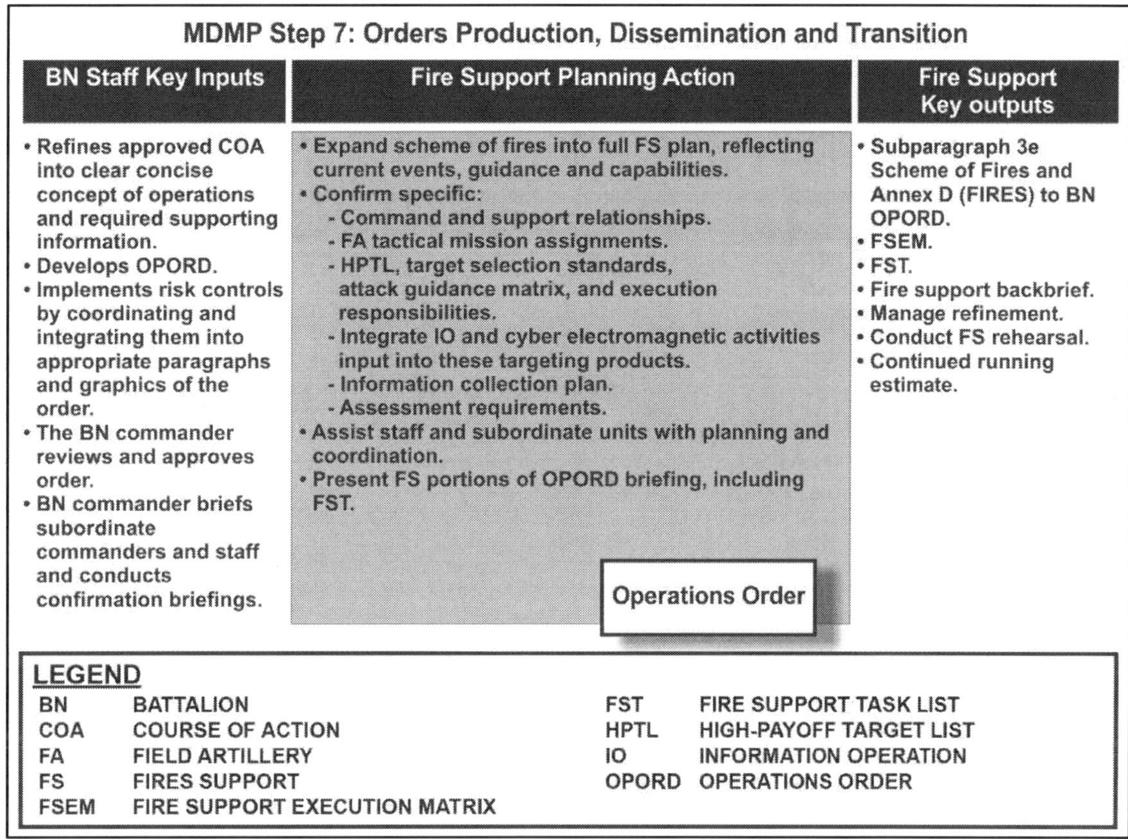

Figure C-1. Fire support planning during the military decisionmaking process (continued)

LETHAL AND NONLETHAL EFFECTS ELEMENT

C-12. The lethal and nonlethal effects element (generally established in the IBCT fire support cell) plans lethal and nonlethal effects for future operations and targeting. The IBCT main command post (CP) facilitates collaboration of lethal and nonlethal effects with the other warfighting functions and subordinate units through the targeting working group. The lethal and nonlethal effects element prepares inputs and products used in targeting at IBCT level and below. The lethal and nonlethal effects element prepares recommendations for the IBCT targeting working group sessions and implements the resulting decisions. Leveraging the reconnaissance and surveillance assets available to the IBCT, the element plans and synchronizes the fires and nonlethal effects of IBCT operations. Targeting functions of the lethal and nonlethal effects element include—

- Providing lethal fires input to the information collection plan.
- Providing nonlethal effects input to the information collection plan.
- Developing inputs to the no-strike list and identifying fire support coordination measures.
- Developing and refining targeting guidance for each course of action.
- Developing target criteria for input into computer systems for each course of action.
- Producing the high priority target list, targeting synchronization matrix and lethal effects tasks for the operation.
- Preparing products for the targeting working group at IBCT level and below.
- Developing MOPs and MOEs for assessment.
- Implementing targeting guidance.
- Updating and purging targeting files.

C-13. Other nonlethal effects targeting team members at IBCT level include the information officer, electronic attack officer, military information support operations officer, civil affairs officer, public affairs officer, and brigade judge advocate. Depending on the mission and the situation, a targeting work group can be established at battalion level though resourced to a lesser extent. (Refer to ATP 3-60 for additional information.)

AIR-GROUND OPERATIONS

C-14. *Air-ground operations* (AGO) are the simultaneous or synchronized employment of ground forces with aviation maneuver and fires to seize, retain, and exploit the initiative (FM 3-04). Employing the combined and complimentary effects of air and ground maneuver and fires through AGO presents the enemy with multiple dilemmas and ensures that aviation assets are position to support ground maneuver. AGO increase the overall combat power, mission effectiveness, agility, flexibility, and survivability of the entire combined arms team. AGO ensure that all members of the combined arms team, whether on the ground or in the air, work toward common and mutually supporting objectives to meet the higher commander's intent.

C-15. Air action by fixed- and rotary-wing aircraft against hostile targets that are in close proximity to friendly forces require detailed integration of each air mission with the fire and movement of ground forces. This section focuses on mortar and artillery, Army attack aviation, and CAS fires in support of the ground maneuver force, and the request processes. Refer to ATP 3-04.64 for discussion on unmanned aircraft system operations to provide surveillance capabilities and to enhance situational awareness during planning, coordination, and execution of combat operations.

FUNDAMENTAL CONSIDERATIONS

C-16. Air and ground forces must integrate effectively to conduct operations and to minimize the potential for fratricide and civilian casualties. *Integration*—the arrangement of military forces and their actions to create a force that operates by engaging as a whole (JP 1)—maximizes combat power through synergy of both forces. The integration of air operations into the ground commander's scheme of maneuver may also require integration of other services or multinational partners. Integration continues through planning, preparation, execution, and assessment. The commander and staff must consider the following framework fundamentals to ensure effective integration of air and ground maneuver forces:

- Understanding capabilities and limitations of each force.
- Standard operating procedures.
- Habitual relationships.
- Regular training events.
- Airspace control.
- Maximizing and concentrating effects of available assets.
- Employment methods.
- Synchronization.

ORGANIZATION FOR FIRE SUPPORT COORDINATION

C-17. The IBCT main CP fire support cell is generally organized with an FSO and assistants, an air defense airspace management/brigade aviation element (known as ADAM/BAE), an electronic warfare element, a targeting element, and digital systems operators. References throughout this document to fire support planners include not only members of the fire support cell but also the IBCT's cannon field artillery battalion commander as the IBCT's fire support coordinator and members of the targeting working group and targeting board. These personnel are all members of the targeting team. The IBCT main CP fire support cell works closely with the battalion FSO, battalion fire support cell, and company fire support teams. These organizations ensure responsive and effective fire support is provided to their respective maneuver commanders and actions are closely coordinated through the IBCT main CP fire support cell. (Refer to FM 3-96 for additional information.)

Note. The IBCT fire support cell uses the division's joint air-ground integration center to ensure continuous collaboration with unified action partners to integrate fires and to use airspace effectively. The IBCT fire support cell sends requests for division-level Army and joint fires to the joint air-ground integration center in the current operations integrating cell of the division. Upon receipt of the request for fire or joint tactical airstrikes, the joint air-ground integration center fire support cell personnel make attack recommendations and, if required, provide target coordinate mensuration. Additionally, the joint air-ground integration center conducts collateral damage estimation and reviews available ground and air component fires capabilities to determine the most effective attack method. Refer to ATP 3-91.1 for additional information on the joint air-ground integration center.

Battalion Fire Support Cell

C-18. Fire support personnel manning the battalion's main CP fire support cell, company fire support teams, and platoon forward observers are assigned to the IBCT's cannon field artillery battalion. These fire support personnel habitually associate with supported battalions and companies or platoons for training, but for combat operations will be deployed by the IBCT commander and the fire support coordinator when and where needed based on the mission variables of METT-TC. The battalion fire support cell provide a fire support coordination capability to the commander and are organized with an FSO and NCO, an electronic warfare NCO, and digital systems operators. The fire support cell may also have an Air Force TACP attached to the battalion. (Refer to ATP 3-09.42 for additional information.)

Fire Support Team and Observers

C-19. Company fire support team headquarters personnel and platoon forward observers provide support to each Infantry rifle company to plan and coordinate all available supporting fires, including mortars, field artillery, naval surface fire support, Army attack aviation, and Air Force CAS integration. Attached fire support teams provide maneuver companies with fire support coordination, precision targeting, type 2 and 3 terminal attack control (see paragraph C-55), and effects assessment capabilities. The use of precision target location tools is the preferred method of establishing accurate target location. These tools include a targeting device or a precision targeting device, a forward entry device, and imagery based mensuration tools. The observer may have an optical device using a laser range finder for distance and an Azimuth Vertical Angle Module to acquire direction and vertical angle. Each fire support team's fire support vehicle, if provided, possesses a target acquisition and communications suite with the capability for laser range finding and designation for laser-guided munitions. (Refer to ATP 3-09.42 for additional information.)

Qualified Observers

C-20. Effective fires require qualified observers to call for and adjust fires on located targets. Forward observers, forward air controllers (FAC), naval gunfire spotter teams, joint fires observers, and joint terminal attack controllers (JTAC) train together and work effectively as a team to request, plan, coordinate, and place accurate fires on targets that create the effects desired by the commander. (Refer to JP 3-09 for additional information.)

Forward Observer

C-21. A *forward observer* is an observer operating with front line troops and trained to adjust ground or naval gunfire and pass back battlefield information (JP 3-09). In the absence of a *forward air controller,* an officer (aviator/pilot) member of the tactical air control party who, from a forward ground or airborne position, controls aircraft in close air support of ground troops (JP 3-09.3), the observer may control CAS strikes. Platoon forward observers are assigned to the fire support team supporting each rifle company in the Infantry battalion. Forward air controllers (airborne), JTACs, and naval gunfire spotter teams may not always be available when and where their support is required. Field artillery observer teams must be proficient in planning and executing CAS when a JTAC is not available. With additional training and certification, the forward observer can qualify as a joint fires observer. (Refer ATP 3-09.42 for additional information.)

Joint Fires Observer

C-22. A *joint fires observer* is a trained Service member who can request, adjust, and control surface-to-surface fires, provide targeting information in support of Type 2 and 3 close air support terminal attack controls, and perform autonomous terminal guidance operations (JP 3-09.3). The joint fires observer is not an additional Soldier in the Army fire support organization, but rather an individual who has received the necessary training and certification to be awarded the joint fires observer's additional skill identifier. A joint fires observer is not a certified JTAC.

- *Terminal attack control* is the authority to control the maneuver of and grant weapons release clearance to attacking aircraft (JP 3-09.3).
- *Terminal guidance operations* are actions using electronic, mechanical, voice or visual communications that provide approaching aircraft and/or weapons additional information regarding a specific target location (JP 3-09).
- *Joint terminal attack controller is* a qualified (certified) Service member who, from a forward position, directs the action of combat aircraft engaged in CAS and other offensive air operations (JP 3-09.3). A qualified and current joint terminal attack controller is recognized across the Department of Defense as capable and authorized to perform terminal attack control.

C-23. Air Force JTACs, if available from the battalion Air Force TACP can deploy forward with a maneuver company and position where they can best support the operation. Tactical air control party JTACs provide the commander and the subordinate and supporting units with recommendations on the use of CAS and its integration with ground maneuver and other attack resources. JTACs also perform terminal attack control of individual CAS missions.

C-24. A *forward air controller (airborne)* is a specifically trained and qualified aviation officer, normally an airborne extension of the tactical air control party, who exercises control from the air of aircraft engaged in close air support of ground troops. (JP 3-09.3). A qualified and current forward air controller (airborne) is recognized across the Department of Defense as capable and authorized to perform terminal attack control. (Refer to JP 3-09.3 for additional information.)

Untrained Observers

C-25. Occasionally the cannon field artillery battalion of the IBCT may need to process fire missions from untrained observers. An untrained observer is anyone not military occupational specialty qualified in requesting and adjusting indirect fire. Often these are critical requests where the requestor is under fire. Field artillery battalion and battery fire direction centers should be identified as the primary handlers of untrained observer missions. For more information on untrained observer procedures, see TC 3-09.81. For more on observers see the ATP 3-09.42 and in ATP 3-09.30.

Scheme of Fires

C-26. *Scheme of fires* is the detailed, logical sequence of targets and fire support events to find and engage targets to accomplish the supported commander's intent (FM 3-09). The battalion commander and staff integrate and synchronize indirect fires, Army attack aviation, and CAS to operations. The commander ensures targets and fire support events are planned for each unit to counter likely enemy obstacles and ambushes and support planned engagement areas. Depending on what other mission is being supported, the unit may not have priority of fires during operations. Internal fire support means are always planned for regardless of external fire support. The scheme of fires is rehearsed to ensure coverage throughout the operation.

Indirect Fire

C-27. The majority of fire support to an Infantry battalion is provided by indirect fire support systems. Indirect fire support systems include mortars and field artillery cannon and rocket systems. (See ATP 3-09.32 for a detailed listing of indirect fire system capabilities and characteristics). Indirect fire support systems may be under direct command of the maneuver battalion or may be in a supporting role. Indirect fire targets during movement are planned on probable locations of enemy attempts to attack the movement. Call for fire (table C-1) is the request for fire containing data necessary for obtaining the required mortar and artillery fire on a target. The ability for mortars and artillery to engage targets from reverse-slopes and areas of defilade is a tremendous advantage, especially in adverse terrain. As with other operations, employing indirect fires in adverse terrain and climate does have its challenges. (Refer to FM 3-09 for additional information.) Unique challenges include—

- Unpredictable weather conditions affecting accuracy of rounds.
- Targets located on peaks and steep terrain making adjustments difficult.
- Intervening crests requiring placement of observers on dominating heights for overwatch.
- Limited terrain suitable for firing positions to cover a particular movement.
- Mortar and artillery locations ideal for range and coverage unsuitable due to intervening adverse terrain features.
- Locations tactically positioned but in an area with difficult or limited access.
- Shifting mortar and artillery assets to alternate locations requiring significant time and engineering and logistical efforts.

Table C-1. Artillery and mortar call for fire

1st Transmission	*3rd Transmission*
1. OBSERVERS IDENTIFICATION (Call signs) 2. WARNING ORDER • Adjust fire • Fire for effect • Suppress • Immediate suppression/immediate smoke	4. TARGET DESCRIPTION • Type • Activity • Number • Degree of protection • Size and shape (length/width or radius) 5. METHOD OF ENGAGEMENT • Type of adjustment • Danger close • Mark • Ammunition • Distribution 6. METHOD OF FIRE AND CONTROL • Method of fire • Method of control
2nd Transmission	
3.TARGET LOCATION • Grid coordinate • Shift from a known point • Polar plot	

C-28. Mortar indirect fires are organic to Infantry battalions. The battalion mortar platoon provides the most responsive indirect fire available to the battalion. These assets provide the commander with close and immediate responsive fires in support of the maneuver companies. These fires harass, suppress, neutralize, or destroy enemy attack formations and defenses; obscure the enemy's vision; or otherwise inhibit their ability to acquire friendly targets.

C-29. The three primary types of mortar fires are high explosive, obscuration, and illumination (visible and infrared). Mortars also can be used for final protective fires and smoke. (Refer to ATTP 3-21.90 for more information.)

C-30. On the battlefield, mortars act as both a killer of enemy forces and as an enhancer of friendly mobility. Mortar fires inhibit enemy fire and movements and allow friendly forces to maneuver to a position of tactical advantage. Effective integration of mortar fires with the overall fire support plan and with maneuver units is critical to successful combat. Listed below are some of the key capabilities of mortar units:

- Mortar units are organic to Infantry battalions and Infantry companies that make them always available and responsive regardless of whether or not the battalion has allocated supporting artillery.
- Organic mortar fires do not have to be externally cleared when firing missions inside the battalion area of operation.
- Mortars provide obscuration and suppression to protect the battalion during the attack or to support it while breaking contact with the enemy in the defense or movement to contact.
- Mortars provide the commander with responsive fires to support the scout platoon's infiltration and exfiltration and the counterreconnaissance force during security operations.

- Maneuver commander can continue to use mortars for indirect fire support in one part of the battle and divert field artillery fires to assist in the critical fight elsewhere.
- Mortars contribute to the battalion's direct fire fight by forcing the enemy to button up, by obscuring their ability to employ supporting fires, and by separating their dismounted Infantry from its armored personnel carriers and accompanying tanks.
- Heavy mortars penetrate buildings and destroy enemy field fortifications, preparing the way for the dismounted assault force. Precision guided mortar munitions can destroy selected high payoff targets.
- Mortars provide battalion and company commanders with the ability to cover friendly obstacles with indirect fire, regardless of the increasing calls for artillery fire against deep targets or elsewhere on the battlefield.
- Mortar fires combine with the final protective fires of company machine guns to repulse the enemy's dismounted assault. Frees artillery to attack and destroy follow-on echelons, which are forced to slow down and deploy, as the ground assault is committed.
- Mortars can use the protection of defilade to continue indirect fires and effects even when subjected to intense counterfire.

Army Attack Aviation

C-31. During the planning process, the battalion commander and staff integrate the employment of Army aviation attack and reconnaissance units into the scheme of maneuver to ensure their responsiveness, synergy and agility during actions on the objective or upon contact with the enemy. Pre-mission development of control measures provides a foundation for the successful integration of Army aviation into the unit's operations. Among these control measures are engagement criteria; the triggers and conditions for execution; fire coordination measures, such as target reference points; engagement areas and target reference points; and airspace coordinating measures, such as aerial ingress and egress routes and *restricted operations zone*, which is airspace reserved for specific activities in which the operations of one or more airspace users is restricted (JP 3-52).

Call for Fire

C-32. Army attack aviation targets are planned on probable enemy locations. Army attack aviation call for fire is a coordinated attack by Army attack aircraft against enemy forces in close proximity to friendly units. Army attack aviation call for fire (table C-2) is not synonymous with CAS flown by Joint and multinational aircraft. Terminal control from ground units or controllers is not required due to aircraft capabilities and enhanced situational understanding of the aircrew. Depending on the enemy situation, Army attack aviation can be on station during times when contact is most likely to occur. Air-ground integration ensures frequencies are known and markings are standardized to prevent fratricide. (Refer to ATP 3-04.1 for additional information)

Table C-2. Army attack aviation call for fire format

1. Observer and Warning Order
"__J27__, **this is** __041__, **fire mission, over"**
(aircraft call sign) (observer call sign)

2. Friendly Location and Mark
"My position __AL78241638__, **marked by** __Strobe__ "
(TRP, grid, etc.) (strobe, beacon, IR strobe, etc.)

3. Target Location
"Target Location __AL82781942__ "
(bearing [magnetic] and range [meters], TRP, grid, etc.)

4. Target Description and Mark
"__Dismounted Infantry__, **marked by** __Tracer__ "
(target description) (IR pointer, tracer, etc.)

5. Remarks: "__At my command__, **over"**
(threats, danger close clearance, restriction, at my command, etc.)

Notes:
1. Clearance: If airspace has been cleared between the employing aircraft and the target, transmission of this brief *is* clearance to fire unless **"danger close"** or **"at my command"** is stated.
2. Danger Close: For danger close fire, the observer or commander must accept responsibility for increased risk. State **"cleared danger close"** in line 5 and pass the initials of the on-scene ground commander. This clearance may be preplanned.
3. At My Command: For positive control of the aircraft, state **"at my command"** on line 5. The aircraft will call **"ready to fire"** when ready.

LEGEND

IR	INFRARED	TRP	TARGET REFERENCE POINT

C-33. During call for fire, the flight lead must have direct communication with the on-scene ground commander to provide direct fire support. After receiving the call for fire brief from ground forces, pilots must be able to positively identify friendly location before engagement. Once the crew has identified both enemy and friendly locations, flight leads formulate an attack plan and brief the supported commander and their other attack team members.

Note. Army attack aviation may be used as a show of force to discourage enemy forces from performing offensive actions. Coordination can be made with aircraft conducting nearby operations to simply fly over or near a planned unit movement to deter aggressive actions by the enemy.

Limitations

C-34. Major limitations for use of attack aircraft include—

- Number of aircraft available. Sorties are often limited and in high demand in combat operations.
- Time needed to get aircraft on station. Available aircraft may be too far away or have to take a lengthy indirect route to be effective.
- Weather conditions. Current or pending adverse weather conditions may ground aircraft.
- Elevation restrictions. High mountain ridges may be at an elevation restricting movement of rotary-wing aircraft across them. Simply getting aircraft to target areas may be restricted if available aircraft are on the other side of mountains with ridges above certain altitudes.

- Rearming and refueling. Travel time to locations may be lengthy and use substantial amounts of fuel. This reduces time on station for the aircraft and requires refueling. Locations for rearming and refueling may be some distance away.

Marking and Identifying Locations and Targets

C-35. Ground units must ensure aircraft have positive identification of friendly unit locations and enemy targets. There are various ways to mark locations or targets. The effectiveness of vision systems on helicopters compared to those found on ground vehicles may differ. During day, vision systems of the AH-64 allow accurate identification of targets.

C-36. During periods of limited visibility, resolution is greatly degraded, requiring additional methods of verification. This situation requires extra efforts from both ground units and aviation elements. Thermal, optical, and radar acquisition devices enable positive identification. Both aviation and ground forces might become overloaded with tasks in battle. Simple, positive identification must be established and known to all.

Marking of Friendly Forces

C-37. A method of target identification is direction and distance from friendly forces. Friendly forces can mark their own positions with infrared strobes, infrared tape, night vision goggle lights, smoke, signal panels, body position, meal, ready to eat heaters, chemical lights, and mirrors. Marking friendly positions is the least desirable method of target location information because it can reveal friendly positions to the enemy.

Marking of Enemy Targets

C-38. Target marking aids aircrews in locating targets the unit in contact desires to attack. The ground commander provides target marking whenever possible. Methods for marking targets include, but are not limited to laser handover, tracer fire, marking rounds (flares or mortars), or laser target marker. To be effective, marking must be timely, accurate, and easily identifiable. Target markings might be confused with other fires on the battlefield, suppression rounds, detonations, and marks on other targets. Although marking is not mandatory, it improves aircrew accuracy, enhances situational awareness, and reduces risk of fratricide.

Target Handover

C-39. The rapid and accurate marking of a target is essential to a positive target handover. Aircraft conducting attacks develop an attack plan that is METT-TC dependent and meets the ground commander's task and purpose. The aircrew generally has an extremely limited amount of time to acquire both the friendly and enemy locations. It is essential that the ground unit has the marking ready and turned on when requested by the aircrew. Attack reconnaissance aircrews use both thermal sight and night vision goggles to fly with and acquire targets. After initially engaging the target, the aircrew generally approaches from a different angle for survivability reasons if another attack is required. The observer makes adjustments using the eight cardinal directions and distance (meters) in relation to the last round's impact and the actual target. At the conclusion of the attack, the aircrew provides its best estimate of battle damage assessment to the unit in contact.

Battle Damage Assessment and Reattack

C-40. After the attack aircraft complete the requested attack mission, the aircrew provides a battle damage assessment to the ground commander. Based on the assessment, the ground maneuver commander determines if another attack is required to achieve the desired end state. The Army attack aviation operation can continue until the aircraft have expended all available munitions or fuel. However, if the air mission commander receives a request for another attack, he must carefully evaluate the unit's ability to extend the operation. If not able, the air mission commander calls for relief on station by another attack team if available. It is unlikely that the original team has enough time to refuel, rearm, and return to station.

Clearance of Fires

C-41. During close combat with numerous aircraft in the vicinity of an area of operation, it is critical to deconflict airspace between aircraft and established indirect fires, to include the following:

- Ensure aircrews have the current and planned indirect fire positions (to include mortars) supporting the ground tactical plan.
- Plan for informal airspace coordination areas and check firing procedures and communications to ensure artillery and mortars firing from within the area of operation do not endanger subsequent serials landing or departing, Army attack aviation, or CAS.
- Ensure that at least one of the aviation team members monitors the fire support net for situational awareness.
- Advise the aviation element if the location of indirect fire units changes from that planned.
- Ensure all participating units are briefed daily on current airspace control order or air tasking order changes and updates that may affect air mission planning and execution.
- Ensure all units update firing unit locations, firing point origins, and final protective fire lines as they change for inclusion in current airspace control order.

C-42. The commander can establish an airspace coordination area. For example, the commander can designate that all indirect fires be south of and all aviation stay north of a specified gridline for a specific period. This is one method for deconflicting airspace while allowing indirect fires and attack aviation to attack the same target. The ground commander then can deactivate the informal airspace coordination area when the situation permits.

Close Air Support

C-43. Infantry battalions generally are allocated CAS sorties. *Close air support* is air action by fixed-wing and rotary-wing aircraft against hostile targets that are in close proximity to friendly forces and that require detailed integration of each air mission with the fire and movement of those forces (JP 3-0). CAS can be employed to blunt an enemy attack; to support the momentum of the ground attack; to help set conditions for battalion and IBCT operations as part of the overall counterfire fight; to disrupt, delay and destroy enemy second echelon forces and reserves; and to provide cover for friendly movements. The effectiveness of CAS is related directly to the degree of local air superiority attained. Until air superiority is achieved, competing demands between CAS and counterair operations may limit sorties apportioned for the CAS role. CAS is the primary support given to committed battalions and IBCT by Air Force, Navy, and Marine aircraft. IBCTs and battalions can request air reconnaissance and battlefield air interdiction missions through the next higher headquarters, but these missions normally are planned and executed at that higher unit level, with the results provided to the Infantry battalion commander and staff.

Missions

C-44. The IBCT normally plans and controls CAS. However, this does not preclude the battalion from requesting CAS, receiving immediate CAS during an operation, or accepting execution responsibility for a planed CAS mission. CAS is another means of indirect fire support available to the battalion. In planning CAS missions, the commander must understand the capabilities and limitations of CAS and synchronize CAS missions with both the battalion fire plan and scheme of maneuver. CAS capabilities and limitations such as windows for use, targets, observers, and airspace coordination present some unique challenges, but the commander and staff must plan CAS with maneuver the same way they do indirect artillery and mortar fires. When executing a CAS mission, the battalion must have a plan that synchronizes CAS with maneuver and the scheme of fires.

Preplanned Close Air Support

C-45. Battalion planners must forward CAS requests as soon as they can be forecasted. These requests for CAS normally do not include detailed timing information because of the lead time involved. Preplanned CAS requests involve any information about planned schemes of maneuver, even general information, which can be used in the apportionment, allocation, and distribution cycle. This includes estimates of weapons effects needed by percentage, such as 60 percent antiarmor and 40 percent antipersonnel; sortie time flows; peak need times; and anticipated distribution patterns. All are vital to preparing the air tasking order. Air liaison officers (ALOs) and S-3s at all planning echelons must ensure that this information is forwarded through higher echelons according to the air tasking order cycle.

Categories

C-46. Preplanned CAS may be categorized as either scheduled or on-call missions.

- A scheduled mission is a CAS strike on a planned target at a planned time.
- An on-call mission is a CAS strike on a planned target or target area executed when requested by the supported unit. Usually, this mission is launched from a ground on-call, but it may be flown from an airborne on-call status. On-call CAS allows the ground commander to designate a general target area within which targets may need to be attacked. The ground commander designates a conditional period within which he later determines specific times for attacking the targets.

Request Channels

C-47. There are specific request channels (figure C-2) for preplanned CAS. Requests for preplanned joint tactical air strike request missions are submitted to the fire support team. The commander, ALO, and S-3 at each echelon evaluate the request; coordinate requirements such as airspace, fires, and intelligence; consolidate; and, if approved, assign a priority or precedence to the request. The S-3 air then forwards approved requests to the next higher echelon. To plan CAS, the S-3 air must work closely with the S-3, FSO, and ALO.

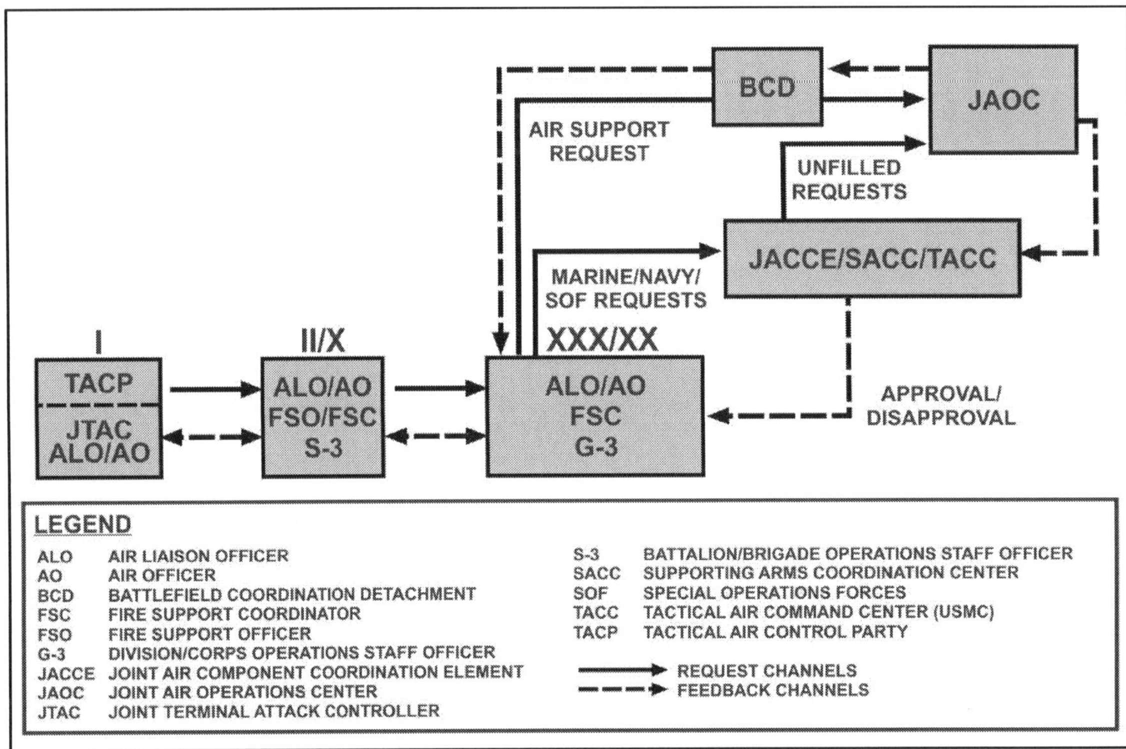

Figure C-2. Preplanned close air support request channels

Engagement Alternatives

C-48. CAS aircraft assigned to attack preplanned targets may be diverted to higher priority targets; therefore, the FSO should plan options for the engagement of CAS targets by other fire support assets.

Immediate Close Air Support

C-49. Immediate requests (figure C-3) are used for air support mission requirements identified too late to be included in the current air tasking order. Those requests initiated below battalion level are forwarded to the battalion main CP by the most rapid means available.

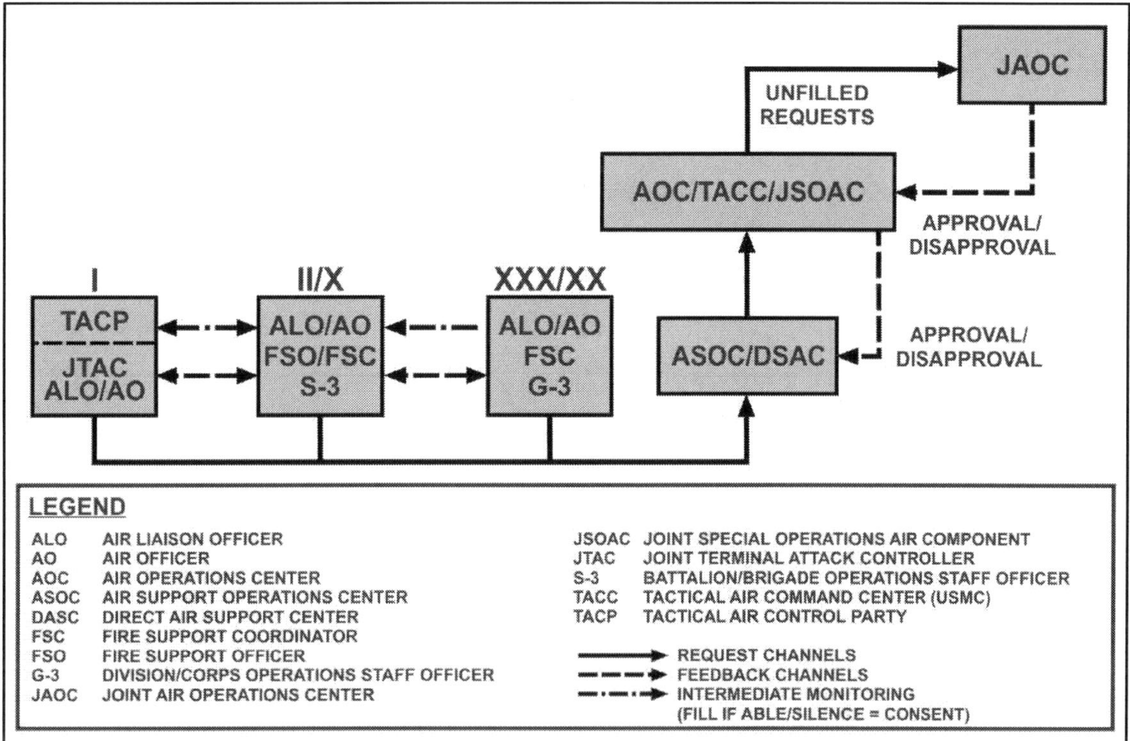

Figure C-3. Immediate close air support request channels

C-50. At battalion level, the commander, FSO, ALO, and S-3 consider each request. Approved immediate CAS requests are transmitted by the TACP over the Air Force air request net (figure C-3) directly to the air support operations center collocated with the corps or separate division tactical operations center.

C-51. The TACP at each intermediate headquarters monitors and acknowledges receipt of the request. Silence by an intermediate TACP indicates approval by the associated headquarters unless disapproval is transmitted.

C-52. The air support operations center coordinates with the G-3 Air at echelons above the IBCT for all air support requests initiated by that unit. Meanwhile, intermediate TACPs pass the request to the associated headquarters G-3 or S-3 for action and coordination.

C-53. All echelons coordinate simultaneously. If any Army echelon above the initiating level disapproves a request or substitutes another support means, such as Army aviation, or field artillery, the TACP at that headquarters notifies the air support operations center at the coordinating unit and the originating TACP, which notifies the requestor.

C-54. When the coordinating unit commander or their representative approves the request, the air support operations center initiates the necessary action to satisfy the request. If all distributed sorties are committed, the coordinating unit commander can request additional sorties from the next higher echelon, when appropriate. If the air support operations center has no CAS missions available, it can, with Army concurrence, divert sorties from lower priority targets or request support from lateral or higher commands.

C-55. Units having a reasonable expectation of conducting terminal attack control need to have certified JTAC available. In rare circumstances, the ground movement commander might require CAS when a JTAC or forward air controller airborne (FAC [A]) is not available, but detailed integration with friendly forces fire and movement is still required. Aircrews executing CAS under these circumstances must be in contact with the ground commander (or the commander's designated representative) and bear increased responsibility for the detailed integration required to minimize fratricide normally done by a JTAC/FAC (A). In these circumstances, the CAS aircrew assist the ground movement commander to the greatest extent possible to bring fires to bear.

Note. Although Army aviation does not consider its aircraft a CAS system, they can conduct attacks employing CAS joint tactics, techniques, and procedures when operating in support of non-U.S. Army forces.

C-56. The flow and prosecution of CAS targets normally begins with a check-in briefing, a situation update briefing followed by a CAS 9-Line and ending with a battle damage assessment report. A game plan is a concise and situational awareness enhancing tool to inform all players of the flow of the following attack (table C-3). At a minimum, the game plan will contain the type of control and method of attack. The method of attack and type of terminal attack control are separate and independent constructs. Method of attack conveys the JTAC's/FAC (A)'s intent for the aircraft prosecution of the target; either the aircraft will be required to acquire the target (bomb on target) or not (bomb on coordinate). The method of attack is broken down into two categories, bomb on target and bomb on coordinate. These two categories define how the aircraft will acquire the target or mark the target. Any type of control can be utilized with either method of attack and no type of control is attached to one particular method of attack.

Table C-3. Game plan and 9-line close air support brief

Do not transmit the numbers. Units of measure are standard unless briefed. Lines 4, 6 and any restrictions are mandatory readbacks.The joint terminal attack controller (JTAC) may request an additional readback. JTAC: "__J27__, **advise when ready for game plan."** JTAC **"Type (1,2,3) control (method of attack, effects desired or ordnance, interval). Advise when ready for 9-line."**
1. Initial Point / Battle Position "__AL78241638__" 2. Heading: "__280 DEGREES__" *(degrees magnetic, initial point or battle position-to-target)* *Offset:* "__LEFT__." *(left or right, when requested)* 3. Distance: "__4000 METERS__" *(initial point-to-target in nautical miles, battle position-to-target in meters)* 4. Target elevation: "__794 FEET__" *(in feet, mean sea level)* 5. Target description "__DISMOUNTED INFANTRY__" 6. Target location: "__AL82781942__" *(latitude and longitude or grid coordinates, or offsets or visual)* 7. Type mark / terminal guidance: "__VIPER 27, CODE 888__" *(description of the mark, if laser handoff, call sign of lasing platform and code)* 8. Location of friendlies: "__4000 METERS EAST OF TARGET__" *(from target, cardinal direction and distance in meters)* *Position marked by:* "__SMOKE__" 9. **" Egress** __SOUTHWEST__ "
Remarks / *Restriction: • Laser to target line (LTL) / pointer target line (PTL). • Desired type and number of ordnance or weapons effects (if not previously coordinated). • Surface-to-air threat, location, and type of suppression of enemy air defense (SEAD). • Additional remarks (e.g., gun-to-target line, weather, hazards, friendly marks). • Additional calls requested. • *Final attack headings or attack direction. • *Airspace coordination areas (ACAs). • *Danger close and initials (if applicable). • *Time on target (TOT) / time to target (TTT). • *Post launch abort restriction (if applicable).
• NOTE: For off-axis weapons, the weapons final attack heading may differ from the aircraft heading at the time of release. The aircrew should inform JTAC when this occurs and ensure weapon final attack headings comply with given restrictions.

C-57. In accordance with JP 3-09.3 and as stated earlier in the appendix, terminal attack control is the authority to control the maneuver of and grant weapons release clearance to attacking aircraft. A certified and qualified JTAC or FAC (A) is recognized across the Department of Defense as capable and authorized to perform terminal attack control. There are three types of terminal attack control: Type 1, 2, and 3. The commander considers the situation and issues guidance to the JTAC/FAC (A) based on recommendations from the staff and associated risks identified in the tactical risk assessment. The intent is to offer the lowest-level supported commander, within the constraints established during risk assessment, the latitude to determine which type of terminal attack control best accomplishes the mission. (Refer to JP 3-09.3 for additional information.) The three types of control are not ordnance specific, but are based on the following factors:

- Type 1 control is used when the JTAC/FAC (A) requires control of individual attacks and the situation requires the JTAC/FAC (A) to visually acquire the attacking aircraft and visually acquire the target for each attack.
- Type 2 control is used when the JTAC/FAC (A) requires control of individual attacks and any or all of the conditions below exist:
  - JTAC/FAC (A) is unable to visually acquire the attacking aircraft at weapons release.
  - JTAC/FAC (A) is unable to visually acquire the target.
- Type 3 control is used when the JTAC/ FAC (A) requires the ability to provide clearance for multiple attacks within a single engagement subject to specific attack restrictions, and any or all of the following conditions exist:
  - JTAC is unable to visually acquire the attacking aircraft at weapons release.
  - JTAC is unable to visually acquire the target.
  - The attacking aircraft is unable to acquire the mark/target prior to weapons release.
  - The JTAC/FAC (A) requires the ability to provide clearance for multiple attacks within a single engagement subject to specific attack restrictions.

Planning Considerations

C-58. CAS mission success directly relates to thorough mission planning based on the following factors and considerations. The S-3 air is responsible for working with the battalion air liaison officer or a joint terminal attack controller (JTAC) before and during tactical air operations. Since there are no digital links with supporting aircraft, the S-3 air must consistently keep the ALO apprised of the ground tactical situation through digital and conventional means.

C-59. When operating in the battalion's AO, CAS aircraft are under the positive control of one of the battalion's TACP. *Positive control* is a method of airspace control that relies on positive identification, tracking, and direction of aircraft within an airspace, conducted with electronic means by an agency having the authority and responsibility therein (JP 3-52). TACPs monitor the ground tactical situation, review the COP, and monitor conventional voice radio nets of the supported ground or maneuver commander to prevent fratricidal air-to-ground or ground-to-air engagements. Other planning factors include time available for planning, procedures, communications, and terrain.

Air Force Support

C-60. Air Force units are attached to the battalion to plan, control, and direct CAS. The ALO and the TACP are the typical air force assets attached to the battalion. ALOs are provided to maneuver units down to battalion. The ALO is responsible for supervising the TACP and coordinating CAS with the fire support team and S-3 air. The ALO is the senior USAF representative for the TACP supporting the battalion. The ALO normally is located with the tactical CP during operations. TACPs are provided to maneuver unit headquarters down to and including the battalion level. TACPs provide direct interaction with the supported maneuver units and should be highly visible to commanders and readily available to assist in the integration and synchronization of air power with land-force fire and movement.

C-61. The supported unit's ALO is the commander of the TACP. TACPs, at higher echelons through the IBCT level, function primarily in an advisory role. These sections provide Air Force operational expertise for the support of conventional planning and operations. They are the point of contact to coordinate local air defense and airspace management activities. Their function is to specifically assist planners in the preparation of the battalion's plan to integrate CAS into the overall scheme of fires and maneuver. They coordinate preplanned and immediate air requests and assist in coordinating air support missions with appropriate airspace control elements. Battalion TACPs have the added responsibility of terminal attack control.

C-62. TACPs coordinate activities through the joint air request net. The TACP performs the following functions:

- Serves as the Air Force commander's representative.
- Advises battalion commander and staff on capabilities, limitations, and employment of air support, airlift, and reconnaissance.
- Coordinates with respective fire support team and airspace control cells.
- Helps synchronize air and surface fires.
- Helps prepare the air support plan.
- Coordinates local air defense and airspace management activities.
- Integrates into the staff for air support planning for future operations.
- Advises on the joint suppression of enemy air defenses.
- Provides appropriate final attack control for CAS and operates the Air Force air request net.

Close Air Support Planning Duties and Responsibilities

C-63. The ALO and members of the battalion TACP provide the necessary expertise for the control and application of tactical air power. The ALO serves as the primary tactical air power advisor for the battalion, while TACP JTACs provide final control for CAS missions executed in the battalion's AO. Their collaborative working relationship established with the IBCT and its maneuver battalion provides a working knowledge of ground operations and enhances their ability to integrate tactical air operations with ground schemes of maneuver effectively.

Forward Air Controller

C-64. The primary responsibility of TACP JTACs includes the positive control of CAS aircraft flying missions in support of brigade operations. Using their knowledge of ground operations, they also are better able to provide the troop safety necessary to avoid fratricidal engagements. The following paragraphs discuss JTAC procedures and responsibilities.

Troop Safety

C-65. The safety of ground forces is a major concern during day and night CAS operations. Fratricidal engagements normally are caused by the incorrect identification of friendly troops operating in an AO or a failure to mark the boundaries of the friendly unit adequately. The use of proper authentication and ground marking procedures assures that a safe separation exists between the friendly forces and the impact area of aerial delivered munitions. Proper radio procedures and markings assist the JTACs and the strike aircraft in the positive identification of ground forces and their operational boundaries.

Identification of Friendly Forces

C-66. As digital technology continues to emerge and digital systems are fielded throughout combat units, the disposition and location of friendly units will become more accurate. These systems will provide enhancements to safety margins and help reduce the potential of fratricidal engagements during joint air attack team or tactical air operations. Friendly unit locations and boundaries can be marked using flash mirrors, marker panels, and direction and distance from prominent land features or target marks. Strobe lights are good markers at night and in overcast conditions. They can be used with blue or infrared filters and made directional using any opaque tube. Any light that can be filtered or covered and uncovered can be used for signaling aircraft or marking friendly locations.

C-67. *Target acquisition* is the detection, identification, and location of a target in sufficient detail to permit the effective employment of weapons (JP 3-60). Targets that are well camouflaged, small and stationary, or masked by hills or other natural terrain are difficult for fast-moving aircraft to detect. Marking rounds or rockets fired from aerial platforms or artillery can enhance target acquisition and help ensure first-pass success.

Target Identification

C-68. ***Target identification* is the accurate and timely characterization of a detected object on the battlefield as friend, neutral, or enemy.** Strike aircraft must have a precise description of the target and know the location of friendly forces in relation to terrain features that are easily visible from the air. Airborne joint terminal attack controllers generally are assigned and area of operation and become intimately familiar with its geographical features as well as the unit operating within the area of operation.

Final Attack Heading

C-69. Choice of the final attack heading depends upon the considerations of troop safety, aircraft survivability, enemy air defense locations, and optimum weapons effects. Missiles or bombs are effective from any angle. Cannons, however, are more effective against the sides and rears of armored vehicles.

Responsibilities

C-70. The following are the responsibilities of key personnel:

- S-3 air: The S-3 air plans for and requests the use of CAS and attack helicopters to support the commander's concept of the operation.
- S-2. The S-2 provides information on the avenues of approach, target array, terrain, and weather as it applies to the time and location of the joint air attack team operation.
- Attack helicopter liaison officer. The Army aviation liaison officer (when attached)—
  - Provides status of Army aviation assets available.
  - Begins planning the air corridors and air battle positions to support the operation.
  - Coordinates with the FSO and the air defense officer to deconflict air corridors.
  - Coordinates for the planned airspace coordination areas.
- Fire support officer. The FSO—
  - Determines the need, availability, and positioning of artillery and battalion mortars, commensurate with the enemy update, to support the joint air attack team.
  - Coordinates with the aviation representative to provide call signs and frequencies to the supporting fire direction center.
  - Helps the TACP deconflict the initial points from artillery positions and develop airspace coordination areas to support the mission.
  - Determines the need for suppression of enemy air defense.
  - Plans and coordinates, in conjunction with the battalion staff, the use of nonlethal attack assets to complement the joint air attack team.
  - Determines when and how priorities of fires shift.
  - Recommends fire support coordination measures to enhance the success of the mission.
  - Establishes a quick fire channel if necessary.
- Air defense officer. The air defense officer—
  - Coordinates to ensure that the air defense assets know the location of air corridors, friendly locations, initial points, and airspace coordination areas.
  - Ensures these assets are informed of friendly air operations and their integration into the battle.
- Tactical air control party: The TACP—
  - Develops contact points, initial points, and airspace coordination areas in coordination with the FSO and the air defense officer.
  - Disseminates contact points, initial points, and airspace coordination areas to the air support operations center for dissemination to the ground liaison officer and wing operations center for preflight briefing.
  - Helps coordinate aircraft forward to the appropriate contact point or initial point) and then hands them off to the aviation commander conducting the joint air attack team operation.

Suppression of Enemy Air Defenses

C-71. Suppression of enemy air defense (SEAD) operations target all known or suspected enemy air defense artillery sites that cannot be avoided and that are capable of engaging friendly air assets and systems, including suppressive fires. The fire support team integrates SEAD fires into an overall fire plan that focuses fires according to the commander's guidance. Synchronization of SEAD fires with the maneuver plan is accomplished using procedural control (an H-hour sequence), positive control (initiating fires on each target as the lead aircraft passes a predetermined reference point or trigger), or a combination of the two. Regardless of the technique, the FSO planning the SEAD must conduct detailed planning and close coordination with the ALO, liaison officer, S-3 air, S-2, air defense officer, field artillery battalion S-3 or fire direction officer, and fire support team.

Note. A *procedural control* is a method of airspace control which relies on a combination of previously agreed and promulgated orders and procedures (JP 3-52).

Weather

C-72. Weather is one of the most important considerations when visually employing aerial-delivered weapons. Weather can hinder target acquisition and identification, degrade weapon accuracy and effectiveness, or negate employment of specific aerial munitions types. The S-3 air can request Integrated Meteorological System data from the division G-2. This will give the S-3 air highly predictive and descriptive weather information for specific times and locations in the battalion's AO. This data improves the S-3 air's ability to determine when to use CAS. The Integrated Meteorological System provides weather data based on inputs from the air weather services and meteorological sensors. This currently system is located at echelons above the IBCT. It predicts weather effects on a specific mission, desired AO, or particular system. It also provides weather hazards for different elevations, surface temperatures in a specific AO, and wind conditions. Meteorological satellite data also may be obtained to show regional cloud cover with high and low pressure systems annotated.

Echelonment of Fires

C-73. Understanding echelonment of fires is critical for the fire support plan to be synchronized effectively with the maneuver plan. The purpose of echeloning fires is to maintain constant and overlapping fires on an objective while using the optimum delivery system up to the point of its risk estimate distance (RED) in combat operations or minimum safe distance in training. Echeloning fires provides protection for friendly forces as they move to and assault an objective, which allows them to get in close with minimal casualties. It prevents the enemy from observing and engaging the assault by forcing the enemy to take cover, which allows the friendly force to continue the advance unimpeded.

Concept of Echeloning Fires

C-74. The concept of echeloning fires begins with attacking targets on or around the objective using the weapons system with the largest RED. As the maneuver unit closes the distance en route to the objective, the fires cease or shift. This triggers the engagement of the targets by the delivery system with the next largest RED. The length of time to engage the targets is based on the rate of the friendly force's movement between the RED trigger lines. The process continues until the system with the smallest RED ceases or shifts fires and the maneuver unit is close enough to eliminate the enemy with direct fires or make its final assault and clear the objective.

C-75. The RED takes into account the bursting radius of particular munitions and the characteristics of the delivery system and associates this combination with a percentage for the probability of incapacitation of Soldiers at a given range. The munitions delivery systems include mortars, field artillery, helicopter, and fixed wing aircraft. The RED is defined as the minimum distance friendly soldiers can approach the effects of friendly fires without suffering appreciable casualties of 0.1 percent or higher probability of incapacitation. Commanders may maneuver their units within the RED area based on the mission; however, in doing so, they are making a deliberate decision to accept the additional risk to friendly forces. Before the commander accepts this risk, he should try to mitigate the probability of incapacitation. For example, maneuvering units in a defilade that provides some protection from the effects of exploding munitions.

> **WARNING**
>
> **Risk estimate distances are for combat use and do not represent the maximum fragmentation envelopes of the weapons listed. Risk estimate distances are not minimum safe distances for peacetime training use.**

C-76. The casualty criterion is the five-minute assault criterion for a prone Soldier in winter clothing and helmet. Physical incapacitation means that a Soldier is physically unable to function in an assault within a five-minute period after an attack. A probability of incapacitation value of less than 0.1 percent can be interpreted as being less than or equal to one chance in one thousand.

C-77. Using echelonment of fires within the specified RED for a delivery system requires the unit to assume some risks. The maneuver commander determines, by delivery system, how close fire will be delivered in proximity to forces. The maneuver commander makes the decision for this risk level, but relies heavily on the FSO's expertise. While this planning normally is accomplished at the battalion level, the company FSO has input and should be familiar with the process because the FSO must execute the same process with the company mortars. (Refer to ATP 3-09.32 appendix H for information on risk estimate distances and Appendix I for information on minimum safe distances.)

Echeloning a preparation

C-78. Echelonment of fires is accomplished when the maneuver commander wishes to conduct preparation fires on an objective. *Preparation fire* is normally a high-volume of fires delivered over a short period of time to maximize surprise and shock effect. Preparation fire can include electronic attack and should be synchronized with other electronic warfare activities (FM 3-09). Not all maneuver tasks warrant preparation fires. Some considerations for conducting preparation fires are—

- Will the loss of surprise from the preparation be offset by the damage done to the enemy?
- Are there enough targets and means to warrant a preparation?
- Can the enemy recover before the preparation fires can be exploited?

C-79. Echeloning a preparation is a 9-step process. The process is outlined and described in detail in ATP 3-09.42. The outline follow the following nine steps for echeloning a preparation:

- Determine what assets, to include ammunition, are required and what assets are currently available or allocated.
- Verify risk estimate distances and attack criteria with the commander.
- Plan targets.
- Develop a communications plan.
- Determine what the rate of movement will be.
- Develop the schedule of fires and decide how the preparation schedule will be initiated.
- Brief the plan and confirm the method with the commander.
- Complete the scheduling worksheet(s) within AFATDS or manually using DA Form 4656 (*Scheduling Worksheet*).
- Rehearse and refine the plan.

Echelonment of Fires, Example

C-80. When the lead elements of the battalion approach the designated phase line en route to the objective, the FSO begins the preparation. Lead element observers and company fire support teams track movement rates and confirm them for the battalion FSO. The battalion FSO may need to adjust the plan during execution based on unforeseen changes to anticipated movement rates (figures C-4 through C-8, page C-24 through C-28.)

C-81. As the unit continues its movement toward the objective, the first delivery system engages its targets. It maintains fires on the targets until the unit crosses the next phase line that corresponds to the RED (in combat) of the weapon.

C-82. To maintain constant fires on the targets, the unit starts the next asset before the previous asset ceases or shifts. This ensures no break in fires, enabling the friendly forces' approach to continue unimpeded. However, if the unit rate of march changes, the fire support system must remain flexible to the changes.

C-83. The FSO shifts and engages with each asset at the prescribed triggers, initiating the fires from the system with the largest RED to the smallest. Once the maneuver element reaches the final phase line to cease all fires on the objective, the FSO shifts to targets beyond the objective.

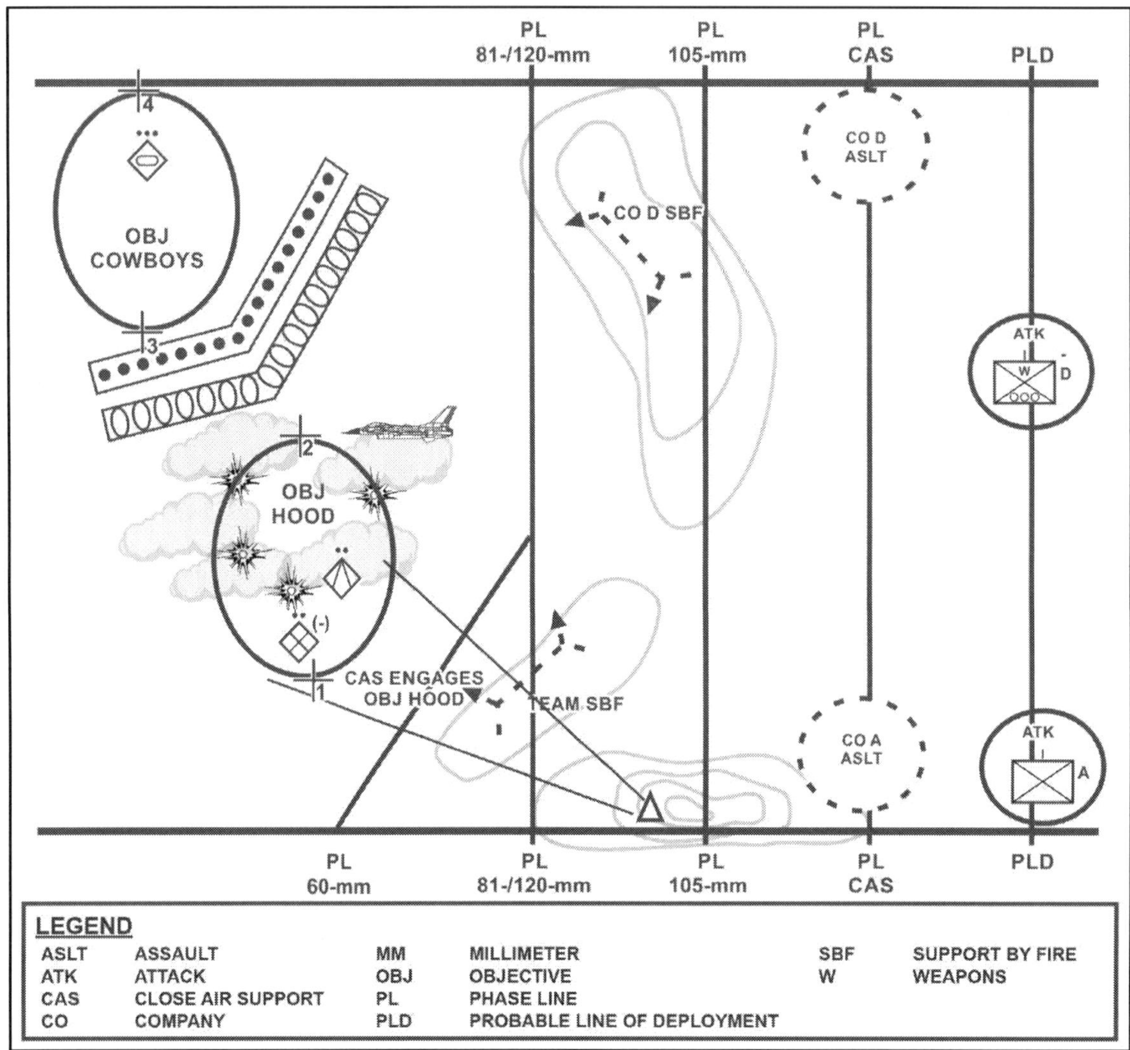

Figure C-4. Beginning of close air support

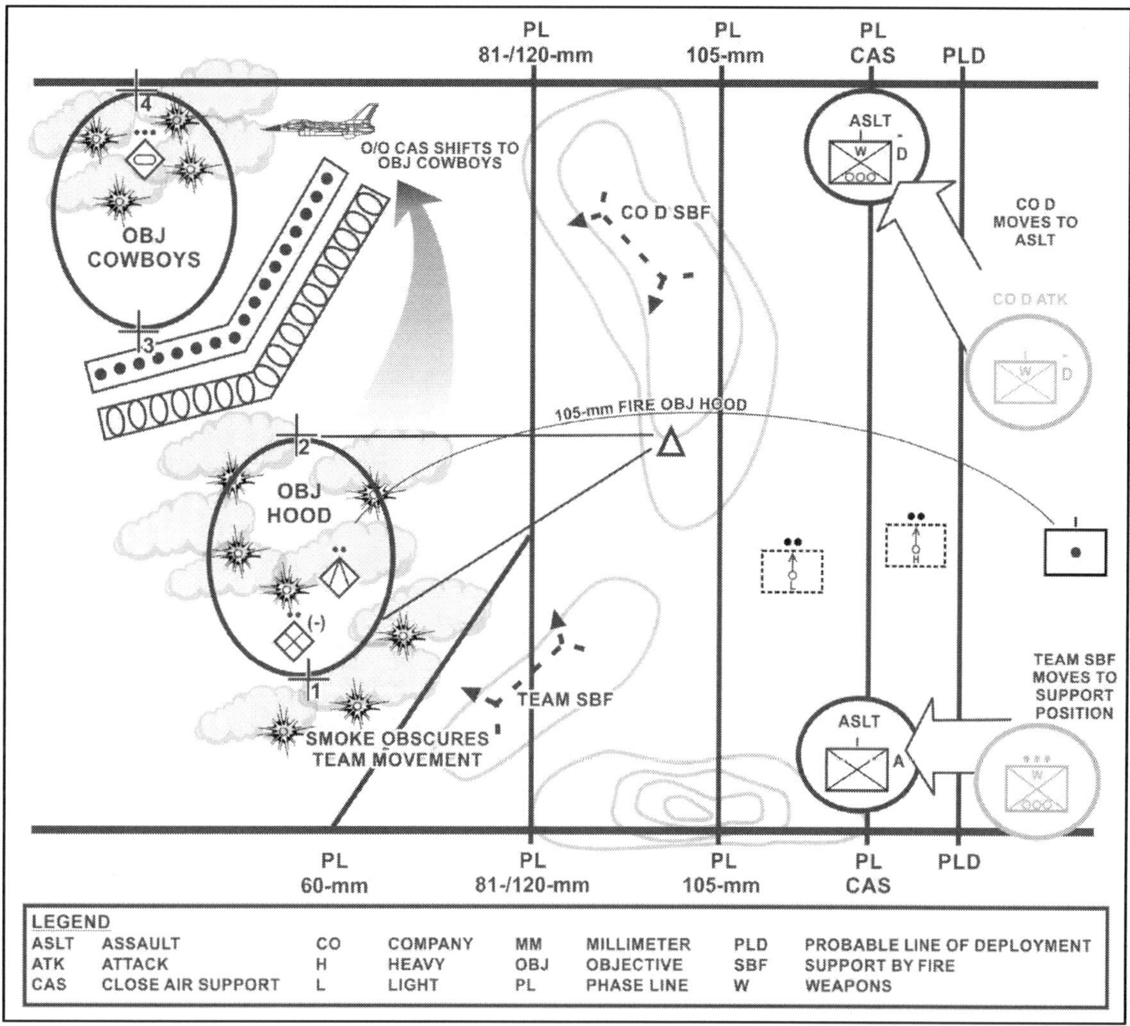

Figure C-5. Execution of 105-mm shaping fires; shifting of close air support

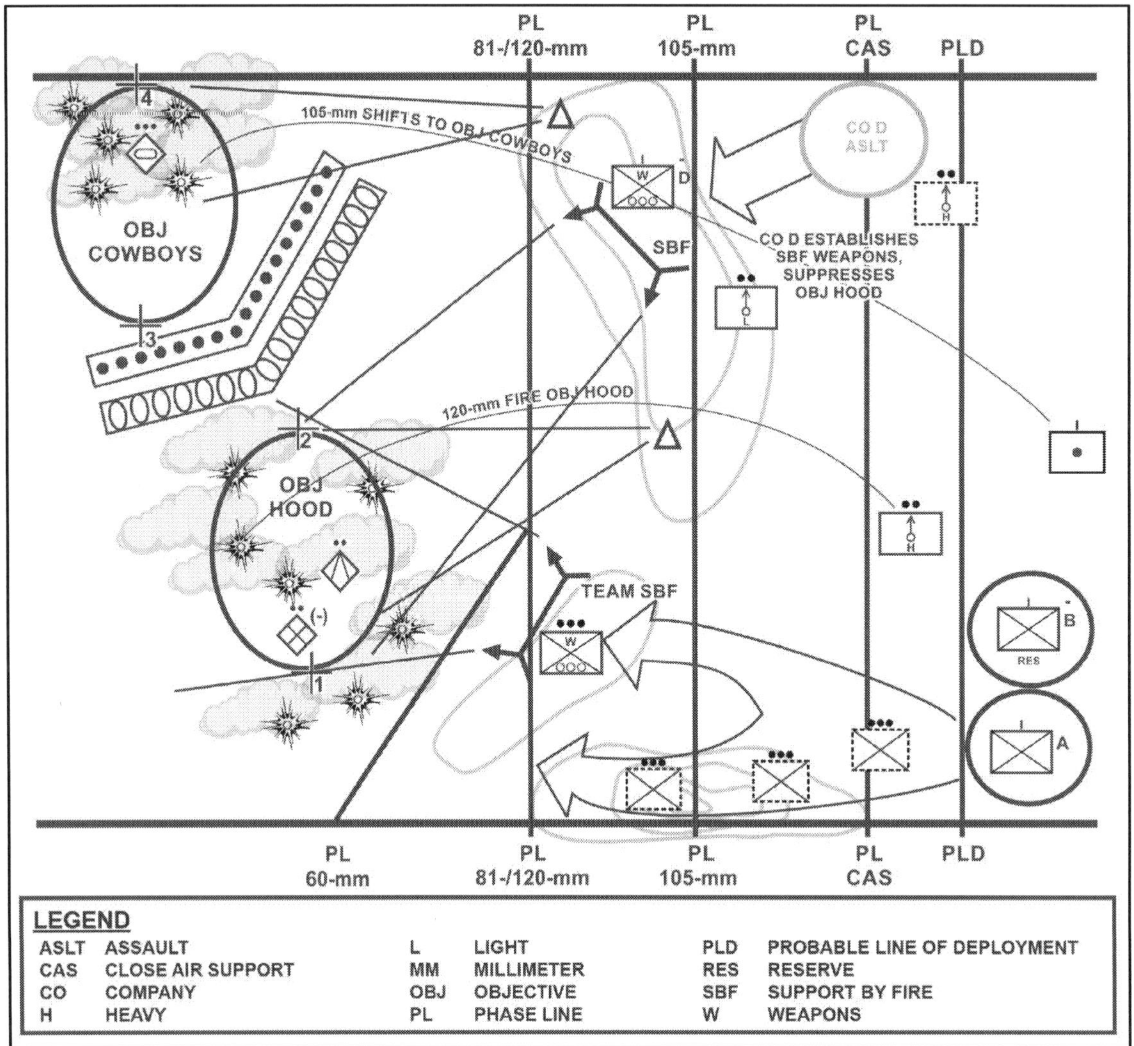

Figure C-6. Beginning of 120-mm and supporting fires; shifting of 105-mm fires

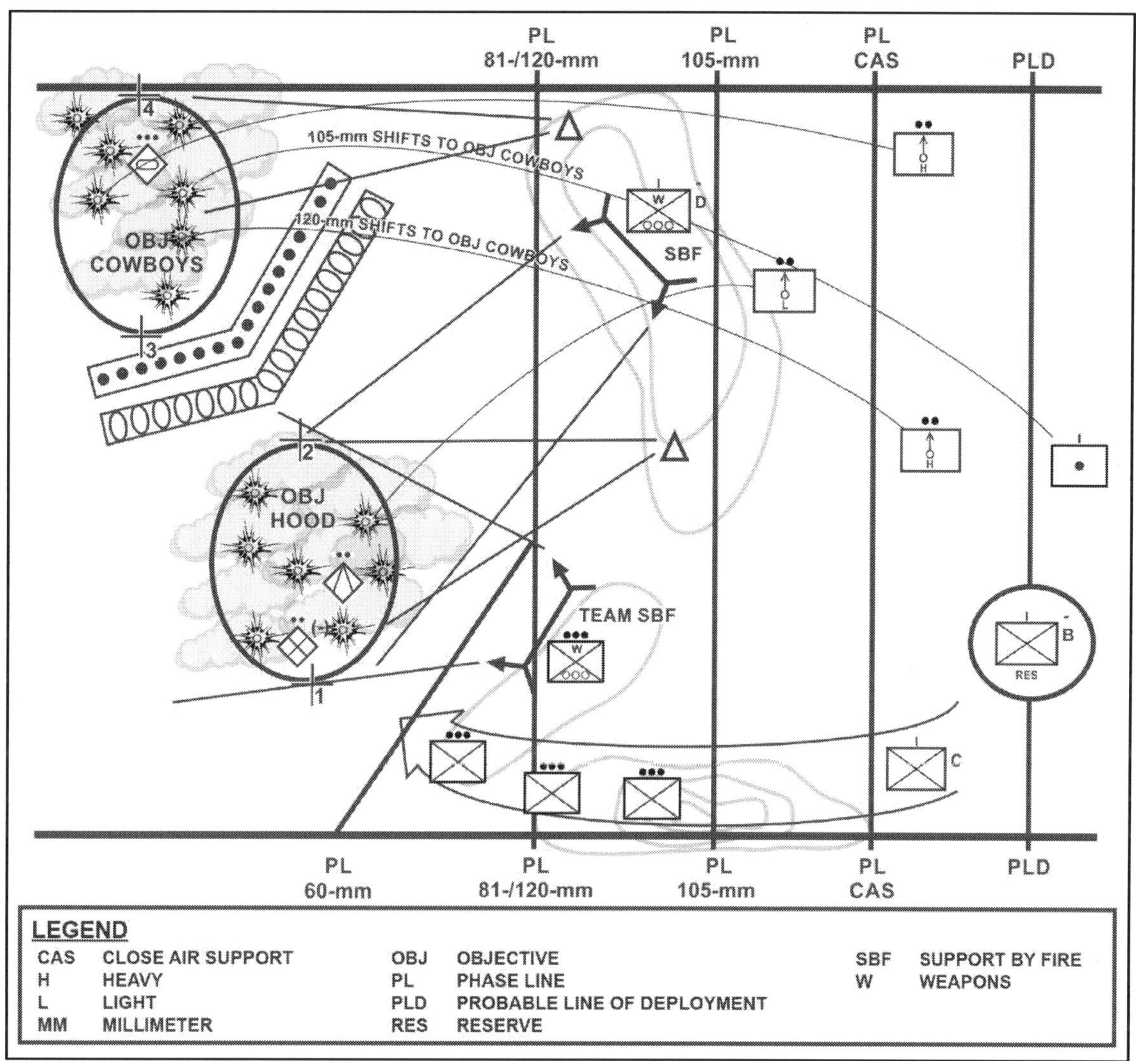

Figure C-7. Beginning of 60-mm fires; shifting of 120-mm fires

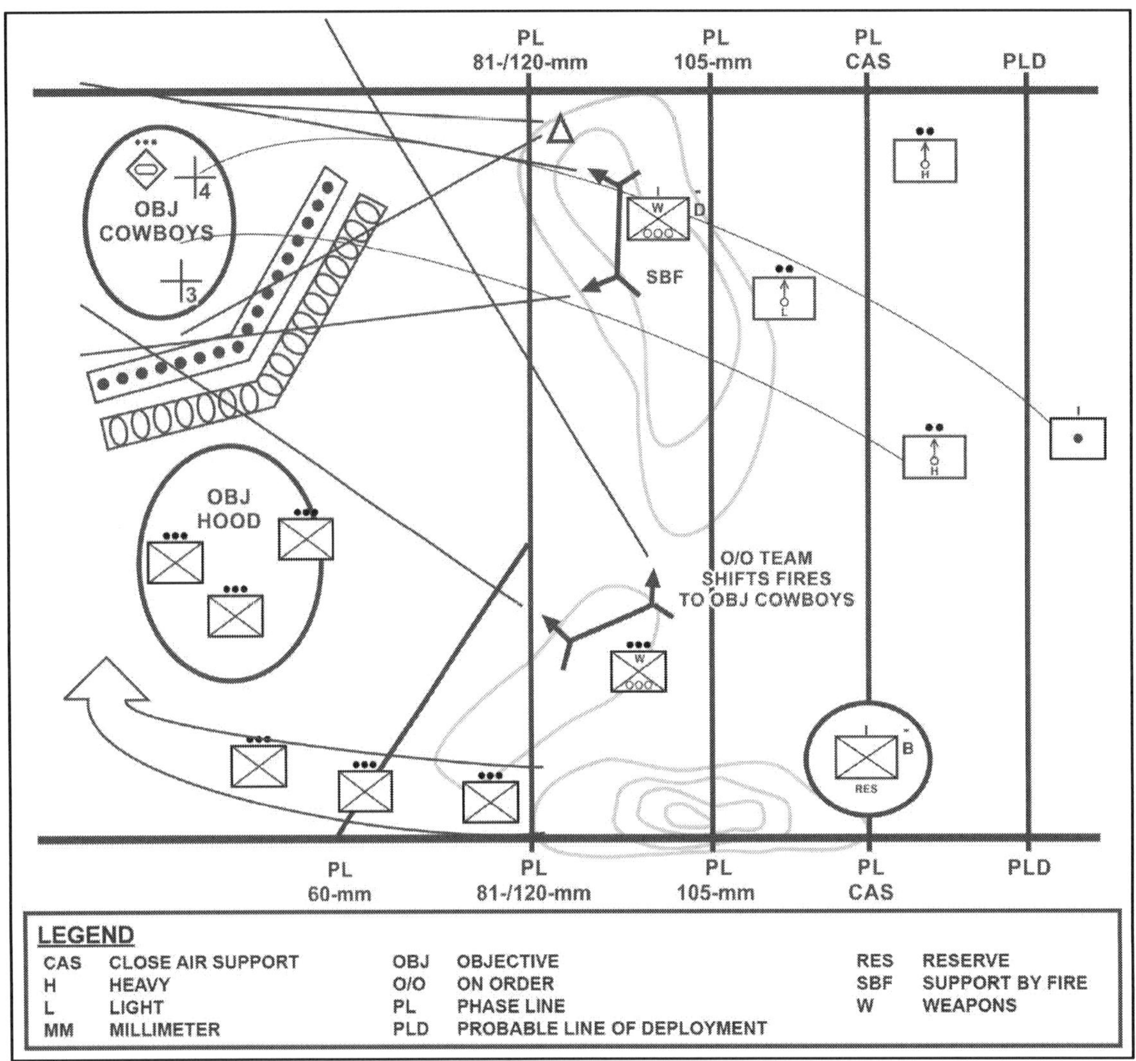

Figure C-8. Cessation of 60-mm fires; shifting of supporting fires

Appendix D

Infantry Weapons Company

Appendix D, in conjunction with the other appendixes and chapters in this publication, provides the doctrinal framework for the Infantry weapons company assigned to Infantry battalions in the IBCT. This appendix is designed to work in conjunction with and complement FM 3-21.10. Among the topics covered in FM 3-21.10 but omitted here are duties and responsibilities, troop leading procedures, precombat checks and inspections, sustainment, armored and Stryker employment, improvised explosive devices, and operations in a CBRN environment. Characteristics and fundamentals of combined arms operations in urban terrain are address in ATTP 3-06.11. Appendix D focuses on the unique characteristics of the Infantry weapons company, including principles, tactics, techniques, procedures, and terms and symbols.

Note. In the near future, ATP 3-21.20 will replace FM 3-21.12.

OVERVIEW

D-1. The Infantry weapons company is uniquely equipped to provide the Infantry battalion with additional capabilities. The organization structure and equipment provide the battalion additional heavy weapons firepower, maneuverability, and long-range communications. This section focuses on the employment of the Infantry weapons company while fighting as a pure company or combined arms team under the command of an Infantry weapons company commander. This framework will help Infantry weapons company leaders effectively—

- Exploit weapons company-unique capabilities.
- Employ the company using unit weapon fundamentals.
- Reduce the vulnerability of the unit.
- Plan for and accomplish missions within the range of military operations.

Note. Though organization of forces may require the detachment of individual weapons company platoons to other units, this appendix does not cover detailed operations of detached platoons. This appendix only provides a general discussion of coordination and operating issues pertaining to detachments. Chapters 2, 3, and 4 of this publication cover detailed operations of detached platoons.

D-2. This section provides an overview of the mission, organizational structure, characteristics, and weapon systems of the Infantry weapons company found in the Infantry battalions of the IBCT. Also referred to as Company D or Delta Company.

ORGANIZATION

D-3. The Infantry weapons company, organic to an Infantry battalion, contains a company headquarters company and four assault platoons (figure D-1, page D-2). Each assault platoon has two sections consisting of two squads each and a leader's vehicle. Each squad contains four Soldiers and a vehicle mounting the heavy weapons. The weapons company is a fully mobile unit consisting of weapons carrier vehicles and a variety of heavy weapons systems.

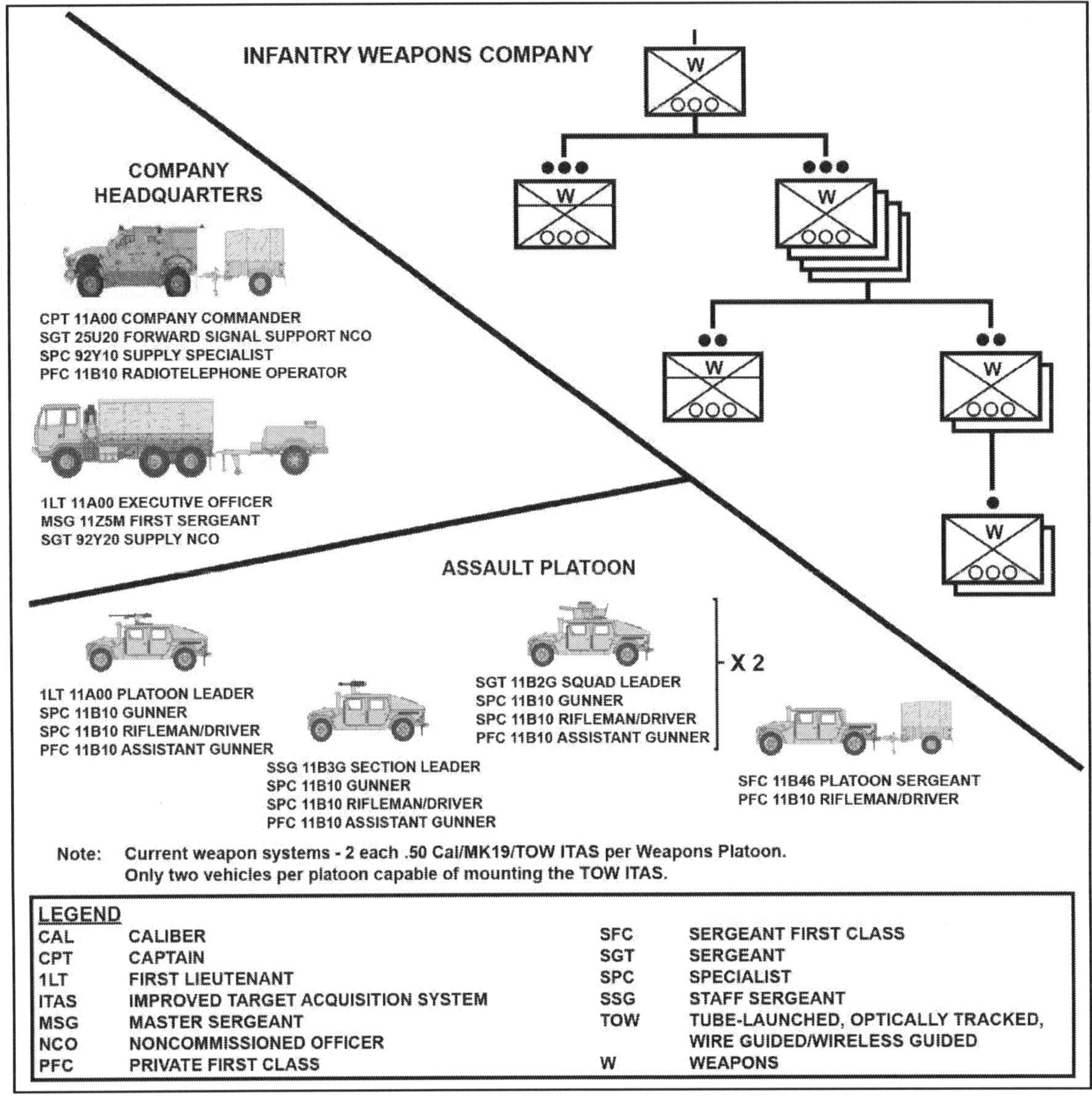

Figure D-1. Weapons company organization and equipment

D-4. The Infantry weapons company is uniquely equipped with heavy weapons to support the maneuver of the rifle companies within the Infantry battalion. The heavy weapons contained in the weapons company include a mix that can be tailored to a particular mission based on mission variables of METT-TC. The company maneuvers in all types of terrain, climates and visibility conditions. Infantry weapons companies are currently equipped with four types of heavy weapons. The selection and employment of weapon system or systems to use for a particular mission is termed the "Arms Room Concept."

D-5. The Infantry weapons company is equipped with the following weapons: the TOW, Improved Target Acquisition System (ITAS), the MK 19 40-mm grenade machine gun, the M2 series heavy machine gun, and the M240 series machine gun. Each vehicle-mounted system is equipped with a tripod for ground mount operations. Only one of these systems can be mounted on each vehicle at a time. While all of the weapons vehicles can mount the MK 19 and the M2, only two vehicles per platoon are equipped to mount the ITAS. Heavy weapons, within the weapons company, can be tailored to a mission based on METT-TC.

D-6. TOW missiles are accurate and lethal; but missile flight time is long and obstacles may interfere with the flight path. The slow rate of fire and the visible launch signature of the TOW missile increase the weapons squad's vulnerability; especially if a vehicle mounted, system engages within an enemy's effective direct-fire range. Units can reduce their vulnerability by employing standoff, displacing often, integrating direct and indirect fires, using obstacles to slow or halt the enemy within an engagement area, and incorporating Javelins and shoulder launched munitions from Infantry rifle platoons.

D-7. Mass, integration, and depth are key components to employing heavy weapons assets. Because of their relatively long range, its fires can be massed to destroy and enemy attack or defense. The weapons company fires should be integrated with the fires from the rifle companies. For example, the fires from the TOW systems can be combined with the fires from the rifle company's Javelins to destroy the enemy within an engagement area. Depth is assured by echeloning systems and by moving units to threatened areas.

D-8. The company's heavy weapon systems provides direct fires against personnel, armored and unarmored vehicles, and other hard targets to support the maneuver of supported units. Vehicular mounted communications systems are used to permit communications between widely separated units and may conduct communications relay if necessary. Optic systems for the heavy weapons system also may aid in surveillance and other information collection tasks.

D-9. The Infantry weapons company receives attachments based on an analysis of the mission variables of METT-TC. The battalion commander determines whether medics and a fire support team are attached. Other attachments may include Infantry, scout, and engineer elements.

Missions

D-10. The mission of the Infantry weapons company is to provide mobile heavy weapons and long range close combat missile fires to the Infantry battalion. The inherent versatility of the weapons company as part of the Infantry battalion makes it well suited for employment regardless of which element of decisive action currently dominants. *Massed fire*—fire from a number of weapons directed at a single point or small area (JP 3-02)—and depth are key components to employing heavy weapons assets. During tactical operations, heavy weapons units can suppress, fix, or destroy enemy at long ranges, allowing other Infantry units or combined arms teams to maneuver. The weapons company provides the Infantry battalion with a highly mobile, multifunctional element that can—

- Deliver precision long-range, large-caliber direct fires to destroy enemy armored vehicles and fortifications.
- Deliver massed heavy machinegun and grenade launcher fires to engage enemy personnel, destroy light vehicles, and provide area suppression.
- Move rapidly on the battlefield to shift combat power where it is needed.
- Communicate over longer distances than units using man-packed radios.
- Employ long-range thermal weapons sights to detect and engage enemy forces during hours of darkness.
- Conduct moving or stationary observation, reconnaissance, screen, and guard missions.
- Provide security and control for armed convoy escorts.
- Coordinate, mass, and shift long-range direct fires.
- Control and execute mounted combat and reconnaissance patrols.
- Support the assaults of other units with massed supporting fires.
- Provide responsive and flexible over-watch of moving elements.
- Provide effective and wide-ranging outer cordon forces.
- Integrate indirect and aerial fires with the unit's direct fire plan.
- Task organize with one or more rifle platoons, or attached armored forces, into a powerful and flexible combined arms team.
- Detach one or more assault platoons to augment rifle companies within the battalion.
- Conduct unit self-sustainment and maintenance within its capability.

Capabilities

D-11. The Infantry weapons company is uniquely equipped with heavy weapons to support the maneuver of rifle companies during the conduct of decisive action through the depth of the Infantry battalion's AO. The heavy weapon systems available to the weapons company provide direct fire against personnel, vehicles, armored or other hard targets to support the battalion's maneuver. Vehicular mounted communications systems may be used to enhance long range communications for the battalion offering communications relay if necessary. Optic systems supplied with the heavy weapons also aid in surveillance, reconnaissance, and security operations.

D-12. In the offense, the weapons company provides the base of fire in a battalion attack in order to suppress, fix, or destroy the enemy in position. The company can also engage enemy in planned engagement areas, isolate objectives by destroying enemy counterattacks, or destroy withdrawing enemy forces. The company is well suited to protect the battalion's flanks. Within the confines of the rules of engagement, the company's heavy weapons may also be useful in combined arms operation in urban terrain with tasks such as creating entry points into buildings, engagement of snipers, and destruction of reinforced structures. (For a further discussion, see chapter 2 of this publication.)

D-13. In the defense, the weapons company may be positioned forward of the defensive area to participate in security operations or to overwatch reconnaissance units or obstacles. As the enemy closes, they can displace to positions that provide direct fires into an engagement area. The company can position throughout the depth of the area of operation to cover likely armor avenues of approach, conduct surveillance and reconnaissance and security operations or provide overwatch to assist in security. During counterattacks, the weapons company can provide overwatch (specifically by support-by-fire and/or attack by fire) for the maneuver element(s). (For a further discussion, see chapter 3 of this publication.)

D-14. When conducting tasks in support of stability-focused operations the weapons company brings with a host of capabilities including transportation, mobility, enhanced optics, and communications assets. These capabilities can be creatively employed to support battalion area security operations. The weapons company can also provide mobile security elements to protect convoys and secure supply routes and other missions, such as cordon and search and search and attack missions (both are subordinate forms of movement to contact discussed in chapter 2, section II) that require the rapid emplacement of forces. (For a further discussion, see chapter 4 of this publication.)

D-15. During movement, un-mounted weapon systems along with any additional equipment, are carried in trailers. When trailers are not feasible or desirable, non-mission essential equipment is left in a stay-behind position. During planning and preparation, the commander determines which weapons systems are best suited for the mission and configure vehicles appropriately.

Limitations

D-16. Although the weapons company's vehicles provide limited protection against small arms and fragmentation, these vehicles lack protection against large caliber direct and indirect fires and are still vulnerable to enemy antiarmor weapons. The weapons carrier vehicle provides increased mobility and can maneuver the weapons quickly to a position of advantage on the battlefield. However, certain types of terrain such as steep slopes, thick vegetation, mud and other restrictive terrain, may restrict vehicular travel. Weather may also prevent vehicles from operating at full capacity. The inherent maintenance requirement for vehicles and heavy weapons are greatly increased for the weapons company over the rifle company. This requirement creates an increase in logistics support to the company.

D-17. The ITAS TOW missile is accurate, but missile flight time is long and obstacles may interfere with the flight path. The slow rate of fire and the visible launch signature of the TOW missile increase the weapons squad's vulnerability especially if a vehicle mounted ITAS engages within an enemy's effective direct-fire range (no standoff). Units can reduce this vulnerability by displacing often and by integrating their fires with those of automatic weapon systems and with other antiarmor weapons in the platoon. Countermobility obstacles, indirect fires, and aviation fires also reduce vulnerability.

Heavy Weapon System Employment Fundamentals

D-18. Heavy weapons system employment fundamentals increase the probability of destroying targets and enhance survivability. Fundamentals include mutual support, security, flank engagements, standoff, dispersion, cover and concealment, employment in-depth, and employment as part of a combined arms team.

Mutual Support

D-19. Mutual support is a condition that exists when heavy weapons systems elements are able to support each other by direct fire in order to prevent the enemy from attacking one position without being subject to direct fire from one or more adjacent positions. To establish mutual support heavy weapons systems elements employ with overlapping primary and secondary sectors of fire (figure D-2). When one element is attacked or forced to displace, other elements continue covering the assigned area. In order to achieve this effect, heavy weapons elements position so that fires directed at one element suppress only that element.

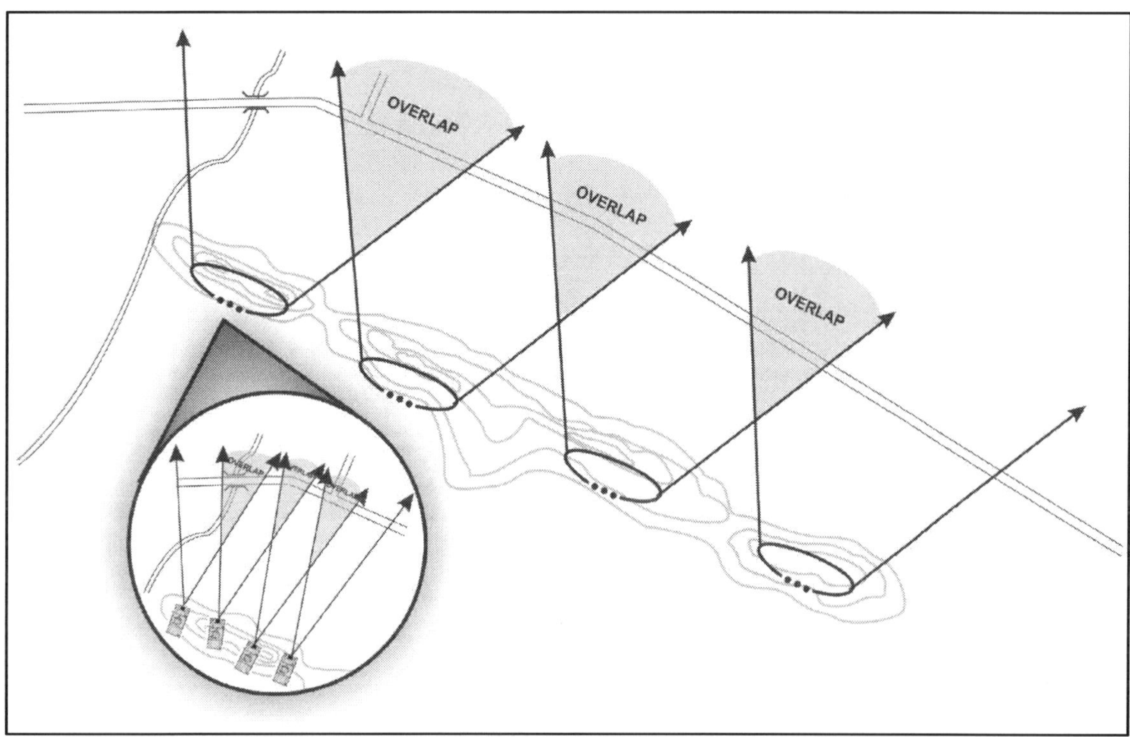

Figure D-2. Overlapping fires

Security

D-20. Because most weapons company personnel are needed to man the weapons systems and operate the vehicles, security against a dismounted threat may be a challenge in certain situations. When the weapons company elements position near other friendly Infantry units, those units may assist in protection against possible enemy attack as part of the overall scheme of maneuver. Though the weapons company element are not always collocated with other units, Infantry units positioned nearby can cover dismounted enemy avenues of approach into weapons positions when coordinated. Weapons company elements are responsible for insuring their own local security whether friendly units position near them or not. Weapons company elements moving with Infantry, again, provide for their own local security. During halts, the driver or assistant gunner may dismount to assist in securing the flank and rear sectors.

Flank and Rear Engagements

D-21. Heavy weapons system elements position to engage tank or armored vehicle threats from the flank or rear. Frontal engagements at enemy armor are less desirable for the following reasons:

- Armored vehicle protection is greatest to the front.
- Armored vehicle firepower and crew are normally oriented to the front.
- Frontal engagements increase the chance of detection and suppression by enemy armored vehicles.
- Armored vehicles provide a smaller target from the front.

Standoff

D-22. Standoff is the difference between a friendly weapon's maximum range and an enemy weapon's maximum effective range. The ITAS maximum range provides it with a standoff advantage over many other weapon systems. Despite this advantage, engaging enemy armored vehicles at greater standoff ranges may not always be tactically feasible. The additional tracking time required to fire an ITAS missile beyond 2000 meters gives a frontal target more time to maneuver against the friendly position and provides a flanking target more time to reach cover. Additionally, the terrain may not provide the fields of fire to support standoff distance engagements.

Dispersion, Cover, and Concealment

D-23. The use of dispersion, cover, and concealment is critical to the effective employment and survivability of heavy weapons systems. Mission analysis is a key step to ensure mission success and to overcome the following inherent heavy weapon system weaknesses:

- Gunner vulnerability when employing the heavy weapon system.
- ITAS missile long flight time.
- Continuous tracking of the missile until impact.
- ITAS slow rate of fire.

D-24. When employing heavy weapon systems leaders should avoid conspicuous terrain and disperse weapons laterally and in-depth so that no single enemy weapon can suppress two weapons systems. Leaders disperse weapons systems to reduce casualties and equipment damage resulting from enemy mortar and artillery fires. These same considerations for weapons system employment apply to route selection, and movement and maneuver.

D-25. During mission analysis, the commander identifies where cover and concealment along a route is good and where cover and concealment is poor. Obscuration and limited visibility (night movement) can enable concealment. Covered and concealed locations with good fields of fire are critical to the employment and survivability of heavy weapons systems. Looking at it from the enemy's view, both in daylight and at night, the commander determines how the enemy can use available cover and concealment.

Employment In-Depth

D-26. Heavy weapon systems elements often employed in-depth. In the offense, routes and firing positions are selected to support the forward movement of attacking units. In the defense, weapon systems elements position forward then moved to positions in-depth as the enemy closes or the elements position in-depth initially.

Employment as a Combined Arms Team

D-27. Skillful integration of combined arms improves the survivability and lethality of the heavy weapons systems of the Infantry weapons company. Infantry rifle platoons and squads can assist in providing local security for the heavy weapons systems. The heavy weapons systems elements support maneuver forces (Infantry, Stryker, and armored) as part of a combined arms team. For example, heavy weapons systems may focus on destroying lightly armored enemy vehicles and dismounted Soldiers at long ranges, allowing tanks to focus on destroying enemy tanks.

D-28. As part of the combined arms team, combat engineers shape the battlefield by enhancing mobility, countermobility, and survivability. General and geospatial engineers augment that effort providing other specialized capabilities. Mobility of the combined arms team is enhanced as engineers provide expertise and assistance in breaching, clearing, gap crossing, and other aspects of mobility support. Countermobility support focuses on the emplacement of tactical obstacles that reduce the enemy's ability to maneuver, mass, or reinforce, and increase their vulnerability to direct and indirect fires. To accomplish this, obstacles must disrupt, fix, turn, or block the enemy. To be effective, obstacles are covered by both direct and indirect fire. Survivability includes supporting the construction of fighting positions, other types of survivability positions, and those aspects of protection related to hardening as well as camouflage, concealment, and deception.

D-29. The commander ensures heavy weapons elements are integrated into the combined arms team and higher headquarters fire support plan and the targeting process. The fire support officer (FSO) habitually attached to the weapons company from the battalion fire support team assists in this coordination and the overall fire support plan. Frequencies, call signs, and priorities of fire must be coordinated. Fires are used to—

- Destroy or neutralize the enemy. Slow the enemy rate of advance.
- Destroy or disrupt enemy formations. Cause enemy vehicles to button up.
- Suppress accompanying enemy artillery and antitank guided missile support by fire positions.
- Conceal weapon system firing signatures and to cover the movement of heavy weapons elements between positions.

Note. When using obscurants, the weapons company commander and subordinate leaders consider the degrading effects these obscurants have on friendly units. For example, covering smoke may alert the enemy to friendly movement and/or reduce the leader's ability to control fires.

PLANNING CONSIDERATIONS

D-30. The unique characteristics of a weapons company requires the commander to factor in additional consideration during mission planning. The company commander plans direct fires as part of the troop-leading procedures. (Refer FM 3-21.10 for information on troop-leading procedures and direct fire planning and control). Determining where and how the company can and will mass fires are essential steps as the commander develops the concept of operation. (See paragraph 3-169 for information on engagement area development.) Along with the task organization for combat, the weapons company commander must also include in mission planning the selection of weapons systems, communications, and vehicle load and modification considerations. As mentioned earlier, commander will not only consider organic assets, but also those of attached personnel and their associated weapons and equipment. The commander bases the selection and employment of the available assets on the mission variables of METT-TC.

COMMANDER'S ROLE IN COURSE OF ACTION DEVELOPMENT

D-31. As the owner of much of the Infantry battalions direct fire heavy weapons, the weapons company commander should be ready to act as a participating member of the battalion-planning cell if requested by the battalion commander. In this role, the commander is the principal advisor on employment of the weapons company heavy weapons in mission support during course of action development. The commander advises on how best to support the mission with heavy weapons and suggest any additional fire control measures to facilitate operations. During course of action development, the commander takes into consideration the fire control principles (see paragraph D-30) that will ultimately increase the effectiveness of direct fires.

DEVELOPING THE CONCEPT OF OPERATION

D-32. After identifying probable (or known) enemy locations, the commander determines points or areas to focus combat power. The commander's situational understanding, and vision of where and how the enemy will attack or defend helps determine the volume of fires to focus at particular points to have a decisive result. When the commander intends to mass the direct fires of more than one platoon, the commander must establish a means for distributing those fires effectively. Based on where and how to focus and distribute direct fires, the commander establishes the weapons ready postures for company elements as well as triggers for initiating fires. The commander evaluates the risk of fratricide and establishes controls to prevent it. Fratricide prevention measures include designation of recognition marking weapons control status, and weapons safety posture.

D-33. After determining where and how to mass and distribute direct fires, the company commander orients platoons to rapidly, and accurately acquire the enemy. The commander anticipates how the enemy will fight. The commander gains this anticipation through a detailed war-game of the selected course(s) of action. During war-gaming, the commander determines probable requirements for refocusing and redistributing fires and for establishing other necessary controls. During the troop-leading procedures, the company commander plans and rehearses direct fires (and the fire control process) based on mission analysis. The company commander continues to apply planning procedures and considerations throughout execution.

Task Organization

D-34. The battalion operations order will contain the task organization for combat depicting attachments and detachments to or from the weapons company for a particular mission. Attachments and/or detachments may be habitual or temporary for a specific operation. If the weapons company receives any attachments, the commander may further task organize within the company itself to accomplish the mission. To provide for mission command and to optimize heavy weapons unit's capabilities, heavy weapons platoons are not normally task organized below platoon level.

Attachments

D-35. Weapons companies may have habitual attachments such as a FSO and a combat medic. Other sustainment support may or may not consist of a medical evacuation or field maintenance team.

D-36. Weapons companies may also receive various other attachments with one of the more common being an Infantry platoon. A typical mission for an Infantry platoon attached to a weapons company would be to provide security for the company while the company performs its primary mission such as support by fire for another maneuvering element.

Detachments

D-37. Elements of the weapons company may also be detached out to other units. Detachments often include an assault platoon attached to an Infantry rifle company. Specific missions given to detached units are the responsibility of the gaining unit commander or subordinate leader and are not covered in detail in this manual (see FM 3-21.10). Typical missions may include, but are not limited to, establishing support by fire in an attack engaging the enemy in a planned engagement area, or security of flanks. Special consideration needs to be given to the maintenance requirements of the detached unit due to the gaining unit not having a habitual requirement for maintenance of their specific weapons or vehicles.

Planning Checklist

D-38. The following example checklist shows several items for consideration for units attached to the weapons company or for elements of the weapons company detached to other units:

- Radio communications between units.
- Command and support relationship.
- Communications requirements.
- Unit tactical standing operating procedures.
- Unit situation report, tactical situation, nature of mission.
- Current operation orders with graphics.
- Signal operating instructions (current frequencies, call signs, challenge, and password).
- Digital communications.
- Special instructions.
- Special equipment.
- Location of units.
- Reporting times.
- Duration of mission.
- Link up information and location.
- Coordination and contact points.
- Precombat checks and precombat inspections.
- Support and sustainment requirements.

Weapon Selection Considerations

D-39. The mix of weapons in a weapons company includes systems that can effectively engage troops, field fortifications, lightly armored and armored vehicles. The arms room concept allows for flexibility in weapons configuration for specific missions given to the weapons company. Considerations during selection of weapons systems must include an analysis of the terrain and threat in conjunction with the characteristics and capabilities of each weapon system. Refer to the appropriate source publication (listed below) for weapon systems characteristics to assist in weapons planning.

Improved Target Acquisition System

D-40. The ITAS is a multipurpose weapon used for long-range engagement of targets. It can be employed in all weather conditions. It fires a TOW missile that provides a long-range capability against armored vehicles, heavily fortified bunkers, buildings, and dug-in or fortified enemy positions. The ITAS optics system, the Target Acquisition System, can also be used to increase visibility for reconnaissance, surveillance, and security operations. For planning purposes, TOW missiles have a maximum range of 3,750 meters and a minimum range of 200 meters for the TOW 2B. TOW missiles have the ability to defeat all known armor units they may encounter during combat operations. For detailed information on all ITAS characteristics and TOW munitions, refer to TC 3-22.32 and FM 3-22.34.

M2, .50 Caliber, Machine Gun

D-41. The M2 can be used against personnel and light armored vehicles with accurate fires past 2000 meters. It is effective in restrictive terrain such as wooded areas. For detailed information on M2 characteristics and munitions, refer to TC 3-22.50.

MK-19, 40mm, Grenade Machine Gun

D-42. The MK-19 is capable of laying down a heavy volume of close, accurate, and continuous fire. As a point weapon, it can penetrate up to 2 inches of steel armor at ranges out to 1,500 meters. As an area weapon, it can inflict personnel casualties out to 15 meters from impact at ranges out to 2,000 meters. Like the M2, the MK-19 can be employed in restrictive terrain conditions. It may also be used to cover dead space. For detailed information on all MK-19 characteristics and munitions, refer to TC 3-22.19.

Javelin Close Combat Missile System

D-43. The Javelin (when support by Javelin teams from the Infantry rifle company) is a dual-mode, man-portable missile with the capability to engage and defeat all known armor including tanks and other armored vehicles. When there is no armored vehicle threat, the Javelin can be employed in a secondary role of providing fire support against point targets such as bunkers and crew-served weapons positions. The Javelin command launch unit can be used as an aid to reconnaissance, security operations and surveillance. The Javelin supports the fires of ITAS and can cover secondary armor avenues of approach and provide observation posts with an antiarmor capability. The Javelin has a maximum effective range of 2000 meters. For detailed information on all Javelin characteristics and munitions, refer to TC 3-22.37.

Optics

D-44. All heavy weapons within the weapons company have optics systems. These systems provide the company with the ability to acquire targets at long range during daylight or limited visibility. All of the systems have magnification and thermal imaging capability allowing thermal acquisition of targets at night or in dense forest or brush areas during daylight. Besides target acquisition, these systems can be used for both day and night observation. Some environmental conditions that affect all optics include limited visibility, night, infrared clutter, and infrared crossover.

Standard Operating Procedures

D-45. When necessary, the commander applies direct fire standard operating procedures (SOPs). A well-rehearsed direct fire SOP enhances direct fire planning and ensures quick, predictable actions by all members of the company. The commander bases the various parts of the SOP on the capabilities of the force and on anticipated conditions and situations. SOP should include standard means for focusing fires, distributing their results, orienting forces, and preventing fratricide.

D-46. The commander adjusts the direct fire SOP whenever changes to the anticipated and actual situation become apparent. A technique to establish a standard respective position for target reference points (TRPs) in relation to friendly elements and to consistently number the TRPs such as from left to right. This allows leaders to quickly determine and communicate the location of the TRPs.

D-47. Two means of distributing the results of the company's direct fires are engagement priorities and target array. Engagement priorities, by type of enemy vehicle or weapon, are assigned for each type of friendly weapon system. The target array technique helps in distribution by assigning specific friendly elements to engage enemy elements of approximately similar capabilities.

D-48. A standard means of orienting friendly forces is to assign a primary direction of fire, using a TRP, to orient each element on a probable (or known) enemy position or likely avenue of approach. To provide all-round security, the SOP can supplement the primary direction of fire with sectors using a friendly-based quadrant. The following sample SOP elements show the use of these techniques:

- The front (center) platoon's primary direction of fire is TRP 2 (center) until otherwise specified; the platoon is responsible for the front two quadrants.
- The left flank platoon's primary direction of fire is TRP 1 (left) until otherwise specified; the platoon is responsible for the left two friendly quadrants (overlapping with the center platoon).
- The right flank platoon's primary direction of fire is TRP 3 (right) until otherwise specified; the platoon is responsible for the right two friendly quadrants (overlapping with the center platoon).

D-49. The company SOP addresses the most critical requirements of fratricide prevention. The SOP directs subordinate leaders to inform the commander, adjacent elements, and subordinates whenever a friendly force is moving or preparing to move. One technique is to establish a standard weapons control status of WEAPONS TIGHT, which requires positive enemy identification prior to engagement. The SOP covers the means for identifying dismounted Infantry squads and other friendly dismounted elements. Techniques include using armbands, medical heat pads, or an infrared light source, as well as detonating a smoke grenade of a designated color at the appropriate time. (Refer to FM 3-21.10 for additional information.)

PREPARATION

D-50. Upon receipt of an order, weapons company units must ensure they are prepared for the mission. This includes personnel, equipment, vehicle, movement preparations. Once the decision is made on heavy weapons configurations for the vehicles they must be mounted for the mission.

VEHICLE LOAD CONSIDERATIONS

D-51. Load configurations for vehicles and trailers will vary between units. During mission planning and preparation leaders will need to plan for what equipment will be taken and where it will be carried. See FM 3-21.10 for additional information on vehicle load considerations. At a minimum leaders should consider—

- How much of what type of ammunition will be carried where?
- Will trailers be taken or left in a separate location?
- Where will non-mounted weapon systems be carried or stored?
- Where will any special equipment taken be stored?
- How will the nature and duration of the mission alter the standard load configuration?
- Will any vehicle modifications alter the load plan?

D-52. Vehicle commanders must ensure that any externally stowed items are secured from theft and do not constitute a fire hazard if the vehicle is attacked by an IED, rocket-propelled grenade, or other flammable device. External stowage should be minimized or modified to lessen the threat of vehicle fire and not restrict the view or movement of gunners or passengers providing security. All loose items stored inside the vehicle must be secured to prevent theft or becoming secondary missiles in the event of a mine or IED strike or a roll over. Commanders should consider stowing flammable items that are mission essential inside the vehicle behind armored portions of the vehicle, and securing nonmission essential and nonflammable items outside the vehicle. Other considerations are listed below:

- Use on-board ammunition storage containers such as 60-mm mortar ammunition cans. These hold several types of ammunition. This saves the crew a lot of time when they have to switch between ammunition for crew-served weapons.
- Carry complete spare wheel and tire assemblies rather than just spare tires. This reduces the time needed to change a flat, and will often allow a crew to repair a vehicle after a mine strike.
- Consider equipping every vehicle or every other vehicle with wheeled vehicle tow bars, so that vehicles can recover or tow each other. Tow bars are better than cables, since no driver is needed in the towed vehicle.
- Consider emplacing civilian or military fire extinguishers in fixed positions inside the vehicle. Normally, locate them to protect the crew rather than the vehicle. This helps ensure crew survivability. Carry additional loose fire extinguishers to fight vehicle fires.

D-53. Commanders should establish load plan SOP for sensitive items. They should account for ammunition and additional special equipment such as breach kits, demolitions, and first aid equipment. They should also account for any additional weapons.

Vehicle Weight, Observation, and Survivability

D-54. Protection is directly linked to mission success and must always be an important consideration in the planning and execution of missions that employ soft-skinned vehicles. The balance between the protection of vehicles and crews, observation, and the employment of weapons is critical. The additional weight of additional armor also places a strain on other vehicle components, such as the engine and the suspension. Normally, heavily armored vehicles, especially wheeled vehicles with extra armor such as the up-armored high mobility multipurpose wheeled vehicle, severely limit crew and passenger observation in restrictive and urban terrain. They can also limit weapons employment at close ranges. At times, insurgent and terrorist enemy forces target vehicles with poor security, because they seem easier to destroy and less likely to respond effectively. Commanders must analyze enemy trends and events in their AO before deciding on the appropriate levels of armor versus offensive capabilities, mission demands, and crew survivability. Other considerations might include—

- Can the vehicle suspension support additional armor and still carry the payload?
- Can the vehicle crew and passengers provide all-round security for themselves?
- Will additional armor affect vehicle mobility over rough terrain or in restrictive urban areas?
- Does the vehicle have sufficient power, acceleration, and speed?
- Can the vehicle crew and passengers quickly and safely mount or dismount? Can they do so under fire?

Movement and Maneuver

D-55. The commander uses tactical movement to position units on the battlefield and prepare them for contact. The process by which units transition from tactical movement to maneuver is actions on contact (see paragraph 2-170). The commander determines the combat formation and movement techniques used based on the mission variables of METT-TC and the likelihood of enemy contact. Platoons employ combat formations and movement techniques within a company movement. In addition to using combat formations and movement techniques, the company employs two stationary formations during temporary halts, the coil and the herringbone. Below is a brief description and diagram of each type formation and technique.

Combat Formations

D-56. The commander can use seven different combat formations depending on the mission variables of METT-TC: column, line, echelon (left or right), box, diamond, wedge, and vee. Terrain characteristics and visibility determine the actual arrangement and location of the unit's personnel and vehicles within a given formation. Mounted combat formations describe the specific locations of the company's elements in relation to each other. They are guides on how to arrange the unit for movements. Each formation aids control, security, and firepower to varying degrees. The following factors should be considered in determining the best formation to use:

- Mission.
- Enemy situation.

- Terrain.
- Weather and visibility conditions.
- Speed of movement desired.
- Degree of flexibility desired.
- Mission command.

Column Formation

D-57. The company uses the column (figure D-3), when moving fast, when moving through restricted terrain on a specific route, or when it does not expect enemy contact. Each platoon normally follows directly behind the platoon to its front. If the situation dictates, platoons can disperse laterally to enhance security.

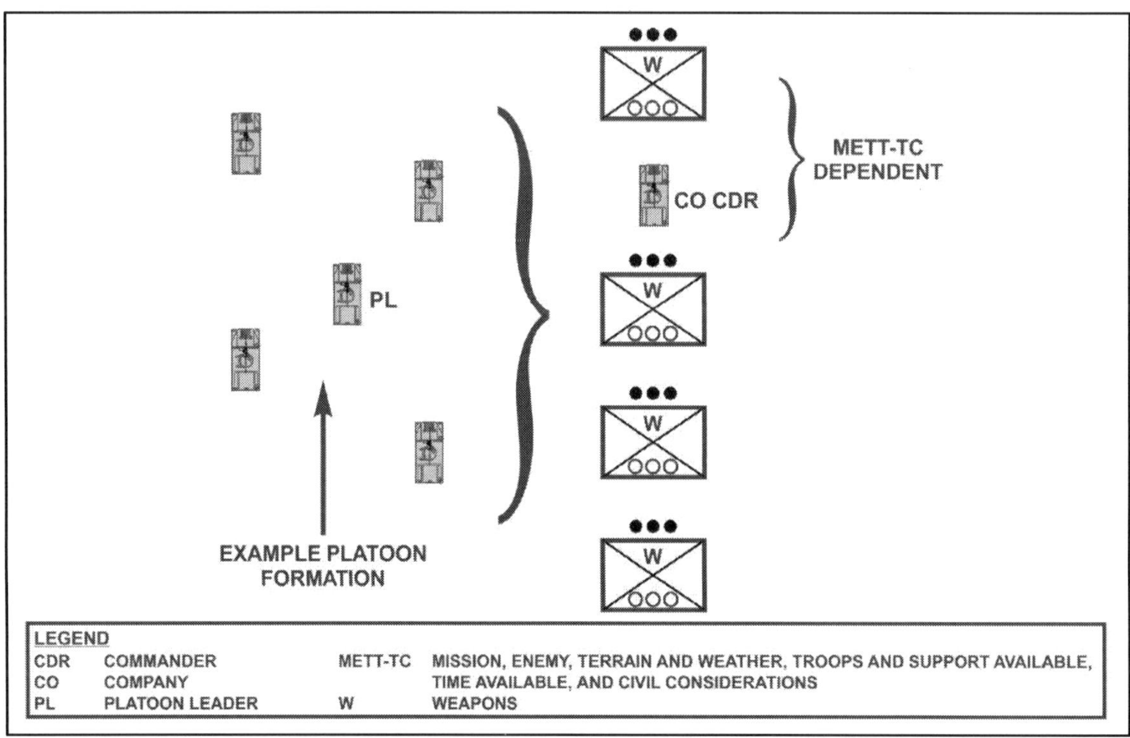

Figure D-3. Column formation

Wedge Formation

D-58. When the enemy situation seems unclear or when contact might occur, leaders often use the wedge formation (figure D-4). In the company wedge, the center platoon is located to the front of the formation, while the flank platoons are to the rear of and outside the center platoon.

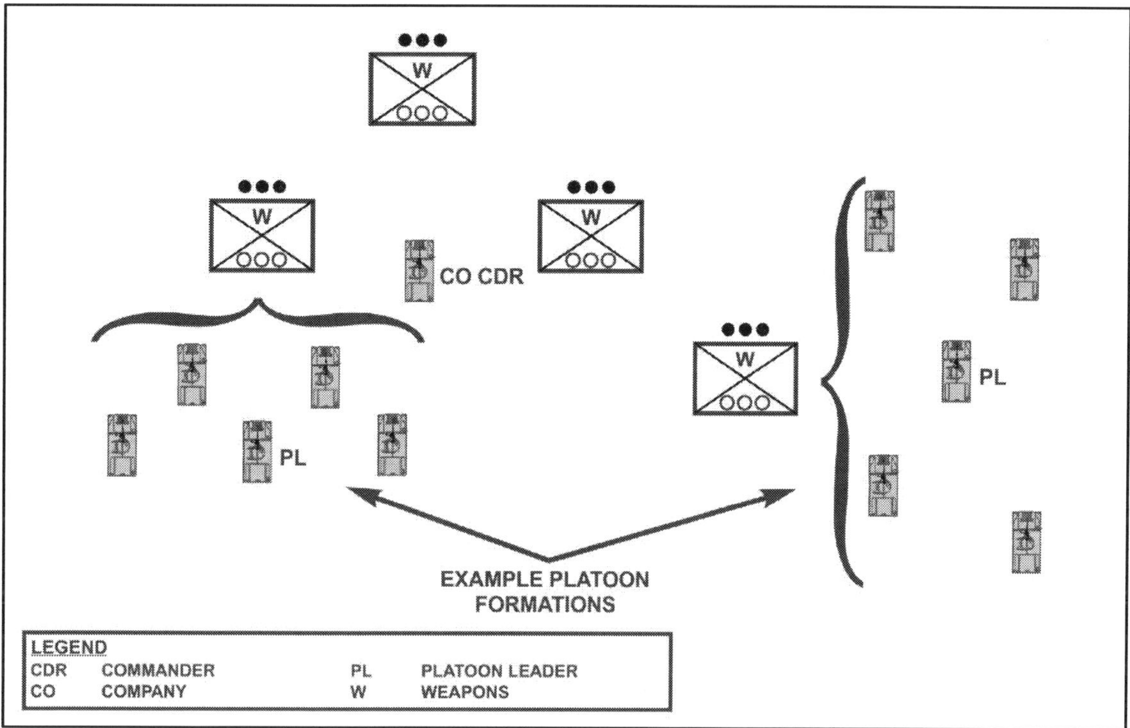

Figure D-4. Wedge formation

Line Formation

D-59. When crossing open areas or occupying a support-by-fire position, the company may use the line formation (figure D-5). Normally, the line formation is used when no terrain remains between it and the enemy, when the enemy's antitank weapons have been suppressed, or when the company is vulnerable to artillery fire and must move fast.

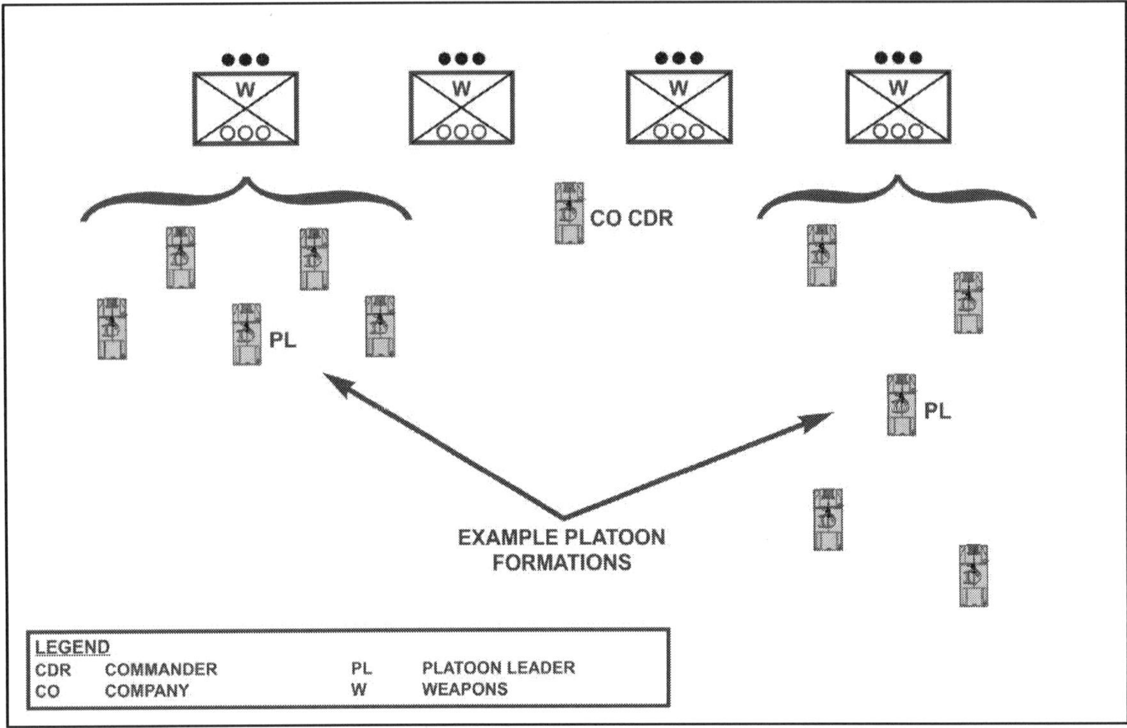

Figure D-5. Line formation

Vee Formation

D-60. The vee formation (figure D-6) is used when enemy contact is possible. In the company vee, the center platoon is located in the rear of the formation, while the flank platoons are to the front of and outside the center platoon.

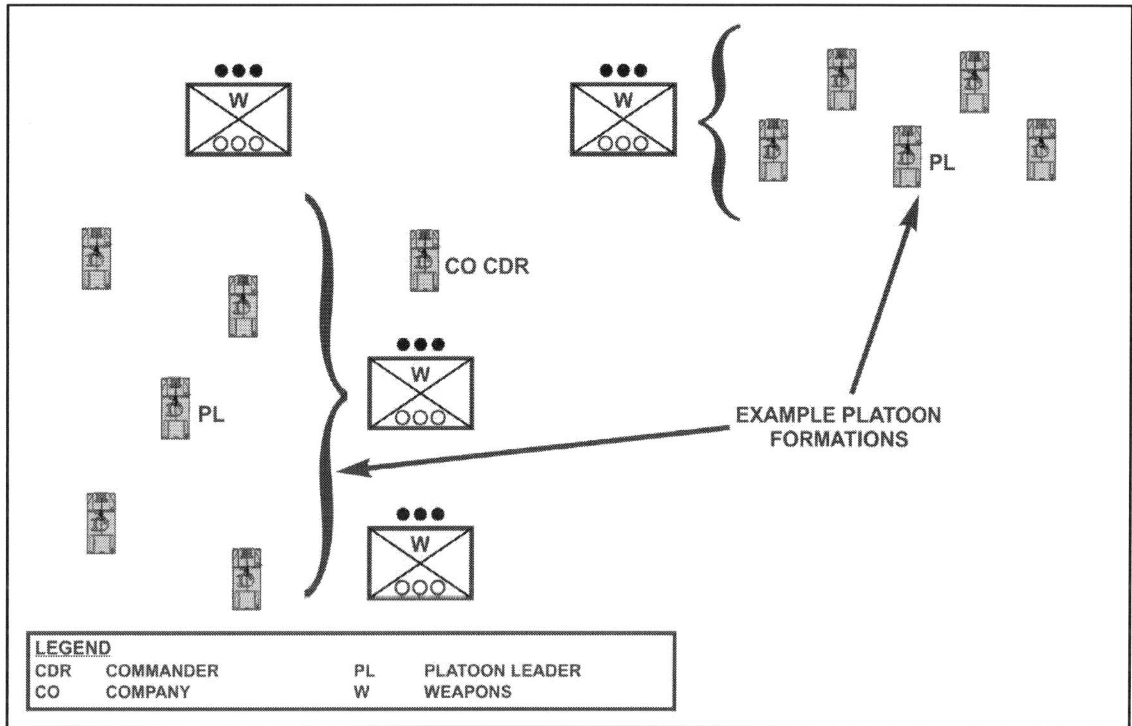

Figure D-6. Vee formation

Diamond Formation

D-61. The company uses the diamond formation (figure D-7) when they want to maintain all around security, and enemy contact is not expected. The company leads with a platoon with two platoons to the flanks and the fourth platoon in the rear.

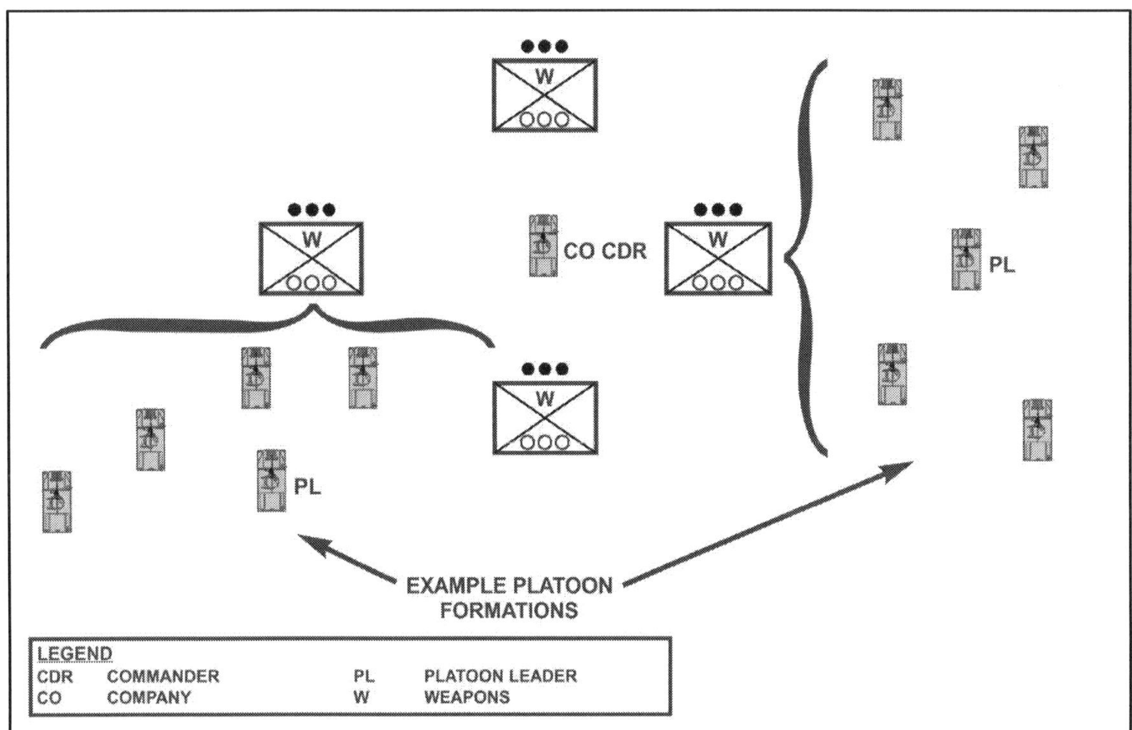

Figure D-7. Diamond formation

Box Formation

D-62. The box formation (figure D-8) arranges the unit with two forward and two trail platoons. A weapons company with only three platoons would have to adopt a vee or another formation. It is often used when executing an approach march, an exploitation, or a pursuit when the commander has only general knowledge about the enemy.

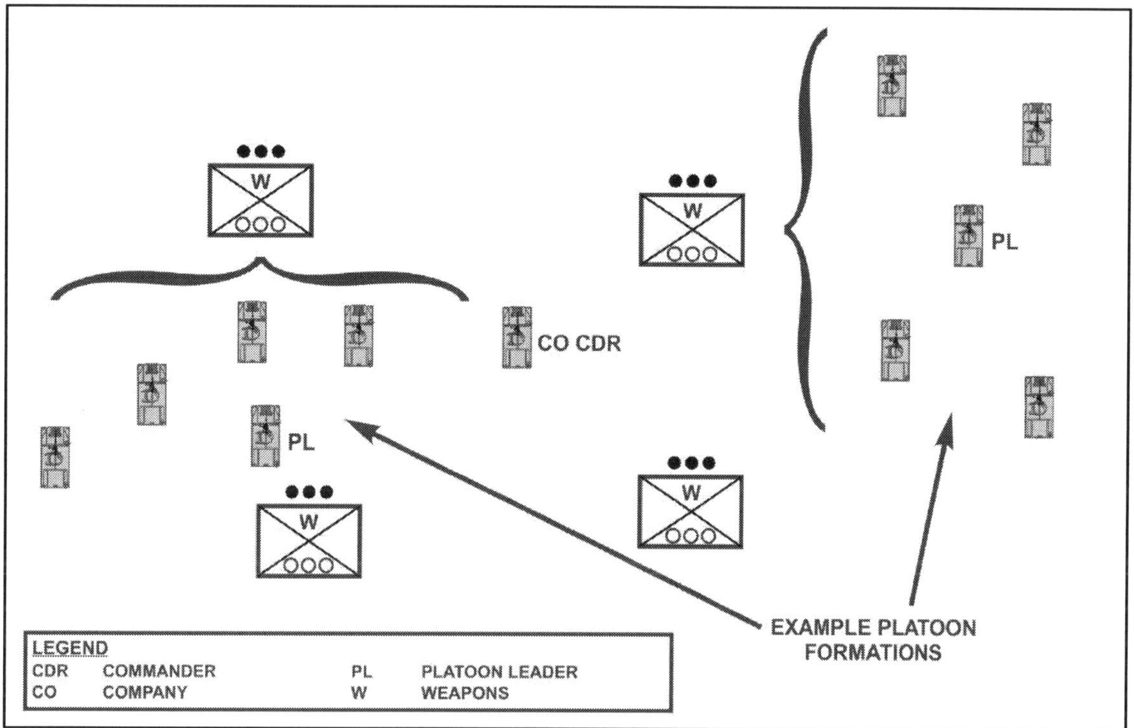

Figure D-8. Box formation

Echelon Formation

D-63. The echelon formation (figure D-9) is used when the company wants to maintain security and/or observation of one flank and enemy contact is not likely. The company echelon formation (either echelon left or echelon right) has the lead platoon positioned farthest from the echeloned flank, with each subsequent platoon located to the rear of and outside the platoon to its front.

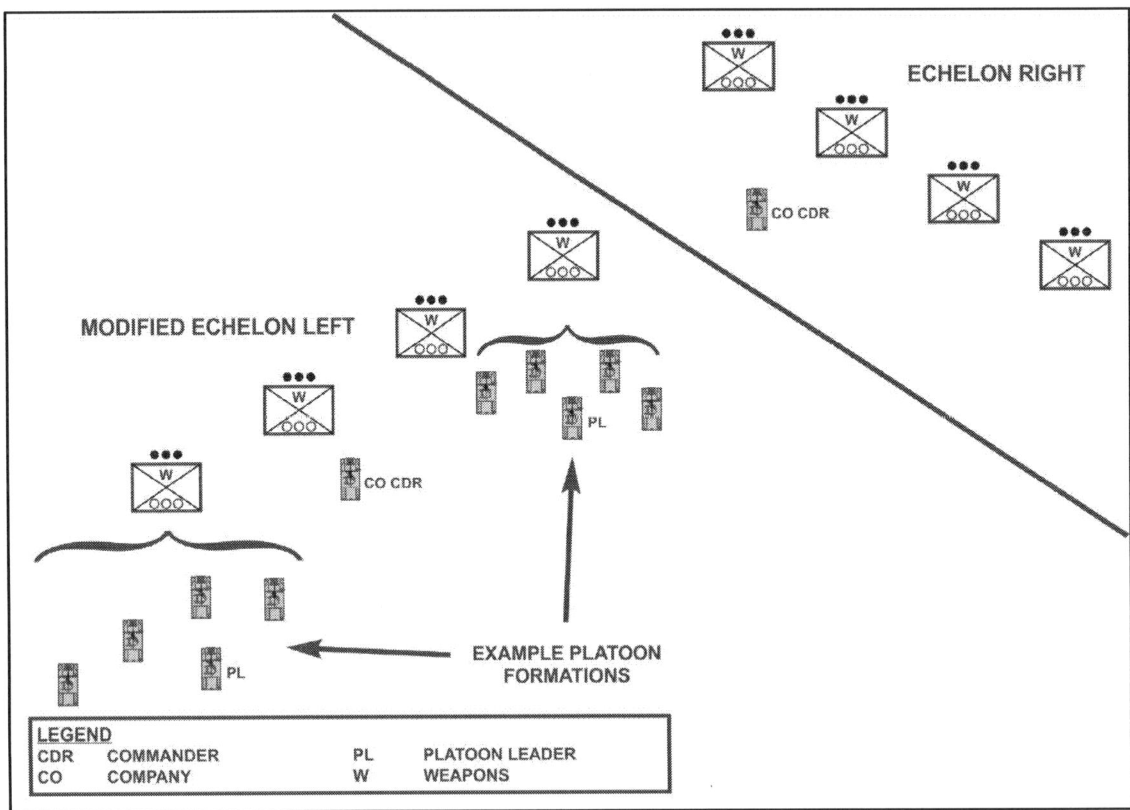

Figure D-9. Echelon left and right formation

Stationary Formations

D-64. The commander employs stationary formations when halted. The coil and herringbone formations are provided as examples below.

Coil

D-65. The coil formation (figure D-10) is used to provide all-round security and observation when the company is stationary. It is also useful for tactical refueling, resupply, and issuing orders. Security is posted to include air guards and dismounted rifleman.

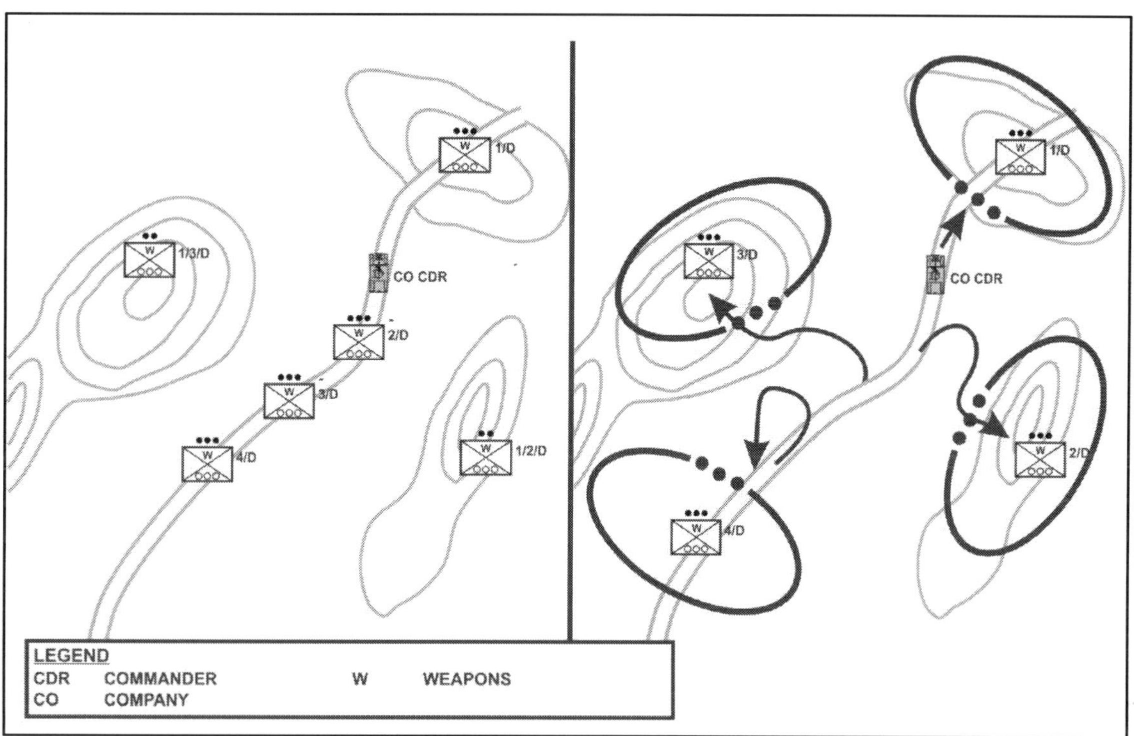

Figure D-10. Coil formation during (left) and after (right)

Herringbone

D-66. The company uses the herringbone formation (figure D-11) to disperse when traveling in column formation. The commander can use this formation during air attacks or when the company must stop during movement. When order, units disperse and move to concealed and covered positions off a road or from an open area and set up all-round security without detailed instructions. The commander repositions the vehicles as needed to take advantage of dispersion, concealment, cover, and fields of fire.

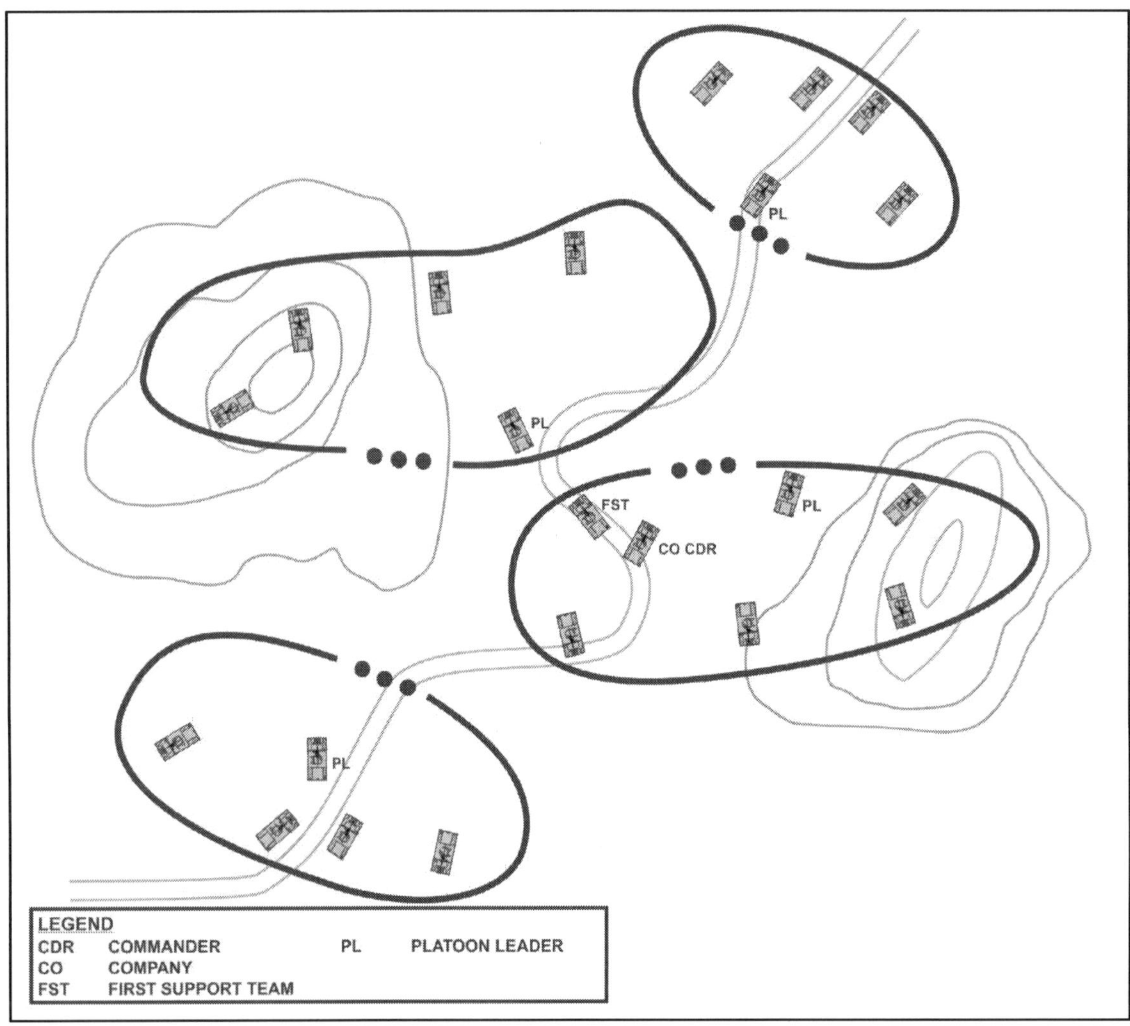

Figure D-11. Herringbone formation

Formation Selection and Comparion

D-67. The commander selects the formation that provides the proper security, fires, control, and speed. Table D-1 compares the four most commonly used combat and temporary halt formations by the weapons company.

Table D-1. Comparison of combat formations

Formation	Security	Fires	Control	Speed
Column	Good dispersion. Limited all-round security.	Limited to front and rear. Excellent to the flanks.	Easy to control. Flexible formation.	Fast.
Line	Excellent to the front. Poor to the flank and rear.	Excellent to the front. Poor to the flank and rear.	Difficult to control. Inflexible formation.	Slow.
Wedge	Good all-round security.	Excellent to the front and good to flanks.	Easy to control but more difficult than the column. Flexible formation.	Slower than the column.
Echelon	Good to the echeloned flank and front.	Excellent to the echeloned flank and front.	Difficult to control.	Slow.
Coil	Excellent all around security.	Excellent to front rear and flanks.	Easy to control.	Used while stationary.
Herringbone	Great dispersion. Good all-round security	Good to front rear and flanks	Easy to implement. Control more difficult after dispersion.	Used to disperse to cover and concealment while traveling.

Movement Techniques

D-68. The company commander selects from the three mounted movement techniques (traveling, traveling overwatch, and bounding overwatch) largely based on the likelihood of enemy contact and other METT-TC considerations. As the probability of enemy contact increases, the commander adjusts the movement technique to provide greater security. For example, if an enemy update received from higher headquarters states that the enemy has moved much closer to the company than the commander anticipated, the commander immediately switches the technique from traveling overwatch to bounding overwatch. Other factors that may influence the commanders decision include—

- The type of contact expected.
- The availability of an overwatch element.
- The terrain over which the moving element will pass.
- The balance of speed and security required during movement.

Traveling

D-69. Continuous movement characterizes the traveling technique by all company elements. The technique is best suited for situations in which enemy contact is unlikely and speed is important. When the commander analyzes the latest information on the enemy and determines that contact with the enemy is unlikely, often the traveling techniques will be used for movement. Figure D-12 shows an example of the traveling technique for the Infantry weapons company.

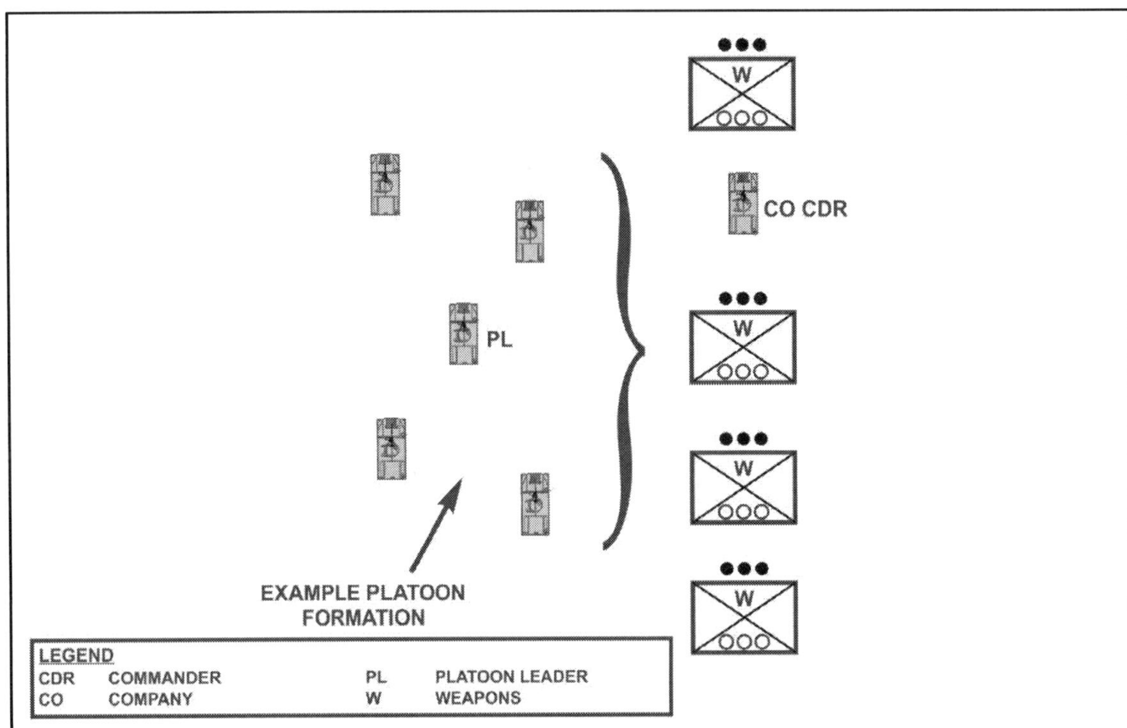

Figure D-12. Traveling technique

Traveling Overwatch

D-70. Traveling overwatch is an extended form of traveling that provides additional security when speed is desirable but contact is possible. The lead element moves continuously and provides security forward of the main body. The commander tracks the movement of forward security elements. Platoons move at various speeds and may halt periodically to overwatch movement of the lead platoon(s). Dispersion between the lead platoons must be based on the second platoon's ability to see the lead platoon and to provide immediate suppressive fires in case the lead platoon is engaged. The intent is to maintain depth, provide flexibility, and maintain the ability to maneuver even if contact occurs. However, if contact is made, ideally a unit should be moving in bounding overwatch rather than traveling overwatch. Figure D-13 shows a traveling overwatch technique for an Infantry weapons company.

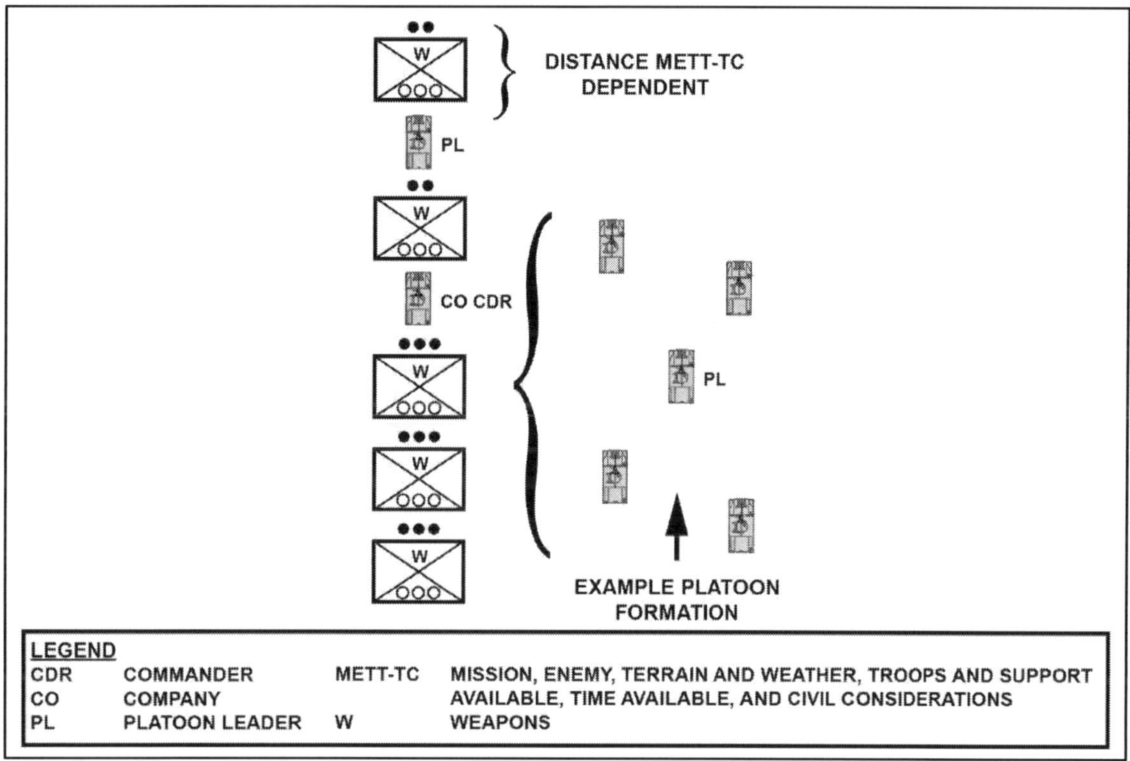

Figure D-13. Traveling overwatch technique

Bounding Overwatch

D-71. Bounding overwatch is used when contact is expected. It is the most secure, but slowest, movement technique. The purpose of bounding overwatch is to deploy prior to contact, giving the unit the ability to protect a bounding element by immediately suppressing an enemy force. In all types of bounding, the overwatch element is assigned sectors to scan while the bounding element uses terrain to achieve cover and concealment. The bounding element avoids masking the fires of the overwatch element; it never bounds beyond the range at which the overwatch element can effectively suppress likely or suspected enemy positions. Ideally, the overwatch element keeps the bounding element in sight. Before bounding, the leader shows the bounding element the location of the next overwatch position. Once the bounding element reaches its overwatch position, it signals READY by voice or visual means to the element that overwatched its bound. The company can employ either of two bounding methods: alternate or successive.

Alternate Bounds

D-72. Covered by the rear element, the lead element moves forward, halts, and assumes overwatch positions. The rear element advances past the lead element and takes up overwatch positions. This sequence continues as necessary with only one element moving at a time. This method is usually more rapid than successive bounds. Figure D-14, shows a bounding overwatch (alternate bounds) technique for an Infantry weapons company.

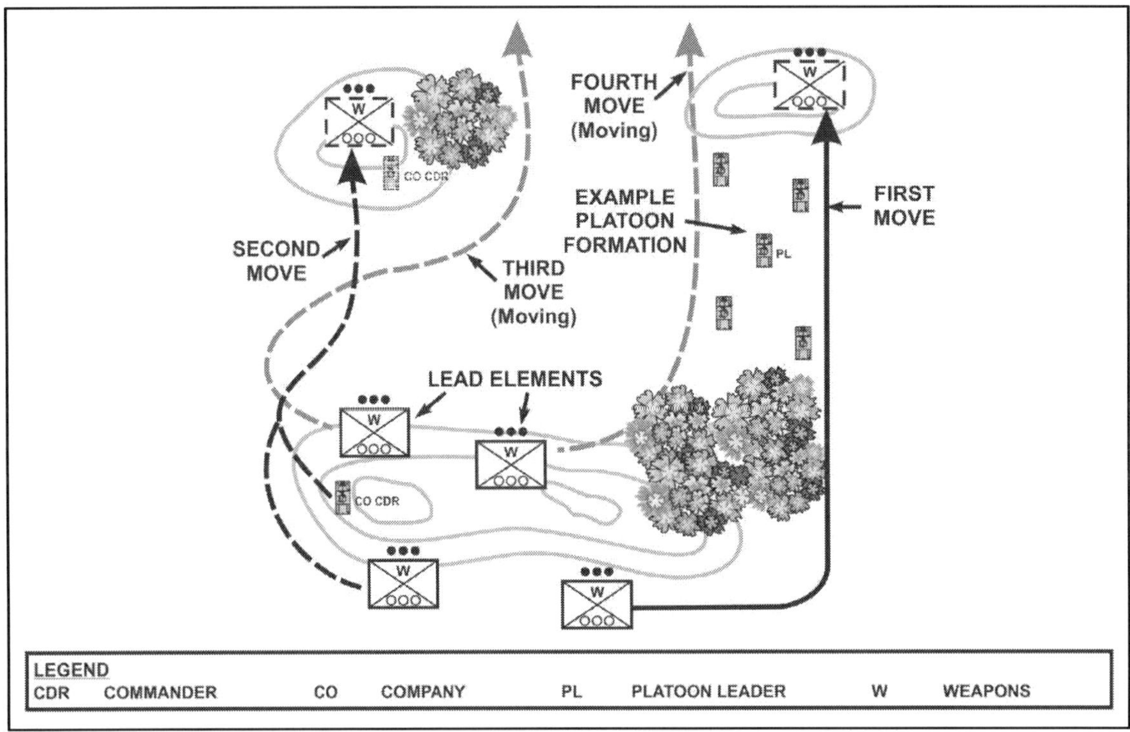

Figure D-14. Bounding overwatch technique (alternate bounds)

Successive Bounds

D-73. In the successive bounding method the lead element, covered by the rear element, advances and takes up overwatch positions. The rear element then advances to an overwatch position roughly abreast of the lead element and halts. The lead element then moves to the next position, and so on. Only one element moves at a time, and the rear element avoids advancing beyond the lead element. This method is easier to control and more secure than the alternate bounding method, but it is slower. Figure D-15 shows a bounding overwatch (successive bounds) technique for an Infantry weapons company.

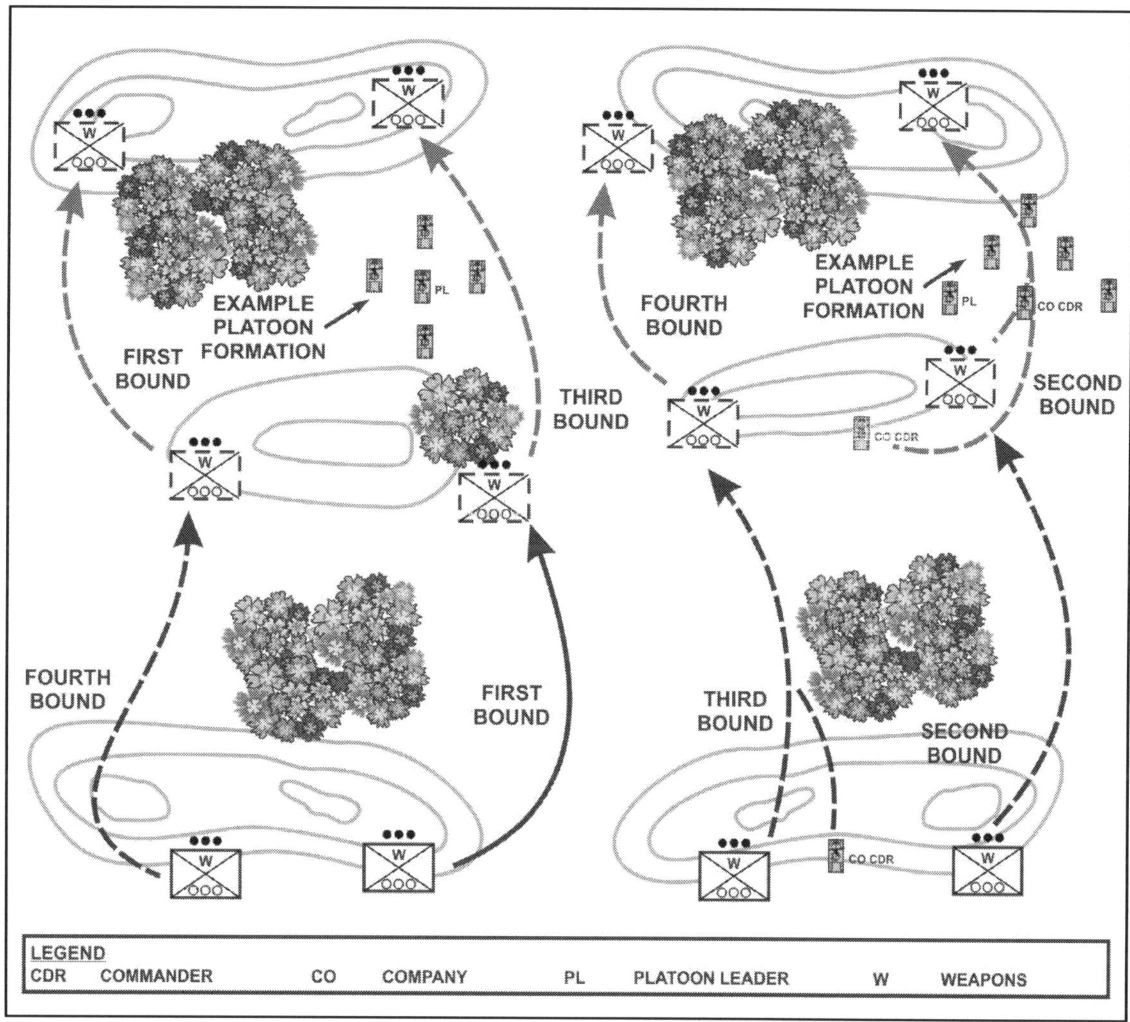

Figure D-15. Bounding overwatch technique (successive bounds)

EXECUTION

D-74. The weapons company commander protects the company during movement and maneuver by ensuring the use of proper combat formations and/or movement techniques. Maneuver should be as rapid as the terrain, mobility of the force (to include dismounted forces), and enemy situation permits. The ability to gain and maintain the initiative often depends on movement being undetected by the enemy. The Infantry weapons company depends heavily upon terrain, mobility and standoff for protection from enemy fire.

EMPLOYMENT CONSIDERATIONS

D-75. The Infantry weapons company commander's mission analysis helps decide how to move efficiently, and with the greatest security. The commander ensures the unit is moving in a way that supports a rapid transition to maneuver. Once in contact with the enemy, squads and platoons execute the appropriate actions on contact, and leaders begin to maneuver their units. The commander uses countermeasures during movement such as suppressive fires, direct and indirect, to degrade the enemy's ability to observe fires upon the unit by use of terrain or obscurants.

D-76. All echelons reconnoiter. The enemy situation and the available planning time may limit the unit's reconnaissance, but leaders at every level seek information about the terrain and enemy. If sufficient information is still lacking, an effective technique is to send a reconnaissance element forward of the lead platoon. Even if this unit is only 15 minutes ahead of the company, it can still provide valuable information and reaction time for the commander.

D-77. Unlike the Infantry rifle company with the ability to move across almost any terrain, the Infantry weapons company is restricted to areas trafficable by vehicles. This restriction is largely counteracted by the speed at which the weapons company can move. To the greatest extent possible, the company moves on covered and concealed routes. Moving in limited visibility may provide better concealment, and the enemy might be less alert during these periods. Leaders should plan to avoid identified danger areas or kill zones such as large open areas surrounded by covered or concealed areas.

D-78. The advantages to moving the company by sections or platoons include—

- Faster movement. The company's faster movement can mean less exposure time in some instances.
- Better security. A smaller unit is less likely detected because it requires less cover and concealment.
- More dispersion. The dispersion gained by moving the company by sections and platoons makes it more difficult for the enemy to concentrate fires against the company, especially indirect fires, CAS, and chemical agents. Subordinate units also gain room to maneuver.
- Better operations security. It is harder for the enemy to determine what the friendly force is doing with only isolated platoon, section or individual (least preferred method) vehicle spot reports.

D-79. When moving by sections and platoons, the commander should also consider the following disadvantages:

- Numerous linkups are required to regroup the company.
- May take longer to mass combat power to support a hasty attack or disengage in the event of enemy contact.

D-80. Security is critical during company and subordinate unit movements and halts. The company commander achieves security by applying the following:

- Use the appropriate formations and movement techniques for the conditions.
- Move as fast as the situation allows to degrade the enemy's ability to detect movements and the effectiveness of enemy fires once detected.
- Orient main weapons systems and vehicle crewmembers to cover the front, flanks, and rear during movements to establish 360-degree security.
- Assigned sectors of fire to include aerial security.
- During halts, dismount as many Soldiers as possible for security. The gunner maintains their position behind the main weapon while the leader, assistant gunner, and driver dismount if possible. The driver stays within arm's reach of the vehicle.
- Enforce noise and light discipline.
- Enforce camouflage discipline (Soldiers and their equipment).

D-81. The battalion, in coordination with the company commander, assigns graphic control measures to synchronize company movement into the battalion's movement or scheme of maneuver. The company commander may establish additional company control measures as required. These may include boundaries, routes, command posts, start points, checkpoints, release points, target reference points, and objectives on known or likely enemy positions to control direct fires. The company commander updates graphic control measures and ensures they are easily recognizable during ground movement.

D-82. Prior reconnaissance and surveillance (R&S) activities enable control during movement. These actions provide the commander with a better idea of where movement is more difficult and where graphic control measures are needed. Elements from the company may perform R&S, however, the battalion scout platoon is more likely to conduct these activities. Either way this information is provided to the battalion intelligence staff officer (S-2) for dissemination to other organizations within the battalion and higher headquarters. The use of unmanned aircraft systems (see FM 3-21-10) can be a valuable surveillance asset. The Improved Target Acquisition System target acquisition system and Javelin's command launch unit are also valuable systems that can aid in R&S.

D-83. Guides who have already seen the terrain are the best way to provide control. When guides are not available for the entire movement, they should reconnoiter the difficult areas along the route and guide the company through those areas. Examples: complex road interchanges, obstacle lanes, and another unit's area of operation.

D-84. The measures listed above are equally useful in limited visibility conditions. However, to aid movement in during limited visibility, the use of night vision devices is a key enabler to control movement. Subordinate leaders may consider closing vehicle intervals in the formation while still maintaining the most dispersion possible at all times and reducing the rate of march or movement to improve control.

INTEGRATION OF DIRECT FIRES

D-85. The integration of direct fires into the battalion's scheme of maneuver is fundamental to success in close combat. While individual elements may be well rehearsed in their particular tasks relating to the mission, effective direct fire engagements do not happen without careful planning, coordination and control. The integration of direct fires in a particular engagement area must be synchronized to enhance combat effectiveness and protect friendly units. Effective direct fires are the unique contribution of maneuver forces to the combined arms team. Although the weapons company is a maneuver unit, a large portion of the weapons company mission is directed toward the direct fire support of other maneuvering friendly forces. The weapons company commander should be well aware of all direct fire control measures and how the direct fires of their subordinate elements fit into the battalion's direct fire plan. The following paragraphs address direct fires considerations specific to the Infantry weapons company during execution. (Refer to FM 3-21.10 for a detail discussion of direct fire planning and control.)

Principles of Direct Fire Control

D-86. When executing direct fires, the weapons company commander and subordinate leaders apply several fundamental principles. The purpose of these principles is not to restrict the actions of subordinates, but to help the company accomplish the primary goal of any direct fire engagement: to eliminate the enemy by acquiring first and shooting first. Applied correctly, these principles give subordinates the freedom to respond rapidly upon acquisition of the enemy. This discussion focuses on the following principles:

- Mass the effects of fire.
- Destroy the greatest threat first.
- Avoid target overkill.
- Employ the best weapon for the target.
- Minimize friendly exposure and fratricide avoidance.
- Plan for limited visibility conditions.
- Develop contingencies.

Mass the Effects of Fire

D-87. The Infantry weapons company must mass its direct fires to achieve decisive results. Whether a battalion or company mission, or whether in a primary or supporting role, the principle of massing fires remains the same. Massing entails focusing direct fires at critical points and distributing the effects. Random application of fires is unlikely to have a decisive result. For example, concentrating the company's fires at a single target may ensure its destruction or suppression; however, that fire control option will fail to achieve the decisive result on the remainder of the enemy formation or position. The weapons company commander will often have assault platoons in a supporting role such as in a support by fire position. The commander integrates fires with those of the maneuvering unit to achieve a combined mass effect of direct fires.

Note. When assault platoons operate separately from the weapons company, for example if attached to a rifle company, the platoon leader integrates assault platoon fires with those of the maneuver element.

D-88. The weapons company masses its fires by positioning elements so that more than one element can fire into an engagement area. Using control measures, such as target reference points, the commander can distribute fires and, in turn, platoon leaders distribute the fires within their assigned area. Using engagement criteria (see paragraph 3-171) and control measures such as engagement lines, the commander destroys the enemy with sudden, distributed, and simultaneous fires from multiple elements. Careful analyses of time-distance factors between positions can also allow leaders to displace units in time to reinforce a threatened area and affect the outcome of the engagement.

Destroy Greatest Threat First

D-89. The order in which the weapons company engages enemy forces is in direct relation to the danger these forces present. The threat posed by the enemy depends on weapons, ranges, and positioning. Presented with multiple targets, subordinate units must initially concentrate direct fires to destroy the greatest threat, and then distribute fires over the remainder of the enemy force. The weapons company with long-range direct fire capabilities will often be the first to engage an approaching enemy unit and will concentrate on attacking the greatest threat first such as tanks and other heavy weapons vehicles.

D-90. The commander establishes a priority for targets. This may be part of the tactical standing operating procedure or revised based on the mission and threat. For example, bridging equipment may become a priority target when the Infantry battalion is defending a river crossing.

Avoid Target Overkill

D-91. Use only the amount of fire required to achieve necessary results. Target overkill wastes ammunition and is not tactically sound. To the other extreme, the company cannot have every weapon engage a different target because the requirement to destroy the greatest threats first remains paramount. To help avoid target overkill, the commander and subordinate leaders use robust fire control measures (see paragraph D-45). If the target is not burning or showing obvious signs of destruction however, it is often difficult to determine if a target is destroyed.

Employ the Best Weapon for a Specific Target

D-92. Using the appropriate weapon for the target increases the probability of rapid enemy destruction or suppression; at the same time, it conserves ammunition. The Infantry weapons company has a variety of weapons with which to engage the enemy. Target type, range, and exposure are key factors in determining the weapon and ammunition employed, as are weapons and ammunition availability and desired target effects. Careful planning and analysis enables the weapons company commander to select the best weapons to mount on the vehicles and to array forces based on the terrain and enemy, and to achieve the desired effects from all direct fire engagements.

D-93. A given weapons system can have its own priority of fires and engage different targets in sequence. Although the specific mission and threat dictates the use of specific weapons, Javelin close combat missile systems, grenade machine guns, and heavy machine guns are most often employed using the guidelines in the following paragraphs:

D-94. Javelin close combat missile systems provide long-range direct fires capable of destroying armored vehicles and fortifications. Javelins have a limited high explosive effect on Infantry. Their primary disadvantages are a relatively slow flight time for the missile and the number of missiles carried are limited. Javelins have a fire-and-forget missile with a top attack capability.

D-95. Grenade machine guns and heavy machine guns are very effective against dismounted Infantry and lightly armored vehicles. They can provide area suppression or reconnaissance by fire. Disadvantages include limited ammunition and effectiveness against more heavily armored vehicles.

Minimize Friendly Exposure and Avoid Fratricide

D-96. Subordinate units of the weapons company increase their survivability by exposing themselves to the enemy only to the extent necessary to engage the enemy effectively. Natural or manmade defilade provides the best cover from antitank guided missiles and other large caliber direct fire munitions. Although armored, weapons company vehicles are vulnerable to direct fires from weapons larger than small arms and to indirect fires. The weapons company commander and subordinate leaders select positions that minimize exposure by constantly seeking effective available cover, trying to engage the enemy from the flank, remaining dispersed, displacing to and firing from multiple positions, and limiting engagement times.

D-97. The weapons company commander works proactively with subordinate leaders to reduce the risk of fratricide and noncombatant casualties. The commander plans and use numerous tools to assist in this effort: identification training for combat vehicles and aircraft, the weapons safety posture, the weapons control status, and recognition markings. Knowledge and employment of applicable rules of engagement are the primary means of preventing noncombatant casualties. Digital tracking systems and control measures also decrease the chance of fratricide.

Plan for Limited Visibility Conditions

D-98. The Infantry weapons company is uniquely equipped to adapt to limited visibility conditions. The night of the TOW missile system not only allows for engagements during limited visibility conditions, but also may be used as a night observation device for security. At night, limited visibility fire control systems enable the weapons company to engage enemy forces at nearly the same ranges that are applicable during the day. However, obscurants such as dense fog, heavy rain, heavy smoke, and blowing sand can reduce the capabilities of thermal and infrared systems to acquire targets. Although a decrease in acquisition capabilities has little effect on area fire, point target engagements are likely to occur at decreased ranges. The commander adjusts firing positions, whether offensive or defensive, closer to the area or point to focus fires. The use of visual or infrared illumination when there is insufficient ambient light for passive light intensification devices can enable target acquisition.

Develop Contingencies

D-99. The commander and subordinate leaders initially develop plans based on their units' maximum capabilities; they make backup plans for implementation in the event of casualties, weapon damage or failure. While the commander and subordinate leaders cannot anticipate or plan for every situation, they develop plans for what they view as the most probable occurrences. Building redundancy into these plans, such as having two systems observe the same sector, is an invaluable asset when the situation (and the number of available systems) permits. Designating alternate sectors of fire and supplementary firing positions provides a means of shifting fires if adjacent elements become unable to fire.

Fire Control Process

D-100. To bring direct fires against an enemy force successfully, the commander and subordinate leaders continuously apply the four steps of the fire control process. At the heart of this process are two critical actions: rapid, accurate target acquisition and the massing of fires to achieve decisive results on the target. Target acquisition consists of detecting, identifying, and locating the enemy in sufficient detail to permit the effective employment of weapon systems. Massing entails focusing fires at critical points and then distributing the fires for optimum effect. The four steps are—

- Identify probable enemy locations and determine the enemy scheme of maneuver.
- Determine where and how to mass (focus and distribute) fires.
- Orient forces to speed target acquisition.
- Shift fires to refocus or redistribute.

D-101. Planning and coordination at the battalion and between the weapons and rifle company commanders is required for an effective operation. For example, if a rifle company is attacking an objective it is vitally important for the weapons company commander in a support by fire role to understand where and when the rifle company commander plans to focus and mass direct fires. The weapons company commander must be aware of all control measures and know when and where to refocus or redistribute fires in synchronization with the rifle company's operation. For a complete discussion of the four steps in the fire control process, refer to FM 3-21.10.

Direct Fire Control

D-102. Acquiring the enemy is a precursor to direct fire engagement. The weapons company commander and subordinate leaders expect the enemy to use covered and concealed routes effectively when attacking and to make best use of flanking and concealed positions in the defense. As a result, the weapons company may not have the luxury of a fully exposed enemy that can easily be see. The acquisition of the enemy often depends on visual recognition of very subtle indicators such as exposed antennas, reflections from the vision blocks of enemy vehicles, small dust clouds, or smoke from vehicle engines or antitank guided missiles or tank fires. Because of the difficulty of target acquisition, the commander develops surveillance plans to assist the company in acquiring the enemy.

Fire Control Measures

D-103. Fire control measures are the means by which leaders control direct fires. Application of these concepts, procedures, and techniques help the unit acquire the enemy, focus fires on the enemy, distribute the results of the fires, and prevent fratricide. At the same time, no single measure is enough to control fires effectively. Weapons company fire control measures are effective only if the entire unit has a common understanding of what they mean and how to employ them. Table D-2 lists terrain based and threat-based fire control measures.

Table D-2. Common fire control measures

TERRAIN-BASED FIRE CONTROL MEASURES	*THREAT-BASED FIRE CONTROL MEASURES*
Target reference point	Rules of engagement
Engagement area	Weapons ready posture
Sector of fire	Weapons safety posture
Direction of fire	Weapons control status
Terrain-based quadrant	Engagement priorities
Friendly-based quadrant	Trigger
Maximum engagement line; restrictive fire line; final protective line	Engagement techniques; fire patterns; target array

D-104. The commander and subordinate leaders use terrain-based fire control measures to focus and control fires on a particular point, line, or area rather than on a specific enemy element. They use threat-based fire control measures to focus and control direct fires by directing the unit to engage a specific enemy element rather than to fire on a point or area. (Refer to FM 3-21.10 for a detailed discussion of each of these measures.)

Fire Commands

D-105. Fire commands are oral orders issued by leaders to focus and distribute fires as required achieving decisive effects against the enemy. Fire commands allow leaders to rapidly, and concisely articulate firing instructions using a standard format. (Refer to TC 3-20.31-4 and FM 3-21.10 for additional information.) Unit fire commands include these elements:

- Alert.
- Weapon or ammunition (optional).
- Target description.
- Direction.
- Range (optional).
- Method.
- Control (optional).
- Execution.
- Termination.

INFANTRY WEAPONS COMPANY RAID, ILLUSTRATION

D-106. This section introduces a fictional raid scenario illustrating one of many ways to employ the Infantry weapons company as a combined arms team under the command of the Infantry weapons company commander. As in all raids, success is based on accurate, timely, and detailed planning, preparation, execution, and assessment. The key elements in determining the level of detail and the opportunities for rehearsal before mission execution are time, operations security, and military deception requirements. During the raid, enemy forces are overcome in a violently executed surprise attack using all available firepower based on the operational environment for shock effect. The design for this company level raid exploits the unique organic capabilities of the Infantry weapons company within the Infantry battalion to rapidly maneuver and maintain long-range communications during the conduct of operations. The Infantry rifle company when similarly organized and equipped, vehicle mounted for example is equally capable of performing this operation. (See chapter 2, paragraphs 2-440 to 2-245, for general information on the conduct of a raid.)

OPERATIONAL OVERVIEW

D-107. In this scenario, used for discussion purposes, the Infantry weapons company conducts a raid to capture enemy personnel, documents, and electronic storage media. The operational environment within this scenario is one dominated by stability. The operation is tailored to the situation within the operational environment and to the number of targets within the weapons company's AO. The weapons company can be task organized, with or without augmentation to raid a single objective or multiple objectives simultaneously or sequentially within its AO. When augmented with joint, interagency, and multinational partners (civilian and military) and host nation partners, trusting relationships must be maintain though there will always be an underlying level of operations security (see paragraph 4-81 for additional information) concerns when working with other partners. For this reason, the commander can decide, for example, to withhold specific operation details from certain partners in order to maintain surprise when the situation warrants. Within this scenario, the weapons company is organized with joint and interagency partners (forensics exploitation and analysis enablers) to conduct a raid on a single objective.

PLANNING CONSIDERATIONS AND PREPARATIONS ACTIVITIES

D-108. Given the inherent complex and uncertain nature of operations dominated by stability, this raid differs from raids conducted in operations dominated by offense and defense (see FM 3-21.10 for information on operations dominated by offense and defense) since the requirement for minimizing collateral damage can be a significant factor. The company commander, supported by the battalion staff, carefully assesses the negative effects of violence on the populace and strictly adheres to the rules of engagement. Operations in support of stability tasks, in general, reflect and promote the host-nation government's authority and legitimacy, thus undermining enemy attempts to influence the narrative and to establish an alternative authority. Planning considerations and preparations activities specific to this raid scenario are address below by warfighting function.

Mission Command

D-109. The commander establishes control measures to allow for maximum decentralized actions and small-unit initiative. Control measures facilitate rapid consolidation and concentration of combat power to seize the objective and accomplishes its assigned task in the objective area before the enemy has time to react. The minimum control measures for the raid are an area of operation, objectives, checkpoints, phase lines, limits of advance, and contact points. The use of TRPs facilitates responsive fire support after contact is made with the enemy. The commander uses objectives and checkpoints to orient forces and guide the movement of subordinate elements. The commander uses other control measures as needed such as restrictive fire lines and marking systems.

D-110. Redundancy in the communications plan is essential to enable the ability to communicate throughout the operation. This is especially true in areas where communications may be degrade due to enemy actions or due to the complexity of the environment (specifically terrain and weather). To ensure redundancy, the commander implements a solid PACE plan (see appendix B); for example, primary means tactical satellite, alternate means fires net (via aerial or ground platform), contingency means digital message, emergency means host nation phone or satellite phone.

D-111. When developing the communications plan, the commander can utilize an execution checklist (commonly referred to as an X check) to develop a COP and to facilitate battle tracking of the operation. Another method to facilitate battle tracking and to decrease the length of transmissions is the use of an operation schedule (commonly referred to as an OPSKED) identifying a sequential list of events designated by numbers. An operation schedule differs from a brevity code in that each number is a cue for several events, even if the number is often a report as opposed to an order. For example, platoon 2 reaches their support position and reports 101; the company FSO, when assigned, automatically calls for suppressive fires, and platoon 1 automatically begins movement to the assault position. Operation schedules can be used separately or in conjunction with execution checklist and matrixes.

D-112. Within this scenario, key radio calls include: leaving the assembly area (may include a slant report [known as a SLANTREP]), arriving at the vehicle drop off (VDO), security set, assault initiated, objective cleared, site exploitation (SE) and tactical questioning (TQ) started and complete, number enemy on the objective (may include enemy and noncombatant number of men, woman, and children) and exfiltration. A slant report is used to provide accurate and routine information regarding the status of critical personnel and equipment necessary for the unit's operation success, submitted when necessary or as directed. The commander designates the information to report during planning or by unit SOPs.

D-113. The commander uses the military grid reference system (commonly referred to as MGRS) and/or graphic control measures to aid participants in maintaining a common operational picture. These aids identify and help communicate an event in time and space and facilitate battle tracking during the operation. Military grid reference system and graphic control measures are disseminated between the ground forces and air assets, and fire support assets and cells to maintain a common operational picture during the mission.

Movement and Maneuver

D-114. Speed and surprise are key considerations when planning and preparing for the raid. The sudden and unexpected delivery of forces into an enemy-held or contested area provides significant advantages to forces conducting a raid. The commander can achieve speed and surprise by air assaulting forces into a target area or surprise through ground (mounted or dismounted) movement to infiltrate forces undetected into the target area. Although a raid is generally an operation to temporarily seize an area culminating in a planned withdrawal, the time on the objective prior to withdrawal may be greater due to the requirement to conduct SE. Within this scenario, the commander utilizes a combination of these techniques to deploy forces into the target area and to withdraw forces from the target area.

D-115. Due to the inherent nature of a raid and the intelligence driven time sensitive targets within this operational environment, the weapons company maintains a high state of readiness throughout the operations process. The commander organizes forces and develops contingency plans to respond to a target (single or multiple target mission) of opportunity on short notice. The commander emphasizes the plan for prisoner of war handling during execution. Civil affairs assets enable the planning of time sensitive targets and the handling of prisoners of war through the facilitation of relationships between local police forces and/or host nation military forces. These relationships help facilitate the operations process throughout all phases of the raid.

D-116. The commander integrates combat engineers into the raiding force to help breach obstacles, keep ground forces maneuvering, and to provide countermobility during actions on the objective to protect the force. Combat engineers can be incorporated into assault teams when required to conduct a breach. When partnering with host nation forces with combat engineering capabilities, these forces may conduct the breach and initiate entry to the target to reinforce host nation legitimacy.

D-117. When working with partner forces and civilian agencies, battalion planners and the weapons company employ operations security specific to operating in a politically sensitive area(s) and/or when in contact with the civilian populous. The commander weighs the requirement for operations security versus the need to integrate host nation and multinational forces and civilian agencies to alleviate points of fiction within the scheme of maneuver. For example, the commander can work with the host nation to obtain written orders for access thru all checkpoints controlled by that host nation force. The company can utilize interpreters to help navigate checkpoints especially in a political sensitive area of operation. During movement interpreters can be attached to each subordinate element (command, assault, support, security, and quick response) when available.

D-118. Army aviation attack reconnaissance units and Air Force CAS enhance ground maneuver and fires, reconnaissance and surveillance, security, and mission command throughout all phases of the raid. Army aviation assault units may conduct infiltration and extraction of dismounted reconnaissance and surveillance elements to conduct detailed reconnaissance and surveillance of the designated target area(s). Army aviation units may insert and extract support and/or assault forces. Extraction assets can evacuate captured enemy personnel for intelligence exploitation, within the target area. Air Force assets, when available, provide preplanned and immediate CAS to operations. (See appendix C.)

Intelligence

D-119. Intelligence analysis for the raid includes considerations of the AO's distinguishing attributes—terrain, society, infrastructure, and the threat. The commander uses these categories to understand the intricacies of the environment that affect operations and assimilates this information into clear mental images. The commander synthesizes these images of the AO with the current status of friendly and threat forces, and develops a desired end state. The commander determines the most decisive sequence of activities that will move forces from the current state to the end state.

D-120. Identifying and understanding environmental characteristics from a counterinsurgent, insurgent, and host nation population's perspective allows the commander to establish and maintain situational understanding. With this understanding, the weapons company commander, in coordination with the battalion commander and support by the staff, develops appropriate courses of action and rules of engagement to accomplish the mission.

D-121. The company commander develops a list of initial information requirements based on higher headquarters tasks, commander's guidance, staff assessments, and subordinate and adjacent unit requests for information. Initial information requirements identify requirements for each potential course of action, any civil considerations, and the potential friendly course of action. During course of action analysis each course of action must meet ethical, legal, political, and technical feasibility criteria. Planning requires precise, time-sensitive, all-source intelligence.

D-122. Information requirements will utilize a host of collection assets in order to develop target packages and networks. When available during the operations process, information collection assets may include signals intelligence, HUMINT, and geospatial intelligence.

Signal Intelligence

D-123. *Signal intelligence* is a category of intelligence comprising either individually or in combination all communications intelligence, electronic intelligence, and foreign instrumentation signals intelligence, however transmitted (JP 2-0). Signals intelligence collectors analyze and report information obtained through intercept of foreign language communications to support the commander's information requirements. Signals intelligence, when available, can include assessment of enemy communications in the vicinity of the objective, including communication devices, scanners, or jammers in the objective area. (Refer ATP 2-22.6 for additional information [this ATP is a classified document]).

Human Intelligence

D-124. *Human intelligence* is the collection by a trained human intelligence collector of foreign information from people and multimedia to identify elements, intentions, composition, strength, dispositions, tactics, equipment, and capabilities (FM 2-22.3). HUMINT uses human sources and a variety of collection methods, both passive and active, to collect information to satisfy the commander's requirements and cue other information collection assets. The human intelligence collection teams are task-organized and employed based on the situation and requirements in the information collection plan. Human intelligence collection teams can collect information on the enemy's composition, strength, disposition, tactics, equipment, personnel, personalities, capability, and intention. (Refer ATP 2-22.31 for additional information [this ATP is a classified document]).

Geospatial Intelligence

D-125. *Geospatial intelligence* is the exploitation and analysis of imagery and geospatial information to describe, assess, and visually depict physical features and geographically referenced activities on the Earth. Geospatial intelligence consists of imagery, imagery intelligence, and geospatial information (JP 2-03). Geospatial intelligence provides correlation and analysis of imagery, imagery intelligence, and geospatial information to create products or display timely intelligence to support the operations process. The geospatial intelligence cell, located with the BCT, combines current information, combat information, and intelligence from other disciplines to produce geospatially referenced products and assessments that support all phases of the operation. (Refer ATP 2-22.7 for additional information.)

Fires

D-126. Effective fire support planning and preparations ensure enemy forces at or near the objective are overcome by surprise and violence of action. When fires are used, support elements either provides a volume of direct or indirect fires (to include precision fires) dictated by civil considerations. Fire support planning and preparations identify targeting options, both lethal and nonlethal, to achieve effects that support the raid's objectives. Lethal assets are normally employed against targets with operations to capture or kill (see appendix C). Nonlethal assets are normally employed against targets that are best engaged through electronic warfare consisting of three functions: electronic attack, electronic protection, and electronic warfare support (see appendix B).

D-127. Fires are closely controlled to ensure precision using fire control measures, marking, and signaling. On order or as planned, fires are adjusted to support the assault element as it enters the target area. As elements withdraw from the target area, support forces provide overwatch or suppressive fires for withdrawal from the target area. During planning, the commander considers the use of stay behind measures [for example a sniper team(s)] or devices to monitor backfill of enemy forces or populous support for the enemy.

Protection

D-128. The weapons company commander, in coordination with the battalion staff, protects elements of the raiding force to deny the enemy the capability to interfere with ongoing operations. To help protect the force, the commander ensures that all protection tasks are addressed during planning, preparation, and execution, while constantly assessing the effectiveness of those protection tasks. The company commander plans and implements survivability and other protection measures to prevent observation of the raiding force thereby reducing the enemy's ability to engage or otherwise interfere with the scheme of maneuver for the raid. During the raid the commander normally plans to secure only the terrain required to protect the elements of the raiding force.

D-129. The commander denies the enemy an effective response to offensive actions through speed and surprise to ensure the survivability of the raid force. Techniques for maintaining speed and surprise include using multiple covered and concealed routes and dispersion, and the wise use of terrain. The exact techniques employed in a specific situation must reflect the mission variables of METT-TC. Other protection measures include the types of tactical movement used and the use of electronic warfare systems—such as counter-radio controlled improvised explosive device and other electronic warfare systems. Protection measures may also include the conduct of countermobility missions to deny the enemy the capability to maneuver against raiding elements within the objective area and during withdrawal from the objective area.

D-130. The IPB process contributes to protection by developing products that help the commander protect subordinate forces, including identification of key terrain features, man-made and natural obstacles, trafficability and cross-country mobility analysis, line of sight overlays, and situation templates. (See appendix B.) For example if an enemy cannot observe the friendly force (line of sight overlay), the enemy cannot engage the friendly force with direct or indirect-fire weapons. The commander can sequence the movement of raid elements to reduce element sizes and movement signatures while executing movements at times and places where the enemy cannot respond effectively. Situation templates can enable the commander in knowing how fast an enemy force can respond once assault elements reach the objective area. Situation templates are developed through determining enemy indirect fire range fans, movement times between enemy locations and the objective area, and other related intelligence items.

D-131. Civil considerations within the IPB involves the staff (lead by the battalion S-2) identifying all significant civil considerations within the objective area so that the interrelationship of threat, friendly forces, and population activities is portrayed. Civil affairs operations can assist in identifying and evaluating civil considerations on raid and evaluating the effects of the raid on civilian populations. A key determination made by the commander is how to minimize the interference of civilians within the objective area with the elements of the raiding force while at the same time protecting these civilians from hostile actions. The commander considers the threat posed to the raiding force and its operations, if enemy agents or saboteurs are part of the civilian population.

D-132. When an enemy air threat is present, the commander establishes air defense priorities based on the concept of operations, scheme of maneuver, air situation, and the air defense priorities established by higher headquarters. The commander weights Army air defense systems, when available, in support of the decisive operation during the assault then to protective corridors over the terrain traversed by elements conducting the raid. Command of all air defense assets is coordinated through the air defense airspace management cell within the IBCT headquarters and requires complete and timely communications to ensure proper weapon status for the protection of friendly air support assets. Passive air defense measures are an essential part of air defense planning and are rehearsed to reduce the effectiveness of the enemy air threat. (See appendix C.)

D-133. Security operations prevent or inhibit the enemy from acquiring accurate information about the raid. However, contact (see paragraph 2-171 for information on the eight forms of contact) with enemy forces before the decisive operation may be deliberate (for example early entry reconnaissance and surveillance assets and/or joint fires observer, forward observer, forward air controller, and forward air controller [airborne]) when designed to shape conditions for the decisive operation. The ultimate goal of security operations (for example screen, guard, and/or local security) is to protect the force from surprise and reduce the unknowns in any situation.

D-134. Operations security and information protection support the raiding force to prevent or inhibit the enemy from acquiring accurate information about the raid. Operations security and any military deception or survivability efforts should conceal the location of the raid, the decisive operation, and the disposition of forces. The commander conceals the timing of the offensive tasks from the enemy or misleads the enemy regarding this information to prevent the enemy from launching effective spoiling attacks. The battalion S-6 continues to refine the information protection plan throughout the operations process. The weapons company works with the battalion's protection cell (when established) to provide staff supervision of the implementation of information system intrusion and attack detection devices.

Note. Information protection is accomplished (generally within the IBCT or higher headquarters) by monitoring protection tools and devices to identify activities that constitute violations of the information protection plan and security policy. Selected events are monitored to detect unauthorized access and inadvertent modification or destruction of data. Network managers react to counter the effects of an incident on the network. Reaction to a network or information system intrusion incorporates restoring essential information services, as well as initiating attack response processes. Disaster recovery requires stopping the breach and restoring the network.

Sustainment

D-135. During planning, the commander ensures the raiding force has the sustainment assets necessary to conduct the raid and maintain momentum through the extraction phase of the operation. By the inherent nature of a raid, sustainment requires deliberate planning due to intermittent or no ground lines of communication between the echelon support area and the raiding force. However, the commander emphasizes carrying only those supplies required to meet the immediate needs of the raiding force.

D-136. During preparation, the weapons company conducts equipment maintenance and rehearses the plans for casualty and medical evacuation and possible engagements with the indigenous civilian within the objective area and along infiltration and extraction routes. Contingency resupply (see paragraph H-37) prepackaging of company-sized resupply sets, as determined during course of action analysis (war gaming branches and sequels to the plan), can ease the execution of sustaining operations when sustainment assets must push supplies due to unexpected changes to the concept of operation during execution.

Execution

D-137. During execution, the raid conducted by the weapons company follows the same five phase process (see paragraphs 2-440 to 2-445) as in operations dominated by offense and defense. In phase I, the raiding force inserts and/or infiltrates into the objective area. In phase II, the objective area is sealed off from outside support or reinforcement, to include enemy air assets. In phase III enemy forces at or near the objective are overcome in a violently executed surprise attack using all available firepower for shock effect. In phase IV, the raiding force seizes the objective and accomplishes assigned tasks quickly before any enemy in the objective area can recover or be reinforced. In phase V, the raiding force withdraws from the objective area and is extracted usually using a different route than what was used for movement to the objective. In this scenario, the weapons company is organized with a quick reaction force (QRF), a security force, an assault force, and a command element.

Phase I – Raid Force Infiltration

D-138. The QRF (assault platoon 1) conducts infiltration, stages along Route Dodge prior to the objective area (figure D-16 on page D-37.) Considerations for QRF best practices and example include the following:

- Share knowledge of the objective area and linkup points to enable the response when call upon to assist the assault force.
- Stage the QRF, when required, as an independent force that travels on different infiltration and exfiltration routes.
- Stage the QRF, when possible (not illustrated), in a secure location(s) in the vicinity of the target area. (For example, a host nation police station, military compound, combat outpost, or other security base.)
- Establish local security (see paragraph 4-61) to protect the QRF at this location. (***Note.*** Drivers do not normally dismount to allow for quick maneuver if contact is made with the enemy at this location.)
- Use the QRF, when required (not illustrated), to secure the ground assault force (GAF) during infiltration to the objective stopping short of the objective area at a staging position. The GAF then continues to the objective area. Upon exfiltration, the GAF can pass by the staging position, to pick up the QRF and travel out as one force.
- Position QRF to respond in a timely manner. When possible do not position checkpoints or obstacles between the staging position(s) and the objective(s).
- Use Army aviation attack and rcconnaissancc units, when available, to conduct manned-unmanned teaming (see paragraph 3-202) during all phases of the operation.
- Use electronic warfare platforms, when available, to conduct electronic attacks (see appendix B) to disrupt enemy cell phones during infiltration and actions on the objective to maintain surprise and to keep the enemy from signal each other that an attack is pending. (***Note.*** Higher echelon electronic warfare activities ensure disruption patterns are not identified by conducting random attacks throughout continuous operations.)
- Use empty seats in QRF vehicles, as a contingency, for casualty evacuation, extra medics and interpreters depending on the situation.
- Preplan primary and alternate pickup and landing zones for casualty and medical evacuation and ground and air routes to trauma centers (not illustrated).
- Establish near and far recognition signals to enable contact with assault and security forces.
- Use contingency resupply (for example, speed ball resupply method [see appendix H]), incorporated into QRF load plans, to resupply extra water, extra ammunition and explosive devises, and medical supplies. (***Note***. Contingency resupply packages can be quickly thrown from the vehicle to the dismounted personnel within the objective area.)

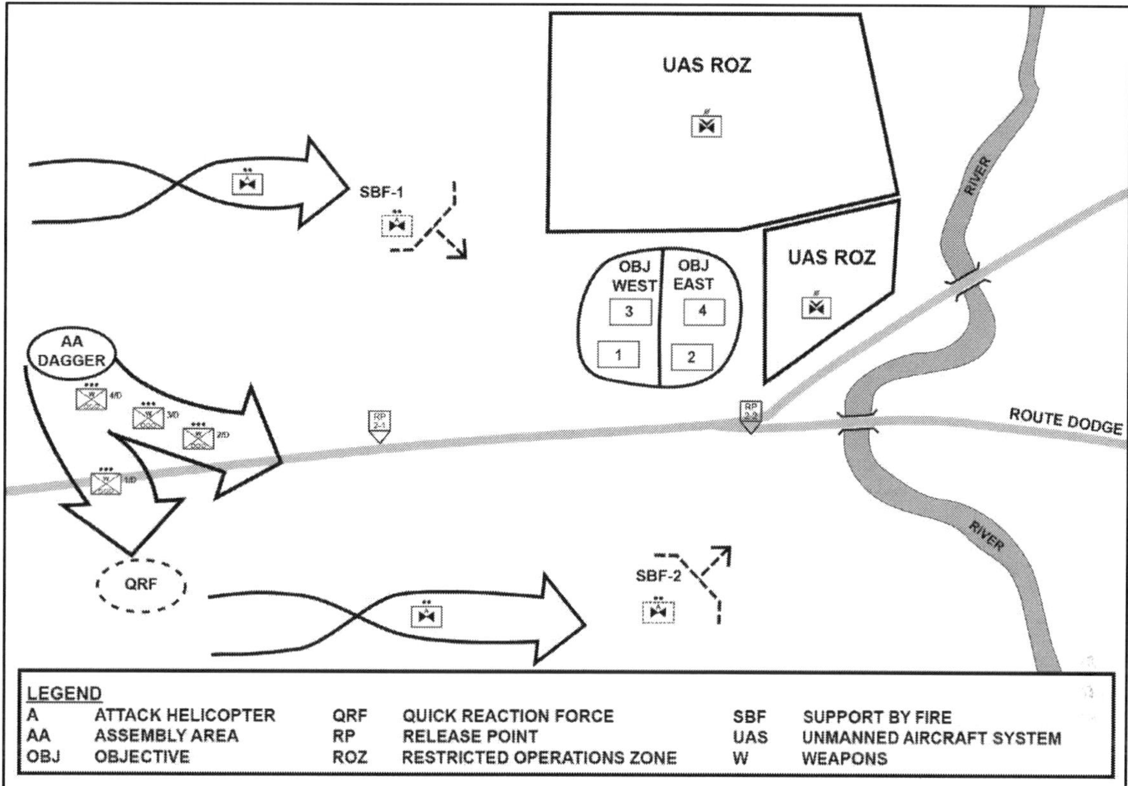

Figure D-16. Initial infiltration and movements to the objective area, example

D-139. The security force (assault platoon 2) follows the QRF, conducts infiltration along Route Dodge. The weapons company commander moves initially with the security force, then with the GAF. The GAF (assault platoons 3 and assault platoon 4) follows the security force, conducts infiltration to vehicle drop off locations 3 and 4. (***Note.*** In this example, each assault platoon is responsible for their own VDO site to ensure elements are effectively deployed to rapidly move to their security or assault position.) Platoon sergeants from each assault platoon are the NCO in charge of respective vehicle drop off locations. As the GAF establishes local security at its vehicle drop off location, security forces (assault platoon 2) under the control of the assault platoon leader, continue infiltration to seal off the objective area (figure D-17, page D-38). Considerations for vehicle drop off best practices and example include the following:

- Establish vehicle drop off as close as possible to the objective area while still maintaining the maximum amount of surprise.
- Maintain, due to the vulnerability of this location, noise and light discipline so as to not alert neighboring areas to the impending raid.
- Establish local security (see paragraph 4-61) to protect the vehicle drop off location. (***Note.*** Drivers do not normally dismount at the vehicle drop off to allow for quick maneuver if contact with an enemy force is made prior to movement from this location.)
- Position the vehicle drop off, especially within an urban terrain, in a manner in which vehicle radio platforms can be used effectively. (***Note.*** When communications are lost, the element[s] occupying the vehicle drop off location moves to a position where communications can be reestablished, then notifies all elements involved in the operation of the new location[s].)
- Information collection assets (internal and external to the battalion) continue to monitor the objective and relay mission essential information to the raiding force.
- Establish, as required, ground and aerial relay transmissions to meet communications requirements for all raid forces and higher command and support elements throughout operations.
- Plan for relays between the higher headquarters, vehicle drop off and staging locations, and element positions within the objective area.

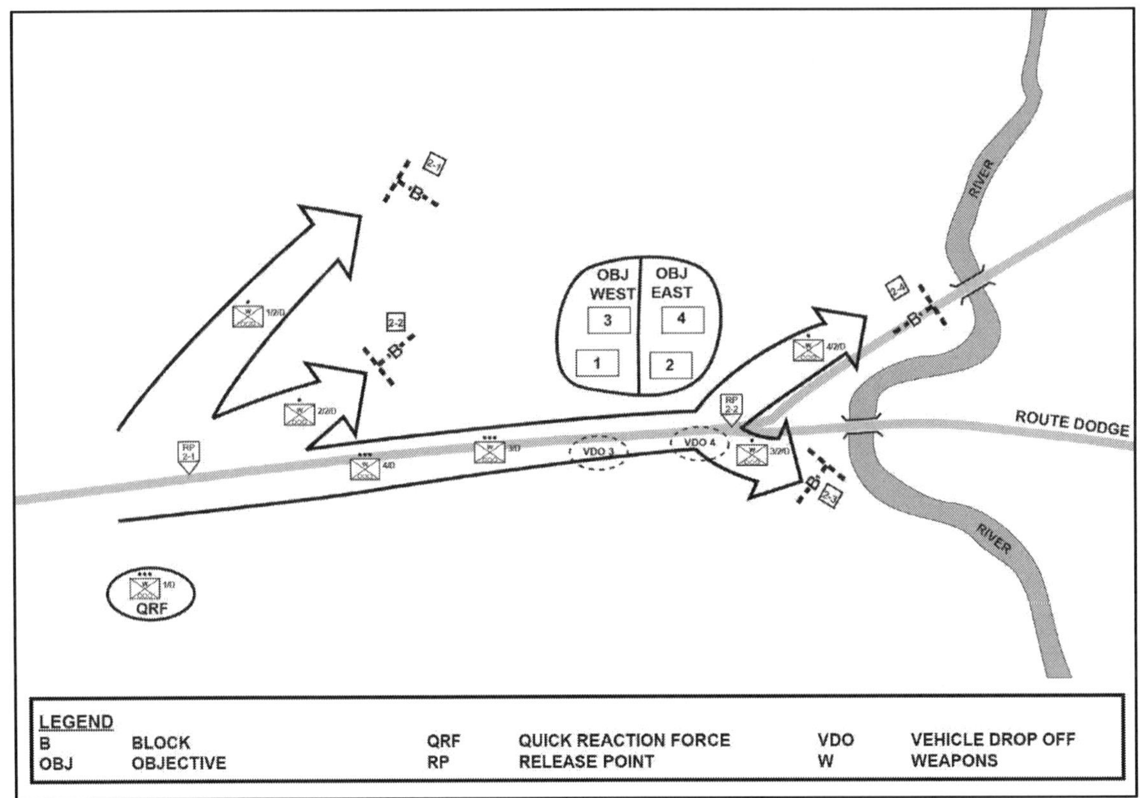

Figure D-17. Security and assault force infiltration, example

Phase II–Security Force Seals off Objective

D-140. Within the scenario, as security forces (assault platoon 2,) reach Release Points 2-1 and 2-2 they deploy to establish two security positions, identified as Blocking Positions 2-1 and 2-2; and 2-3 and 2-4 respectively (figure D-18). The platoon leader, assault platoon 2, controls blocking position 2-3 and 2-4. The platoon sergeant, assault platoon 2, controls blocking position 2-1 and 2-2.

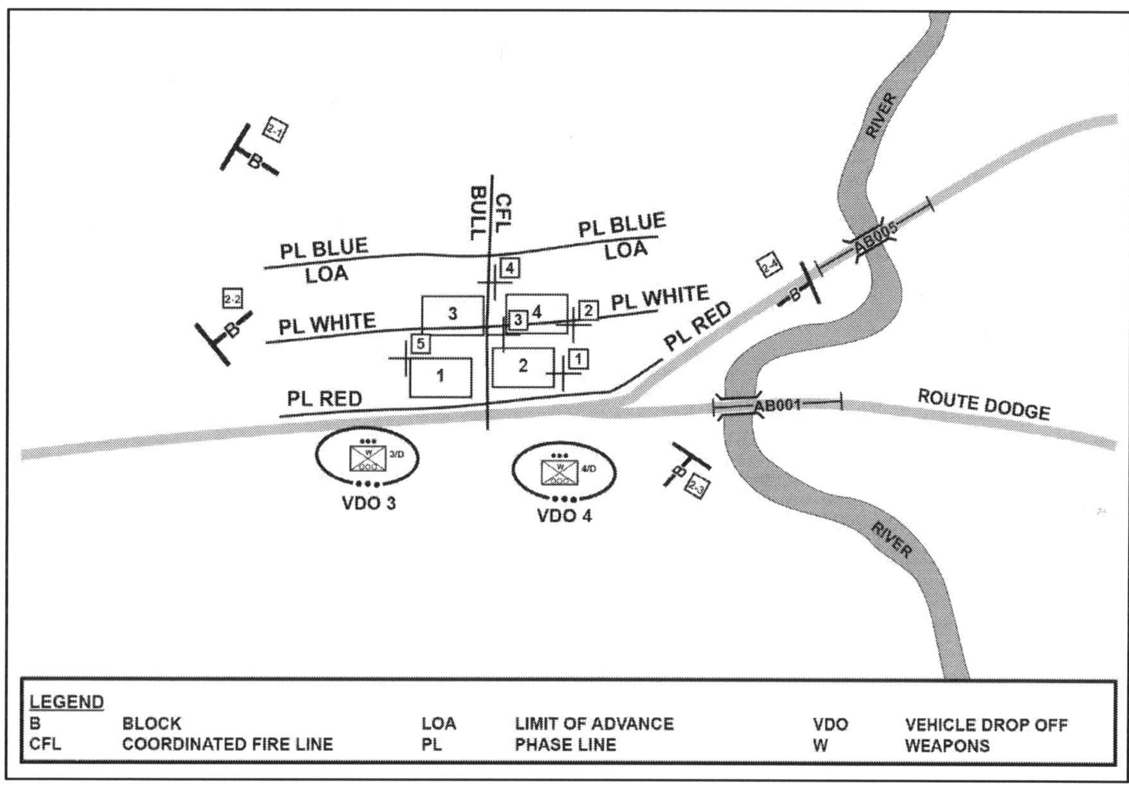

Figure D-18. Blocking position established, example

Phase III – Overcome Enemy Forces at or Near the Objective (Using all Available Firepower)

D-141. The actions executed by the raiding force in phase III are measured dependent on the constraints inherent within the complex and uncertain nature of operations dominated by stability. Within this scenario, a violently executed surprise attack using all available firepower for shock effect is not possible due to the requirement to limit collateral damage within the objective area.

Phase IV – Assault Force Seizes the Objective and Accomplishes Assigned Tasks

D-142. Assault platoon 3 clears Objective East (buildings 1 and 3), and assault platoon 4 clears Objective West (buildings 2 and 4) (figure D-19, page D-40). Moving from each platoons respective VDO (VDO 3 and VDO 4), and subsequent assault positions (in some instances this may be the same location) assault platoons 3 and 4 breach and clear buildings 1 and 2 simultaneously. After both platoons have cleared the first set of buildings the assault platoon leaders ensure they report to the commander (located vicinity VDO 3) in accordance with the previously provided operation schedule, to ensure synchronization of forces prior to their advance to buildings 3 and 4 sequentially. In this scenario assault platoon 4 provides support by fire on both buildings 3 and 4 prior to assault platoon 3's advance. (*Note.* The commander ensures that assault forces know the location of the coordinated fire line and have a thorough understanding of the triggers for firing across Coordinated Fire Line Bull.) Once assault platoon 4 shifts fires off of building 3 and west of TRP 4, the weapons company commander orders assault platoon 3 to assault and clear building 3. (*Note.* The commander ensures both platoons understand the location of TRPs, and who is responsible for each of them during the assault). Directly after building 3 is reported clear, assault platoon 4 clears building 4, establishes local security and reports that they have reached the limit of advance (Phase Line Blue). Additional considerations for initiating the assault, and partnering with host-nation forces and other agencies and other best practices within urban terrain (see ATTP 3-06.11 for additional information) include the following:

- Assault forces may initiate the assault with either a dynamic entry (illustrated) to the objective or in some cases with a soft knock (not illustrated).
- Assault forces may conduct a tactical call out where the assault force gives the target the opportunity to come out on their own accord.
- Assault forces conducting a dynamic breach to enter the objective, can mitigating risk by the use of flash bang or concussion grenades.
- When the assault force receives effective enemy fire from a target house the commander can employ force escalation methods, dependent on the rules of engagements, for example—
  - Place well-aimed shots thru the windows.
  - Move vehicles forward from the VDO to employ heavy weapons.
  - Breach wall(s) in objective buildings.
  - Gain access from the rear or top of the building.
  - Lastly, destroy the building with direct or indirect fires.
- Host-nation assault forces, when partnering with the U.S. assault force, can initiate the assault to reflect and promote the host-nation government's authority and legitimacy.
- Assault forces use assault ladders to gain access to single and multiple story building rooftops, enabling the force to—
  - Prevent the enemy from escaping to the roof then fleeing across adjacent rooftops.
  - Conduct simultaneous breaches at ground and roof top level.
  - Establish security over courtyard walls.
  - Overwatch ground forces as they move to assault an objective.

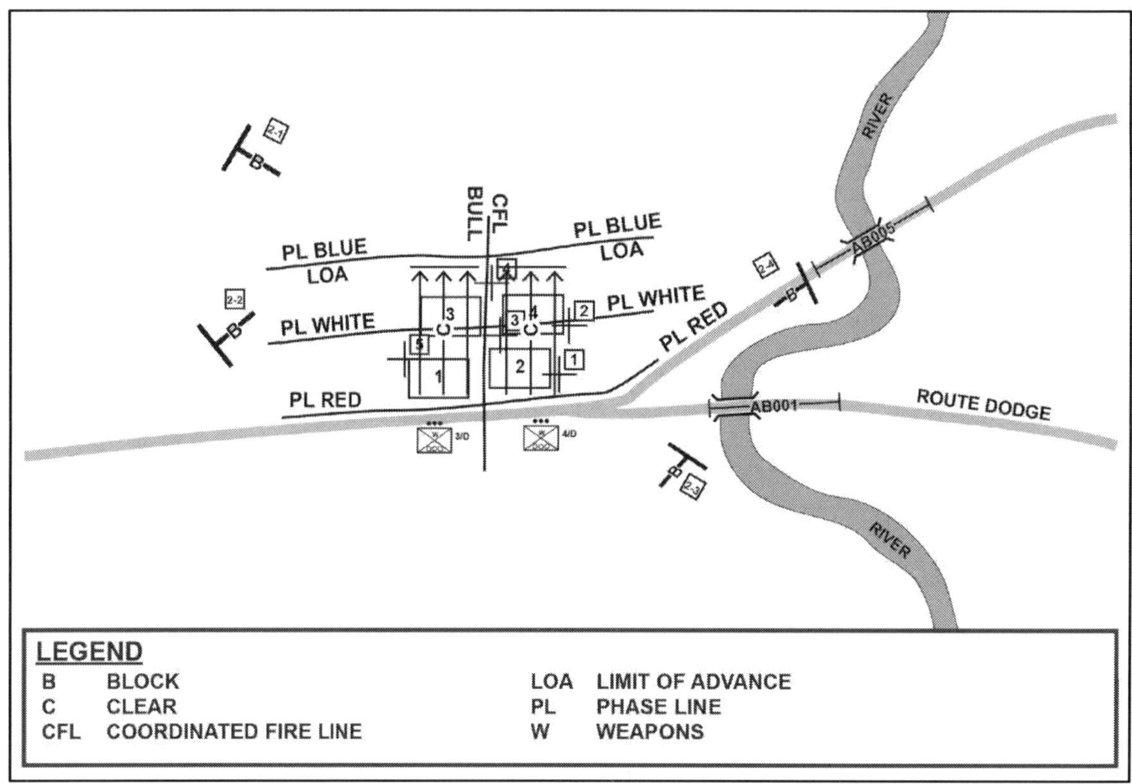

Figure D-19. Seize the objective, example

D-143. During actions on the objective, assigned tasks can include tactical SE, possibly within a sensitive site, and TQ. The commander manages SE and TQ on the objective with subordinate leaders, along with attached interpreters when available, leading the effort. Considerations for tactical SE and TQ best practices (see ATP 3-90.15 for additional information) include—

- Separate all military aged males in different parts of the target building or area.
- Separate the men, woman, and children and then question each individual in a bathroom or other small room in the building.
- Package SE equipment kits to include: zip ties, sealable plastic bags, sand bags, evidence bags, and camera(s).
- Attached law enforcement professionals, when available, to tactical SE and TQ teams to perform more specialized collect methods (latent prints (such as the palm and fingers), biometric collection, and deoxyribonucleic acid (commonly referred to as DNA collection).
- Separately categorize target vehicles within the target area.

Phase V–Withdrawal from the Objective Area

D-144. Upon completion of SE and TQ tasks, the QRF, security forces, and the GAF bring preparations for exfiltration. During SE and TQ, platoon sergeants within VDO positions 3 and 4 reposition vehicles outside the target house(s) for exfiltration. When leaving the objective area assault and security forces take different routes from the one used during infiltration. When this is not possible, the commander re-task ground and aerial reconnaissance assets along the infiltration route for the withdrawal of the raiding force. The QRF can move to the objective area for withdrawal with security and assault forces, or it can take an alternative withdrawal route alone depending on the situation.

Note. Upon completion of the mission, raid leaders are debriefed by the battalion S-2 to provide all relevant information for analysis. When established, detainees and evidence is turned over to the detainee holding facility. The weapons company then refits for follow-on missions.

This page intentionally left blank.

Appendix E
Infantry Battalion Sniper Squad

Appendix E, in conjunction with the other appendixes and chapters in this publication, provides the doctrinal framework for the Infantry sniper squad assigned to Infantry battalions in the IBCT. This appendix is designed to work in conjunction with and complement TC 3-22.10, *Sniper.* The battalion sniper squad provides the commander with precise, long-range, and discriminatory fires. The sniper squad is especially valuable when fighting an enemy that tries to blend in with the local population. Sniper targets designated by the commander can include enemy leaders, specialists, and snipers; enemy personnel and other individuals as specified; and key enemy equipment and vehicles. The sniper squad's second mission is to collect and report accurate and detailed battlefield information.

Note. TC 3-22.10 distribution is restricted. Point of contact for release is the United States Army Sniper School located at Fort Benning, Georgia.

OVERVIEW

E-1. During the conduct of decisive action, through the depth of the Infantry battalion's AO, the primary mission of the Infantry battalion's sniper squad is to deliver precise long-range fire on selected targets. Sniper fires create a marked effect on enemy troops; creates casualties, slows movement, instills fear and influences their decisions and actions, lowers morale, and adds confusion to their operations.

E-2. The sniper squad's observational and navigational skills and specialized equipment help them see the terrain in detail and observe changes. Snipers provides the commander with details about the terrain, obstacles, likely avenues of approach, or other pertinent battlefield information.

E-3. Snipers infiltrate enemy areas to engage the enemy from unexpected directions. Snipers employed in advance of a unit's movement allows them to move at their own pace so to remain undetected. During movement or from static positions snipers attempt to identify enemy positions or movements and provide overwatch and/or engage enemy targets that threaten the protected element's movement or position.

Note. Though organization of forces may require the sniper squad or team to task organize to battalion control or detach to subordinate company control on occasion, this appendix does not cover detailed operations of the squad or teams. This appendix only provides a general discussion of coordination and operating issues pertaining to the employment of the sniper squad or individual teams. Chapters 2, 3, and 4 of this publication cover the detailed operations of snipers within the Infantry battalion.

E-4. This section provides an overview of the mission, organizational structure, characteristics, and weapon systems of the sniper squad found in the Infantry battalions of the IBCT. (Refer to TC 3-22.10 for additional information.)

ORGANIZATION AND EQUIPMENT

E-5. The sniper squad is assigned to the headquarters and headquarters company of the Infantry battalion. The sniper squad is composed of the squad leader, three senior snipers (team leaders), and six snipers. All should be school-qualified snipers. The squad usually operates in three three-Soldier sniper teams composed of a senior sniper and two snipers. Sniper teams, however, can be specifically configured to meet METT-TC conditions. Within the team, the senior sniper is usually the observer with a primary and alternate sniper. The alternate sniper usually provides security but may be assigned a sniper mission (Figure E-1).

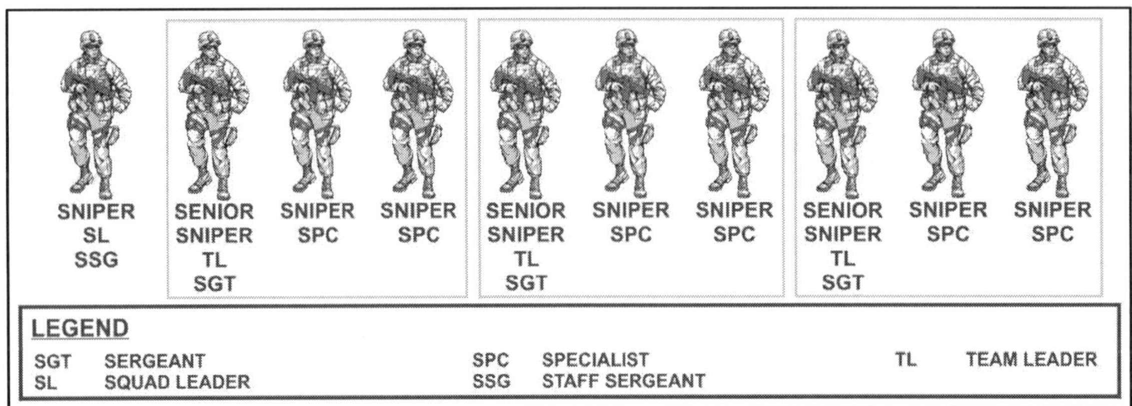

Figure E-1. Infantry battalion sniper squad

E-6. The three sniper teams carry standard equipment along with specialized sniper equipment. The team leader or observer normally carries an M4 with the M320 grenade launcher. Dependent on the mission variables of METT-TC, the team leader carries the observation telescope to determine wind speed and direction. The primary sniper carries the sniper weapon. The alternate sniper or security Soldier carries the M249 squad automatic weapon or another sniper weapon. The sniper squad uses the M4, M320, and M249 to provide security and to disengage when required. Other equipment carried by the sniper squad can include image intensification, infrared devises, ghillie suits, and field expedient directional antennas.

E-7. Using the arms room concept, the sniper squad uses the best weapon for the mission. Sniper weapons assigned to the sniper squad include the M110 (7.62-mm) semiautomatic sniper system and the M107 (.50 caliber) long-range sniper rifle. The sniper employment officer (SEO) or squad leader normally designates the weapon(s) to carry. This may include using the M4, one or both of the sniper rifles, or some other weapon for the mission. Refer to TC 3-22.10 for additional information on special equipment and weapons performance data specific to the employment of snipers.

SNIPER EMPLOYMENT OFFICER

E-8. The Infantry battalion's SEO is responsible for advising the commander, subordinate commanders, and staff on the employment and control of assigned snipers in the battalion. The sniper employment officer can be the headquarters and headquarters company commander, the executive officer, the scout platoon leader, or any other officer. In the SEO's absence, either the sniper squad leader or the individual sniper team leader can represent. SEO duties and responsibilities are address below and in the following four paragraphs. The memory aid—knowledge, advise, coordinate, training, issue, supervise, and debrief (also known as the mnemonic, KACTIS-D)—lists the specific duties of the SEO below:

- Knowledge of sniper capabilities. The SEO understands employ considerations and the capabilities and limitations of sniper teams, to include movement and camouflage techniques, observation techniques, hide site operations, and marksmanship.
- Advise the supported unit commander. The SEO advises the supported commander, operations staff officer (S-3), and intelligence staff officer (S-2) on how to employ the sniper units (by squad, section, or team). In addition to sniper capabilities and limitations, the SEO ensures the commander sees the sniper as the eyes, ears, and trigger finger on the battlefield for the battalion.

- Coordinate all aspects of the sniper mission. Coordination begins during mission analysis, and is continuous throughout the operations process. The SEO coordinates the assignment of sniper teams to missions to support units or as an integrated part of a sniper mission. To prevent fratricide and possible compromise of position and mission, the SEO coordinates terrain management with units in the AO.
- Training should be realistic, varied, challenging, and mission-oriented. The SEO assembles the battalion sniper squad, to include any snipers assigned to company sniper teams, in order to standardize training and standard operating procedures (SOPs). Every skill required of a sniper (to include information collection) is perishable and requires continual practice. The SEO ensures that snipers are allocated the proper time and resources needed to maintain their effectiveness. An experienced sniper squad leader serves as the SEO's primary trainer. Concurrent sniper training at the IBCT level ensures SEO and sniper training and SOPs are standardized across the IBCT.
- Issue orders to the sniper squad/team leader(s). The SEO issues the order, formally or informally, to the sniper squad/team leader(s) using an OPORD or fragmentary order.
- Supervise planning and preparation (specific to rehearsals and backbriefs). The SEO should provide supervision for the planning of the mission, the preparation of the mission, and mission rehearsals. A good briefback indicates the sniper squad's/team's readiness for the mission.
- Debrief all members of the team upon completion of the mission. The SEO, in coordination with the battalion S-2 should conduct detailed debriefings after completion of the mission. The sniper squad/ team(s) brings all pertinent information (for example, their data books, sniper logs, field sketches, range cards, and digital photos) to the debriefing.

E-9. Knowing sniper capabilities and limitations is important. The SEO participates in sniper training at every opportunity, and ensures that snipers are trained in reconnaissance and surveillance (R&S), and security operations as well as sniper skills. An important aspect of the sniper team employment is the time required to plan, rehearse, travel, and recover from a mission. The SEO needs to convey to the commander that sniper teams cannot be continuously on missions. Furthermore, if all three teams are employed at the same time, then there will be a period when no teams are available.

E-10. The SEO must know how to employ snipers effectively in support of various operations to advise the commander, S-3, and S-2 on their employment. When assigned to a supported unit other than the sniper team's parent unit, the SEO represents the sniper team and advises the supported unit commander on what the assigned team can do for commander. Coordinating with the supported commander gives the SEO an opportunity to explain proper sniper employment tactics and techniques, and to clarify misconceptions about sniper assets.

E-11. The SEO can assign teams to missions to support units or to serve as a part of the overall scout platoon mission. When assigned to a subordinate unit for employment, the SEO and unit commander meet face-to-face, so the SEO can advise the commander on sniper employment. Discussing who occupies what terrain or area of operation(s) ensures that all parties understand each other's mission to prevent fratricide, and to protect the integrity of the mission. The arrangement for the insertion, resupply, and extraction of the sniper squad or individual teams operating independently is always a key planning concern to the SEO.

E-12. The OPORD or FRAGO provided by the SEO can assign sniper teams to specific companies or the teams can operate as a squad with missions and teams assigned accordingly. Missions assigned one at a time, with succeeding missions issued as fragmentary orders lets the sniper squad or teams focus on planning and executing each mission. Once the sniper squad or team is in receipt of the order, the SEO provides supervision during mission planning and preparation. The SEO leaves the detailed supervision of mission preparation to the squad leader and team leaders. Detailed preparation includes precombat checks and inspections per SOPs, movement techniques and battle drill rehearsals, route reconnaissance, load tailoring, time management, and cross loading equipment.

SNIPER SQUAD TASKS

E-13. Sniper squad tasks are specific in certain operations, though many tasks performed apply to most operations. The SEO or squad leader develop sniper tasks based on the unit commander's intent and mission. Though not all inclusive, tasks specific to sniper squad employment include:

Long-Range Precision Fire

E-14. The sniper supports operations by delivering long-range precision fire on key selected targets and targets of opportunity. Specific missions include—

- Accurate fires on enemy:
  - Command posts (CPs) and key leaders.
  - Crew-served weapons and crews.
  - Bunkers and embrasures.
  - Key weapons and equipment.
  - Selected targets just prior to an attack.
- Security, overwatch, and covering fires for friendly:
  - Leader's reconnaissance.
  - Patrols (combat and reconnaissance).
  - Observation posts and combat outposts.
  - Engineers, demolition guards, and supply columns.
  - Checkpoints, contact points, and linkup points.
  - Key terrain and defiles.
  - Cordons and routes.
  - Flanks and rear.
- Countersniper.
- Ambushes or harassment of withdrawing enemy.
- Observation and control of fires.

Information Collection and Reporting

E-15. The advanced optics of the sniper team(s) enable collection and reporting of detailed battlefield information from concealed positions supports the battalion's R&S efforts and security operations. ATP 3-21.8 provides information on the fundamentals of Infantry patrolling (reconnaissance and combat), and information collection (including surveillance and security operations) and reporting.

PLANNING CONSIDERATIONS

E-16. During initial planning the commander and staff's mission analysis determines whether to employ snipers or not. The SEO along with the sniper squad leader assists the commander, in coordination with the staff, in making this decision and in determining the exact sniper tasks and the number of teams to deploy. During sniper employment planning the commander considers—

- Rules of engagement.
- Collateral damage.
- Potential mines and unexploded ordinance.
- Shoot-on-command capability.
- Surveillance and reconnaissance environment.
- Response force availability.
- Sustainment.
- Communication.

PLANNING FOR SNIPER EMPLOYMENT

E-17. Planning for sniper employment should account for all events from departure from friendly lines or insertion to reentry of friendly lines or extraction. Prior to sniper employment, the SEO uses information collected from R&S efforts by ground or air or, at the least, with maps, ground and air photos, and patrol debriefings to develop the plan. Due to its small size, a sniper team must request and coordinate the use of many mission-essential items such as transportation, special equipment, and staging areas. The team coordinates closely with forces in contact and with units to move the team. Unit SOPs and mission checklists are valuable in helping planners concentrate on the unique aspects of the operation.

Selection of Targets

E-18. Targets are selected based on—

- Tactical value of the target.
- Nature and type of the armor or cover.
- Active defensive measures employed by the enemy to protect the target.
- Potential collateral destruction.
- Spot on the target with the best balance of vulnerability and high payoff.
- Angle of fire relative to target.
- Maximum range from the target that will ensure penetration and the effect of the round to the area behind the target.
- Selection of ammunition that will achieve the desired effect.

Assignment of Targets

E-19. When assigning targets, the sniper squad or team must stay within the commander's intent, the Law of Land Warfare, and the rules of engagement (ROE). The sniper squad or team makes decisions based on the above criteria and its own survival. The commander can use different methods to designate and prioritize targets. The commander may or might—

- Describe the affects or results expected and allow the snipers to select key targets.
- Prescribe specific types of targets. (For example, if the commander wants to disrupt an enemy's defensive preparation, the commander might task snipers to engage equipment operators and vehicle drivers.)
- Assign specific or key targets. (These can include specific personnel, leaders, radiotelephone operators, antitank guided missile gunners, armored vehicle commanders, or weapons crews.)

Target Intelligence

E-20. The decision to employ a sniper team is based on accurate and up-to-date intelligence of the target area. The commander considers the current threat and whether suitable targets can be identified within the target area. Leaders must provide the sniper team with the most current target intelligence available. Target intelligence include specific details about the target or target complex to include enemy locations, equipment, strengths, vulnerabilities, capabilities, composition, and relative importance.

E-21. In addition to target descriptions, the SEO also must have other current information on the area such as aerial photographs. This helps the sniper leader determine the type of terrain and identify indigenous vegetation. It also helps the sniper leader identify suitable positions that offer the maximum standoff range while allowing the sniper team to destroy the target. The sniper leader needs to know the unit's planned routes to and from the target area. Data on the target area's meteorological and environmental conditions, such as prevailing winds, is important. The sniper leader decides on the direction of approach that offers—

- The best fields of fire.
- An effective range.
- A good angle of attack.
- Cover, concealment and security.
- Quick exfiltration route.

E-22. Based on target and area information received, the sniper leader/team chooses the best sniper weapon for the job. For example, a heavy sniper team armed with the M107 would be overkill if a conventional sniper detachment armed with the M110 could accomplish the mission effectively.

E-23. The sniper leader/team must select a tentative final fire position that meets all the requirements to ensure a successful engagement. The sniper team considers meteorological and environmental conditions such as wind speed and direction. If the ground and tactical situation permit, the commander should place the team where the sniper can shoot at a minimum angle to the wind.

E-24. The sniper leader/team develops contingency plans for exfiltrating and extracting the sniper team. The leader/team confirms or adjusts these plans once the team is in position. During planning, the leader/team can seldom confirm the exfiltration routes.

Indirect Fires on the Target

E-25. The commander considers using indirect fire in and around the target area in conjunction with the sniper leader/team. Indirect fires can—

- Augment the direct engagement of the team on specific targets; saturate the area and inflict collateral damage on non-priority targets.
- Disguise the sniper fire and reduce the chance that the enemy can identify the team's location.
- Divert enemy attention while the team extracts. This requires detailed coordination with indirect-fire elements to ensure that the team's intended target is clear of smoke and dust before the sniper engages the target.

PREPARATION

E-26. Regardless of what type of mission a sniper is conducting, preparation, inspections, and rehearsals are vital for success. The squad or team leader specifies the following elements during preparation:

- Security.
- Control. (Control measures include communications, emergency actions, and specific control measures.)
- Routes. (Routes include primary and alternate routes to and from the objective.)
- Navigation and navigational aids.
- Weather.
- Intelligence.
- Coordination with other units.
- Contingency plans.

E-27. Before the squad/team departs friendly lines, leaders conduct precombat checks, precombat inspections, and rehearsals. (Refer to ATP 3-21.8 for a detailed discussion of small unit checks, inspections, and rehearsals.) Leaders conduct rehearsals during the same hours of the day or night as the mission.

EXECUTION

E-28. During execution, the control of sniper teams is complicated by their isolation from other units and their locations. ROEs may require direct communications between the commander and sniper teams. Prior to employment the SEO, in coordination with the maneuver commander's plan, formulates policies and procedures to aid in the control of sniper teams. (Refer to TC 3-22.10 additional information.) Policies and procedures, though not all inclusive, may include—

- Centrally controlling the sniper squad and teams, tasking them to support subordinate units as required.
- Avoiding the temptation to use snipers on pure reconnaissance missions, although they might or might not form part of the scout platoon.
- Ensuring sniper squad or team leader have situational understanding, through prior backbriefs and rehearsals, to make independent decisions when communications are degraded.

Multiteam Operations

E-29. For targets that the commander designates as high priority, and for targets dispersed throughout a large, defined target area, the commander may assign multiple sniper teams. When the target area has multiple high priority and well-defended targets, the commander can deploy multiple sniper teams. Multiteam employment may be required to ensure complete coverage and destruction or disabling of all key targets simultaneously. On the ground, coordination and communication between the teams is vital.

E-30. For multiteam operations to succeed, sniper teams and resources must be coordinated as thoroughly as possible. The SEO divides and assigns target area responsibilities to each sniper team. For control on the ground, the squad leader or senior team leader in the target area is generally the mission leader. This mission leader assesses the tactical situation in the target area(s) and determines the best time to initiate the engagement. Depending on the tactical situation, the leader might grant the teams the authority to engage their respective targets in a set period of time, unless the targets appear at different times than initially expected. This way the teams can divert, confuse, and delay the enemy's response, and then exfiltrate back to a linkup point for extraction.

E-31. Within each team, the individual team leader is responsible for selecting, prioritizing, and destroying targets in specific sectors of fire. Each team must communicate its location and intended targets to each other to reduce the chance of fratricide. All teams might engage an extremely high-value target at once to ensure its destruction. Depending on the mission and size of the objective area, teams might insert together or individually at designated positions that offer the best approach to their target area.

Sniper Employment Mission

E-32. Sniper employment missions support the accomplishment of the battalion's mission. Commanders and subordinate leaders at all levels must know the value of employing friendly snipers and the threat posed by enemy snipers. They must understand the effects a sniper can have on unit operations, and how the enemy could counter friendly sniper's operations. The four phases of a sniper employment mission are—

- Insertion phase.
- Execution phase.
- Extraction phase.
- Recovery/debrief phase.

E-33. All phases are subordinate to the battalion's mission, and planned accordingly. Each phase may involve multiple steps. The techniques used for these phases may be limited to the type of unit to which the sniper team is assigned and depending on the unit's resources. (Refer to TC 3-22.10 for a detail discussion of the four phases.)

Offensive Tasks

E-34. Sniper tasks during the office assists the commander in accomplishing the mission by: depriving the enemy of resources, deceiving or diverting the enemy from the main effort, fixing the enemy in place, disrupting enemy plans, and obtaining information. The fluidity occurring during support to offensive tasks presents good opportunities for the employment of snipers. In a movement to contact for example, snipers can infiltrate enemy areas and engage the enemy from unexpected directions. Sniper teams should move out well in advance of the projected main body movement. This allows snipers to move at their own pace so they remain undetected. It also allows sniper teams to engage any targets that threaten the advance. The teams may use normal stalking methods, or insert by ground vehicle, helicopter, parachutes, or boats. Maintaining communication between the battalion and the team is critical.

E-35. The coordination and planning stages of a deliberate attack give commanders enough time to take full advantage of the unit's sniper capabilities. In a deliberate attack, snipers can be effectively employed near the fire support team and/or support positions. Sniper accuracy and optics allow them to continue to reduce enemy targets in the midst of friendly forces. The unit should take care to avoid drawing enemy attention to the sniper team's position. For example, the battlefield noise generated by the fire-support element can interfere with sniper team communication. The team also may be deployed forward of the fire support team to support the attack with accurate selective rifle fire; or deployed with a cutoff force with the same task. If time permits, snipers can infiltrate behind the enemy positions to disrupt counterattacks or withdrawal, and to harass enemy reinforcements. During reorganization, sniper teams can be deployed forward of the forward edge of the battle area on likely counterattack routes.

E-36. During hasty operations, when planning time is constrained, one way to employ snipers is to have them operation and move directly with the supported maneuver unit. As the maneuver unit deploys to attack sniper precision fires can reduce delays or stalemates during the hasty attack.

Defensive Tasks

E-37. Snipers play an important role in the defense. Snipers missions are assigned within the security area as well as the main battle area. Sniper missions in the forward security area are generally controlled by the battalion but may be assigned to subordinate companies operating in the main area or security area for specific missions. Their presence and mission has to be coordinated with the unit responsible for the AO in which they operate.

E-38. The sniper team can perform the following tasks during the defense:

- Cover obstacles, minefields, roadblocks, and demolitions.
- Perform counterreconnaissance.
- Engage enemy observation posts, armored-vehicle commanders exposed in turrets, and antitank guided missile teams.
- Damage enemy vehicles' optics to degrade their movement.
- Suppress enemy crew-served weapons.
- Disrupt follow-on units with long-range small-arms fire.
- Observe named areas of interest and target areas of interest.
- Collect battlefield information.

E-39. Snipers in an area defense often are able to establish primary, alternate and supplementary firing positions, prepare hide positions, prepare range cards, and become familiar with their assigned area. Snipers can be given missions anywhere in the battalion's AO. They may initially locate and be given missions within the security area and then displace back to the main battle area. Because of likely enemy fire, sniper teams normally are not assigned to positions close to friendly Infantry forces. They should however, be close enough to Infantry forces for protection.

E-40. As part of a mobile defense, the Infantry battalion normally is be part of the fixing force and conducts a area defense within the larger mission of a mobile defense. Sniper teams normally receive the same type of missions as they do in an area defense. If assets are available to provide equal or greater mobility than the enemy, such as Army aviation, sniper teams and other battalion elements may be assigned missions to the battalion's front or flank to deceive and delay the enemy.

E-41. During a retrograde, the sniper team is employed with and moves with the rear security elements. Snipers can delay and reduce the momentum of the enemy advance. They also can cover obstacles and be employed in an economy of force mission. Missions and considerations for the employment of sniper teams during a retrograde include—

- Dominate key terrain.
- Cover obstacles.
- Cover primary and secondary avenues of approach.
- Confuse the enemy.
- Gather and report detailed information on the terrain, route, and enemy.
- Control fire support.

E-42. Sniper teams also can conduct deliberate or ad hoc stay behind missions to conduct operations and to conduct surveillance. Communications must be maintained so they can pass along information and intelligence, to control their movement, to arrange for their extraction, and allow the sniper team to call fires on large enemy groups.

E-43. Sniper teams may be assigned missions to cover intervals between units, flanks, and the rear of friendly positions where regular patrols and forward edge of the battle area observation activities cannot. When possible, the commander should provide the snipers with Infantry protection to their rear. This protection should be close enough, usually within 1 kilometer, to help the snipers extract should they have to, but not close enough to compromise the snipers.

E-44. Sniper teams must coordinate with units responsible for the AO in which they operate. To avoid fratricide, the sniper team must make sure that it is included in that unit's defensive plan. Fire control measures such as no fire areas are used for the sniper team's position to further prevent the possibility of fratricide. Coordination also may include the sniper team providing information.

SUPPORT TO OPERATIONS FOCUSED ON STABILITY TASKS

E-45. During stability, sniper teams are effective primarily because they can deliver selective, precision fires against specific targets in accordance with the ROE. Due to their skills in observation and familiarity with the AO, snipers play a vital role in providing information. Support to operations focused on stability places an additional strain on the sniper to kill without the motivational stimulus normally associated with the battlefield. During support to operations focused on stability tasks, the sniper's ROE may differ from the supported Infantry unit.

E-46. Based on the current ROE and as authorized by the commander, sniper teams may be assigned the following:

- Engage dissidents involved in hijacking, kidnapping, and holding hostages.
- Engage dissident snipers as opportunity targets or as part of a deliberate clearance operation.
- Covertly occupy concealed positions to observe selected areas.
- Record and report all suspicious activity in the area of observation.
- Assist in the coordination of activities by other elements from hidden observation positions.
- Protect other elements, including auxiliaries such as fire fighters, and repair crews.

E-47. Commanders should avoid using sniper teams when other units are adequate. They should not over-commit sniper teams but rather employ them where their specialized skills are required.

E-48. As the enemy may specifically target snipers, commanders must carefully protect their anonymity. Ideally, snipers are held in a central location and employ only as required. If needed, snipers may deploy in hidden observation posts.

E-49. The sniper team must understand its responsibilities and the correct authority to authorize its use of deadly force. The ROEs spell out these responsibilities and provide the sniper team with the command authority to carry out those responsibilities. The sniper team must understand how to determine when its fire constitutes reasonable force.

E-50. Ideally, a sniper team should deploy where it can receive the order to fire from the appropriate commander. This is often difficult. Therefore, all orders, to include targets and ROEs, must be clear to the sniper team before it deploys. The sniper team should rehearse its actions during all possible scenarios.

COUNTERSNIPER OPERATIONS

E-51. Countersniper operations eliminate enemy snipers. The sniper squad must thoroughly plan any countersniper operation. (Refer to TC 3-22.10 for a detailed discussion on countersniper operations.) During countersniper operations, snipers perform the following steps:

- Determine the threat.
- Gather information.
- Determine patterns.
- Determine best location and time to engage the enemy.
- Engage the enemy.

E-52. The SEO and sniper squad leader should be informed when any of the following are reported:

- Sightings of enemy Soldiers wearing special camouflage uniforms.
- Sightings of enemy Soldiers carrying weapons with long barrels, mounted scopes, or bolt-action receivers, or carried in weapon cases or drag bags.
- Key personnel casualties such as commanders, senior noncommissioned officers, or weapons crewmembers, and a simultaneous reduction in enemy patrol activities.
- Reports of reflections off optical lenses.
- Reports from intelligence or reconnaissance units of small groups of enemy personnel (one to three Soldiers).
- Finding single spent cartridges in sizes used by enemy snipers.

E-53. When an enemy sniper is operating in the unit area, the sniper force or any other force uses active measures such as the following to protect the unit against sniper fire:

- Gather information.
- Make a plan.
- Observe likely locations.
- Locate the enemy through observation. Stalking and tracking expose the team to higher risk.
- Kill the enemy sniper.

E-54. Examples of unit-level passive countersniper measures follow:

- Avoid consistent routines, such as meal times, ammunition resupply times, assembly area procedures, patrol routes, routes to the objective rally point, or any consistent day-to-day activities.
- Gather, meet, and brief under cover or in limited visibility.
- Cover or conceal equipment, such as maps, radios, and antennas.
- Remove rank and do not salute.
- Leaders avoid behaving authoritatively.
- Increase the unit's observation capabilities through such means as observation posts.
- Inform patrols to look for signs of the presence of a sniper such as single spent rounds and camouflage materials other than those your unit uses.
- Do not dismiss the potential for a woman to be the sniper.

Appendix F
Combined Arms Breaching Operations

A combined arms breach is a complex operation with many moving parts. Combined arms breaching requires detailed planning, preparation, and execution. Breaching is an inherent part of maneuver. Effective breaching operations allow friendly maneuver in the face of obstacles. Combined arms breaching techniques cannot be discussed in isolation. There is continuity between tactics and procedures, covered in field manuals, and techniques covered in Army techniques publications. This appendix provides a brief overview of combined arms breaching operation types, areas, and tenets, and includes a discussion on deliberate and hasty breaching operations.

OVERVIEW

F-1. A *breach* is a tactical mission task in which the unit employs all available means to break through or establish a passage through an enemy defense, obstacle, minefield, or fortification (FM 3-90-1). As a tactical mission task, a breach is an action by a friendly force conducted to allow maneuver despite the presence of obstacles. During maneuver, the commander attempts to bypass and avoid obstacles and enemy defensive positions to the maximum extent possible to maintain tempo and momentum. Breaching enemy defenses and obstacle systems is normally the last choice. A breach is a synchronized operation under the control of the maneuver commander in contact with the obstacle or enemy defense. The breach begins when friendly forces detect an obstacle and begin to apply the breaching fundamentals (see paragraph F-18) and ends when battle handover has occurred between follow-on forces and a unit conducting the breach. Refer to ATP 3-21.8 for information on Battle Drill 8-Breach a Mined Wire Obstacle (07-3-D9412).

Note. A *tactical mission task* is the specific activity performed by a unit while executing a form of tactical operation or form of maneuver. It may be expressed in terms of either actions by a friendly force or effects on an enemy force (FM 3-90-1).

F-2. As defined in ATP 3-90.4, a *breach* is a synchronized combined arms activity under the control of the maneuver commander conducted to allow maneuver through an obstacle (ATP 3-90.4). A breach is a synchronized combined arms mission under the control of the maneuver commander. When breaching operations are required to support an attack along the continuum from a deliberate to a hasty attack, regardless of where the attack falls along the continuum, the breaching tenets (see paragraphs F-14) apply when conducting breaching operations in support of an attack. A breaching activity includes the reduction of minefields, other explosive hazards, and other obstacles. Generally, breaching requires significant combat engineering support to accomplish.

F-3. *Reduction* is the creation of lanes through a minefield or obstacle to allow passage of the attacking ground force (JP 3-15). A *lane* is a route through, over, or around an enemy or friendly obstacle that provides passage of a force (ATP 3-90.4). The route may be reduced and proofed as part of breaching, constructed as part of the obstacle, or marked as a bypass. The number and width of lanes vary depending on the enemy situation, the size and composition of the assaulting force and the scheme of movement and maneuver.

F-4. *Proof* is the verification that a lane is free of mines or explosive hazards and that the width and trafficability at the point of breach (POB) are suitable for the passing force (ATP 3-90.4). Proofing can be conducted visually, electronically, or mechanically. Some mines are resistant to reduction assets and may require a combination of breaching techniques; for example, magnetic and double impulse mines may resist a mine clearing line charge (known as a MICLIC) blast. Proofing is an important component of breaching considering the wide variety of explosive obstacle threats in use today.

F-5. Most combined arms breaching is conducted by a IBCT or a battalion-size task force as a tactical mission, but higher echelons may also execute operational-level combined arms breaching tasks. Significant engineer augmentation from echelon above brigade is typically required to enable an IBCT breach or a battalion task force hasty or deliberate breach. (Refer to ATP 3-90.4 for additional information.)

TYPES OF BREACH

F-6. Army forces are task organized specifically for an operation to provide a fully synchronized combined arms team. Most operations lie somewhere along a continuum between two extremes—deliberate operations and hasty operations. (See paragraph 1-36.) Attacks take place along this continuum (commonly referred to as a—deliberate attack or hasty attack) based on the knowledge of enemy capability and disposition and the intentions and details of friendly force planning and preparation. Deliberate attack and hasty attack refer to the opposite ends of that continuum and describe characteristics of the attack. As breaching may be required to support an attack anywhere along this continuum. Breaching activities must be adapted to exploit the situation. The level and type of planning distinguish which of the three general types of breaching (deliberate, hasty, and covert) are used to meet mission variables. (Refer to ATP 3-90.4 for additional information.)

DELIBERATE BREACH

F-7. A deliberate breach is the creation of a lane through a minefield or a clear route through a barrier or fortification, which is systematically planned and carried out. A deliberate breach is used against a strong defense or complex obstacle system. It is similar to a deliberate attack, requiring detailed knowledge of the defense and obstacle systems. It is characterized by the planning, preparation, and buildup of combat power on the near side of obstacles. Subordinate units are task-organized to accomplish the breach. The breach often requires securing the far side of the obstacle with an assault force before or during reduction. Amphibious breaching is an adaptation of the deliberate breach intended to overcome antilanding defenses to allow a successful amphibious landing.

HASTY BREACH

F-8. *Hasty breach* (land mine warfare) is the creation of lanes through enemy minefields by expedient methods such as blasting with demolitions, pushing rollers or disabled vehicles through the minefields when the time factor does not permit detailed reconnaissance, deliberate breaching, or bypassing the obstacle (JP 3-15). A hasty breach is an adaptation to the deliberate breach and is conducted when less time is available. It may be conducted during a deliberate or hasty attack due to lack of clarity on enemy obstacles or changing enemy situations, to include the emplacement of scatterable mines (SCATMINEs).

F-9. An in-stride breach is a type of hasty breach used to describe the situation when a subordinate unit is expected to be able to organize for and conduct a hasty breach with its organic or task-organized assets, without affecting the higher unit scheme of movement and maneuver or commander's intent. For example, an IBCT is considered to be conducting an in-stride breach when a subordinate battalion is able to organize for the breach (support, breach, assault forces) and breach an obstacle without affecting the IBCT scheme of movement and maneuver or the commander's intent. In-stride breach is generally not used below the company level.

COVERT

F-10. A covert breach is the creation of lanes through minefields or other obstacles that is planned and intended to be executed without detection by an adversary. Its primary purpose is to reduce obstacles in an undetected fashion to facilitate the passage of maneuver forces. A covert breach is conducted when surprise is necessary or desirable and when limited visibility and terrain present the opportunity to reduce enemy obstacles without being seen. A covert breach uses elements of deliberate and hasty breaching, as required.

F-11. A covert breach is characterized by using stealth to reduce obstacles, with support and assault forces executing their mission only if reduction is detected. Through surprise, the commander conceals their capabilities and intentions and creates the opportunity to position support and assault forces to strike the enemy while unaware or unprepared. The support force does not usually provide suppressive fire until the initiation of the assault or in the event that the breach force is detected. Covert breaches are usually conducted during periods of limited visibility. A battalion is the principal unit to conduct a covert breach. The covert breach requires a level of detailed planning, information collection, and mission command that is normally beyond the capability of a company. The IBCT is usually too large to maintain the level of stealth necessary to conduct a covert breach. The covert breach is ideally suited for foot-mobile forces.

BREACH AREA

F-12. The *breach area* is a defined area where a breach occurs (ATP 3-90.4). It is established and fully defined by the higher headquarters of the unit conducting the breach. Within the breach area is the POB, the reduction area, the far side objective, and the point of penetration (POP). Their definitions follow:

- Point of breach is the location at an obstacle where the creation of a lane is being attempted (ATP 3-90.4). Initially, POBs are planned locations only. Normally, the breach force determines the actual POBs during the breach.
- Reduction area is a number of adjacent points of breach that are under the control of the breaching commander (ATP 3-90.4). The commander conducting the attack determines the size and location of the reduction area that supports the seizure of a POP. The reduction area is indicated by the area located between the arms of the control graphic for breach. (See FM 3-90-1). The length and width of the arms extend to include the entire depth of the area that must be reduced.
- Far side objective is a defined location oriented on the terrain or on an enemy force that an assaulting force seizes to eliminate enemy direct fires to prevent the enemy from interfering with the reduction of obstacles and allows follow-on forces to move securely through created lanes (ATP 3-90.4). A far side objective can be oriented on the terrain or on an enemy force. The higher headquarters assigns the objective; however, the attacking unit normally subdivides the objective into smaller objectives to assign responsibilities and to control and focus the assault of subordinate forces. When breaching as part of a larger force, seizing the far side objective provides the necessary maneuver space for the higher unit follow-on forces to move securely through the lanes, assemble or deploy, and continue the attack without enemy interference.
- Point of penetration is the location, identified on the ground, where the commander concentrates their efforts at the enemy's weakest point to seize a foothold on the farside objective (ATP 3-90.4). This is achieved along a narrow front through maneuver and direct and indirect fires that are accurately placed against enemy forces. A commander conducting a breach establishes a POP that supports planning locations for the reduction area and the seizure of the far side objective.

F-13. The breach area must be large enough to allow the attacking unit to deploy its support force and extend far enough on the far side of the obstacle to allow follow-on forces to deploy before leaving the breach area. One technique is to establish the breach area using phase lines (PLs) or unit boundaries. The PL defining the far side of the breach area may be established as a battle handover line (figure F-1, page F-4).

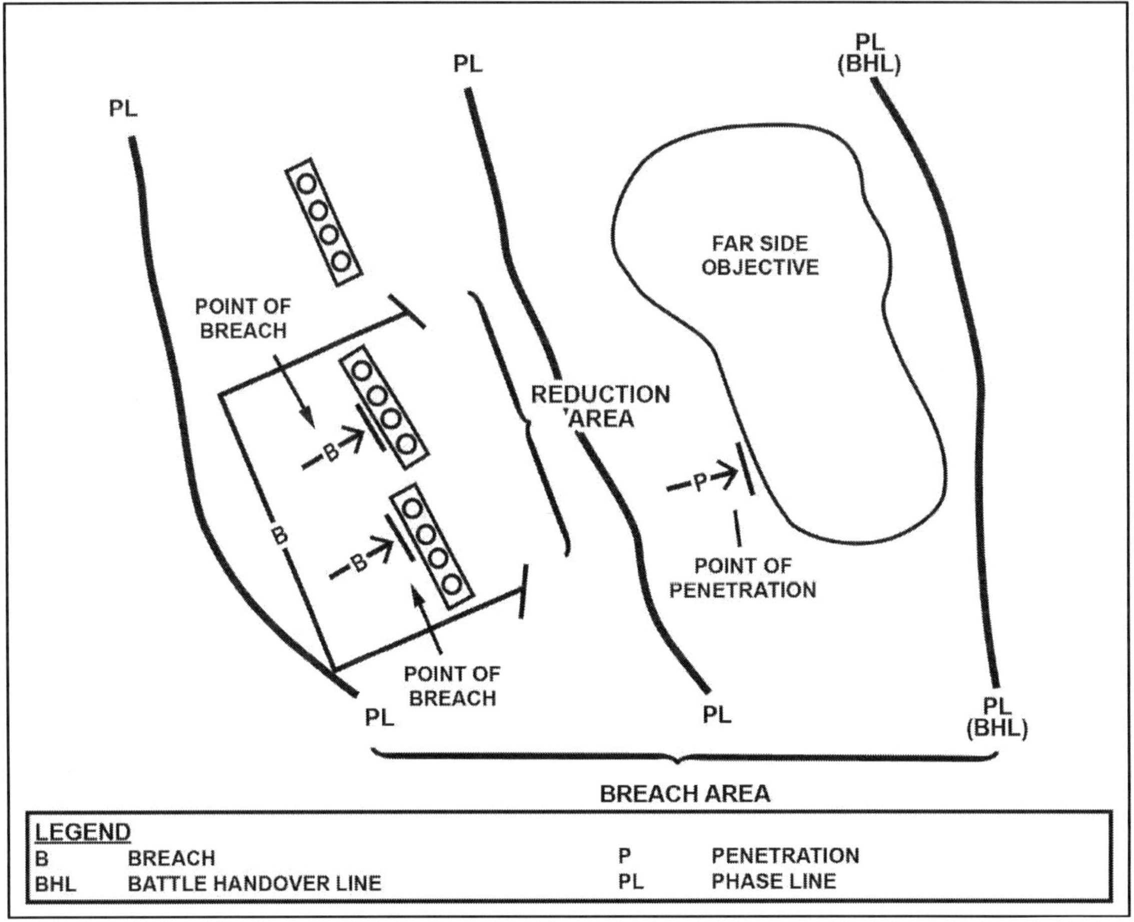

Figure F-1. Breach area

BREACHING TENETS

F-14. Breaching activities are planned by incorporating the breaching tenets within the planning steps of the military decisionmaking process (MDMP). The need to conduct a breach is determined based on the identification of specified, implied, and essential tasks for mobility as part of mission analysis, step 2 of the MDMP. A unit may be tasked to conduct a breach in support of the higher headquarters mission, the commander's intent, and the scheme of movement and maneuver; or it may be implied based on the enemy situation, the terrain (mobility corridors), and the commander's intent. (ATP 3-90.4, table 3-4 provides a detailed listing of breach planning considerations in relation to the MDMP.)

F-15. Breaching missions are characterized by applying breaching tenets. Breaching tenets apply whenever an obstacle is encountered, whether friendly forces are conducting an attack or route clearance operations. These tenets are integrated during planning. The tenets are—

- Intelligence.
- Breaching fundamentals (suppress, obscure, secure, reduce, and assault [known as SOSRA]).
- Breaching organization.
- Mass.
- Synchronization.

INTELLIGENCE

F-16. The ability to identify how the enemy applies obstacles to the terrain is critical to a commander's success. The commander and staff conduct IPB to develop initial situational templates and priority intelligence requirements. Intelligence gathered by reconnaissance forces is essential to developing a finalized situational template and final POB locations. Unverified enemy situational templates may cause friendly forces to deploy to reduce obstacles early, waste time trying to locate nonexistent obstacles, develop courses of action using ineffective obstacle reduction methods, and fail to locate bypasses or become surprised by an obstacle.

F-17. The engineer staff planner shows template enemy obstacles on the situational templates based on—

- Threat patterns based on past operations and emerging tactics, techniques, and procedures.
- Enemy countermobility capabilities (based on manpower, equipment, materials, and time available), to include SCATMINEs.
- Terrain and weather effects.
- The range of enemy weapon systems covering obstacles and emplacing SCATMINEs.

F-18. Augmentation of reconnaissance forces by engineer squads or sections may be used as part of the overall information collection effort. Examples of information used to produce obstacle intelligence products include—

- Location of existing or reinforcing obstacles.
- Orientation and depth of obstacles.
- Soil conditions. (Determines ability to use mine plows, if available.)
- Lanes or bypass locations.
- Composition of minefields (buried or surface laid antitank and antipersonnel mines).
- Types of mines and fuses. (Determines effectiveness of mechanical or explosive reduction techniques.)
- Composition of complex obstacles.
- Location of direct and indirect fire systems overwatching obstacle.

BREACHING FUNDAMENTALS

F-19. Breaching fundamentals are integrated into the planning process and always apply when reducing a defended obstacle. This includes breaching, gap crossing, and route clearance missions. The breach fundamentals: suppress, obscure, secure, reduce, and assault are described by the memory aid SOSRA.

Suppress

F-20. Suppress is a tactical mission task that results in the temporary degradation of the performance of a force or weapons system below the level needed to accomplish its mission. Suppression protects friendly forces reducing and maneuvering through an obstacle. Successful suppression generally triggers the rest of the actions at the POB. Fire control measures ensure that all fires are synchronized with other actions at the POB. The mission of the support force is to suppress the enemy overwatching the obstacle. The breach force also provides additional suppressive fires as the situation dictates; however, its primary focus is on reducing the obstacle. In many situations, the weapons company may be ideal to provide suppression (see figure F-2, page F-8).

Obscure

F-21. Obscuration degrades observation and target acquisition of the enemy forces while concealing friendly force reduction and assault activities. Obscuration planning factors include wind direction, type of obscuration systems available (mechanical smoke, artillery delivered, mortar delivered, smoke pots), and the capabilities and limitations of these systems. In urban areas, indirect delivered obscuration and suppressive fires will be more restricted. In some situations, using mortars (because of the ability to fire high-level trajectory), smoke pots, and smoke grenades rather than artillery-fired obscurants may be more effective. Normally, obscuration starts with smoke delivered by indirect fire that builds quickly, followed by mechanical or smoke pots that have a longer duration but take more time to place and build. Typically, the most effective placement of obscuration is between the obstacle and the overwatching enemy forces. (See ATP 3-11.50 for additional information on obscuration.)

Secure

F-22. *Secure* is a tactical mission task that involves preventing a unit, facility, or geographical location from being damaged or destroyed as a result of enemy action (FM 3-90-1). Identifying the extent of the enemy defense is critical in selecting the appropriate technique to secure the POB. The POB must be secured before reducing the obstacle. Friendly forces secure the POB to prevent enemy forces from interfering with the reduction of lanes and passage of assault forces. The breach force must be resourced with sufficient maneuver assets to provide local security against the enemy that the support force cannot adequately engage. Elements within the breach force that secure the reduction area may also be used to suppress the enemy once reduction is complete.

Reduce

F-23. *Reduce* is a mobility task to create and mark lanes through, over, or around an obstacle to allow the attacking force to accomplish its mission (ATP 3-90.4). Reduction cannot be accomplished until effective suppression and obscuration is achieved and the POB secured. The breach force will reduce, proof, and mark the required number of lanes to pass the assault force through the obstacle. The number and width of lanes needed depend on the enemy situation, terrain, size and composition of the assault force, and scheme of movement and maneuver. Follow-on forces will continue to improve and reduce the obstacle when required. When possible, the breach force also should try to secure a foothold to assist in the passage of the assault force.

Assault.

F-24. The assault force's primary mission is to seize terrain on the far side of the obstacle in order to prevent the enemy from placing or observing direct and indirect fires on the reduction area. If planned, the battle handover with follow-on forces occurs.

Breaching Organization

F-25. Establishing the breach organization facilitates the application of the breaching fundamentals. The commander and staff develop courses of action that organize friendly forces into a support force, a breach force, and an assault force to quickly, and effectively execute the breach fundamentals. Table F-1 shows the relationship between the breach organization as well as the responsibilities of each force.

Table F-1. Breaching organization and responsibilities

Breach Organization	*Responsibilities*
Support force	•Suppress an enemy capable of placing direct fires on the reduction area to protect the breach force as it reduces the obstacle and the assault force as it passes through the created lane. •Fix enemy forces to isolate the reduction area. •Control obscuration.
Breach Force	•Reduce, proof, and mark the necessary number of lanes through the obstacle. •Report the status and location of created lanes. •Provide local security on the near side and far side of the obstacle. •Provide additional suppression of enemy overwatching the obstacle. •Provide additional obscuration in the reduction area. •Assist the passage of the assault force through created lanes.
Assault Force	•Seize the far side objective. •Reduce the enemy protective obstacles. •Provide clear routes from the reduction area to the battle handover line for follow-on forces. •Prevent the enemy from placing direct fires on follow-on forces as they pass through the created lanes. •Conduct battle handover with follow-on forces. •Provide reinforcing fires for the support force. •Destroy the enemy on the obstacle far side that is capable of placing direct fires on the reduction area.

Support Force

F-26. Support force responsibilities are to isolate the reduction area with direct and indirect fires and suppress enemy's direct and indirect fire at the POB. The support force controls friendly direct and indirect fires and obscuration within the breach area.

Breach Force

F-27. The breach force must have sufficient combat power to secure the POB as well as sufficient reduction assets to reduce required number of lanes through the obstacle. Critical friendly zones should be activated at the POB before commitment of the breach force to protect it from enemy indirect fires.

Assault Force

F-28. The assault force's primary mission is the destruction of enemy forces on the far side of the obstacle to prevent the enemy from placing direct fires on the breach lanes. In complex or restrictive terrain, the assault force may be constrained to a single lane and the assault force commander must ensure that the sequencing of forces through the lane is appropriate to achieve the mission.

Reverse Planning

F-29. The size and composition of the support, breach, and assault forces (breach organization) are determined during course of action development using reverse planning. Reverse planning begins with actions on the objective and moves backward to the line of departure, since seizing an objective is typically the decisive point and directly tied to mission accomplishment. Battalion planners use reverse planning and force ratios to determine the size and composition of the maneuver forces that will perform the tasks that support the decisive and shaping operations for each course of action. Reverse planning for breaching is performed using the following steps:

- Step 1. Identify available reduction assets.
- Step 2. Template enemy obstacles.
- Step 3. Understand the scheme of movement and maneuver.
- Step 4. Identify the number of required breach lanes.
- Step 5. Identify the assets required to reduce, proof, and mark lanes.
- Step 6. Task-organize reduction assets within the maneuver force.

F-30. Detailed reverse planning initiates during the IPB and the development of enemy situational template. The scheme of maneuver, engineer operations, fires, air defense, and actions at the obstacle are based upon this common situational template. The situation template, developed by the intelligence staff officer (S-2) depicting enemy direct- and indirect-fire coverage of templated enemy obstacles, determines the size and composition of the support force.

F-31. The enemy's ability to interfere with the breach force at the POB determines size and composition of the security element within the breach force. The enemy's ability to mass fires on the POB determines the amount of suppression required as well as the size and composition of the breach force. Lane requirements and the composition of obstacles drive the amount and type of reduction assets needed by the breach force. The engineer staff planner focuses on the allocation of reduction assets.

F-32. Actions on the objective drive the size and composition of the force that conducts the final assault onto the objective as part of an attack, which dictates lane requirements (the number and location of required lanes). The engineer staff planner for the battalion and other planners determine how best to allocate reduction assets within the arrayed forces to facilitate the scheme of movement and maneuver for each course of action (figure F-2, page F-8).

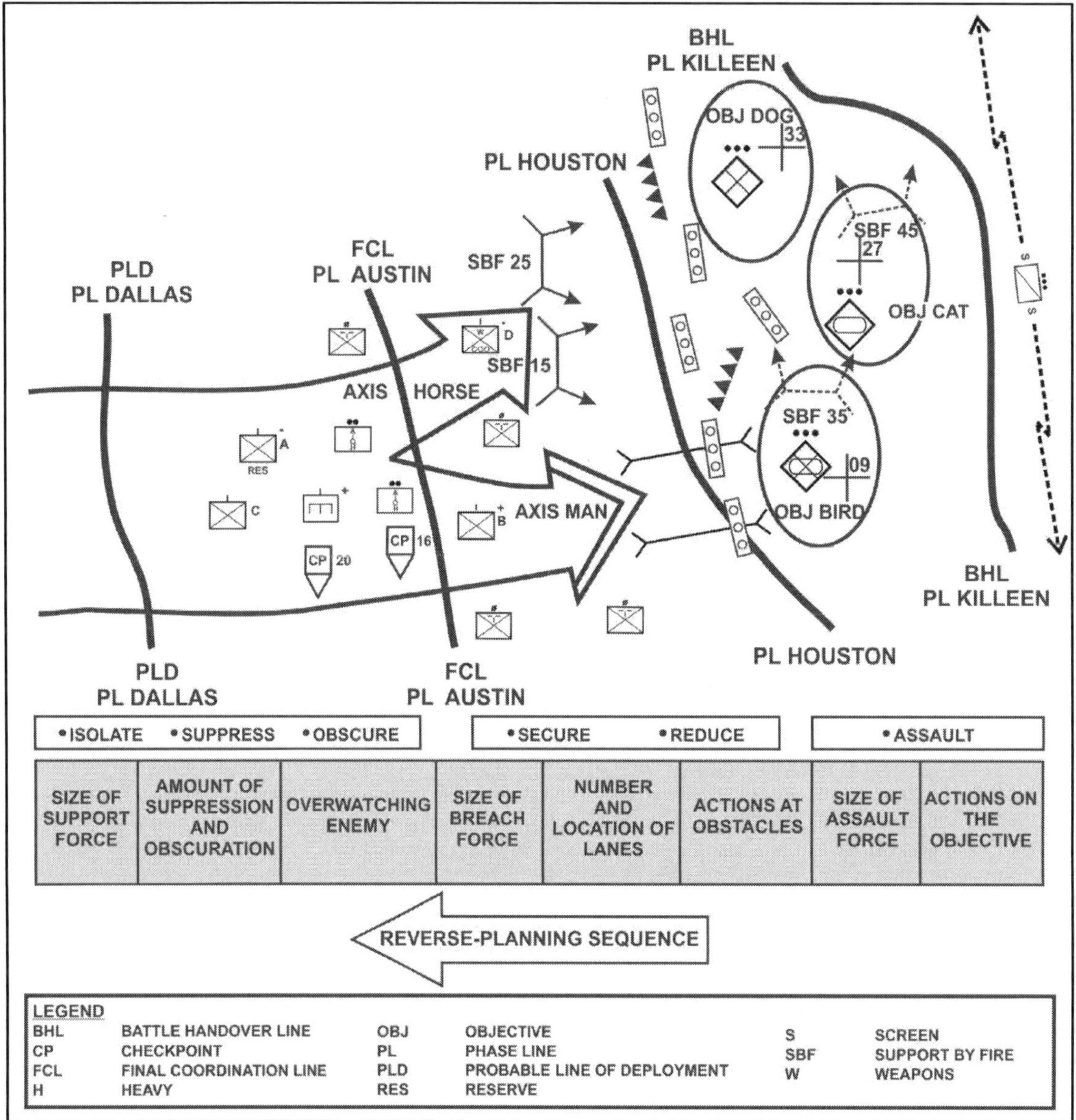

Figure F-2. Breach organization course of action, example

Massed Combat Power

F-33. Breaching is conducted by rapidly applying concentrated efforts at a POB to reduce the obstacle and penetrate the defense. A POB is planned where the enemy can be isolated, fixed, or disrupted. The location selected for breaching depends largely on enemy weakness, where its covering fires are minimized. If friendly forces cannot find a natural weakness, they create one by fixing most of the enemy force and isolating a small portion of it for attack. Denying the enemy's ability to mass combat power against the breach is achieved by isolating, fixing, or disrupting the defending forces; synchronizing the application of friendly combat power; and simultaneously breaching at separate locations to prevent the enemy from concentrating fires and defeating a breaching force in detail.

Synchronization

F-34. Synchronization of combined-arms elements to achieve the breach fundamentals is essential. The commander achieves synchronization through detailed reverse planning of offensive tasks from the objective back to the assembly area, by issuing clear subordinate unit instructions, planning effective mission command, and ensuring their forces are well rehearsed. The commander may use an execution matrix to synchronize forces. An execution matrix lists subunit instructions sequentially in relation to key events or the sequence of the attack. It provides subordinate commanders with an understanding of how the mission and those of adjacent units fit into the overall plan. It allows subordinates to better track the battle and coordinate their own maneuver with that of adjacent units. See ATP 3-90.4 for a sample execution matrix for a breaching mission.

BATTALION DELIBERATE BREACH

F-35. The following paragraphs discuss the detailed planning, preparation, and execution necessary in conducting a combined-arms breach during deliberate operations. (Refer to ATP 3-90.4 for additional information.)

PLANNING CONSIDERATIONS

F-36. Planning a breaching operation begins with the command and engineer estimates. The battalion S-2 templates the enemy's order of battle, and the engineer planner assesses its engineer capabilities. The enemy's tactical and protective obstacles are doctrinally templated by the S-2 and engineer. The staff develops courses of action using the templates, and the engineer develops their scheme of engineer operations for each course of action. After selecting a course of action, the commander allocates available assets to the breach, assault, and support forces to ensure that they can accomplish their assigned tasks.

F-37. Identifying the enemy's vulnerability is important so that the force can mass direct and indirect fires and maneuver against that weakness. The battalion isolates a portion of the enemy to achieve the desired combat ratio at the point of assault. It achieves mass by directing fire on the enemy from multiple directions and by narrowing attack zones to concentrate its force against a smaller defending element.

F-38. When the attack requires breaching two or more complex obstacle systems, the commander must retain enough engineers and sufficient breaching assets to reduce subsequent obstacles. The commander must not commit all the engineers to breach the first obstacle system unless they are willing to risk their capability to breach follow-on obstacles. Depleted engineer forces need significantly more time to conduct follow-on breaches.

F-39. The breach and assault forces may require fires and obscurants under their control in addition to that controlled by the support force. The support, breach, and assault forces place direct fires on enemy positions. This makes synchronization of direct and indirect fires extremely complex. Fire control is planned in detail using surface danger zones, risk estimate distances, minimum safe distances, well-understood control measures, and triggers that are carefully rehearsed.

F-40. When a battalion conducts a combined-arms breach during a deliberate operation or plans to conduct a passage of lines of a large force after a breach, breach plans must include detailed planning for the staging and movement of follow-on forces and equipment. The plan should consider improvements to—

- The breach lanes.
- Markings of the lanes.
- Contact points and guides.
- Preparation for an enemy counterattack.
- Repositioning of indirect fire assets to provide extended coverage.
- Control measures for follow-on forces to continue the attack.

PREPARATION

F-41. The battalion continues an intelligence collection plan using the scout platoon, snipers, unmanned aircraft system, attached engineers, patrols, and aerial reconnaissance. The S-2 and the battalion engineer planner continually refine the template based on intelligence. The battalion may have to adjust task organization as it uncovers more details of the defense and obstacle system. It also uses this information during the combined-arms rehearsals.

Rehearsals

F-42. The battalion meticulously plans, manages, and controls rehearsals. The battalion S-3 allocates time for each unit to perform a combined-arms rehearsal. When possible, the force rehearses the operation under the same conditions expected during the actual engagement, including battlefield obscuration, darkness, CBRN posture, and inclement weather. The battalion chooses terrain as similar as possible to that of the operational area and constructs a practice obstacle system based on obstacle intelligence products. Rehearsals include a leader and key personnel walk-through as well as individual rehearsals by support, breach, and assault forces.

F-43. When the force commander rehearses the breaching operation, the commander rehearses several contingency plans. The contingencies should include possible enemy counterattacks and attack by enemy indirect fire systems

Generating Obstacle Intelligence

F-44. The success of combined-arms breaching during a deliberate operation depends heavily on the success of the information collection plan. The S-2 develops the collection plan, with the scout platoon and snipers concentrating on confirming enemy locations. The engineers focus on gathering intelligence on obstacle orientation and composition as well as on the types of fortifications the battalion may encounter. Unmanned aircraft systems can gather information on approaches to and composition of the obstacles, minefields, and enemy reserve forces locations and composition. Intelligence is used to refine the task organization of support, breach, and assault forces and the scheme of maneuver.

Execution

F-45. The force crosses the line of departure or probable line of deployment organized to conduct the combined-arms breach. If the battalion encounters obstacles en route, it executes the breach with this organization. On arrival, the battalion's scout platoon and snipers adjusts artillery fires on the enemy positions to cover deployment of the support force. Snipers engage key targets to reduce enemy command and control, and destroy antitank and crew-served weapons. The support force moves into position and establishes its support by fire position. Breach and assault forces move into position and prepare to execute their tasks. The battalion commander continues to incorporate last-minute information into their plan and make final adjustments of positions and locations.

F-46. The support force occupies its support by fire position and immediately begins suppressing the enemy. The support force fire support team and battalion FSO execute group targets planned on enemy positions. Mortar and artillery obscurants are adjusted to provide initial obscuration of the breaching site from enemy target acquisition. Depending on the wind conditions, the support force or the breach force will provide mechanical obscurants or emplace smoke pots to continue obscuration. The breach force begins movement once suppression and obscurants are effective, based on clearly defined commitment criteria. Timing is critical since the high volume of suppressing fires and smoke can be sustained only for a short duration. Support by fire positions have interlocking sectors of fires and are positioned to ensure suppression of the enemy's positions.

F-47. Once suppression and obscuration have built to effective levels, the breach force moves forward to the breaching site. The engineers create the lanes while the combined-arms breach force provides for local security. As they finish the lanes, engineers mark the lanes to assist the assault and following forces in maneuvering to the lanes. The assault force penetrates the objective after receiving the order from the battalion commander. The assault force must seize and clear the objective, prepare for counter attacks and be prepared to pass additional forces for attacks beyond the objective. Due to the complexity of the breach, the mission command systems spread out to ensure synchronization. For example, the battalion S-3 may control a multi-company team support force while the battalion commander positions where best to control the entire breaching operation.

F-48. The obstacle system acts as a chokepoint and is dangerous even after the battalion has overcome the defenses. The battalion constructs additional lanes to speed the passage of follow-on forces. Next, it widens the lanes to allow two-lane traffic through the obstacles and constructs switch lanes to prevent blocking by disabled vehicles or artillery fires. Deliberate marking and fencing systems are installed, and military police, when assigned, establish the necessary traffic control. Eventually, rear-area engineer or Infantry forces clear the obstacles and eliminate the chokepoint. After passage through the lanes, the maneuver force continues its mission.

F-49. Both the breaching and follow-on force must be aware of the potential for the enemy to reseed breached obstacles with remotely delivered SCATMINEs or other rapidly emplaced obstacles. The breaching commander may develop a response plan and position remaining reduction assets near the breach lane or lanes to rebreach, repair, or improve lanes as necessary. In addition, the commander may develop a reaction plan for maneuver or other forces that encounter a reseeded portion of the obstacle while passing through the lane. The commander of the follow-on force, regardless of the reported status of the breach lanes the commander is about to pass through, should organize reduction assets forward in their formation that are prepared to rebreach, repair, or improve these lanes as necessary.

BATTALION HASTY BREACH

F-50. Hasty breaching operations are conducted when the enemy situation is vague and the commander may be required to execute the combined-arms breach with their current task organization. Therefore, the battalion commander must either task-organize their subordinate company teams with sufficient combat power to conduct company team-level breaching operations or have a plan that allows for the flexible application of combat power necessary to execute breaching operations. When conducting offensive tasks such as a movement to contact, or while when en route during an approach march or attack, and when conducting a passage of lines and movements through defiles, the battalion commander must address breaching operations. The battalion breach planning considerations and process discussed previously apply to combined-arms breach planning during hasty operations as well. The only difference is the organizational echelon at which the breach is planned, prepared, and executed.

F-51. Subsequent to course of action development, the commander and staff anticipate where units are most likely to encounter obstacles based on the scheme of maneuver and situational template. From this analysis, the commander refines the task organization, if necessary, in order to apply the combat power required for executing the templated breach. The battalion engineer planner recommends a task organization of engineer platoons and critical breaching equipment to create enough lanes for the breaching unit. The commander maintains a mobility reserve that can create additional lanes for follow-on forces. This mobility reserve can also mass reduction assets if the battalion must transition to a deliberate operation. The battalion FSO designs their fire plan to provide priority of fires and obscurants to company teams likely to conduct a breach. Above all, the commander task-organizes company teams for the mission first, then modifies the task organization where necessary to provide company teams with the additional forces needed to conduct independent breaching operations as part of the battalion effort.

This page intentionally left blank.

Appendix G
CBRN Defense and Countering Weapons of Mass Destruction

Chemical, biological, radiological, and nuclear defense are measures taken to minimize or negate the vulnerabilities to, and effects of, a chemical, biological, radiological, or nuclear hazard or incident (JP 3-11). The battalion commander integrates chemical, biological, radiological, and nuclear (CBRN) defense considerations into mission planning depending on the CBRN threat. This includes CBRN defense activities, such as contamination avoidance, individual and collective protection, and decontamination. CBRN protection may slow the tempo, degrade combat power, and increase logistics requirements. CBRN reconnaissance and surveillance (R&S) consumes resources, especially time. Personnel wearing individual protective equipment find it difficult to work or fight for an extended period. (Refer to FM 3-11 and ATP 3-11.37 for additional information.)

Countering weapons of mass destruction (CWMD) is the efforts against actors of concern to curtail the conceptualization, development, possession, proliferation, use, and effects of weapons of mass destruction, related expertise, materials, technologies, and means of delivery (JP 3-40). At the tactical level combined arms teams conduct specialized activities to understand the environment, threats, and vulnerabilities; control, defeat, disable, and dispose of WMD; and safeguard the force and manage consequence. (Refer to JP 3-40 for additional information.)

This appendix provides an overview of CBRN defense and CWMD activities. The appendix introduces a fictional scenario used as a discussion vehicle for illustrating one of many ways the Infantry battalion can support CWMD activities.

OVERVIEW

G-1. The use of WMD in future conflict is inevitable. Many threat organizations already possess WMD, (chemical, biological, radiological, or nuclear weapons) and their delivery systems (for example, rockets and artillery). Enemies employ these WMD to obtain a relative advantage over United States forces to achieve their objectives. Threat organizations that do not currently possess WMD consistently seek opportunities to acquire them.

G-2. The possibility of the use of improvised chemical, biological, and nuclear as well as radiological dispersal devices by terrorist groups cannot be overlooked. Planning must routinely address the use of each of these as well as protective measures against conventional CBRN weapons. The potential catastrophic effects associated with the threat or use of WMD adds greater uncertainty to an already complex environment.

G-3. The intentional or unintentional release of CBRN material, including toxic industrial materials, and nontraditional agents can seriously challenge military operations. Chemical weapons could be used early in an operation or from its onset to hinder U.S. and partner nation's movement; disrupt its command, control, and communications; produce casualties; destroy or disable equipment; and disrupt operations. Biological weapons could target rear area objectives such as food supplies, water sources, troop concentrations, convoys, and urban and rural population centers, rather than frontline forces. Any of these materials may be employed separately or together and may supplement conventional weapons. The IBCT commander and subordinate commanders must anticipate and plan for CBRN defense and CWMD activities.

Chemical, Biological, Radiological, and Nuclear Environment

G-4. *Chemical, biological, radiological, and nuclear environment* is an operational environment that includes chemical, biological, radiological, and nuclear threats and hazards and their potential resulting effects. (JP 3-11). CBRN environment conditions can be the result of deliberate enemy or terrorist actions or the result of an industrial accident. CBRN threats include the intentional employment of, or intent to employ, weapons or improvised devices to produce CBRN hazards. CBRN hazards include those created from accidental or intentional releases of toxic industrial materials, biological pathogens, or radioactive matter. Toxic industrial material is a generic term for toxic or radioactive substances in solid, liquid, aerosolized, or gaseous form that may be used or stored for industrial, commercial, medical, military, or domestic purposes. Toxic industrial material may be chemical, biological, or radiological. (Refer to FM 3-11 for more information on CBRN hazards)

G-5. *A chemical, biological, radiological, and nuclear hazard* is chemical, biological, radiological, and nuclear elements that could create adverse effects due to an accidental or deliberate release and dissemination. (JP 3-11). CBRN hazards may result from WMD employment. The key distinction between WMD and CBRN hazards is that WMD refers to the actual weapon, while CBRN refers to the contamination or effects resulting from the employment of WMD and from the dispersal of CBRN materials. When Department of Defense capabilities are called upon to conduct CBRN response activities, they essentially will be responding to CBRN hazards or contamination. such as—

- The deposit, absorption, or adsorption of radioactive material or a biological or chemical agent on or near a structure, area, person, or object.
- Food or water that is unfit for consumption.

Chemical Hazards

G-6. A *chemical hazard* is any chemical manufactured, used, transported, or stored that can cause death or other harm through toxic properties of those materials, including chemical agents and chemical weapons prohibited under the Chemical Weapons Convention as well as toxic industrial chemicals (JP 3-11). This includes—

- Chemical weapons, toxic chemicals specifically designed as a weapon.
- Chemical agents, chemical substance that is intended for use in military operations to kill, seriously injure, or incapacitate, mainly through physiological effects.
- Toxic industrial chemicals, chemicals that are developed or manufactured for use in industrial operations or research.

G-7. A chemical weapon is a munition or device, specifically designed to cause death or other harm through the toxic properties of specified chemicals. Chemicals are released as a result of the employment of such munition or device; and any equipment specifically designed for use directly in connection with the employment of munitions or devices.

G-8. A *chemical agent* is a chemical substance that is intended for use in military operations to kill, seriously injure, or incapacitate mainly through its physiological effects (JP 3-11). Chemical agents (see FM 3-11.9), cause casualties, degrade performance, slow maneuver, restrict terrain, and disrupt operations (table G-1). They can cover large areas and may be delivered as liquid, vapor, or aerosol. Chemical agents can be delivered by artillery, mortars, rockets, missiles, aircraft spray, bombs, land mines, and covert means.

Table G-1. Characteristics of chemical agents

Agent	Nerve	Blister	Blood	Choking
Protection	Mask and IPE	Mask and IPE	Mask	Mask
Detection	JCAD, M256A2, CAM, and M8 and M9 paper	JCAD, M256A2, CAM, and M8 and M9 paper	JCAD, M256A2	Odor (freshly mowed hay)
Symptoms	Difficult breathing, drooling, nausea, vomiting, convulsions, and blurred vision	Burning eyes, stinging skin, irritated nose	Convulsions and coma	Coughing, nausea, choking, headache, and tight chest
Effects	Incapacitates	Blisters skin, damages respiratory tract	Incapacitates	Floods and damages lungs
First Aid	ATNAA and CANA DECON	As for 2nd and 3rd degree burns	None	Keep warm and avoid movement
Decontamination	RSDL and flush eyes with water	RSDL and flush eyes with water	RSDL	RSDL

Legend:

ATNAA	antidote treatment nerve agent auto-injector	IPE	individual protection equipment
DECON	decontamination	JCAD	joint chemical agent detector
CAM	chemical agent monitor	RSDL	reactive skin decontamination lotion
CANA	convulsive antidote nerve agent		

Biological Hazards

G-9. *Biological hazard*—an organism, or substance derived from an organism, that poses a threat to human or animal health. (JP 3-11). This can include medical waste or samples of a microorganism, virus, or toxin from a biological source that can impact human health. Biological agents (see FM 3-11.9) are microorganisms that can spread disease through humans and agriculture. They are categorized as pathogens, disease-producing microorganisms, or toxins. A *biological agent is* a microorganism (or a toxin derived from it) that causes disease in personnel, plants, or animals or causes the deterioration of materiel (JP 3-11).

Radiological Hazards

G-10. Radiological hazards include any electromagnetic or particulate radiation that is capable of producing ions to cause damage, injury, or destruction. Radiological hazards also include toxic industrial materials. Enemies could disperse radioactive material in a number of ways, such as—

- Arming the warhead of a missile with radioactive material from a nuclear reactor.
- Releasing low-level radioactive material intended for use in industry or medicine.
- Disseminating material from a research or power-generating nuclear reactor.

Nuclear Hazards

G-11. Nuclear weapon effects are qualitatively different from biological or chemical weapon effects. The nature and intensity of nuclear detonation effects are determined by the type of weapon, its yield, and the physical medium in which the detonation occurs. The effects of a nuclear detonation include—

- Blast produces shockwaves that can cause critical injuries to personnel and destroy material.
- Thermal radiation causes severe burns and secondary fires.
- Electromagnetic pulse can cause widespread disruption or electrical and electronic equipment.
- Ionizing radiation is a significant threat to personnel and materiel.
- Fallout is residual radiation and may be a lingering, widespread hazard that limits military operations.

G-12. Cover and shielding offer the best protection from the immediate effects of a nuclear detonation; this includes cover in fighting positions with 18 inches overhead cover, culverts, and ditches. Soldiers should cover exposed skin and stay down until the blast wave passes and debris stops falling. Immediately after a nuclear detonation, continuous radiation monitoring should begin.

G-13. Operations in a nuclear environment are complicated by the necessity to control exposure of personnel to nuclear radiation. An operation exposure guide determines the maximum radiation dose to which units may be exposed and still accomplish a mission. Determination of this dose is based on the accumulated dose or radiation history of the unit.

Chemical, Biological, Radiological, and Nuclear Passive Defense

G-14. Operationally, CBRN passive defense maintains the commander's ability to continue military operations in a CBRN environment while minimizing or eliminating the vulnerability of the force to the degrading effects of those CBRN threats and hazards. Tactical-level doctrine has traditionally segregated CBRN passive defense into the distinct principles of contamination avoidance, protection, and decontamination. While these principles remain valid, they are now recognized to be components of the more expansive concepts of hazard awareness and understanding and contamination mitigation. Since hazard awareness and understanding largely focuses strategic aspects of operations in a CBRN environment, tactical level doctrine is organized around the key activities (figure G-1) of CBRN protection and contamination mitigation.

- *Chemical biological, radiological, and nuclear protection* consists of measures taken to keep chemical, biological, radiological, and nuclear threats and hazards from having an adverse effect on personnel, equipment, and facilities (ATP 3-11.32). CBRN protection encompasses the following activities: protect personnel, equipment, and facilities.
- *Contamination mitigation* is described as the planning and actions taken to prepare for, respond to, and recover from contamination associated with all chemical, biological, radiological, and nuclear threats and hazards in order to continue military operations (JP 3-11). The two subsets of contamination mitigation are contamination control and decontamination. (Refer to ATP 3-11.32 a detailed discussion of CBRN passive defense activities.)

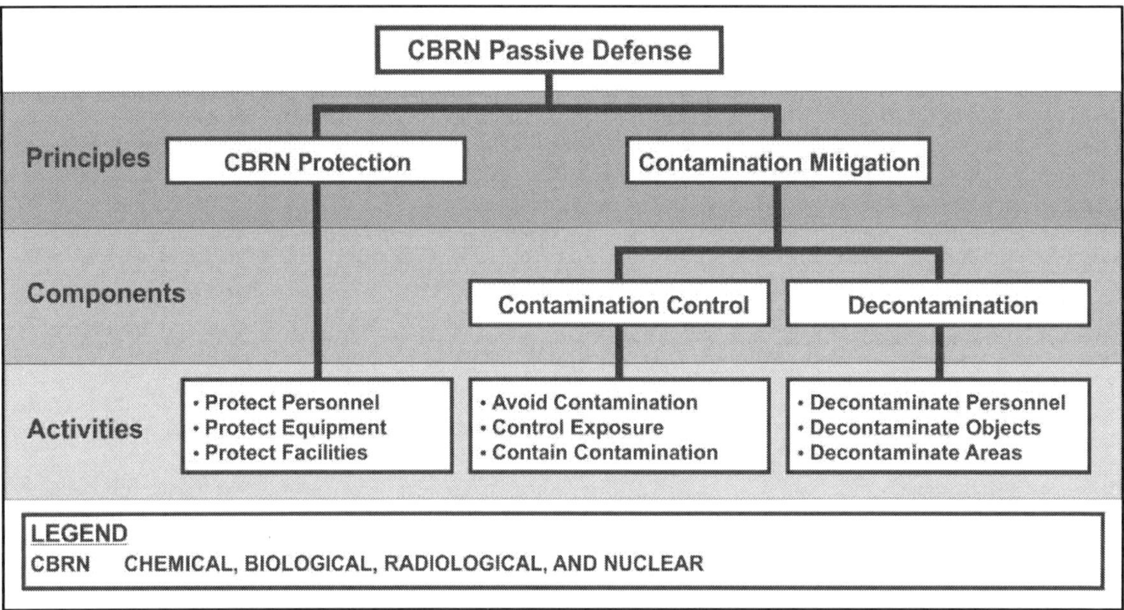

Figure G-1. Chemical, biological, radiological, and nuclear passive defense architecture

CHEMICAL, BIOLOGICAL, RADIOLOGICAL, AND NUCLEAR DEFENSE TASKS

G-15. CBRN passive defense includes measures taken to minimize or negate the vulnerability to, and effects of, CBRN attacks. (See ATP 3-11.32.) Passive defense focuses on maintaining the battalion's ability to continue operations in a CBRN environment. Success depends on the effective integration of equipment; CBRN training; and CBRN tactics, techniques, and procedures. Passive defense measures by the battalion designed to mitigate the immediate effects of a CBRN incident enable and protect the force conducting the operation. The application of CBRN defense task address the hazards created by CBRN incidents or accidents and help minimize vulnerabilities, protect friendly forces, and maintain the battalion's operational tempo.

CBRN Protection

G-16. CBRN protection measures are taken to keep chemical, biological, radiological, and nuclear threats and hazards from having an adverse effect on personnel, equipment and facilities. Tasks that enable CBRN protection include the following:

- Employ IPE and other CBRN protective equipment.
- Establish CBRN alarm conditions.
- Exercise personal hygiene and force health protection programs.
- Utilizing shielding or protective cover.

G-17. CBRN protection is an integral part of all operations. Techniques that work for avoidance also work for protection such as shielding Soldiers and units and shaping the battlefield. Other forms of protection involve sealing or hardening positions, protecting Soldiers, assuming appropriate mission-oriented protective posture (MOPP) levels (table G-2), reacting to attack, and using collective protection. Individual protective items include the protective mask, joint service lightweight integrated suit technology, overboots, and gloves. The higher-level commander above the IBCT establishes the minimum level of protection. Subordinate units may increase this level as necessary but may not decrease it. The joint service lightweight integrated suit technology may be worn for 45 days with up to six launderings or up to 120 days with no launderings. The joint service lightweight integrated suit technology can be worn for 24 hours once contaminated. The overboots provide 60 days of durability and 24 hours of protection against liquid chemical agents.

Table G-2. Mission oriented protective posture levels

Level/Equipment	MOPP Ready	MOPP0	MOPP1	MOPP2	MOPP3	MOPP4	Mask Only
Mask	Carried	Carried	Carried	Carried	Worn	Worn	Worn***
JSLIST	Ready*	Available**	Worn	Worn	Worn	Worn	
Overboots	Ready*	Available**	Available**	Worn	Worn	Worn	
Gloves	Ready*	Available**	Available**	Available**	Available**	Worn	
Helmet Cover	Ready*	Available**	Available**	Worn	Worn	Worn	

Legend:
* Items available to Soldier within two hours with replacement available within six hours.
** Items must be positioned within arms-reach of the Soldier.
*** Never "mask only" if nerve or blister agent is used in area of operation.
JSLIST joint service lightweight integrated suit technology.
MOPP mission-oriented protective posture.

G-18. Passive measures can be used to monitor for the presence of CBRN hazards. Depending on the threat and probability of use, periodic or continuous techniques are used. An area array of CBRN detectors and/or monitors can be positioned within a given area for detection and early warning of a CBRN incident. ATP 3-11.32 describes specific techniques for monitoring radiological exposure including determining correlation factor data, radiation exposure status and recording exposure. All units initiate continuous monitoring when they receive a fallout warning, when a unit is on an administrative or tactical move, when a nuclear burst occurs, when radiation levels above one centigray per hour are detected by periodic monitoring, and on order of the commander. Except for units on the move, continuous monitoring stops on instructions from the commander or higher headquarters or when the dose rate falls below one centigray per hour.

Contamination Mitigation

G-19. Contamination mitigation describes how the commander should conduct mitigation of hazardous contamination in support of operations to achieve assigned objectives. Contamination mitigation provides the framework for the commander and staff to estimate the proficiency and sufficiency of future contamination mitigation capabilities for passive defense and CBRN response. The commander can control contamination by avoiding contaminated areas, controlling exposure, and containing contamination. Example tasks for contamination control include—

- Marking contaminated areas.
- Establishing control zones.
- Controlling run off contamination.
- Implementing warning and reporting.
- Documenting exposures.

G-20. Use of CBRN weapons creates unique residual hazards that may require decontamination. In addition to the deliberate use of these weapons, collateral damage, natural disasters, and industrial emitters may require decontamination. Contamination forces units into protective equipment that degrades performance of individual and collective tasks. Decontamination restores combat power and reduces casualties that may result from exposure, allowing commanders to sustain combat operations. Use the three principles of decontamination listed below when planning decontamination operations:

- Decontaminate as soon as possible.
- Decontaminate only what is necessary.
- Decontaminate as far forward as possible, which is METT-TC dependent.

G-21. When a CBRN incident occurs, the battalion quickly responds to and initially mitigates the effects of contamination. The battalion performs only those actions required to allow continuation of the mission and, within mission constraints, save lives. To recover, the commander decides whether decontamination is required to restore combat power, and if so, what level of decontamination is required. The levels of decontamination are immediate, operational, thorough, and clearance. (Refer to ATP 3-11.32 for a detailed discussion of decontamination levels.)

G-22. Immediate decontamination minimizes casualties and limits the spread or transfer of contamination. The contaminated individual to save lives and reduce penetration of agent into surfaces carries out immediate decontamination. This may include decontamination of personnel, clothing, and equipment. Immediate decontamination should help prevent casualties and permit the use of individual equipment and key systems.

G-23. Operational decontamination sustains operations by reducing the contact hazard, limiting the spread of contamination, and eliminating or reducing the duration that MOPP equipment should be used. The contaminated unit carries out operational decontamination with possible assistance from an organic decontamination organization. Operational decontamination is restricted to the specific parts of contaminated, operationally essential equipment, material, and work areas to minimize contact and transfer hazards and to sustain operations. This may include individual decontamination beyond the scope of immediate decontamination, decontamination of mission-essential equipment, and limited terrain decontamination. Operational decontamination reduces the level of contamination, thus lessening the chance of spread and transfer.

G-24. Thorough decontamination reduces contamination to the lowest detectable level by the use of tactical-level capabilities. The intent of thorough decontamination is to reduce or eliminate the level of MOPP. This is accomplished by units (with or without external support) when operations and resources permit. Detailed equipment decontamination, detailed troop decontamination and detailed aircraft decontamination are conducted as part of a reconstitution effort during breaks in combat operations. These operations require immense logistic support and are manpower-intensive. Thorough decontamination is carried out to reduce contamination on personnel, equipment, materiel, and work areas. This permits the partial or total removal of IPE and maintains operations with minimum degradation. While conducting thorough decontamination, contaminated units are non-mission capable. The resulting decrease in MOPP should allow the unit to operate with restored effectiveness.

G-25. Clearance decontamination provides decontamination to a level that allows unrestricted transportation, maintenance, employment, and disposal. Clearance decontamination of equipment and personnel allows the operation to continue unrestricted. Decontamination at this level will probably be conducted at or near a shipyard, advanced base, or other industrial facility. Clearance decontamination involves factors such as suspending normal activities, withdrawing personnel, and having materials and facilities not normally present. During clearance decontamination, resource expenditures are documented, force health protection measures are conducted, and after action reviews are prepared.

G-26. Additional decontamination considerations include the following:

- Plan decontamination sites throughout the width and depth of the area of operation.
- Tie decontamination sites to the scheme of maneuver and templated CBRN incidents.

- Plan for contaminated routes.
- Plan logistics and resupply of mission-oriented protective posture, mask parts, water, and decontamination supplies.
- Consider medical concerns, including treatment and evacuation of contaminated casualties.
- Plan for site security.

CBRN Reconnaissance and Surveillance

G-27. CBRN R&S is the detection, identification, reporting, and marking of CBRN hazards. CBRN reconnaissance consists of search, survey, surveillance, and sampling operations. Due to limited availability and number of the CBRN reconnaissance vehicles within the IBCT, consider alternate means of conducting CBRN reconnaissance such as reconnaissance elements, engineers, and maneuver units. (See ATP 3-11.37.) At a minimum, consider the following actions when planning and preparing for CBRN reconnaissance:

- Use the intelligence preparation of the battlefield process to orient on CBRN enemy named areas of interest.
- Pre-position R&S assets to support requirements.
- Establish command and support relationships.
- Assess the time and distance factors for the conduct of CBRN R&S.
- Report all information rapidly and accurately.
- Plan for resupply activities to sustain CBRN R&S operations.
- Determine possible locations for post-mission decontamination.
- Plan for fire support requirements.
- Plan fratricide prevention measures.
- Establish medical evacuation procedures.
- Identify CBRN warning and reporting system procedures and frequencies.

COUNTERING WEAPONS OF MASS DESTRUCTION TASKS

G-28. CWMD is described as actions undertaken in a hostile or uncertain environment to systematically locate, characterize, secure, and disable, or destroy WMD programs and related capabilities. Collecting forensic evidence from the WMD program during CWMD is a priority for ascertaining the scope of a WMD program and for follow-on attribution. Many technical chemical, biological, radiological, nuclear, and explosives (CBRNE) forces have the capability to conduct some activities within CWMD; however, no single technical CBRNE force can accomplish the entire CWMD mission alone. CBRNE response teams conduct exploitation and destruction. They also have the capability to provide field confirmatory identification of CBRN hazards. Nuclear disablement teams (specialized forces) perform site exploitation and disable critical radiological and nuclear infrastructure during CWMD. (Refer to ATP 3-11.24 for additional information.)

G-29. CWMD missions require extensive collaborative planning, coordination, and execution oversight by IBCT commander and staff, and subordinate commanders. CWMD will likely involve teams of experts to include both technical forces (but are not limited to, CBRN reconnaissance teams, hazardous response teams, CBRN dual-purpose teams, and explosive ordnance disposal elements) and specialized forces (but are not limited to, technical escort units, nuclear disablement teams, and chemical analytical remediation activity elements). Associated planning will begin at echelons above the IBCT characterized by centralized planning and decentralized execution of CWMD missions to ensure that the right assets are provided. (Refer to FM 3-94, ATP 3-91, and ATP 4-32.2 for additional information.)

G-30. CWMD operations may be lethal or nonlethal as indicators are identified that meet the commander's critical information requirements and priority intelligence requirements suggesting that a site contains sensitive information. CWMD operations may develop intelligence that feeds back into the planning process to include the intelligence preparation of the battlefield and targeting process. The priority for CWMD activities is to reduce or eliminate the threat. CWMD operations may be conducted under two circumstances—planned and opportunity. While planned operations are preferred, some operations involving WMD sensitive sites may occur because the opportunity presents itself during operations to accomplish another mission. Not every operation requires destruction tasks—tactical isolation or exploitation may be the only elements executed. Nonetheless, the IBCT commander and staff, and subordinate commanders always consider each element of CWMD operations (isolation, exploitation, destruction, and monitoring and redirection) and its relevance to the situation. A particular element may be unnecessary, but making that judgment is the appropriate level commander's responsibility. (Refer to ATP 3-11.23 for additional information.)

G-31. An explosive ordnance disposal company, when tasked, provides explosive ordnance disposal, protection planning, and operations support to the IBCT and subordinate battalions. The explosive ordnance disposal company supporting the IBCT may provide an operations officer and noncommissioned officer to the IBCT to provide appropriate explosive ordnance disposal planning and to perform liaison officer duties that include facilitating cooperation and understanding among the IBCT commander, staff, subordinate battalions, and explosive ordnance disposal battalion and company commanders. The explosive ordnance disposal company coordinates tactical matters to achieve mutual purpose, support, and action. In addition, the company ensures precise understanding of stated or implied coordination measures to achieve synchronized results.

G-32. Explosive ordnance disposal elements supporting subordinate maneuver units can neutralize hazards from conventional unexploded ordnance, explosives and associated materials, improvised explosive devices, booby traps containing both conventional explosives and CBRN explosives that present a threat to those units. These elements may dispose of hazardous foreign or U.S. ammunition, unexploded ordnance, individual mines, booby-trapped mines, and chemical mines. Breaching and clearance of minefields is primarily an engineer responsibility. (Refer to ATP 4-32.2 for additional information about unexploded ordnance procedures.)

BATTALION TASKS AND ACTIVITIES WITHIN A CHEMICAL, BIOLOGICAL, RADIOLOGICAL, AND NUCLEAR ENVIRONMENT

G-33. Effective mission command requires the integration with hazards planning, preparation, and execution—along with continuous assessment activities—to prevent CBRN incidents from occurring; to protect personnel, equipment, and information during a CBRN incident response; and to mitigate/recover from a CBRN incident that involves casualties and/or contamination. This section addresses key CBRN process and assessment tasks and activities the commander and staff, and subordinate commanders use to measure performance and the MOE (paragraph 4-17) for unit activities.

INFANTRY BATTALION COMMANDER

G-34. The battalion commander, supported by the staff, characterizes and manages through understanding the CBRN threats and hazards in a particular operational environment. (See ATP 3-11.36.) The commander applies this situational understand to the military decision-making process to shape the operational environment involving CBRN threats and hazards and to better understand where and when to expect CBRN hazards.

G-35. The battalion commander ensures the units and personnel prepare to operate in a CBRN environment. To do this, the commander ensures the battalion takes the proper protective measures including—

- CBRN vulnerability analysis and assessment.
- Dispersion and use of terrain as shielding.
- Continuous CBRN monitoring with detection equipment.
- Assumption of the appropriate mission-oriented protective posture level.

INFANTRY BATTALION STAFF

G-36. The staff collects, processes, displays, stores, assesses, and disseminates relevant information for creating the common operational picture and using information—primarily from collection assets external to the battalion—to synchronize operations through the operations process. The battalion CBRN officer (see ATP 3-11.36) collaborates with CBRN staff members at each echelon, with the commander and other staff sections to support the commander's intent to direct units and control resource allocations. As with any operation, the staff is alert for enemy or friendly situations that require command decisions and advises the commander concerning them.

G-37. Additionally, the battalion CBRN officer, assisted by the battalion CBRN NCO, provides technical advice to the commander and the remainder of the battalion staff. (see ATP 3-11.36 for a complete listing of CBRN core staff functions) The CBRN staff officer and NCO—

- With the intelligence staff officer (S-2), template strikes and predict the effects of enemy CBRN attacks.
- Disseminates information received via the CBRN warning and reporting system.
- Recommends reconnaissance, monitoring, and surveying requirements (see ATP 3-11.37).
- Recommends MOPP and operation exposure guide based on S-2's threat analysis and higher headquarters guidance.
- Maintains records of unit contamination to include radiological dose records.
- Conducts vulnerability analysis of unit positions.
- Plans battalion decontamination operations in conjunction with the S-3.
- Coordinates for nonorganic CBRN assets (decontamination, reconnaissance, and surveillance) support.
- Monitors and plans for passive defense measures (ATP 3-11.32).
- Acts as the liaison between attached CBRN units and the S-3.

Chemical, Biological, Radiological, and Nuclear Workgroup

G-38. The CBRN working group, generally established at the IBCT main command post and led by the IBCT CBRN officer, includes members from the protection-working group, higher headquarters elements, host-nation agencies, unified action partners, and other representatives (CBRN representatives specifically) from subordinate battalions and other units. The CBRN working group—disseminates CBRN operations information, including trend analysis, defense best practices and mitigating measures, operations, the status of equipment and training issues, CBRN logistics, and management consequence and remediation efforts and refines the CBRN threat, hazard, and vulnerability assessments. The working group helps to develop, train, and rehearse a CBRN defense plan to protect personnel and equipment from an attack or incident involving CBRN threats or hazards. CBRN threat and hazard assessments made by the working group help determine initial, individual protective equipment levels and the positioning of decontaminants. Force health personnel maintain the medical surveillance of personnel strength information for indications of force contamination, epidemic, or other anomalies apparent in force health trend data. (Refer to FM 3-11 and ADRP 3-37 for additional information.)

ILLUSTRATION OF A COUNTER WEAPONS OF MASS DESTRUCTION MISSION

G-39. In this scenario, used for discussion purposes, Infantry Battalion 2 conducts a search and attack (refer to chapter 2 for additional information on search and attack missions) as a shaping operation to support the identification and disposal of WMD material within the IBCT's area of operation. During the battalion's mission, a search team located an area once occupied by an enemy artillery unit. Within the area were several ammunition caches containing artillery shells. One of the caches contained shell cases with different markings than the others. The battalion chemical officer investigated and determined that these shells, subject to further investigation, may contain an extremely dangerous nerve agent. Exact type of nerve agent is unknown. Figure G-2, page G-10 represents the battalion's area of operation.

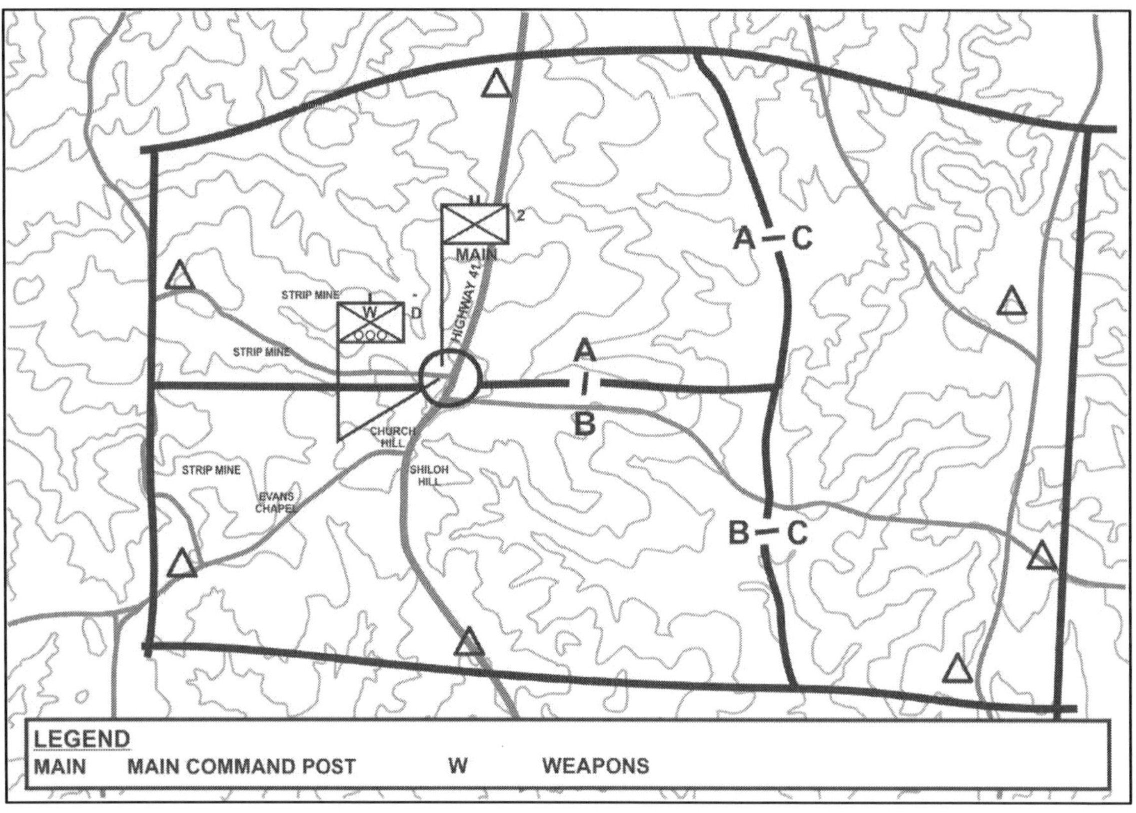

Figure G-2. Battalion area of operation, example

Immediate Actions

G-40. The battalion commander immediately—

- Avoids contamination by—
  - Pulling all Soldiers out of the immediate area and established a cordon 200 meters from the suspected chemical site.
  - Conducted CBRN R&S by establishing monitoring teams with chemical detectors around the suspected site.
  - Initiated CBRN R&S within the rest of the battalion AO.
- Protects unit by initiating MOPP4 protective posture level. After conducting CBRN reconnaissance and monitoring, the commander may decide to reduce the MOPP level.
- Conducts decontamination by—
  - Decontaminating the Soldiers who had the initial contact with the suspected chemical agent, medically evaluated, and evacuated if required.
  - Establishing decontamination sites for the search team Soldiers and vehicles.
- Begins sending the appropriate CBRN reports to the IBCT.
- Requests additional trained CBRN personnel and units.

G-41. The IBCT commander directed that the battalion—

- Establish an outer and inner cordon around the site with the inner cordon no less than 200 meters from the cache (figure G-3).
- Establish monitoring sites throughout the AO.
- Conduct a search of the site after receiving elements of a chemical company.

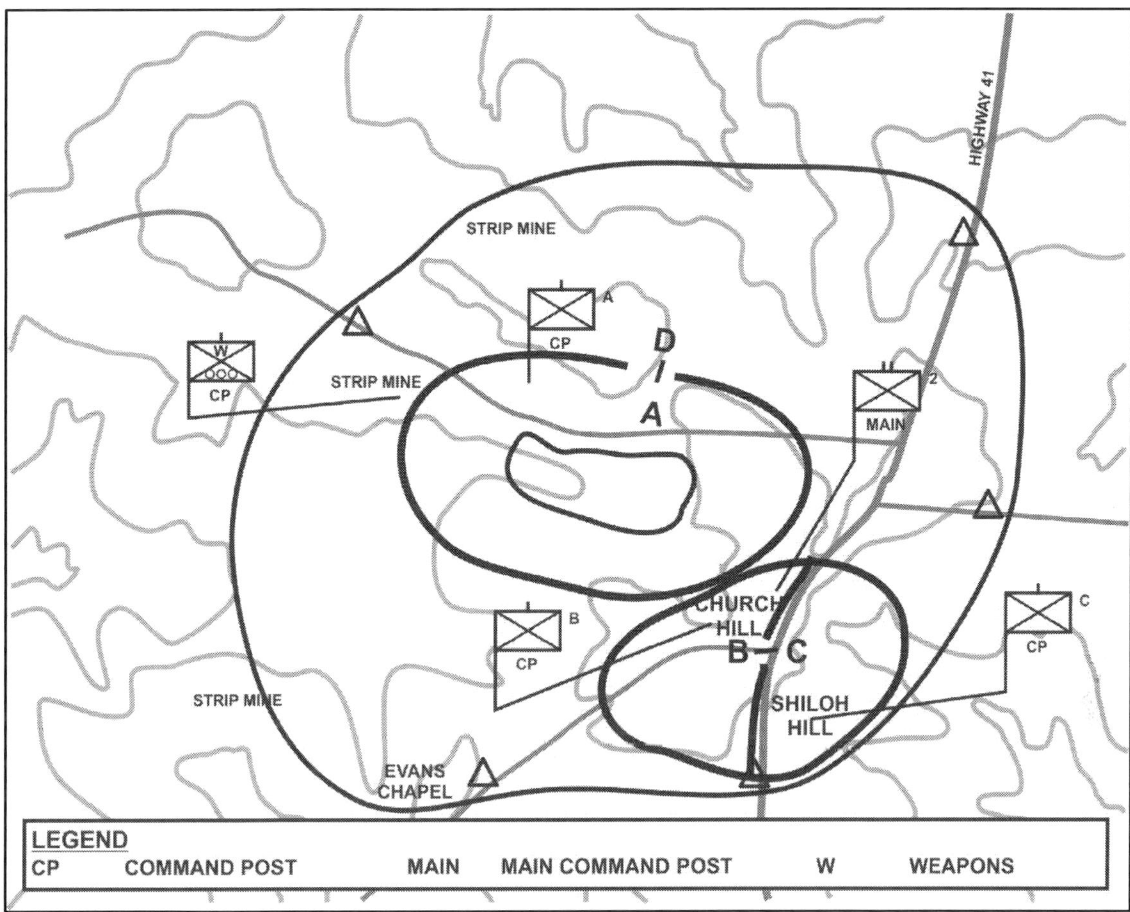

Figure G-3. Cordon and search (countering weapons of mass destruction), example

Battalion Subordinate Unit Task Organization and Missions

G-42. The battalion conducts a cordon and search (refer to chapter 2 for additional information on cordon and search missions) with—

- Company A, security element, responsible for the inner cordon (approximately 2000 meters in length). Company A has one additional Infantry rifle platoon attached.
- Company D (weapons company), security element, responsible for the outer cordon. Attached sniper squad establishes three observations posts within the outer cordon.
- Company B, support element, with attached scout platoon (and decontamination platoon) supports search element (CBRN R&S platoon) within the target area.
- Company C, battalion reserve. Attaches one platoon to Company A.
- Mortar platoon, section 1 establishes fire support position vicinity battalion main command post, section 2 establishes battalion quick response force vicinity Church Hill to assist Company B.

Preparation for the Search

G-43. From the IBCT commander, the battalion commander received a CBRN R&S platoon, a CBRN decontamination platoon, and the necessary equipment to conduct a search of the target area. After being briefed by the two attached platoon leaders, the battalion commander directs the following:

- The CBRN R&S platoon leader is in charge of the search element.
- The CBRN decontamination platoon is attached to Company B, support element.
- The CBRN decontamination platoon is provided any required personnel and equipment and establishes a decontamination site as close to the target area as possible and upwind from it.

- The medical platoon establishes a battalion aid station within the support area and is prepared to treat CBRN contaminated Soldiers.
- The CBRN reconnaissance personnel and attached search teams from the battalion conduct rehearsals.
- Vehicles from the forward support company are prepared to remove the artillery shells when and if the reconnaissance platoon leader determines that removal is the best method of disposal. Bags and other equipment to contain contaminated material also are prepared.
- Earth-moving engineer equipment are requested in case the contaminated material has to be buried at the site.

EXECUTION OF THE SEARCH

G-44. Once preparations are complete, the search element searches the target area. Search teams, with designated areas to search, find only the one cache of artillery shells. The CBRN R&S platoon leader confirms the type of chemical contained in the shell. The platoon leader confirms that the shells do not have fuzes, the bodies are intact, and the contents are stable.

G-45. Once the technical CBRNE forces arrive, the shells are sealed and moved to the designated vehicles and are, accompanied by the scout platoon for security, moved out of the area to a location designated by the IBCT commander.

G-46. All members of the search teams, their equipment, and their vehicles go through a thorough decontamination conducted by the CBRN decontamination platoon. All clothing and other contaminated material are properly disposed of.

G-47. The battalion continues to monitor the AO until the threat is eliminated. The battalion reverts to MOPP1 and continues its search and attack mission.

Appendix H
Sustainment

Sustainment is the provision of logistics, personnel services, and health service support necessary to maintain operations until successful mission completion (ADP 4-0). Sustainment operations provide support and services to ensure freedom of action, extend operational reach, and prolong endurance. Within the Infantry battalion, the sustainment planners synchronize sustainment operations and the forward support company executes sustainment operations in support of the Infantry battalion under all conditions to allow the Infantry battalion to seize, retain, and exploit the initiative. The Infantry battalion sustainment staffs anticipate future needs to retain freedom of movement and action at the end of extended and contested lines of operation. The forward support company (FSC) commander is the Infantry battalion's senior logistician. The FSC commander is responsible for sustainment synchronization and execution across the Infantry battalion's area of operation. This appendix describes sustainment operations in support of the Infantry battalion, specifically the functions, command and staff roles and responsibilities, functions, and unit relationships throughout high operating tempo decentralized operations. This appendix illustrates, through a fictional scenario used as a discussion vehicle, the provision of continuous support during the IBCT's conduct of offensive actions.

SUSTAINING THE INFANTRY BATTALION

H-1. Sustainment based on an integrated process, (people, systems, materiel, health services, and other support) inextricably links sustainment to operations. Sustaining the Infantry battalion in austere environments, often at the ends of extended lines of communications, requires a logistics network capable of projecting and providing the support and services necessary to ensure freedom of action, extend operational reach, and prolong endurance. Success will require deployment and distribution systems capable of delivering and sustaining the battalion from strategic bases to points of employment within and throughout the operational area at the precise place and time of need.

SUSTAINMENT STAFF

H-2. The battalion commander and staff integrate forces, the operational plan, and existing and available logistics and services to ensure that the battalion can win across the range of military operations. The sustainment staff plans, directs, controls and coordinates sustainment, with unrelenting endurance in support of those operations. The following proponents make up the sustainment staff. (Refer to FM 3-96 and FM 6-0 for additional information.)

Executive Officer

H-3. The Infantry battalion executive officer (XO) provides oversight of operations and sustainment planning for the battalion commander. He directs, coordinates, supervises and synchronizes the work of the staff to ensure the staff is integrated and aligned with the Infantry battalion commander's priorities. The XO's primary sustainment duties and responsibilities in relation to sustainment operations include the following:

- Responsible for sustainment within the battalion.
- Ensuring the concept of support is synchronized with the scheme of maneuver in-depth.
- Providing oversight over the maintenance status of the battalion.
- Setting priorities for the battalion staff sustainment cell (personnel staff officer [S-1], logistics staff officer [S-4], surgeon, and chaplain).

- Managing contract operations for the battalion.
- Leads the battalion's sustainment rehearsal in cooperation with the battalion S-4.

Logistics Staff Officer

H-4. The battalion S-4 is the coordinating staff officer for logistical operations and plans. The S-4 provides staff oversight to battalion units in the areas of supply, maintenance, transportation, and field services. The S-4 is the battalion staff integrator between the battalion commander and the forward support company (FSC) commander who executes sustainment operations for the battalion. (Refer to FM 6-0 for additional information.) Primary duties and responsibilities include, but are not limited to—

- Developing the logistics plan to support battalion's operations and determining support requirements necessary to sustain battalion's operations.
- Developing associated logistics annexes to the plans and orders, anticipates and forecasts requirements for support and maintains the logistics running estimate.
- Developing the battalion concept of support.
- Coordinating support requirements with the IBCT S-4 and support operations officer on current and future support requirements and capabilities.
- Conducting logistics preparation of the battlefield.
- Managing the logistics status report (LOGSTAT) for the battalion and ensure accurate LOGSTAT reporting from company level to include expenditure reports.
- Monitoring and analyzing equipment readiness status of all battalion units.
- Planning transportation to support special transportation requirements such as casualty evacuation.
- Coordinating for all classes of supply, food preparation, water purification, mortuary affairs, aerial delivery, laundry, shower, and clothing/light textile repair. (See FM 4-95.)
- Recommending sustainment priorities and controlled supply rates to the commander.
- Monitoring and enforcing the battalion command supply discipline program throughout all phases of the operation.
- Managing organizational and theater provided equipment assigned to the battalion.
- Planning for inter-theater movement and the deployment of battalion personnel and equipment.
- Leads the sustainment rehearsal to further synchronize the sustainment plan.

Personnel Staff Officer

H-5. The battalion S-1 is the principle staff advisor to the battalion commander for all matters concerning human resources support. The function of the battalion S-1 section is to plan, provide, and coordinate the delivery of human resources support, services, or information to all assigned and attached personnel within the battalion and subordinate units. The battalion S-1 may coordinate the staff efforts of the battalion equal opportunity, Inspector General, and morale support activities. (Refer to FM 6-0 for additional information.) The S-1's primary duties and responsibilities include, but are not limited to—

- Maintaining unit strength and personnel accountability statuses.
- Preparing personnel estimates and annexes.
- Planning casualty replacement operations.
- Assisting the support operations officer plan enemy prisoner of war and displaced civilian movement.
- Planning the battalion postal operation plan.
- Conducting essential personnel services for the battalion.

Surgeon

H-6. The battalion surgeon is normally the medical platoon leader and serves as the personal staff officer responsible for health service support. The surgeons the advisor to the commander on the physical and mental health of the battalion. The surgeon manages health service support activities and coordinates implementation through the battalion S-3 and the battalion S-4 that affect battalion medical operations. The surgeon provides health service support and force health protection mission planning to support battalion operations. (Refer to

FM 6-0 or ATP 4-02.3 for additional information.) Primary duties and responsibilities include, but are not limited to the following:

- Provide Role I medical care from point of injury through the battalion aid station and on to higher roles of care, as required.
- Planning casualty care and area support medical treatment.
- Planning medical evacuation (ground and air).
- Planning dental care (operational dental care and emergency dental care).
- Coordinating medical logistics (class VIII, medical supplies, blood management, and field level and sustainment support medical maintenance).
- Planning for battalion behavioral health and neuropsychiatric treatment.
- Treating patients contaminated with CBRN hazards.
- Planning and coordinating force health protection activities (preventive medicine, medical surveillance, occupational and environmental health, and field sanitation).
- Planning and coordinating for combat and operational stress control.
- Planning and coordinating veterinary services, dental services, and laboratory services.
- Advising on medical humanitarian assistance.
- Advising the command on the battalion health status, and the occupied or friendly territory's health situation within the command's assigned area of operation.
- Identifying potential medical hazards associated with the geographical locations and climatic conditions with the battalion's area of operation.

Chaplain

H-7. The battalion chaplain and unit ministry team provides religious support to the command group and battalion staff. Chaplains personally deliver religious support. They have dual roles: religious leader and religious staff advisor. The chaplain as a religious leader executes the religious support mission to ensure the free exercise of religion for Soldiers, families, and authorized civilians. As a personal staff officer, the chaplain advises the commander and staff on religion, morals, morale, and ethical issues, both within the command and throughout the area of operations. (Refer to FM 1-05 for additional information.) Primary duties and responsibilities include, but are not limited to—

- Developing plans, policies, and programs for religious support.
- Coordinating and synchronizing area and denominational religious support coverage.
- Coordinating and synchronizing all tactical, logistical, and administrative actions for religious support operations.

Medical Support

H-8. The Infantry battalion has organic medical resources within the unit headquarters to include a battalion surgeon and medical platoon. Role 1 (also referred to as unit-level medical care) is the first medical care a Soldier receives. Nonmedical personnel performing first-aid procedures assist the combat medic in their duties. An individual (self-aid and buddy-aid) administers first aid and combat lifesavers administer enhanced first aid. If needed, the Soldier is evacuated to the Role 1 medical treatment facility (battalion aid station) at the battalion or squadron, or the Role 2 medical treatment facility (brigade support medical company) in the BSB of the IBCT. (Refer to ATP 4-02.3 for additional information.)

Battalion Aid Station

H-9. The mission of the medical platoon is to provide Role 1 Army Health System support to the maneuver battalion. A medical treatment platoon is organic to each and is the unit level Role 1 medical treatment facility, usually referred to as the battalion aid station. The medical platoon is dependent upon the maneuver elements to which it is assigned for all logistic support, with the exception of Class VIII (medical) supplies. For information on Class VIII coordination, synchronization, and execution of medical logistics support see paragraph 9-100.

H-10. Medical platoons within the Infantry battalion configure with a headquarters section, medical treatment squad, ambulance squad (ground), and combat medic section. The treatment squad consists of two teams (treatment team alpha and team bravo). The treatment squad operates the battalion aid station and provides Role 1 medical care and treatment (to include sick call, tactical combat casualty care, and advance trauma management). Team alpha is clinically staffed with the battalion surgeon while team bravo is clinically staffed with the physician assistant.

H-11. Medical platoon ambulances provide medical evacuation and en route care from the Soldier's point of injury, the casualty collection point, or an ambulance exchange point to the battalion aid station. The ambulance squad is four teams of two ambulances composed of one emergency care sergeant and two ambulance aide/drivers assigned to each ambulance.

Combat Lifesavers

H-12. The combat lifesaver is a nonmedical Soldier trained to provide enhanced first aid and lifesaving procedures beyond the level of self-aid or buddy-aid. He is usually the first person on the scene of a medical emergency. He provides enhanced first aid to wounded and injured personnel. The squad leader is responsible for ensuring that an injured Soldier receives immediate first aid and is responsible for informing the commander of the casualty.

Combat Medic

H-13. The combat medic is the first individual in the medical chain that makes medical decisions based on medical specialty-specific training. The platoon combat medic goes to the casualty and initiates tactical combat casualty care or the casualty may be brought to the combat medic at the casualty collection point. The medic makes an assessment; administers initial medical care; initiates the DD Form 1380, *Tactical Combat Casualty Care (TCCC) Card*, or other documents; requests evacuation; or returns the Soldier to duty.

Forward Support Company

H-14. The Forward Support Company (FSC) is the link from the BSB to the Infantry battalion and provides the Infantry battalion the greatest flexibility for logistics support. The FSC is organic to the brigade support battalion and is organized to provide direct support to the Infantry battalion. The FSC provides field feeding, bulk fuel, general supply, ammunition, and field maintenance support to the Infantry battalion. FSCs are structured similarly with the most significant differences in the maintenance capabilities.

H-15. The FSC commander assists the battalion S-4 and battalion XO with the battalion logistics' planning and is responsible for executing the logistics plan according to the BSB and supported maneuver commanders' guidance. Integrating the logistics plan early into the supported battalions or squadron S-3's operational plan will help to mitigate logistic shortfalls, and support the commander to seize, retain, and exploit gains.

H-16. The FSC receives technical logistic directions from the BSB commander. This allows the BSB commander and the BSB support operations officer to task organize the FSC and cross-level assets amongst FSCs when it is necessary to weight logistics support to the IBCT. The task organization of the FSCs is a collaborative, coordinated effort that involves analysis by the staff and consensus amongst all commanders within the IBCT. The BSB provides administrative support, some logistic support, and technical oversight to the FSC.

H-17. The IBCT commander may attach or place a forward support company under operational control of the Infantry battalion. The FSC attachment or operational control to its supported battalion is generally limited in duration and may be for a specific mission or phase of an operation. Regardless of what command relationship is determined for the FSC, the FSC must retain their technical relationship with the BSB commander.

H-18. The forward support company normally operates in close proximity to the Infantry battalion. The location of the FSC commander and the distance separating the FSC and the battalion is METT-TC dependent, with mission command, logistics asset protection, and required resupply turn-around times being key considerations.

H-19. The FSC may be divided with some elements collocated with the Infantry battalion forward in the combat trains and some elements located in the field trains usually located within the brigade support area (BSA. The FSC commander in collaboration with the BSB commander and supported unit commander determines the task organization for the mission. Thorough mission analysis, accurate and continuous logistics running estimates, and an understanding of the FSC's capabilities can assist the planning staff in the emplacement of the forward support company to best support the Infantry battalion. The sustainment planners (in conjunction with the BSB commander) must have the flexibility to adjust and move the FSC across the combat trains and field trains. FSC employment considerations include—

- Location, time, and distance of the FSC in relation to the supported battalion.
- Decision to separate elements of the FSC by platoon or other sub-elements into multiple locations.
- Benefits of locating FSC elements in the BSA or the combat trains command post.
- Benefits of collocating battalion staff sections with the FSC.
- Benefits of collocating battalion medical elements with the FSC.
- Security of the FSC locations and during movement.
- Establishment and location of a maintenance collection point.
- Security needed to protect the FSC conducting replenishment operations.

H-20. A technique for FSC employment at the field trains is to have a team of logistics personnel from the FSC under supervision of the FSC executive officer. The logistics team could include a food operations sergeant, an ammo handler, additional vehicle operators, and a supply specialist that can provide expertise for commodity management needed to assist field trains personnel to include company supply sergeants or representatives. The logistics team could prepare commodities and assets requested on the LOGSTAT and can configure loads for the battalion field trains to pick-up if utilizing service station distribution at the BSA or for the BSB distribution company transportation platoon to deliver to a logistics resupply point or combat trains. This team could also be used to receive and direct all of the FSC LOGPACs arriving or departing from the BSA and could serve as a direct liaison to the brigade support operations section.

H-21. A technique for FSC employment at the combat trains if METT-TC allows, is to have a majority of the distribution platoon, maintenance control, field maintenance, and service and recovery sections placed at the combat trains to provide distribution and maintenance support to the Infantry battalion. This technique would allow the distribution platoon to receive the battalion configured loads from the field trains and have the capability to break the battalion configured loads into company configured loads and delivered via combat trains to company trains at an established logistics release point (LRP). (Refer to ATP 4-90 for additional information.)

Headquarters Section

H-22. The FSC's headquarters food service section provides the Class I support to the Infantry battalion. The food service section provides food service and food preparation for the Infantry battalion and prepares, serves and distributes the full range of operational rations.

Distribution Platoon

H-23. The FSC's distribution platoon is the primary distribution hub of the battalion. It provides the supply and transportation components of logistics support to the battalion. The distribution platoon leader leads the platoon, oversees LOGPAC operations, and manages the distribution of supplies for the battalion. The distribution platoon consists of a platoon headquarters, class III section, general supply section, and class V section that can be tasked organized to provide Class II, III, IV, V, and VII distribution to the Infantry battalion. The distribution platoon can be tasked organized into company support squads, a technique to provide all supplies for a supported company if the Infantry battalion is occupying a widely dispersed geographical area.

H-24. The class III section is capable of receiving, mobile storage, and issuing petroleum for the Infantry battalion. The class III section provides retail class III bulk fuel distribution to the Infantry battalion and provides refueling options in support of the IBCT units passing through the Infantry battalion's area of operations.

H-25. The general supply section provides supply support to the battalion directly. General supply operations include the requisition, receipt, storage, protection, maintenance, issue, distribution, redistribution and retrograde of supplies. The general supply section provides Class II, III packaged, IV, V, VII and IX support to the Infantry battalion.

H-26. The Class V Section provides direct transportation support to include management and distribution of to the Infantry battalion. The distribution platoon is responsible for transporting Class V from the BSB distribution company, ammunition transfer holding point (ATHP) to the Infantry battalion and supported units as appropriate.

The FSC commander and distribution platoon leader must coordinate with the battalion S-4, battalion S-3 and the master gunner in order to ensure timely and accurate munitions distribution.

H-27. The distribution platoon may provide limited transportation support to the Infantry battalion. When troop transport is required that is not within the FSC's capability, the S-4 must coordinate the requirement with the IBCT S-4.

Field Maintenance platoon

H-28. The maintenance platoon performs field maintenance support to the Infantry battalion as well as all maintenance functions, dispatching, and scheduled service operations. The maintenance platoon consists of a platoon headquarters section, the maintenance control section, maintenance section, service and recovery section, and the field maintenance teams. The FSC's maintenance priorities are determined by the Infantry battalion chain of command with recommendations from the FSC commander and the maintenance control officer.

H-29. The maintenance control section manages all maintenance actions in the FSC and the Infantry battalion. This section performs management functions, dispatching operations, and tracks scheduled services for the Infantry battalion and the FSC. This section also provides CLIX support and exchange of reparable items.

H-30. After operator level maintenance, the Infantry battalion's next level of maintenance support comes from the field maintenance teams, which provides field maintenance support to all combat platforms in the Infantry battalion. The field maintenance teams perform repairs as far forward as possible to return equipment to the battle quickly. The field maintenance teams operates under the operational control of the maneuver company and is supervised by the maintenance non-commissioned officer in charge.

RESUPPLY OPERATIONS

H-31. The Infantry battalion S-4 is the principal staff officer responsible for synchronizing resupply operations for all units assigned or attached to the Infantry battalion. The battalion S-4 identifies requirements through daily logistic status reports and logistics running estimates conducted during the operations process. The battalion's forward support company commander applies capabilities to the battalion's requirements and the distribution platoon leader executes resupply operations in support of the Infantry battalion. The following paragraphs discuss resupply techniques and delivery methods to sustain the battalion during operations.

Routine Resupply

H-32. Whenever possible, routine resupply by LOGPAC is conducted on a regular basis and is the preferred method for the distribution of supplies. Routine resupply, conducted ideally during hours of limited visibility, through LOGPAC covers all classes of supply, mail, and any other items usually requested. The LOGPAC, a grouping of multiple classes of supply and supply vehicles under the control of a single ground convoy commander (see ATP 4-01.45) or through aerial delivery under certain situations (see ATP 4-48), is an efficient method to accomplish routine resupply operations. The key feature is a centrally organized resupply operation carrying all items needed to sustain the force for a specific period, usually 24 hours or until the next scheduled LOGPAC.

H-33. The battalion S-4 within and the Infantry battalion tries to standardize a LOGPAC (commonly referred to as a push package) as much as possible while still providing subordinate units with sufficient quantities of each supply item in anticipation of their requirements. Together with the commander's guidance for issuance of scarce, but heavily requested supply items, accurate reporting allows the S-4 to quickly forecast supply constraints and submit requisitions to alleviate projected shortages. Inaccurate or incomplete reporting can severely handicap efforts to balance unit requirements and available supplies. As a result, some units may go into combat without enough supplies to accomplish their mission while others may have an excess of certain items.

H-34. The distribution platoon leader oversees LOGPAC operations and manages the distribution of supplies within the FSC. Replenishment operations are conducted by the distribution platoon of the FSC in support of the Infantry battalion.

Immediate Resupply

H-35. Immediate resupply also referred to as emergency or urgent resupply is the least preferred method for the distribution of supplies. While resupply may be required when combat losses occur, requests for immediate resupply not related to combat loss indicates a breakdown in coordination and collaboration between sustaining and operating forces. Immediate resupply that extends beyond the Infantry battalion's echelons of support capabilities requires immediate intervention of the brigade support battalion or next higher sustainment echelon capable of executing the support mission.

H-36. When a unit has an immediate need for resupply that cannot wait for a routine LOGPAC an immediate resupply may involve Class III (petroleum, oils, and lubricants), Class V (ammunition), and Class VIII (medical), and, on occasion, Class I (rations). In this situation the battalion might use its forward support company distribution platoon located in its unit, field, or combat trains (when established) to conduct the resupply. An immediate resupply, by aerial delivery (see below) is dependent the availability of aviation assets. The fastest and most appropriate means of delivery is normally used, although, procedures may have to be adjusted when in contact with the enemy.

Contingency Resupply

H-37. Contingency resupply is the on call delivery of prepackaged supplies during the execution phase of an operation. This type of on call delivery of a prepackaged resupply is generally used to support an operation of limited duration, such as an airborne or air assault or other limited engagement of short duration. Contingency resupply operations are identified during the military decision-making process, normally during war gaming as each course of action is analyzed. Contingency resupply differs from a routine LOGPAC or immediate resupply, in that, prior to execution, triggers for delivery are developed to tie contingency resupply operations to the ground tactical plan. During the planning and preparation phases of the operations process units develop menus for prepackaged classes of supply to ensure their availability for expedited delivery as needed. A contingency resupply package can be as simple as a container or bag filled with a small amount of supplies or a unit basic load prepackaged for delivery when needed. Delivery methods vary between rotary wing, fixed wing, and ground delivery assets.

Aerial Delivery

H-38. Aerial delivery, by airland, airdrop, and sling-load operations, provides additional capability to resupply the Infantry battalion when the terrain or enemy situation limits access by ground transportation. Aerial delivery of routine, (resupply by LOGPAC), immediate resupply (emergency or urgent resupply), and contingency resupply, provides an effective means to by-pass enemy activities and reduces the need for route clearance of ground lines of communications. When planning aerial delivery operations, the commander considers the enemy's ability to locate the delivery and receiving unit by observing the delivery aircraft. Drop zones and landing zones are located away from the main unit in an area that can be defended for a short time unless the resupply is conducted in an area under friendly control and away from direct enemy observation. When delivered, supplies are immediately transported away from the drop zone or landing zone.

Note. In order for aerial delivery to be effective, friendly forces must control the airspace and neutralize enemy ground-based air defenses along the aerial delivery route. FM 3-99 addresses planning considerations for the suppression of enemy air defenses along aerial routes and guidance for selecting landing zones and drop zones. ATP 4-48 describes the planning, preparation, execution process for aerial delivery; and identifies responsibilities in the conduct of aerial delivery.

H-39. Six aerial delivery means, common to battalion resupply operations, are discussed in the following paragraphs. The six include internal and external (sling) loading, speedball, kicker box, Container Delivery System, Low Cost Low Altitude, and Joint Precision Airdrop System.

Internal and External (Sling) Loading

H-40. Internal and external (sling) loading, conducted by rotary-wing assets, provides the Infantry battalion with an aerial delivery capability to load internally or to rig and sling external cargo loads with nets, bags or sling legs. Prior to the resupply, the commander determines whether internal or external (sling) loading is the best delivery means for situation. The method used depends largely on mission and load requirements and the availability of sling loading and rigging equipment.

H-41. Aircraft internal loads utilizing the airland method are the preferred method of aerial delivery. This method of delivery can be used by short-take-and-landing fixed-wing aircraft as well as rotary wing aircraft. Aircraft carrying internal loads have better flight characteristics, including greater speed and maneuverability, and require less rigging equipment to deliver the supplies.

H-42. Sling Loading operations are performed by rotary wing aircraft. Sling loaded equipment requires trained personnel to prepare the cargo whether using cargo nets, bags or sling legs. Sling loading allows for larger cargo loads and quicker unloading but degrades the aircrafts flight characteristics. Aircraft normally have to fly at lower altitudes at lower speeds when carrying a sling-loaded cargo. Sling load operations are normally coordinated at the battalion level and require special training for landing zone personnel.

Speedball

H-43. Speedball, conducted by rotary-wing assets (known as utility helicopter [UH] 60), is simply a small prepacked amount of supplies (for example, ammunition, water, and/or food) put in a bag (human remains pouch, duffle bag, aviator's kit bag, or other suitable container) packaged in bubble wrap or other shock-absorbing material to minimize damage. This small-prepacked amount of supplies is dropped as close to the unit drop point as possible then the rotary wing asset leaves the area quickly to reduce exposure times to the delivery asset and receiving unit.

Kicker Box

H-44. Kicker box, conducted by rotary-wing assets (typically CH-47), generally involves delivering larger, or higher quantity items (for example, construction or barrier materials required for constructing obstacles and fighting positions), again, as close to the unit drop point as possible then leaves the area quickly to reduce exposure times to the delivery asset and receiving unit. Recovery of kicker box supplies will require more time due to the bulk of delivered items.

Container Delivery System

H-45. The container delivery system (CDS) is a commonly used airdrop via either low or high-velocity airdrops for the insertion of supplies quickly for airborne and air assault operations. CDS bundles are used as a means of delivering additional equipment or other resupply item by airdrop. CDS loads are heavy, yet they are ideal for commodities such as water and larger munitions. The battalion S-4, in coordination with the battalion S-3 air, request air assets for the CDS bundles through the BSB support operations officer to the brigade S-3 and brigade aviation element. The supporting FSC or trained Infantry battalion personnel can build CDS bundles from 501-2200 pound configurations. Depending on the location and unit standard operating procedures, the CDS material A-22 bag and parachutes may require retrograding.

Low Cost Low Altitude

H-46. Low cost low altitude is a one-time use, stand-along airdrop system consisting of a modular suite of low cost airdrop items, comprised of parachutes, containers, skid boards, and other air items configured for high-velocity and low velocity drops. All components are simple in design and operation, require no maintenance, and have low production and lifecycle costs. This aerial delivery system is comparable to the current Container Delivery System performance and cargo delivery capability of 2,200 pounds and was designed to address the persistent lack of retrograde of aerial delivery hardware. Coordination and requests for this delivery system are similar to those mentioned in paragraph H-45.

Joint Precision Airdrop System

H-47. Joint Precision Airdrop System uses the Global Positioning System, steerable parachutes, and an onboard computer to steer loads to a designated point of impact on a drop zone. The Joint Precision Airdrop System integrates the Army's Precision and Extended Glide Airdrop System (known as PEGASYS) and the Air Force's Precision Airdrop System program. Army's Precision and Extended Glide Airdrop System consists of several precision airdrop systems, ranging from extra light to heavy payloads. These systems are particularly valuable to units operating in contested areas where traditional aerial delivery methods are too dangerous. A notable disadvantage is the sensitive items that require retrograding. The Airborne Guidance Unit System is in excess of 25 pounds and must be recovered and turned in. Coordination and requests for this delivery system are similar to those mentioned in paragraph H-45.

Prepositioned

H-48. Prepositioning of supplies must be carefully planned and executed at every level when utilized. All leaders must know the exact locations of prepositioned sites, which they verify during reconnaissance and rehearsals. The commander take measures to ensure survivability. These measures may include digging in prepositioned supplies and selecting covered and concealed positions. The commander must also have a plan to remove or destroy prepositioned supplies if required.

Cache

H-49. A cache is a prepositioned and concealed supply point. Caches are an excellent tool for reducing the Soldier's load and can be set up for a specific mission or as a contingency measure. Cache sites have the same characteristics as an objective rally point or patrol base, with the supplies concealed above or below ground. An above ground cache is easier to get to but is more likely to be discovered by the enemy, civilians, or animals. A security risk always exists when returning to a cache. A cache site should be observed for signs of enemy presence and secured before being used as it may have been booby-trapped and may be under enemy observation.

Operational Contract Support

H-50. Operational contract support is the process of planning for and obtaining supplies, services, and construction from commercial sources in support of joint operations. While varying in scope and scale, operational support contracting, and its subset of theater contact support capability, is a critical force multiplier in unified land operations, especially long-term stability operations. (Refer to ATP 4-10 for additional information.) Theater support contracting and purchasing will likely be a method of sustainment that helps to round out the Infantry battalion's concept of support. The Infantry battalion must have trained and ready contracting officer representatives, field ordering officers, and pay agents. These designated personnel must be carefully selected, as they will make up the acquisition team within the Infantry battalion. They must work closely together as these personnel are part of a larger acquisition team that includes the contract and financial management experts who will provide the guidance and direction to each contracting officer representative, field-ordering officer, and pay agent to meet unit needs. (Refer to ATP 4-10 for additional information.)

H-51. The contracting officer representative (COR), sometimes referred to as a contracting officer technical representative, is an individual appointed in writing by a contracting officer. (See ATP 4-10.) Responsibilities include monitoring contract performance and performing other duties as specified by their appointment letter. The requiring unit or designated support unit normally nominates a COR. All CORs must complete mandatory training requirements and have the expertise to review the Quality Assurance Surveillance Plan. The Infantry battalion commander must ensure CORs are allowed sufficient time to execute their quality surveillance tasks. Additional information on COR responsibilities can be located in the Defense Contingency COR handbook at the Office of the Under Secretary of Defense for Acquisition, Technology, and Logistics Website.

H-52. A field-ordering officer is an individual who is trained to make micro-purchases within established thresholds (normally with local vendors) and places orders for goods or services. A pay agent is an individual who is trained to account for government funds and make payments in relatively small amounts to local vendors. While performing as field ordering officers or pay agents, individuals work for and must respond to guidance from their appointing contracting official. One individual cannot serve as both field ordering officer and pay agent. Property book officers cannot serve as field ordering officers or pay agents. Field ordering officers and pay agents must be careful when dealing with local nationals because field ordering officers and paying agents have a ready source of cash, local nationals may overestimate the influence of field ordering officers and pay agent teams. (Refer to ATP 1-06.1 for additional information.) Considerations for field ordering officers and pay agents include—

- Security (personal and cash).
- Unauthorized purchases:
  - Type of purchase.
  - Number of items purchased.
  - Single item or extended dollar amount.
- Not splitting purchases to get around limits.
- Poor record keeping.
- Accepting gifts of any kind and not reporting gifts.

MAINTENANCE SUPPORT

H-53. The purpose of the Army maintenance system is to generate/regenerate combat power and to preserve the equipment to enable mission accomplishment. The Army employs field and sustainment levels of maintenance as described in the following paragraphs.

H-54. The forward support company provides field maintenance support to the Infantry battalion. Field maintenance is generally characterized by on (near) system maintenance, often-using line replaceable unit and component replacement, in the owning unit, using tools and test equipment found in the unit. Field maintenance is not limited to remove and replace actions, but also allows for repair of components or end items on (near) system. Field maintenance includes adjustment, alignment, service, applying approved field-level modification work orders as directed, fault/failure diagnoses, battle damage assessment, repair, and recovery.

H-55. The FSC's maintenance platoon establishes the maintenance collection point and provides vehicle and equipment evacuation, and maintenance support to the field maintenance teams. The maintenance collection point is normally located near or collocated with the combat trains for security, and should be on or near a main axis or supply route. Field maintenance teams evacuate vehicles and equipment that require evacuation for repair and return, have an extended repair time, or when the vehicle or equipment exceeds its maintenance capabilities and augmentation is necessary. Maintenance priorities are determined by the Infantry battalion chain of command with recommendations from the FSC commander and the maintenance control officer. Additionally, the BSB commander with the IBCT commander maintains the flexibility to task organize and move field maintenance teams and other personnel across the formation to weight the battle where needed.

H-56. Sustainment maintenance is off-system component repair and/or end item repair and entails operations employing job shops, bays, or production lines. Sustainment maintenance is performed by Department of Defense civilians and contractors, who return equipment to a national standard, after which the equipment is placed back into the overall supply system. When a unit sends equipment to a sustainment maintenance organization, the owning unit removes the equipment from its hand receipt. In rare exceptions, for example a unit reset, the equipment is returned to the owning unit.

OPERATIONS PROCESS

H-57. Sustainment planning is fully integrated throughout the operations process, with the sustainment concept of support synchronized with other areas within the concept of operations. Planning is continuous and concurrent with ongoing support preparation, execution, and assessment. Key sustainment planners at all levels actively participate in the military decision-making process, to include war-gaming. Through a logistics running estimate, sustainment planners continually assess the current situation to determine if the current operation is proceeding according to the commander's intent and if planned future operations are supportable. (Refer to FM 3-96 and FM 6-0 for additional information.)

PLANNING CONSIDERATIONS

H-58. Sustainment planning supports operational planning (including branch and sequel development) and the targeting process. Sustainment planning is a collaborative function primarily performed by key members of the battalion staff (executive officer, S-4, S-1, surgeon, and chaplain) Sustainment planners and operators must understand the mission statement, the commander's intent, and the concept of operations to develop a viable and effective concept of support. The goal is to ensure support during all phases of an operation.

H-59. The battalion S-4 is the lead planner for sustainment within the battalion staff. The battalion S-1, the surgeon, and chaplain assist the S-4 in developing the battalion concept of support. Representatives from these and other sections form a sustainment planning cell at the battalion main CP to ensure sustainment plans are integrated fully into all operations planning. Sustainment standard operating procedures within the battalion should be the basis for sustainment operations, with planning conducted to determine specific requirements and to prepare for contingencies. Battalion and subordinate unit orders should address only specific support requirements for the operation and any deviations from standard operating procedures. The battalion S-4 is responsible for producing the sustainment paragraph and annexes of the operations order. (Refer to FM 6-0 for additional information.)

Operations Logistics Planner or Online Planning Tools

H-60. When the battalion S-4 is planning logistics for the battalion, the S-4 should have access to the operations logistics planner. The operations logistics planner is a web-based interactive tool designed to assist logistics planners in developing a logistic estimate based off of requirements for supply of class I, class II, class III (package), class IV, class VI, class VII, class X, including water, ice, and mail in support of operations. The S-4 will use the Tables of Organization to create the battalion in operations logistics and will create the units into task organizations. The tasks organizations can then be assigned to a multi-phase order and mission parameters can be set. The operations logistics planner will then generate reports that provide supply consumption by unit, task organization, phase and order. The operations logistics planner can be downloaded at http://www.cascom.army.mil/g_staff/g3/SUOS/site-operational/pages/bn.htm#S4-log-loc under sustainment estimation resources. For more information on the operations logistics planner, refer to FM 4-95.

H-61. Quick logistics estimation tools are also available for the S-4 to use to calculate initial class of supply requirements. These spreadsheets can be downloaded at http://www.cascom.army.mil/g_staff/g3/SUOS/site-sustainment/pages/other_units.htm under sustainment estimation tools.

H-62. The logistics estimate workbook is another automated planning tool the battalion S-4 can use to determine logistics requirements. Requirements can be determined by phase of the operation for the Infantry battalion and provide information to create the logistics running estimate. The logistics estimate workbook can be found at https://www.us.army.mil/suite/doc/47273756.

Logististcs Running Estimate

H-63. The S-4, with planning guidance is responsible for the battalion's logistics running estimate, an analysis of how logistics support factors can affect mission accomplishment. The logistics estimate will address maintenance, supply and services, transportation, medical, and support to enemy prisoner of war operations. It will contain the staff's comparisons of requirements and capabilities, conclusions and recommendations about the feasibility of supporting a specified course of action. The logistics running estimate begins with the receipt of the WARNORD and will continue through the orders production process. To determine requirements, the S-4 should take into consideration historical data if available, utilize the operations logistics planner and coordinate with the battalion staff.

H-64. The battalion S-4 will determine the requirements for the Infantry battalion and the FSC commander will apply capabilities to requirements. The logistics plan should be integrated into the S-3's operational plan early on to mitigate shortfalls. This will allow the S-4 to identify possible solutions and recommendations and better plan a course of action to support the mission. The FSC commander is the senior logistician to the battalion and is responsible for executing the logistics support according to the BSB and Infantry battalion commander's guidance. The key to successful logistical operations is creating a logistics estimate of the tactical operation through phases and the S-4 must take into consideration through each phase of the operation the distance traveled by the battalion, the time the battalion needs to travel that distance, and the consumption rate of all classes of supply. The logistics estimate and METT-TC will assist in determining the logistics task organization and the placement of FSC assets between the combat trains and field trains. The logistics estimate is used to create the concept of support. The concept of support and the logistics task organization will be discussed during the battalion sustainment rehearsal.

Planning Requirements

H-65. Supply commodities are used in support of operations and are separated into ten supply classes. Supply provides the material and life support that gives the troop the combat power and prolonged endurance to accomplish the mission (see FM 4-95). When the battalion S-4 is planning logistics for the battalion, the S-4 should have access to the automated tools. When digital systems are not available, the S-4 should know how to estimate requirements by class of supply and will take the following into consideration when planning for logistics:

Class I - Perishable and Semi-Perishable Subsistence Items, Water, and Gratuitous Health and Comfort Items

H-66. Class I includes rations that are packaged as individual or group meals. The S-4 will forecast rations based upon the headcount of the battalion multiplied by the ration cycle or type of meal and multiplied by the issue cycle (how often the battalion will receive the bulk rations.) There are three categories of meals: meal, ready to eat (known as MREs), unitized group rations–A Option, or unitized group rations–heat option. Ration cycle and issue cycle will be provided in the brigade order. To simplify, the headcount X Ration Cycle X Issue Cycle = the number of meals needed for the battalion. For example, the S-4 will use this number to calculate how many cases of MRE meals per company and how many pallets each company will receive. An MRE case contains 12 MREs per case and there are 48 cases per pallet. The weight of the pallet should be taken into consideration; however, the FSC commander will plan transportation requirements. Further information on planning for Class I operations can be found in ATP 4-41.

H-67. Bulk water planning is calculated on a per-person, per-day cycle and water for drinking, personal hygiene, heat injury treatment and field feeding must be potable. To plan for minimum water consumption that is required to maintain the battalion, refer to ATP 4-44, appendix A for water planning tables.

Class II Individual Equipment and General Supplies

H-68. Class II items consists of common consumable items such as clothing, individual equipment, tentage, tool sets and kits, maps, administrative/housekeeping supplies, and chemical, biological radiological and nuclear protective equipment. The battalion typically deploys with a minimum load of Class II. The battalion S-4 must keep up with battalion shortages and ensure company supply sergeants are ordering shortages in enough time to receive prior to mission. The S-4 should be aware of the requirement for Class II and work in close coordination with the BSB Supply Support Activity (SSA) to place on order and to determine transportation requirements for the requests of Class II items.

Class III Packaged Fuel - Petroleum, Oils, and Lubricants

H-69. Class III packaged consists of packaged petroleum, oils and lubricants that can be handled similarly to dry cargo. The S-4 must anticipate requirements based of the mission and coordinate with the supporting maintenance officer from FSC for Class IIIP forecasting assistance. Environmental considerations such as dust, snow, and rain will affect the consumption rates of Class IIIP and should also be taken into consideration.

Class III Bulk Fuel - Gasoline, Diesel and Aviation Fuel

H-70. Class III bulk fuel consists of gasoline, diesel, and aviation fuel. The S-4 should know how to determine the consumption rate of bulk fuel as well as be able to forecast requirements based off of the mission assigned and requires a detailed analysis of the maneuver concept of the operation. To calculate the estimated fuel usage, take the vehicle or number of vehicles and multiply consumption rate stated in gallons per hour, multiplied by the number of hours that the equipment will be in operation. The battalion S-4 will have to take into consideration if the vehicles will be idle, cross-country, or traveling across roads. The S-4 should also take into consideration historical data and actual consumption when planning for bulk fuel usage. This process will be used with each vehicle type in the battalion and will provide an accurate estimate of Class III consumption that will assist the S-4 in identifying and mitigating shortfalls and ensure mission success.

Class IV Construction and Barrier Materials

H-71. Class IV consists of fortifications, barrier and constructions materials. Typically, a battalion will not have a requirement for Class IV materials unless in the defense. In the defense, sustainment units at higher levels consider prepositioning Class IV as far forward as METT-TC allows. Class IV is configured into combat configured loads and delivered to the supporting FSCs normally within 48 hours to allow maneuver unit's enough time to construct their fighting positions and improve their defensive positions. The S-4 should take into consideration all materials needed to handle Class IV material. For example, gloves and picket pounders for handling of concertina wire and placement of pickets. The S-4 ensures FSC is tracking requirement for Class IV and configuration of combat configured loads to each company to plan transportation requirements. If transportation requirements are beyond FSC capabilities, the S-4 will submit a request to the IBCT S-4/ BSB support operations officer for additional transportation support.

Class V Ammunition

H-72. Battalion unit basic loads are determined by the brigade ammunition office and validated through the Total Ammunition Management Information System. Unit basic load is determined by the weapon density, number of soldiers, and specific mission requirements over time. The battalion will receive their combat configured load from the BSB.

H-73. The planning factor for unit basic loads for a battalion is one with the company, one with the FSC, and one stored at the brigade's ammunition transfer and holding point. The S-4 will account for the basic loads and the FSC and battalion should be able to transport all combat configured loads with organic assets. The S-4 will determine how additional ammunition will be replenished and will ensure units are submitting accurate expenditure reports daily through the LOGSTAT. Expenditure reporting starting at the lowest level ensures accurate replenishment of the unit basic load. Replenishment of Class V is requested on a DA Form 581, *Request for Issue and Turn-In of Ammunition.*

Class VI Sundry, Personal Demand Items

H-74. Class VI supplies are personal demand items such as toiletry, hygiene, and small recreational items. In most cases, a Soldier deploys with a 30-day supply of health and comfort items. After the first 30 days, health and comfort packages are centrally funded and provided at 30-day intervals through Class I channels at the request of the unit commander and until Army and Air Force Exchange Service support can be established (ATP 4-42). The S-4 should plan for each Soldier to deploy with a 30 day of supply of health and comfort items and it should be included in the Service Support paragraph of the operations order.

Class VII Major End Items

H-75. Class VII Supplies include major end items. This class of supply is intensely managed and controlled through higher command channels. See FM 4-40 for detailed information on Class VII operations and their inherent critical property accountability issues. The S-4 should anticipate combat power loss and be prepared to conduct recovery operations and battle loss/battle damage replacement operations.

Class VIII Medical Supplies

H-76. Typically, the medical platoon of the Infantry battalion will deploy with a three-day supply of Class VIII to support the battalion. The S-4 in coordination with the medical platoon leader or sergeant request their Class VIII supplies from the Brigade Medical Support Officer. When forecasting Class VIII requirements, the mission, location, projected casualty rates, and available medical assets are taken into consideration. See ATP 4-02.3 for more information.

Class IX Repair Parts

H-77. Class IX repair parts include individual repair parts and major assemblies such as engines, transmissions, and final drives which are required to maintain battalion equipment and operational readiness. The S-4 and the FSC maintenance officer, with oversight by the battalion executive officer, tracks all repair parts ordered and received and ensures equipment is repaired in a timely manner. The S-4 should be aware of the requirement for Class IX and work in close coordination with the BSB SSA to place on order and to determine transportation requirements for the requests of Class IX items. Retrograde operations. The S-4 will plan for backhaul of unserviceable parts during LOGPAC operations and discuss during the sustainment rehearsal.

Concept of Support

H-78. The battalion S-4 is responsible for developing the battalion concept of support which is nested with the IBCT concept of support. The battalion concept of support describes how sustainment support will be executed during the operation. Once approved by the battalion commander, the battalion S-4 briefs the concept of support to all commanders and staff to ensure a shared understanding across the battalion. The FSC commander executes the battalion concept of support. The concept of support ultimately tells the FSC how the Infantry battalion is going to be supported.

H-79. The concept of support establishes priorities of support (by phase or before, during, and after) for the operation and gives the commander the authority to weight support organizations and task organize accordingly. The commander sets these priorities for each level in the intent statement and in the concept of operations. Priorities include such items as personnel replacements; maintenance and evacuation by unit and by system (air and surface systems are given separate priorities); fuel and ammunition; road network use by unit and commodity; and any resource subject to competing demands or constraints. To establish the concept of support, sustainment planners must know—

- Subordinate units' missions.
- Times missions are to occur.
- Desired end states.
- Schemes of movement and maneuver.
- Timing of critical events.
- Number and type of personnel and equipment.
- Supply consumption history.
- When attachments are effective and for how long.
- The support assets that will accompany attached elements.

Logistics Common Operational Picture

H-80. Accurate and timely reporting of the LOGSTAT and other key logistics systems can assist with developing the logistics COP for the Infantry battalion. Digitized mission command systems provide data and facilitate situational understanding of the mission. The logistics COP is a single display of relevant information within a commander's area of interest tailored to the user's requirements and based on common data and information shared by more than one command. The S-4 develops the logistics COP for the Infantry battalion and the development is ongoing throughout the operation. The Infantry battalion standard operating procedures often will provide detailed unit instruction on how to configure the logistics COP.

Synchronization of Battle Rhythm and Sustainment Operations

H-81. Sustainment operations are fully integrated with the battalion battle rhythm through integrated planning and oversight of ongoing operations. Sustainment and operational planning, and the targeting process occur simultaneously rather than sequentially. Incremental adjustments to either the maneuver or the sustainment plan during its execution must be visible to all battalion elements. The sustainment synchronization matrix and LOGSTAT initiate and maintain synchronization between operations and sustainment functions. (Refer to ATP 4-90 for additional information.)

Fusion of Sustainment and Maneuver Situational Understanding

H-82. Effective sustainment operations by the FSC depend on a high level of situational understanding. Situational understanding enables the battalion S-4 and FSC commander to maintain visibility of current and projected requirements; to synchronize movement and materiel management; and to maintain integrated visibility of transportation and supplies. The Command Post of the Future, analog systems, Movement Tracking System, Joint Capabilities Release, and Joint Capabilities Release-Logistics are some of the fielded systems the battalion S-4 or FSC commander uses to ensure effective situational understanding and logistics support. These systems enable sustainment commanders and staffs to exercise mission command, anticipate support requirements, and maximize battlefield distribution.

Reports

H-83. The LOGSTAT is an internal status report that identifies logistics requirements, provides visibility on critical shortages, allows commanders and staff to project mission capability, and informs the logistics COP. Accurate reporting of logistics and Army Health System support status is essential for keeping units combat ready. Brigade standard operating procedures establish report formats, reporting times, redundancy requirements, and radio voice brevity codes to keep logistic nets manageable. Data collection for the LOGSTAT is based upon operational and mission variables and should not overwhelm subordinate units with submission requirements.

H-84. LOGSTAT reporting begins at the lowest level. The company first sergeant or executive officer compiles reports from subordinate elements, and completes the unit's LOGSTAT report. Once completed, reports are forwarded from a unit to its higher headquarters and its supporting logistics headquarters, to include the FSC. Normally, LOGSTATs flow through S-4 channels. The FSC reports on hand supply and supply point on hand quantities. The battalion staff has an interest in both reports, as does the supporting sustainment unit.

H-85. The frequency of a LOGSTAT varies and is dependent on the operational tempo of the battalion or subordinate units. LOGSTATs should be completed at least daily, but may be required more frequently during periods of increased intensity or high operational tempo. As long as automation is available, logistics status relayed via near-real time automation provides the commander with the most up to date information, ultimately improving the supporting unit's ability to anticipate requirements.

H-86. The LOGSTAT can be completed through any means of communication to include written reports, radio, email or Joint Capabilities Release. Army Health System status is typically reported through the Medical Communications for Combat Casualty Care (MC4) System. The Joint Capabilities Release System helps lower level commanders automate the sustainment data-gathering process. The system does this through logistics situation reports, personnel situation reports, logistics call for support, logistics task order messaging, situational understanding, and task management. This functionality affects the synchronization of all logistics support in the area of operation between the supported and the supporter.

H-87. The command relationship of the FSC to the battalion will determine who reports the LOGSTAT and to whom for the FSC. The higher command mission order must delineate relationships and reporting requirements. If the FSC is attached to the Infantry battalion, the attachment orders could state the FSC submits its LOGSTAT to the supported battalion S-4. The battalion S-4 would have the responsibility to cross-level supplies within the Infantry battalion, adjust the battalion LOGSTAT and forward to the BDE S-4.

H-88. The battalion S-4 collects the LOGSTAT from the Infantry battalion's supported units. The battalion S-4 with the battalion XO's approval, determines which units receive designated supplies from the FSC and when. This decision is based on which unit has mission priority and the battalion's commander's guidance.

H-89. In addition to Joint Capabilities Release, sustainment leaders utilize the Global Combat Support System-Army (GCSS-Army) to track supplies, spare parts, and the operational readiness of organizational equipment. GCSS-Army Enterprise Resource Planning Solution is an automated information system that serves as the primary tactical logistics enabler supporting Army and joint transformation for sustainment. The program re-engineers current business processes to achieve end-to-end logistics and integration with applicable mission command and joint systems.

H-90. The sustainment staff must proactively identify and solve sustainment issues. This includes—

- Using Joint Capabilities Release, GCSS-Army, and other Army mission command systems to maintain sustainment situational understanding.
- Working closely with higher headquarters staff to resolve sustainment problems.
- Recommending sustainment priorities that conform to mission requirements.
- Recommending sustainment-related commander's critical information requirement.
- Ensuring the commander is kept aware of critical sustainment issues.
- Coordinating as required with key automated system operators and managers to assure focus and continuity of support.

H-91. The S-6 and the information systems technician work together to ensure that Joint Capabilities Release and GCSS-Army have interconnectivity. The battalion S-4, S-1, surgeon, and FSC commander monitor the functionality of these systems and implement alternate means of reporting during degraded communications or as required. The MC4 system supports information management requirements for the battalion surgeon's section and the battalion medical platoon to higher echelons of care. The Joint Capabilities Release, GCSS-Army, and the MC4 systems are used to support mission planning, coordination of orders and subordinate tasks, and to monitor and ensure mission execution.

PREPARATION

H-92. Preparation for the sustainment consists of activities performed by units to improve their ability to execute an operation. Preparation includes but is not limited to plan refinement, rehearsals, information collection, coordination, inspections, and movements. Sustainment preparation of the operational environment identifies friendly resources (host-nation support, contractible, or accessible assets) or environmental factors (endemic diseases, climate) that affect sustainment. Factors to consider, although not inclusive, include geography information and the availability of supplies and services, facilities, transportation, maintenance, and general skills (such as translators, laborers).

H-93. Sustainment preparation of the operational environment assists the sustainment planning staff to refine the logistics running estimate and concept of support. Sustainment planners forecast and build operational stocks as well as identify endemic health and environmental factors. Integrating environmental considerations will sustain vital resources and help reduce the logistics footprint. Sustainment planners take action to optimize means (force structure and resources) for supporting the commander's plan. These actions include, resupplying, maintaining, and issuing supplies or equipment along with any repositioning of sustainment assets. Additional considerations may include identifying and preparing bases, host-nation infrastructure and capabilities, contract support requirements, and lines of communications.

H-94. Sustainment rehearsals help synchronize the sustainment warfighting function with the Infantry battalion's overall operation. These rehearsals are led by the battalion S-4 and typically involve coordination and procedure drills for transportation support, resupply, maintenance and vehicle recovery, and medical and casualty evacuation. Throughout preparation, the FSC and staffs rehearse battle drills and standard operating procedures. Leaders place priority on those drills or actions they anticipate occurring during the operation. For example, the distribution platoon may rehearse a battle drill on reacting to an ambush while waiting to begin movement. Sustainment rehearsals and combined arms rehearsals complement preparations for the operation. Units may conduct rehearsals separately and then combine them into full-dress rehearsals. Although these rehearsals differ slightly by warfighting function, they achieve the same result. (Refer to chapter 3 and FM 6-0 for additional information.)

EXECUTION

H-95. Sustainment plays a key role in enabling decisive action. Sustainment determines the depth and duration of the Infantry battalion operation and is essential to retaining and exploiting the initiative to provide the support necessary to maintain operations until mission accomplishment. Failure of sustainment operations could cause a pause or culmination of an operation resulting in the loss of the initiative. Sustainment planners and operation planners work closely to synchronize all of the warfighting function, in particular sustainment, to allow the commander the maximum freedom of action.

Support to Offensive Tasks

H-96. Support to offensive tasks is by nature a high-intensity operation that requires anticipatory support as far forward as possible. The Infantry commander and staff ensure adequate support as they plan and synchronize the operation. Plans should include flexible sustainment capabilities to follow exploiting forces and continue support. Considerations during execution include the following:

- Establish protection for sustainment units from bypassed enemy forces in a fluid, noncontiguous area of operations.
- Recover damaged vehicles from the main supply route.
- Preposition essential supplies far forward to minimize lines of communication interruptions.

- Plan increased consumption of petroleum, oils, lubricants, and ammunition.
- Anticipate longer lines of communications as the offensive moves forward.
- Anticipate poor trafficability for sustainment vehicles across fought-over terrain.
- Consider preconfigured LOGPACs of essential items.
- Anticipate increased vehicular maintenance especially over rough terrain.
- Maximize field maintenance teams forward.
- Request distribution at forward locations.
- Increase use of meals, ready to eat.
- Use captured enemy supplies, equipment, support vehicles, and petroleum, oils, and lubricants. Test for contamination before use.
- Suspend most field service functions except airdrop and mortuary affairs.
- Prepare for casualty evacuation and mortuary affairs requirements.
- Select potential and/or projected supply routes, logistics release points, drop zones, landing zones and/or pickup zones, and support areas based on map reconnaissance.
- Plan and coordinate enemy prisoner of war operations.
- Plan replacement operations based on known and/or projected losses.
- Ensure that sustainment preparations do not compromise tactical plans such as excess stockpiles of vehicles and supplies and operational security.
- Builds contingency resupply packages to be delivered based off triggers of the ground mission.

Support to Defensive Tasks

H-97. The Infantry battalion commander with guidance from the BSB commander and FSC commander positions sustainment assets to support the forces in the defense. Sustainment requirements in the defense depend on the type of defense. Increased quantities of ammunition and decreased quantities of fuel characterize most area defenses. Barrier and fortification materiel to support the defense often has to move forward, placing increased demands on the transportation system. The following sustainment considerations will apply during operations:

- Pre-position ammunition, petroleum, oil, and lubricants, and barrier materiel in centrally located position well forward.
- Make plans to destroy stocks if necessary.
- Resupply during limited visibility to reduce the chance of enemy interference.
- Plan to reconstitute lost sustainment capability.
- Use maintenance support teams from the maintenance collection point to reduce the need to recover equipment to the BSA.
- Consider and plan for the additional transportation requirements for movement of CL IV barrier materiel, mines, and pre-positioned ammunition.
- Consider and plan for sustainment requirements of additional engineer units assigned for preparation of the defense.
- Plan for pre-positioning and controlling ammunition on occupied and prepared defensive positions.

Support to Stability-Focused Tasks

H-98. Sustainment while conducting operations focused on stability often involves supporting United States and multinational forces in a wide range of missions for an extended period. Tailoring supplies, personnel, and equipment to the specific needs of the task is essential for the Infantry battalion commander to accomplish the mission.

H-99. The Infantry battalion may utilize to a greater extent sustainment support from host-nations, contractors, and local entities. This can reduce dependence on the logistics system, improve response time and free airlift and sealift for other priority needs. Support may include limited classes of supplies and services (catering, maintenance and repair, sanitation, laundry, and transportation).

H-100. The logistics civil augmentation program provides the ability to contract logistics support requirements in a theater of operations. The Infantry battalion commander should expect contractors to be involved in operations focused on stability after the initial response phase. The terms and conditions of the contract establish relationships between the military and the contractor. The commander and staff planners must assess the need for providing operational area security to a contractor and designate forces to provide security when appropriate. The mission of, threat to, and location of the contractor determines the degree of protection needed. (Refer to ATP 4-92 for additional information.)

ECHELON SUPPORT

H-101. How the Infantry battalion, including external and attached organizations and BSB, array in echelon varies widely based METT-TC. The FSC, in support of the Infantry battalion's concept of support, plans and synchronizes *echelon support*, which is the method of supporting an organization arrayed within an area of operation (ATP 4-90). *Area support* is the method of logistics, medical support, and personnel services in which support relationships are determined by the location of the units requiring support. Sustainment units provide support to units located in or passing through their assigned areas (ATP 4-90). Current mission, task organization, mission command, concept of support, capability and capacity, and terrain influence how support is echeloned.

ECHELON OF SUPPORT

H-102. Echeloning support within the Infantry battalion is a carefully planned and executed process. The method employed to echelon support is a deliberate, collaborative decision based upon a thorough mission analysis within the military decision-making process. During this analysis, there must be an understanding at all levels of the capabilities of the support organization within and supporting the Infantry battalion. As the Infantry battalion's primary sustainment organization, the FSC's organization facilitates echeloned support. Common echelon of support at the lowest level of sustainment is executed at the battalion, and company echelons.

Battalion Echelons

H-103. As discussed earlier, a FSC from the BSB supports the Infantry battalion. The FSC performs the logistics function within the battalion echelon of support, referred to as unit trains in one location, or echeloned trains within an area of operation. Unit trains at the battalion level are appropriate when the unit is consolidated in an assembly area (AA), during reconstitution, major movements, or when terrain or distances restrict movement causing the unit to depend on aerial resupply and evacuation for support. The battalion normally operates in echeloned trains where subordinate unit trains employ into multiple locations.

H-104. Echeloned trains at the battalion can be organized into combat trains and field trains. Battalion trains are used to array subordinate sustainment elements (unit personnel, vehicles, and equipment) including the FSC. The battalion commander and staff, and the FSC commander collaborate to determine the best method of employment commensurate with the Infantry battalion's concept of support. Echeloning of support can include the battalion aid station, elements of the S-1 section and S-4 section, and elements of the FSC (figure H-1).

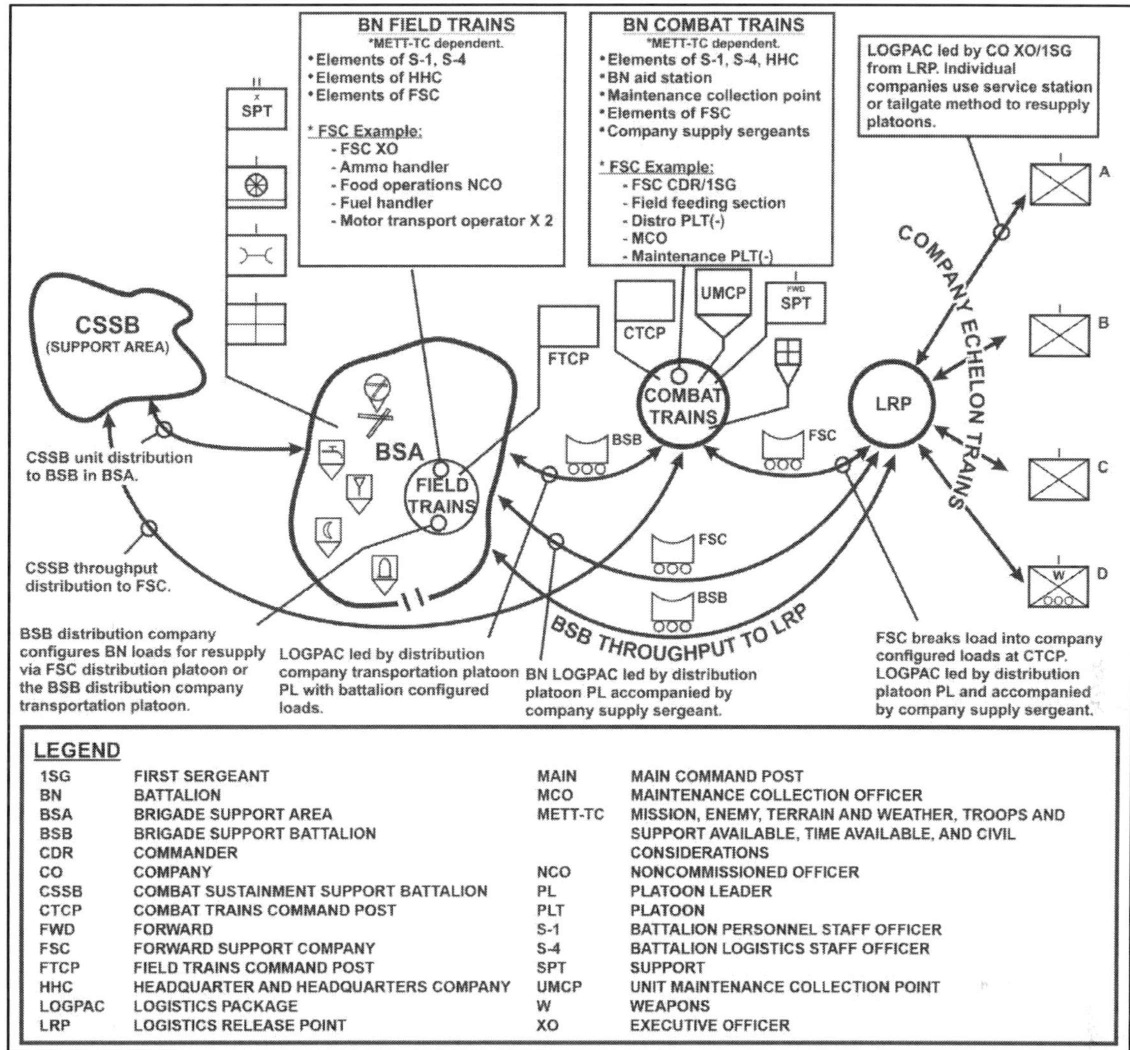

Figure H-1. Battalion concept of support, example

Combat Trains

H-105. Combat trains usually consist of elements of the battalion S-1 section, S-4 section, and battalion aid station, the maintenance collection point and other selected elements of the forward support company. Units consider the mission variables of METT-TC when locating combat trains in a battalion support area. The maintenance collection point should be positioned where recovery vehicles have access, or where major or difficult maintenance is performed. The combat trains must be mobile enough to support frequent changes in location, time and terrain permitting, under the following conditions when—heavy use or traffic in the area may cause detection, area becomes worn by heavy use such as in wet and muddy conditions, security is compromised. Refer to appendix A for information on the combat CP.

Field Trains

H-106. Field trains are positioned based on METT-TC considerations and often will be located in the BSA. The field trains include battalion sustainment assets not located with the combat trains. Field trains can provide direct coordination between the battalion and the BSB. When organized the field trains usually consist of the elements of the headquarters and headquarters company and the battalion S-1 and S-4 sections, and may include FSC elements not located in the combat trains. Field trains personnel help facilitate the coordination and movement of support from the BSB to the battalion. The battalion S-4 coordinates all unit supply requests with the IBCT S-4 and BSB. The BSB fills orders with on-hand-stocked items through unit distribution to the FSC, typically located at the combat trains. Refer to appendix A for information on the field trains CP.

Company Echelons

H-107. Echeloning of support begins at the company level. Companies within the Infantry battalion have no organic logistics organizations. Echeloning support within these units, if required, must be done with internal personnel and equipment used to facilitate or expedite logistics support within these units.

H-108. The commander determines the composition of echeloned support, often referred to as company trains, and may consist of the first sergeant, supply sergeant, and medic. Maintenance teams from the FSC may be included. This echeloned support expedites replenishment of subordinate elements using either the supply point distribution or the unit distribution method. The operations order must described the method used.

H-109. Supply point distribution requires unit representatives to move to a supply point to pick up their supplies. Supply point distribution is commonly executed by means of a logistics release point. The logistics release point may be any place on the ground where unit vehicles return to pick up supplies and then take them forward to their unit. In unit distribution, supplies are configured in unit sets and delivered to one or more central locations. Depending on the distribution method used, the first sergeant may send unit personnel and vehicles to a logistics release point designated by the FSC (supply point distribution) or the first sergeant may coordinate for the forward support company to deliver supplies to a location (unit distribution).

H-110. Within the company, the first sergeant will replenish company elements using various methods depending on the situation. Unit elements may move from their positions to the designated site to feed, resupply, or turn in damaged equipment. This is often referred to as a service station technique. This method is normally used in assembly areas and when contact is not likely. This method takes the least amount of time for the sustainment operators.

H-111. Conversely, the first sergeant may use unit or support personnel and vehicles to go to each element to replenish them. Soldiers can remain in position when using this method. This method is the most lengthy resupply method and may compromise friendly positions. This is often referred to as the tailgate technique or the in-position resupply.

Brigade Support Area

H-112. The *brigade support area* is a designated area in which sustainment elements locate to provide support to a brigade (ATP 4-90). The BSA is the sustainment (logistics, medical, personnel, and administrative) node for the IBCT, and is the BSB's terrain from which to conduct sustainment operations. It consists of the BSB main CP (which can also serve as an IBCT alternate CP if required), the brigade engineer battalion, signal assets, and other sustainment units from echelons above brigade. The BSB commander is responsible for the mission command of all support organizations within the BSA for terrain management and security unless otherwise stated by the operations or fragmentary order. The IBCT commander, with the support of their staff and upon the advice of the BSB commander, determine the control exercised by the BSB commander in governing the authority and limitations of the BSB to execute area security within the BSA. Considerations used in determining the authority and limitations of the BSB commander to execute area security within the BSA are—threat levels and situation; utility of different locations; and civil considerations. The Infantry battalion can place their field trains within the BSA. (Refer to FM 3-96 for additional information.)

Locations for Echelon Support Areas

H-113. The battalion S-4, assisted by the battalion XO and battalion S-3, recommends to the battalion commander the layout of the echelon support areas for the field trains and combat trains, when established. The echelon support areas should be located so that support from the BSA (if not collocated) can be maintained, but does not interfere with the tactical movement of battalion units or with units that must pass through the battalion's area, while still maximizing security. The echelon support area's size varies with terrain and how the forward support company is organized between the field trains and combat trains. Usually the echelon support areas are on a main supply route and out of the range of the enemy's medium artillery. The echelon support areas should be positioned away from the enemy's likely avenues of approach and entry points into the battalion's main battle area.

H-114. In determining the location for the support areas, there is a constant balancing of support and security, which ultimately determines the best placement of support areas. The commander or responsible officer for the support area integrates both supporting and security activities so as to not degrade the battalion's combat effectiveness. The commander or responsible officer for the echelon support areas must ensure logistics missions and associated activities continue without restriction and that all units within or transiting the support area are capable of conducting self-protection against a Level I threat. (See appendix I for a discussion of threat levels).

H-115. Once positioned, echelon support areas should not be considered permanent or stationary. Echelon support areas (specifically echeloned field trains and combat trains) must be mobile to support the units when they move, and should change locations frequently depending on available time and terrain. A change of location may occur with a change of mission or change in a unit's area of operation. Movement to a new location may be required to avoid detection caused by heavy use or traffic in the area or an area becomes worn by heavy use (wet and muddy conditions). Echeloned trains locations may need to change when security becomes lax or complacent due to familiarity. (Refer to ATP 4-90 for additional information.) Support area location considerations include the following:

- Cover and concealment (natural terrain or man-made structures).
- Room for dispersion.
- Level, firm ground to support vehicle traffic and sustainment operations.
- Suitable helicopter landing sites.
- Distance from known or templated enemy indirect fire assets.
- Good road or trail networks.
- Good routes in and out of the area (preferably separate routes going in and going out).
- Access to lateral routes.
- Good access or positioned along the main supply route.
- Positioned away from likely enemy avenues of approach.

SUPPLY ROUTES AND CONVOYS

H-116. The IBCT S-4, in coordination with the BSB support operations officer and the IBCT S-3, select supply routes between echeloned support areas and is discussed in the IBCT concept of support during the sustainment rehearsal. Main supply routes are designated within the battalion's area of operation. A main supply route is selected based on the terrain, friendly disposition, enemy situation, and scheme of maneuver. Alternate supply routes are planned in the event that a main supply route is interdicted by the enemy or becomes too congested. In the event of CBRN contamination, either the primary or the alternate main supply route can be designated as the dirty main supply route to handle contaminated traffic. Alternate supply routes should meet the same criteria as the main supply route. (Refer to FM 3-96 for additional information.)

H-117. Security of supply routes in a noncontiguous environment may require the battalion commander to commit combat units to the field trains and combat trains. The security and protection of supply routes along with lines of communications are critical to military operations since most support traffic moves along these routes. The security of supply routes presents one of the greatest security challenges in an area of operation. Route security operations are defensive in nature and are terrain-oriented. A route security force may prevent an enemy or adversary force from impeding, harassing, or destroying traffic along a route or portions of a route by establishing a movement corridor. Units conduct synchronized operations (mobility and information collection) within the movement corridor. A movement corridor may be established in a high-risk area to facilitate the movement of a single element, or it may be an enduring operation. (Refer to FM 3-90-2 for additional information.)

ILLUSTRATION OF CONTINUOUS SUPPORT DURING OFFENSIVES ACTIONS

H-118. Conduct of offensive actions requires large amounts of supplies, and the provision of continuous support along open and secure lines of communications. Those lines of communications lengthen during the conduct of the offense, which in turn requires the forward movement of stocks and sustainment units and the establishment of forward support areas and/or bases. The forward movement of sustainment units and stocks must be timed to minimize the impact on support to subordinate maneuver forces. This section illustrates, through a fictional scenario used as a discussion vehicle, the provision of continuous support during the IBCT's conduct of offensive actions.

Operational Overview for the Illustration

H-119. The IBCT currently occupies an intermediate staging base in the corps area of responsibility (not illustrated). The IBCT conducts reorganization and reconfiguration of capabilities to meet evolving divisional offensive actions and logistical rhythms. The IBCT moves from the intermediate staging base to occupy multiple AAs within the division's AO to prepare for future operations (not illustrated). Advanced force operations, long-range surveillance and special operations forces, assists IBCT entry forces. (See FM 3-99.) Advanced force operations conduct shaping operations to assist air assault operations and ground tactical movements, including reconnaissance and surveillance of proposed helicopter landing zones (not illustrated). Insurgent and terrorist groups encountered during the operation are fixed until sufficient combat power can be brought to bear to destroy them and capture or kill the personnel associated with these groups.

IBCT Concept of Operation

H-120. The IBCT conducts operations to seize Objective Airfield and to retain key terrain and lines of communication to the south (illustrated in figures H-2 to H-5) in support of the division's main effort (not illustrated) to the east. The IBCT conducts an air assault operation north of Objective Airfield then clears to the south to seize Objective Airfield in concert with follow-on ground movement and maneuver to expand the lodgment within the IBCTs area of operation. The IBCT conducts a zone reconnaissance [cavalry squadron minus (two mounted cavalry troops)] to the along Route Dodge. The IBCT continues to defend and prepare for follow on forces. IBCT subordinate unit task and order of movement—

Cavalry Squadron

H-121. The squadron, first in the order of movement, conducts zone reconnaissance south along Route Dodge. On order, conducts screen in the south and southeast portion of the IBCT's AO. The squadron confirms or denies the presence of enemy forces south of Objective Airfield. Upon reaching Phase Line (PL) Orange, the squadron reports their front line trace, triggering the IBCT commander to order battalion 1 to begin movement. The Squadron attaches one cavalry troop (dismounted) to battalion 1. Cavalry troop conducts air assault operation, on order conducts guard mission (dismounted) to protect battalion 1's right flank during its mission to clear Objective Airfield (see figures H-1 and H-2, pages H-24 and H-26).

Infantry Battalion 1

H-122. Battalion 1 conducts air assault, on order, maneuvers south to clear Objective Airfield. The battalion seizes key terrain and open lines of communication vital to future operations. Battalion 1, reinforced with elements from the brigade engineer battalion, clears Objective Airfield to limit of advance, PL Red. Battalion 1 then retains cleared area of operation. Attached cavalry troop conducts air assault, then moves to the battalions west flank to protect battalion's clearing mission to the south (see figures H-1 and H-2, pages H-24 and H-26).

Infantry Battalion 2

H-123. Battalion 1 reaches PL Orange. On order, battalion 2 (with brigade engineer battalion, attached sustainment assets, field artillery battery 2, and the IBCT tactical CP) begins tactical movement to Objective Airfield. Lead elements of Battalion 2, with attached engineer assets, clears along Route Ford to Objective Airfield. Upon arrival, battalion 2 conducts link up with battalion 1. Once linkup completed, battalion 2 conducts a forward passage of lines with battalion 1. Once forward passage of lines completed, battalion 2 moves to conduct movement to contact to the south to prevent the enemy from influencing IBCT operations within and surrounding Objective Airfield (see figure H-2, page H-25).

Infantry Battalion 3

H-124. On order, battalion 3 (with field artillery battery 3, field artillery headquarters, the IBCT Main CP and the remaining element of the BSB) conducts ground tactical movement along Route Ford. Lead elements of Battalion 3, with attached engineer assets, clear along Route Ford. Battalion 3 conducts linkup and forward passage of lines with battalion 1. Upon completion of the forward passage of lines, battalion 3 conducts a movement to contact, seizing key terrain to prevent the enemy from influencing IBCT operations within and surrounding Objective Airfield (see figure H-3, page H-27).

Brigade Engineer Battalion

H-125. The brigade engineer battalion, minus detachments, follows battalion 2, conducts ground tactical movement to Objective Airfield. The brigade engineer battalion provides engineer, intelligence, and signal support through all phases of the operation.

Brigade Support Battalion

H-126. The BSB follows battalion 3 during Phase III, conducts tactical convoy operations to Objective Airfield. The BSB establishes and repositions the brigade support area throughout all phases of the IBCT's operation.

Field Artillery Battalion

H-127. The field artillery battalion moves via ground tactical movement and rotary wing air movement, sling-loading equipment during all phases of the operation as required. Movement by phase ensures the most effective employ of continuous fires in support of the IBCT's operation. Battery 3 establishes a position area for artillery (PAA) vicinity in AA 1 and provides suppression of enemy air defense and preparation fires for battalion I during air assault operations into the designated helicopter-landing zone. Battery 1 conducts artillery air assault raid, to the northeast of Objective Airfield, to provide direct support fire to battalion 1 during phases I, II, and III. Battery 1 establishes PAA 1, positioned ready to fire within 15 minutes of occupation. On order, Battery 1 moves via rotary wing air movement, sling-loading equipment to support battalion 3 during Phase IV of the operation. Battery 2 conducts ground tactical movement with battalion 2. Battery 2 establishes a PAA between PL Red and PL White during Phase III and provides fire to support battalion 2 and 3's objectives. The field artillery battalion TAC CP conducts a movement with battery 2 and establishes the battalion TAC CP within the same vicinity. Battery 3's trigger to conduct movement to the Objective Airfield is both batteries in position ready to fire. Battery 3 follows battalion 3 ground tactical movement to Objective Airfield along Route Ford. Upon arrival to Objective Airfield, battery 3 establishes a PAA and provides indirect fires for follow on missions.

Concept of Sustainment

H-128. Support relationship for FSCs: Each FSC is attached to support maneuver battalions. Scenarios are based on units submitting LOGSTATs and expenditure reports twice daily from lowest level (company). Example concept of sustainment may include the following:

- Class I, Field Feeding. Issue/Ration Cycle 3-3-3/Meals, Ready to Eat.
- Class I, Bulk water. Units deploy with M149s (commonly referred to as a water buffalo) and 1 X Load Handling System Compatible Water Tank Rack (commonly referred to as a Hippo) per FSC and 6 X hippos with the BSB distribution company. The combat service sustainment battalion (CSSB) will distribute bulk water with hippo exchange with the distribution company. The distribution company will conduct hippo exchange with the forward support companies (FSCs). The consumption rate will dictate the resupply.
- Class II. Units deploy with 10 days of supply (DOS) and request resupply via Global Combat Support System (GCSS) Army system.
- Class III, bulk fuel ([II (bulk)]. The distribution company will provide FSCs replenishment of CLIII (bulk). Type of resupply is based off what phase the operation is in.
- Class III, packaged [III (package)]. Units will deploy with 10 days of supply of Class III (package) and resupply based off LOGSTAT.
- Class IV. Each unit deploys with mission-configured loads for the air assault mission. If a requirement arises for additional Class IV, the BSB Support Operations Officer will coordinate requirements with the supporting CSSB.
- Class V. Units enter the AA with one basic load per Soldier, one combat load per weapons platform, and one sustainment load for each battalion maintained by the FSC, and one sustainment load for the

IBCT maintained at the BSB's distribution company ATHP. Class V resupply is based off of expenditure reports turned in twice daily with the LOGSTAT.

- Class VI. Units deploy with 30 days of supply of health and comfort items. Health and comfort packs can be ordered through supply system if operation is longer than 30 days.
- Class VII. There is currently no replacements for Class VII items during the early phases of this operation. Units will report damaged equipment and evacuate to closest maintenance collection point.
- Class VIII/Medical. Units deploy with 6 days of supply on hand and request resupply via brigade medical supply office or through MC4. Emergency resupply packages available upon request. Casualties will be evacuated from point of injury to the respective battalion aid station. Casualties requiring role II or higher will be evacuated according to priority. Primary means of evacuation is medical evacuation and secondary means is casualty evacuation. Units will use predesignated ambulance exchange point (not shown in illustration) for ground casualty evacuation.
- Class IX Maintenance. Units deployed with full Class IX shop stock and authorized stockage list. Battalions will receive resupply from their supported FSC with secondary resupply through the distribution company SSA. Units will requisition Class IX parts to the SSA via GCSS-Army. For damaged equipment, units will secure all equipment until the equipment can be recovered. Units will evacuate battle-damaged items to designated maintenance collection point for repair. All units provide their own recovery assets. Additional requests for recovery will be submitted to the support operations officer for further coordination for evacuation. All units deploy with battle damage assessment, and repair (known as BDAR) kits and tow bars.

H-129. Mortuary affairs. The BSB receives a mortuary affairs collection section to provide direct support to the IBCT and to establish the brigade mortuary affairs collection point. Each unit will have a search and recovery team to conduct initial recovery operations. Each unit will recover remains to the brigade mortuary affairs collection point. Remains will be stored in accordance with unit mortuary affairs standard operating procedures and will backhaul remains utilizing quickest method available, ground or air mode. The support operations mortuary affairs NCO will coordinate further transport of remains to theater mortuary affairs collection point.

H-130. Information systems. Units deploy with all logistics information systems. Logistics status reports and expenditure reports will be turned in twice daily. The primary, alternate, contingency, and emergency plan for logistics reporting is Joint Battle Command–Platform/Joint Capabilities Release-Logistics, command post of the Future Secret Internet Protocol Router Network, and Tactical Nonclassified Internet Protocol Router.

H-131. Transportation. For this vignette, transportation movement requests were coordinated with the BSB transportation officer (resides in BSB support operations) prior to mission execution. Any requirement outside of the BSB's capability were coordinated through the supporting CSSB.

Phase I – Zone Reconnaissance/Air Assault—Concept of Support

H-132. During phase I, the squadron conducts zone reconnaissance along Route Dodge. Upon reaching PL Orange, battalion 1 task force conducts an air assault north of Objective Airfield. Squadron continues zone reconnaissance operations further south along Route Dodge. In AA 1, the BSB continues sustainment operations and prepares for ground tactical movement (figure H-2).

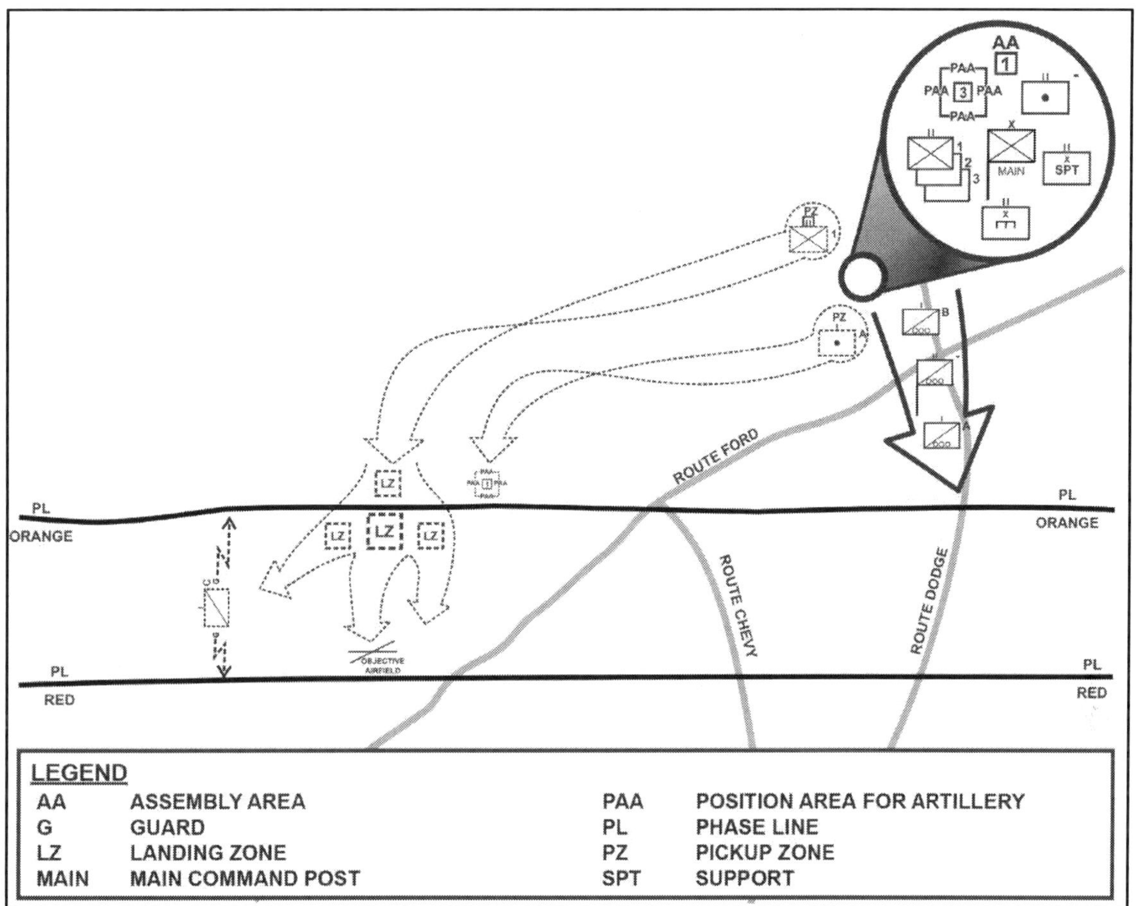

Figure H-2. Phase I—zone reconnaissance/air assault concept of support

H-133. The BSB has established supply points for Class I, III, IV, limited SSA operations for units to receive Class II, III, IV, V, and Class IX support prior to the mission. The BSB support operations officer/S-3 coordinates with aviation brigade elements to synchronize sustainment efforts and rotary wing assets to prepare to execute resupply of preplanned contingency packages to support the battalion 1 task force during the air assault. All fuel assets for the BSB and FSCs were full prior to AA occupation, The FSC holds 1 DOS for their supported battalion and the BSB holds 1 DOS for the IBCT. Primary means of resupply during phase I will be contingency resupply (see paragraphs H-37) packages delivered via aerial delivery based off triggers set by the battalion during planning. Secondary means of resupply is via LOGPAC. LOGPAC operations can be executed once Battalion 2 clears Route Ford. Contingency resupply consists of CDSs, speedballs, or kicker boxes prepared by each supporting FSC and staged at designated pick-up zone. The contingency resupply packages are trigger based, if a trigger does not happen, the contingency resupply package is not used and FSC simply reloads the package on designated flat racks. The priority of support during Phase I is cavalry squadron then to Battalion 1. Priority of supply for phase I is Class V, IV, III and VIII. Contingency resupply packages are called forward based off triggers set by each battalion.

> ***Note.*** The BSB also coordinates and plans in concert with the air liaison officer and brigade aviation officer for fixed and rotary wing delivery of sustainment assets when applicable. In this scenario, the BSB has coordinated for fixed wing and rotary wing assets to deliver contingency resupply packages. The BSB also coordinates with the supporting combat sustainment support battalion (CSSB) to receive 10 Hippos for water support to the battalion, executed via Hippo exchange. The CSSB has the ability to provide throughput distribution when needed. The brigade engineer battalion FSC E deploys with 2 X 55 gallon fuel drums of aviation gas to support unmanned aircraft system operations. Distance along Route Ford from AA 1 to the airfield is about 40 kilometers and time to get there for battalion 2 depends on enemy contact during movement.

H-134. The BSB movement plan to support the IBCT is further broken down to move key elements within the BSB forward at the appropriate trigger and timing. The quartering party moving within the first ground tactical movement for the IBCT may be further broken down into serials, commonly referred to as the torch party (first to move), and advanced echelon (ADVON). A forward logistics element (FLE) is pushed forward with the ADVON to be able to provide initial sustainment support if aerial resupply goes black. The FLE will provide Class V support, Class III support, and limited Class IV and maintenance support. The main body would then follow with the bulk of sustainment forces, followed by the trail party with remaining sustainment assets.

Note. A *forward logistics element* is comprised of task-organized multifunctional logistics assets designed to support fast-moving offensive operations in the early phases of decisive action. Also called FLE (ATP 4-90).

H-135. Battalion 2, the brigade engineer battalion, the IBCT TAC, and elements of the BSB (BSA quartering party and ADVON to include FLE) prepares for tactical movement during phase II. The BSB provides maintenance support, Class III bulk fuel support, and is prepared to push FSC's contingency resupply packages or speedballs via aerial delivery from AA 1.

H-136. FSC D, supporting the cavalry squadron prepares contingency resupply packages and stages them in designated pick-up zones. If the trigger set by squadron is met, contingency resupply packages are dropped at designated landing zones that are secured by the squadron. FSC D will be on one of the first ground tactical movements after battalion 1 moves to seize the airfield during Phase I. A small logistics team is left at AA 1 to conduct contingency resupply operations as needed during phase I and part of phase II of the operation. This team is pulled forward with the rest of the BSB in phase III.

H-137. FSC G supports battalion 1. Prior to battalion 1 departure to conduct air assault mission, FSC G provides logistics support to battalion 1 and builds contingency resupply packages of Class IV and V that are staged at designated pick-up zone to be delivered to battalion 1 via aerial delivery. Battalion 1 deploys with 1 day of supply (DOS) on hand. During phase I air assault, FSC G resupplies Battalion 1 utilizing contingency resupply. Several contingency resupply packages of Class V and configured loads of Class IV were pushed out once battalion 1 reached PL Red and battalion 1's transition to retain area surrounding Objective Airfield, triggers set by Battalion 1. The FSC prepares for tactical convoy operations during Phase III secured by battalion 3.

H-138. FSC F supports the field artillery battalion. While in AA 1, FSC F prepares battery 1 for movement via sling load. Equipment for battery is staged for sling load and executed with the initial air assault. Additional sling load equipment will be pushed to battery 1 for follow on mission, sling load of battery 1 to support battalion 3 during Phase III. FSC F prepares contingency resupply packages of Class V to be pushed out once the PAA 1 is in position and ready to fire.

H-139. All other FSCs are executing logistics operations in order to prepare their supported battalion for ground tactical convoy operations movement. Each FSC, excluding FSC J supporting battalion 3, builds contingency resupply packages for aerial delivery during phase I and II of this operation. Each battalion establishes triggers to execute the delivery of contingency packages to call forward. The contingency resupply packages consists of Classes V, IV, and VIII.

Phase II – First Ground Assault Convoy Movement to Objective Airfield - Concept of Support

H-140. Once battalion 1 reaches PL Orange, battalion 2, brigade engineer battalion, IBCT TAC, field artillery battery B, and elements of the BSB begin tactical movement to the airfield. Battalion 2 clears Route Ford to Objective Airfield, conducts linkup with battalion 1 north of PL Red along Route Ford. Battalion 2 conducts a movement to contact and seizure of key terrain south of PL Red to prevent enemy forces from influencing the IBCT (figure H-3).

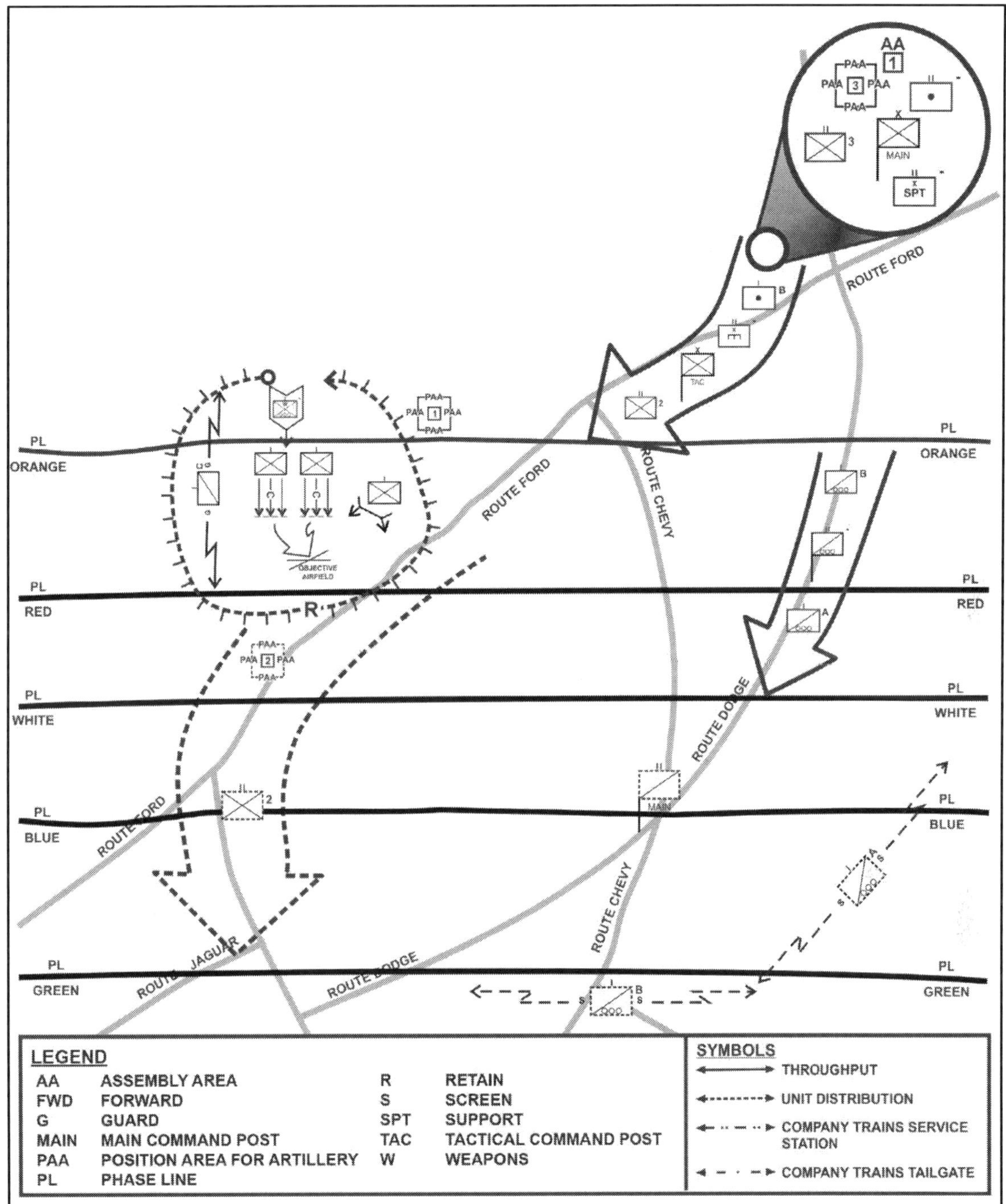

Figure H-3. Phase II—first ground tactical movement—concept of support

H-141. Once Battalion 1 reaches PL Orange, battalion 2, brigade engineer battalion, IBCT TAC, and BSB quartering party begins ground tactical movement in serials along Route Ford from AA 1 to Objective Airfield (distance about 40 km). Battalion 1 continues to limit of advance, vicinity PL Red. The BSB places a logistics team consisting of a small maintenance team from the maintenance company and fuel support from the distribution company and stages at designated rest area and fuel stop at checkpoint 2 with a security element from battalion 2. The logistics team provides fuel support and limited maintenance support to the IBCT during the first ground tactical movement as needed and will be picked up by the trail party of the IBCT. Each battalion will provide medical support to each serial. The BSB quartering party will be a part of first few serials with the IBCT and the BSB ADVON to include the FLE positioned in the last serial with FSC D.

H-142. Once the first ground tactical movement reaches the north side of the Objective Airfield, the BSB quartering party will begin to establish key locations within BSA to set-up and prepare to receive follow-on sustainment serials metering into the BSA. The quartering party will begin preparing for transfer of mission command by establishing initial communications. Quartering party sustainment planners continue coordination's of appropriate security elements and if applicable begin making changes to proposed BSA defense plan. Party prepares to conduct security if battalion 1 is called forward, based on METT-TC as needed and establishes each unit's AO to establish local security. Key to this phase is communication with sustainment units within AA 1 to report any changes to BSA proposed set-up and priorities of work.

H-143. BSB ADVON, consisting of elements of the headquarters and headquarters company (HHC), distribution company, maintenance company, and brigade medical support company reaches the Northside of the airfield. The BSB ADVON continues set-up of BSA operations to include initial set up of the battalion main, distribution, maintenance, and initial medical support operations. The ADVON begins establishing the BSA entry control point (ECP) and BSA defense operations. The FLE is on standby to provide sustainment support to battalion 1, elements of the brigade engineer battalion, and battalion 2 as needed. The squadron establishes the field trains within the BSA and upon arrival of FSC D, the FSC will receive an AO within the BSA (while residing in the BSA) to support the squadron. When the FSC is pulled forward with field trains or combat trains due to mission requirements, the BSA defense plan has to be addressed immediately to retask units to secure that portion of the BSA. In this vignette, Battalion 1 provides area security to include elements of the brigade engineer battalion securing the BSA while the BSA sets up the BSA. The remainder of the BSB locates in the AA and prepares for the second ground tactical movement comprised of battalion 3, IBCT main, and remaining sustainment assets.

Note. Location of the field trains CP is METT-TC dependent but is generally located within the brigade support area. For more information on field trains CP and combat trains CP operations, refer to appendix A.

H-144. Priority of supply during phase II is Class V, IV, VIII, and III. One DOS is with each unit, 1 DOS with the FSC, and 1 DOS with the BSB. Contingency resupply packages are staged and ready to be sent forward. Priority of support is battalion 1, cavalry squadron, then battalion 2. Primary means of resupply is contingency resupply packages via aerial delivery and alternate means of resupply for units forward will be provided by the FLE once established in the BSA.

H-145. The remaining BSB elements at the AA 1 prepares for the second ground tactical movement while tracking the operation and providing logistics support. BSB staff continues to receive LOGSTATs, expenditure reports, and ensures contingency resupply packages are pushed out when triggers are met, and ensures the IBCT has continuous, uninterrupted sustainment support.

H-146. FSC D supports the squadron via contingency resupply and the squadron field trains. During the second ground tactical movement, contingency resupply packages planned off squadron triggers are pushed forward via sling load operations by the logistics team left with the BSB in the AA 1. Secondary means of resupply is by LOGPAC via echelon trains. Once battalion 2 moves to vicinity of PL Blue, squadron field trains move forward with company configured loads to the designated LRP to resupply the Squadron. At the LRP, troop 1SGs pick up the LOGPAC and move forward to resupply the troops A and B. The squadron communicates with battalion 1 that the squadron field trains are moving through battalion 1 AO.

H-147. FSC G and FSC E continues to support battalion 1 and elements of the brigade engineer battalion attached to battalion 1 via aerial resupply if needed and prepares for movement during Phase III. FSC J continues to prepare battalion 3 for ground tactical movement with the trail party. Secondary means of resupply for forward units is the FLE located in the BSA.

H-148. FSC F continues to support the field artillery battalion from the AA 1. Class V is provided through contingency resupply packages called forward and delivered via sling load to battery 1. FSC F prepares for movement during Phase III but leaves a logistics team with battery 3 that will provide limited fuel, distribution and maintenance support to trail party ground tactical movement with battery 3 as needed.

Phase III – Second Ground Tactical Movement to Objective Airfield - Concept of Support

H-149. Battalion 3 and the remaining elements of BSB and field artillery battalion headquarters and battery C conducts a ground tactical movement to Objective Airfield. Once battalion 3 reaches the Route Chevy, it moves to conduct a movement to contact along Route Chevy and seizure of key terrain to the southeast of the lodgment to prevent the enemy from influencing IBCT future operations. Battalion 2 continues movement to contact and seizure of key terrain to the southwest of the lodgment to prevent the enemy from influencing IBCT future operations. Infantry Battalion 1 retains to secure lodgment for future operations (figure H-4).

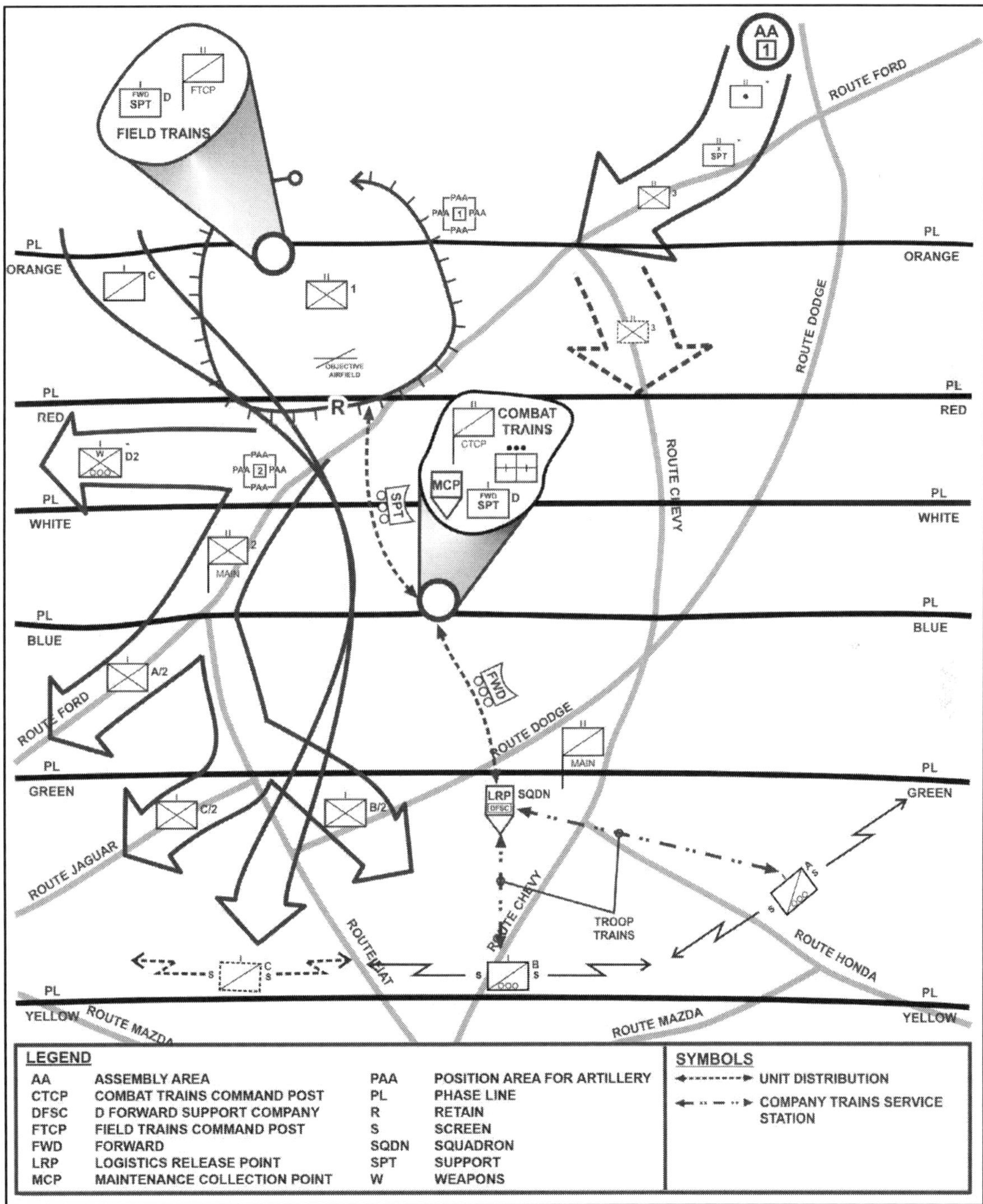

Figure H-4. Phase III—second ground tactical movement—concept of support

H-150. During this phase, the final element of the BSB conducts movement with the second ground tactical movement to assist in the establishment of the BSA. Upon arrival, the BSB HHC sets up the BSB main CP, establishes ECP operations, and begins BSA operations. Key to occupation is the continuous strength of the base defense or the triggers associated with assuming BSA defense AOs when IBCT forces move forward from the BSA. Each company within the BSA, to include FSCs currently located within the BSA, are assigned an area of operation to establish security and communications within the assigned AO. During this phase, elements of the brigade engineer battalion currently located within the BSA are part of the BSA defense plan. The FLE provides logistics support to units within the BSA and prepares to send logistics elements forward for resupply as needed. Priority of supply is Class V, VIII, III, and I. Priority of support in this phase is battalion 2, battalion 3, squadron, and battalion 1. BSB will receive resupply from the CSSB every two days. Primary means of resupply is LOGPAC via tactical convoy operations and alternate means of resupply is via aerial resupply.

H-151. During phase III, the HHC finalizes the set-up of mission command for the BSB command group and staff, establishes support operations, field feeding for the battalion, ECP operations for the BSA, and plans and coordinates for mortuary affairs support. The BSB staff continues overseeing logistics operations to include maintenance management, commodity management, and transportation operations for all units assigned to the IBCT.

H-152. The BSB distribution company establishes security in assigned AOs and establishes the SSA operations, the ATHP, Class III and Class I points and prepares for distribution operations to the FSCs. For Class I bulk water, the BSB retained six Hippos of the 12 requested from the CSSB. Class I bulk water support will be conducted via hippo exchange with the FSCs. The distribution company establishes a designated landing zone for sling load operations on the airfield. The BSB support operations officer, in coordination with the BSB S-3, coordinates with the aviation brigade elements to synchronize sustainment efforts for rotary wing assets. The distribution company conducts replenishment operations via LOGPAC to FSC D at the squadron's combat trains located north of PL Green.

H-153. The field maintenance company establishes security in assigned AOs and establishes field maintenance operations to the BSB, units located within the BSA, and provides limited maintenance support to the FSC for low density commodities/electronics and armament equipment.

H-154. The brigade support medical company establishes security in assigned AOs and establishes a Role II medical facility within the BSA augmented with a forward surgical team.

H-155. FSC D: The squadron established the squadron combat trains vicinity of Route Dodge, south of PL Blue. FSC D pushed sustainment assets forward to augment the combat trains. FSC D receives routine resupply from the distribution company at the combat trains. The FSC left a team at the squadron field trains to ensure squadron configured loads are built off LOGSTAT reports and verified during the IBCT logistics synchronization meeting. (The squadron S-4 forecasts what is needed off LOGSTATs and submits requests to the IBCT S-4 and BSB support operations officer.) Upon arrival at the combat trains, the FSC begins breaking down the squadron configured loads into troop-configured loads with the assistance of each troop supply sergeant. The combat trains are ready to conduct replenishment operations to LRPs at next designated resupply times during this phase. The LRP will be established in vicinity of PL Yellow between Route Fiat and Route Chevy. Once LRPs are established, the troop first sergeant or designated representative of each troop will pick up their designated resupply and continue forward to resupply the troop. If resupply via tactical convoy cannot be accomplished, the FSC team in the BSA has resupply packages to be pushed via aerial delivery to designated drop zones.

H-156. Each FSC arrives at the BSA and establishes security in assigned AO and establishes the logistics footprint within their supported battalion's field trains. Each FSC prepares to support its battalion via routine resupply operations.

H-157. The trail party consisting of a security element of battalion 3 and battery 3 of the field artillery battalion begin movement to BSA and picks up logistics team that halted at checkpoint 2.

Phase IV – IBCT Continues Operations/BSA Fully Operational - Concept of Support

H-158. During Phase IV of the operation, the squadron continues operations further to the south vicinity PL Brown. Troop C conducted air assault during Phase III, reattaches to squadron. Battalion 2 occupies key terrain in southwest of lodgment, and Battalion 3 occupies key terrain southeast of lodgment. Battalion 1 continues to occupy key terrain surrounding lodgment (figure H-5, page H-32).

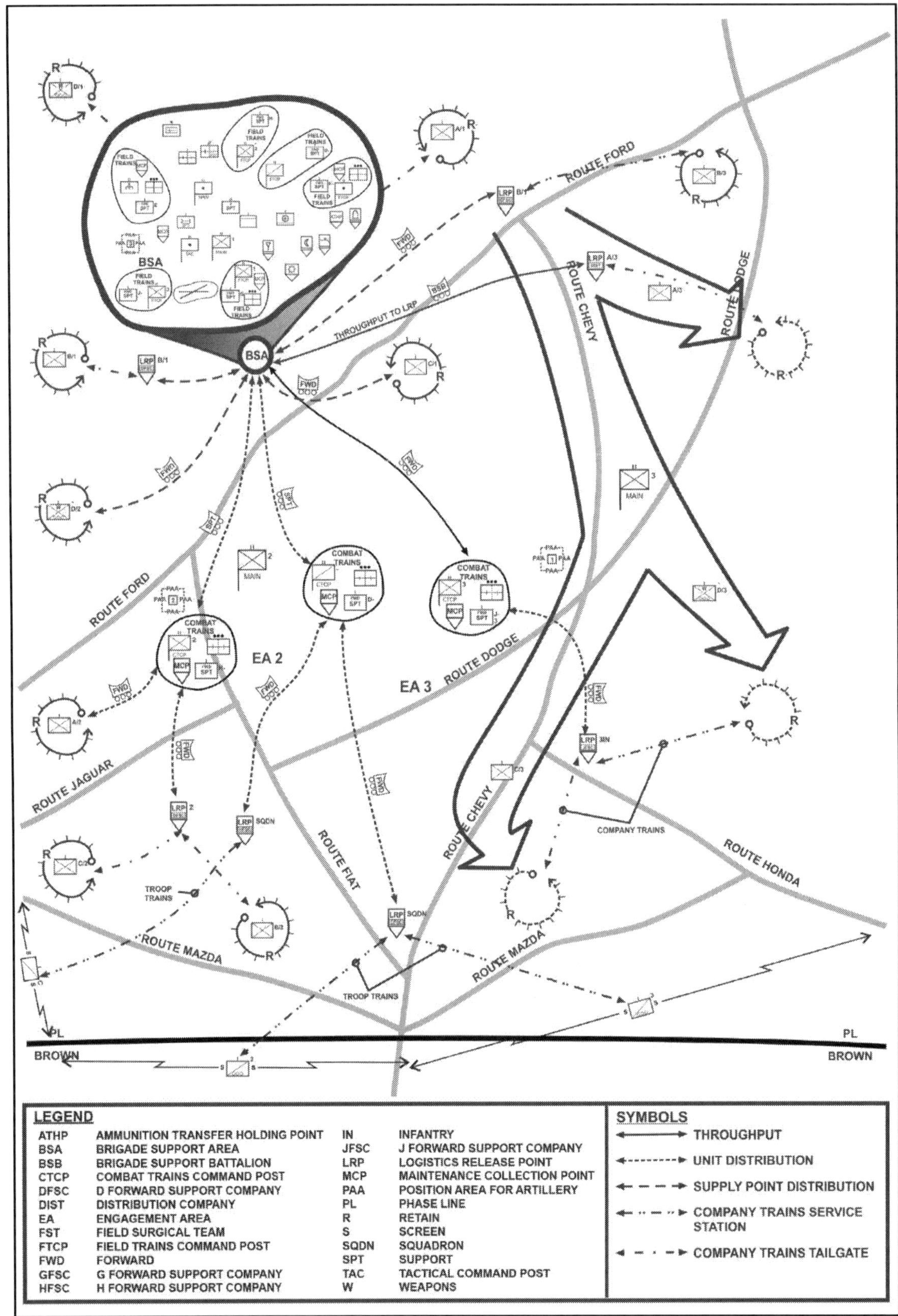

Figure H-5. Phase IV—IBCT continuous operations—BSA fully operational

H-159. The BSB is operational in the BSA and is providing logistics support to the IBCT through routine resupply. The support operations officer continues synchronizing BSB operations for the IBCT. The support operations officer applies BSB capabilities against IBCT's requirements, and coordinates logistic support from the CSSB support operations officer in coordination with the IBCT S-4. The support operations officer monitors and updates the logistics COP, monitors support operations, and continues to make adjustments to ensure support requirements are met. The support operations officer sets the schedule for the brigade logistics synchronization meeting that will include the supporting CSSB coordinating staff, IBCT S-4, FSC commanders, BSB support operations officer staff and battalion S-4s. (Physical representations from each battalion is dependent on how the battalion positions personnel within the field trains and combat trains. For example, if the FSC commander is forward the commander could listen in via communications established, or the FSC XO could attend the meeting in the commander's place). The BSA is prepared to jump if needed and a FLE will be established forward if the BSA does jump to provide logistics support while BSA is moving forward.

H-160. Priority of supply is Class V, VIII, III and I. Priority of support in this phase is squadron, battalion 1, Battalion 2 and battalion 3. BSB receives resupply from CSSB every two days. Primary means of resupply during this operation is routine resupply by LOGPAC via tactical convoy operations. Alternate means of resupply is routine resupply packages pushed via aerial resupply.

H-161. The headquarters company continues providing support to the BSB commander and staff. The BSB headquarters continues overseeing logistics operations for all units assigned and attached to the IBCT. The headquarters company continues ECP operations with details rotating from the distribution company, the field maintenance company, and the brigade support medical company, and continues field-feeding operations for the BSB.

H-162. The distribution company provides supply point distribution to FSCs E, G, and F and unit distribution to FSCs D, H, and J at the designated combat trains. Battalion 3 S-4 requested throughput distribution to LRP A/3 for company A, battalion 3. The distribution company continues to push SSA, ATHP, fuel, and water operations. Distribution company pushes routine resupply LOGPACs via tactical convoy operations at times coordinated by the support operations officer and listed on the BSB Synchronization Matrix.

H-163. Field maintenance company continues providing field maintenance operations to units located within the BSA and provides limited maintenance support to the FSC for low density commodities/electronics and armament equipment. The brigade support medical company continues providing Role II support to the IBCT.

H-164. Squadron continues screening operations south along PL Brown. The squadron's combat trains are positioned to jump once battalion 2 and battalion 3 clear pass Route Mazda. The combat trains will move near intersection of Route Fiat and Route Chevy and will be prepared to conduct routine resupply to designated LRP.

H-165. Battalion 1 occupies key terrain surrounding lodgment. FSC G establishes replenishment operations within the battalion field trains location within the BSA. The FSC receives supply point distribution (also known as Service Station Distribution) from the BSB distribution company. The FSC then builds company configured loads at the BSA and field trains push LOGPAC forward to the designated LRP(s).

H-166. Once battalion 2 clears past Route Jaguar, sustainment assets from FSC H are called forward with first resupply to establish sustainment footprint within the combat trains. FSC H leaves a logistics team behind in the BSA. This team builds battalion configured loads for the BSB distribution company to push to the combat trains. Once sustainment assets from FSC H are received at the combat trains, the combat trains personnel break battalion configured loads into company configured loads. The combat trains accompanied by company supply sergeants move from the combat trains to established LRP. The company first sergeant or executive officer meets the combat trains at LRP and pushes forward to resupply the company utilizing either service station distribution or the tailgate technique.

H-167. Once Battalion 3 clears pass route Honda, sustainment assets from FSC J are called forward with the combat trains with first resupply and push directly to the LRP established by battalion 3. The combat trains move along established supply routes to LRP to resupply company A and company B. Company executive officers pick up their portion of the combat trains and move forward to resupply each company. Each company will be resupplied Classes I, III, V, and VIII. Class VIII resupply is requested through the battalion surgeon further to the brigade medical officer located within the IBCT S-4. After the resupply is complete, the distribution elements returns to the combat trains. The battalion establishes another LRP located along route dodge to resupply company A of battalion 3 and requests the distribution company push resupply to this LRP. Prior to moving forward, the

FSC built company configured loads at the BSA. The distribution company then pushes the company-configured load to the established LRP to resupply company A. Company A of Battalion 3 with utilize the tailgate technique for resupply.

H-168. The brigade engineer battalion continues to support the BSA defense plan. The FSC sets up sustainment operations within the BSA and is prepared to provide resupply to any asset pulled forward to assist the squadron, battalion 2 or battalion 3. The brigade engineer battalion field trains will conduct replenishment operations via LOGPAC to support the brigade engineer battalion to designated combat trains (squadron, battalion 2 or battalion 3) if called forward. The FSC receives resupply via supply point distribution from the distribution company.

H-169. The field artillery battalion continues to provide fire in support of squadron, battalion 2 and 3 objectives. Battery 3 establishes PAA north of the airfield and is ready to provide fire for follow on missions. The FSC has established operations within the BSA and receives resupply via supply point distribution from the distribution company. The FSC then builds company configured loads and conducts service station distribution from the BSA to Battery 3 positioned to the east of the airfield. Field trains resupply via LOGPAC to LRP to support established PAAs. Class V routine resupply packages are built to be pushed via aerial delivery to PAA 1 and PAA 2 when called forward.

Appendix I

Base Operations

A base is a locality from which operations are projected or supported (JP 4-0). The Infantry battalion requires a secure area to prepare for future operations, to recover and refit, and conduct sustainment. It uses perimeters and other measures to protect units during offensive and defensive tasks, especially within a noncontiguous AOs. While supporting primarily stability-focused tasks however, Infantry battalions use base camps to protect sustainment operations and provide a secure area for its units to recover from and prepare for operations. This appendix provides an overview of base and base camp operations, and includes a discussion of base camp operations in support of stability-focused tasks. (Refer to ATP 3-37.10 for additional information.)

OVERVIEW

I-1. A base camp is an evolving military facility that supports the military operations of a deployed unit and provides the necessary support and services for sustained operations (ATP 3-37.10). Base camps provide a protected location from which to project and sustain combat power. Operating from base camps is a fundamental tactic of ground-based forces while conducting primarily missions, tasks, and activities to stabilize an operational environment in crisis or vulnerable state. The level of protection afforded by a base camp is based on the threat.

I-2. A base cluster, in base defense operations, a collection of bases geographically grouped for mutual protection and ease of command and control (JP 3-10). Units located within the base or base camp are under the tactical control of the base or base camp for base security and defense. Within large echelon support areas, controlling commanders may designate base clusters for mutual protection and accomplishment of mission objectives.

I-3. Base camps are characterized by four principles that are incorporated throughout the life cycle. These principles are—

- Scalability.
- Sustainability.
- Standardization.
- Survivability.

I-4. Commanders and staffs use the base camp principles as a guide for analytical thinking. These principles are not a set of rigid rules, nor do they apply in every situation. They should be applied with creativity, insight, and boldness.

CLASSIFICATION

I-5. Base camps are broadly classified by duration, purpose, and size. This classification system provides common terminology and a framework that aids in the conduct of all base camp life cycle activities.

Base Camp Duration

I-6. A base camp may be classified according to its expected duration as shown in table I-1, page I-2. A contingency base camp is expected to operate 2–10 years or less, while an enduring base camp is expected to operate more than 5 years or longer. Facilities should transition from contingency to enduring standards when appropriate, typically any time within a 6-month to 5-year period. These timelines provide a framework to plan for the transition of standards. The actual trigger for transition is based on conditions and other factors.

Table I-1. Bases camp duration

Phase	Construction Standard	Expected Duration
Contingency	Organic	Up to 90 days
	Initial	Up to 6 months
	Temporary	Up to 5 years
	Semipermanent	2-10 years
Enduring	Permanent	5 years or greater

I-7. Expected base camp duration affects the construction standards used for facilities and infrastructure. While enduring construction standards are not typically used during the contingency phase of an operation, at times semipermanent construction standards may sometimes be used in place of initial (completed with organic equipment) or temporary construction when site considerations require or mission parameters lead to their use. The combatant commander, in coordination with Service components and the Services, specifies the construction standards for facilities in the theater to optimize the engineer effort expended on any given facility while assuring that the facilities are adequate for health, safety, and mission accomplishment.

BASE CAMP PURPOSE

I-8. Each base or base camp is unique, based on mission requirements and the theater-specific facility allowances and construction standards that apply. Bases and base camps are developed to serve a specific purpose such as to serve as an intermediate staging base, a forward operating base, or a logistic base; support reception, staging, onward movement, integration, training, and detention operations; or they may be multifunctional. The designated purpose and the operational requirements of tenant units serve as the primary guide in designing the base or base camp.

BASE CAMP SIZE

I-9. There are five sizes of base camps: platoon, company, battalion, brigade combat team, and support area. Table I-2 shows base camp sizes and the populations associated with each. The base camp population typically includes tenant and transient units and organizations, including U.S., multinational, and host nation personnel, units, and organizations to include contractors authorized to accompany the force and selected non-contractors authorized to accompany the force. Transient units and organizations are those that come to the base camp for specified services and support. This may not necessarily include staying overnight. Determining the number of transients that a base camp can serve and understanding service and support relationships with other base camps are critical factors in accurately identifying requirements for base camp facilities and infrastructure, services, and support.

Table I-2. Bases camp sizes and approximate populations

Base Camp Size	Approximate Population
Platoon	50
Company	3000
Battalion	1000
Brigade Combat Team	3000
Support Area	6000 or greater

I-10. The base camp population includes both tenant and transient units and organizations, which can include U.S., multinational, and host nation personnel, units, and organizations to include contractors. Transient units and organizations are those that come to the base camp for specified services and support, which may not necessarily include remaining overnight. Determining the number of transients that a base camp will serve and understanding service and support relationships with other base camps are critical factors in accurately identifying requirements for base camp facilities and infrastructure, services, and support.

CHARACTERISTICS

I-11. Levels of capabilities describe the characteristics of a base camp in terms of support and services provided and the nature of the construction effort applied that are commensurate with the anticipated duration of the mission. Base camps in support of short-duration missions are more austere and require fewer resources to establish and operate, while those for longer-duration missions generally require greater resources. Not all similar sized base camps will have the same level of capability, and the implementation of these capabilities is not directly linked to operational phases. There are three levels of capabilities for base camps: basic, expanded, and enhanced.

I-12. Basic capabilities are established as part of initial entry and are implemented primarily using organic capabilities and prepositioned stocks. Basic capabilities are—

- Activities and services that are essential for sustaining operations for a minimum of 60 days.
- Characterized by rapid deployment and emplacement.
- Highly flexible and moveable.

I-13. Expanded capabilities are basic capabilities that have been improved to increase efficiencies in the provision of base camp support and services. Expanded capabilities are expanded to sustain operations for a minimum of 180 days. Engineer units or contracted support may be used to achieve the desired results.

I-14. Enhanced capabilities are expanded capabilities that have been improved to operate at optimal efficiency and support operations for an unspecified duration. Many of the activities, facilities, and services and support resemble those of a permanent base or installation.

CONSTRUCTION STANDARDS

I-15. There are three construction standards for base camps: initial, temporary, and semi-permanent. The time periods for each standard are derived from the expected design life, not how long a facility may actually be used. Units use their organic construction capabilities to the fullest extent possible to construct base camps to the directed standard. Initial construction standards are characterized by austere facilities requiring minimal engineer effort that take full advantage of a unit's organic capabilities and are intended for immediate operational use by units upon arrival for up to 6 months. Temporary construction standards are characterized by austere facilities requiring additional engineer effort above that required for initial construction standards and are intended to serve a life expectancy of 5 years or less. Extending the life of temporary construction is generally not cost effective. Semipermanent construction refers to buildings and facilities designed and constructed to serve a life expectancy of less than 10 years.

TRANSFER AND CLOSURES

I-16. All or portions of a base camp may be closed when no longer needed or transferred to another service, multinational force, governmental or nongovernmental organization, or the host nation. As the operation progresses and mission objectives are achieved, base camps are often realigned and closed to consolidate resources and reduce the overall logistic footprint in support of the basing strategy. Proper transfer and closure procedures facilitate the timely withdrawal of forces.

BASE CAMP ACTIVITIES

I-17. Base camp activities provide useful constructs to aid in describing general areas of knowledge and in visualizing and characterizing the base camp-operating environment. Base camp activities are interrelated and interdependent; each activity provides an action that mutually supports the others. The foundation of all activities is master planning. The base camp activities are—

- Master planning.
- Operations and maintenance.
- Protection.
- Sustainment

I-18. During mission planning, the base camp activities help the commander and staff to organize the broad range of base camp requirements and the supporting information and tasks required for execution. Base camp activities are used in organizing people and equipment within base operations centers, base cluster operations centers, base camp working groups, and base camp management centers to facilitate the exercise of authority and direction and the management of base camps (see paragraphs I-19 to I-24).

BASE CAMP OPERATIONS

I-19. A base or base camp can contain one or more units from one or more services and will typically support both U.S. and multinational forces, as well as other unified action or inter-organizational partners operating anywhere along the range of military operations. A base camp must be viewed through a life cycle construct that includes the development of base camps from pre-establishment through transfer or closure, with levels of increasing base camp capabilities.

I-20. A commander designates an area or facility as a base camp, and often designates a single commander as the base or base camp commander responsible for protection, terrain management, and day-to-day operations of the base or base camp. This allows other units to focus on their primary function. Units located within the base or base camp are under the tactical control of the base or base camp commander for base security and defense. Within large echelon support areas, controlling commanders may designate base clusters for mutual protection and accomplishment of mission objectives.

I-21. The base operations center is the command post for the base camp commander. It is a centralized facility that operates and manage the base camp. It is the base camp commander's primary means for monitoring the situation and managing the performance of base camp activities and the provision of services and support. Base operations centers are organized into functional areas that generally reinforce the base camp to help focus efforts. The size, composition, and configuration of the base operations center may vary between base camps based on the—

- Base camp size.
- Level of base camp services and support.
- Real property asset management requirements.
- Complexity of facility and infrastructure operations and management requirements.
- Span of control based on the number and echelon of tenant and transient units or subordinate base camps for base clusters.

I-22. Base cluster operations centers are established to control several subordinate base camps that may be grouped together in a cluster for mutual support for either sustainment or protection. Base cluster operations centers, commonly found at the IBCT level or higher headquarters, are similar to the base camp management center in both organization and function.

I-23. Commanders at all levels may form base camp working groups by grouping select staff members who meet to focus on base camp planning or problem-solving. Base camp working groups may be used to conduct the initial base camp development planning until the necessary augmentation needed for adequate base camp development becomes available. When a base camp working group is established, the commander normally designates a group facilitator to focus the group's efforts and prevent duplication of effort. The group facilitator should brief the commander and staff on a recurring basis to maintain visibility and command emphasis on base camps.

I-24. Base camp management centers coordinate, monitor, direct, and synchronize actions needed for establishing, operating, sustaining, and managing base camps within an echelon's AO. On smaller camps, and/or when augmentation is unavailable, base camp commanders must rely on reachback to technical expertise residing in higher headquarters base camp management centers or other support agencies and centers.

BASE SECURITY AND DEFENSE

I-25. A primary task of base camps is to protect the combat power of the units assigned to it. Ensuring that base camps provide the necessary protection requires an application of the protection tasks that are detailed in ADRP 3-37 and FM 3-90-1. Base camps will typically protect their personnel and assets through application of area security. Area security is a security task conducted to protect friendly forces, installations, routes, and actions within a specific area (ADRP 3-90). Operations in noncontiguous AOs require commanders to emphasize area security. Area security operations focus on the protected force, installation, route, or area. (See chapter 4, section III for information on area security operations.)

General Planning Considerations

I-26. Numerous threats to operations security exist in stability-focused operations beyond those encountered during the conduct of combat-focused operations. The establishment of semi-permanent bases and the life support area provide locations where insurgent, terrorist, and organized criminal elements focus their data collection and data corruption efforts. The presence of local-hire and third-nation contract civilians to support the battalion's operations gives increased opportunities for computer viruses and worms to be inserted into battalion and higher headquarters information systems. This results not only in restrictions on when, where, and under what supervision civilian janitorial staff and other support personnel perform their duties but also to how higher-level partners of all types are integrated into the staff's operations. The mix of other U.S. governmental, host nation, and multinational agencies along with international and private volunteer organizations and the media, all of which may have totally different cultures and standards when it comes to safeguarding information further complicates the situation. IBCT and battalion signal staff planners consider these additional factors when addressing information protection.

I-27. Base camp defense (see paragraph I-32) includes the activities needed to defeat Level I and Level II threats to a base camp or base cluster, and shape or delay Level III threats until they can be defeated by a tactical combat force or other available response forces that is part of the higher commander's area security efforts. The three levels of threat categories are—

- Level I threat—a small enemy force that can be defeated by those units normally operating in the echelon support area or by the perimeter defenses established by friendly bases and base clusters (ATP 3-91).
- Level II threat—an enemy force or activities that can be defeated by a base or base cluster's defensive capabilities when augmented by a response force (ATP 3-91).
- Level III threat—an enemy force or activities beyond the defensive capability of both the base and base cluster and any local reserve or response force (ATP 3-91).

I-28. In most cases, base camps will be placed in locations where the risk of Level III threats have been eliminated or effectively mitigated by the area commander; however, base camps often become focal points for hostile actions. Because of the uncertainty in contingency operations and the acknowledgement of hybrid threats, base camp commanders must be prepared to conduct defensive tasks to repel a Level III attack when the threat assessment indicates the possibility of a Level III threat in the AO, regardless of which element of decisive action/simultaneous activities is currently dominant. This may involve significant increases in area denial measures; offensive actions; hardening, dispersal, and other protection measures; and immediate reaction to hostile actions. While hardening of facilities and maintaining a response force is the responsibility of the base camp commander, area denial actions and offensive tasks to reduce the risks of Level III threats are the responsibility of the area commander.

I-29. Base camps are designed and constructed to be resistant to attack and recover quickly after an attack so that they can continue to operate. The ability to quickly recover from an attack is enhanced through detailed planning and rehearsals of procedures. Base camps must be prepared to defend in any direction through flexible base defense plans, including the use of dedicated response forces positioned to respond to the widest possible range of contingencies. Base camp commanders apply the principles of protection described in ADRP 3-37 as well as the fundamentals of security and characteristics of defense identified in FM 3-96 and FM 3-90-1 in preparing base camp defense plans.

I-30. Base camp defense is based on the characteristics of the defense discussed in FM 3-96 and FM 3-90-1. One of the key characteristic, and one that is difficult for a base camp to achieve, is depth. The base camp commander uses a variety of techniques to increase the time a threat has to take to achieve a penetration and their exposure to friendly fires. These can include—

- Patrolling outside the perimeter.
- Wire, concrete, or other barriers used to reinforce the perimeter.
- Entry control points and associated obstacle and countermobility plans used to canalize and control incoming personnel or vehicles.
- Barriers employed to block high-speed avenues of approach, both externally on approaches to the perimeter and internally to protect high-risk targets.
- Perimeter guard towers and observation posts.

- Ditches, berms, or other earthen obstacles.

I-31. Base camps are vulnerable to indirect fires because of the concentration of forces within a small area. The base camp commander mitigates this threat by dispersal, use of protected buildings and barriers, and active counterbattery techniques.

Base Camp Security Framework

I-32. The framework for base camp security and defense consists of three primary areas (figure I-1). This structuring provides a means for organizing protection and defense information and requirements and focusing efforts. These three areas are—

- Outer security area. This is the area outside the perimeter that extends out to the limit of the base camp commander's AO. Commanders establish an outer security area to provide early warning and reaction time, and deny enemy reconnaissance efforts and vantage points for conducting standoff attacks. The outer security area is typically patrolled by security elements.
- Perimeter zone. This zone includes the base camp perimeter and area immediately in front or behind it that is needed for observation posts, fighting positions, and entry control points. Selected base camps may have designated inner and outer perimeters. Larger base camps will seldom employ this double layer of perimeters, and will rely more on a single perimeter supplemented with inner barriers and access control measures around critical facilities. Creation of a double perimeter is extremely resource intensive.
- Inner security area. This is the area inside the base camp perimeter. Interior barrier plans can be used around individual unit locations, critical assets, and as traffic control measures to add depth to the base camp security plan and to halt or impede the progress of threat penetrations of the perimeter zone.

I-33. Collectively, these three areas form the base camp AO. Commanders assigned an AO have inherent responsibilities that are described in FM 3-90-1. Not all commanders that may serve as base camp commanders will have the organic capabilities within their units to perform all of these responsibilities. In those situations, the higher commander must clearly articulate in the order which AO responsibilities will not be performed by the base camp commander (and who will perform them), or provide the necessary augmented capabilities to perform them.

I-34. Base camp commanders and their staffs apply the framework for base security and defense, much like a perimeter defense, (see chapter 3, section I for a discussion of perimeter defenses) to focus planning activities and ensure all critical elements of base security and defense are addressed. The framework is not intended as an all-inclusive solution to base security and defense, but is intended to provide a general template for planning.

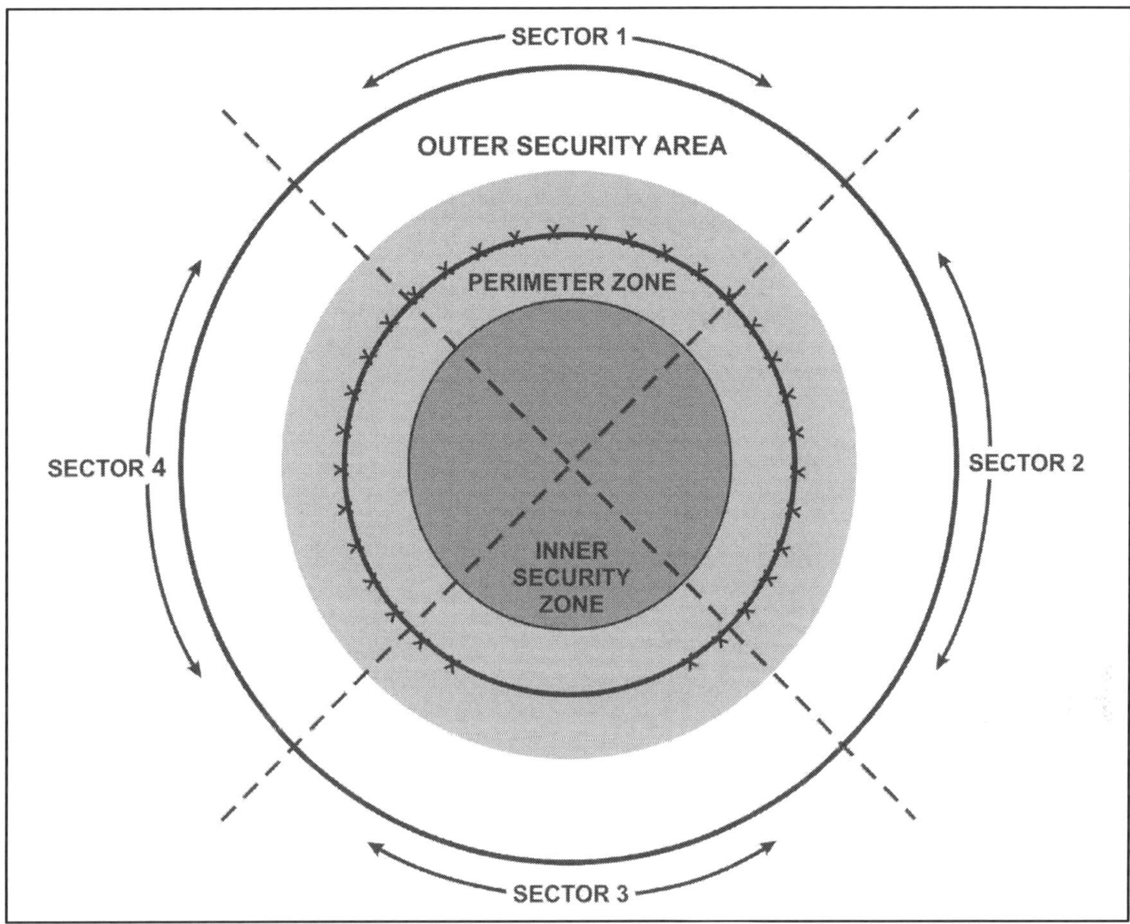

Figure I-1. Framework for base camp security and defense, example

BASE CAMP PROTECTION FORCES

I-35. Base defense is the local military measures, both normal and emergency, required to nullify or reduce the effectiveness of enemy attacks on, or sabotage of, a base, to ensure that the maximum capacity of its facilities is available to United States forces (JP 3-10). The base defense force consists of all the personnel and equipment needed to perform both security and defense tasks on the base camp. (For more information, refer to ATP 3-37.10.) The base defense force is capable of—

- Defeating Level I threats and defeating Level II threats with a response force.
- Defending against Level III threats until they can be defeated by a tactical combat force.
- Conducting reconnaissance patrols within its AO for detecting and reporting the location, strength, and capabilities of the enemy.

I-36. The major components of the base defense force is generally composed of the following:

- Base defense operations center.
- Perimeter security force.
- Response force.

I-37. The base defense force may consist of tenant and transient units. The force may also consist of contractor support. Tenant or transient unit may be tasked with specific security requirements while residing on the base. Base camp commanders must consider the risks of using contractors on the base camp. The use of contractors from the local population may incur additional risks based on the attitudes of the local population.

Base Defense Operations Center

I-38. The base defense operations center serves as the focal point for base security and defense for planning direction, coordination, integration, and control of all base defense efforts. The base defense operations center consists of two primary sections, the command section and the plans and operations section, with additional sections as required. The center is staffed with representatives of the intelligence, movement and maneuver, and fires warfighting functions and other functional areas. The center's composition reflects the base camp population.

Perimeter Security Force

I-39. The perimeter security force occupies observation posts and firing positions along the perimeter and conducts area security tasks within the base camp's security area. The perimeter security force is prepared to integrate the response force and other reinforcements and/or response forces that are committed as part of the perimeter defense.

Response Force

I-40. A response force is a dedicated force on a base with adequate tactical mobility and fire support designated to defeat Level I and Level II threats and shape Level III threats until they can be defeated by a tactical combat force or other available response forces. The response force provides the base camp/cluster commander with a depth for security and defense. Once committed the commander will be prepared to reconstitute a response force. The base camp commander may assign the response force a wide variety of tasks both within the base camp security area and within the base camp perimeter to—

- Reinforce a threatened area or respond to a penetration of the perimeter.
- Establish contact with potential threats and engage those threats as required within the base camp's security area, defeating Level II threats, and delaying Level III threats until the threat can be defeated by a tactical combat force.

I-41. The size and composition of the response force is based on a threat assessment and the levels of uncertainty and risk; and is adjusted based on changes in the situation. Its level of responsiveness (readiness condition) is adjusted based on threat conditions. The response force should be mounted to ensure adequate protection and tactical mobility. Typically, Infantry, armor, military police, and combat engineers are capable of performing response force operations.

Tactical Combat Force

I-42. A tactical combat force (TCF) is a rapidly deployable, air-ground mobile combat unit, with appropriate combat support and combat service support assets assigned to and capable of defeating Level III threats including combined arms (JP 3-10). The TCF is established, when required, at the division and corps level. The Infantry battalion can be assigned the TCF mission for the division or corps. The TCF size, composition, and response time is based on the TCF's operations order. Depending on the threat, the tactical combat unit mission may be on call or a unit's only mission. At a minimum, the tactical combat force must—

- Be positioned so that it can respond within the required time.
- Be mounted (generally), but can be foot-mobile when transported by helicopters or conditions warrant.
- Have sufficient ammunition and supplies to accomplish the mission to include engineer support as required.
- Be able to communicate with the supported base defense operations center to include monitoring the center's command net if required.
- Understand and rehearse the requisite parts of the supported base defense plan.

I-43. TCF assigned missions can include the following:

- Reinforce engaged units outside the perimeter.
- Conduct reconnaissance and surveillance activities.
- Respond to threats on critical assets, infrastructure, or high-risk personnel.
- Conduct security checks and random patrolling within the base camp perimeter.

BATTALION BASE CAMPS

I-44. The Infantry battalion makes extensive use of base camps during operations to support to stability-focused tasks Each AO however, has unique combinations of mission variables that make it difficult to discuss a standard method that Infantry battalions incorporate bases camps into the conduct of missions. Each subordinate commander evaluates current situation within the AO. (See chapter 4, section III for an illustration of base operations within an area security mission.) The following considerations assist the commander in developing techniques to incorporate base camps into operations.

I-45. Base camps provide an area to—

- Protect the force.
- Conduct sustainment.
- At least partially conduct mission command.
- Refit and prepare units for movement and maneuver.
- Prepare intelligence.
- Secure fires assets.
- Conduct administrative activities.
- Plan missions and conduct rehearsals.

I-46. Base camps can be a liability. They provide a fixed point for the enemy to observe friendly activity as well as the start and endpoint for friendly missions. Base camps also provide a concentrated target for enemy indirect fires. The enemy can also—

- Observe the entry and exit of units from the base camp.
- Identify defensive positions and obstacles.
- Identify patterns of movement, sustainment.
- Emplace improvised explosive devices in high-use areas.
- Plan ambushes and other attacks.

I-47. The battalion commander can mitigate these liabilities by—

- Not developing patterns.
- Conducting missions for extended periods with less physical contact with the base camp.
- Using Army aviation to move and resupply units.
- Identifying enemy direct and indirect firing points and developing plans to suppress them.
- Minimizing exposure to enemy attacks and fires by—
  - Conducting patrols.
  - Developing good relations with the local population.
  - Building protective structures.
  - Splitting mortars and field artillery assets.
  - Separating critical assets.

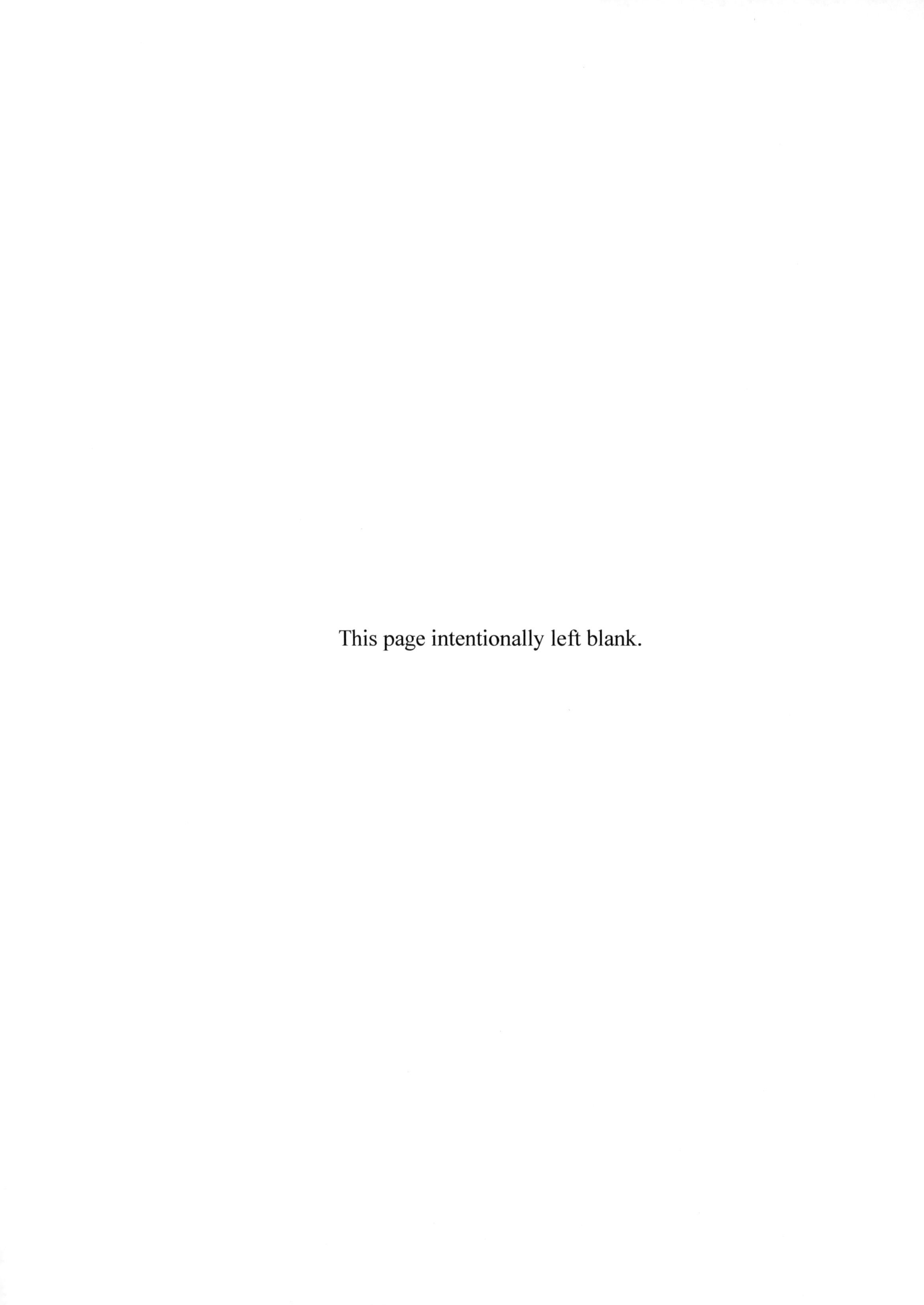

This page intentionally left blank.

Glossary

The glossary lists acronyms and terms with Army or joint definitions. Where Army and joint definitions differ, (Army) precedes the definition. Terms and acronyms for which ATP 3-21.20 is the proponent are marked with an asterisk (*). The proponent publication for other terms is listed in parentheses after the definition.

SECTION I – ACRONYMS AND ABBREVIATIONS

Acronym	Definition
	A
AA	assembly area
ABCT	armored brigade combat team
	B
BAE	brigade aviation element
BCT	brigade combat team
BSA	brigade support area
BSB	brigade support battalion
	C
CAS	close air support
CBRN	chemical, biological, radiological, and nuclear
CBRNE	chemical, biological, radiological, nuclear, and explosive
CCIR	commander's critical information requirement
CDS	container delivery system
COMSEC	communications security
COP	common operational picture
COR	contracting officer representative
CP	command post
CWMD	countering weapons of mass destruction
	D
D3A	decide, detect, deliver, and assess
DA	Department of the Army
DD	Department of Defense
DOS	days of supply
	E
EA	engagement area
ECP	entry control point
EMS	electromagnetic spectrum
EP	electronic protection
EW	electronic warfare
	F
FAC	forward air controller

FAC(A) forward air controller (airborne)

FLE forward logistics element

FOB forward operating base

FPF final protective fire

FSC forward support company

FSF foreign security forces

FSO fire support officer

G

GAF ground assault force

H

HUMINT human intelligence

I

IBCT Infantry brigade combat team

IED improvised explosive device

IPB intelligence preparation of the battlefield

ITAS Improved Target Acquisition System

J

JTAC joint terminal attack controller

L

LOGPAC logistics package

LOGSTAT logistics status report

LRP logistics release point

M

METT-TC mission, enemy, terrain and weather, troops and support available, time available, civil considerations (mission variables)

MDMP military decisionmaking process

MI military intelligence

MOE measure of effectiveness

MOP measure of performance

MOPP mission-oriented protective posture

N

NAI named area of interest

NCO noncommissioned officer

O

OPORD operations order

OTERA-A organize, train, equip, rebuild and build, advise and assist, and assess

P

PAA position area for artillery

PACE primary, alternate, contingency, and emergency

PL phase line

POB point of breach

POP point of penetration

	Q
QRF	quick reaction force
	R
R&S	reconnaissance and surveillance
RED	risk estimate distance
	S
S-1	battalion or brigade personnel staff officer
S-2	battalion or brigade intelligence staff officer
S-3	battalion or brigade operations staff officer
S-4	battalion or brigade logistics staff officer
S-6	battalion or brigade signal staff officer
S-9	battalion or brigade civil affairs operations staff officer
SC	security cooperation
SCATMINE	scatterable mine
SE	site exploitation
SEAD	suppression of enemy air defense
SEO	sniper employment officer
SFA	security force assistance
SOP	standard operating procedure
SSA	supply support activities
STANAGs	standardization agreements
	T
TOW	tube launched, optically tracked, wire guided
	U
U.S.	United States
	V
VDO	vehicle drop off
	W
WARNORD	warning order
WMD	weapons of mass destruction
	X
XO	executive officer

SECTION II – TERMS

actions on contact

A series of combat actions, often conducted simultaneously, taken upon contact with the enemy to develop the situation. (ADRP 3-90)

administrative movement

A movement in which troops and vehicles are arranged to expedite their movement and conserve time and energy when no enemy ground interference is anticipated. (FM 3-90-2)

adversary

A party acknowledged as potentially hostile to a friendly party and against which the use of force may be envisaged. (JP 3-0)

air-ground operations

The simultaneous or synchronized employment of ground forces with aviation maneuver and fires to seize, retain, and exploit the initiative. (FM 3-04)

alternate position

A defensive position that the commander assigns to a unit or weapon for occupation when the primary position becomes untenable or unsuitable for carrying out the assigned task. (ADRP 3-90)

ambush

An attack by fire or other destructive means from concealed positions on a moving or temporarily halted enemy. (FM 3-90-1)

approach march

The advance of a combat unit when direct contact with the enemy is intented. (ADRP 3-90)

area defense

A defensive task that concentrates on denying enemy forces access to designated terrain for a specific time rather than destroying the enemy outright. (ADRP 3-90)

area of influence

A geographical area wherein a commander is directly capable of influencing operations by maneuver or fire support systems normally under the commander's command or control. (JP 3-0)

area of interest

That area of concern to the commander, including the area of influence, areas adjacent thereto, and extending into enemy territory. (JP 3-0)

area of operations

An operational area defined by a commander for land and maritime forces that should be large enough to accomplish their missions and protect their forces. (JP 3-0)

area reconnaissance

A form of reconnaissance that focuses on obtaining detailed information about the terrain or enemy activity within a prescribed area. (ADRP 3-90)

area security

A security task conducted to protect friendly forces, installations, routes, and actions within a specific area. (ADRP 3-90)

area support

Method of logistics, medical support, and personnel services in which support relationships are determined by the location of the units requiring support. Sustainment units provide support to units located in or passing through their assigned areas. (ATP 4-90)

Army design methodology

A methodology for applying critical and creative thinking to understand, visualize, and describe unfamiliar problems and approaches to solving them. (ADP 5-0)

art of command

The creative and skillful exercise of authority through timely decision making and leadership. (ADP 6-0)

assault position

A covered and concealed position short of the objective, from which final preparations are made to assault the objective. (ADRP 3-90)

assembly area

An area a unit occupies to prepare for an operation. (FM 3-90-1)

assessment

The determinatin of the progress toward accomplishing a task, creating a condition, or achieving an objective (JP 3-0).

assured mobility

A framework—of processes, actions, and capabilities—that assures the ability of a force to deploy, move, and maneuver where and when desired, without interruption or delay, to achieve the mission. (ATP 3-90.4)

attack

An offensive task that destroys or defeats enemy forces, seizes and secures terrain, or both. (ADRP 3-90)

attack by fire

A tactical mission task in which a commander uses direct fires, supported by indirect fires, to engage an enemy force without closing with the enemy to destroy, suppress, fix, or deceive that enemy. (FM 3-90-1)

attack position

The last position an attacking force occupies or passes through before crossing the line of departure. (ADRP 3-90)

backbrief

A briefing by subordinates to the commander to review how subordinates intend to accomplish their mission. (FM 6-0)

base

A locality from which operations are projected or supported. (JP 4-0)

base camp

An evolving military facility that supports that military operations of a deployed unit, and provides the necessary support and services for sustained operations. (ATP 3-37.10)

base cluster

In base defense operations, a collection of bases, geographically grouped for mutual protection and ease of command and control. (JP 3-10)

base defense

The local military measures, both normal and emergency, required to nullify or reduce the effectiveness of enemy attacks on, or sabotage of, a base, to ensure that the maximum capacity of its facilities is available to Uunited States forces. (JP 3-10)

battle handover line

A designated phase line on the ground where responsibility transitions from the stationary force to the moving force and vice versa. (ADRP 3-90)

battle position

A defensive location oriented on a likely enemy avenue of approach. Also called a BP. (ADRP 3-90)

battle rhythm

A deliberate cycle of command, staff, and unit activities intended to synchronize current and future operations. (FM 6-0)

biological agent

A microorganism (or a toxin derived from it) that causes disease in personnel, plants, or animals or causes the deterioration of materiel. See also chemical agent. (JP 3-11)

biological hazard

An organism, or substance derived from an organism, that poses a threat to human or animal health. (JP 3-11)

block

A tactical mission task that denies the enemy access to an area or prevents their advance in a direction or along an avenue of approach. Block is also an engineer obstacle effect that integrates fire planning and obstacle efforts to stop an attacker along a specific avenue of approach or to prevent the attacking force from passing through an engagement area. (FM 3-90-1)

bounding overwatch

A movement technique used when contact with enemy forces is expected. The unit moves by bounds. One element is always halted in position to overwatch another element while it moves. The overwatching element is positioned to support the moving unit by fire or fire and movement. (FM 3-90-2)

breach

1. A tactical mission task in which the unit employs all available means to break through or establish a passage through an enemy defense, obstacle, minefield, or fortification. (FM 3-90-1) 2. A synchronized combined arms activity under the control of the maneuver commander conducted to allow maneuver through an obstacle. (ATP 3-90.4)

breach area

A defined area where a breach occurs. (ATP 3-90.4)

breakout

An operation conducted by an encircled force to regain freedom of movement or contact with friendly units. It differs from other attacks only in that a simultaneous defense in other areas of the perimeter must be maintained. (ADRP 3-90)

brigade support area

A designated area in which sustainment elements locate to provide support to a brigade. Also called BSA. (ATP 4-90).

bypass criteria

Measures during the conduct of an offensive operation established by higher headquarters that specify the conditions and size under which enemy units and contact may be avoided. (ADRP 3-90)

canalize

A tactical mission task in which the commander restricts enemy movement to a narrow zone by exploiting terrain coupled with the use of obstacles, fires, or friendly maneuver. (FM 3-90-1)

casualty evacuation

(Army) Nonmedical units use this to refer to the movement of casualties aboard nonmedical vehicles or aircraft without en route medical care. (FM 4-02)

***checkpoint**

A predetermined point on the ground used to control movement, tactical maneuver, and orientation. Also called a CP.

chemical agent

A chemical substance that is intended for use in military operations to kill, seriously injure, or incapacitate mainly through its physiological effects. See also chemical warfare; riot control agent. (JP 3-11)

chemical, biological, radiological, and nuclear defense

Measures taken to minimize or negate the vulnerabilities to, and/or effects of, a chemical, biological, radiological, or nuclear hazard or incident. Also called CBRN defense. (JP 3-11)

chemical, biological, radiological, and nuclear environment

An operational environment that includes chemical, biological, radiological, and nuclear threats and hazards and their potential resulting effects. Also called CBRN environment. (JP 3-11)

chemical, biological, radiological, and nuclear hazard

Chemical, biological, radiological, and nuclear elements that could create adverse effects due to an accidental or deliberate release and dissemination. Also called CBRN hazard. (JP 3-11)

chemical hazard

Any chemical manufactured, used, transported, or stored that can cause death or other harm through toxic properties of those materials, including chemical agents and chemical weapons prohibited under the Chemical Weapons Convention as well as toxic industrial chemicals. (JP 3-11)

clear

A tactical mission task that requires the commander to remove all enemy forces and eliminate organized resistance within an assigned area. (FM 3-90-1)

close air support

Air action by fixed-wing and rotary-wing aircraft against hostile targets that are in close proximity to friendly forces and that require detailed integration of each air mission with the fire and movement of those forces. (JP 3-0)

close area

The portion of a commander's area of operations assigned to subordinate maneuver forces. (ADRP 3-0)

close combat

Warfare carried out on land in a direct-firefight, supported by direct and indirect fires, and other assets. (ADRP 3-0)

collaborative planning

Commanders, subordinate commanders, staffs, and other partners sharing information, knowledge, perceptions, ideas, and concepts regardless of physical location throughout the planning process. (ADRP 5-0)

collateral damage

Unintentional or incidental injury or damage to persons or objects that would not be lawful military targets in the circumstances ruling at the time. (JP 3-60)

combat formation

A combat formation is an ordered arrangement of forces for a specific purpose and describes the general configuration of a unit on the ground. (ADRP 3-90)

combat identification

The process of attaining an accurate characterization of detected objects in the operational environment sufficient to support an engagement decision. Also called CID. (JP 3-09)

combat outpost

A reinforced observation post capable of conducting limited combat operations. (FM 3-90-2)

combat power

(Army) The total means of destructive, constructive, and information capabilities that a military unit or formation can apply at a given time. (ADRP 3-0)

combined arms

The synchronized and simultaneous application of all elements of combat power that together achieve an effect greater than if each element was used separately or sequentially. (ADRP 3-0)

command post

A unit headquarters where the commander and their staff perform their activities. Also called CP. (FM 6-0)

command post cell

A grouping of personnel and equipment organized by warfighting function or by planning horizon to facilitate the exercise of mission command. (FM 6-0)

commander's critical information requirement

An information requirement identified by the commander as being critical to facilitating timely decision making. Also called CCIR. (JP 3-0)

commander's intent

A clear and concise expression of the purpose of the operation and the desired military end state that supports mission command, provides focus to the staff, and helps subordinate and supporting commanders act to achieve the commander's desired results without further orders, even when the operation does not unfold as planned. (JP 3-0)

complex terrain

A geographical area consisting of an urban center larger than a village and/or of two or more types of restrictive terrain or environmental conditions occupying the same space. (ATP 3-34.80)

concealment

Protection from observation or surveillance. (FM 3-96)

concept of operations

(Army) A statement that directs the manner in which subordinate units cooperate to accomplish the mission and establish the sequence of actions the force will use to achieve the end state. (ADRP 5-0)

confirmation brief

A briefing subordinate leaders give to the higher commander immediately after the operation order is given. It is the leader's understanding of the commander's intent, their specific tasks, and the relationship between their mission and the other units in the operation. (ADRP 5-0)

consolidate gains

The activities to make permanent any temporary operational success and set the conditions for a sustainable stable environment allowing for a transition of control to legitimate civil authorities. (ADRP 3-0)

consolidation

The organizing and strengthening a newly captured position so that it can be used against the enemy. (FM 3-90-1)

contain

A tactical mission task that requires the commander to stop, hold, or surround enemy forces or to cause them to center their activity on a given front and prevent them from withdrawing any part of their forces for use elsewhere. (FM 3-90-1)

contamination mitigation

The planning and actions taken to prepare for, respond to, and recover from contamination associated with all chemical, biological, radiological, and nuclear threats and hazards in order to continue military operations. (JP 3-11)

contiguous area of operations

Where all of a commander's subordinate forces' areas of operations share one or more common boundaries. (FM 3-90-1)

control measure

A means of regulating forces or warfighting functions. (ADRP 6-0)

cordon and search

A technique of conducting a movement to contact that involves isolating a target area and searching suspected locations within that target area to capture or destroy possible enemy forces and contraband. (FM 3-90-1)

counterattack

An attack by part or all of a defending force against an enemy attacking force, for such specific purposes as regaining ground lost or cutting off, or destroying enemy advance units, and with the general objective of denying to the enemy the attainment of the enemy's purpose in attacking. In sustained defensive operations, it is undertaken to restore the battle position and is directed at limited objectives. (FM 3-90-1)

countering weapons of mass destruction

The efforts against actors of concern to curtail the conceptualization, development, possession, proliferation, use, and effects of weapons of mass destruction, related expertise, materials, technologies, and means of delivery. (JP 3-40)

countermobility operations

Those combined arms activities that use or enhance the effects of natural and man-made obstacles to deny enemy freedom of movement and maneuver. (ATP 3-90.8)

counterreconnaissance

A tactical mission task that encompasses all measures taken by a commander to counter enemy reconnaissance and surveillance efforts. Counterreconnaissance is not a distinct mission, but a component of all forms of security operations. (FM 3-90-1)

cover

Protection from the effects of fires. (FM 3-96)

cyberspace electromagnetic activities

The process of planning, integrating, and synchronizing cyberspace and electronic warfare operations in support of unified land operations. (ADRP 3-0)

cyberspace operations

The employment of cyberspace capabilities where the primary purpose is to achieve objectives in or through cyberspace. (JP 3-0)

decision point

A point in space and time when the commander or staff anticipates making a key decision concerning a specific course of action. (JP 5-0)

decisive action

The continuous, simultaneous combinations of offensive, defensive, and stability or defense support of civil authorities tasks. (ADRP 3-0)

decisive operation

The operation that directly accomplishes the mission. (ADRP 3-0)

decisive terrain

Decisive terrain, when present, is key terrain whose seizure and retention is mandatory for successful mission accomplishment. (FM 3-90-1)

deep area

The portion of the commander's area of operations that is not assigned to subordinate units. (ADRP 3-0)

defeat

A tactical mission task that occurs when an enemy force has temporarily or permanently lost the physical means or the will to fight. The defeated force's commander is unwilling or unable to pursue that individual's adopted course of action, thereby yielding to the friendly commander's will and can no longer interfere to a significant degree with the actions of friendly forces. Defeat can result from the use of force or the threat of its use. (FM 3-90-1)

defeat in detail

Concentrating overwhelming combat power against separate parts of a force rather than defeating the entire force at once. (ADRP 3-90)

defeat mechanism

A method through which friendly forces accomplish their mission against enemy opposition. (ADRP 3-0)

defensive task

A task conducted to defeat an enemy attack, gain time, economize forces, and develop conditions favorable for offensive or stability tasks. (ADRP 3-0)

delay line

A phase line where the date and time before which the enemy is not allowed to cross the phase line is depicted as part of the graphic control measure. (FM 3-90-1)

delaying operation

An operation in which a force under pressure trades space for time by slowing down the enemy's momentum and inflicting maximum damage on the enemy without, in principle, becoming decisively engaged. (JP 3-04)

deliberate operation

An operation in which the tactical situation allows the development and coordination of detailed plans, including multiple branches and sequels. (ADRP 3-90)

demonstration

In military deception, a show of force in an area where a decision is not sought that is made to deceive an adversary It is similar to a feint but no actual contact with the adversary (Army uses the term enemy instead of adversary) is intended. (JP 3-13.4)

denial operations

Actions to hinder or deny the enemy the use of space, personnel, supplies, or facilities. (FM 3-90-1)

***deny**

A task to hinder or prevent the enemy from using terrain, space, personnel, supplies, or facilities.

depth

The extension of operations in time, space, or purpose to achieve definitive results. (ADRP 3-0)

destroy

A tactical mission task that physically renders an enemy force combat-ineffective until it is reconstituted. Alternatively, to destroy a combat system is to damage it so badly that it cannot perform any function or be restored to a usable condition without being entirely rebuilt. (FM 3-90-1)

detachment left in contact

An element left in contact as part of the previously designated (usually rear) security force while the main body conducts its withdrawal. (FM 3-90-1)

direct fire

Fire delivered on a target using the target itself as a point of aim for either the weapon or the director. (JP 3-09.3)

directed obstacle

An obstacle directed by a higher commander as a specified task to a subordinate unit. (ATP 3-90.8)

disengage

A tactical mission task where a commander has the unit break contact with the enemy to allow the conduct of another mission or to avoid decisive engagement. (FM 3-90-1)

disengagement line

A phase line located on identifiable terrain that, when crossed by the enemy, signals to defending elements that it is time to displace to their next position. (ADRP 3-90)

dismounted march

Movement of troops and equipment, mainly by foot, with limited support by vehicles. Also called foot march. (FM 3-90-2)

double envelopment

Results from simultaneous maneuvering around both flanks of a designated enemy force. (FM 3-90-1)

dynamic targeting

Targeting that prosecutes targets identified too late, or not selected for action in time to be included in deliberate targeting. (JP 3-60)

echelon support

The method of supporting an organization arrayed within an area of operation. (ATP 4-90)

electronic warfare

Military action involving the use of electromagnetic and directed energy to control the electromagnetic spectrum or to attack the enemy. (JP 3-13.1)

encirclement operations

Operations where one force loses its freedom of maneuver because an opposing force is able to isolate it by controlling all ground lines of communication and reinforcement. (ADRP 3-90)

enemy

A party identified as hostile against which the use of force may be envisaged. (ADRP 3-0)

engagement area

Where the commander intends to contain and destroy an enemy force with the massed effects of all available weapons and supporting systems. (FM 3-90-1)

engagement criteria

Protocols that specify those circumstances for initiating engement with an enemy force. (FM 3-90-1)

engagement priority

The order in which the unit engages enemy systems or functions. (FM 3-90-1)

envelopment

A form of maneuver in which an attacking force seeks to avoid the principal enemy defenses by seizing objectives behind those defenses that allow the targeted enemy force to be destroyed in their current positions. (FM 3-90-1)

essential element of friendly information

A critical aspect of a friendly operation that, if known by the enemy, would subsequently compromise, lead to failure, or limit success of the operation and therefore should be protected from enemy detection. (ADRP 5-0)

execution

Putting a plan into action by applying combat power to accomplish the mission. (ADP 5-0)

exfiltrate

A tactical mission task where a commander removes Soldiers or units from areas under enemy control by stealth, deception, surprise, or clandestine means. (FM 3-90-1)

exploitation

(Army) The removal of personnel or units from areas under enemy control by stealth, deception, surprise, or clandestine means. See also special operations; unconventional warfare. (ADRP 3-90)

far side objective

A defined location oriented on the terrain or on an enemy force that an assaulting force seizes to eliminate enemy direct fires to prevent the enemy from interfering with the reduction of obstacles and allows follow-on forces to move securely through created lanes. (ATP 3-90.4)

feint

In military deception, an offensive action involving contact with the adversary conducted for the purpose of deceiving the adversary as to the location and/or time of the actual main offensive action. (JP 3-13.4)

final coordination line

A phase line close to the enemy position used to coordinate the lifting or shifting of supporting fires with the final deployment of maneuver elements. (ADRP 3-90)

final protective fire

An immediately available prearranged barrier of fire designed to impede enemy movement across defensive lines or areas. Also called FPF. (JP 3-09.3)

fire and movement

The concept of applying fires from all sources to suppress, neutralize, or destroy the enemy, and the tactical movement of combat forces in relation to the enemy (as components of maneuver, applicable at all echelons). At the squad level, fire and movement entails a team placing suppressive fire on the enemy as another team moves against or around the enemy. (FM 3-96)

fire superiority

That degree of dominance in the fires of one force over another that permits that force to conduct maneuver at a given time and place without prohibitive interference by the enemy. (FM 3-90-1)

fire support

Fires that directly support land, maritime, amphibious, and special operations forces to engage enemy forces, combat formations, and facilities in pursuit of tactical and operational objectives. (JP 3-09)

fire support plan

A plan that addresses each means of fire support available and describes how Army indirect fires, joint fires, and target acquisition are integrated with maneuver to facilitate operational success. (FM 3-09)

fire support planning

The continuing process of analyzing, allocating, and scheduling fires to describe how fires are used to facilitate the actions of the maneuver force. (FM 3-09)

fires

The use of weapon systems or other actions to create specific lethal or nonlethal effects on a target. (JP 3-09)

fires warfighting function

The related tasks and systems that provide collective and coordinated use of Army indirect fires, air and missile defense, and joint fires through the targeting process. (ADRP 3-0)

fix

A tactical mission task where a commander prevents the enemy from moving any part of the force from a specific location for a specific period. Fix is also an obstacle effect that focuses fire planning and obstacle effort to slow an attacker's movement within a specified area, normally an engagement area. (FM 3-90-1)

follow and assume

(Army) A tactical mission task in which a second committed force follows a force conducting an offensive task and is prepared to continue the mission if the lead force is fixed, attrited, or unable to continue. (FM 3-90-1)

follow and support

A tactical mission task in which a committed force follows and supports a lead force conducting an offensive task. (FM 3-90-1)

force tailoring

The process of determining the right mix of forces and the sequence of their deployment in support of a joint force commander. (ADRP 3-0)

forms of maneuver

Distinct tactical combinations of fire and movement with a unique set of doctrinal characteristics that differ primarily in the relationship between the maneuvering force and the enemy. (ADRP 3-90)

forward air controller

An officer (aviator/pilot) member of the tactical air control party who, from a forward ground or airborne position, controls aircraft in close air support of ground troops. Also called FAC. (JP 3-09.3)

forward air controller (airborne)

A specifically trained and qualified aviation officer, normally an airborne extension of the tactical air control party, who exercises control from the air of aircraft engaged in close air support of ground troops. (JP 3-09.3)

forward logistics element

(Army) Comprised of task-organized multifunctional logistics assets designed to support fast-moving offensive operations in the early phases of decisive action. (ATP 4-90)

forward observer

An observer operating with front line troops and trained to adjust ground or naval gunfire and pass back battlefield information. (JP 3-09)

forward operating base

An airfield used to support tactical operations without establishing full support facilities. (JP 3-09.3)

friendly force information requirement

Information the commander and staff need to understand the status of friendly and supporting capabilities. (JP 3-0)

frontal attack

A form of maneuver in which an attacking force seeks to destroy a weaker enemy force or fix a larger enemy force in place over a broad front. (FM 3-90-1)

geospatial intelligence

The exploitation and analysis of imagery and geospatial information to describe, assess, and visually depict physical features and geographically referenced activities on the Earth. Geospatial intelligence consists of imagery, imagery intelligence, and geospatial information. (JP 2-03)

guard

A security task to protect the main force by fighting to gain time while also observing and reporting information and preventing enemy ground observation of and direct fire against the main body. Units conducting a guard mission cannot operate independently because they rely upon fires and functional and multifunctional support assets of the main body. (ADRP 3-90)

hasty breach

The creation of lanes through enemy minefields by expedient methods such as blasting with demolitions, pushing rollers or disabled vehicles through the minefields when the time factor does not permit detailed reconnaissance, deliberate breaching, or bypassing the obstacle. (JP 3-15)

hasty operation

An operation in which a commander directs immediately available forces, using fragmentary orders, to perform activities with minimal preparation, trading planning, and preparation time for speed of execution. (ADRP 3-90)

high-value target

A target the enemy commander requires for the successful completion of the mission. (JP 3-60)

human intelligence

The collection by a trained human intelligence collector of foreign information from people and multimedia to identify elements, intentions, composition, strength, dispositions, tactics, equipment, and capabilities. (FM 2-22.3)

hybrid threat

The diverse and dynamic combination of regular forces, irregular forces, terrorist forces, or criminal elements unified to achieve mutually benefitting threat effects. (ADRP 3-0)

indirect fire

Fire delivered at a target not visible to the firing unit. (TC 3-09.81)

infiltration

A form of maneuver in which an attacking force conducts undetected movement through or into an area occupied by enemy forces to occupy a position of advantage behind those enemy positions while exposing only small elements to enemy defensive fires. (FM 3-90-1)

infiltration lane

A control measure that coordinates forward and lateral movement of infiltrating units and fixes fire planning responsibilities. (FM 3-90-1)

information collection

An activity that synchronizes and integrates the planning and employment of sensors and assets as well as the processing, exploitation, and dissemination systems in direct support of current and future operations. (FM 3-55)

information environment

The aggregate of individuals, organizations, and systems that collect, process, disseminate, or act on information. (JP 3-13)

information management

(Army) The science of using procedures and information systems to collect, process, store, display, disseminate, and protect data, information, and knowledge products. (ADRP 6-0)

information operations

The integrated employment, during military operations, of information-related capabilities in concert with other lines of operation to influence, disrupt, corrupt, or usurp the decision-making of adversaries and potential adversaries while protecting our own. (JP 3-13)

information-related capability

A tool, technique, or activity employed within a dimension of the information environment that can be used to create effects and operationally desirable conditions. (JP 3-13)

information system

Equipment that collects, processes, stores, displays, and disseminates information. This includes computers—hardware and software—and communications, as well as policies and procedures for their use. (ADP 6-0)

integration

(DOD) 2. The arrangement of military forces and their actions to create a force that operates by engaging as a whole. (JP 1)

intelligence analysis

The process by which collected information is evaluated and integrated with existing information to facilitate intelligence production. (ADRP 2-0)

intelligence operations

(Army) The tasks undertaken by military intelligence units and Soldiers to obtain information to satisfy validated requirements. (ADRP 2-0)

intelligence preparation of the battlefield

The systematic process of analyzing the mission variables of enemy, terrain, weather, and civil considerations in an area of interest to determine their effect on operations. (ATP 2-01.3)

intelligence synchronization

The "art" of integrating information collection and intelligence analysis with operations to effectively and efficiently support decision making. (ADRP 2-0)

intelligence warfighting function

The related tasks and systems that facilitate understanding the enemy, terrain, weather, civil considerations, and other significant aspects of the operations environment. (ADRP 3-0)

interdict

A tactical mission task where the commander prevents, disrupts, or delays the enemy's use of an area or route. (FM 3-90-1)

interdiction

An action to divert, disrupt, delay, or destroy the enemy's military surface capability before it can be used effectively against friendly forces, or to achieve enemy objectives. (JP 3-03)

joint fires observer

A trained Service member who can request, adjust, and control surface-to-surface fires, provide targeting information in support of Type 2 and 3 close air support terminal attack controls, and perform autonomous terminal guidance operations. (JP 3-09.3)

joint terminal attack controller

A qualified (certified) Service member who, from a forward position, directs the action of combat aircraft engaged in close air support and other offensive air operations. Also called JTAC. (JP 3-09.3)

key terrain

Any locality, or area, the seizure or retention of which affords a marked advantage to either combatant. (JP 2-01.3)

kill zone

That part of an ambush site where fire is concentrated to isolate, fix, and destroy the enemy. (FM 3-90-1)

knowledge management

The process of enabling knowledge flow to enhance shared understanding, learning, and decision making. (ADRP 6-0)

lane

A route through, over, or around an enemy or friendly obstacle that provides passage of a force. (ATP 3-90.4)

land mine

A munition on or near the ground or other surface area that is designed to be exploded by the presence, proximity, or contact of a person or vehicle. (ATP 3-90.8)

leadership

The process of influencing people by providing purpose, direction, and motivation to accomplish the mission and improve the organization. (ADP 6-22)

Level I threat

A small enemy force that can be defeated by those units normally operating in the echelon support area or by the perimeter defenses established by friendly bases and base clusters. (ATP 3-91)

Level II threat

An enemy force or activities that can be defeated by a base or base cluster's defensive capabilities when augmented by a response force. (ATP 3-91)

Level III threat

An enemy force or activities beyond the defensive capability of both the base and base cluster and any local reserve or response force. (ATP 3-91)

limit of advance

A phase line used to control forward progress of the attack. The attacking unit does not advance any of its elements or assets beyond the limit of advance, but the attacking unit can push its security forces to that limit. (ADRP 3-90)

line of contact

A general trace delineating the locations where friendly and enemy forces are engaged. (FM 3-90-1)

line of departure

A phase line crossed at a prescribed time by troops initiating an offensive operation. (ADRP 3-90)

linkup

A meeting of friendly ground forces, which occurs in a variety of circumstances. (ADRP 3-90)

linkup point

A point where two infiltrating elements in the same or different infiltration lanes are scheduled to meet to consolidate before proceeding on with their missions. (FM 3-90-1)

local security

A security task that includes low-level security activities conducted near a unit to prevent surprise by the enemy. (ADRP 3-90)

main battle area

The area where the commander intends to deploy the bulk of the unit's combat power and conduct decisive operations to defeat an attacking enemy. (ADRP 3-90)

main command post

A facility containing the majority of the staff designed to control current operations, conduct detailed analysis, and plan future operations. (FM 6-0)

main effort

A designated subordinate unit whose mission at a given point in time is moct critical to overall mission success. (ADRP 3-0)

maneuver

(DOD) 4. The employment of forces in the operational area through movement in combination with fires to achieve a position of advantage in respect to the enemy. (JP 3-0)

manned-unmanned teaming

The integrated maneuver of Army Aviation rotary wing and unmanned aircraft systems to conduct movement to contact, attack, reconnaissance, and security tasks. (FM 3-04)

march column

A march column consists of all elements using the same route for a single movement under control of a single commander. (FM 3-90-2)

march serial

A major subdivision of a march column that is organized under one commander who plans, regulates, and controls the serial. (FM 3-90-2)

march unit

A subdivision of a march serial. It moves and halts under the control of a single commander who uses voice and visual signals. (FM 3-90-2)

massed fire

(DOD) 2. Fire from a number of weapons directed at a single point or small area. (JP 3-02)

meeting engagement

A combat action that occurs when a moving force, incompletely deployed for battle, engages an enemy at an unexpected time and place. (FM 3-90-1)

military decision-making process

An iterative planning methodology to understand the situation and mission, develop a course of action, and produce an operation plan or order. (ADP 5-0)

mission

(DOD) 1. The task, together with the purpose, that clearly indicates the action to be taken and the reason therefore. (JP 3-0)

mission command

The exercise of authority and direction by the commander using mission orders to enable disciplined initiative within the commander's intent to empower agile and adaptive leaders in the conduct of unified land operations. (ADP 6-0)

mission command warfighting function

The related tasks and systems that develop and integrate those activities enabling a commander to balance the art of command and the science of control in order to integrate the other warfighting functions. (ADRP 3-0)

mission orders

Directives that emphasize to subordinates the results to be attained, not how they are to achieve them. (ADP 6-0)

mission variables

The categories of specific information needed to conduct operations. (ADP 1-01)

mobile defense

A defensive task that concentrates on the destruction or defeat of the enemy through a decisive attack by a striking force. (ADRP 3-90)

mobility

A quality or capability of military forces which permits them to move from place to place while retaining the ability to fulfill their primary mission. (JP 3-17)

mounted march

The movement of troops and equipment by combat and tactical vehicles. (FM 3-90-2)

movement and maneuver warfighting function

The related tasks and systems that move and employ forces to achieve a position of relative advantage over the enemy and other threats. (ADRP 3-0)

movement to contact

An offensive task designed to develop the situation and establish or regain contact. (ADRP 3-90)

mutual support

That support which units render each other against an enemy, because of their assigned tasks, their position relation to each other and to the enemy, and their inherent capabilities. (JP 3-31)

named area of interest

A geospatial area or systems node or link against which information that will satisfy a specific information requirement can be collected. Named areas of interest are usually selected to capture indications of adversary courses of action, but also may be related to conditions of the operational environment. (JP 2-01.3)

networked munitions

A remotely controlled, interconnected, weapons system designed to provide rapidly emplaced ground-based countermobility and protection capability through scalable application of lethal and nonlethal means. (JP 3-15)

neutral

A party identified as neither supporting nor opposing friendly, adversary, or enemy forces. (ADRP 3-0)

neutralize

A tactical mission task that results in rendering enemy personnel or materiel incapable of interfering with a particular operation. (FM 3-90-1)

noncontiguous area of operations

Where one or more of the commander's subordinate force's areas of operation do not share a common boundary. (FM 3-90-1)

objective area

A geographical area, defined by competent authority, within which is located an objective to be captured or reached by the military forces. Also called OA. (JP 3-06)

objective rally point

A rally point established on an easily identifiable point on the ground where all elements of the infiltrating unit assemble and prepare to attack the objective. (ADRP 3-90)

observation post

A position from which military observations are made, or fire directed and adjusted, and which possesses appropriate communications. While aerial observers and sensor systems are extremely useful, those systems do not constitute aerial observation posts. (FM 3-90-2)

obstacle

Any natural or man-made obstruction designed or employed to disrupt, fix, turn, or block the movement of an opposing force, and to impose additional losses in personnel, time, and equipment on the opposing force. (JP 3-15)

obstacle belt

A brigade-level command and control measure, normally given graphically, to show where within an obstacle zone the ground tactical commander plans to limit friendly obstacle employment and focus the defense. (JP 3-15)

obstacle control measures

Specific measures that simplify the granting of obstacle-emplacing authority while providing obstacle control. (FM 3-90-1)

obstacle groups

One or more individual obstacles grouped to provide a specific obstacle effect. (FM 3-90-1)

obstacle restricted area

A command and control measure used to limit the type or number of obstacles within an area. (JP 3-15)

obstacle zone

A division-level command and control measure, normally done graphically, to designate specific land areas where lower echelons are allowed to employ tactical obstacles. (JP 3-15)

offensive task

Task conducted to defeat and destroy enemy forces and seize terrain, resources, and population centers. (ADRP 3-0)

operational approach

A description of the broad actions the force must take to transform current conditions into those desired at end state. (JP 5-0)

operational area

An overarching term encompassing more descriptive terms (such as area of responsibility and joint operations area) for geographic areas in which military operations are conducted. Also called OA. (JP 3-0)

operational area security

A form of security operations conducted to protect friendly forces, installations, routes, and actions within an area of operations. (ADRP 3-37)

operational environment

A composite of the conditions, circumstances, and influences that affect the employment of capabilities and bear on the decisions of the commander. Also called OE. (JP 3-0)

operational framework

A cognitive tool used to assist commanders and staffs in clearly visualizing and describing the application of combat power in time, space, purpose, and resources in the concept of operations. (ADP 1-01)

operational variables

A comprehensive set of information categories used to define an operational environment. (ADP 1-01)

operations process

The major mission command activities performed during operations: planning, preparing, executing, and continuously assessing the operation. (ADP 5-0)

parallel planning

Two or more echelons planning for the same operation sharing information sequentially through warning orders from the higher headquarters prior to the higher headquarters publishing their operation plan or operation order. (ADRP 5-0)

passage of lines

An operation in which a force moves forward or rearward through another force's combat positions with the intention of moving into or out of contact with the enemy. (JP 3-18)

penetration

A form of maneuver in which an attacking force seeks to rupture enemy defenses on a narrow front to disrupt the defensive system. (FM 3-90-1)

personnel recovery

The sum of military, diplomatic, and civil efforts to prepare for and execute the recovery and reintegration of isolated personnel. Also called PR. (JP 3-50)

phase

A planning and execution tool used to divide an operation in duration or activity. (ADRP 3-0)

planning

The art and science of understanding a situation, envisioning a desired future, and laying out effective ways of bringing that future about. (ADP 5-0)

planning horizon

A point in time commanders use to focus the organization's planning efforts to shape future events. (ADRP 5-0)

point of breach

The location at an obstacle where the creation of a lane is being attempted. (ATP 3-90.4)

point of departure

The point where the unit crosses the line of departure and begins moving along a direction of attack. (ADRP 3-90)

point of penetration

The location, identified on the ground, where the commander concentrates their efforts at the enemy's weakest point to seize a foothold on the farside objective. (ATP 3-90.4)

position of relative advantage

A location or the establishment of a favorable condition within the area of operations that provides the commander with temporary freedom of action to enhance combat power over an enemy or influence the enemy to accept risk and move to a position of disadvantage. (ADRP 3-0)

positive control

A method of airspace control that relies on positive identification, tracking, and direction of aircraft within an airspace, conducted with electronic means by an agency having the authority and responsibility therein. (JP 3-52)

preparation

Those activities performed by units and Soldiers to improve their ability to execute an operation. (ADP 5-0)

preparation fire

Normally a high-volume of fires delivered over a short period of time to maximize surprise and shock effect. Preparation fire can include electronic attack and should be synchronized with other electronic warfare activities. (FM 3-09)

primary position

The position that covers the enemy's most likely avenue of approach into the area of operations. (ADRP 3-90)

priority intelligence requirement

An intelligence requirement stated as a priority for intelligence support, that the commander and staff need to understand the adversary or other aspects of the operational environment. Also called PIR. (JP 2-01)

probable line of deployment

A phase line that designates the location where the commander intends to deploy the unit into assault formation before beginning the assault. Also called PLD. (ADRP 3-90)

procedural control

A method of airspace control which relies on a combination of previously agreed and promulgated orders and procedures. (JP 3-52)

procedures

Standard, detailed steps that prescribe how to perform specific tasks. (CJCSM 5120.01)

proof

The verification that a lane is free of mines or explosive hazards and that the width and trafficability at the point of breach are suitable for the passing force. (ATP 3-90.4)

protection warfighting function

The related tasks and systems that preserve the force so the commander can apply maximum combat power to accomplish the mission. (ADRP 3-0)

pursuit

An offensive task designed to catch or cut off a hostile force attempting to escape, with the aim of destroying it. (ADRP 3-90)

quartering party

A group of unit representatives dispatched to a probable new site of operations in advance of the main body to secure, reconnoiter, and organize an area before the main body's arrival and occupation. (FM 3-90-2)

radio silence

The status on a radio network in which all stations are directed to continuously monitor without transmitting, except under established criteria. (ATP 6-02.53)

raid

An operation to temporarily seize an area to secure information, confuse an adversary, capture personnel or equipment, or to destroy a capability culminating with a planned withdrawal. (JP 3-0)

***rally point**

(Army) 2. An easily identifiable point on the ground at which units can reassemble and reorganize if they become dispersed.

reconnaissance

A mission undertaken to obtain, by visual observation or other detection methods, information about the activities and resources of an enemy or adversary, or to secure data concerning the meteorological, hydrographic, or geographic characteristics of a particular area. (JP 2-0)

reconnaissance objective

A terrain feature, geographic area, enemy force, adversary, or other mission or operational variable, such as specific civil considerations, about which the commander wants to obtain additional information. (ADRP 3-90)

***reconstitution**

Actions that commanders plan and implement to restore units to a desired level of combat effectiveness commensurate with mission requirements and available resources.

reduce

A mobility task to create and mark lanes through, over, or around an obstacle to allow the attacking force to accomplish its mission. (ATP 3-90.4)

reduction

The creation of lanes through a minefield or obstacle to allow passage of the attacking ground force. (JP 3-15)

reduction area

A number of adjacent points of breach that are under the control of the breaching commander. (ATP 3-90.4)

rehearsal

A session in which a staff or unit practices expected actions to improve performance during execution. (ADRP 5-0)

release point

A location on a route where marching elements are released from centralized control. (FM 3-90-2)

relief in place

An operation in which, by direction of higher authority, all or part of a unit is replaced in an area by the incoming unit and the responsibilities of the replaced elements for the mission and the assigned zone of operations are transferred to the incoming unit. (JP 3-07.3)

reorganization

All measures taken by the commander to maintain unit combat effectiveness or return it to a specified level of combat capability. (FM 3-90-1)

reserve

That portion of a body of troops, which is withheld from action at the beginning of an engagement, in order to be available for a decisive movement. (ADRP 3-90)

reserved obstacle

Obstacles of any type, for which the commander restricts execution authority. (ATP 3-90.8)

restricted operations zone

Airspace reserved for specific activities in which the operations of one or more airspace users is restricted. Also called ROZ. (JP 3-52)

retirement

A form of retrograde in which a force out of contact moves away from the enemy. (ADRP 3-90)

retrograde

A defensive task that involves organized movement away from the enemy. (ADRP 3-90)

route reconnaissance

A directed effort to obtain detailed information of a specified route and all terrain from which the enemy could influence movement along that route. (ADRP 3-90)

running estimate

The continuous assessment of the current situation used to determine if the current operation is proceeding according to the commander's intent and if planned future operations are supportable. (ADP 5-0)

scheme of fires

The detailed, logical sequence of targets and fire support events to find and engage targets to accomplish the supported commander's intent. (FM 3-09)

science of control

Systems and procedures used to improve the commander's understanding and support accomplishing missions. (ADP 6-0)

screen

A security task that primarily provides early warning to the protected force. (ADRP 3-90)

search

(DOD) 1. A systematic reconnaissance of a defined area, so that all parts of the area have passed within visibility. (JP 3-50)

search and attack

A technique for conducting a movement to contact that shares many of the characteristics of an area security mission. (FM 3-90-1)

sector of fire

That area assigned to a unit, a crew-served weapon, or an individual weapon within which it will engage targets as they appear in accordance with established engagement priorities. (FM 3-90-1)

secure

A tactical mission task that involves preventing a unit, facility, or geographical location from being damaged or destroyed as a result of enemy action. (FM 3-90-1)

security

Measures taken by a military unit, activity, or installation to protect itself against all acts designed to, or which may, impair its effectiveness. (JP 3-10)

security area

That area that begins at the forward area of the battlefield and extends as far to the front and flanks as security forces are deployed. Forces in the security area furnish information on the enemy and delay, deceive, and disrupt the enemy and conduct counterreconnaissance. (ADRP 3-90)

security cooperation

All Department of Defense interactions with foreign defense establishments to build defense relationships that promote specific U.S. security interests, develop allied and friendly military capabilities for self-defense and multinational operations, and provide U.S. forces with peacetime and contingency access to a host nation. Also called SC. (JP 3-22)

security operations

Those operations undertaken by a commander to provide early and accurate warning of enemy operations, to provide the force being protected with time and maneuver space within which to react to the enemy, and to develop the situation to allow the commander to effectively use the protected force. (ADRP 3-90)

security sector reform

A comprehensive set of programs and activities undertaken by a host nation to improve the way it provides safety, security, and justice. Also called SSR. (JP 3-07)

seize

A tactical mission task that involves taking possession of a designated area by using overwhelming force. (FM 3-90-1)

shaping operation

An operation that establishes conditions for the decisive operation through effects on the enemy, other actors, and the terrain. (ADRP 3-0)

signal intelligence

A category of intelligence comprising either individually or in combination all communications intelligence, electronic intelligence, and foreign instrumentation signals intelligence, however transmitted. (JP 2-0)

single envelopment

A form of maneuver that results from maneuvering around one assailable flank of a designated enemy force. (FM 3-90-1)

site exploitation

(DOD) A series of activities to recognize, collect, process, preserve, and analyze information, personnel, and/or materiel found during the conduct of operations. Also called SE. (JP 3-31) (Army) The synchronized and integrated application of scientific and technological capabilities and enablers to answer information requirements, facilitate subsequent operations, and support host-nation rule of law. (ATP 3-90.15)

situational obstacle

An obstacle that a unit plans and possibly prepares prior to starting an operation, but does not execute unless specific criteria are met. (ATP 3-90.8)

situational understanding

The product of applying analysis and judgment to relevant information to determine the relationship among the operational and mission variables to facilitate decision making. (ADP 5-0)

special reconnaissance

Reconnaissance and surveillance actions conducted as a special operation in hostile, denied, or politically sensitive environments to collect or verify information of strategic or operational significance, employing military capabilities not normally found in conventional forces. Also called SR. (JP 3-05)

spoiling attack

A tactical maneuver employed to seriously impair a hostile attack while the enemy is in the process of forming or assembling for an attack. (FM 3-90-1)

stability mechanism

The primary method through which friendly forces affect civilians to attain conditions that support establishing a lasting, stable peace. (ADRP 3-0)

stability tasks

Tasks conducted as part of operations outside the United States in coordination with other instruments of national power to maintain or reestablish a safe and secure environment and provide essential governmental services, emergency infrastructure reconstruction, and humanitarian relief. (ADP 3-07)

start point

A location on a route where the marching elements fall under the control of a designated march commander. (FM 3-90-02)

stay-behind operation

An operation in which the commander leaves a unit in position to conduct a specified mission while the remainder of the forces withdraw or retire from an area. (FM 3-90-1)

striking force

A dedicated counterattack force in a mobile defense constituted with the bulk of available combat power. (ADRP 3-90)

strong point

A heavily fortified battle tied to a natural or reinforcing obstacle to create an anchor for the defense or to deny the enemy decisive or key terrain. (ADRP 3-90)

subsequent position

A position that a unit expects to move to during the course of battle. (ADRP 3-90)

supplementary position

A defensive position located within a unit's assigned area of operations that provides the best sectors of fire and defensive terrain along an avenue of approach that is not the primary avenue where the enemy is expected to attack. (ADRP 3-90)

support area

The portion of the commander's area of operations that is designated to facilitate the positioning, employment, and protection of base sustainment assets required to sustain, enable, and control operations. (ADRP 3-0)

support by fire

A tactical mission task in which a maneuver force moves to a position where it can engage the enemy by direct fire in support of another maneuvering force. (FM 3-90-1)

supporting distance

The distance between two units that can be traveled in time for one to come to the aid of the other and prevent its defeat by an enemy or ensure it regains control of a civil situation. (ADRP 3-0)

supporting effort

A designated subordinate unit with a mission that supports the success of the main effort. (ADRP 3-0)

supporting range

The distance one unit may be geographically separated from a second unit, yet remain within the maximum range of the second unit's weapons systems. (ADRP 3-0)

suppress

A tactical mission task that results in the temporary degradation of the performance of a force or weapon system below the level needed to accomplish its mission. (FM 3-90-1)

suppression

Temporary or transient degradation by an opposing force of the performance of a weapons system below the level needed to fulfill its mission objectives. (JP 3-01)

surveillance

The systematic observation of aerospace, surface, or subsurface areas, places, persons, or things, by visual, aural, electronic, photographic, or other means. (JP 3-0)

survivability

A quality or capability of military forces which permits them to avoid or withstand hostile actions or environmental conditions while retaining the ability to fulfill their primary mission. (ATP 3-37.34)

survivability operations

Those military activities that alter the physical environment to provide or improve cover, concealment, and camouflage. (ATP 3-37.34)

sustaining operation

An operation at any echelon that enables the decisive operation or shaping operations by generating and maintaining combat power. (ADRP 3-0)

sustainment

(Army) The provision of logistics, personnel services, and health service support necessary to maintain operations until successful mission completion. (ADP 4-0)

sustainment warfighting function

The related tasks and systems that provide support and services to ensure freedom of action, extend operational reach, and to prolong endurance. (ADRP 3-0)

tactical combat force

A rapidly deployable, air-ground mobile combat unit, with appropriate combat support and combat service support assets assigned to and capable of defeating Level III threats including combined arms. Also called TCF. (JP 3-10)

tactical command post

A facility containing a tailored portion of a unit headquarters designed to control portions of an operation for a limited time. (FM 6-0)

tactical mission task

The specific activity performed by a unit while executing a form of tactical operation or form of maneuver. It may be expressed in terms of either actions by a friendly force or effects on an enemy force. (FM 3-90-1)

tactical road march

A rapid movement used to relocate units within an area of operations to prepare for combat operations. (ADRP 3-90)

tactics

The employment and ordered arrangement of forces in relation to each other. (CJCSM 5120.01)

target acquisition

The detection, identification, and location of a target in sufficient detail to permit the effective employment of weapons. (JP 3-60)

target area of interest

The geographical area where high-value targets can be acquired and engaged by friendly forces. Also called TAI. (JP 2-01.3)

***target identification**

The accurate and timely characterization of a detected object on the battlefield as friend, neutral, or enemy.

target reference point

A predetermined point of reference normally a permanent structure of terrain feature that can be used when describing a target location. Also called TRP. (JP 3-09.3)

targeting

The process of selecting and prioritizing targets and matching the appropriate response to them, considering operational requirements and capabilities. (JP 3-0)

task organization

(Army) A temporary grouping of forces designed to accomplish a particular mission. (ADRP 5-0)

task-organizing

The act of designing a force, support staff, or sustainment package of specific size and composition to meet a unique task or mission. (ADRP 3-0)

techniques

Nonprescriptive ways or methods used to perform missions, functions, or tasks. (CJCSM 5120.01)

tempo

The relative speed and rhythm of military operations over time with respect to the enemy. (ADRP 3-0)

terminal attack control

The authority to control the maneuver of and grant weapons release clearance to attacking aircraft. (JP 3-09.3)

terminal guidance operations

Actions using electronic, mechanical, voice or visual communications that provide approaching aircraft and/or weapons additional information regarding a specific target location. (JP 3-09)

terrain management

The process of allocating terrain by establishing areas of operation, designating assembly areas, and specifying locations for units and activities to deconflict activities that might interfere with each other (ADRP 5-0).

threat

Any combination of actors, entities, or forces that have the capability and intent to harm United States forces, United States national interests, or the homeland (ADRP 3-0).

traffic control post

A manned post that is used to preclude the interruption of traffic flow or movement along a designated route (FM 3-39).

trail party

The last march unit in a march column and normally consists of primarily maintenance elements in a mounted march (FM 3-90-2).

***traveling**

A movement technique used when speed is necessary and contact with enemy forces is not likely. All elements of the unit move simultaneously. The commander or small-unit leader locates where best to control the situation. Trailing elements may move in parallel columns to shorten the column and reaction time.

traveling overwatch

A movement technique used when contact with enemy forces is possible. The lead element and trailing element are separated by a short distance, which varies with the terrain. The trailing element moves at variable speeds and may pause for short periods to overwatch the lead element. It keys its movement to terrain and the lead element. The trailing element overwatches at such a distance that, should the enemy engage the lead element, it will not prevent the trailing element from firing or moving to support the lead element (FM 3-90-2).

***trigger line**

A phase line located on identifiable terrain that crosses the engagement area—used to initiate and mass fires into an engagement area at a predetermined range for all or like weapon systems.

troop movement

The movement of troops from one place to another by any available means (ADRP 3-90).

turning movement

A form of maneuver in which the attacking force seeks to avoid the enemy's principle defensive positions by seizing objectives behind the enemy's current positions thereby causing the enemy force to move out of their current positions or divert major forces to meet the threat. (FM 3-90-1)

unified action

The synchronization, coordination, and/or integration of the activities of governmental and nongovernmental entities with military operations to achieve unity of effort. (JP 1)

unified action partners

Those military forces, governmental and nongovernmental organizations, and elements of the private sector with whom Army forces plan, coordinate, synchronize, and integrate during the conduct of operations. (ADRP 3-0)

unity of effort

Coordination and cooperation toward common objectives, even if the participants are not necessarily part of the same command or organization, which is the product of successful unified action. (JP 1)

***vehicle distance**

the clearance between vehicles in a column, which is measured from the rear of one vehicle to the front of the following vehicle

vertical envelopment

A tactical maneuver in which troops that are air-dropped, air-landed, or inserted via air assault, attack the rear and flanks of a force, in effect cutting off or encircling the force. (JP 3-18)

warfighting function

A group of tasks and systems united by a common purpose that commanders use to accomplish missions and training objectives. (ADRP 3-0)

withdrawal operation

A planned retrograde operation in which a force in contact disengages from an enemy force and moves in a direction away from the enemy. (JP 3-17)

zone reconnaissance

A form of reconnaissance that involves a directed effort to obtain detailed information on all routes, obstacles, terrain, and enemy forces within a zone defined by boundaries. (ADRP 3-90)

This page intentionally left blank.

References

URLs accessed on 1 September 2017.

REQUIRED PUBLICATIONS

These documents must be available to intended users of this publication.

DOD Dictionary of Military and Associated Terms. August 2017.

ADRP 1-02. *Terms and Military Symbols*. 16 November 2016.

RELATED PUBLICATIONS

These documents contain relevant supplemental information.

JOINT AND DEPARTMENT OF DEFENSE PUBLICATIONS

Most joint publications are available online: http://www.dtic.mil/doctrine/new_pubs/jointpub.htm.

CJCSM 5120.01. *Joint Doctrine Development Process.* 29 December 2014.

DODI 5000.68, *Security Force Assistance (SFA)*. 27 October 2010. http://www.esd.whs.mil/DD/

JP 1. *Doctrine for the Armed Forces of the United States.* 25 March 2013.

JP 2-0. *Joint Intelligence.* 22 October 2013.

JP 2-01. *Joint and National Intelligence Support to Military Operations.* 5 July 2017.

JP 2-01.3. *Joint Intelligence Preparation of the Operational Environment.* 21 May 2014.

JP 2-03. *Geospatial Intelligence Support to Joint Operations*. 5 July 2017.

JP 3-0. *Joint Operations.* 17 January 2017.

JP 3-01. *Countering Air and Missile Threats.* 21 April 2017.

JP 3-02. *Amphibious Operations*. 18 July 2014.

JP 3-03. *Joint Interdiction.* 9 September 2016.

JP 3-04. *Joint Shipboard Helicopter and Tiltrotor Aircraft Operations.* 6 December 2012.

JP 3-05. *Special Operations.* 16 July 2014.

JP 3-06. *Joint Urban Operations.* 20 November 2013.

JP 3-07. *Stability.* 3 August 2016.

JP 3-07.3. *Peace Operation.* 1 August 2012.

JP 3-09. *Joint Fire Support.* 12 December 2014.

JP 3-09.3. *Close Air Support.* 25 November 2014.

JP 3-10. *Joint Security Operations in Theater.* 13 November 2014.

JP 3-11. *Operations in Chemical, Biological, Radiological, and Nuclear Environments.* 4 October 2013.

JP 3-13. *Information Operations.* 27 November 2012.

JP 3-13.1. *Electronic Warfare.* 8 February 2012.

JP 3-13.4. *Military Deception.* 14 February 2017.

JP 3-15. *Barriers, Obstacles, and Mine Warfare for Joint Operations.* 6 September 2016.

JP 3-17. *Air Mobility Operations*. 30 September 2013.

JP 3-18. *Joint Forcible Entry Operations*. 11 May 2017.

JP 3-22. *Foreign Internal Defense*. 12 July 2010.

JP 3-31. *Command and Control for Joint Land Operations.* 24 February 2014.

JP 3-40. *Countering Weapons of Mass Destruction.* 31 October 2014.

JP 3-50. *Personnel Recovery.* 2 October 2015.

JP 3-52. *Joint Airspace Control.* 13 November 2014.

JP 3-60. *Joint Targeting.* 31 January 2013.

JP 4-0. *Joint Logistics.* 16 October 2013.

JP 5-0. *Joint Operations Planning.* 16 June 2017.

ARMY PUBLICATIONS

Most Army publications are available online: http://armypubs.army.mil

ADP 1-01. *Doctrine Primer.* 2 September 2014.

ADP 3-0. *Operations*. 6 October 2017.

ADP 3-07. *Stability*. 31 August 2012.

ADP 4-0. *Sustainment.* 31 July 2012.

ADP 5-0. *The Operations Process.* 17 May 2012.

ADP 6-0. *Mission Command.* 17 May 2012.

ADP 6-22. *Army Leadership.* 1 August 2012.

ADRP 1. *The Army Profession.* 14 June 2015.

ADRP 2-0. *Intelligence*. 31 August 2012.

ADRP 3-0. *Operations*. 6 October 2017.

ADRP 3-07. *Stability*. 31 August 2012.

ADRP 3-09. *Fires*. 31 August 2012.

ADRP 3-28. *Defense Support of Civil Authorities*. 14 June 2013.

ADRP 3-37. *Protection*. 31 August 2012.

ADRP 3-90. *Offense and Defense*. 31 August 2012.

ADRP 4-0. *Sustainment*. 31 July 2012.

ADRP 5-0. *The Operations Process*. 17 May 2012.

ADRP 6-0. *Mission Command.* 17 May 2012.

ADRP 6-22. *Army Leadership.* 1 August 2012.

AR 27-1. *Judge Advocate Legal Services.* 24 January 2017.

ATP 1-0.1. *G-1/AG and S-1 Operations.* 23 March 2015.

ATP 1-06.1. *Field Ordering Officer (FOO) and Pay Agent (PA) Operations.* 10 May 2013.

ATP 2-01. *Plan Requirements and Assess Collection.* 19 August 2014.

ATP 2-01.3. *Intelligence Preparation of the Battlefield/Battlespace.* 10 November 2014.

ATP 2-19.4. *Brigade Combat Team Intelligence Techniques*. 10 February 2015.

ATP 2-22.6. *(U) Signals Intelligence Techniques (TS).* 17 December 2015.

ATP 2-22.7. *Geospatial Intelligence.* 26 March 2015.

ATP 2-22.31. *(U) Human Intelligence Military Source Operations Techniques (S//NF).* 17 April 2015.

ATP 3-01.4. *Multi-Service Tactics, Techniques, and Procedures for Joint Suppression of Enemy Air Defense (J-SEAD) MCRP 3-22.2A; NTTP 3-01.42; AFTTP 3-2.28.* 15 December 2015.

ATP 3-01.7. *Air Defense Artillery Brigade Techniques.* 16 March 2016.

ATP 3-01.8. *Techniques for Combined Arms for Air Defense.* 29 July 2016.

ATP 3-01.15. *IADS Multi-Service Tactics, Techniques, and Procedures for an Integrated Air Defense System. ATP 3-01.15 [FM 3-01.15], MCRP 3-25E, NTTP 3-01.8, AFTTP 3-2.3.* 9 September 2014.

ATP 3-01.50. *Air Defense and Airspace Management (ADAM) Cell Operation.* 5 April 2013.

ATP 3-01.60. *Counter-Rocket, Artillery, and Mortar Operations*. 10 May 2013.

ATP 3-04.1. *Aviation Tactical Employment.* 13 April 2016.

ATP 3-04.64. *Multi-Service Tactics, Techniques, and Procedures for the Tactical Employment of Unmanned Aircraft Systems, ATP 3-04.64, MCRP 3-42.1A, NTTP 3-55.14, AFTTP 3-2.64.* 22 January 2015.

ATP 3-05.2. *Foreign Internal Defense*. 19 August 2015.

ATP 3-06.20. *Multi-Service Tactics, Techniques, and Procedures for Cordon and Search Operations, ATP 3-06.20, MCRP 3-30.5 [MCRP 3-31.4B], NTTP 3-05.8, AFTTP 3-2.62.* 18 August 2016.

ATP 3-09.12. *Field Artillery Target Acquisition*, 24 July 2015.

ATP 3-09.30. *Techniques for Observed Fire*, 28 September 2017.

ATP 3-09.32. *JFIRE Multi-Service Tactics, Techniques, and Procedures for Joint Application of Firepower, ATP 3-09.32, MCRP 3-16.6A, NTTP 3-09.2, AFTTP 3-2.6.* 21 January 2016.

ATP 3-09.42. *Fire Support for the Brigade Combat Team.* 1 March 2016.

ATP 3-11.23. *Multi-Service Tactics, Techniques, and Procedures for Weapons of Mass Destruction Elimination Operations ATP 3-11.23, MCWP 3-37.7, NTTP 3-11.35, AFTTP 3-2.71.* 1 November 2013.

ATP 3-11.24. *Technical Chemical, Biological, Radiological, Nuclear, and Explosives Force Employment.* 6 May 2014.

ATP 3-11.32. *Multi-Service Tactics, Techniques, and Procedures for Chemical, Biological, Radiological, and Nuclear Passive Defense, ATP 3-11.32, MCWP 3-37.2, NTTP 3-11.37.* 13 May 2016.

ATP 3-11.36. *Multi-Service Tactics, Techniques, and Procedures for Chemical, Biological, Radiological, and Nuclear Aspects of Command and Control, ATP 3-11.36, MCRP 3-37B, NTTP 3-11.34, AFTTP 3-2.70.* 1 November 2013.

ATP 3-11.37. *Multi-Service Tactics, Techniques, and Procedures for Chemical, Biological, Radiological, and Nuclear Reconnaissance and Surveillance, ATP 3-11.37, MCWP 3-37.4, NTTP 3-11.29, AFTTP 3-2.44.* 25 March 2013.

ATP 3-11.50. *Battlefield Obscuration.* 15 May 2014.

ATP 3-20.98. *Reconnaissance Platoon.* 5 April 2013.

ATP 3-21.8. *Infantry Platoon and Squad.* 12 April 2016.

ATP 3-21.18. *Foot Marches.* 17 April 2017.

ATP 3-28.1. *Multi-Service Tactics, Techniques, and Procedures for Defense Support of Civil Authorities (DSCA), ATP 3-28.1, MCWP 3-36.2, NTTP 3-57.2, ATTP 3-2.67.* 25 September 2015.

ATP 3-34.22. *Engineer Operations—Brigade Combat Team and Below.* 5 December 2014.

ATP 3-34.80. *Geospatial Engineering.* 22 February 2017.

ATP 3-34.81. *Engineer Reconnaissance.* 1 March 2016.

ATP 3-36. *Electronic Warfare Techniques.* 16 December 2014.

ATP 3-37.2. *Antiterrorism.* 3 June 2014.

ATP 3-37.10. *Base Camps.* 27 January 2017.

ATP 3-37.34. *Survivability Operations.* 28 June 2013.

ATP 3-39.30. *Security and Mobility Support.* 30 October 2014.

ATP 3-39.32. *Physical Security.* 30 April 2014.

ATP 3-39.35. *Protective Services.* 31 May 2013.

ATP 3-52.1. *Multiservice Tactics, Techniques, and Procedures for Airspace Control, ATP 3-52.1 [FM 3-52.1], MCWP 3-25.13, NTTP 3-56.4, AFTTP 3-2.78.* 9 April 2015.

ATP 3-55.4. *Techniques for Information Collection During Operations Among Populations.* 5 April 2016.

ATP 3-57.60. *Civil Affairs Planning.* 27 April 2014.

ATP 3-60. *Targeting.* 7 May 2015.

ATP 3-60.1. *Dynamic Targeting, Multi-Service Tactics, Techniques, and Procedures for Dynamic Targeting.* 10 September 2015.

ATP 3-90.4. *Combined Arms Mobility.* 8 March 2016.

ATP 3-90.8. *Combined Arms Countermobility Operations.* 17 September 2014.

ATP 3-90.15. *Site Exploitation.* 28 July 2015.

ATP 3-91. *Division Operations.* 17 October 2014.

ATP 3-91.1. *The Joint Air Ground Integration Center.* 18 June 2014.

ATP 4-01.45. *Multi-Service Tactics, Techniques, and Procedures for Tactical Convoy Operations, ATP 4-01.45, MCRP 3-40F.7[MCRP 4-11.3H], AFTTP 3-2.58..* 22 February 2017.

ATP 4-02.3. *Army Health System Support to Maneuver Forces.* 9 June 2014.

ATP 4-02.8. *Force Health Protection.* 9 March 2016.

ATP 4-10. *Multi-Service Tactics, Techniques and Procedures for Operational Contract Support, ATP 4-10, MCRP 4-11H, NTTP 4-09.1, AFMAN 10-409-O.* 18 February 2016.

ATP 4-32.2. *Multi-Service Tactics, Techniques, and Procedures for Explosive Ordnance, ATP 4-32.2 [ATTP 4-32.2], MCRP 3-17.2B, NTTP 3-02.4.1, AFTTP 3-2.12.* 15 July 2015.

ATP 4-41. *Army Field Feeding and Class I Operations.* 31 December 2015.

ATP 4-42. *General Supply and Field Services Operations.* 14 July 2014.

ATP 4-44. *Water Support Operations.* 2 October 2015.

ATP 4-48. *Aerial Delivery.* 21 December 2016.

ATP 4-90. *Brigade Support Battalion.* 2 April 2014.

ATP 4-92. *Contracting Support to Unified Land Operations.* 15 October 2014.

ATP 5-0.1. *Army Design Methodology.* 1 July 2015.

ATP 5-19. *Risk Management.* 14 April 2014.

ATP 6-0.5. *Command Post Organization and Operations.* 1 March 2017.

ATP 6-01.1. *Techniques for Effective Knowledge Management.* 6 March 2015.

ATP 6-02.53 (FM 6-02.53). *Tactical Radios Operations.* 7 January 2016.

ATP 6-02.70. *Techniques for Spectrum Management Operations.* 31 December 2015.

ATP 6-02.75. *Techniques for Communications Security (COMSEC) Operations.* 17 August 2015.

ATTP 3-06.11. *Combined Arms Operations in Urban Terrain.* 10 June 2011.

ATTP 3-21.50. *Infantry Small-Unit Mountain Operations.* 28 February 2011.

ATTP 3-21.90. *Tactical Employment of Mortars.* 4 April 2011.

FM 1-0. *Human Resources Support.* 1 April 2014.

FM 1-04. *Legal Support to the Operational Army.* 18 March 2013.

FM 1-05. *Religious Support.* 5 October 2012.

FM 2-0. *Intelligence Operations.* 15 April 2014.

FM 2-22.3. *Human Intelligence Collector Operations.* 6 September 2006.

FM 3-04. *Army Aviation.* 29 July 2015.

FM 3-06. *Urban Operations.* 26 October 2006.

FM 3-07. *Stability.* 2 June 2014.

FM 3-09. *Field Artillery Operations and Fire Support.* 4 April 2014.

FM 3-11. *Multi-Service Doctrine for Chemical, Biological, Radiological, and Nuclear Operations, FM 3-11, MCWP 3-37.1, NWP 3-11, AFTTP 3-2.42.* 1 July 2011.

FM 3-11.9. *Potential Military Chemical/Biological Agents and Compounds, FM 3-11.9, MCRP 3-37.1B, NTRP 3-11.32, AFTTP (I) 3-2.55.* 10 January 2005.

FM 3-12 (FM 3-38). *Cyberspace and Electronic Warfare Operations.* 11 April 2017.

FM 3-13. *Information Operations.* 6 December 2016.

FM 3-21.10. *The Infantry Rifle Company.* 27 July 2006.

FM 3-21.12. *The Infantry Weapons Company.* 1 July 2008.

FM 3-22. *Army Support to Security Cooperation.* 22 January 2013.

FM 3-22.34. *TOW Weapon System.* 28 November 2003.

FM 3-24.2. *Tactics in Counterinsurgency.* 21 April 2009.

FM 3-34. *Engineer Operations.* 2 April 2014.

FM 3-39. *Military Police Operations.* 26 August 2013.

FM 3-50. *Army Personnel Recovery.* 2 September 2014.

FM 3-52. *Airspace Control*. 20 October 2016.

FM 3-55. *Information Collection*. 3 May 2013.

FM 3-57. *Civil Affairs Operations*. 31 October 2011.

FM 3-61. *Public Affairs Operations*. 1 April 2014.

FM 3-90-1. *Offense and Defense, Volume 1*. 22 March 2013.

FM 3-90-2. *Reconnaissance, Security, and Tactical Enabling Tasks, Volume 2*. 22 March 2013.

FM 3-94. *Theater Army, Corps, and Division Operations*. 21 April 2014.

FM 3-96. *Brigade Combat Team*. 8 October 2015.

FM 3-99. *Airborne and Air Assault Operations*. 6 March 2015.

FM 4-02. *Army Health System*. 26 August 2013.

FM 4-40. *Quartermaster Operations*. 22 October 2013.

FM 4-95. *Logistics Operations*, 1 April 2014.

FM 6-0. *Commander and Staff Organization and Operations*. 5 May 2014.

FM 6-02. *Signal Support Operations*. 22 January 2014.

FM 6-02.71. *Network Operations*. 14 July 2009.

FM 6-05. *CF-SOF, Multi-Service Tactics, Techniques, and Procedures for Conventional Forces and Special Operations Forces Integration, Interoperability, and Interdependence, FM 6-05 [FM 6-03.05], MCWP 3-36.1, NTTP 3-05.19, AFTTP 3-2.73, USSOCOM Pub 3-33*. 13 March 2014.

FM 7-0. *Train to Win in a Complex World*. 5 October 2016.

FM 27-10. *The Law of Land Warfare*. 18 July 1956.

FM 90-3. *Desert Operations*. 24 August 1993.

FM 90-5. *Jungle Operations*. 16 August 1982.

TC 2-91.4. *Intelligence Support to Urban Operations*. 23 December 2015.

TC 3-09.81. *Field Artillery Manual Cannon Gunnery*. 13 April 2016.

TC 3-20.31-4. *Direct Fire Engagement Process (DIDEA)*. 23 July 2015.

TC 3-22.10. *Sniper*. 7 December 2017.

TC 3-22.19. *Grenade Machine Gun MK 19 MOD 3*. 10 May 2017.

TC 3-22.32. *M41 Improved Target Acquisition System (ITAS) and Tube-Launched, Optically Tracked, Wire Guided/Wireless (TOW) Missile*. 18 November 2015.

TC 3-22.37. *Javelin-Close Combat Missile System, Medium*. 13 August 2013.

TC 3-22.50. *Heavy Machine Gun M2 Series*. 19 May 2017.

OTHER PUBLICATIONS

STANAGs are available at https://nso.nato.int/protected/nsdd/ListPromulg.html
(Note: you need to login to have access.)

STANAG 2020, *Operational Situation Reports*, 3 April 1967.

PRESCRIBED FORMS

This section contains no entries.

REFERENCED FORMS

Forms are available online: https://armypubs.army.mil.

DA Form 581. *Request for Issue and Turn-In of Ammunition*.

DA Form 2028. *Recommended Changes to Publications and Blank Forms*.

DA Form 4656. *Scheduling Worksheet*.

DD Form 1380. *Tactical Combat Casualty Care (TCCC) Card*.

DD Form 2977. *Deliberate Risk Assessment Worksheet*.

Index

Entries are by paragraph number.

Made in the USA
Middletown, DE
07 September 2022